A Differential Quadrature Hierarchical Finite Element Method

A Differential Quadrature Hierarchical Finite Element Method

Bo Liu
Beihang University, China

Cuiyun Liu
Beihang University, China

Yang Wu
Institute of Applied Physics and
Computational Mathematics, China

Yufeng Xing
Beihang University, China

国防工業出版社
National Defense Industry Press

World Scientific

Published by

World Scientific Publishing Co. Pte. Ltd.

5 Toh Tuck Link, Singapore 596224

USA office: 27 Warren Street, Suite 401-402, Hackensack, NJ 07601

UK office: 57 Shelton Street, Covent Garden, London WC2H 9HE

Library of Congress Cataloging-in-Publication Data
Names: Liu, Bo, 1981– author. | Liu, Cuiyun, author. |
Wu, Yang (Physicist), author. | Xing, Yufeng, author.
Title: A differential quadrature hierarchical finite element method / Bo Liu, Beihang University, China; Cuiyun Liu, Beihang University, China; Yang Wu, Institute of Applied Physics and Computational Mathematics, China; and Yufeng Xing, Beihang University, China.
Description: [Beijing Shi] : Guo fang gong ye chu ban she ; New Jersey : World Scientific, [2022] | Includes bibliographical references.
Identifiers: LCCN 2021009806 | ISBN 9789811236754 (hardcover) |
ISBN 9789811236761 (ebook for institutions) | ISBN 9789811236778 (ebook for individuals)
Subjects: LCSH: Finite element method. | Quadrature domains. | Differential equations--Numerical solutions. | Structural analysis (Engineering)--Mathematical models.
Classification: LCC QC20.7.F56 L54 2022 | DDC 518/.25--dc23
LC record available at https://lccn.loc.gov/2021009806

British Library Cataloguing-in-Publication Data
A catalogue record for this book is available from the British Library.

微分求积升阶谱有限元方法
Originally published in Chinese by National Defense Industry Press

For any available supplementary material, please visit
https://www.worldscientific.com/worldscibooks/10.1142/12271#t=suppl

Desk Editors: Vishnu Mohan/Steven Patt

Typeset by Stallion Press
Email: enquiries@stallionpress.com

Preface

The differential quadrature hierarchical finite element method (DQHFEM) combines the hierarchical finite element method (HFEM) with the weak form of the differential quadrature method (DQM) as the name indicates. It is also called the hierarchical quadrature element method (HQEM) for simplicity. I came to know about the DQM around 2007 and I was surprised of its accuracy levels that can almost be exact. I was working on the analytical method at that time. I had a profound understanding that analytical methods need complex derivations and can only be used in very simple domains like square or circular disk with simple boundary conditions. So, I began to work on the DQM in 2008. I was pleased that the DQM can solve problems in regular domains with arbitrary boundary conditions. When I tried to solve problems in irregular domains using the DQM, I found it had numerical instability problems and the boundary condition imposition of the DQM is not simple. This led to the development of the differential quadrature finite element method (DQFEM) in 2009. During the time I worked on DQM, I read a book *The Hierarchical Finite Element Method* by D.C. Zhu, a senior Professor in our group, and learned that the HFEM can only be applied to rectangular and hexahedral domains. The HFEM has an embedding property, due to which the local refinement will not propagate from patch to patch. I tried the HFEM but found that the orthogonal polynomials in this method cannot be higher than 25 degrees in two-dimensional problems. It is not until 2014 when I began to work on isogeometric analysis (IGA) that I found that there are several ways in computer-aided geometric design (CAGD) to write polynomials to avoid numerical instability. This inspired me and I tried the recursion formula of orthogonal polynomials. It turned out that the recursion formula worked well and the HQEM was developed. I received support from the Natural Science Foundation of China (NSFC) and the investigation on the HFEM continued until now. Finally, we now have this book which is a collection of my publications on this topic.

The hierarchical concept of FEM was proposed by Zienkiewicz *et al.* in 1971 for three-dimensional stress analysis. At that time, the objective of the hierarchical functions was simply the introduction of p-graded (degrees of freedom) meshes in an *a priori* chosen manner when advancing to a region of high stress concentration, then some of the total degrees of freedom (and therefore the cost) can be saved. Such elements were called mixed, super-parametric, "hierarchical" elements. In the context of such elements, it was believed to be more convenient to assign *edge variables* rather than create actual nodes along any specific edge. Therefore, the degrees of the edge bases are usually of second or cubic order. Three-dimensional HFEM elements of order higher than 3 were not suggested. Then Zhu proposed in 1986 to use Jacobi orthogonal polynomials or integrated Legendre orthogonal polynomials as bases for HFEM, which elevated the maximum order to seventh degree. D.C. Zhu used the series form of orthogonal polynomials. This influenced many researchers and established numerical instability problems as a common feature of the HFEM. Although Szabó and Babuška began to use the recursion formula in 1991, many people were not aware of this fact before my first paper on HQEM. After the fast development of the HFEM in the 1970s and 1980s, the first commercial codes based on p-version FEM, StressCheck, was developed by ESRD Company created by Szabó, Myers and Babuška. In the 1990s, the HFEM was applied to nonlinear problems and thin-walled structures. In more recent applications, the HFEM has been combined with the finite cell method, IGA, DQFEM, and other methods.

The investigations on the HFEM significantly reduced after the 1990s. This would have been caused by the limitation that automatic generator of coarse high-order mesh was not available at that time. Without coarse high-order mesh, the p-version FEM may not be competitive compared with its h-version counterpart as summarized by

Szabó, Düster and Rank in 2004 in "The p-version of the Finite Element Method," which is a subsection in *The Encyclopedia of Computational Mechanics* edited by Stein, Borst and Hughes. Simplex HFEM elements were developed by Tong in 1977. Dubiner proposed multivariate orthogonal polynomials on simplexes in 1991. Webb and Abouchacra introduced modified Dubiner's triangular bases by splitting them into interior and boundary modes in 1995. However, the application of the HFEM to structural mechanics has been limited to tensor product domains as can be seen from the summary by Szabó *et al.* in 2004. Significant development of high-order methods and high-order mesh generation was achieved after the ADIGMA (A European Initiative on the Development of Adaptive Higher-Order Variational Methods for Aerospace Applications) and IDIHOM (Industrialization of High-Order Methods) projects funded by the European Union from 2006 to 2014. Such development mainly takes place in fluid dynamics where a small geometric error may induce a great fluctuation in the final results in problems like boundary layer analyses.

The FEM uses interpolative nodal bases, which makes simulation intuitive and simple. The hierarchical bases of HFEM do not have clear physical meaning, which makes boundary condition imposition and element assembly in the HFEM not as simple as in the FEM. The HQEM included interpolative bases constructed by DQ nodes on the boundaries of elements to overcome the limitations of the HFEM. The interpolation bases of the HQEM on quadrilateral and hexahedral domains were constructed by blending the function together with Lagrange interpolation. The interpolation bases of HQEM on simplex domains were constructed by using modified Dubiner's bases together with Fekete points. The Fekete points are known as optimized interpolation nodes on arbitrary domains. They are the Gauss–Lobatto points on tensor product domains. The HFEM on pyramid domains is special. Rational vertex bases are required on pyramid domains so that the requirements of C^0 continuity can be satisfied at all faces.

This book is organized into 12 chapters. The first chapter provides a comprehensive review of high-order methods, especially the HFEM, on structural mechanics. A chronological review of the HFEM, applications of the HFEM to nonlinear problems, the computation of Fekete points, and high-order mesh generation are presented in this chapter. A brief review of the DQM and its weak form, and the application of the p-version FEM to plates and shells are also covered in this chapter. Finally, an outlook of high-order method is presented in this chapter.

Chapter 2 introduces the DQFEM. The HQEM is developed to overcome the limitation of the DQFEM in local order elevation and the HQEM uses some techniques of the DQFEM. Therefore, this chapter is helpful in understanding the HQEM.

Chapters 3 and 4 introduce the application of the HQEM to Mindlin plates and plane problems. C^0 quadrilateral and triangular HQEM elements are developed in the two chapters. The method used to construct the two types of elements in Chapter 4 is slightly different from that in Chapter 3. The method in Chapter 3 can be understood easily, but the method in Chapter 4 is simpler in coding. Examples for statics and free vibration of isotropic and composite plates are presented in these two chapters. In Chapters 3 and 4, the HQEM is still called the DQHFEM as it was proposed.

Chapter 5 introduces the HQEM for 3D solids. Tetrahedral, wedge, hexahedral and pyramid HQEM elements are presented in this chapter. These elements are then applied to various structural problems to test their performance.

Chapter 6 introduces the HQEM for Kirchhoff plates. C^1 quadrilateral and triangular HQEM elements are developed in this chapter. Six vertex degrees of freedom are required for C^1 problems to reach high accuracy, and the DOFs of rotations on element boundaries should be at least one more than those for deflection to ensure numerical stability. Bending and vibration analyses in the chapter validates the exponential convergence rate of the developed elements.

Chapters 7 and 8 respectively introduce the HQEM for shells in orthogonal and general curvilinear coordinate systems. The two chapters adopt the C^0 quadrilateral HQEM elements and analyze various structural problems for isotropic materials as well as composite laminates.

Chapters 9 and 10 introduce the application of the HQEM to geometrically nonlinear and incremental elasto-plastic analyses. Two-dimensional as well as three-dimensional structures are analyzed in the two chapters. The HQEM showed fast convergence rate in all examples. Chapter 11 applies the C^1 HQEM to finite deformation analyses of isotropic and composite laminated thin shells.

Chapter 12 introduces the incorporation of the HQEM with IGA. The IGA can naturally provide high accuracy mesh for HQEM. The HQEM can also be used to express geometry in high accuracy, which can overcome the limitation of IGA on geometries with polar singularity. The generation of high-order mesh with non-equally spaced nodes on faces through Gmsh and Open CASCADE is also covered in this chapter.

This book is the first one that systemically introduces the construction of various HFEM elements of commonly used geometric shapes such as triangles, tetrahedrons, pyramids and has overcome most of the computational limitations of the elements. The HQEM is also developed by incorporating the advantages and the latest research achievements of the DQM and the IGA. So, this book can also be referenced by researchers work on the two fields. The examples provided include statics and dynamics, isotropic materials and composites, linear and nonlinear problems, plates as well as shells and solid structures. The results can be used as benchmarks for low-order schemes as the HQEM is of extremely high accuracy. High-order mesh generation has been regarded as the major bottleneck for the wide application of high-order FEM. This book provides an introduction to some latest advancements on this topic.

Prospective readers of the book include graduate students in mechanical, civil, aerospace, biomedical, and industrial engineering and engineering mechanics as well as researchers and design engineers from the aforementioned fields.

I would like to thank Professor Yufeng Xing with whom I worked for many years. I was his PhD student when I began to work on the DQM and finally proposed the DQFEM. During the past five years of work on the HFEM, I discussed various topics with Prof. Xing and he always helped me with all sincerity. I would like also to thank Dr. Yang Wu who worked with me on thin-walled structures in recent years. Dr. Wu spent much time in preparing the Chinese version of this book. Chapters 6, 8 and 11 are based on collaborative papers with him. The review on the application of p-version FEM to plates and shells in Chapter 1 was also prepared by him. I am thankful to Prof. Cuiyun Liu who spent much time in proofreading the Chinese and English versions of this book. I am also thankful to my students, Liang Zhao, Shuai Lu, Mao Guo, Jiajia Song,

Tao Shi, who worked with me on the topics of Chapters 2–5, Chapters 7, 9 and 10, and Hao Sun and Zeyu Peng, who worked with me on the topics of Chapter 12. Finally, the financial support of the National Natural Science Foundation of China (Grant Nos. 11772031, 11972004, 12002018, 11402015) on the research, and the National Defense Industry Press and World Scientific Publishing Co Pte Ltd on the publication of this book is gratefully acknowledged.

Bo Liu
The Solid Mechanics Research Centre
School of Aeronautic Science and Engineering
Beihang University (BUAA)
Beijing 100191, China

December 2020

About the Authors

Dr. Bo Liu is an associate professor of Beihang University (BUAA), Beijing, China. His research currently mainly focuses on high accuracy numerical methods in solid mechanics and the software development. Dr. Liu has about 40 peer-reviewed publications on international journals and 3 monographs (two of the books are in Chinese).

Dr. Cuiyun Liu is an assistant professor of Beihang University (BUAA), Beijing, China. Her research currently mainly focuses on high accuracy numerical methods in solid mechanics and property characterization of composites. Dr. Liu has about 10 peer-reviewed journal papers.

Dr. Yang Wu current works at Institute of Applied Physics and Computational Mathematics, CAEP Software Center for High Performance Numerical Simulation, Beijing, China. His research currently mainly focuses on high accuracy numerical methods in thin-walled structures. Dr. Wu has about 10 peer-reviewed publications on international journals and 2 monographs.

Dr. Yufeng Xing is a professor of Beihang University (BUAA), Beijing, China. His research currently mainly focuses on analytical and numerical methods for structure dynamics and multiscale simulation of composites. Dr. Xing has won two national defense science and technology progress awards. He has about 120 peer-reviewed publications on international journals, one monograph and two text books.

Contents

Chapter 1

An Overview of High-Order Methods for Structural Mechanics

High-order methods are rapidly increasing in popularity due to their favorable numerical characteristics and ability to more effectively use of modern computing hardware compared to traditional low-order methods [1, 2]. Therefore, various higher-order methods on unstructured/hybrid grids have been developed in the last five decades. From the latest review on strong formulation finite element method based on the differential quadrature method (DQM) [3], one can see that only the DQM on one-dimensional, quadrilateral or hexahedral domains were considered. Similar is the case of the review paper [4] two years later on weak form Quadrature Element Method (QEM). It was summarized that in order to efficiently analyze problems in science and engineering, including analyzing large-scale problems in structural mechanics and performing multiscale analysis, "various transition elements should be proposed. These problems offer challenges to future researchers of the weak form quadrature element method." The hierarchical finite element method (HFEM) was proposed by Zienkiewicz [5] in 1971 and has a history of about half a century. However, one can see from some recent books [6–9] that its applications are still limited to one-dimensional, quadrilateral or hexahedral domains. The HFEM on simplex and pyramid domains actually have already been formulated, but this has been scattered over many papers in the fields of mathematics, computational electromagnetics, fluid dynamics and structural mechanics. This chapter presents a state-of-the-art review of the DQM and the HFEM on simplex and pyramid domains. In what follows, the basic mathematical concepts underlying the DQM, QEM and HFEM on one-dimensional, quadrilateral or hexahedral domains are first presented in Sections 1.1–1.3.2. This is done because the concepts are also needed in simplex and pyramid domains. An introduction of the method of formulating orthogonal polynomials on simplex and pyramid domains are discussed thereafter.

After that, a state-of-the-art review of the HFEM for nonlinear problems is presented in Section 1.4. Both geometric nonlinearities and material nonlinearities are considered. Some publications using methods such as the differential quadrature method and the finite cell method that relate with the HFEM are also reviewed. It is shown that early publications on the applications of the HFEM to nonlinear problems mainly focused on the performances or properties of the HFEM on the problems. Later publications focused on taking full advantage of the HFEM on the targeted problems.

Thin-walled structures like plates and shells are the most common construction elements in nature and technology. This is independent of the specific scale; they might be small like cell membranes and tiny machine parts or very large like fuselages and cooling towers. This preference to apply walls as thin as possible is a natural optimization strategy to reduce dead load and to minimize construction material. In addition to the slenderness, the advantageous effect of curvature is utilized in shell structures allowing one to carry transverse loading in an optimal way, a design principle already known to the ancient master builders. A review of the hierarchical finite element method for thin walled structures is presented in Section 1.5.

Because of the importance in high-order simulations and the impact of the geometric accuracy on the convergence of high-order discretization methods [10], significant progress on high-order curvilinear mesh generation techniques has been made in the last decade. Generally speaking, there are two kinds of approaches to generate high-order mesh. The first one is to generate high-order elements directly from the CAD model using a high-order version of classical meshing algorithm. The other one is to convert the initial straight-edge (P1) elements generated by some traditional grid generation software into the high-order curvilinear version. The second one is more popular than the first because the initial grids can be generated easily by available grid generation software. In order to convert the straight-edge P1 elements into high-order curvilinear ones, many techniques have been presented in the literature. The first of these focuses around the concept of solid body deformation, whereby the mesh is treated as a solid body which is deformed to incorporate curvature at the boundary. The second theme follows a different route, whereby the mesh is equipped with an associated functional that denotes its energy. A nonlinear optimization problem is then solved in order to minimize this functional and yield a valid mesh. Again, most studies in this area have focused around the choice of functional. A review of the method of high-order mesh generation is presented in Section 1.6.

Finally, an outlook of high-order methods is presented in Section 1.7. It was concluded that with the fast development of high-order mesh generation techniques and the strong requirements of highly accurate simulation methods in some fields, the wide application of the p-FEM may come true in a few decades.

1.1 The Differential Quadrature Method

1.1.1 *The quadrature rules*

Instead of showing the chronological development of the DQM that can be found in [3, 11], this section is aimed at presenting some main techniques of the DQM that will be used throughout the paper. However, some related important contributions are reviewed. The DQM was proposed by Bellman [12] in 1971 for the solution of initial- and/or boundary-value problems. The term "quadrature," as ordinarily used, applies to the approximate evaluation of an integral

$$\int_0^1 f(x)\mathrm{d}x \cong \sum_{i=1}^{N} w_i f(x_i) \tag{1.1}$$

He extended the approximation technique to far more general linear functionals and presented the following approximate relation:

$$f'(x_i) \cong \sum_{j=1}^{N} A_{ij}^{(1)} f(x_j), \quad i = 1, 2, \ldots, N \tag{1.2}$$

with the coefficient matrix ($A_{ij}^{(1)}$) determined in various fashions. He called this procedure "differential quadrature." This method requires the solution to be sufficiently smooth. A year later, he presented the method of determining weighting coefficients [13]. This was done by analogy with the classical quadrature case, demanding that Eq. (1.2) be exact for all polynomials of degree less than or equal to $N-1$. The test function $p_k(x) = x^{k-1}, k = 1, \ldots, N$, for arbitrary distinct x_i leads to the set of linear algebraic equations

$$\sum_{j=1}^{N} A_{ij}^{(1)} x_j^{k-1} = (k-1)x_i^{k-2}, \quad i = 1, 2, \ldots, N \tag{1.3}$$

which has a unique solution since the coefficient matrix is a Vandermonde matrix.

Rather than solving a set of linear algebraic equations, $A_{ij}^{(1)}$ were readily determined explicitly once and for all under the condition that x_i are properly selected. In Bellman's work [13], x_i were chosen to be the roots of the shifted Legendre polynomial of degree N, $P_N^*(x)$, with orthogonality range of [0, 1]. Note that the zeros of the shifted Legendre polynomials do not include boundary points ($x = 0$ and 1). The polynomials $P_N^*(x)$ are defined in terms of the usual Legendre polynomials by the relation

$$P_N^*(x) = P_N(1-2x) \tag{1.4}$$

The method of solving x_i will be discussed later. By analogy with Lagrange's interpolation formula, the test function is taken to be of the form

$$p_k(x) = \frac{P_N^*(x)}{(x-x_k)P_N^{*\prime}(x_k)} \tag{1.5}$$

It follows that $p_k(x)$ is a polynomial of degree $(N-1)$ such that $p_k(x_j) = \delta_{kj}$. Using the fact that Eq. (1.1) is to be exact for $f(x) = p_k(x)$, we see that

$$A_{ik}^{(1)} = \frac{P_N^{*\prime}(x_i)}{(x_i - x_k)P_N^{*\prime}(x_k)}, \quad i \neq k \tag{1.6}$$

For the case $k = i$, use of L'Hospital's rule plus the fact that $P_N^*(x)$ satisfies the differential equation

$$x(1-x^2)P_N^{*\prime\prime}(x) + (1+2x)P_N^{*\prime}(x) + N(1+N)P_N^*(x) = 0 \tag{1.7}$$

gives

$$A_{kk}^{(1)} = \frac{1 - 2x_k}{2x_k(x_k - 1)} \tag{1.8}$$

Using Eqs. (1.4), (1.6) and (1.8), the constant coefficients $A_{ij}^{(1)}$ for any value of N, say $N = 3, \ldots, 15$, can be easily calculated. These values are then used as input data for a given problem once N has been chosen.

Bert *et al.* [14] first applied the DQM to structural problems or to problems involving partial differential equations (PDE) higher than second order. He extended Eq. (1.3) to nth derivative. This leads to

$$\sum_{j=1}^{N} A_{ij}^{(n)} x_j^{k-1} = (k-1)(k-2)\cdots(k-n)x_i^{k-n-1}, \quad i, k = 1, 2, \ldots, N \tag{1.9}$$

or

$$\begin{bmatrix} 1 & 1 & 1 & \cdots & 1 \\ x_1 & x_2 & x_3 & \cdots & x_N \\ x_1^2 & x_2^2 & x_3^2 & \cdots & x_N^2 \\ \vdots & \vdots & \vdots & \ddots & \vdots \\ x_1^{N-1} & x_2^{N-1} & x_3^{N-1} & \cdots & x_N^{N-1} \end{bmatrix} \begin{Bmatrix} A_{i1}^{(n)} \\ A_{i2}^{(n)} \\ A_{i3}^{(n)} \\ \vdots \\ A_{iN}^{(n)} \end{Bmatrix} = \begin{Bmatrix} g_{i1} \\ g_{i2} \\ g_{i3} \\ \vdots \\ g_{iN} \end{Bmatrix} \tag{1.10}$$

where

$$g_{ik} = (k-1)(k-2)\cdots(k-n)x_i^{k-n-1}, \quad i = 1, 2, \ldots, N \tag{1.11}$$

The weighting coefficients $A_{ij}^{(n)}$ were solved from the N sets of N linear algebraic equations [15] given by Eq. (1.10). It is noted that the sets have a unique solution for the weighting coefficients $A_{ij}^{(n)}$, since the matrix of elements x_j^{k-1} represents a Vandermonde matrix which always has an inverse. Equation. (1.2) is correspondingly extended as

$$f^{(n)}(x_i) \cong \sum_{j=1}^{N} A_{ij}^{(n)} f(x_j), \quad i = 1, 2, \ldots, N \tag{1.12}$$

The weighting coefficients $A_{ij}^{(n)}$ are then used in Eq. (1.12) to express the derivatives of a function at a discrete point in terms of all the discrete function values. It is emphasized that the quadrature must be of higher order than the order of any partial derivative, i.e., $N > n$.

Multidimensional problems in more than one space variable can be treated in essentially the same way by using linear transformations with respect to the space variables for the partial derivatives. For example [11], the rth-order partial derivatives of $f(x, y)$ with respect to x, the sth-order partial derivatives of $f(x, y)$ with respect to y, and $(r+s)$th-order partial derivatives of $f(x, y)$ with respect to both x and y can be expressed at point (x_i, y_j) as

$$\left.\frac{\partial^r f}{\partial x^r}\right|_{ij} = \sum_{m=1}^{M} A_{im}^{(r)} f_{mj}, \quad \left.\frac{\partial^s f}{\partial y^s}\right|_{ij} = \sum_{n=1}^{N} B_{jn}^{(s)} f_{in}, \quad \left.\frac{\partial^{r+s} f}{\partial x^r \partial y^s}\right|_{ij} = \sum_{m=1}^{M} A_{im}^{(r)} \sum_{n=1}^{N} B_{jn}^{(s)} f_{mn} \tag{1.13}$$

where M and N are respectively the numbers of grid points in x-direction and y-direction (Fig. 1.1). $A_{ij}^{(r)}$ and $B_{ij}^{(s)}$ are the weighting coefficients associated with the rth-order and sth-order partial derivatives of $f(x, y)$ with respect to x and y at the discrete points x_i and y_i, respectively. Further, noting the definition of the differential

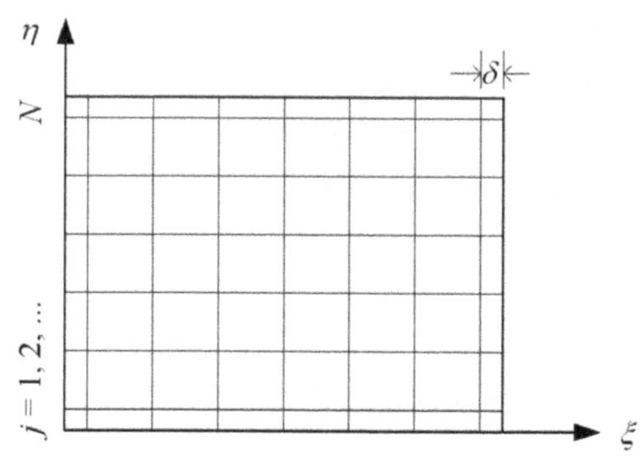

Fig. 1.1. Nodal distribution of the DQM and the δ method.

operators,

$$\frac{\partial^r f}{\partial x^r} = \frac{\partial}{\partial x}\left(\frac{\partial^{r-1} f}{\partial x^{r-1}}\right) = \frac{\partial^{r-1}}{\partial x^{r-1}}\left(\frac{\partial f}{\partial x}\right) \tag{1.14}$$

Using Eqs. (1.13) and (1.14), one may easily obtain the following recurrence relationships for the weighting coefficients:

$$\boldsymbol{A}^{(r)} = \boldsymbol{A}^{(1)}\boldsymbol{A}^{(r-1)} = \boldsymbol{A}^{(r-1)}\boldsymbol{A}^{(1)} \tag{1.15}$$

It may be seen that having the matrix $\boldsymbol{A}^{(1)}$ of first-order derivative weighting coefficients, one can obtain the weighting coefficients of the higher-order derivatives by successive multiplications of the $\boldsymbol{A}^{(1)}$ matrix by itself. Equations (1.14) and (1.15) are given for the x-partial derivatives; the equations for the y-partial derivatives follow in an identical manner.

Equation (1.13) can be reformulated in the same compact form as Eq. (1.12) by defining the following vector and matrixes [16]:

$$\bar{\mathbf{f}} = \begin{bmatrix} f_{11} & \cdots & f_{M1} & f_{12} & \cdots \\ f_{M2} & \cdots\cdots & f_{1N} & \cdots & f_{MN} \end{bmatrix}^{\mathrm{T}} \tag{1.16}$$

$$\bar{\boldsymbol{A}}^{(r)} = \begin{bmatrix} \boldsymbol{A}^{(r)} & \boldsymbol{O} & \cdots & \boldsymbol{O} \\ \boldsymbol{O} & \boldsymbol{A}^{(r)} & \cdots & \boldsymbol{O} \\ \vdots & \vdots & \ddots & \vdots \\ \boldsymbol{O} & \boldsymbol{O} & \cdots & \boldsymbol{A}^{(r)} \end{bmatrix}, \quad \bar{\boldsymbol{B}}^{(s)} = \begin{bmatrix} \boldsymbol{B}_{11}^{(s)} & \boldsymbol{B}_{12}^{(s)} & \cdots & \boldsymbol{B}_{1N}^{(s)} \\ \boldsymbol{B}_{21}^{(s)} & \boldsymbol{B}_{22}^{(s)} & \cdots & \boldsymbol{B}_{2N}^{(s)} \\ \vdots & \vdots & \ddots & \vdots \\ \boldsymbol{B}_{N1}^{(s)} & \boldsymbol{B}_{N2}^{(s)} & \cdots & \boldsymbol{B}_{NN}^{(s)} \end{bmatrix} \tag{1.17}$$

where $\bar{\boldsymbol{A}}^{(r)}$ and $\bar{\boldsymbol{B}}^{(s)}$ are $(M \times N) \times (M \times N)$ matrices, $\boldsymbol{B}_{ij}^{(s)} = \mathrm{diag}(B_{ij}^{(s)}, \ldots, B_{ij}^{(s)})_{M\times M}$, and $\boldsymbol{A}^{(r)} = (A_{ij}^{(r)})_{M\times M}$. Thus, Eq. (1.13) becomes

$$\begin{aligned} \left.\frac{\partial^r f}{\partial x^r}\right|_k &= \sum_{m=1}^{M\times N} \bar{A}_{km}^{(r)} \bar{f}_m, \quad \left.\frac{\partial^s f}{\partial y^s}\right|_k \\ &= \sum_{m=1}^{M\times N} \bar{B}_{km}^{(s)} \bar{f}_m, \quad \left.\frac{\partial^{r+s} f}{\partial x^r \partial y^s}\right|_k = \sum_{m=1}^{M\times N} \bar{F}_{km}^{(r+s)} \bar{f}_m \end{aligned} \tag{1.18}$$

where $k, m = (j-1)M + i$, $(i = 1, 2, \ldots, M; j = 1, 2, \ldots, N)$. It is noteworthy that the two matrices defined by Eq. (1.17) can also be obtained by the standard tensor product as

$$\bar{\boldsymbol{A}}^{(r)} = \boldsymbol{E} \otimes \boldsymbol{A}^{(r)} \tag{1.19}$$

$$\bar{\boldsymbol{B}}^{(s)} = \boldsymbol{B}^{(s)} \otimes \boldsymbol{E} \tag{1.20}$$

The matrix of weighting coefficients for cross-differentials can be given as

$$\bar{\boldsymbol{F}}^{(r+s)} = \bar{\boldsymbol{A}}^{(r)} \bar{\boldsymbol{B}}^{(s)} = \bar{\boldsymbol{B}}^{(s)} \bar{\boldsymbol{A}}^{(r)} = \boldsymbol{B}^{(s)} \otimes \boldsymbol{A}^{(r)} \tag{1.21}$$

where $\boldsymbol{B}^{(s)} = (\boldsymbol{B}_{ij}^{(s)})_{N \times N}$. $\boldsymbol{E}$ is an $N \times N$ unit matrix for Eq. (1.19) and an $M \times M$ unit matrix for Eq. (1.20).

Outside Bellman's group, the DQM seems to have been first employed by Mingle [17]. A very interesting idea in Mingle's work was the incorporation of the boundary conditions into the two weighting coefficient matrices of orders 1 and 2 as needed in the problem. It is noted that since the boundary conditions are applied both at the boundaries and also at very small distances δ from the boundaries (Fig. 1.1), an attempt to make polynomial approach points that are not on the polynomial will cause the polynomial to oscillate [11]. These oscillations get larger as the degree of the polynomial is increased. Thus, the cure for a poor polynomial fit seems to be to stay with lower-order polynomials.

Bert applied the DQM to spatially free vibration problems [14]: bars, beams, membranes and plates. All of the DQM results were computed with a grid of no more than eight points in one direction. In problems governed by differential equations involving up to second derivatives with respect to position, the grid points were chosen to be equally spaced. However, in problems governed by differential equations involving up to fourth derivatives with respect to position, *equal spacing of grid points would lead to appreciable error* in the application of one of the two boundary conditions at each edge unless an excessive number of grid points is used. Thus, in these problems, it is advantageous to use non-uniform grid point spacing such that, for each boundary point, there is an interior point in close proximity and lying along the normal to the boundary at the boundary point. In this way, the two respective boundary conditions can be applied on and very close to each edge. Despite the difficulties mentioned, the DQM was found to require much less computational effort than the finite-element and finite-difference methods to achieve comparable accuracy.

From then on, the DQM is wellknown in structural analyses as a method approximating the partial derivative of a function at a given discrete point as a weighted linear sum of function values at all given discrete points. Using the DQM, a partial differential equation is reduced to a set of algebraic equations. This method, as any polynomial approach, is dependent on the order of the approximating polynomial, i.e., with an increase in the order of the polynomial, the accuracy of the solution increases. It has been applied to mechanical, civil and aerospace engineering due to its simple and straightforward implementation and great versatility.

1.1.2 *Weighting coefficients and sampling points*

Two extensively decisive factors in the accuracy of the differential quadrature solutions are as follows: (1) the accuracy of the weighting coefficients and (2) the choice of sampling points [11]. In order to obtain the weighting coefficients, one may solve the Vandermonde system of equations, such as Eqs. (1.9)–(1.11), using the usual linear equation solvers. However, Vandermonde matrices are known to be inherently illconditioned and, in fact, it is experienced that *the weighting coefficients obtained by a direct solution of the Vandermonde equations become increasingly inaccurate with an increasing number of sampling points.* The weighting coefficients for the derivatives may be obtained directly, and most accurately, irrespective of the number and positions of the sampling points, from the explicit formulae [18–20]. These extremely useful formulae, taken from Shu and Richards [19, 20], are given here for the interest of the readers. These are given with respect to the x coordinate only; the formulae with respect to the y coordinate would follow in an identical manner.

Quan and Chang [18] found the general collocation method is equivalent to the method of differential quadrature. They studied the relations between Eqs. (1.9) and (1.5), and presented explicit formulae for the weighting coefficients. The weighting coefficients of the first-order derivative are given by

$$A_{ij}^{(1)} = \begin{cases} \prod_{k=1,k\neq i,j}^{N} (x_i - x_k) \Big/ \prod_{k=1,k\neq j}^{N} (x_j - x_k), & (i \neq j) \\ \sum_{k=1,k\neq i}^{N} \dfrac{1}{(x_i - x_k)}, & (i = j) \end{cases} \tag{1.22}$$

Shu and Richards [20] derived the method of computing weighting coefficients for high-order derivatives. The off-diagonal terms of a weighting coefficient matrix of the second- and higher-order derivatives may be obtained through the following recurrence relationship:

$$A_{ij}^{(n)} = n\left(A_{ii}^{(n-1)} A_{ij}^{(1)} - \frac{A_{ij}^{(n-1)}}{x_i - x_j}\right) \tag{1.23}$$

where $i, j = 1, 2, \ldots, N$ and $i \neq j$, $2 \leq n \leq N - 1$. The diagonal terms of a weighting coefficient matrix are given by

$$A_{ii}^{(n)} = -\sum_{j=1,j\neq i}^{N} A_{ij}^{(n)} \tag{1.24}$$

where $i = 1, 2, \ldots, N$ and $1 \leq n \leq N - 1$. It should be noted that having the weighting coefficients of the first-order derivative from Eq. (1.22), the weighting coefficients of higher-order derivatives may also be obtained using the recurrence relationship of Eq. (1.15). However, the number of arithmetic operations in Eqs. (1.23) and (1.24) is substantially smaller than that in Eq. (1.15). Consequently, with increasing number of sampling points, one should use Eqs. (1.23) and (1.24) rather than Eq. (1.15) to reduce computational costs and round-off errors.

A significant factor for the accuracy of the differential quadrature (DQ) solution is the choice of the sampling or grid points. A natural and often convenient choice for the grid points are the equally spaced points. However, better accuracy can be obtained in most cases by using unequally spaced points as

$$x_j = \frac{1}{2}\left(1 - \cos\frac{j-1}{N-1}\pi\right) \quad (j = 1, 2, \ldots, N) \tag{1.25}$$

They are known as the Chebyshev–Gauss–Lobatto points and well accepted in the DQM. It should be mentioned that in the quadrature solutions, the sampling points in the various coordinate directions may be different in number as well as in type. In fact, one may even use different types of test functions in the various coordinate directions.

1.1.3 *The method of imposing initial/boundary conditions*

It is well known that the implementation of the initial/boundary conditions in DQM can be very tricky [11]. Recently, some researchers have investigated the problems of imposing multiple boundary conditions [21, 22]. Here, the method in [21, 23] for imposing the initial/boundary conditions by modifying the coefficient matrixes of the DQ analog equations of the differential equations is exploited, which is capable of implementing inhomogeneous multiboundary conditions with higher-order derivatives simply. The following boundary value problem [24] is used to illustrate this method:

$$\begin{aligned} &\text{D.E.} \quad \frac{\partial^2 u}{\partial x^2} = h(x), \quad 0 \le x \le L \\ &\text{B.C.} \quad \left.\frac{\partial u}{\partial x}\right|_{x=0} = c_1, \quad u(L) = c_2 \end{aligned} \tag{1.26}$$

To replace the DQ analog equations of the governing equations at grid points on or near the boundaries by those of the boundary conditions, the DQ analog equations of Eq. (1.26) are

$$\text{D.E.}\quad \begin{bmatrix} A_{21}^{(2)} & A_{22}^{(2)} & \cdots & A_{2M}^{(2)} \\ A_{31}^{(2)} & A_{32}^{(2)} & \cdots & A_{3M}^{(2)} \\ \vdots & \vdots & \ddots & \vdots \\ A_{M-1,1}^{(2)} & A_{M-1,2}^{(2)} & \cdots & A_{M-1,M}^{(2)} \end{bmatrix} \begin{bmatrix} u_1 \\ u_2 \\ \vdots \\ u_M \end{bmatrix} = \begin{bmatrix} h(x_2) \\ h(x_3) \\ \vdots \\ h(x_{M-1}) \end{bmatrix} \tag{1.27}$$

$$\text{B.C.}\quad \begin{bmatrix} A_{11}^{(1)} & A_{12}^{(1)} & \cdots & A_{1,M-1}^{(1)} & A_{1M}^{(1)} \\ 0 & 0 & \cdots & 0 & 1 \end{bmatrix} \begin{bmatrix} u_1 \\ u_2 \\ \vdots \\ u_M \end{bmatrix} = \begin{bmatrix} c_1 \\ c_2 \end{bmatrix} \tag{1.28}$$

The DQ analog equations (1.28) of the boundary conditions can be rewritten as

$$\begin{bmatrix} u_1 \\ u_M \end{bmatrix} = \begin{bmatrix} A_{11}^{(1)} & A_{1M}^{(1)} \\ 0 & 1 \end{bmatrix}^{-1} \times \left\{ - \begin{bmatrix} A_{12}^{(1)} & \cdots & A_{1,M-1}^{(1)} \\ 0 & \cdots & 0 \end{bmatrix} \begin{bmatrix} u_2 \\ \vdots \\ u_{M-1} \end{bmatrix} + \begin{bmatrix} c_1 \\ c_2 \end{bmatrix} \right\} \tag{1.29}$$

The inhomogeneous boundary conditions can be imposed by substituting Eq. (1.29) into Eq. (1.27). The imposing of multiboundary conditions with higher-order derivatives can be done in the same way as shown in Eq. (1.29), using the DQ rules of Eq. (1.13).

1.1.4 *Strong formulation finite element method*

Strong formulation finite element method (SFEM) refers to the strong form-based differential quadrature element method (DQEM) or the strong form spectral element method (SEM) [3]. The DQM and spectral method (SM) combined with domain decomposition methods are respectively the DQEM and the SEM. Both DQM and SM use an interpolation approach in solving governing equations in their strong form. Therefore, the two methods are essentially the same.

As can be seen from the comprehensive review paper [3], the main difficulties of the SFEM include the imposition of boundary conditions and inter-element compatibility conditions, especially at corners [3]. The difficulties can be easily understood from the procedures shown in Section 1.1.3. However, boundary condition imposition and element assembly are trivial in the weak formulation. Since this work mainly focuses on the weak formulation, the main skills of the SFEM will not be discussed here.

1.1.5 *A few comments on the DQM*

Here, a few comments based on the review paper by Tornabene *et al.* [3] and comparison with the weak formulation are presented for the DQM and the SFEM.

(1) It has been recognized that the DQM has the capability to produce highly accurate solutions with minimal computational effort [11]. This fact can be understood from the application of the DQM in a wide range of initial- and/or boundary-value problems of physical and engineering sciences [3, 11].
(2) Although "some aspects are still in a developing stage (e.g., boundary conditions in a domain decomposition approach)" [3], the DQM does not need a weak formulation and can be naturally used together with the meshfree methods as can be seen from the reviews in [3, 25].
(3) To avoid ill-conditioning problem and reduce computational cost, it was recommended that "when a physical problem is decomposed into several domains, the number of collocation points should be limited" [3]. This can be taken as a skill of using the DQM and can also be interpreted as its limitations.

The DQM also has some limitations that are the same as the weak formulation. They will be discussed in Section 1.2.

1.2 Weak Form Quadrature Element Method

1.2.1 *Weak form quadrature rules*

The term "quadrature element method (QEM)" was first used by Striz *et al.* [26], and refers to the differential quadrature element method (DQEM). It currently includes both the strong formulations based on the governing as well as boundary equation and the weak formulations based on the principle of minimum potential energy [4]. "The weak form quadrature element method is more flexible than the strong form quadrature element method in dealing with complex geometry since it is essentially a higher order FEM. The method combines the generality of the finite element method with the accuracy of spectral techniques" [4]. The first work using the DQ rules in weak form was presented by Striz *et al.* [27], and this was done by direct combination of Eqs. (1.9)–(1.11) with the weak form. They formulated the weak form QEM for static analyses of plane stress problems and Kirchhoff thin plates in rectangular domains. In their works [27–29], any kind of grid points can be used as the element nodes. During formulations, however, it requires computing the inverse of a full Vandermonde matrix, and thus the number of elemental nodes should be fixed and cannot be greater than 15. In the work by Franciosi and Tomasiello in 2004 for Euler–Bernoulli structure, they used Eqs. (1.22)–(1.24) to evaluate the weighting coefficients and used non-uniformly distributed points as the element nodes. "To remove the deficiency existing in the early version of the QEM and raise the computational efficiency, Gauss–Lobatto–Legendre (GLL) points are used as the element nodes and GLL quadrature is used to obtain the element stiffness matrix and mass matrix [16, 30, 31]; thus the number of element nodes is not limited and the derivatives at nodal points can be calculated explicitly by using the available formulations in DQM. This reduces the difficulty in calculations of the derivatives for high-order elements, the round-off error, and the programming effort. To use the DQ law, however, the element nodal points should be the same as the quadrature points; thus, only GLL points can be used as the element nodes" [4].

The weak form QEM for various elasticity problems can be found in [4, 32]. For convenience of discussions in the following, the weak form QEM for Kirchhoff thin plates in irregular domains [16] are presented. First, the GLL quadrature rule defined at $[-1, 1]$ is presented as follows:

$$\int_{-1}^{1} f(x)\mathrm{d}x = \sum_{j=1}^{n} C_j f(x_j) \tag{1.30}$$

where the weights C_j of Gauss–Lobatto integration are given by

$$C_1 = C_n = \frac{2}{n(n-1)}, C_j = \frac{2}{n(n-1)[P_{n-1}(x_j)]^2} \quad (j \neq 1, n) \tag{1.31}$$

where x_j is the $(j-1)$th zero of $P'_{n-1}(x)$, and $P_n(x)$ is the Legendre polynomial of degree n. The GLL points and corresponding weights for $n < 22$ can be found in [33]. A short program calculating the GLL points and corresponding weights if $n > 21$ can be found in [4]. A detailed description of the method of solving GLL nodes with arbitrary n that reaches more than 2000 can be found in [34, 35]. For thin plate, the strain energy U and work potential W have the forms as

$$U = \frac{D}{2} \iint_S \left[\left(\frac{\partial^2 w}{\partial x^2} \right)^2 + \left(\frac{\partial^2 w}{\partial y^2} \right)^2 + 2v \frac{\partial^2 w}{\partial x^2} \frac{\partial^2 w}{\partial y^2} \right.$$
$$\left. + \, 2\,(1-v) \left(\frac{\partial^2 w}{\partial x \partial y} \right)^2 \right] \mathrm{d}x \, \mathrm{d}y \tag{1.32}$$

$$W = -\iint_S w \left(-\rho h \frac{\partial^2 w}{\partial t^2} \right) \mathrm{d}x \mathrm{d}y - \iint_S wq \mathrm{d}x \, \mathrm{d}y \tag{1.33}$$

where $D = Eh^3/12(1-v^2)$. Using Lagrange function as the bases, we have the trial function as

$$w(x, y) = \sum_{i=1}^{M} \sum_{j=1}^{N} l_i(x) l_j(y) w_{ij} \tag{1.34}$$

An additional relation as follows is needed to discretize the potential functional:

$$\mathrm{d}x \mathrm{d}y = |\boldsymbol{J}| \mathrm{d}\xi \, \mathrm{d}\eta \tag{1.35}$$

Substituting the trial function into Eqs. (1.32) and (1.33), and using the DQ rules in conjunction with the Gauss-Lobatto quadrature rule, one can have

$$U = \frac{D}{2} \sum_{k=1}^{K} C_k J_k \left\{ [\tilde{\boldsymbol{A}}_k^{(2)} \boldsymbol{w}]^2 + [\tilde{\boldsymbol{B}}_k^{(2)} \boldsymbol{w}]^2 + 2v [\tilde{\boldsymbol{A}}_k^{(2)} \boldsymbol{w}][\tilde{\boldsymbol{B}}_k^{(2)} \boldsymbol{w}] \right.$$
$$\left. + \, 2(1-v) [\tilde{\boldsymbol{F}}_k^{(1+1)} \boldsymbol{w}]^2 \right\} \tag{1.36}$$

$$W = \sum_{k=1}^{K} C_k J_k \left[w_k \left(\rho h \frac{\partial^2 w_k}{\partial t^2} \right) - w_k q_k \right] \tag{1.37}$$

where $K = M \times N$, $C_k = C_i^x C_j^y$, $J_k = |\boldsymbol{J}|_{ij}$, $\tilde{\boldsymbol{A}}_k^{(2)}$, $\tilde{\boldsymbol{B}}_k^{(2)}$ and $\tilde{\boldsymbol{F}}_k^{(1+1)}$ are respectively the kth rows of $\tilde{\boldsymbol{A}}^{(2)} = \tilde{\boldsymbol{A}}^{(1)}\tilde{\boldsymbol{A}}^{(1)}$, $\tilde{\boldsymbol{B}}^{(2)} = \tilde{\boldsymbol{B}}^{(1)}\tilde{\boldsymbol{B}}^{(1)}$ and $\tilde{\boldsymbol{F}}^{(1+1)} = \tilde{\boldsymbol{A}}^{(1)}\tilde{\boldsymbol{B}}^{(1)}$. $\boldsymbol{w} = (w_m)_{K\times 1}$ is a column vector, where $m = (j. - 1)M + i, i = 1, \ldots, M, j = 1, \ldots, N$. For brevity, let $\boldsymbol{A} = \tilde{\boldsymbol{A}}^{(2)}$, $\boldsymbol{B} = \tilde{\boldsymbol{B}}^{(2)}$, $\boldsymbol{F} = \tilde{\boldsymbol{F}}^{(1+1)}$, $\boldsymbol{C} = \mathrm{diag}[C_k J_k]$, then Eqs. (1.36) and (1.37) can be rewritten as

$$U = \frac{1}{2}\boldsymbol{w}^{\mathrm{T}}\boldsymbol{K}\boldsymbol{w}, \quad W = \boldsymbol{w}^{\mathrm{T}}\boldsymbol{M}\ddot{\boldsymbol{w}} + \boldsymbol{w}^{\mathrm{T}}\boldsymbol{R} \tag{1.38}$$

where

$$\begin{aligned} \boldsymbol{K} &= D[\boldsymbol{A}^{\mathrm{T}}\boldsymbol{C}\boldsymbol{A} + \boldsymbol{B}^{\mathrm{T}}\boldsymbol{C}\boldsymbol{B} + 2v\boldsymbol{A}^{\mathrm{T}}\boldsymbol{C}\boldsymbol{B} + 2(1-v)\boldsymbol{F}^{\mathrm{T}}\boldsymbol{C}\boldsymbol{F}] \\ \boldsymbol{M} &= \rho h \boldsymbol{C}, \quad \boldsymbol{R} = \boldsymbol{C}\boldsymbol{q} \end{aligned} \tag{1.39}$$

are respectively the stiffness matrix, the mass matrix and the force vector. For convenience of boundary condition imposition and element assembly, the elements matrices need to be modified [16]. Using mixed Lagrange–Hermite-type Kirchhoff plate elements [4, 31, 36], modification of the boundary degrees of freedoms (DOFs) is still needed for element assembly but can be simplified and the transformation matrix can be better conditioned.

In the literatures, the Lobatto-type QEM is often regarded as the time-domain SEM or SFEM [4]. The SEM is popular for simulating wave propagations and is an established technique in the wave and fluid fields. The main differences between weak form QEM and SEM seem to be that QEM uses DQ rules and can use various types of nodes and bases [4]. Therefore, the weak form QEM was also named as a DQ finite element method (DQFEM) [16, 32]. Since the SEM is a special case of the QEM [37], the quadrature element method also possesses the high accuracy of spectral methods with the flexibility of finite element method. For more details about the similarities and differences between QEM and SEM, see Ref. [4].

1.2.2 *A few comments on the weak form QEM*

Here, a few comments based on the review paper by Wang *et al.* [4] and comparison with the strong formulation are presented for the weak form QEM.

(1) The weak form QEM has the capability to produce highly accurate solutions with minimal computational effort as the DQM. It also combines the generality of the finite element method and is more flexible than the DQM in dealing with complex geometry, boundary conditions in a domain decomposition approach.
(2) The weak form QEM alleviates the ill-conditioning problem in DQM since the order of differentiations in a weak form is half of that in the corresponding strong form. Therefore, the polynomial order of weak form QEM elements can be as high as needed [30, 34].
(3) In order to analyze large-scale problems in structural mechanics and doing multiscale analysis [4], "various transition elements should be proposed. These problems offer challenges to future researchers of the weak form quadrature element method." This is also true for the DQM.
(4) Another limitation of the weak or strong form QEM is that the number of nodes on two opposite edges of a QEM element should be the same. Thus, order elevation ($p-$refinement) will propagate from patch to patch.

1.3 The Hierarchical Finite Element Method

1.3.1 *The HFEM on 1D and its tensor product domains*

Hierarchical functions were introduced simply in a p-type context for three-dimensional stress analysis by Zienkiewicz *et al.* [5] as early as 1971. The number of elements and of the degrees of freedom in three-dimensional elastic stress analysis as compared with two-dimensional situations increases rapidly (assuming that similar subdivisions are needed for accuracy) and the bandwidth of assembled equations puts excessive demands on computer storage. To allow practical engineering solutions to be achieved within a reasonable computer cost,

Zienkiewicz *et al.* put much effort into the subject. At that time, the objective of the hierarchical functions was simply the introduction of p-graded (degrees of freedom) meshes in an *a priori* chosen manner when advancing to a region of high stress concentration, then some of the total degrees of freedom (and therefore the cost) can be saved. Such elements were called mixed, super-parametric, "hierarchical" elements [5]. In the context of such elements, it was believed to be more convenient to assign *edge variables* rather than create actual nodes along any specific edge.

Consider the variation of the unknown (displacement), u, along an edge placed in ξ direction, where $\xi \in [-1, 1]$. Let this be given by a linear summation of interpolation basis on the two ends and the hierarchical bases inside the domain as

$$u(\xi) = \frac{1-\xi}{2}u(\xi_1) + \frac{1+\xi}{2}u(\xi_2) + \sum_{n=1}^{N} \varphi_{n+2}(\xi)u_{n+2} \tag{1.40}$$

where $\varphi_1 = (1-\xi)/2$, $\varphi_2 = (1+\xi)/2$, $\xi_1 = -1$ and $\xi_2 = 1$, N is the number of hierarchical basis in the ξ-direction, a_n, $n = 1, 2, \ldots, N$ are coefficients of the hierarchical basis, and $\varphi_{n+2}(\xi)$ are the integrated Legendre polynomials as follows:

$$\varphi_{n+2}(\xi) = \int_{-1}^{\xi} P_n(\xi)d\xi = \frac{1}{2^n n!}\frac{\mathrm{d}^n}{\mathrm{d}\xi^n}(\xi^2-1)^{n+1}, \quad \xi \in [-1, 1] \tag{1.41}$$

where $n = 1, 2, \ldots, P_n(\xi)$ is the Legendre polynomial of degree n. The first few bases are

$$\begin{aligned} &\varphi_3 = \xi^2 - 1, \varphi_4 = 2\xi(\xi^2 - 1) \\ &\varphi_5 = (15\xi^4 - 18\xi^2 + 3)/4 \\ &\varphi_6 = 7\xi^5 - 10\xi^2 + 3\zeta \end{aligned} \tag{1.42}$$

One can see from Eq. (1.41) that the hierarchical shape functions have zero function values at the two endpoints. They present simply a new way of writing a complete polynomial in ξ of appropriate order, they have no influence on inter-element compatibility. The interpolation basis and the hierarchical basis are respectively shown in Figs. 1.2(a) and (b).

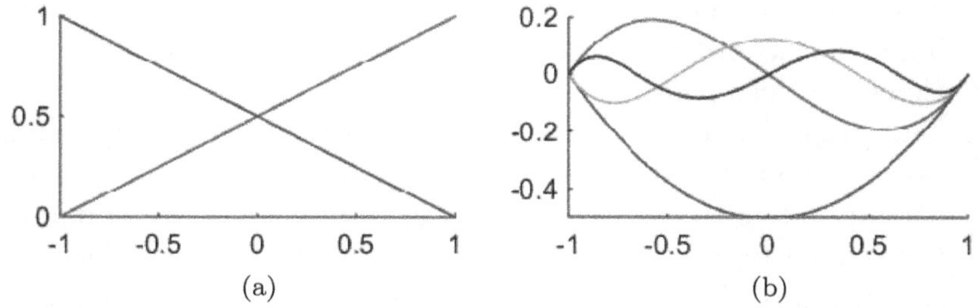

Fig. 1.2. 1D bases of the HFEM: (a) interpolation basis at the two ends, (b) hierarchical basis inside the domain.

Proceeding now to the actual shape functions of a 3D element, we define the response to parameter u_m as N_m in the usual manner. This will now be written as

$$N_m = \varphi_i(\xi)\varphi_j(\eta)\varphi_k(\zeta) \tag{1.43}$$

with i, j and k being integers suitably chosen. For 3D analyses, if two adjacent elements share an edge, then, provided these higher parameters are shared by both elements, continuity will ensue. The only difficulty is that along a common edge, the variables in both elements must be measured in the same direction.

One can see from Eq. (1.42) that the hierarchical bases can be odd and even. Thus, before assembly of an element, one needs to examine the sense of its global nodal numbers and the sense of its local nodal numbers and make the necessary alternation of the signs of some components of the element stiffness matrix [38].

A few years after the pioneering work of HFEM by Zienkiewicz *et al.* [5], new and useful families of hierarchical p-type elements were introduced by Peano [39], who presented hierarchies of C^0 and C^1 interpolations over triangles. The triangular elements will be discussed in the following section. Here, a summary of the main features of their finite element program conducted at Washington University before that time is given as follows. These features are typical for HFEM and have been accepted by many researchers [8]:

(1) It is based on assumed displacements.
(2) The assumed displacements are approximated by complete polynomial sequences of arbitrary order.

(3) The inter-element continuity conditions are satisfied exactly, however the order of polynomial interpolation may vary from element to element.
(4) The shape functions corresponding to a polynomial approximation of order p constitute a subset of the set of shape functions corresponding to the $(p+1)$th-order polynomial approximation. Consequently, the stiffness matrix of the element of order p is a submatrix of the stiffness matrix of the $(p+1)$-order element.
(5) Since the order of the approximating polynomials can be increased in just a few elements where it is needed merely by bordering the global stiffness matrix by the rows and columns corresponding to the new variables, improved solutions may be computed by very efficient numerical techniques.
(6) The capability of including, at a low cost, additional degrees of freedom where needed in order to improve the accuracy is likely to give automated control on the discretization error.

The distinguishing feature of the proposed finite element computer program is that it permits the user to exercise control over both the number of finite elements and the order of approximation over each element. Consequently, it will not be necessary to define more finite elements than needed for describing the geometry of the structure. This will lead to substantial savings in mesh generation, particularly around stress concentrations where automated mesh grading is difficult to obtain, and in three-dimensional problems where perspective views of the mesh are difficult to scrutinize.

Szabó and Mehta [40] carried out p-convergent finite element approximations study in fracture mechanics. The numerical results presented there clearly indicated that p-convergent approximations require substantially fewer degrees of freedom than h-convergent approximations to achieve similar levels of precision.

After the full advantages of HFEM have been realized, Zienkiewicz *et al.* [41] summarized the HFEM concept. Its advantages include the following in addition:

(1) improved conditioning;
(2) ease of introducing error indicators if successive refinement is sought.

Further, it is possible to use the ideas to construct a range of error estimators which compare well with alternatives and are ideally suited for adaptive refinements of analysis.

As the hierarchical elements are equally simple to implement as "standard" fixed-order elements, it is assumed that more programs will, in the future, turn to utilize their advantages. This is especially true in the field of nonlinear analysis where even today computational economies are necessary.

Peano *et al.* [42] further modified the bases shown in Eq. (1.41) for convenience of element assembly. The alternative convenient form for the hierarchical functions is defined by

$$\varphi_{n+1}(\xi) = \begin{cases} \frac{1}{n!}(\xi^n - 1) & n \quad \text{even} \\ \frac{1}{n!}(\xi^n - \xi) & n \quad \text{odd} \end{cases} \tag{1.44}$$

where $n \geq 2$. This yields the set of shape functions

$$\varphi_3 = \frac{1}{2}(\xi^2 - 1), \quad \varphi_4 = \frac{1}{6}(\xi^3 - \xi)$$
$$\varphi_5 = \frac{1}{24}(\xi^4 - 1), \quad \varphi_6 = \frac{1}{120}(\xi^5 - \xi), \quad \text{etc.} \tag{1.45}$$

One can see that the first two bases in Eq. (1.45) are similar to those of Eq. (1.42). We observe that all derivatives of φ_{n+1} of second or higher order have the value zero at $\xi = 0$, apart from $\mathrm{d}^n N_{n+1}/\mathrm{d}\xi^n$, which equals unity at that point, and hence, when shape functions of the form given by Eq. (1.45) are used, we can identify the parameters in the approximation as

$$u_{n+1} = \left.\frac{\mathrm{d}^n u}{\mathrm{d}\xi^n}\right|_{\xi=0} \quad n \geq 2 \tag{1.46}$$

This identification gives a general physical significance but is by no means necessary, so Eq. (1.44) was adopted in Zienkiewicz's book *The Finite Element Method* [43].

In practice, three-dimensional HFEM elements of order higher than 3 were seldom used [38]. In the summarized paper on HFEM by Zienkiewicz *et al.* [41], they inserted a word of caution on the use of very high-order polynomials which may lead to local oscillation. For this reason, they suggest to limit in general $p \leq 4$ and would recommend a start from a reasonable mesh not calling for a higher refinement. In future automatic serf adaptive programs, it may well be necessary to introduce both h and p refinements simultaneously, but computation problems may become over complex.

Zhu [44] suggested improved shape functions for one-dimensional hierarchal finite element, such that the condition numbers of the corresponding stiffness matrices for bars can greatly be reduced as compared with the traditional shape functions. Therefore, it is beneficial to avoid numerical trouble during calculation when high-order elements are used. This was done by defining polynomials of which the zeros are located at

$$\xi = \pm 1, 0, \pm\frac{1}{2}, \pm\frac{1}{4}, \pm\frac{3}{4}, \cdots \tag{1.47}$$

namely,

$$\begin{aligned} \varphi_3 &= \xi^2 - 1, \varphi_5 = (4\xi^2 - 1)\varphi_3 \\ \varphi_7 &= (16\xi^2 - 1)\varphi_5, \varphi_9 = (16\xi^2 - 9)\varphi_7 \\ &\cdots \\ \varphi_{2n}(\xi) &= \xi\varphi_{2n-1}(\xi) \end{aligned} \tag{1.48}$$

In order to examine their applicability from the viewpoint of numerical computation, the eigenvalues and condition numbers of the corresponding stiffness matrices up to the ninth-order polynomials of bars are calculated. The condition numbers of Eq. (1.48) were shown to be much smaller than those of Eq. (1.45).

Though a number of papers as cited above have shown that the p-convergence is superior to the h-convergence in many cases, the inevitable numerical trouble encountered during the use of high-order hierarchal finite elements must be overcome in order to bring this method into full play. To do so, Zhu [45] developed unified and explicit formulas for evaluating the relevant element matrices, such that the numerical error during computation to the lowest level can be reduced. However, the method is useful for rectangular or brick elements. It cannot deal with the numerical trouble over arbitrary quadrilateral or hexahedral domains with high performance.

In the excellent work by Zhu [45], he also presented three kinds of hierarchal shape functions, based on the polynomials of progressively ascending power in explicit and the neatest form, were presented and their merits and demerits were compared with one another:

(1) The Rodrigues formula of Jacobi orthogonal polynomials $J_n(x)$ which satisfy the following orthogonality relation

$$\int_{-1}^{1} g(\xi)J_m(\xi)J_n(\xi)\mathrm{d}\xi = 0, \quad m \neq n \tag{1.49}$$

is

$$J_n(\xi) = \frac{1}{(-2)^n n!}\frac{1}{g(\xi)}\frac{\mathrm{d}^n}{\mathrm{d}\xi^n}[g(\xi)(1-\xi^2)^n] \tag{1.50}$$

where $g(\xi) = (1-\xi)^r(1+\xi)^s$. Obviously, $J_n(\xi)$ degenerate into Legendre orthogonal polynomials $P_n(\xi)$ when $r = s = 0$. By expanding Eq. (1.50) in accordance with the binomial theorem and then by induction, the most convenient expression for Legendre orthogonal polynomials can be written as

$$P_n(\xi) = \sum_{k=0}^{[n/2]} \frac{(-1)^k(2n-2k)!}{2^n k!(n-k)!(n-2k)!}\xi^{n-2k} \tag{1.51}$$

where $[n/2]$ denotes its own integer part only. Let $P_n^s(\xi)$ denotes the following s-multiple integral of $P_{n_{-s}}(\xi)$:

$$P_n^s(\xi) = \int_{-1}^{\xi}\cdots\int_{-1}^{\xi} P_{n-s}(\xi)d\xi\cdots d\xi \tag{1.52}$$

with

$$\begin{aligned} P_n^1(\xi) &= \int_{-1}^{\xi} P_n(\xi)d\xi = \frac{(\xi^2-1)}{n(n+1)}\frac{dP_n(\xi)}{d\xi} \\ P_n^2(\xi) &= \int_{-1}^{\xi}\int_{-1}^{\xi} P_{n+1}(\xi)d\xi = \frac{(\xi^2-1)^2}{n(n+1)(n+2)(n+3)}\frac{d^2P_{n+1}(\xi)}{d\xi^2} \end{aligned} \tag{1.53}$$

It is noted that $P_n^s(\xi)$ can be used as hierarchal shape functions of $C_{s_{-l}}$ continuity since all their derivatives of order lower than s vanish at $\xi = \pm l$. Furthermore, due to the orthogonality relation Eq. (1.49), $P_n^s(\xi)$ which make the relevant stiffness submatrix diagonal for elements with uniform rigidity distribution may be called $C_{s_{-l}}K$-orthogonal hierarchal shape functions.

(2) Let $Q_n^t(\xi)$ denote $J_n(\xi)$ when $r = s = 2t$. Thus, the $C_{t_{-l}}$ hierarchal shape functions which satisfy the $C_{t_{-l}}$ continuity requirements at $\xi = \pm l$ can be expressed as $(l - \xi^2)^t Q_{n-2t}^t(\xi)$. Due to the orthogonality relation Eq. (1.49), such functions may be called M-orthogonal hierarchal shape functions because the relevant mass submatrix for elements with uniform mass distribution is diagonal.
(3) The simplest C_0 hierarchal shape functions can be defined as Eq. (1.44) or

$$\varphi_{n+2}(\xi) = (1 - \xi^2)\xi^{n-1}, n \geq 1 \tag{1.54}$$

The C_1 hierarchal shape functions can be expressed as the products of $(1 - \xi^2)$ and $\varphi_{n+2}(\xi)$.
(4) The vibration eigenmodes of the element with fixed boundary conditions can be used as the biorthogonal hierarchal shape functions because both the corresponding mass and stiffness submatrices are diagonal. Actually, such method is just the substructure technique in the finite element method. This method needs to compute vibration mode shapes in advance, and therefore it is convenient only for some special cases.

Generally, the mass, rigidity, tension and load distributions of commonly used elements can all be expanded into power series. The merits and demerits of the hierarchal shape functions mentioned above can be evaluated and listed in Table 1.1. It can be seen that the K-orthogonal hierarchal shape functions (type 1) have the best synthetic properties.

Table 1.1. The synthetic comparison of four kinds of 1D hierarchal shape functions.

Type		1	2	3	4
Condition of submatrix	K	very well	well	ill	very well
	G	well	well	ill	well
	M	well	very well	ill	very well
Semi-bandwidth of submatrix	K	diagonal	full	full	diagonal
	G	2 for C_0, 3 for C_1	full	full	full
	M	3 for C_0, 5 for C_1	diagonal	full	diagonal

After the work by Zhu [45], the K-orthogonal hierarchal shape functions have been used in many works as can be seen from the recent books [6–9]. The p-version finite element method has also been added to a number of well-known commercial codes [46]. Besides polynomials, trigonometric functions can also be used as hierarchal shape functions [8, 47]. Advantages of the trigonometric functions are that they are numerically more stable than polynomials and so higher-order functions can be used. However, trigonometric functions also have the limitation that they need more integration nodes than polynomials.

As discussed earlier, the p-method approximates the unknown variation using high-order trial solutions. The number of elements required when using the p-version is noticeably less than these when using the h-version because these high-order approximations can model more complex variations. Thus, p-elements are larger than h-elements, and consequently, geometry mapping is more demanding for p-elements. The standard h-version approach of isoparametric mapping does not suffice. Thus, Campion [46] proposed to use the blending function method to model curved geometries accurately in the p-version finite element method. For an arbitrary quadrilateral domain, the geometric mapping has the form

$$\begin{aligned} V(\xi,\eta) &= \frac{1-\xi}{2}M(-1,\eta) + \frac{1+\xi}{2}M(1,\eta) + \frac{1-\eta}{2}M(\xi,-1) \\ &\quad + \frac{1+\eta}{2}M(\xi,1) - \frac{(1-\xi)(1-\eta)}{4}M(-1,-1) \\ &\quad - \frac{(1-\xi)(1+\eta)}{4}M(-1,1) - \frac{(1+\xi)(1-\eta)}{4}M(1,-1) \\ &\quad - \frac{(1+\xi)(1+\eta)}{4}M(1,1) \end{aligned} \tag{1.55}$$

where $-1 \leq \xi,\ \eta \leq 1$. The functions $M(\xi_i,\ \eta)$ and $M(\xi,\ \eta_j)$ represent the four parametric curves that constitute the boundary of the quadrilateral domain. The function $M(\xi_i,\ \eta_j)$ represents the x–y coordinates of the points corresponding to the non-dimensional coordinates $(\xi_i,\ \eta_j)$. The blending function method can provide exact geometric mapping. Therefore, it is more accurate than the serendipity-family interpolation functions.

1.3.2 *The HQEM on quadrilateral and hexahedral domains*

Literatures used p-version meshes in HFEM enriching the auxiliary degrees of freedom both inside and on the boundary of an HFEM element by orthogonal polynomials [46, 48–50]. However, the HFEM in this case is much like the meshfree method since most of the freedoms do not have clear physical meaning, and boundary conditions imposition and element assembly in HFEM become difficult [38]. Computed quantities defined on boundaries are usually the most important ones in engineering applications [51, 52], and boundary conditions are usually imposed on the boundary of elements. Therefore, interpolation shape functions on the boundary of HFEM elements are necessary. Liu *et al.* [35] proposed a hierarchical quadrature element method (HQEM) by constructing nodal interpolation basis functions on the boundary of HFEM elements using blending function and Lagrange functions, which made the imposing of boundary conditions and the assembling of elements more convenient than before. The HQEM in 1D case is the same as the HFEM. The HQEM in 2D and 3D cases is equivalent to the fixed interface mode synthesis method to some extent.

The method of constructing boundary bases for C^0 quadrilateral elements in the HQEM [35] is briefly introduced here. Assume ξ_i ($i = 1, 2, \ldots, M$) and $\eta_j (j = 1, 2, \ldots, N)$ are the sampling points along the edges of a square domain in the natural $\xi - \eta$ plane. The preferable nodes are non-uniformly distributed nodes such as Legendre–Gauss–Lobatto nodes, which lead to much lower condition numbers than uniformly distributed nodes in the p-method [53]. Using the blending function as Eq. (1.55), the nodal shape function at $\xi_1 = \eta_1 = -1$ can be constructed from Lagrange functions as

$$S_1(\xi, \eta) = \frac{1-\eta}{2} L_1^M(\xi) + \frac{1-\xi}{2} L_1^N(\eta) - \frac{(1-\xi)(1-\eta)}{4} \tag{1.56}$$

The nodal shape functions on edge $\eta_1 = -1$ are

$$S_i(\xi, \eta) = \frac{1-\eta}{2} L_i^M(\xi), \quad i = 2, \ldots, M-1 \tag{1.57}$$

where

$$L_i^M(\xi) = \prod_{j=1, j\neq i}^{M} \frac{\xi - \xi_j}{\xi_i - \xi_j} \tag{1.58}$$

is the Lagrange function. The serendipity shape functions at other corners and edges are of the same form as Eqs. (1.56) and (1.57). The subscripts $k = 1, 2, \ldots, K$ of shape functions are numbered along the edge of a domain. The shape functions on the boundary of elements constructed herein make the imposition of boundary conditions of HFEM more convenient than before [46, 49]. Moreover, besides two neighboring edges, the number of nodes on the two opposite edges of an element can also be different.

As can be seen from the review above, the hierarchical shape functions constructed from Taylor basis functions were used in HFEM [5, 44] in the beginning. The corresponding HFEM matrices may become ill conditioned even for the seventh-order hierarchical shape functions [44, 45]. The hierarchical shape functions derived from Legendre polynomials [45] can guarantee monotonous convergence of HFEM results to exact solutions. The resulting HFEM matrices have narrow bandwidth, good numerical stability and high computational efficiency. The HFEM matrices using orthogonal polynomials are theoretically well conditioned. However, HFEM matrices computed from high-order orthogonal polynomials through numerical integration have non-ignorable computational error. Explicit integration [45, 54] was used to solve this problem in regular domains, meanwhile the computational cost can be saved. Numerical integration and differentiation have to be used in irregular domains. The interior bases of the HQEM still used tensor product of integrated Legendre polynomials. However, instead of using the Tylor series expression as Eq. (1.51), it was found that the following recursion formula should be used:

$$P_{n+1}(\xi) = \frac{2n+1}{n+1}\xi P_n(\xi) - \frac{n}{n+1}P_{n-1}(\xi) \tag{1.59}$$

where $P_0(\xi) = 1, P_1(\xi) = \xi$. The mth-order derivatives of Legendre polynomials can be obtained from the following recursion formula:

$$P_{n+1}^{(m)}(\xi) = \xi P_n^{(m)}(\xi) + (n+m)P_n^{(m-1)}(\xi) \tag{1.60}$$

It was shown that by using the recursion formulas above, the order n can reach several thousand without numerical problem. It was found that Szabó and Babuška [55] have used these properties in HFEM very early. However, many researchers still used the Taylor series expression of Legendre polynomials and the maximum order can only reach 45th order even if symbolic computation is used

[8, 47, 49, 54, 56–60]. Therefore, the application of HFEM needs to take account of many factors and is not as simple and straightforward as h-version FEM.

It was summarized in [35] that, in order to achieve p-adaptive analysis with high accuracy, the following points should be noted:

(1) If sampling points are needed, they should be the roots of the orthogonal polynomials that satisfy the boundary conditions. Lagrange polynomials using GLL points provide the best accuracy for C^0 problem because the jth Lagrange polynomial has maximum value at jth GLL point and the maximum value is 1.
(2) If high-order orthogonal polynomials are needed, they should be computed from the recursion formula that has the minimum round-off errors. Many publications still use symbolic computation to overcome the computational stability problem of HFEM just because they were not aware of this fact.
(3) The derivatives in HFEM should be computed analytically whenever possible to save the computational cost. For a 2D or 3D problem, the separation variable feature of the bases should be used so that 2D or 3D differentiation can be computed from 1D cases and computational cost can be saved greatly.

The three points can be found in the literatures, but they were usually discussed separately in different papers. Successful application of the HQEM needs to consider the three points simultaneously.

1.3.3 *The HFEM on simplex domains*

In the pioneering works on HFEM by Zienkiewicz *et al.* [5, 61], triangular and tetrahedral elements are respectively degenerated from quadrilateral (2D) and brick (3D) elements. In general [62], a corrective factor must be introduced to modify the shape functions in order to get proper quadratic response of the elements. The problem of what are the required modifications for higher-order responses in the degenerated elements has not been investigated. Peano [39] first formulated hierarchies of C^0 interpolations over triangles using "area" coordinate shape functions corresponding to polynomials of arbitrary degree. Tong [38] used a hierarchical approach to construct the simplex elements which permits easy merging of elements of different orders. The explicit interpolation functions and the algorithm

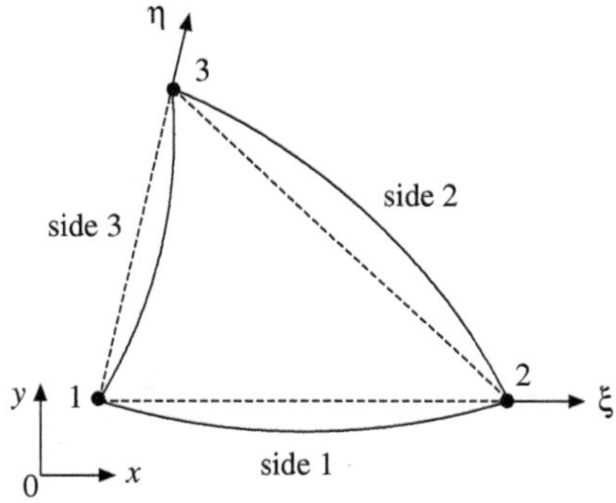

Fig. 1.3. A curved triangular element.

of using such elements are presented. In terms of the area coordinates, the displacement field u of an HFEM element on triangles was given as

$$u[x(\xi,\eta),y(\xi,\eta)] = \sum_{v=1}^{3} V^v(\xi,\eta)u_v + \sum_{e=1}^{3}\sum_{i=0}^{n_e-1} E_i^e(\xi,\eta)a_i^e + \sum_{i=0}^{n}\sum_{j=0}^{n-i} F_{ij}(\xi,\eta)c_{ij} \tag{1.61}$$

where $v = 1,2,3$ denotes the vertexes, and $e = 1,2,3$ denotes the edges (Fig. 1.3), n_e ($e = 1,2,3$) denotes the number of functions on each edge (Fig. 1.4) and n denotes the order of bases on one direction of the triangular face (Fig. 1.4). The coefficient a_i^e denotes the hierarchical bases on the edges of the element and c_{ij} denotes the hierarchical bases on the face of the triangle. These coefficients do not have a clear physical meaning. The bases V^v defined on the three vertexes of the triangle can be given as

$$V^1 = 1-\xi-\eta, V^2 = \xi, V^3 = \eta \tag{1.62}$$

The bases E_i^e and F_{ij} are respectively defined on the three edges and on the face of the triangle. In Tong's work, the edge function for edge 1, for example, is taken as

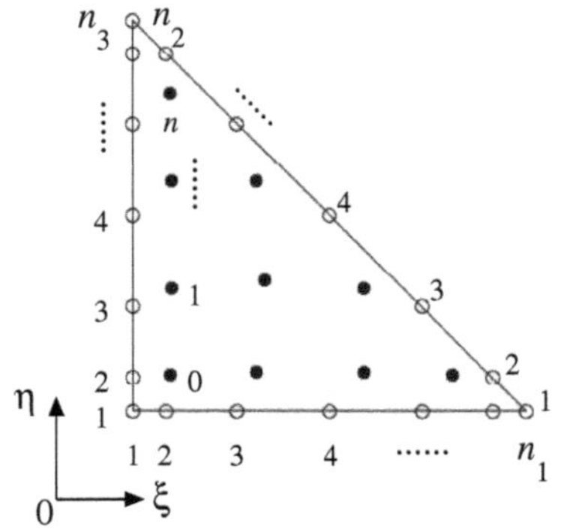

Fig. 1.4. Nodes on a unit triangular domain.

$$E_i^1(\xi,\ \eta) = V^1V^2\left(\xi - \frac{1}{2}\right)^i \tag{1.63}$$

The face function is taken as

$$F_{pi}(\xi,\eta) = V^1V^2V^3\xi^p\eta^i \tag{1.64}$$

The displacement field u of a tetrahedral element was given as

$$\begin{aligned} &u[x(\xi,\eta,\zeta),y(\xi,\eta,\zeta),z(\xi,\eta,\zeta)] \\ &= \sum_{v=1}^{4} V^v(\xi,\eta,\zeta)u_v + \sum_{e=1}^{6}\sum_{i=0}^{n_e-1} E_i^e(\xi,\ \eta,\zeta)a_i^e \\ &\quad + \sum_{f=1}^{4}\sum_{i=0}^{n_f}\sum_{j=0}^{n_f-i} F_{ij}^f(\xi,\eta,\zeta)c_{ij}^f + \sum_{i=0}^{m}\sum_{j=0}^{m-i}\sum_{k=0}^{m-i-j} B_{ijk}(\xi,\eta,\zeta)d_{ijk} \end{aligned} \tag{1.65}$$

where $v = 1, 2, 3, 4$ denotes the vertexes, $e = 1, 2, \ldots, 6$ denotes the edges (Fig. 1.5), n_e ($e = 1, 2, \ldots, 6$) denotes the number of functions on each edge, n_f denotes the number of bases on one direction of a triangular face of the tetrahedron, and m denotes the number of bases on one direction of tetrahedral body. The coefficients a_i^e refer to the hierarchical bases on the edges of the element, the coefficients c_{ij}^f refer to the hierarchical bases on each face of the tetrahedron, and the coefficients d_{ijk} refer to the hierarchical bases inside the body of the

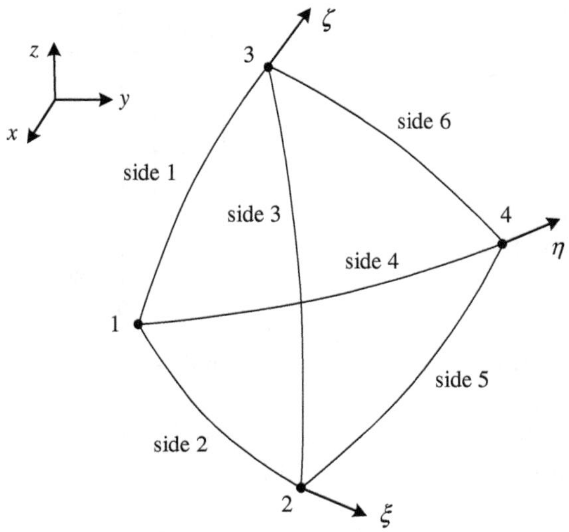

Fig. 1.5. A curved tetrahedral element.

tetrahedron. The bases V^v defined on the vertexes of the tetrahedral element can be given as

$$V^1 = 1 - \xi - \eta - \zeta, \quad V^2 = \xi, \quad V^3 = \zeta, \quad V^4 = \eta \tag{1.66}$$

The bases E_i^e, F_{ij}^f and B_{ijk} are respectively defined on the six edges, four faces and inside the body of the tetrahedral element. In Tong's work [38], these bases were defined in the same way as above for the triangular element. Tong hoped that, with the hierarchical shape functions explicitly constructed for the simplex elements, one may investigate Irons' problem [62] of what is required to modify the higher-order shape functions of a more complex (e.g., a brick-type) element when it is degenerated into a simplex element.

Dubiner [63] showed how to obtain multidimensional spectral methods as a warped product of one-dimensional spectral methods, thus generalizing direct (tensor) products. Applications were given for spectral approximation on a disk and on a triangle. A rectangular domain can be mapped into a triangular domain (Fig. 1.6) by Duffy's transformation [64] as

$$\xi = \frac{(1+s)(1-t)}{4}, \quad \eta = \frac{(1+s)(1+t)}{4} \tag{1.67}$$

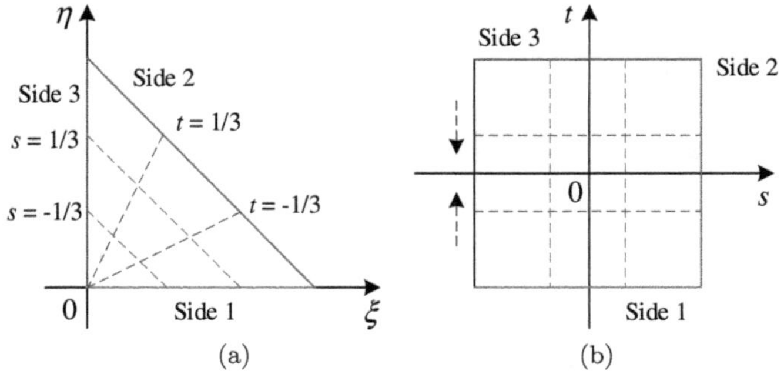

Fig. 1.6. A unit triangle and a unit square domain.

The triangular domain can be similarly mapped into the rectangular domain by the inverse transformation

$$s = 2(\xi + \eta) - 1, \quad t = \frac{\eta - \xi}{\eta + \xi} \tag{1.68}$$

Dubiner [63] proposed two types of polynomial expansion bases for use on triangular domains. The first expansion basis is orthogonal in the Legendre inner product, namely,

$$g_{ij}(\xi, \eta) = P_i[2(\xi + \eta) - 1]P_j\left(\frac{\eta - \xi}{\eta + \xi}\right) \tag{1.69}$$

The expansion is truly independent and a direct product. But it is not smooth at the corner $(\xi, \eta) = (0, 0)$ and cannot easily be extended to form a C^0 continuous basis from the union of multiple triangular domains or elements. The second expansion basis can be extended to form a C^0 continuous basis from such a union. Let us denote $P_n^{(\alpha,\beta)}(x)$ as the nth-order Jacobi polynomial in the $[-1, l]$ interval with the orthogonality relationship,

$$\int_{-1}^{1} (1-x)^\alpha (1+x)^\beta P_m^{(\alpha,\beta)}(x) P_n^{(\alpha,\beta)}(x) \mathrm{d}x = \delta_n^m C_n^{(\alpha,\beta)} \tag{1.70}$$

where δ_n^m is the Kronecker delta and

$$C_n^{(\alpha,\beta)} = \frac{2^{\alpha+\beta+1}}{2n + \alpha + \beta + 1} \frac{\Gamma(n + \alpha + 1)\Gamma(n + \beta + 1)}{n!\Gamma(n + \alpha + \beta + 1)} \tag{1.71}$$

then Dubiner's orthogonal expansion bases satisfy the following orthogonal condition:

$$\int_S g_{ij}g_{pq}\mathrm{d}S = \int_0^1\int_0^{1-\xi} g_{ij}g_{pq}\mathrm{d}\xi\mathrm{d}\eta = \delta_i^p\delta_j^q C_{ij} \tag{1.72}$$

where C_{ij} is a constant. The determinate of the Jacobian of coordinates $(\xi,\ \eta)$ with respect to $(s,\ t)$ is

$$|\boldsymbol{J}| = \left|\frac{\partial(\xi,\ \eta)}{\partial(s,\ t)}\right| = \begin{vmatrix} \frac{1-t}{4} & -\frac{1+s}{4} \\ \frac{1+t}{4} & \frac{1+s}{4} \end{vmatrix} = \frac{1+s}{2^3} \tag{1.73}$$

Thus, the substitution of Eq. (1.67) into Eq. (1.72) leads to

$$\int_S g_{ij}g_{pq}\mathrm{d}S = \int_{-1}^1\int_{-1}^1 \frac{1+s}{2^3} g_{ij}(s,t)g_{pq}(s,t)\mathrm{d}s\mathrm{d}t \tag{1.74}$$

Using the orthogonal property of Jacobi polynomials given by Eq. (1.70), the bases $g_{ij}(s,\ t)$ can be determined as

$$g_{ij}(s,t) = [(1-s)^a(1+s)^b P_i^{(2a,\ 2b+1)}(s)][(1-t)^c(1+t)^d P_j^{(2c,\ 2d)}(t)] \tag{1.75}$$

where a, b, c and d should be determined according to convenience in applications and the requirements of the Jacobi polynomials. In order to eliminate $(\xi+\eta)$ in the denominator of t in Eq. (1.75), one needs to take $b = j$. For the C^0 problem considered in this work, we can take $a = c = d = 0$. So, we have

$$g_{ij}(s,t) = (1+s)^j P_i^{(0,2j+1)}(s)P_j^{(0,0)}(t) \tag{1.76}$$

or

$$g_{ij}(\xi,\eta) = 2^j P_i^{(0,2j+1)}[2(\xi+\eta)-1]\left[(\xi+\eta)^j P_j^{(0,0)}\left(\frac{\eta-\xi}{\eta+\xi}\right)\right] \tag{1.77}$$

The constant coefficient 2^j in Eq. (1.77) is usually omitted in application. Dubiner refers to Eq. (1.77) as a warped product to differentiate it from the standard tensor product associated with quadrilateral domains.

Note that Koornwinder [65] has constructed orthogonal bases on triangles, disks, etc. before Dubiner [63] using methods of constructing two-variable analogues of the Jacobi polynomials. Therefore, some researchers refer to the orthogonal bases derived through Dubiner's approach as Koornwinder-Dubiner bases.

The orthogonal bases proposed by Dubiner would be ideal in both two- and three-dimensions except that we want to discretize a complex-geometry solution domain into many triangular or tetrahedral subdomains, or elements. We are interested in discretizing Second-order problems and for this we need to enforce a C^0 continuity between each subdomain. C^0 continuity is not easily enforced but might be achieved by a constraint on all modes. A constraint of this sort would introduce a non-diagonal contribution to the matrices thereby destroying the most desirable property of the expansion. To overcome this problem, multidomain triangular and tetrahedral expansions can be constructed using a more standard finite element approach. In this approach, we need to decompose the expansion into vertex, edge and interior modes.

Carnevali *et al.* [66] used the Gram–Schmidt procedure to obtain edge, face and body basis functions. Szabó and Babuška [67] presented edge, face and body basis functions using Legendre polynomials but without the rational term, as shown in Eq. (1.69). For example,

$$E_i^1(\xi,\eta) = (1-\xi-\eta)\xi P_i(2\xi-1) \tag{1.78}$$

Sherwin and Karniadakis [68, 69] and Webb and Abouchacra [54] independently modified Dubiner's triangular bases by splitting them into interior and boundary modes where all interior modes are zero on the triangle boundary similar to the bubble modes used above on quadrilateral domains. The modified edge bases are

$$\begin{aligned} E_i^1(\xi,\eta) &= (1-\xi-\eta)\xi P_i^{(2,3)}[2(\xi+\eta)-1] \\ E_j^2(\xi,\eta) &= \xi\eta(\eta+\xi)^j P_j^{(2,2)}\left(\frac{\eta-\xi}{\eta+\xi}\right) \\ E_i^3(\xi,\eta) &= (1-\xi-\eta)\eta P_i^{(2,3)}\left[2(\xi+\eta)-1\right] \end{aligned} \tag{1.79}$$

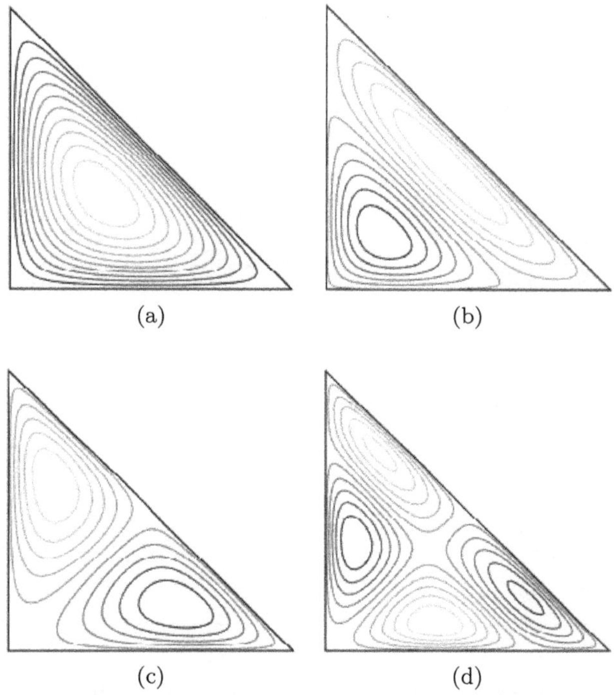

Fig. 1.7. Face functions inside a triangular element: (a) F_{00}, (b) F_{01}, (c) F_{10}, (d) F_{11}.

The modified face bases are

$$F_{ij}(\xi,\eta) = (1-\xi-\eta)\xi\eta P_i^{(2,2j+5)}[2(\eta+\xi)-1] \times \left[(\eta+\xi)^j P_j^{(2,2)}\left(\frac{\eta-\xi}{\eta+\xi}\right)\right] \quad (1.80)$$

The vertex bases are the same as Eq. (1.62). Figure 1.7 shows some face functions of a triangular element computed from Eq. (1.80). Some edge functions are plotted in Fig. 1.8.

According to the derivations above, the general form of the warped bases, and the edge and face shape functions presented above can be given as

$$N_{ij}(\xi,\eta) = (1-\xi-\eta)^h \xi^l \eta^m H_{ij}(\xi,\eta) \quad (1.81)$$

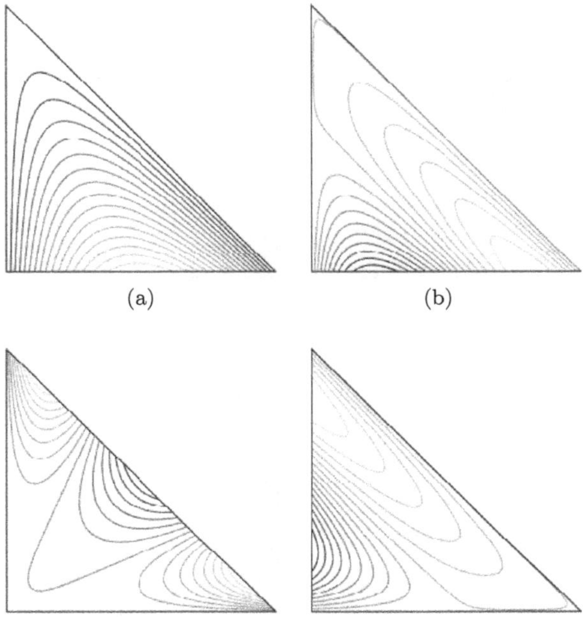

Fig. 1.8. Edge functions on a triangular element: (a) E_0^1, (b) E_1^1, (c) E_2^2, (d) E_1^3.

which should satisfy the following orthogonal condition:

$$\begin{aligned}\int_0^1\int_0^{1-\xi} & N_{ij}N_{pq}\mathrm{d}\xi\mathrm{d}\eta \\ &= \int_{-1}^1\int_{-1}^1 \frac{(1-s)^{2h}(1+s)^{2l+2m+1}(1-t)^{2l}(1+t)^{2m}}{2^{3+2h+4l+4m}} \\ &\quad \times H_{ij}(s,t)H_{pq}(s,t)\mathrm{dsdt} = \delta_i^p\delta_j^q C_{ij} \end{aligned} \tag{1.82}$$

where H_{ij} can be determined through the same way as Eqs. (1.71)–(1.77):

$$H_{ij}(s,t) = (1+s)^j J_i^{(2h,2l+2m+1+2j)}(s)J_j^{(2l,2m)}(t) \tag{1.83}$$

or

$$\begin{aligned} H_{ij}(\xi,\eta) = 2^j(\xi+\eta)^j P_i^{(2h,2j+2l+2m+1)} \\ \times[2(\xi+\eta)-1]P_j^{(2l,2m)}\left(\frac{\eta-\xi}{\eta+\xi}\right) \end{aligned} \tag{1.84}$$

where h, l, m are equal to 0 or 1. The constant coefficient 2^j in Eq. (1.84) is omitted in application. The constant C_{ij} in Eq. (1.82) is

$$C_{ij} = \frac{C_i^{(2h,2j+2l+2m+1)} C_j^{(2l,2m)}}{2^{3+2h+4l+4m}} \tag{1.85}$$

One can see that Eq. (1.81) becomes the face shape function given by Eq. (1.80) when $h = l = m = 1$. The edge shape functions above can be obtained in similar ways. Dubiner's bases given by Eq. (1.77) can be obtained by taking $h = l = m = 0$.

Dubiner [63] also presented warped orthogonal basis on tetrahedrons. Sherwin and Karniadakis [68, 69] have also modified Dubiner's tetrahedral bases by splitting them into interior and boundary modes. They are also given in the general form in the following:

$$N_{ijk}(\xi, \eta, \zeta) = (1 - \xi - \eta - \zeta)^h \xi^l \eta^m \zeta^n H_{ijk}(\xi, \eta, \zeta) \tag{1.86}$$

where h, l, m and n are non-negative integers. In order to apply Jacobi polynomials to the tetrahedron element, coordinate transformation is necessary. Duffy's transformation from a unit hexahedral domain to a unit tetrahedral domain, as shown in Fig. 1.9, can be given as

$$\begin{aligned} \xi &= \frac{(1+r)(1+t)}{4} \\ \eta &= \frac{(1-r)(1+s)(1+t)}{8}, \quad \zeta = \frac{(1-r)(1-s)(1+t)}{8} \end{aligned} \tag{1.87}$$

The corresponding inverse transformation is

$$r = 1 - \frac{2(\eta+\zeta)}{(\xi+\eta+\zeta)}, \quad s = \frac{\eta-\zeta}{\eta+\zeta}, \quad t = 2(\xi+\eta+\zeta) - 1 \tag{1.88}$$

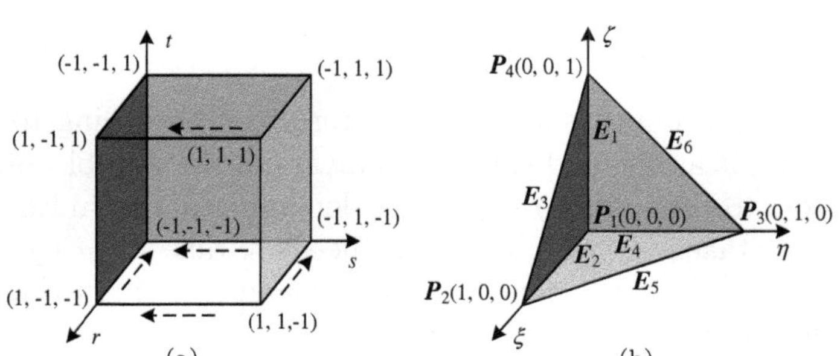

Fig. 1.9. A unit hexahedron (a) and tetrahedron (b).

By the transformation given by Eq. (1.87), the face $r = -1$ of the unit hexahedron corresponds to the face $\xi = 0$ of the unit tetrahedron, $s = -1$ corresponds to $\eta = 0$, $s = 1$ corresponds to $\zeta = 0$, $t = 1$ corresponds to $\xi + \eta + \zeta = 1$, the face $t = -1$ is degenerated to be a point and the face $r = 1$ is degenerated to be a line. According to the orthogonal requirements, the following formulation should hold for the shape functions on the unit tetrahedron:

$$\int_V N_{ijk} N_{pqw} \mathrm{d}V = \int_0^1 \int_0^{1-\xi} \int_0^{1-\xi-\eta} N_{ijk} N_{pqw} \mathrm{d}\zeta \mathrm{d}\eta \mathrm{d}\xi = \delta_i^p \delta_j^q \delta_k^w C_{ijk} \tag{1.89}$$

The substitution of Eqs. (1.86) and (1.87) into Eq. (1.89) leads to

$$\int_0^1 \int_0^{1-\xi} \int_0^{1-\xi-\eta} N_{ijk} N_{pqw} \mathrm{d}\zeta \mathrm{d}\eta \mathrm{d}\xi = \int_{-1}^1 \int_{-1}^1 \int_{-1}^1 \tilde{J} H_{ijk} H_{pqw} \mathrm{d}r \mathrm{d}s \mathrm{d}t \tag{1.90}$$

where

$$\tilde{J} = \frac{\begin{array}{c}(1+r)^{2l}(1-r)^{2m+2n+1}(1+s)^{2m} \\ \times (1-s)^{2n}(1+t)^{2l+2m+2n+2}(1-t)^{2h}\end{array}}{2^{2h+4l+6m+6n+6}} \tag{1.91}$$

Using the orthogonal property of Jacobi polynomials as shown by Eq. (1.70), $H_{ijk}(r,\ s,\ t)$ can be determined as

$$\begin{aligned} H_{ijk}(r,s,t) = {} & [(1-r)^a(1+r)^b P_i^{(2m+2n+1+2a,2l+2b)}(r)] \\ & \times [(1-s)^c(1+s)^d P_j^{(2n+2c,2m+2d)}(s)] \\ & \times [(1-t)^e(1+t)^f P_k^{(2h+2e,2l+2m+2n+2+2f)}(t)] \end{aligned} \tag{1.92}$$

where a, b, c, d, e and f should be determined according to convenience of applications and the requirements of the Jacobi polynomials. In order to eliminate $(\eta + \zeta)$ in the denominator of s in Eq. (1.92), one needs to take $a = j$. Similarly, one needs to take $f = i + j$ to eliminate $(\xi + \eta + \zeta)$ in the denominator of r and $(1 - r)$ in Eq. (1.92). The indices i, j, k should have the relation $i + j + k \leq \gamma$–4, where $\gamma \geq 4$ is the expected maximum order of the bases if one wants to keep the bases below a certain maximum order. For the C^0 problem considered in this work, we can take $b = c = d = e = 0$, whereas we have

$$H_{ijk}(r,s,t) = [(1-r)^j P_i^{(2m+2n+1+2j,2l)}(r)] P_j^{(2n,2m)}(s) \times [(1+t)^{i+j} P_k^{(2h,2(l+m+n+1)+2(i+j))}(t)] \quad (1.93)$$

The substitution of Eq. (1.88) into Eq. (1.93) leads to

$$H_{ijk}(\xi,\eta,\zeta) = \left[2^{i+2j}(\xi+\eta+\zeta)^i P_i^{(2m+2n+1+2j,2l)}\left(\frac{\xi-\eta-\zeta}{\xi+\eta+\zeta}\right)\right] \times \left[(\eta+\zeta)^j P_j^{(2n,2m)}\left(\frac{\eta-\zeta}{\eta+\zeta}\right)\right] \times P_k^{(2h,2(l+m+n+1)+2(i+j))}[2(\xi+\eta+\zeta)-1] \quad (1.94)$$

The constant 2^{i+2j} in Eq. (1.94) will not be considered in computation. The constants C_{ijk} in Eq. (1.89) is given by

$$C_{ijk} = \frac{C_i^{(2m+2n+1+2j,2l)} C_j^{(2n,2m)} C_k^{(2h,2(l+m+n+1)+2(i+j))}}{2^{2h+4l+6m+6n+6}} \quad (1.95)$$

Dubiner's warped orthogonal bases can be obtained by taking $h = l = m = n = 0$. The modified bases presented by Sherwin and Karniadakis can be obtained by taking at least one of h, l, m, n to be non-zero. Figure 1.10 plots a few bases on a unit tetrahedral domain.

The method of deriving orthogonal bases on simplex domains proposed by Dubiner [63] and developed by Sherwin and Karniadakis [68, 69] has wide recognition and applications [70, 71]. There are also some other bases on simplexes that have some advantages in some applications. They are briefly reviewed in the following.

Shephard *et al.* [72] presented a geometry-based methodology to support the construction of variable p-order meshes employing elements of various topological forms. Key to the effectiveness of this approach is the decomposition of higher-order mesh entity shape functions into a function over the mesh entity multiplied by a blending function over the domain of the element. The shape function decomposition process is efficient and easily accomplished when an appropriate set of mesh entity parametric coordinates are used. In fact, all the shape functions given there are decomposed versions of previously presented p-version shape functions [66–68].

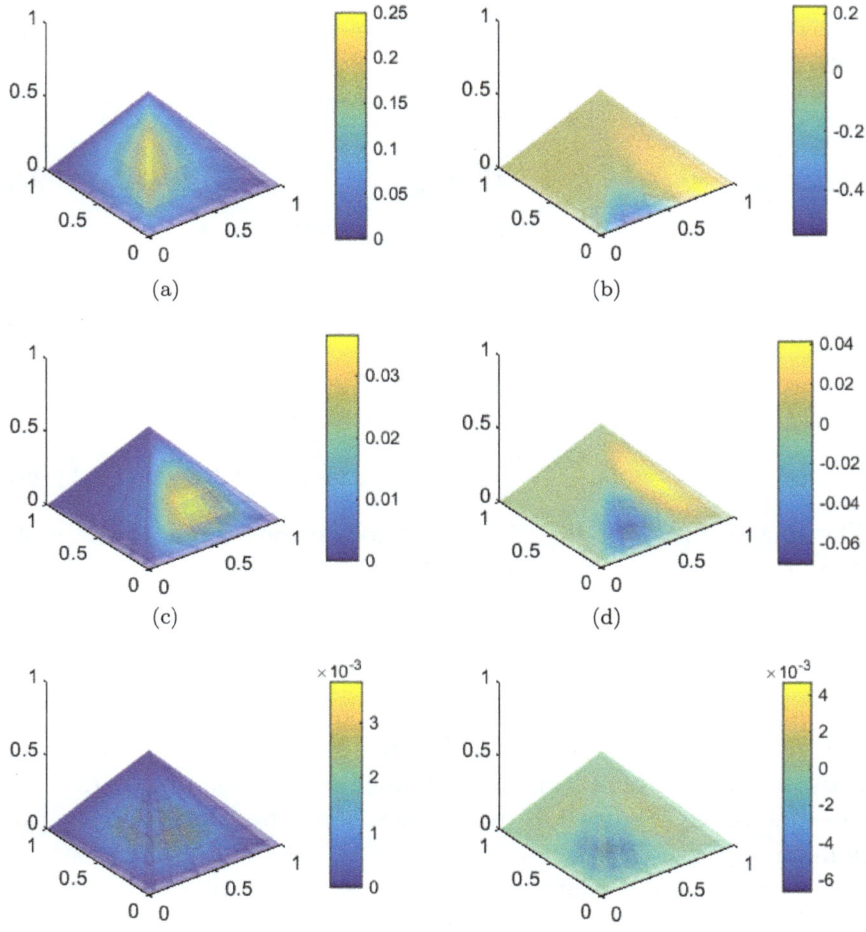

Fig. 1.10. Bases on edges 1 and 2 (a, b), face $\eta = 0$ (c, d) and body (e, f) of a unit tetrahedron.

Adjerid *et al.* [73] indicated how to construct hierarchical bases on triangles and tetrahedrons that have a quadratic growth in the condition number with polynomial degree p. This is substantially better than the exponential growth associated with the hierarchical bases of Carnevali *et al.* [66] and Szabó and Babuška [67]. The strategy for improving the condition number involves extracting an orthogonal set of shape functions and proper basis representation for the Gram–Schmidt procedure. A further improvement of the bases by Carnevali *et al.* [66] and Adjerid *et al.* [73] was presented by Bittencourt [74, 75] by fully constructing tensorial nodal and modal shape functions.

Houmat [48] developed a trigonometrically enriched curved triangular p-element through rational blending function interpolation. Express the curved sides of a triangle (Fig. 1.3) in the following parametric forms:

$$\begin{aligned} &x = x_1(\xi), \quad y = y_1(\xi), \quad 0 \leq \xi \leq 1 \\ &x = x_i(\eta), \quad y = y_i(\eta), \quad 0 \leq \eta \leq 1 \quad (i = 2, 3) \end{aligned} \tag{1.96}$$

where the subscripts denote the side numbers. The element rational blending mapping functions are

$$\begin{aligned} x(\xi, \eta) = &\frac{1-\xi-\eta}{1-\xi} x_1(\xi) + \frac{\xi}{1-\eta} x_2(\eta) \\ &+ \frac{1-\xi-\eta}{1-\eta} x_3(\eta) - (1-\xi-\eta)X_1 - \frac{\xi(1-\xi-\eta)}{1-\xi} X_2 \\ y(\xi, \eta) = &\frac{1-\xi-\eta}{1-\xi} y_1(\xi) + \frac{\xi}{1-\eta} y_2(\eta) \\ &+ \frac{1-\xi-\eta}{1-\eta} y_3(\eta) - (1-\xi-\eta)Y_1 - \frac{\xi(1-\xi-\eta)}{1-\xi} Y_2 \end{aligned} \tag{1.97}$$

where X_1, X_2 and Y_1, Y_2 are respectively the x and y coordinates of vertexes 1 and 2 shown in Fig. 1.6. The first derivatives of $x(\xi, \eta)$ in Eq. (1.97) with respect to ξ and η are

$$\begin{aligned} \frac{\partial x(\xi, \eta)}{\partial \xi} = &\frac{1-\xi-\eta}{1-\xi} \frac{dx_1(\xi)}{d\xi} + \frac{x_2(\eta) - x_3(\eta)}{1-\eta} \\ &+ \frac{\eta[X_2 - x_1(\xi)]}{(1-\xi)^2} + X_1 - X_2 \\ \frac{\partial x(\xi, \eta)}{\partial \eta} = &\frac{\xi}{1-\eta} \frac{\partial x_2(\eta)}{\partial \eta} + \frac{1-\xi-\eta}{1-\eta} \frac{\partial x_3(\eta)}{\partial \eta} \\ &+ \frac{\xi X_2 - x_1(\xi)}{1-\xi} + \frac{\xi[x_2(\eta) - x_3(\eta)]}{(1-\eta)^2} + X_1 \end{aligned} \tag{1.98}$$

The derivatives of $y(\xi, \eta)$ with respect to ξ and η are similar. One can see that this method is complex and the generalization of the method to tetrahedron is not an easy task.

Recently, Liu *et al.* [76] discussed how to compute the rational term in the orthogonal bases on simplex. Take Eq. (1.80) as an example. For convenience of derivation, the recursion formula of Jacobi polynomials is given in the following:

$$a_{1n}P_{n+1}^{(\alpha,\beta)}(\xi) = (a_{2n} + a_{3n}\xi)P_n^{(\alpha,\beta)}(\xi) - a_{4n}P_{n-1}^{(\alpha,\beta)}(\xi) \tag{1.99}$$

where n is a non-negative integer, .$\alpha > -1$, $\beta > -1$ and

$$P_0^{(\alpha,\beta)}(\xi) = 1, \quad P_1^{(\alpha,\beta)}(\xi) = \frac{1}{2}[\alpha - \beta + (\alpha + \beta + 2)\xi] \tag{1.100}$$

and

$$\begin{aligned}
a_{1n} &= 2(n+1)(n+\alpha+\beta+1)(2n+\alpha+\beta) \\
a_{2n} &= (2n+\alpha+\beta+1)(\alpha^2-\beta^2) \\
a_{3n} &= (2n+\alpha+\beta)(2n+\alpha+\beta+1)(2n+\alpha+\beta+2) \\
a_{4n} &= 2(n+\alpha)(n+\beta)(2n+\alpha+\beta+2)
\end{aligned} \tag{1.101}$$

In order to eliminate the computation of $\xi + \eta = 0$ that is in the denominator, make the following variable substitution:

$$\tilde{P}_n^{(\alpha,\beta)}(\tilde{s},\tilde{t}) = \tilde{t}^p P_n^{(\alpha,\beta)}(\tilde{s}/\tilde{t}), \quad \tilde{s} = \eta - \xi, \quad \tilde{t} = \eta + \xi \tag{1.102}$$

The substitution of Eq. (4.36) into Eq. (4.19) leads to

$$a_{1n}\tilde{P}_{n+1}^{(\alpha,\beta)}(\tilde{s},\tilde{t}) = (a_{2n}\tilde{t} + a_{3n}\tilde{s})\tilde{P}_n^{(\alpha,\beta)}(\tilde{s},\tilde{t}) - a_{4n}\tilde{t}^2\tilde{P}_{n-1}^{(\alpha,\beta)}(\tilde{s},\tilde{t}) \tag{1.103}$$

where

$$\tilde{P}_0^{(\alpha,\beta)}(\tilde{s},\tilde{t}) = 1, \quad \tilde{P}_1^{(\alpha,\beta)}(\tilde{s},\tilde{t}) = \frac{1}{2}[(\alpha-\beta)\tilde{t} + (\alpha+\beta+2)\tilde{s}] \tag{1.104}$$

This technique can also be used for tetrahedrons and pyramids.

1.3.4 *The DQM and HQEM on simplex domains*

The DQM has been primarily used in quadrilateral and hexahedral domains. In order to overcome these limitations, Zhong [77–80] proposed a triangular DQM (TDQM) using uniform nodes on triangles.

The TDQM has numerical instability problem, so Zhong [81, 82] proposed a localized triangular differential quadrature method to solve the problem, which may sacrifice some accuracy. The numerical instability problem of DQM on quadrilateral and hexahedral domains was solved by using tensor product of 1D non-uniformly distributed nodes that are roots of orthogonal polynomials [11]. Thus, the difficulty of applying the DQM on triangles and tetrahedrons can also be solved by using non-uniformly distributed nodes on triangles and tetrahedrons. It is wellknown that the optimized interpolation nodes on tensor product domain are tensor product of 1D Gauss–Lobatto nodes [83]. The Gauss–Lobatto nodes generalized to non-tensor-product domains like triangles and tetrahedrons are known to be the Fekete points [84]. Thus, the difficulty of applying the DQM to triangles and tetrahedrons becomes the efficient computation of the Fekete points on triangles and tetrahedrons.

An open question concerns the spatial distribution of nodes that are suitable for high-order Lagrange interpolation on the triangle and tetrahedron. Bos [85] first introduced an approach of choosing nodal locations that maximize the determinant of the classical Vandermonde matrix defined in terms of a bivariate monomial basis up to the seventh order. The resulting nodes are referred to as Fekete nodes [86]. The method is briefly revisited in the following.

Denote the space of polynomials of degree n in m variables by $\boldsymbol{P}_n^m$ and let $\boldsymbol{X} \subset \boldsymbol{R}$ be compact. It is well known that the dimension of the approximation space

$$\dim \boldsymbol{P}_n^m = C_n^{m+n} = \frac{(m+n)!}{m!n!} =: N_n^m \tag{1.105}$$

given there is no ambiguity present for N. Hence, N_n^m conditions are needed to determine a polynomial uniquely.

The interpolation problem is then as follows: given N_n^m distinct points, $\boldsymbol{x}_i \in \boldsymbol{X}$, and a function $f \in L(\boldsymbol{X})$, find that $p \in \boldsymbol{P}_n^m$ satisfying $p(\boldsymbol{x}_i) = f(\boldsymbol{x}_i)$, $1 \leq i \leq N$. If we let $\{m_i(\boldsymbol{x}) : 1 \leq i \leq N\}$ be a basis for $\boldsymbol{P}_n^m$, then this is equivalent to finding N_n^m numbers, $a_j \in \boldsymbol{R}$, so that

$$p(\boldsymbol{x}_i) = \sum_{j=1}^{N} a_j m_j(\boldsymbol{x}_i) = f(\boldsymbol{x}_i), \quad 1 \leq i \leq N \tag{1.106}$$

or

$$\begin{bmatrix} m_1(\boldsymbol{x}_1) & \cdots & m_N(\boldsymbol{x}_1) \\ \vdots & & \vdots \\ m_1(\boldsymbol{x}_N) & \cdots & m_N(\boldsymbol{x}_N) \end{bmatrix} \begin{bmatrix} a_1 \\ \vdots \\ a_N \end{bmatrix} = \begin{bmatrix} f(\boldsymbol{x}_1) \\ \vdots \\ f(\boldsymbol{x}_N) \end{bmatrix} \tag{1.107}$$

or VDM$\boldsymbol{a} = \boldsymbol{f}$, a linear system solvable for all right-hand sides if its determinant is non-zero. We will refer to this determinant again and give it a name VDM$(\boldsymbol{x}_1, \boldsymbol{x}_2, \ldots, \boldsymbol{x}_N)$, in recognition of the fact that in one dimension, if the basis is chosen to be $m_i(x) = x^{i-1}$, it is the Vandermonde determinant.

An equivalent geometrical condition for all such interpolating polynomials to exist is that the points do not lie on the zero set of any non-trivial $q \in \boldsymbol{P}_{n^m}$.

Now, consider a configuration of points with VDM$(\boldsymbol{x}_1, \boldsymbol{x}_2, \ldots, \boldsymbol{x}_N) \neq 0$. A very useful thing to do is to form the associated Lagrange polynomials l_i of degree n defined by the conditions $l_i(\boldsymbol{x}_j) = \delta_i^j$, $1 \leq i, j \leq N$. Then for $f \in L(\boldsymbol{X}), (\Pi_n f)(\boldsymbol{x}) := \sum_{i=1}^N f(\boldsymbol{x}_i) l_i(\boldsymbol{x})$ is its interpolating polynomial. Consequently, the interpolating Lagrange polynomials may in general be found as the solution to the dual interpolation problem, $\Pi_n(p) =$ for all $p \in \boldsymbol{P}_n^m$, or

$$\begin{bmatrix} m_1(\boldsymbol{x}_1) & \cdots & m_N(\boldsymbol{x}_1) \\ \vdots & & \vdots \\ m_1(\boldsymbol{x}_N) & \cdots & m_N(\boldsymbol{x}_N) \end{bmatrix} \begin{bmatrix} l_1(\boldsymbol{x}) \\ \vdots \\ l_N(\boldsymbol{x}) \end{bmatrix} = \begin{bmatrix} m_1(\boldsymbol{x}) \\ \vdots \\ m_N(\boldsymbol{x}) \end{bmatrix} \tag{1.108}$$

through which, using Cramer's rule, we obtain the direct solution

$$l_i(\boldsymbol{x}) = \frac{|\text{VDM}(\boldsymbol{x}_1, \ldots, \boldsymbol{x}_{i-1}, \boldsymbol{x}, \boldsymbol{x}_{i+1}, \ldots, \boldsymbol{x}_N)|}{\text{VDM}(\boldsymbol{x}_1, \ldots, \boldsymbol{x}_N)} \tag{1.109}$$

The solution of Vandermonde systems, Eq. (1.108), is notoriously difficult due to exponential-like growth of the condition number.

When both spaces are equipped with the supremum norm, the norm of this projection is well known and is easily seen to be

$$\left\| \prod_n \right\| = \max_{\boldsymbol{x}} \sum_{i=1}^N |l_i(\boldsymbol{x})| \tag{1.110}$$

The function $\Lambda(\boldsymbol{x}) = \sum_i |l_i(\boldsymbol{x})|$ is usually referred to as the Lebesgue function of the interpolation process. Consequently, by computing the Lebesgue constant, we obtain a measure of how close the approximation is to the best polynomial approximation.

This norm is of interest for the following two reasons:

(1) If $\boldsymbol{P}_n^*$ is the best uniform approximation to f on $\boldsymbol{X}$, then

$$\left\| f - \prod_n f \right\| \leq \left(1 + \left\| \prod_n \right\| \right) \| f - \boldsymbol{P}_n^* \| \tag{1.111}$$

Thus, $|||\Pi_n|||$ yields an estimate on how close the interpolation is to f.

(2) For $p \in \boldsymbol{P}_n^m$,

$$\|p\| \leq \left\| \prod_n \right\| \max_{1 \leq i \leq N} |p(\boldsymbol{x}_i)| \tag{1.112}$$

Thus, $||\Pi_n||$ yields a simple method of bounding polynomials. It is now clear that problems of interest are to first of all bound $||\Pi_n||$ for a given set of points and secondly to find points for which this norm is small, if not as small as possible.

Bos [87] conjectured that in the triangle (and some other domains) Fekete points will have the maximum number of points on the boundary. Under this assumption, he proved that the boundary points will be the one-dimensional Gauss–Lobatto points.

Mavriplis and Rosendale [88] suggested to restrict the Gauss–Lobatto mesh of the quadrilateral to the triangle in the field of spectral methods.

Chen and Babuška [89] extended the approach of Bos [85] up to 13 order in a triangle with an L^2-norm optimal Lebesgue constant, although these points did not have an edge distribution which can be identified with Gauss–Lobatto–Jacobi points.

Hesthaven [90] proposed an extension to the concept of using the minimization of an electrostatic potential to determine a nodal set of points in a triangle and found that the solution of that electrostatic problem led to nodal sets which are very close to the best known sets at that time. Using the one-dimensional distribution of points to constrain the nodal distribution along the edges, the interior nodal points were then determined as the minimum of an electrostatic potential.

By constraining the edge points in this manner, the nodes along all edges can be constrained to the Gauss–Lobatto–Legendre quadrature points. The nodal distribution, therefore, offers a compatible extension to the quadrilateral expansion where both quadrilateral and triangular regions can be assembled into a global C^0 expansion.

Taylor *et al.* [84] extended the approach of Bos [85] up to 19th order. They used the steepest ascent method to maximize the determinant of the Vandermonde matrix. The method is equivalent to solving the following system of nonlinear equations:

$$\frac{\partial \mathrm{VDM}(\boldsymbol{x}_1, \ldots, \boldsymbol{x}_N)}{\partial \boldsymbol{d}x_i} = 0, \quad 1 \le i \le N \tag{1.113}$$

In order to get the Lagrange polynomials $l_i(\boldsymbol{x})$, the Vandermonde matrix $\mathrm{VDM}(\boldsymbol{x}_1, \boldsymbol{x}_2, \ldots, \boldsymbol{x}_N)$ must be inverted (through Gaussian elimination) at each iteration. Since we must compute this inverse numerically, it is important that VDM be well conditioned. For this, they have found that the Dubiner [63] basis functions in the right triangle are far superior to Legendre polynomials. Using monomials for the basis functions is practically impossible. For their largest case, degree 19, the MATLAB reported condition number of VDM is less than 50 when computed with the Dubiner polynomials.

Pasquetti and Rapetti [91] compared a triangle-based SEM with the classical quadrangle-based SEM and with a standard spectral method. *The requirement of a highly accurate quadrature rule was particularly emphasized.* Then it was shown that the convergence properties of the triangle SEM compare well with those of the classical SEM by solving an elliptic equation with smooth (but steep) analytical solution. When compared to the standard SEM, the main advantage of the TSEM comes from the highest capability of triangles to handle complex geometries. However, the computation of the derivatives involves all collocation point values, so that the computational cost is heavier. The most natural approach will then combine the SEM with the TSEM to take advantage of the features of both methods.

For the Fekete points, the computation of the point distributions for a specified polynomial order, m, requires a fair amount of effort [84, 89, 90, 92]. It is desirable to devise a more practical method without overly compromising on the interpolation accuracy. As a competitive alternative, Blyth and Pozrikidis [92] presented a simple, easy-to-implement interpolation scheme. Edge nodes are positioned at the zeros of the Lobatto polynomials, while the interior nodes are generated by constructing barycentric coordinate lines inside the triangle and then averaging the coordinates of the vertices of particular internal triangles. The approach has been generalized to tetrahedrons [93, 94]. Lebesgue constants computed for the new distributions are competitive with those of the more complicated node configurations.

Heinrichs [95] calculated new sets of points with improved Lebesgue constants in the triangle. Starting with the Fekete points, improved Lebesgue constants were obtained in the triangle. The minimization procedure was based on a combination of the steepest descent algorithm with a damped Newton method. It is quite surprising that the optimal points are not symmetric. The points along the boundary of the triangle are the 1D Gauss-Lobatto points.

Warburton [96] presented a simple and geometrically motivated way of spatially distributing interpolation nodes in a triangle and tetrahedron. The method was inspired by techniques employed in the creation of curvilinear finite element methods and represents a new approach and perspective on the problem of creating these nodal sets. The new *warp and blend* nodes were shown to have Lebesgue constants which are better than or comparable with all existing node sets up to at least 10th-order interpolation.

Briani *et al.* [97] computed, by the MATLAB Optimization Toolbox, Fekete and Lebesgue points for polynomial interpolation on three basic bidimensional compacts: the simplex, the square, and the disk. These point sets can be affinely mapped on any triangle, parallelogram, and ellipse, respectively, and give for almost all degrees the best results known so far in terms of Lebesgue constant.

Zhong *et al.* [98] made an effort to construct non-uniform grids for triangular differential quadrature analogs. It starts from a uniform iso-solid-angle grid in an octant of a unit sphere which is generated with the aid of the evenly divided solid angle in the three directions of the spherical triangle. With an appropriate projection onto the equilateral plane triangle passing the three vertices of the spherical triangle, a non-uniform grid is generated. The triangular differential quadrature analog is implemented in the triangle with the non-uniform grid wherein weighting coefficients are determined in terms of complete monomials of area coordinates.

Liu *et al.* [76] presented a simple formula of direct map 1D non-uniform interpolation nodes to triangles that are close to the Fekete points. Similar to the 1D Gauss–Lobatto–Chebyshev nodes that are transformed from 1D uniformly distributed nodes, this was shown to be also possible for triangles. Let the transformation formula from uniform nodes $(\xi, \eta, 1-\xi-\eta)$ to non-uniform nodes $(\lambda_1, \lambda_2, \lambda_3)$ be

$$\lambda_1 = f(\xi), \quad \lambda_2 = f(\eta), \quad \lambda_3 = f(1-\xi-\eta) \tag{1.114}$$

If $(\lambda_1, \lambda_2, \lambda_3)$ are also area coordinates, then $\lambda_i (i = 1, 2, 3)$ is symmetric with respect to each other and we can let

$$\begin{aligned} s(\lambda_1, \lambda_2, \lambda_3) &= \frac{\lambda_1 + \alpha\lambda_1\lambda_2\lambda_3}{\lambda_1 + \lambda_2 + \lambda_3 + 3\alpha\lambda_1\lambda_2\lambda_3} \\ t(\lambda_1, \lambda_2, \lambda_3) &= \frac{\lambda_2 + \alpha\lambda_1\lambda_2\lambda_3}{\lambda_1 + \lambda_2 + \lambda_3 + 3\alpha\lambda_1\lambda_2\lambda_3} \end{aligned} \tag{1.115}$$

where α is an arbitrary constant. Note that λ_3 is not equal to $1-\lambda_1-\lambda_2$. It is equal to $f(1-\xi-\eta)$, where ξ and η are uniform nodes. If 1D Chebyshev–Lobatto nodes are used, $\lambda_1 = \sin(\pi\xi/2)$, $\lambda_2 = sin(\pi\eta/2)$ and $\lambda_3 = \sin[\pi(1-\xi-\eta)/2]$. If 1D Gauss–Lobatto nodes are used, for example, let $\boldsymbol{\alpha} = [\alpha_1, \alpha_2, \ldots, \alpha_n]$ be the 1D Gauss–Lobatto nodes defined on [0, 1]. Let $\lambda_1 = \alpha_i$ and $\lambda_2 = \alpha_j$, then we have $\lambda_3 = \alpha_{n+2-i-j}$ instead of $\lambda_3 = 1-\lambda_1-\lambda_2$. It was empirically found that $\alpha = 3$ is a good choice. One can easily verify that s and t can also be used as coordinates of a unit triangle (Fig. 1.9(a)) as ξ and η. Especially,

$$s(\lambda_1, 0, \lambda_3) = \lambda_1, \quad t(0, \lambda_2, \lambda_3) = \lambda_2 \tag{1.116}$$

which means if Gauss–Lobatto nodes are used to generate the non-uniform nodes in a unit triangle, the generated nodes along the edges of the triangle are still the Gauss–Lobatto quadrature nodes. This property is important, since the Fekete points on the edge of a triangle have been proved to be the Gauss–Lobatto quadrature points [90]. Using 1D Gauss–Lobatto nodes and indexes of uniform nodes shown in Fig. 1.11(a), non-uniform nodes can be generated by Eq. (4.40), as shown in Fig. 1.11(b). The nodes on an equilateral triangle of Fig. 1.11(c) are transformed from the nodes in Fig. 1.11(b) by taking $\tilde{\eta} = \sqrt{3}\eta$, $\tilde{\xi} = 2\xi + \eta - 1$. They were shown to be very close to the Fekete points.

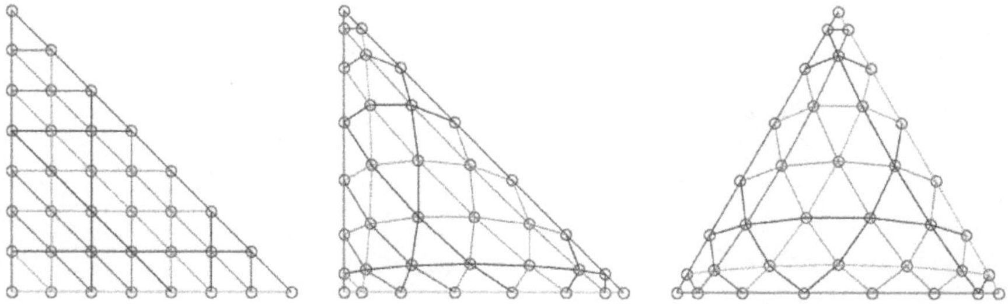

Fig. 1.11. A unit triangle with uniform (a) and non-uniform (b, c) nodes.

Similar to the triangular element, a simple formula of directly mapping 1D Gauss–Lobatto nodes to tetrahedrons was presented by Liu *et al.* [76]. Let the transformation formula from uniform nodes $(\xi, \eta, \zeta, 1-\xi-\eta-\zeta)$ to non-uniform nodes $(\lambda_1, \lambda_2, \lambda_3, \lambda_4)$ be

$$\lambda_1 = f(\xi), \quad \lambda_2 = f(\eta), \quad \lambda_3 = f(\zeta), \quad \lambda_4 = f(1-\xi-\eta-\zeta) \tag{1.117}$$

If the number of nodes on each edge is taken as the same, then $\lambda_i (i = 1, 2, 3, 4)$ is symmetric with respect to each other. Therefore, we can let

$$\begin{aligned} r(\lambda_1, \lambda_2, \lambda_3, \lambda_4) &= \frac{\lambda_1 + \alpha\lambda_1\lambda_2\lambda_3\lambda_4}{\lambda_1 + \lambda_2 + \lambda_3 + 4\alpha\lambda_1\lambda_2\lambda_3\lambda_4} \\ s(\lambda_1, \lambda_2, \lambda_3, \lambda_4) &= \frac{\lambda_2 + \alpha\lambda_1\lambda_2\lambda_3\lambda_4}{\lambda_1 + \lambda_2 + \lambda_3 + 4\alpha\lambda_1\lambda_2\lambda_3\lambda_4} \\ t(\lambda_1, \lambda_2, \lambda_3, \lambda_4) &= \frac{\lambda_3 + \alpha\lambda_1\lambda_2\lambda_3\lambda_4}{\lambda_1 + \lambda_2 + \lambda_3 + 4\alpha\lambda_1\lambda_2\lambda_3\lambda_4} \end{aligned} \tag{1.118}$$

where α is a constant to be determined. Note that λ_4 is not equal to $1 - \lambda_1 - \lambda_2 - \lambda_3$ as has been discussed for triangles. If 1D Gauss–Lobatto nodes are used, for example, let $\boldsymbol{\alpha} = [\alpha_1, \alpha_2, \ldots, \alpha_n]$ be the 1D Gauss–Lobatto nodes defined on [0, 1]. If $\lambda_1 = \alpha_i \lambda_2 = \alpha_j$ and $\lambda_3 = \alpha_k$, then we have $\lambda_4 = \alpha_{n+3-i-j-k}$. It was empirically found that the best choice of α on a unit tetrahedron is 64. Figure. 1.12 shows approximated nodes on a unit tetrahedron generated through the procedure.

The non-uniform interpolation nodes generated through Eqs. (4.40) and (1.118) together with orthogonal polynomials on the

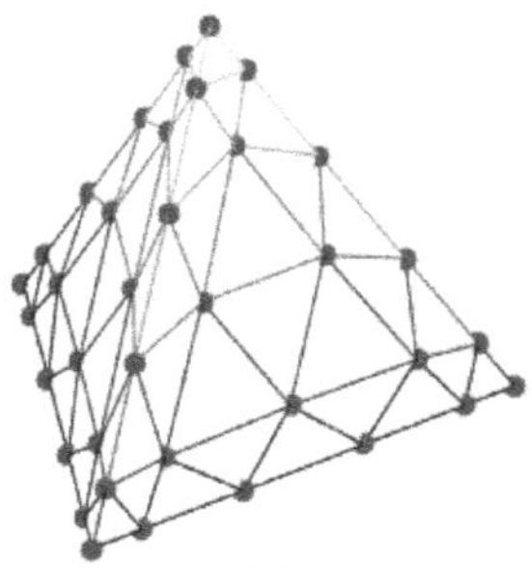
(a)

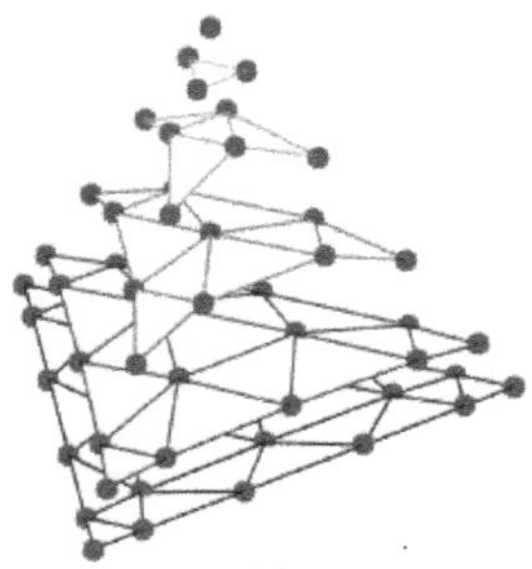
(b)

Fig. 1.12. A unit tetrahedron with non-uniform nodes.

corresponding domains were used to solve the Fekete points through the Newton–Raphson method by Liu *et al.* [76]. Then interpolation bases and DQ rules on triangles and tetrahedrons were formulated using the Fekete points and the orthogonal polynomials. For the triangular element, the displacement field given by Eq. (1.61) can be written as

$$u[x(\xi,\eta),y(\xi,\eta)] = \boldsymbol{\psi}_{N_t}^{\mathrm{T}}(\xi,\eta)\tilde{\boldsymbol{u}} \tag{1.119}$$

where N_t is the total number of bases. Let the number of bases on all edges be $m+2$ (see Fig. 1.7), namely, let $n_1 = n_2 = n_3 = m+2$, where $m = n$ if face functions are included, otherwise, $m = 0$. Use $m+2$ 1D Gauss–Lobatto quadrature points to generate non-uniform nodes on the triangle through Eq. (4.40) and denote the resulting node vector as $(\boldsymbol{\xi},\boldsymbol{\eta})$. So, we can construct a generalized Vandermonde matrix as

$$\boldsymbol{V} = \boldsymbol{\psi}_{N_t}^{\mathrm{T}}(\boldsymbol{\xi},\boldsymbol{\eta}) \tag{1.120}$$

Hence, we have

$$u[x(\xi,\eta),y(\xi,\eta)] = \boldsymbol{\psi}_{N_t}^{\mathrm{T}}(\xi,\eta)\boldsymbol{V}^{-1}u(\boldsymbol{\xi},\boldsymbol{\eta}) = \sum_{i=1}^{N_t} L_i(\xi,\eta)u(\xi_i,\eta_i) \tag{1.121}$$

where $L_i(\xi,\eta)$ is defined as a Lagrange function on the triangle at point (ξ_i,η_i). The condition numbers of the generalized Vandermonde matrix in Eq. (4.43) can be approximately given by $e^{0.45m}$, where

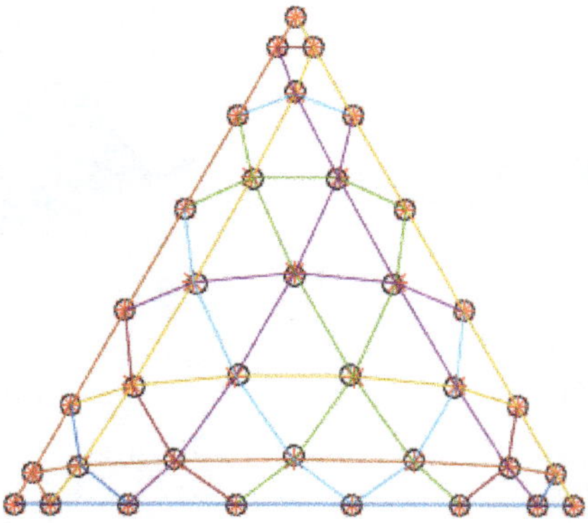

Fig. 1.13. Comparison of the solved Fekete points (red *) and the explicitly generated nodes (black ∘).

$(m+1)$ is the order of the element. The Fekete points $(\boldsymbol{\xi}, \boldsymbol{\eta})$ should make

$$|L_i(\xi, \eta)| \leq 1 \text{ and } L_i(\xi_i, \eta_i) = 1 \tag{1.122}$$

This relation can usually be replaced by

$$\frac{\partial L_i(\boldsymbol{\xi}, \boldsymbol{\eta})}{\partial \xi_i} = 0, \quad \frac{\partial L_i(\boldsymbol{\xi}, \boldsymbol{\eta})}{\partial \eta_i} = 0 \tag{1.123}$$

Since the nodes on the edges of the triangle have already been chosen as Gauss–Lobatto nodes and the Fekete points on the edges of a triangle have been proved to be Gauss–Lobatto quadrature points, only the nodes inside the triangle need to be optimized. Since the initial guess of the Fekete points is very close to the real Fekete points, the Newton–Raphson method that is more efficient than the steepest ascent algorithm can be used to solve the system of nonlinear equations. Therefore, Liu *et al.* [76] extended the approach of Bos [85] up to 13th order with ease. Figure 1.13 compares the solved Fekete points with the explicitly generated nodes through Eq. (4.40). One can see that they are very close to each other. Figure 1.14 plots four Lagrange bases obtained through the Fekete points and the orthogonal polynomials of this work.

Using the Fekete points presented above, one can also construct triangular element with nodes distribution shown in Fig. 1.4, namely, the numbers of nodes on different edges of a triangle can be different and do not relate with (this word only refers to the node numbers)

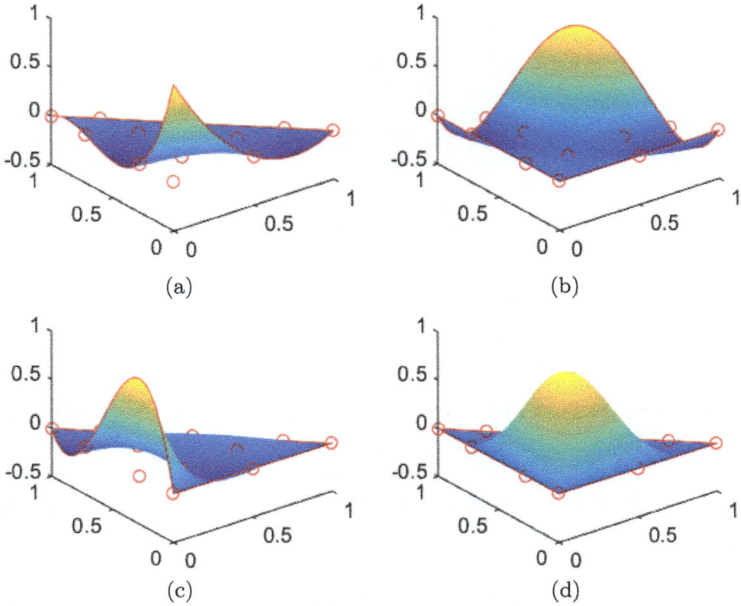

Fig. 1.14. Four interpolation bases on a unit triangle.

the nodes inside the triangle. The bases shown in Fig. 1.14 used the node distribution shown in Fig. 1.4. Take the derivatives of Eq. (4.44) with respect to the parametric coordinates and then substitute the Fekete points into it, one obtains the DQ rules on a unit triangle with non-uniform node distribution, namely,

$$\frac{\partial u(\bar{\boldsymbol{\xi}}, \bar{\boldsymbol{\eta}})}{\partial \xi} = \boldsymbol{A}^{(1)} u(\boldsymbol{\xi}, \boldsymbol{\eta}), \quad \frac{\partial u(\bar{\boldsymbol{\xi}}, \bar{\boldsymbol{\eta}})}{\partial \eta} = \boldsymbol{B}^{(1)} u(\boldsymbol{\xi}, \boldsymbol{\eta}) \tag{1.124}$$

The high-order derivatives can be obtained as

$$\frac{\partial^r u(\bar{\boldsymbol{\xi}}, \bar{\boldsymbol{\eta}})}{\partial \xi^r} = \boldsymbol{A}^{(r)} u(\boldsymbol{\xi}, \boldsymbol{\eta}), \quad \frac{\partial^s u(\bar{\boldsymbol{\xi}}, \bar{\boldsymbol{\eta}})}{\partial \eta^s} = \boldsymbol{B}^{(s)} u(\boldsymbol{\xi}, \boldsymbol{\eta})$$

$$\frac{\partial^{r+s} u(\bar{\boldsymbol{\xi}}, \bar{\boldsymbol{\eta}})}{\partial \xi^r \partial \eta^s} = \boldsymbol{F}^{(r+s)} u(\boldsymbol{\xi}, \boldsymbol{\eta}) \tag{1.125}$$

Note that the evaluation points $(\bar{\boldsymbol{\xi}},\ \bar{\boldsymbol{\eta}})$ and the nodes $(\boldsymbol{\xi}, \boldsymbol{\eta})$ were taken as different. In this case, the following interpolation relation may be needed:

$$u(\bar{\boldsymbol{\xi}}, \bar{\boldsymbol{\eta}}) = \boldsymbol{G}u(\boldsymbol{\xi}, \boldsymbol{\eta}) \quad \text{or} \quad \bar{\boldsymbol{u}} = \boldsymbol{G}\boldsymbol{u} \tag{1.126}$$

If the evaluation points and the nodes are taken as the same, the following recurrence formulations hold:

$$\boldsymbol{A}^{(r)} = \boldsymbol{A}^{(1)}\boldsymbol{A}^{(r-1)}, \quad \boldsymbol{B}^{(s)} = \boldsymbol{B}^{(1)}\boldsymbol{B}^{(s-1)}, \quad \boldsymbol{F}^{(r+s)} = \boldsymbol{A}^{(r)}\boldsymbol{B}^{(s)} \tag{1.127}$$

The direct usage of Eq. (4.48) is usually more efficient and accurate than Eq. (4.50). However, it is obvious that the usage of Eq. (4.50) is simpler than Eq. (4.48). Using the orthogonal polynomials on a quadrilateral domain [35], one can define DQ rules as Eqs. (4.47)–(4.50) on the domain with node distribution on the boundary of the element similar to Fig. 1.4 through the same procedure as Eqs. (4.42)–(4.44). Since the Fekete points on a quadrilateral region are the Gauss–Lobatto integration nodes, the optimization with requirements similar to Eqs. (4.45) and (4.46) is not necessary. Figure 1.15 shows some interpolation bases on a unit quadrangle.

The approximated nodes on a unit tetrahedron (Fig. 1.12) can be optimized by the bases presented above through the same procedure as Eqs. (4.42)–(4.46). With the optimized nodes, interpolation bases can be obtained as shown in Fig. 1.16 and DQ rules similar to Eq. (4.47) can be defined. The condition numbers of the generalized Vandermonde matrix for a unit tetrahedron using the Fekete points can be approximately given by $e^{0.9(n-2)}$, where $(n-1)$ is the order of the element.

The obtaining of orthogonal polynomials on wedge and hexahedral domains is much simpler than on a tetrahedral domain. To get orthogonal polynomials on a wedge domain, one can use tensor product of the bases on a triangular domain and 1D orthogonal polynomials. To get orthogonal polynomials on a hexahedral domain, one can use tensor product of 1D orthogonal polynomials on three directions. Figure 1.17 shown some bases on a unit wedge domain and a unit hexahedral domain.

The DQM for triangles and tetrahedrons formulated by Liu *et al.* [76] using Fekete points and Lagrange functions on the domains is free of numerical instability and is as accurate and stable as the DQM on quadrilateral and hexahedral domains. Moreover, since the

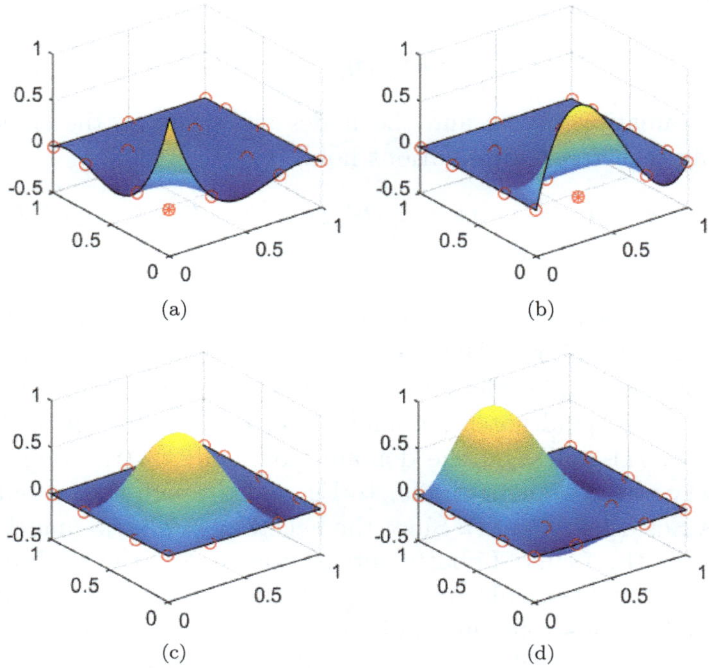

Fig. 1.15. Four interpolation bases on a unit quadrangle.

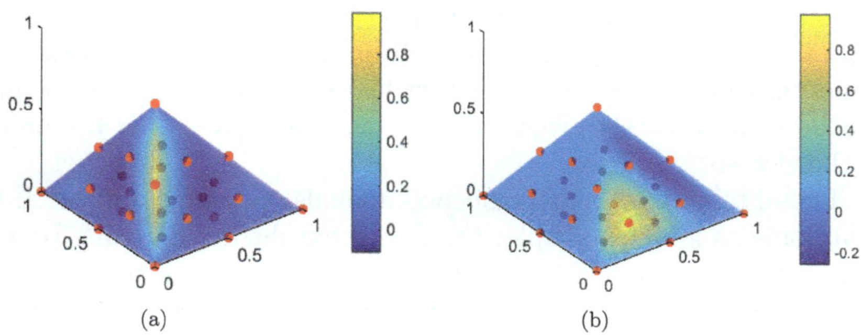

Fig. 1.16. Interpolation bases on an edge (a) and a face (b) of a unit tetrahedron.

orthogonal polynomials are also the hierarchical bases on triangles and tetrahedrons, the number of nodes on different edges of an element and on the boundary and interior of the element were made not to influence each other. Therefore, mesh refinements in the new DQM will not propagate from patch to patch. Similar DQ rules were

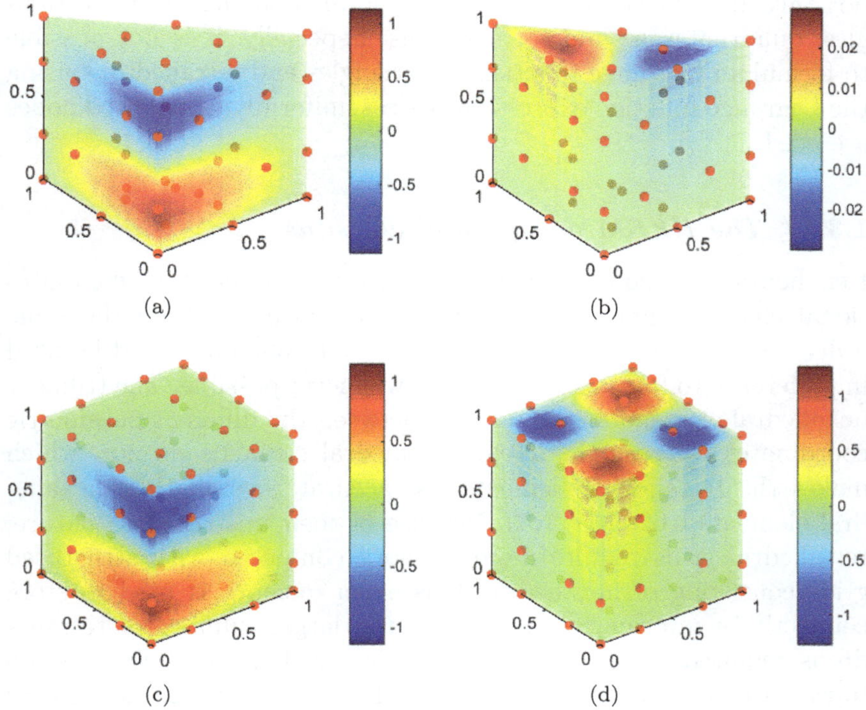

Fig. 1.17. Orthogonal polynomials on a unit wedge (a, b) and a unit hexahedron (c, d).

also constructed for quadrilateral, wedge (triangular prism) and hexahedral domains. Since the DQM presented in the work has some properties of the HFEM, it is named as the hierarchical quadrature method (HQM). Similar HQEM formulations as [35, 99] that only have interpolation bases on the boundary of elements were also constructed in [76].

It was also found in [76] that the best way of formulating Lagrange interpolation functions for HQEM is not always through the Lagrange function expressions as used by [35, 99]. If non-uniformly distributed points that are roots of orthogonal polynomials are used as nodes and high-order bases are needed, the Lagrange function should be transformed from orthogonal polynomials because the evaluation of orthogonal polynomials using the recursion formula can be very efficient and the inverse of the interpolation matrix is free of numerical instability. It was found that this conclusion holds not only for 1D case but also for 2D and 3D domains, such as triangles, quadrilaterals, and tetrahedrons. Especially, it is not possible to formulate Lagrange functions on triangles and tetrahedrons using the Lagrange function expressions unless uniformly distributed nodes are used.

1.3.5 *The HFEM on pyramid domains*

Hexahedral elements are generally superior in terms of computational efficiency and accuracy to tetrahedral elements of the same order for finite element stress analyses [100–106] and would be used in preference to tetrahedral elements whenever possible. The trilinear hexahedral elements have coupling between the different parametric coordinates whereas the linear tetrahedral elements do not, which makes the tetrahedral elements less accurate overall than hexahedral elements [102, 105, 107]. Because of their unstructured nature, tetrahedral grids require more memory compared with structured grids consisting of hexahedra. This extra memory is used to store connectivity information as well as the larger number of tetrahedrons required to fill a specific domain, i.e. five to six times more tetrahedrons than hexahedrons [107, 108]. From the parallel solver point of view, large aspect ratio tetrahedrons can lead to substantial degradation of convergence rate in iterative solvers, and certain topological constraints need to be imposed to maintain a balanced parallel computation [109]. Quadrilateral and hexahedral elements may offer significant benefits over triangular and tetrahedral elements in high-order (p-type) finite element methods [100] as well — exploiting a tensor product structure allows for simplified data structures [35, 76, 99] as well as fast, low memory applications of operators in matrix-free methods [100]. In p-type finite element method, it is crucial that the total number of elements should be as small as possible, while maintaining their geometrical quality [110]. Additionally, high-order discontinuous Galerkin (DG) methods [111] benefit from the use of hexahedral elements by reducing the number of flux calculations and also reducing the number of degrees of freedom compared to tetrahedral elements, while still delivering the same global order of approximation [100]. However, while various robust mesh generation techniques, such as advancing front and Delaunay triangulation, exist for mesh complex geometrical domains with tetrahedral elements, no general technique exists for hexahedral elements [100–103, 110–116].

Often, it is at most possible to construct hex-dominant meshes, which contain primarily hexahedral elements and also minor number of tetrahedral elements to fill the remaining, geometrically more complex or less critical regions of the solid [100–109, 116–118]. In order to avoid hanging nodes, wedge or pyramidal elements can be used to connect hexahedral and tetrahedral elements [100–108, 116–121], or two regions of structured hexahedral elements of different sizes [102, 108, 120]. For above reasons, the pyramidal elements have become increasingly popular for both low-order (h-type) [106] and high-order (p-type) [100] schemes using hex-dominant meshes for reducing DOFs as well as improving computational efficiency and accuracy.

The topic of finite element shape functions has been extensively discussed in Zienkiewicz [122] and other standard reference books on finite elements. The choice of degrees of freedom and the types of functions that basis can be used in the pyramidal elements are constrained by the requirement for the resulting elements to be compatible with neighboring hexahedral and tetrahedral finite elements [123]. Indeed, the need for compatibility between hexahedral pyramids and between tetrahedral pyramids dictates the choice of the degrees of freedom on the faces of the pyramid [123]. The problems start when one seeks to select a basis for the pyramidal elements. In 1971, Zienkiewicz [122] first proposed to collapse all of the nodes of one face of a hexahedral element into one node to obtain a pyramidal element (Fig. 1.18). The shape functions for the uncollapsed nodes are the same as before, while the shape function for the collapsed (degenerated) node is the sum of the original shape functions for that face, i.e. the vertex shape functions are

$$\begin{aligned}
S_1^v &= \frac{1}{4}(1-r)(1-s)(1-t) \\
S_2^v &= \frac{1}{4}(1+r)(1-s)(1-t) \\
S_3^v &= \frac{1}{4}(1+r)(1+s)(1-t) \\
S_4^v &= \frac{1}{4}(1-r)(1+s)(1-t) \\
S_5^v &= t
\end{aligned} \tag{1.128}$$

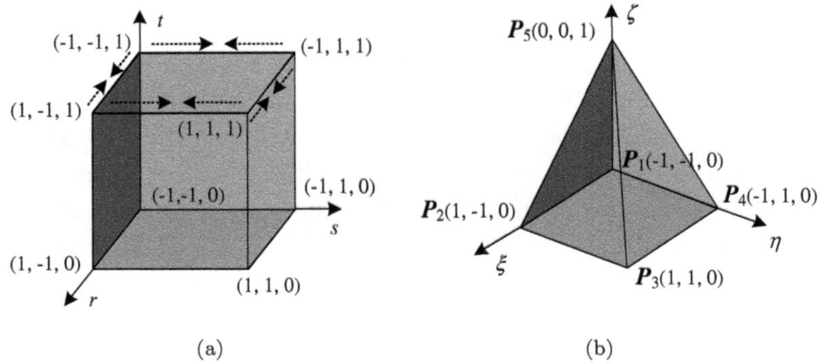

Fig. 1.18. A unit hexahedron (a) and a unit pyramid (b).

The collapsing of the four vertices means that the Jacobian locally tends to zero, making derivatives impossible to compute. The errors introduced seem in practice to be fairly local, so these elements have some applications although their description is accompanied by a warning regarding unreliable local stress results (for example, in the ANSYS code, the degeneration form of the element designated SOLID186) [104]. In 1973, Newton [124] showed that a more extensive modification is required for the shape functions of degenerated brick-type isoparametric elements. In 1974, Irons [62] suggested a method for degenerating brick-type isoparametric elements. It has shown that some of the shape functions for a quadratic brick-type isoparametric element require unexpected modification when the element is degenerated by coalescing the nodes of one edge.

First work on nodal pyramidal elements was released by Bedrosian [108] in 1992 who noted the impossibility of choosing only polynomial shape functions if we want to preserve the conformity with neighboring elements. As a solution, he proposes what he calls "rabbit functions" that are rational basis functions on the pyramids. They are as follows:

$$S_1^v = \frac{1}{4}\left(1 - \xi - \eta - \zeta + \frac{\xi\eta}{1-\zeta}\right)$$

$$S_2^v = \frac{1}{4}\left(1 + \xi - \eta - \zeta - \frac{\xi\eta}{1-\zeta}\right)$$

$$S_3^v = \frac{1}{4}\left(1 + \xi + \eta - \zeta + \frac{\xi\eta}{1-\zeta}\right)$$

$$S_4^v = \frac{1}{4}\left(1 - \xi + \eta - \zeta - \frac{\xi\eta}{1-\zeta}\right)$$
$$S_5^v = \zeta \tag{1.129}$$

It was verified that these shape functions satisfy the requirements of C^0 continuity at all faces. For any face, it suffices to show that the shape functions reduce to a 2D triangular or quadrilateral shape function, since adjacent tetrahedrons, prisms or hexahedrons have the same property. It was further observed that the Jacobian matrix is constant for any pyramid element of this kind with a planar parallelogram base. In the general case of a base of any shape, not necessarily in a plane, the Jacobian will, of course, vary with position, but far less strongly than for the degenerate hexahedron.

The rational "rabbit functions" given by Bedrosian [108] for first- and second-order pyramidal elements have been demonstrated to be at least as good as classic elements by Zgainski *et al.* [119, 125]. Coulomb *et al.* [125] discussed the behavior of the "rabbit functions" and their derivatives. At the top of the pyramid, $\zeta = 1$ and consequently $\xi = \eta = 0$ nullify both numerators and denominators of the rational terms. These local coordinates verify the following inequalities:

$$\begin{aligned} -(1-\zeta) \leq \xi \leq (1-\zeta) \\ -(1-\zeta) \leq \eta \leq (1-\zeta) \\ 0 \leq \zeta \leq 1 \end{aligned} \tag{1.130}$$

As the rational terms $\xi\eta/(1-\zeta) \to 0$ when $\zeta \to 1$, the nodal functions S_1^v, S_2^v, S_3^v and S_4^v are well defined at the summit. Inequalities, Eq. (1.130), in local coordinates can also be written as follows:

$$\begin{aligned} -1 \leq \frac{\xi}{1-\zeta} \leq 1 \\ -1 \leq \frac{\eta}{1-\zeta} \leq 1 \end{aligned} \tag{1.131}$$

which shows that derivatives are indeterminate. However, Bravo *et al.* [105] have proved that the rational term in the vertex bases itself and its derivatives vanish at the top of the pyramid, so it does not cause any inconvenience in computation.

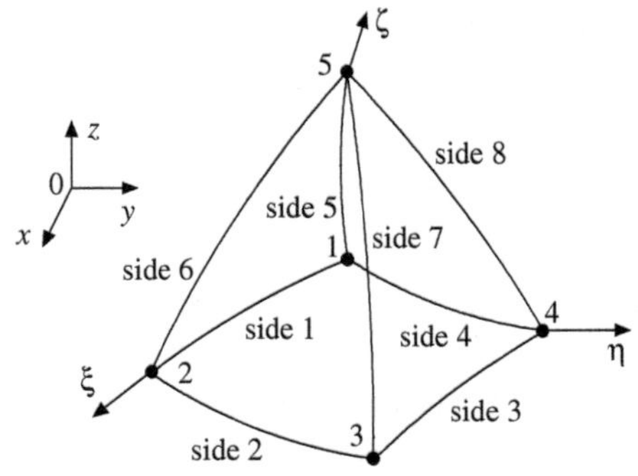

Fig. 1.19. A curved pyramidal element.

Sherwin [126] first derived Dubiner's orthogonal expansion bases on pyramid domains. Then Sherwin *et al.* [109] split the bases into interior and boundary modes (Fig. 1.19). The general form of the displacement field of the three-dimensional hierarchical finite elements (including pyramid elements) can be written as

$$\begin{aligned} & u[x(\xi,\eta,\zeta), y(\xi,\eta,\zeta), z(\xi,\eta,\zeta)] \\ & = \sum_{i=1}^{N_v} S_i^v(\xi,\eta,\zeta)u_i + \sum_{i=0}^{L_k}\sum_{k=1}^{N_e} S_i^{e,k}(\xi,\eta,\zeta)a_i^{e,k} \\ & + \sum_{i=0}^{M_k}\sum_{j=0}^{N_k}\sum_{k=1}^{N_f} S_{ij}^{f,k}(\xi,\eta,\zeta)a_{ij}^{f,k} \\ & + \sum_{i=0}^{L}\sum_{j=0}^{M}\sum_{k=0}^{N} S_{ijk}^{b}(\xi,\eta,\zeta)a_{ijk}^{b} \end{aligned} \tag{1.132}$$

where v, e, f and b respectively denote the vertexes, edges, faces and body, N_v, N_e and N_f respectively denote the total number of vertexes, edges and faces, i, j and k are integer index variables, $S_i^v(\xi,\eta,\zeta)$ are vertex shape functions, $S_i^{e,k}(\xi,\eta,\zeta)$ are edge shape functions, $S_{ij}^{f,k}(\xi,\eta,\zeta)$ are face shape functions, and $S_{ijk}^b(\xi,\eta,\zeta)$ are body shape functions. Although orthogonal bases were derived on pyramid domains, the vertex shape functions shown by Eq. (1.129) were not used.

Nigam and Phillips [121, 127] developed the method of Graglia and Gheorma [120] to define high-order interpolatory curl- and divergence-conforming vector basis functions on pyramidal elements for the numerical solution of 3D electromagnetic field problems. The vector bases were obtained through the combinations of monomials and rational "rabbit functions." Fuentes *et al.* [128] further presented a unified construction of high-order vector bases for elements of "all" shapes including pyramids.

In 2010, Bergot *et al.* [118] constructed highly efficient nodal pyramidal finite elements of any order using rational "rabbit functions" on the pyramid vertices and Dubiner's orthogonal expansion polynomials inside the pyramid domain. Numerical experiments conducted with these elements (up to order 6) exhibit a low phase error, a good stability condition, and a very good behavior in hybrid meshes. Bergot and Duruflé [129, 130] have applied their pyramidal elements [118] to discontinuous Galerkin methods and edge elements for the 3D Maxwell equations.

In 2014, Bravo *et al.* [105] defined the set of rational form functions of degree p in a hierarchical way for pyramidal elements. The concept of side of the pyramidal element was extended to include its vertices, edges, faces and interior. The rational terms are limited within their domains and they go to zero at the element's boundary. The edge, face and body bases were constructed through the rational "rabbit functions" and the Chebyshev polynomials. One of the main tests consisted in interpolating every monomial of degrees 0–7 in three variables, including the rational monomials.

One can see through the above review that orthogonal polynomials using the rational vertex bases shown in Eq. (1.129) have not been derived so far. They will be presented in Section 5.2.4.

1.3.6 *The HQEM on pyramid domains*

In the pioneering work of high-order finite elements defined on pyramids by Bergot *et al.* [118], they placed the degrees of freedom on Legendre–Gauss–Lobatto (LGL) points on the quadrangular base of the pyramid and on Hesthaven points [90] on each triangular face. They added n_i degrees of freedom inside the pyramid

$$n_i = \frac{1}{6}(p-1)(p-2)(2p-3) = \sum_{1\leq k\leq p-2} k^2 \tag{1.133}$$

and they placed them on $(p-2)$ parallel planes of k^2 degrees of freedom.

Chan and Warburton [100] adapted existing procedures for constructing optimized tetrahedral nodal sets for high-order interpolation to the pyramid with constrained face nodes, including two generalizations of the explicit warp and blend construction of nodes on the tetrahedron [96]. Comparisons between nodal sets showed that the lowest Lebesgue constants are given by warp and blend nodes for order $N \leq 7$ and Fekete nodes for $N > 7$, though numerical experiments showed little variation in the conditioning and accuracy of all surveyed non-equidistant points.

A new explicit formula of transforming face Fekete points to the interior of the pyramidal element through blending function interpolation will be presented in Section 5.2.4.

1.4 The *p*-version Finite Element Method for Nonlinear Problems

The higher-order versions of the finite element method (*p* and *hp*-versions) are fairly well established as highly efficient methods for monitoring and controlling the discretization error in linear problems before the 1990s. Holzer and Yosibash [131, 132] are the first ones who discussed which aspects of incremental elasto-plastic finite element analysis are particularly amenable to improvements by the *p*-version. It was demonstrated that the *p*-version performs very well even in cycles of elasto-plastic loading and unloading, not only as compared with the traditional *h*-version but also with respect to the exact solution. An example of considerable practical importance — the analysis of a cold-working lug — is presented, which demonstrates how the modeling tools offered by higher-order finite element techniques can contribute to an improved approximation of practical problems. Szabó *et al.* [133] discussed the solution of small-strain elastic–plastic stress analysis problems by the *p*-version of the finite element method. The formulation of the elastic–plastic problem described in the paper was based on the displacement method. This is possible because *p*-extensions are not susceptible to Poisson ratio locking. It was shown that certain types of nonlinearities, such as material nonlinearities, do not perturb the smoothness of the underlying exact solution significantly. Therefore, the *p*-version is an effective method for solving such problems. Krause *et al.* [134] are the first ones who applied a *p*-version of the finite element method and an *hp*-extension to a geometrically nonlinear model problem. Starting from a standard formulation in finite elasticity, some implementation details were outlined. The robustness and efficiency of the method, as already well known for linear problems, was demonstrated. In two numerical examples, high accuracy and exponential rate of convergence in energy were shown. The above early investigations [131–134] showed numerically that the advantages of the high convergence rate of the *p*-approximation survive in the investigated nonlinear situation.

Han and Petyt [135, 136] applied the HFEM to geometrically nonlinear static analysis of laminated plates and free vibration of isotropic rectangular plates. The von Karman type of geometrically nonlinear strain–displacement relationships and harmonic balance method were used in deriving the equation of motion. It was shown that the use of very high-order polynomial displacement shape functions makes it possible to model the whole rectangular plate with only one element that can greatly reduce the number of degrees of freedom, which is an important feature of nonlinear analysis using the finite element method.

Jeremić and Xenophontos [137] discussed the use of the *p*-version of the finite element method applied to elasto-plastic problems that exhibit sharp (but continuous) deformation gradients. The deformation theory of deviatoric [133], linearly hardening elasto-plasticity with an iterative, displacement-based finite element framework is used. They showed that the use of the *p*-version of the FEM in conjunction with non-uniform but mixed meshes enabled us to accurately predict the shear band formation without the use of enhanced finite element spaces and/or special basis functions. It was concluded that it would be beneficial to extend the use of the *p*-version of the FEM to the incremental, large deformation elasto-plastic theories. The application to problems in failure mechanics of engineering solids would be of great interest.

Ribeiro, Petyt and their collaborators [59, 138–142] applied the hierarchical finite element method (HFEM) to the geometrically nonlinear vibration of beams and isotropic and composite laminated thin plates. The HFEM model is characterized by its small number of DOFs and of elements. This results in very significant time savings when constructing the model, iteratively solving the equations of motion and calculating the characteristic exponents by solving an eigenvalue problem. Ribeiro and Heijden [143] developed a model based on a Timoshenko beam p-version finite element to analyze oscillations that are, simultaneously, elasto-plastic and geometrically nonlinear. The main advantages of the beam p-version hierarchical element were shown to be that it requires fewer DOFs than h-version beam and shell elements and provides a more detailed description of stress and strain fields than the h-version beam elements.

Düster and Rank [144] applied a p-version of the finite element method to the deformation theory of plasticity and the results are compared to a state-of-the-art adaptive h-version. The p-version turned out to be significantly more accurate than the h-version, even when compared to an adaptive h-refinement. They expect from their investigations that the p-version for the deformation theory is superior, as long as the computational domain can be discretized into a small number of elements. Düster *et al.* [145] applied a high-order finite element approach to elasto-plastic problems in two as well as in three dimensions. It was demonstrated that high-order finite elements are indeed competitive to low-order formulations. Moreover, it was shown that a strictly three-dimensional high-order finite element formulation applied to thin-walled structures allows one to consistently compute structures which are commonly still discretized by dimensionally reduced elements. Düster and Rank [146] implemented a two- and three-dimensional p-version approach to the J_2 flow theory with nonlinear isotropic hardening including also cyclic loading for small displacements. Curved boundaries were taken care of with the blending function method, allowing an accurate representation of geometry with only a few p-elements. It is numerically demonstrated that the p-version is an accurate and efficient discretization strategy for the investigated physically nonlinear model problem.

In the past, a lot of effort has been invested in the development of low-order finite elements in order to obtain robust and non-locking 3D finite elements to solve geometrically and physically nonlinear problems of solid mechanics. In this context, mixed finite elements and special element formulations based on stabilization techniques have been proposed to overcome problems occurring in bending dominated structures or for compressible materials. On the other hand, it has been shown by Suri [147] and Szabó and Babuska [67] in the case of small strains that beyond a certain polynomial order of the shape functions locking-free results are obtained. In this context,

Düster *et al.* [148] applied the p-version finite element method to finite strain problems. In particular, the behavior of high-order finite elements was studied for an isotropic hyperelastic material in the case of near incompressibility. It was demonstrated that elements of high order are highly insensitive to distortion, which plays an important role in the field of finite strains. It was shown that the p-version finite element approach is a promising method to compute geometrically highly nonlinear structures.

Nübel *et al.* [149] presented an rp-discretization strategy for physically nonlinear problems based on a high-order finite element formulation. In order to achieve convergence, the p-version leaves the mesh unchanged and increases the polynomial degree of the shape functions locally or globally, whereas the r-method moves nodes and edges of an existing FE-mesh. The rp-adaptive algorithm iteratively detects the boundary of the elastic–plastic interface to subdivide the whole computational domain into areas where smooth solutions exist. Three examples demonstrate that even for this class of problems an exponential rate of convergence can be obtained.

Yosibash *et al.* [150] analyzed axisymmetric pressure boundary loading for finite deformation using p-FEM. Restricting to axisymmetrical applications, they provided analytical and numerical solutions for a set of problems in compressible Neo-Hookean materials so as to serve as benchmark problems for verifying the accuracy and efficiency of various FEMs for follower load applications. Thereafter, the weak formulation for the follower load in 3D domain was reduced to an axisymmetrical setting and, subsequently, consistently linearized in the framework of p-FEMs, exploiting the blending function mapping techniques. Then comparison studies were carried out. As for the efficiency, in terms of DOFs, it was shown that in order to achieve the same accuracy p-FEMs require less than 10% compared to h-FE DOFs, and furthermore, an exponential convergence rate was observed. In terms of CPU time, a direct comparison is difficult because the computer code AdhoC (p-FE solutions) runs under a Linux OS on a PC and ABAQUS runs on a Win/XP OS on a different PC. Nevertheless, both machines are of comparable performance, and the CPU time required for the p-FE runs was shorter overall by about a factor of 10.

Heisserer *et al.* [151] demonstrated the locking-free properties of the displacement formulation of p-finite elements when applied to nearly incompressible Neo-Hookean material under finite deformations. Later, Heisserer and his co-authors [152] considered the simulation of powder compaction problems (die compaction and cold isostatic pressing) by an implicit high-order (p-version) finite element method. In this class of problems, use was made of a finite strain viscoplasticity model with evolution equations for internal variables developed for the highly compressible behavior in powder compaction processes. It was shown that the benefit of the p-version approach for the investigated geometric and material nonlinear applications is a faster convergence in terms of unknowns compared to low-order h-refinement. The return of investment for increasing the polynomial order and using more complicated element stiffness matrices and data structures justifies the effort also with respect to computation time.

Wang and his collaborators [153, 154] investigated the elasto-plastic buckling behavior of thin and thick rectangular plates under biaxial loadings by using the DQ method. Both J_2 flow and deformation plasticity theories were adopted and iteration processes were involved due to the material nonlinearity. DQ results agreed very well with known analytical solutions. This fact indicates that the DQ method can be employed for obtaining elasto-plastic buckling loads of plates subjected to combined distributed loadings. Andakhshideh *et al.* [155] presented nonlinear static analysis of laminated sector plates with any combination of clamped, simply supported and free edges using the generalized differential quadrature (GDQ) method. Particular interest of the study was large deformation of asymmetric sector plates with free edges. Various examples showed predictions of the presented method even with a few number of grid points for various stress and displacement components have very good agreement with other numerical solutions.

He and Zhong [156] applied a weakform quadrature element method to large deflection elasto-plastic analysis of frames. The quadrature element formulation differs from traditional high-order finite elements: displacements at the integration points in the functional of variation principle are designated as DOFs; derivatives at the integration points are directly approximated using differential quadrature analogs. The weakform quadrature beam (QB) element can simply and directly deal with state determination, geometrical nonlinearity and non-uniform distributed loads, contrasting with the force-based element. For both large deflection effect and material nonlinearity, results of the QB model converge with the increase of the number of integration points, and much fewer DOFs are needed in a QB model than those in a classical displacement-based model.

Zhong and his collaborators [157–163] further applied the weak form quadrature element formulation to the geometrically exact beam and shell models undergoing large displacements and rotations. Spatial rotations were represented by quaternion algebras to achieve a total Lagrangian formulation without singularities. Besides high computational efficiency, the formulation retains strain objectivity and avoids locking problems that may occur in low-order finite element formulations. Numerical examples demonstrated the feasibility, accuracy and efficiency of the weak form quadrature element formulation. Fu *et al.* [164] proposed asymptotic formulations incorporating with spectral spatial discretization approach for wrinkling in hyperelastic sheets at finite membrane strain. This is a complex instability problem with highly geometric and material nonlinearities.

Schillinger *et al.* [165, 166] presented the integration of the finite cell method (FCM) into the framework of nonlinear finite element technology. The FCM combines the fictitious domain approach with the p-version of the finite element method and adaptive integration. As a consequence of the weak penalization, nonlinear strain measures provoke excessive stress oscillations in the cells cut by geometric boundaries, leading to a low algebraic rate of convergence. Therefore, the FCM approach was complemented by a local overlay of linear hierarchical basis functions in the sense of the *hp-d* method, which synergetically uses the h-adaptivity of the integration scheme. Numerical experiments show that the *hp-d* overlay effectively reduces oscillations and permits stronger penalization of the fictitious domain by stabilizing the deformation map. The *hp-d*-adaptive FCM is thus able to restore high convergence rates for the geometrically nonlinear case, while preserving the easy meshing property of the original FCM.

Abedian *et al.* [167, 168] compared the finite cell method to h-version finite element method for elasto-plastic problems. The results demonstrated the efficiency of the FCM to solve materially nonlinear problems. Numerical examples showed that the FCM provides efficient and accurate approximations to this class of physically nonlinear problems. It is easier to control the discretization error with the FCM than that with the h-FEM. Therefore, the degree of error can be reduced by increasing the order of elements and not necessarily by a new mesh. Error monitoring and control were achieved automatically, with minimum user interaction and without substantial loss of computational efficiency. Taghipour *et al.* [169, 170] applied *p*-version finite elements and finite cells to finite strain elasto-plastic problems with complex geometries. The performance of the FCM is numerically investigated in the finite strain range under nearly incompressible plastic deformations, and the most important conclusions are as follows: (1) *p*-version cells with pure displacement-based formulation can provide locking-free behavior. Non-overstiff solutions can be obtained when a high enough *p* value is used. (2) It offers high convergence rates and provides easy mesh generation capabilities for highly complex geometries.

1.5 The *p*-version Finite Element Method for Plates and Shells

Shells are common and most efficient structures found in nature. Man-made shell structures are abundant in engineering applications, for example, roofs, bridges, the bodies of cars and airplanes, rockets, ship hulls, and pressure vessels, to name a few. In these cases, a thin structure covers a wide area and holds large external and internal loads, making it possible to build lightweight structures with the minimum amount of material [171, 172]. Shells with composite material often exhibit special performance such as high strength-to-weight and stiffness-to-weight ratios, excellent fatigue strength, and excellent design flexibility [173]. Owing to the wide applications of shell structures, great efforts have been devoted to the development of advanced theoretical models and numerical methods to predict their mechanical behaviors.

For single-layer shell analysis, the classical Kirchhoff–Love (KL) theory and Reissner–Mindlin (RM) theory are the most popular ones. The KL theory, also known as "thin shell theory", assumes that the shell cross-sections remain normal to its mid-surface in the deformed configuration, which implies that the transverse shear strains are zero. By contrast, the RM theory, also known as "thick shell theory," assumes a linear expansion of transverse shear strains along the thickness direction. As a result, additional rotation fields are used to describe the position of normal fibers. In the field of laminated plates and shells, theory models corresponding to the KL and RM theories are respectively known as the classical lamination theory (CPT) and the first shear deformation theory (FSDT) [173]. These two models along with those high-order shear deformation theories (HSDT) [174, 175] are often referred to as the equivalent single-layer (ELS) models because the displacements through all layers are described equivalently by a common smooth function. These models are widely used in the prediction of macro-structural responses such as deflections and vibration frequencies [174, 176, 177]. In addition to the ESL models, layerwise and zigzag models [178–181] were also studied by researchers because of their superiorities in predicting the local zigzag effects that arise frequently in thick laminated plates and shells. For a comprehensive review of different plate and shell models, one can refer to the works of Reddy [173, 182, 183], Carrera [184] and Oñate [185].

Compared with the common h-version finite elements, p-version elements typically have a higher convergence rate and, hence, produce highly accurate results using fewer DOFs [186]. In some engineering problems, the convergence of the results may decline substantially if the domains of the problems are represented inaccurately. An illustrative example can be found in [187] (Chapter 5), where elements that have slightly inaccurate geometries cause the solutions to converge to erroneous results. Thus, for analyzing problems with irregular domains with high accuracy, p-version finite elements that have curved boundaries are essential.

As for analysis using the p-version FEM, with HFEM as its special case, this has been found to be highly accurate, insensitive to mesh distortion and locking effects in many application fields. For these reasons, the application of HFEM to the analysis of plates and shells has been studied by a lot of researchers. Since the implementation of p-version FEM relies heavily on a high-accurate curved mesh generator to ensure the solution error to be controlled by p rather than the errors of geometry discretization [188–190], the p-version plate elements have been studied more extensively in the literatures, compared with the shell elements, because the curved mesh generation for a 2D plane is simpler than that for the 3D surfaces. However, due to the difficulty of satisfying C^1 conformity, most p-version plate elements are based on the RM-like plate theories with only C^0 continuity of the solution space. For example, Ahmed *et al.* [191] developed p-version finite elements for the 2D laminated composite plates based on a high-order kinematical model; Rank *et al.* [192] studied the accuracy of p-elements for Reissner–Mindlin plate problems, and they found that very accurate displacements and bending moments can be obtained using p-elements with very coarse meshes; Woo *et al.* [193] proposed a p-element for the material and geometrically nonlinear analysis of laminated plates using a von Karman-hypothesis-based ESL model; and Yazdani [194] proposed a p-elements for large-deflection analysis of composite plates with curvilinear fibers based on the zigzag layerwise theory.

Compared with the C^0 type elements, the p-elements with C^1 conformity are found to be rare. Earlier literature regarding p-version C^1 elements can be found in [195–197]. These bases were constructed via linear transformation of the monomials of Pascal's triangle or by using rational functions. However, these studies were limited to straight-sided triangular elements. Bardell [198, 199] developed a series of hierarchical bases for rectangular elements using the tensor product of third-order Hermite bases, along with Legendre orthogonal polynomials, and applied the elements to vibration analyses of rectangular and skew plates. However, C^1 continuity of the elements was only satisfied for rectangular elements that had sides that are parallel to the global axis since the starting point of the bases was the BFS elements that have such a difficulty [200]. Beslin and Nicolas [47] developed a type of hierarchical bases using trigonometric functions and applied the bases to the prediction of very high-order plate-bending modes. Their study was limited to analysis using a single patch due to the conformity difficulty, as in Bardell's work. Xing and Liu [16] developed a weak-form DQM that uses tensor product Lagrange bases that are based on non-uniformly distributed Gauss–Lobatto points for thin plate vibration on curvilinear domains. The method was further improved by Zhong and Yue [201] and Jin and Wang [36] via the combinational use of Lagrange and Hermite bases. Recently, Ferreira *et al.* [202] proposed a new set of C^1-conforming hierarchical bases for quadrangular and triangular elements. Their quadrilateral H_3 elements were also based on the BFS elements, like the Bardell elements, and the H_5 elements were derived from the tensor products of Fifth-order Hermite bases that were enriched by high-order terms. The triangular bases were based on the TUBA elements and were enriched by tensor product edge and face functions. Fast convergence of the elements was observed in thin plate analyses. However, elements that had curved boundaries were not contained in their studies and local refinement may propagate from element to element due to the use of tensor product bases. In summary, most of the existing p-version C^1 elements are limited to the straight-side case. As aforementioned, for the finite element discretization using p-version elements with curved boundaries are essential for the domains with irregular shape.

However, the construction of fully C^1 conforming elements that have curved boundaries was found to be difficult. The formulae may become very complicated even when only one curved side is involved [203, 204]. The method of isogeometric analysis (IGA), which was introduced by Hughes *et al.* [205], is a simple approach for tackling the problem of C^1 conformity within a single patch. Using the strategy of k-refinement [205, 206], one can easily obtain the bases of high-order continuity. However, the advantage of adjusting the continuity order is limited to the interior of the patch. Due to the non-interpolative property of NURBS bases and the rotation-free feature of control variables, the enforcement of C^1 conformity on patch boundaries is difficult, as in FEM, and additional conformity-enhancement techniques are often employed. For example, Kiendl and Shojacee *et al.* [207–209] employed the NURBS bases to analyse the Kirchhoff shells and plates. The C^1 conformity between patches was approximately satisfied by using the bending strip method, where a penalty factor for rotation discontinuity was introduced. The penalty factor should be selected carefully such that the global stiffness matrix does not become poorly conditioned [208]. Guo *et al.* [210, 211] used Nitsche's method to enforce the C^1 conformity in an integral sense and applied it to the NURBS-based IGA of Krichhoff shells. However, compared with conventional FEM, this method is complex. To overcome the limitations of NURBS-based IGA, Liu *et al.* [212] constructed non-uniform rational Hermite (NURH) functions for C^0 and C^1 isogeometric analysis. However, they encountered the same difficulties as with the C^1 tensor product Hermite elements [198, 199]. In our recent work [213], a strategy of interpolation of normal derivatives on Gauss–Lobatto nodes had been proposed to deal with the problem of C^1 conformity on curved boundaries, and new triangular and quadrilateral C^1 elements were developed for Kirchhoff plates. Through several numerical examples, it was shown that exponential rate of convergence can be achieved even when curved elements are used in the discretization.

For the analysis of curved shells using p-elements, most elements in the literatures were developed based on the 3D degenerated shell model. For example, Surana *et al.* [214, 215] proposed a series of 3D p-version curved shell elements for the elasto-static analysis of isotropic and laminated composite shells. In their work, the element geometry was described by the nodes located at the middle surface of the element and the nodal vectors. Using the hierarchical bases, the approximation of element displacement can be of arbitrary polynomial order in the in-plane as well as out-plane direction. With some benchmark problems, they demonstrated that their elements are highly accurate, efficient and practical means of simulating linear elastic behavior of plate and shell structures as well as many three-dimensional solid applications. Rank *et al.* [216–218] developed a p-version element for solid shells based on anisotropic Ansatz spaces. The geometry of their element was expressed using the exact 3D blending function interpolation. It was demonstrated numerically that the use of anisotropic Ansatz spaces in combination with the blending function method allows one to efficiently compute the structural behavior of plate and shell-like structures. The flexibility of their elements enables one to switch consistently from solid to shell-like structures without using dimensionally reduced models. Regardless of the feasibility of solid shell elements, in practical CAD modeling a shell-like structure is generally expressed by curved surfaces. Therefore, it is attractive that a shell element is constructed directly based on the curved surfaces, which means curved mesh generation on surfaces is required. Relative research can be found in the work by Payette and Reddy [219], where a seven-parameter spectral/*hp* finite element was proposed for the isotropic, laminated composite and functionally graded shells. However, the limitation of their work is that the shell surfaces are expressed by analytical functions with simple geometry. With one step further, in the authors' recent work [181], a p-version shell element was constructed for the isotropic and laminated shells. The geometries of the elements were based on the NURBS mapping which is a general tool for CAD geometry representation [220]. In this way, the geometry discretization errors can be fully removed, which is beneficial for the p-version discretization with very coarse mesh.

Finally, it should be noted that the aforementioned curved shell elements are all C^0 type elements. However, for thin shell analysis based on the Kirchhoff–Love model, the C^1 conformity of the solution space is required. Compared with the C^1 plate elements, publications on the C^1 shell elements are found to be rare. Early developed thin shell elements were the h-version flat shell elements derived from the composition of flat thin plate element and the membrane elements [221, 222]. In these formulations, the C^1 plate elements with only the first-order derivatives [223, 224] have been widely adopted due to the clear physical significance. By contrast, the application of C^1 elements with high-order derivatives is implemented in a different way. For example, the early developed SHEBA elements, see the works of Argyris [225, 226], used the same interpolation strategy for the displacement field and the geometry representation to achieve the C^1 continuity. In their implementation, a global mapping function is required to produce the geometry interpolation information that is required to define the finite element mesh. More recently, Ivannkov *et al.* [227] generalized the TUBA plate elements to the nonlinear thin shell analysis in a different way. The main feature of their formulation is that the reference configuration of shell body is taken as the flat parametric domain. After that, the C^1 conforming displacement fields of shell elements can be constructed using the flat TUBA plate elements along with the geometric mapping. However, in both formulations of Argyris and Ivannkov, it requires that the entire shell surface is represented by one parametric function, which is inconvenient in dealing with shells defined by multipatches of different parameterizations. Compared with the h-version C^1 elements, literature research shows that no publication has be found in the field of p-version curved C^1 shell elements. The reason is of course the difficulty of C^1 conformity on the surface boundary of shell elements. This problem has been studied in our recent work [228], where the interpolation strategy for our C^1 plate element [213] has been used to solve the C^1 conformity problem of the shell elements, and good convergence performance was verified. In recent years, several new methods for thin shell analysis were developed by invoking the bases for CAD geometry representation. Some of them such as the subdivision surface method [229, 230] and the isogeometric analysis method [207, 210, 211, 231–233] are worthy of mention. These methods greatly promoted the practical application of the thin shell model.

1.6 A Review of High-Order Mesh Generation

The advantages of *p*- and *hp*-adaptive finite element computational strategies have been well established before the 1990s. For properly designed meshes, *p*- and *hp*-adaptive schemes are capable of delivering exponential rates of convergence of the finite element solution error for domains with planar and piecewise smooth curvilinear boundaries. The increased approximation power allows the use of coarse meshes in which relatively few elements are used to cover large portions of the computational domain. Realization of exponential convergence possible with *hp*-adaptive finite element methods requires that the approximation of the domain geometry be carefully controlled. Approximate geometry representation leads to the inexact representation of boundary and initial conditions and to the approximate evaluation of element level integrals. In order to sustain the exponential rate of convergence with *hp*-refinement, these perturbation errors must also converge exponentially. Thus, in a *p*-adaptive environment, the accuracy of geometry approximations and that of geometry-related items must be related to *p*, namely, a curvilinear mesh geometry representation that captures the features of the geometry should be used during the finite element computations. The ability to construct suitable computational meshes was gradually regarded as a significant limiting factor in the development of compact high-order algorithms in very complex geometries. Thus, investigations on this topic appeared in 1997 and grew fast thereafter.

Shephard and his collaborators [72, 234] presented a mesh geometry mapping scheme, based on using model entity parametric coordinates and the boolean sum interpolation theory, which conforms exactly to the shape of the domain boundary as defined in the geometric modeling system. On simplex domains, the linear blending schemes have limited smoothness and hence theoretically only guarantee optimal convergence of the discretization error up to a critical *p*. Dey *et al.* [235] presented issues related to creating curvilinear meshes suitable for *p*-version finite element analysis. These issues include the representation of curved mesh geometry as well as definition of what constitutes a valid curvilinear mesh. Basic requirements for curvilinear meshing were discussed as well as two general approaches for curvilinear mesh generation were presented: (1) *direct generation* of curvilinear meshes of CAD geometry; (2) *a posteriori* curving of straight-sided meshes with respect to the model geometry. They discussed meshing tools required to generate valid curvilinear meshes and included detecting interference between curved mesh entities and defining distortion measures using Jacobians. The use of local mesh modification tools to aid in *correcting invalid curved mesh regions* was then presented. These operations included face and edge swapping, edge deletion, and curving mesh entities. Finally, curvilinear meshes that were the result of applying these techniques to straight-sided meshes were presented. These meshes were curved using quadratic Lagrange interpolation functions.

Traditional unstructured mesh generation techniques can provide a coarse linear discretization of a complex geometry. However, the subdivision of this mesh into a boundary conforming mesh of high-order elements requires careful handling due to the distortion that non-planar surfaces might introduce. Sherwin and Peiró [236] presented three complementary approaches for improving high-order meshes. Firstly, an approach was developed to generate optimal surface discretizations for high-order elements that accounts for surface curvature. This generation process minimizes the element distortion due to the surface curvature by searching for the shortest geodesic line between two points while maintaining a Gaussian quadrature distribution. Secondly, the use of hybrid discretization employing prismatic and tetrahedral elements was investigated. Finally, the use of curvature has proven to be a very helpful tool to enable valid high-order mesh generation. Using the computational reconstruction of the arterial bypass graft as an example computational domain, they demonstrated how the combination of all three strategies leads to a valid high-order discretization of the computational domain.

Luo *et al.* [237] presented a procedure for automatic *p*-version mesh generation for general three-dimensional domains defined in solid modeling systems. The key steps are as follows: (1) automatically isolate the singular edges and vertices in the model;

(2) generate a coarse linear mesh accounting for the isolated singular features; (3) curve the singular feature isolation mesh to maintain a properly curved mesh isolation; (4) curve the remaining mesh entities classified on the curved boundaries; (5) apply local mesh modifications as needed to ensure a valid mesh is produced. Shephard [238] considered the technologies needed to support the creation of adaptively constructed meshes for general curved three-dimensional domains and outlined one set of solutions for providing them. Luo *et al.* [239] presented a procedure to track moving adaptive mesh refinement in curved domains, which is capable of generating suitable curvilinear meshes to enable large-scale accelerator simulations. The procedure combined a general mesh curving tool and size-driven mesh adaptation to produce valid curved meshes with substantially fewer elements.

Remacle *et al.* [240] presented a general method for the post-processing treatment of high-order finite element fields. The method applies to general polynomial fields, including discontinuous finite element fields. The technique uses error estimation and h-refinement to provide an optimal visualization grid. Schemes and algorithms that are discussed in the paper are readily available as part of an open source project that was developed by the authors of the paper, namely, Gmsh [241], which is an open-source 3D finite element grid generator with a built-in CAD engine and post-processor.

Persson and Peraire [242] proposed a method for generating well-shaped curved unstructured meshes using a nonlinear elasticity analogy. The geometry of the domain to be meshed was represented as an elastic solid. The undeformed geometry is the initial mesh of linear triangular or tetrahedral elements. The external loading results from prescribing a boundary displacement to be that of the curved geometry, and the final configuration was determined by solving for the equilibrium configuration. The deformations were represented using piecewise polynomials within each element of the original mesh. When the mesh is sufficiently fine to resolve the solid deformation, this method guarantees non-intersecting elements even for highly distorted or anisotropic initial meshes. They applied their technique on many problems with excellent results. The technique was used by Hindenlang *et al.* [243] to generate curved surface and volume meshes for curved boundary layers.

In the book *ADIGMA — A European Initiative on the Development of Adaptive Higher-Order Variational Methods for Aerospace Applications*, which summarized the results of a collaborative research project funded by the European Union from 2006 to 2009, Kroll *et al.* [244] concluded as follows: "In order to realize the full potential of adaptive higher order methods, further concentrated research effort is required. *Particular research areas to be addressed are generation of coarse higher-order meshes* and memory-efficient solver strategies for large-scale simulations for turbulent flow problems." With realization of "the ability of generating coarse unstructured meshes suitable for high-order methods has been identified as a major bottleneck for industrial type applications" [245], the follow-on project IDIHOM (Industrialization of High-Order Methods) included high-order grid generation as one of the major parts. The works by Remacle *et al.* [240], Geuzaine and Remacle [241], Johnen *et al.* [246–248], and Toulorge *et al.* [249, 250] seem to be supported by this project.

Johnen *et al.* [246–248] presented a way to compute accurate bounds on Jacobian determinants of curvilinear polynomial finite elements. The key feature of the method is to expand the Jacobian determinant using a polynomial basis, built using Bézier functions, that has both properties of boundedness and positivity. The proposed algorithm can either be used to determine the validity or invalidity of curved elements or provide an efficient way to measure their distortion. Triangles, quadrangles, tetrahedrons, prisms, pyramid and hexahedra can be analyzed using the same algorithm, which is available in the open source mesh generator, Gmsh.

Toulorge *et al.* [249] presented a technique that allows one to untangle high-order/curvilinear meshes. The technique makes use of unconstrained optimization where element Jacobians are constrained to lie in a prescribed range by moving logbarriers. The untangling procedure starts from a possibly invalid curvilinear mesh and moves mesh vertices with the objective of producing elements that all have bounded Jacobians. Bounds on Jacobians are computed using the results of Johnen *et al.* [246]. The technique is applicable to any kind of polynomial element for surface, volume, hybrid or boundary layer meshes. All the material presented in the paper is readily available in Gmsh. Later, Toulorge *et al.* [250] further presented a method to generate valid high-order meshes with optimized geometrical accuracy. The optimization procedure is used to both untangle invalid elements and optimize the geometrical accuracy of the mesh.

Xie *et al.* [251] presented an *a posteriori* strategy for obtaining high-order curved meshes suitable for finite element analysis in both two and three dimensions. The strategy adopted is based upon curving a generated initial mesh with planar edges and faces by using a linear elasticity analogy. The analogy employs boundary loads which ensure that nodes representing curved boundaries lie on the true surface. Several examples, in both two and three dimensions, illustrate the performance of the proposed approach, with the quality of the generated meshes being analyzed in terms of a distortion measure. The technique was used by Szałtys *et al.* [252] to generate anisotropic hybrid mesh for high-Reynolds number flows. Hartmann and Leicht [111], and Abgrall *et al.* [253] used the linear elasticity technique to compute curved meshes for fluid dynamics problems. It is clear that when large deformations are induced to produce the deformed curvilinear high-order mesh, a linear elastic model can result in non-valid elements due to the violation of the hypothesis of small deformations. In order to alleviate this problem, Poya *et al.* [254] proposed an incrementally linearized elasticity formulation (wherein the geometry, the tangent operator and the stresses are updated incrementally) to generate curvilinear high-order meshes.

Lu *et al.* [255] presented a parallel adaptive mesh control procedure designed to operate with high-order finite element analysis packages to enable large-scale automated simulations on massively parallel computers. The core of the procedure are two classes of mesh modification: entity geometry modification and local mesh modification for curved meshes. For the entity geometry modification, curved entity reshape operations that explicitly resolve element invalidity and improve the shape quality of curved elements were presented. The local mesh modification operations for curved meshes were extended from the operations for straight-sided meshes with additional consideration and treatment of curve boundary entities and selected curved interior entities.

Xu and Chernikov [256] presented a novel approach for automatically generating a high-order mesh with two main features: (1) the boundary of the mesh is globally smooth; (2) the mesh boundary satisfies a required fidelity tolerance. Invalid elements were eliminated. Example meshes demonstrated the features of the algorithm.

Gargallo-Peiró *et al.* [257–260] presented a robust method for generating and defining distortion (quality) measures for curved high-order tetrahedral and surface meshes. The proposed definition of mesh distortion (quality) is valid for any interpolation degree and allows detecting the validity of a high-order tetrahedron. The mesh generation approach consists of modifying an initial linear mesh by first introducing high-order nodes, second, displacing the boundary nodes to ensure that they are on the computer-aided design surface, and third, smoothing and untangling the mesh obtained after the displacement of the boundary nodes to produce a valid curved high-order mesh. The smoothing algorithm is based on the optimization of a regularized measure of the mesh distortion relative to the original linear mesh. This means that whenever possible, the resulting mesh preserves the geometrical features of the initial linear mesh, such as shape, stretching, and size. They presented several examples to illustrate the performance of the proposed algorithm. Furthermore, the examples showed that the implementation of the optimization problem is robust and capable of handling situations in which the mesh before optimization contains a large number of invalid elements.

Moxey *et al.* [261, 262] proposed an isoparametric approach, whereby a mesh containing a valid coarse discretization comprising high-order triangular prisms near walls was refined to obtain a finer prismatic or tetrahedral boundary-layer mesh. The validity of the prismatic mesh provides a suitable mapping that allows one to obtain very fine mesh resolutions across the thickness of the boundary layer.

Several curvilinear mesh generation techniques above utilized mesh deformation to move the interior nodes of the linear mesh in order to accommodate curvature at the boundary. Many of these were based on elastic models, where the mesh was treated as a solid body and deformed according to a linear or nonlinear stress tensor. However, such methods typically have no explicit control over the validity of the elements in the resulting mesh. Moxey *et al.* [263] presented an extension of this elastic formulation, whereby a thermal stress term was introduced to "heat" or "cool" elements as they deform. They outlined a proof-of-concept implementation and showed that the adoption of a thermo-elastic analogy led to an additional degree of robustness by considering examples in both two and three dimensions.

To overcome the limitations of trivariate NURBS and T-splines, Engvall and Evans [264] proposed the use of rational Bernstein–Bézier tetrahedrons to create analysis suitable volumetric parameterizations for isogeometric analysis. In part one of the two-part series, they presented a method for blending isogeometric analysis with traditional meshing techniques to create geometrically exact 2D meshes. By using rational Bernstein–Bézier triangles as their finite elements, they have limited their basis to be C^0-continuous across element boundaries. However, this was considered to be an acceptable trade-off for the flexibility and robustness offered by triangular mesh generation technology.

Fortunato and Persson [189] proposed a method to generate high-order unstructured curved meshes using the classical Winslow equations. They started with an initial straight-sided mesh in a reference domain and fixed the position of the nodes on the boundary on the true curved geometry. In the interior of the domain, they solved the Winslow equations using a new continuous Galerkin finite element discretization. The formulation appears to produce high-quality curved elements, which are highly resistant to inversion. In addition, the corresponding nonlinear equations can be solved efficiently using Picard iterations even for highly stretched boundary layer meshes. Compared to several previously proposed techniques, such as optimization and approaches based on elasticity analogies, this can significantly reduce the computational cost while producing curved elements of similar quality.

Ruiz-Gironés *et al.* [265, 266] proposed a 3D mesh curving method that converts a straight-sided mesh to an optimal-quality curved high-order mesh that interpolates a CAD boundary representation. They devised the method as follows. First, the boundary of a straight-sided high-order mesh was curved to match the curves and surfaces of a CAD model. Second, the method minimized the volume mesh distortion with respect to the coordinates of the inner nodes and the parametric coordinates of the curve and surface nodes. The proposed minimization features untangling capabilities, and therefore, it repairs the invalid elements that may arise from the initial curving step. Compared with other mesh curving methods, it is less constrained since the boundary nodes are free to slide on the CAD curves and surfaces. Hence, the proposed method is well suited to generate curved high-order meshes of optimal quality from CAD models that contain thin parts or high-curvature entities.

The generation of sufficiently high-quality unstructured high-order meshes remains a significant obstacle in the adoption of high-order methods. However, there is little consensus on which approach is the most robust, fastest and produces the "best" meshes. Turner *et al.* [1, 267] reformulated a number of previous methods for curvilinear mesh generation and compared their relative performance. They found that elasticity methods generally produce higher-quality meshes — at least under the definition of quality given by the scaled Jacobian — but require more iterations to reach this point of optimization. In a problem where cost can be prohibitive, use of the distortion or Winslow formulations will converge faster and could therefore be an advantage. Eichstädt *et al.* [268] investigated how the variational framework introduced by Turner *et al.* [1, 267] can effectively make use of modern heterogeneous architectures.

Bock and Stiller [269] presented a strictly local optimization method to construct high-order triangular surface patches of high quality and accuracy. It combined fitting and energy minimization, in which approximate bending and stretching functionals were minimized by means of an incremental procedure. The method was applied to analytically defined smooth surfaces as well as scattered surface data derived from scanning data. In both cases, the optimization yielded considerable improvements in patch quality while preserving the accuracy of pure least-squares fitting.

Feng *et al.* [270] proposed an extension of Optimal Delaunay Triangulations (ODT) to curved and graded isotropic meshes. By exploiting a continuum mechanics interpretation of ODT instead of the usual approximation theoretical foundations, they formulated a very robust geometry and topology optimization of Bézier meshes based on a new simple functional promoting isotropic and uniform Jacobians throughout the domain. They demonstrated that the resulting curved meshes can adapt to complex domains with high precision even for a small count of elements thanks to the added flexibility afforded by more control points and higher-order basis functions.

Ziel *et al.* [10] presented two curving methods that take a piecewise fine linear mesh as input: a least squares approach and a continuous optimization in the H^1-seminorm. Hierarchic, modal shape functions are used as basis for the geometric approximation. This approach allows one to create very high-order curvilinear meshes efficiently ($q > 4$) without having to optimize the location of non-vertex nodes. The methods were compared on two test geometries and then used to solve a Helmholtz problem at various input frequencies.

Ims and Wang [271] presented an algorithm for the CAD-free conversion of linear unstructured meshes into curved high-order meshes, which are necessary for high-order flow simulations. The algorithm operates via three steps: (1) autonomous detection of feature curves along the mesh surface, (2) reconstruction of the surface curvature from the combination of surface node positions and feature curve positions, and (3) alignment of the mesh interior to the newly curved surface.

The characterization and analysis of interpolation error bounds for curved, high-order finite elements are often desired to efficiently obtain the solution of PDEs when using the FEM. Using a Taylor series expansion approach, Moxey *et al.* [272] derived an interpolation error bound for both straight-sided and curved high-order elements. The availability of this bound facilitates better node placement for minimizing interpolation error compared to the traditional approach of minimizing the Lebesgue constant as a proxy for interpolation error. This is useful for the adaptation of the mesh in regions where increased resolution is needed and where the geometric curvature of the elements is high, e.g., boundary layer meshes.

Li [273] presented a 3D curved mesh generator for converting linear elements to quadratic elements required by high-order methods, which was based on the reconstruction of Cubic Bézier surfaces. Successive curved mesh refinement is also supported by inquiring the middle nodes of the edges and faces of the reconstructed quadratic elements via the Cubic Bézier surface method. Numerical test cases were shown to demonstrate the capability of both mesh curving and refinement around 3D geometries.

Stees and Shontz [274] proposed a heuristic for correcting tangled second- and third-order meshes. For each interior edge, their method minimized an objective function based on the unsigned angles of the pair of triangles that share the edge. They also proposed an optimization-based approach for generating high-order meshes based on affine combinations of nodal positions [275].

Zhao *et al.* [2] developed a high-order curvilinear hybrid mesh generation technique based on an improved radius basic function (RBF) approach by which the straight-edge mesh was converted into high-order curve mesh. Firstly, an initial straight-edge mesh was prepared by traditional grid generation software. Then, high-order interpolation points were inserted into the mesh entities such as edges, faces, and cells according to the final demand of mesh order. To preserve the original geometry, the inserted points on solid wall were then projected onto the CAD model using an open source tool "Open Cascade." Finally, other inserted points in the field near the solid wall were moved to appropriate positions by the improved RBF approach to avoid tangled cells. To avoid possible improper mesh, a weighting based on the local grid aspect ratio between normal direction and tangential direction was introduced into the baseline RBF approach. Three typical configurations were tested to validate the mesh generator.

Mohammadi *et al.* [276] presented a new method of generating high-order curvilinear triangular meshes directly from curved geometries. Their method does not require a post-processing step, such as mesh untangling, since they generated each element as a high-quality valid element. The method can be used to generate curvilinear triangular meshes from patient-specific epicardial and endocardial contours, brain ventricular contours, etc., extracted via segmentation from medical images, such as MRI or CT images.

1.7 Outlook

Although implementations of the p-version are available in a number of commercial finite element computer codes, widespread applications of the p-version in professional practice were shown to be limited by three factors [277, 278]:

(1) The infrastructure of the most widely used FEA software products was designed for the h-version and cannot be readily adapted to meet the technical requirements of the p-version. Existing high-order schemes are generally less robust and more complex to implement than their low-order counterparts.

(2) In typical industrial problems, finite element meshes are generated by automatic mesh generators that produce very large numbers of tetrahedral elements mapped by low-order (linear or quadratic) polynomial mapping functions. When the mapping functions are of low degree, the use of high-order elements is generally not justified. As mesh generators improve and produce fewer elements and more accurate mappings, this obstacle will be gradually removed.
(3) The demand for verified information in industrial applications of FEMs has been generally weak; however, as computed information is becoming an increasingly important part of the engineering decision-making process, the demand for verified data, and hence the importance of the p-version, is likely to increase.

The situations are changing now. In some engineering problems like boundary layer analyses in fluid dynamics and thin-walled structures analyses in solid mechanics, a small geometric error may induce a great fluctuation in the final results. Significant development of high-order methods and high-order mesh generation was achieved after the ADIGMA (A European Initiative on the Development of Adaptive Higher-Order Variational Methods for Aerospace Applications) and IDIHOM (Industrialization of High-Order Methods) projects funded by the European Union from 2006 to 2014. The wide application of p-FEM may come true in a few decades.

Chapter 2

A Differential Quadrature Finite Element Method

This chapter formulates one- to three-dimensional (1D to 3D) element matrices of the differential quadrature finite element method (DQFEM) as a combination of differential quadrature method (DQM) and standard finite element method (FEM). The Lagrange polynomials are used as the trial functions of C^0 and C^1 differential quadrature (DQ) finite elements with regular and/or irregular shapes. The inter-element compatibility requirements for problems with C^1 continuity are implemented by modifying the nodal parameters using the DQ rules. The DQ finite element matrices are computed by simple algebraic operations of the weighting coefficient matrices of the DQ rules and Gauss–Lobatto integral rules, which greatly simplify the computation of element matrices, especially the higher-order elements. The diagonal mass matrices of C^0 finite element in DQFEM can reduce the computational cost for dynamic problems. Numerical comparison studies through 2D and 3D elasticity problems demonstrate the high accuracy and rapid convergence of the DQFEM.

2.1 The Reformulated Differential Quadrature Rule

The DQM is based on the idea that the partial derivatives of a field variable at the ith discrete point in the computational domain are approximated by a weighted linear sum of the field variable along the line that is parallel to the coordinate direction of the derivatives and passes through that point. For example, the nth derivatives of the field variable $f(x)$ at point x_i is approximated as

$$f_i^{(n)} = \sum_{j=1}^{N} A_{ij}^{(n)} f_j \quad (i = 1, 2, \ldots, N) \tag{2.1}$$

where $A_{ij}^{(n)}$ are the weighting coefficients of the nth order derivatives and N is the number of grid points in the x-direction. For the computation of weighting coefficients and more details about the DQM, one may refer to the survey paper by Bert and Malik [11].

The above DQ approximation is shown for a one-dimensional (1D) case, based on which the two- and three-dimensional (2D and 3D) DQ approximations can be easily obtained. The rth-order partial derivatives of $f(x, y)$ with respect to x, the sth-order partial derivatives of $f(x, y)$ with respect to y, and $(r + s)$th-order partial derivatives of $f(x, y)$ with respect to both x and y can be expressed at point (x_i, y_j) as

$$\left.\frac{\partial^r f}{\partial x^r}\right|_{ij} = \sum_{m=1}^{M} A_{im}^{(r)} f_{mj}, \quad \left.\frac{\partial^s f}{\partial y^s}\right|_{ij} = \sum_{n=1}^{N} B_{jn}^{(s)} f_{in}$$
$$\left.\frac{\partial^{r+s} f}{\partial x^r \partial y^s}\right|_{ij} = \sum_{m=1}^{M} A_{im}^{(r)} \sum_{n=1}^{N} B_{jn}^{(s)} f_{mn} \tag{2.2}$$

where M and N are the numbers of grid points in x-direction and y-direction (see Fig. 2.1), respectively. $A_{ij}^{(r)}$ and $B_{ij}^{(s)}$ are the

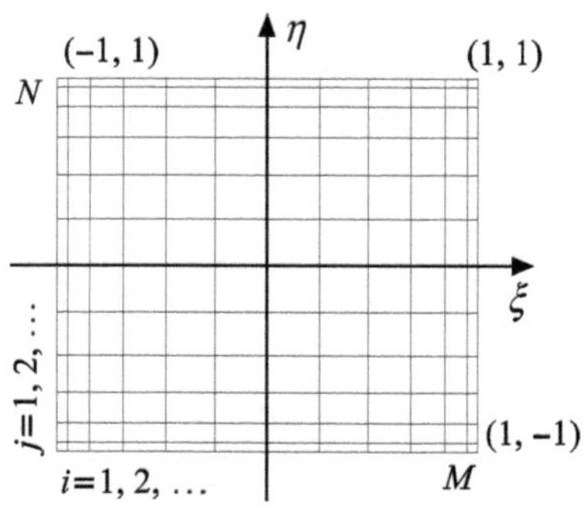

Fig. 2.1. A square region.

weighting coefficients associated with the rth-order and sth-order partial derivative of $f(x, y)$ with respect to x and y at the discrete points x_i and y_i, respectively.

The DQ approximates for the first-order derivatives of function $f(x, y, z)$ are required in the 3D formulation of this paper. They can be written as

$$\left.\frac{\partial f}{\partial x}\right|_{ijk} = \sum_{m=1}^{M} A_{im}^{(1)} f_{mjk}, \quad \left.\frac{\partial f}{\partial y}\right|_{ijk} = \sum_{n=1}^{N} B_{jn}^{(1)} f_{ink}$$

$$\left.\frac{\partial f}{\partial z}\right|_{ijk} = \sum_{l=1}^{L} C_{kl}^{(1)} f_{ijl} \tag{2.3}$$

where $A_{ij}^{(1)}$, $B_{ij}^{(1)}$, and $C_{ij}^{(1)}$ are the weighting coefficients associated with the first-order partial derivative of $f(x, y, z)$ with respect to x, y, and z at the discrete points x_i, y_i, and z_i, respectively.

DQM has been applied to irregular domains with the help of the natural-to-cartesian geometric mapping with the serendipity-family interpolation functions [279] or the blending functions which permit exact mapping [280]. In fact, the mapping using serendipity-family interpolation functions is applicable to arbitrary domain. For an arbitrary quadrilateral domain (see Fig. 2.2), the geometric mapping has the form

$$\begin{cases} x = x(\xi, \eta) = \sum S_k(\xi, \eta) x_k \\ y = y(\xi, \eta) = \sum S_k(\xi, \eta) y_k \end{cases} \quad -1 \le \xi, \eta \le 1 \tag{2.4}$$

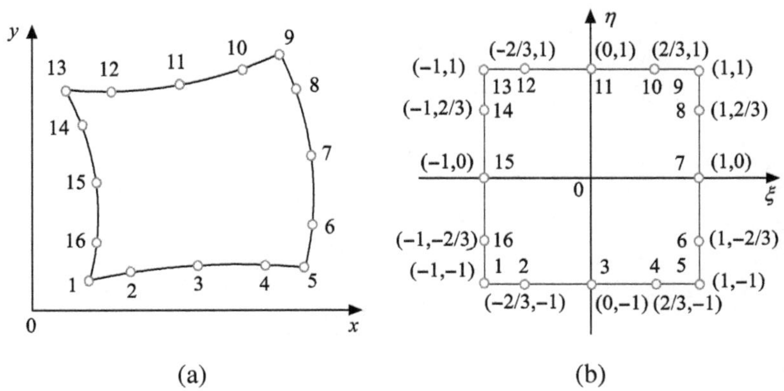

Fig. 2.2. (a) A curvilinear quadrilateral region in cartesian x–y plane; (b) a square parent domain in natural ξ–η plane.

where x_k, y_k; $k = 1, 2, \ldots, N_s$ are the coordinates of N_s boundary grid points in the cartesian $x-y$ plane and $S_k(\xi, \eta)$ is the serendipity interpolations defined in the natural $\xi-\eta$ plane. Since the base function S_k has a unity value at the kth node and zeros at the remaining $(N_s - 1)$ nodes, the domain mapped by Eq. (2.4) and the given quadrilateral domain matches exactly at the nodal points.

Subsequently, we should express the derivatives of a function $f(x, y)$ with respect to x, y coordinates in terms of its derivatives in $\xi - \eta$ coordinates. Consider $f(x, y)$ to be an implicit function of ξ and η as $f = f[x(\xi, \eta), y(\xi, \eta)]$. Using the chain rule of differentiation, we have the following results:

$$\frac{\partial f}{\partial x} = \frac{1}{|\boldsymbol{J}|}\left(\frac{\partial y}{\partial \eta}\frac{\partial f}{\partial \xi} - \frac{\partial y}{\partial \xi}\frac{\partial f}{\partial \eta}\right), \quad \frac{\partial f}{\partial y} = \frac{1}{|\boldsymbol{J}|}\left(\frac{\partial x}{\partial \xi}\frac{\partial f}{\partial \eta} - \frac{\partial x}{\partial \eta}\frac{\partial f}{\partial \xi}\right) \tag{2.5}$$

where the determinant $|\boldsymbol{J}|$ of the Jacobian $\boldsymbol{J} = \partial(x, y)/\partial(\xi, \eta)$ is

$$|\boldsymbol{J}| = \frac{\partial x}{\partial \xi}\frac{\partial y}{\partial \eta} - \frac{\partial y}{\partial \xi}\frac{\partial x}{\partial \eta} \tag{2.6}$$

Then the partial derivatives $\partial f/\partial x$ and $\partial f/\partial y$ at the gird point $x_{ij} = x(\xi_i, \eta_j)$, $y_{ij} = y(\xi_i, \eta_j)$ in the mapped curvilinear quadrilateral domain as shown in Fig. 2.3 can be computed by inserting Eq. (2.2)

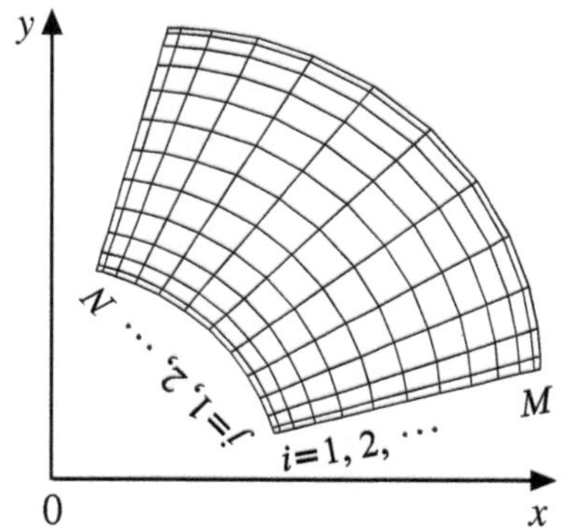

Fig. 2.3. A sectorial region.

into Eq. (2.5) as

$$\left(\frac{\partial f}{\partial x}\right)_{ij} = \frac{1}{|\boldsymbol{J}|_{ij}}\left[\left(\frac{\partial y}{\partial \eta}\right)_{ij}\left(\sum_{m=1}^{M} A_{im}^{(1)} f_{mj}\right) - \left(\frac{\partial y}{\partial \xi}\right)_{ij}\left(\sum_{n=1}^{N} B_{jn}^{(1)} f_{in}\right)\right] \tag{2.7}$$

$$\left(\frac{\partial f}{\partial y}\right)_{ij} = \frac{1}{|\boldsymbol{J}|_{ij}}\left[\left(\frac{\partial x}{\partial \xi}\right)_{ij}\left(\sum_{n=1}^{N} B_{jn}^{(1)} f_{in}\right) - \left(\frac{\partial x}{\partial \eta}\right)_{ij}\left(\sum_{m=1}^{M} A_{im}^{(1)} f_{mj}\right)\right] \tag{2.8}$$

Equations (2.7) and (2.8) define the DQ rules of the first-order partial derivatives with respect to the cartesian x, y coordinates for the irregular domain. Certainly, these rules can also be written in a compact form using a single index notation for grid points as

$$\left.\frac{\partial f}{\partial x}\right|_k = \sum_{m=1}^{M\times N} \bar{A}_{km}^{(1)} \bar{f}_m, \quad \left.\frac{\partial f}{\partial y}\right|_k = \sum_{m=1}^{M\times N} \bar{B}_{km}^{(1)} \bar{f}_m \tag{2.9}$$

where $\bar{A}_{km}^{(1)}$ and $\bar{B}_{km}^{(1)}$ are the weighting coefficients for the first-order derivatives with respect to the x and y coordinates for the irregular domain, respectively:

$$\bar{f}_m = f_{ij} = f(\xi_i, \eta_j); \quad i = 1, 2, \ldots, M; \; j = 1, 2, \ldots, N \tag{2.10}$$

$$k, m = (j-1)M + i \tag{2.11}$$

The general quadrature rules in the mapped region may be written as

$$\begin{gathered} \left.\frac{\partial^r f}{\partial x^r}\right|_k = \sum_{m=1}^{M\times N} \bar{A}_{km}^{(r)} \bar{f}_m, \quad \left.\frac{\partial^s f}{\partial y^s}\right|_k = \sum_{m=1}^{M\times N} \bar{B}_{km}^{(s)} \bar{f}_m \\ \left.\frac{\partial^{r+s} f}{\partial x^r \partial y^s}\right|_k = \sum_{m=1}^{M\times N} \bar{F}_{km}^{(r+s)} \bar{f}_m \end{gathered} \tag{2.12}$$

where the weighting coefficients may be obtained from the recurrence relationships

$$\begin{gathered}[\bar{\boldsymbol{A}}^{(r)}] = [\bar{\boldsymbol{A}}^{(1)}][\bar{\boldsymbol{A}}^{(r-1)}],\ [\bar{\boldsymbol{B}}^{(s)}] = [\bar{\boldsymbol{B}}^{(1)}][\bar{\boldsymbol{B}}^{(s-1)}] \quad (r, s \geq 2) \\ [\bar{\boldsymbol{F}}^{(r+s)}] = [\bar{\boldsymbol{A}}^{(r)}][\bar{\boldsymbol{B}}^{(s)}] \quad (r, s \geq 1)\end{gathered} \tag{2.13}$$

An additional result that will be needed is the relation

$$\mathrm{d}x\mathrm{d}y = |\boldsymbol{J}|\mathrm{d}\xi\, \mathrm{d}\eta \tag{2.14}$$

Its proof may be found in calculus books and will not be given here.

For 3D problems, using the corresponding isoparametric mapping, after processes similar to Eqs. (2.4)–(2.13), the first-order derivatives of function $f(x, y, z)$ in the mapped region may be written as

$$\begin{gathered}\left.\frac{\partial f}{\partial x}\right|_q = \sum_{q=1}^{M\times N\times L} \bar{A}_{qp}^{(1)} \bar{f}_p,\ \left.\frac{\partial f}{\partial y}\right|_q = \sum_{m=1}^{M\times N\times L} \bar{B}_{qp}^{(1)} \bar{f}_p \\ \left.\frac{\partial f}{\partial z}\right|_q = \sum_{n=1}^{M\times N\times L} \bar{C}_{qp}^{(1)} \bar{f}_p\end{gathered} \tag{2.15}$$

where

$$\begin{gathered}\bar{f}_p = f_{ijk} = f(\xi_i, \eta_j, \zeta_k) \\ \text{for } i = 1, 2, \ldots, M;\ j = 1, 2, \ldots, N; \quad k = 1, 2, \ldots, L\end{gathered} \tag{2.16}$$

$$p, q = (k-1) \times L \times N + (j-1) \times N + i \tag{2.17}$$

For regular domains, the weighting coefficients of Eqs. (2.9) and (2.11) can be obtained by assembling those of Eq. (2.2) according to Eq. (2.10); the weighting coefficients of Eq. (2.14) can be obtained by assembling those of Eq. (2.3) according to Eq. (2.16).

2.2 Gauss–Lobatto Quadrature Rule

The well-known Gauss–Lobatto quadrature is the Gauss integration with two endpoints fixed, which can be found in mathematics handbooks, but it is introduced here to make the paper self-contained.

The Gauss–Lobatto quadrature rule with precision degree $(2n-3)$ for function $f(x)$ defined at $[-1,1]$ is

$$\int_{-1}^{1} f(x)\mathrm{d}x = \sum_{j=1}^{n} C_j f(x_j) \tag{2.18}$$

where the weights C_j of Gauss–Lobatto integration are given by

$$C_1 = C_n = \frac{2}{n(n-1)}, \quad C_j = \frac{2}{n(n-1)[P_{n-1}(x_j)]^2} \quad (j \neq 1, n) \tag{2.19}$$

where x_j is the $(j-1)$th zero of $P'_{n-1}(x)$, and the Legendre polynomial $P_n(x)$ of degree n has the form

$$P_n(x) = \sum_{k=0}^{[n/2]} \frac{(-1)^k(2n-2k)!}{2^n k!(n-k)!(n-2k)!} x^{n-2k} \tag{2.20}$$

The zeros of $P'_{n-1}(x)$ are the same as the eigenvalues of its companion matrix. The companion matrix of a polynomial $c_1x^n + c_2x^{n-1} + \cdots + c_nx + c_{n+1}$ is

$$B = \begin{bmatrix} b_1 & b_2 & \cdots & b_{n-1} & b_n \\ 1 & 0 & \cdots & 0 & 0 \\ 0 & 1 & \cdots & 0 & 0 \\ \vdots & \vdots & \ddots & \vdots & \vdots \\ 0 & 0 & \cdots & 1 & 0 \end{bmatrix} \tag{2.21}$$

where $b_j = -c_{j+1}/c_1$ $(j = 1, 2, \ldots, n)$.

2.3 The Differential Quadrature Finite Element Method for Kirchhoff Thin Plates

2.3.1 *The quadrature element method (QEM)*

In general, the convergence of the p-version elements is more rapid than that of the h-version elements in a finite element analysis using the same number of degrees of freedom (DOFs). Using the p-version with higher-order polynomials, one whole plate can be modeled by

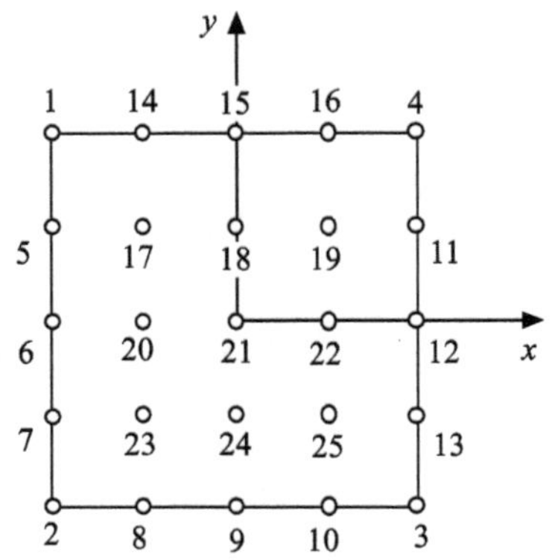

Fig. 2.4. The nodal configuration of a quadrature plate element.

just one element while satisfying the accuracy requirement. In the following, the quadrature plate element, a 25-node rectangular element with 49 DOFs [26, 28] as shown in Fig. 2.4, is recalled. The displacement field of the 49-DOF quadrature plate element is expressed in terms of polynomial-type basis functions such that the displacement of the element is assumed as

$$\begin{aligned} w(x,y) = &\sum_{i=1,2,3,4} [N_{i1}w_i + N_{i2}w_{ix} + N_{i3}w_{iy} + N_{i4}w_{ixy}] \\ &+ \sum_{i=17-25} [N_{i1}w_i] + \sum_{i=5,6,7,11,12,13} [N_{i1}w_i + N_{i2}w_{ix}] \\ &+ \sum_{i=8,9,10,14,15,16} [N_{i1}w_i + N_{i2}w_{iy}] \end{aligned} \tag{2.22}$$

where w_i, $w_{ix} = (\partial w/\partial x)_i$, $w_{iy} = (\partial w/\partial y)_i$, and $w_{ixy} = (\partial^2 w/\partial x \partial y)_i$ are the local degrees of freedom associated with node i. Then one can obtain the ordinary dynamic equations of thin plate by substituting the displacement function into the strain energy U and the work potential W and using the variational principle $\delta(U + W) = 0$. For thin plate, U and W have the forms

$$\begin{aligned} U = \frac{D}{2}\iint_S \Bigg[&\left(\frac{\partial^2 w}{\partial x^2}\right)^2 + \left(\frac{\partial^2 w}{\partial y^2}\right)^2 + 2v\frac{\partial^2 w}{\partial x^2}\frac{\partial^2 w}{\partial y^2} \\ &+ 2(1-v)\left(\frac{\partial^2 w}{\partial x \partial y}\right)^2\Bigg] \mathrm{d}x\,\mathrm{d}y \end{aligned} \tag{2.23}$$

$$W = -\iint_S w\left(-\rho h \frac{\partial^2 w}{\partial t^2}\right) \mathrm{d}x\, \mathrm{d}y - \iint_S wq \mathrm{d}x\, \mathrm{d}y \tag{2.24}$$

where $D = Eh^3/12(1 - \upsilon^2)$.

Using the above QEM, quite accurate results were obtained with relatively small number of DOFs. It deserves to be mentioned that the QEM based on the trial function (2.22) is relevant to DQM due to the use of deflections themselves at inner grid points and to FEM due to the use of nodal shape functions and slopes in trial functions as well as variational principle.

2.3.2 *The differential quadrature finite element method for Kirchhoff thin plates*

The need to improve the computational efficiency and simplify the implementation of boundary conditions of the QEM while maintaining its high accuracy motivates the development of a new method, referred to as the differential quadrature finite element method in this study. In this novel method, the trial function is assumed as

$$w(x, y) = \sum_{i=1}^{M} \sum_{j=1}^{N} l_i(x) l_j(y) w_{ij} \tag{2.25}$$

It should be emphasized that there are no slopes and curvatures in expression (14), it differs from the trial functions used in FEM, QEM and DQEM for thin plate. Substituting the trial function into Eqs. (12) and (13) and using the DQ rules in conjunction with the Gauss–Lobatto quadrature rule, one can have

$$\begin{aligned} U = \frac{D}{2} \sum_{k=1}^{K} C_k \{ [\bar{\boldsymbol{A}}_k^{(2)} \bar{\boldsymbol{w}}]^2 + [\bar{\boldsymbol{B}}_k^{(2)} \bar{\boldsymbol{w}}]^2 + 2\upsilon [\bar{\boldsymbol{A}}_k^{(2)} \bar{\boldsymbol{w}}][\bar{\boldsymbol{B}}_k^{(2)} \bar{\boldsymbol{w}}] \\ + 2(1 - \nu)[\bar{\boldsymbol{F}}_k^{(1+1)} \bar{\boldsymbol{w}}]^2 \} \end{aligned} \tag{2.26}$$

$$W = \sum_{k=1}^{K} C_k \left[w_k \left(\rho h \frac{\partial^2 w_k}{\partial t^2} \right) - w_k q_k \right] \tag{2.27}$$

where $K = M \times N$, $C_k = C_i^x C_j^y$, $\bar{\boldsymbol{A}}_k^{(2)}$, $\bar{\boldsymbol{B}}_k^{(2)}$ and $\bar{\boldsymbol{F}}_k^{(1+1)}$ are the kth rows of $\bar{\boldsymbol{A}}^{(2)}$, $\bar{\boldsymbol{B}}^{(2)}$ and $\bar{\boldsymbol{F}}^{(1+1)}$, respectively. $\bar{\boldsymbol{w}} = (w_m)_{K\times 1}$ is a column vector, where $m = (j - 1)M + i$, $i = 1, \ldots, M$, $j = 1, \ldots, N$.

For brevity, let $\boldsymbol{A} = \bar{\boldsymbol{A}}^{(2)}$, $\boldsymbol{B} = \bar{\boldsymbol{B}}^{(2)}$, $\boldsymbol{F} = \bar{\boldsymbol{F}}^{(1+1)}$, $\boldsymbol{C} = \mathrm{diag}[C_k]$ then Eqs. (2.26) and (2.27) can be rewritten as

$$U = \frac{D}{2}\bar{\boldsymbol{w}}^{\mathrm{T}}[\boldsymbol{A}^{\mathrm{T}}\boldsymbol{C}\boldsymbol{A} + \boldsymbol{B}^{\mathrm{T}}\boldsymbol{C}\boldsymbol{B} + 2\upsilon\boldsymbol{A}^{\mathrm{T}}\boldsymbol{C}\boldsymbol{B} + 2(1-\upsilon)\boldsymbol{F}^{\mathrm{T}}\boldsymbol{C}\boldsymbol{F}]\bar{\boldsymbol{w}} \tag{2.28}$$

$$W = \bar{\boldsymbol{w}}^{\mathrm{T}}(\rho h\boldsymbol{C})\ddot{\bar{\boldsymbol{w}}} + \bar{\boldsymbol{w}}^{\mathrm{T}}\boldsymbol{C}\boldsymbol{q} \tag{2.29}$$

For imposing the inter-element compatibility conditions of any order, some modification on the displacement vector $\bar{\boldsymbol{w}}$ is necessary. One choice of the displacement vectors satisfying the C^1 inter-element compatibility conditions for thin plate element of a square region as shown in Fig. 2.4 can be given by

$$\begin{aligned}\boldsymbol{w} = [&w_m \;\; w_{mx} \;\; w_{my} \;\; w_{mxy} \;\; (i = 1, M; j = 1, N),\\ &w_m \;\; w_{mx}(i = 3, \ldots, M-2; j = 1, N),\\ &w_m \;\; w_{my} \;\; (i = 1, M; j = 3, \ldots, N-2),\\ &w_m \;\; (i = 3, \ldots, M-2; j = 3, \ldots, N-2)]\end{aligned} \tag{2.30}$$

which is analogous to nodal parameters used in trial function (2.22). For a curvilinear quadrilateral element as shown in Fig. 2.3, the slopes on the edge should be defined as the derivatives of w with respect to the normal. The transformation from $\bar{\boldsymbol{w}}$ to $\boldsymbol{w}$ can be finished by using the DQ rules as

$$\boldsymbol{w} = \boldsymbol{Q}\bar{\boldsymbol{w}} \tag{2.31}$$

It is apparent that Q is not singular for any case, and as a diagonal unit matrix for C^0 element, this causes mass matrix given below to be a diagonal matrix, see Eq. (2.33). Substitution of Eq. (2.31) into Eqs. (2.28) and (2.29) yields the stiffness matrix, the mass matrix and the force vector of the new rectangular thin plate as

$$\begin{aligned}\boldsymbol{K} = D\boldsymbol{Q}^{-\mathrm{T}}[&\boldsymbol{A}^{\mathrm{T}}\boldsymbol{C}\boldsymbol{A} + \boldsymbol{B}^{\mathrm{T}}\boldsymbol{C}\boldsymbol{B} + \upsilon(\boldsymbol{A}^{\mathrm{T}}\boldsymbol{C}\boldsymbol{B} + \boldsymbol{B}^{\mathrm{T}}\boldsymbol{C}\boldsymbol{A})\\ &+ 2(1-\upsilon)\boldsymbol{F}^{\mathrm{T}}\boldsymbol{C}\boldsymbol{F}]\boldsymbol{Q}^{-1}\end{aligned} \tag{2.32}$$

$$\boldsymbol{M} = \boldsymbol{Q}^{-\mathrm{T}}(\rho h\boldsymbol{C})\boldsymbol{Q}^{-1}, \quad \boldsymbol{R} = \boldsymbol{Q}^{-\mathrm{T}}(\boldsymbol{C}\boldsymbol{q}) \tag{2.33}$$

Apparently, both $\boldsymbol{K}$ and $\boldsymbol{M}$ are symmetrical matrices implying good numerical properties, but, generally, they are unsymmetrical in

DQEM. Moreover, the above formulations for DQFEM are simpler and more adaptable, for plate with irregular shape, the formulation of the DQFEM can maintain the above simple form as shown in the following.

2.4 The Differential Quadrature Finite Element Method for Elasticity

The differential quadrature finite element method was developed in [281] where the DQ and Gauss–Lobatto quadrature rules were used to discretize the potential energy functional, and the free vibrations of thin plates were investigated extensively by using DQFEM.

Here, we will extend the DQFEM to rod, beam, thick plate, plane and 3D problems. For linear elastic bodies, the total potential energy involves the strain energy (U) and work potential (W). The strain energy U is given by

$$U = \frac{1}{2}\iiint_V \boldsymbol{\varepsilon}^{\mathrm{T}}\boldsymbol{\sigma}\,\mathrm{d}V = \frac{1}{2}\iiint_V \boldsymbol{\varepsilon}^{\mathrm{T}}\boldsymbol{D}\boldsymbol{\varepsilon}\,\mathrm{d}V \tag{2.34}$$

where ε is the strain field vector, σ the stress field vector, and $\boldsymbol{D}$ the material matrix. The work potential is given by

$$W = -\iiint_V \boldsymbol{u}^{\mathrm{T}}\left(-\rho\frac{\partial^2\boldsymbol{u}}{\partial t^2}\right)\mathrm{d}V - \iint_S \boldsymbol{u}^{\mathrm{T}}\boldsymbol{q}\,\mathrm{d}S \tag{2.35}$$

where $\boldsymbol{u}$ is the displacement field vector, $\boldsymbol{q}$ the distributed surface force vector, ρ the volume density and t the time. Then the element matrices of different kinds of structures can be obtained from the discrete quadratic forms of U and W.

2.4.1 *Rod element*

Consider a uniform rod element of length L and cross section area S. Assuming that the longitudinal displacement function is

$$u(x) = \sum_{i=1}^{M} l_i(x)u_i \tag{2.36}$$

where l_i are the Lagrange polynomials, $u_i = u(x_i)$ the displacements of the Gauss–Lobatto quadrature points or the nodal displacements of the differential quadrature finite rod element, x_i the Gauss–Lobatto node coordinates, and M the total node number. Using DQ and Gauss–Lobatto quadrature rules, Eqs. (2.34) and (2.35) can be written as

$$U = \frac{1}{2}\int_0^L ES\left(\frac{\partial u}{\partial x}\right)^2 \mathrm{d}x = \frac{ES}{2}\boldsymbol{u}^T\boldsymbol{A}^{(1)^{\mathrm{T}}}\boldsymbol{C}\boldsymbol{A}^{(1)}\boldsymbol{u} \tag{2.37}$$

$$W = -\int_0^L u\left(-\rho S\frac{\partial^2 u}{\partial t^2}\right)\mathrm{d}x - \int_0^L qu\,\mathrm{d}x = \boldsymbol{u}^{\mathrm{T}}(\rho S\boldsymbol{C})\ddot{\mathbf{u}} - \boldsymbol{u}^{\mathrm{T}}(\boldsymbol{C}\boldsymbol{q}) \tag{2.38}$$

where E is the Young's modulus, $\boldsymbol{u}^{\mathrm{T}} = [u_1\ u_2\ \cdots\ u_M]$ the nodal displacement vector, $\boldsymbol{q}^{\mathrm{T}} = [q(x_1)\ q(x_2)\ \cdots\ q(x_M)]$ the nodal load vector, $\boldsymbol{A}^{(1)\mathrm{T}} = (\boldsymbol{A}^{(1)})^{\mathrm{T}}$ where $\boldsymbol{A}^{(1)}$ indicates the weighting coefficient matrix of DQ rules for the first-order derivatives with respect to the Gauss–Lobatto nodes, and

$$\boldsymbol{C} = \mathrm{diag}(C_1 \quad C_2 \quad \cdots \quad C_M) \tag{2.39}$$

where C_j are the weighting coefficients of Gauss–Lobatto integration. Therefore, the stiffness matrix $\boldsymbol{K}$, mass matrix $\boldsymbol{M}$ and load vector $\boldsymbol{R}$ are

$$\boldsymbol{K} = ES\boldsymbol{A}^{(1)\mathrm{T}}\boldsymbol{C}\boldsymbol{A}^{(1)}, \quad \boldsymbol{M} = \rho S\boldsymbol{C}, \quad \boldsymbol{R} = \boldsymbol{C}\boldsymbol{q} \tag{2.40}$$

It is notable that the finite element matrices in DQFEM can be obtained by simple algebraic operations of the weighting coefficient matrices of the DQ rule and Gauss–Lobatto integral rule, and that the mass matrix $\boldsymbol{M}$ is diagonal. For the comparison of the standard FEM with DQFEM, the 3-DOF rod element matrices of standard FEM for a bar subjected to uniformly distributed load q_0 are given:

$$\boldsymbol{K} = \frac{ES}{3L}\begin{bmatrix} 7 & -8 & 1 \\ -8 & 16 & -8 \\ 1 & -8 & 7 \end{bmatrix}, \quad \boldsymbol{M} = \frac{\rho SL}{30}\begin{bmatrix} 4 & 2 & -1 \\ 2 & 16 & 2 \\ -1 & 2 & 4 \end{bmatrix}$$
$$\boldsymbol{R} = \frac{q_0 L}{6}\begin{bmatrix} 1 \\ 4 \\ 1 \end{bmatrix} \tag{2.41}$$

The corresponding matrices of DQFEM are

$$\boldsymbol{K} = \frac{ES}{3L}\begin{bmatrix} 7 & -8 & 1 \\ -8 & 16 & -8 \\ 1 & -8 & 7 \end{bmatrix}, \quad \boldsymbol{M} = \frac{\rho SL}{6}\begin{bmatrix} 1 & 0 & 0 \\ 0 & 4 & 0 \\ 0 & 0 & 1 \end{bmatrix}$$
$$\boldsymbol{R} = \frac{q_0 L}{6}\begin{bmatrix} 1 \\ 4 \\ 1 \end{bmatrix} \tag{2.42}$$

It is noteworthy that the stiffness matrices and force vectors of the two methods are the same when their nodes are the same and that the mass matrix of the DQFEM is diagonal for C^0 elements, which can reduce the computational cost of dynamic problems.

2.4.2 *Euler beam element*

Consider a uniform Euler beam element with length L and cross-section area S. Assuming that the deflection function is

$$w(x) = \sum_{i=1}^{M} l_i(x) w_i \tag{2.43}$$

where $w_i = w(x_i)$ are the deflections of the Gauss–Lobatto quadrature points of the DQ finite beam element. Similarly, as in Section 2.4.1, using DQ and Gauss–Lobatto quadrature rules, Eqs. (2.34) and (2.35) can be written as

$$U = \frac{1}{2}\int_0^L EI\left(\frac{\partial^2 w}{\partial x^2}\right)\mathrm{d}x = \frac{EI}{2}\bar{\boldsymbol{w}}^T \boldsymbol{A}^{(2)\mathrm{T}}\boldsymbol{C}\boldsymbol{A}^{(2)}\bar{\boldsymbol{w}}$$
$$W = -\int_0^L w\left(-\rho S\frac{\partial^2 w}{\partial t^2}\right)\mathrm{d}x - \int_0^L qw\,\mathrm{d}x = \bar{\boldsymbol{w}}^{\mathrm{T}}(\rho S\boldsymbol{C})\ddot{\bar{\boldsymbol{w}}} - \bar{\boldsymbol{w}}^{\mathrm{T}}\boldsymbol{C}\boldsymbol{q} \tag{2.44}$$

where I is the moment of inertia, and

$$\bar{\boldsymbol{w}}^{\mathrm{T}} = [w_1 \quad w_2 \quad \cdots \quad w_M] \tag{2.45}$$

In order to construct element satisfying C^1 inter-element continuity requirements, the element displacement vector should be

$$\boldsymbol{w}^{\mathrm{T}} = \left[w_1 \quad w_1' \quad w_3 \quad \cdots \quad w_M \quad w_M'\right] \tag{2.46}$$

Using DQ rules one can find the relation between $\boldsymbol{w}$ and $\bar{\boldsymbol{w}}$ as

$$\boldsymbol{w} = \boldsymbol{Q}\bar{\boldsymbol{w}} \tag{2.47}$$

where

$$\boldsymbol{Q} = \begin{bmatrix} 1 & 0 & 0 & \cdots & 0 & 0 \\ A_{1,1}^{(1)} & A_{1,2}^{(1)} & A_{1,3}^{(1)} & \cdots & A_{1,M-1}^{(1)} & A_{1,M}^{(1)} \\ 0 & 0 & 1 & \cdots & 0 & 0 \\ \vdots & \vdots & \vdots & \ddots & \vdots & \vdots \\ 0 & 0 & 0 & \cdots & 0 & 1 \\ A_{M,1}^{(1)} & A_{M,2}^{(1)} & A_{M,3}^{(1)} & \cdots & A_{M,M-1}^{(1)} & A_{M,M}^{(1)} \end{bmatrix} \tag{2.48}$$

Substituting Eq. (2.47) into Eq. (2.44), the stiffness matrix, mass matrix and load vector of the DQ finite Euler beam are obtained as

$$\begin{aligned} &\boldsymbol{K} = EI\boldsymbol{Q}^{-\mathrm{T}}\boldsymbol{A}^{(2)\mathrm{T}}\boldsymbol{C}\boldsymbol{A}^{(2)}\boldsymbol{Q}^{-1}, \quad \boldsymbol{M} = \boldsymbol{Q}^{-\mathrm{T}}(\rho S\boldsymbol{C})\boldsymbol{Q}^{-1}, \\ &\boldsymbol{R} = \boldsymbol{Q}^{-\mathrm{T}}\boldsymbol{C}\boldsymbol{q} \end{aligned} \tag{2.49}$$

It is readily shown that the transformation matrix $\boldsymbol{Q}$ in Eq. (2.48) is well conditioned in general. In the same way as above, the construction of element with C^n continuity is possible. In the following, the comparisons of the beam element matrices of the DQFEM and the FEM are presented. For a beam subjected to uniformly distributed load q_0, the 4-DOF beam element matrices and vector of FEM are

$$\begin{aligned} &\boldsymbol{K} = \frac{EI}{L^3}\begin{bmatrix} 12 & 6L & -12 & 6L \\ 6L & 4L^2 & -6L & 2L^2 \\ -12 & -6L & 12 & -6L \\ 6L & 2L^2 & -6L & 4L^2 \end{bmatrix}, \quad \boldsymbol{R} = \frac{q_0 L}{12}\begin{bmatrix} 6 \\ L \\ 6 \\ -L \end{bmatrix} \\ &\boldsymbol{M} = \frac{\rho SL}{420}\begin{bmatrix} 156 & 22L & 54 & -13L \\ 22L & 4L^2 & 13L & -3L^2 \\ 54 & 13L & 156 & -22L \\ -13L & -3L^2 & -22L & 4L^2 \end{bmatrix} \end{aligned} \tag{2.50}$$

The corresponding matrices and vector of DQFEM are

$$\boldsymbol{K} = \frac{EI}{L^3}\begin{bmatrix} 12 & 6L & -12 & 6L \\ 6L & 4L^2 & -6L & 2L^2 \\ -12 & -6L & 12 & -6L \\ 6L & 2L^2 & -6L & 4L^2 \end{bmatrix}, \quad \boldsymbol{R} = \frac{q_0 L}{12}\begin{bmatrix} 6 \\ L \\ 6 \\ -L \end{bmatrix}$$
$$\boldsymbol{M} = \frac{\rho SL}{420}\begin{bmatrix} 156.8 & 22.4L & 53.2 & -12.6L \\ 22.4L & 4.2L^2 & 12.6L & -2.8L^2 \\ 53.2 & 12.6L & 156.8 & -22.4L \\ -12.6L & -2.8L^2 & -22.4L & 4.2L^2 \end{bmatrix} \tag{2.51}$$

Apparently, the stiffness matrices and force vectors of FEM and DQFEM are the same, but the Lagrange polynomials are used in Eq. (2.43) while the Hermite interpolation functions are used in standard FEM. The mass matrix of the DQFEM is slightly different from that of the FEM.

2.4.3 *Plane element*

Consider a quadrilateral domain with uniform thickness h, as shown in Fig. 2.2(a), the displacement fields have the forms

$$[u(x,y), v(x,y)] = \sum_{i=1}^{M}\sum_{j=1}^{N} l_i(x)l_j(y)[u_{ij}, v_{ij}] \tag{2.52}$$

The strain–displacement relations of plane problems are

$$\begin{bmatrix} \varepsilon_x \\ \varepsilon_y \\ \gamma_{xy} \end{bmatrix} = \begin{bmatrix} \partial/\partial x & 0 \\ 0 & \partial/\partial y \\ \partial/\partial y & \partial/\partial x \end{bmatrix} \begin{bmatrix} u \\ v \end{bmatrix} \tag{2.53}$$

Define the following element displacement vectors:

$$\begin{aligned} \boldsymbol{u}^{\mathrm{T}} &= \begin{bmatrix} u_{11} \cdots u_{M1}\, u_{12} \cdots u_{M2} \cdots u_{1N} \cdots u_{MN} \end{bmatrix} \\ \boldsymbol{v}^{\mathrm{T}} &= \begin{bmatrix} v_{11} \cdots v_{M1}\, v_{12} \cdots v_{M2} \cdots v_{1N} \cdots v_{MN} \end{bmatrix} \end{aligned} \tag{2.54}$$

Then by inserting the DQ rules for the first-order partial derivatives defined by Eq. (2.9) into the strain–displacement relations Eq. (2.53), one can obtain the corresponding nodal strain vector

$$\begin{bmatrix} \varepsilon_x \\ \varepsilon_y \\ \gamma_{xy} \end{bmatrix} = \begin{bmatrix} \bar{\boldsymbol{A}}^{(1)} & \boldsymbol{0} \\ \boldsymbol{0} & \bar{\boldsymbol{B}}^{(1)} \\ \bar{\boldsymbol{B}}^{(1)} & \bar{\boldsymbol{A}}^{(1)} \end{bmatrix} \begin{bmatrix} \boldsymbol{u} \\ \boldsymbol{v} \end{bmatrix} \tag{2.55}$$

where the DQ rule and Gauss–Lobatto rule have been involved, and $\bar{\boldsymbol{A}}^{(1)}$, $\bar{\boldsymbol{B}}^{(1)}$ are the weighting coefficient matrices defined by Eq. (2.9). Thus, we can obtain the matrices of the DQ finite curvilinear quadrilateral plane element; for plane stress problem, they are

$$\boldsymbol{K} = c \begin{bmatrix} \bar{\boldsymbol{A}}^{(1)\mathrm{T}} \boldsymbol{C} \bar{\boldsymbol{A}}^{(1)} + \upsilon_1 \bar{\boldsymbol{B}}^{(1)\mathrm{T}} \boldsymbol{C} \bar{\boldsymbol{B}}^{(1)} & \begin{matrix} \upsilon \bar{\boldsymbol{A}}^{(1)\mathrm{T}} \boldsymbol{C} \bar{\boldsymbol{B}}^{(1)} \\ + \upsilon_1 \bar{\boldsymbol{B}}^{(1)\mathrm{T}} \boldsymbol{C} \bar{\boldsymbol{A}}^{(1)} \end{matrix} \\ \upsilon \bar{\boldsymbol{B}}^{(1)\mathrm{T}} \boldsymbol{C} \bar{\boldsymbol{A}}^{(1)} + \upsilon_1 \bar{\boldsymbol{A}}^{(1)\mathrm{T}} \boldsymbol{C} \bar{\boldsymbol{B}}^{(1)} & \begin{matrix} \bar{\boldsymbol{B}}^{(1)\mathrm{T}} \boldsymbol{C} \bar{\boldsymbol{B}}^{(1)} \\ + \upsilon_1 \bar{\boldsymbol{A}}^{(1)\mathrm{T}} \boldsymbol{C} \bar{\boldsymbol{A}}^{(1)} \end{matrix} \end{bmatrix} \tag{2.56}$$

$$\boldsymbol{M} = \rho h \begin{bmatrix} \boldsymbol{C} & \boldsymbol{0} \\ \boldsymbol{0} & \boldsymbol{C} \end{bmatrix}, \quad \boldsymbol{R} = \begin{bmatrix} \boldsymbol{C}\boldsymbol{q}_u \\ \boldsymbol{C}\boldsymbol{q}_v \end{bmatrix} \tag{2.57}$$

where the corresponding nodal displacement vector is $[\boldsymbol{u}^{\mathrm{T}} \ \ \boldsymbol{v}^{\mathrm{T}}]$, $c = Eh/(1-\upsilon^2)$, $\upsilon_1 = (1-\upsilon)/2$, $\boldsymbol{C} = \mathrm{diag}(J_k C_k)$, $J_k = |\boldsymbol{J}|_{ij}$ is the determinant of the Jacobian defined by Eq. (2.6), $C_k = C_i^x C_j^y$, the integer k is defined by Eq. (2.11); C_i^x and C_j^y, $\boldsymbol{q}_u$ and $\boldsymbol{q}_v$ are the Gauss–Lobatto weights and the nodal load vectors with respect to x- and y-directions, respectively. Replacing E and υ in Eq. (2.56) with $E/(1-\upsilon^2)$ and $\upsilon/(1-\upsilon)$ will yield the stiffness matrix of plane strain element.

2.4.4 *Mindlin plate element*

In Mindlin plate theory, one can choose the deflection w and two rotations θ_x and θ_y of the normal line with respect to the middle surface as the generalized displacements which can be expressed as

$$[w, \theta_x, \theta_y] = \sum_{i=1}^{M} \sum_{j=1}^{N} l_i(x) l_j(y) [w_{ij}, \theta_{xij}, \theta_{yij}] \tag{2.58}$$

Define the nodal displacement vector as $[\boldsymbol{\theta}_x^{\mathrm{T}}\ \boldsymbol{\theta}_y^{\mathrm{T}}\ \boldsymbol{w}^{\mathrm{T}}]$ whose elements are arranged as in Eq. (2.54), similar to the above, one can determine the Mindlin plate element matrices as

$$\boldsymbol{K} = D\begin{bmatrix} \boldsymbol{K}_{11} & & \text{sym} \\ \boldsymbol{K}_{21} & \boldsymbol{K}_{22} & \\ \boldsymbol{K}_{31} & \boldsymbol{K}_{32} & \boldsymbol{K}_{33} \end{bmatrix}, \quad \boldsymbol{M} = \rho\begin{bmatrix} J\boldsymbol{C} & \boldsymbol{0} & \boldsymbol{0} \\ \boldsymbol{0} & J\boldsymbol{C} & \boldsymbol{0} \\ \boldsymbol{0} & \boldsymbol{0} & h\boldsymbol{C} \end{bmatrix}$$
$$\boldsymbol{R} = \begin{bmatrix} \boldsymbol{C}\boldsymbol{m}_x \\ \boldsymbol{C}\boldsymbol{m}_y \\ \boldsymbol{C}\boldsymbol{q}_w \end{bmatrix} \tag{2.59}$$

$$\begin{bmatrix} \boldsymbol{K}_{11} \\ \boldsymbol{K}_{22} \\ \boldsymbol{K}_{33} \end{bmatrix} = \begin{bmatrix} 1 & \upsilon_1 & \upsilon_s \\ \upsilon_1 & 1 & \upsilon_s \\ \upsilon_s & \upsilon_s & 0 \end{bmatrix} \begin{bmatrix} \bar{\boldsymbol{A}}^{(1)\mathrm{T}}\boldsymbol{C}\bar{\boldsymbol{A}}^{(1)} \\ \bar{\boldsymbol{B}}^{(1)\mathrm{T}}\boldsymbol{C}\bar{\boldsymbol{B}}^{(1)} \\ \boldsymbol{C} \end{bmatrix}$$
$$\boldsymbol{K}_{21} = \upsilon\bar{\boldsymbol{B}}^{(1)\mathrm{T}}\boldsymbol{C}\bar{\boldsymbol{A}}^{(1)} + \upsilon_1\bar{\boldsymbol{A}}^{(1)\mathrm{T}}\boldsymbol{C}\bar{\boldsymbol{B}}^{(1)} \tag{2.60}$$
$$\boldsymbol{K}_{31} = -\upsilon_s\bar{\boldsymbol{A}}^{(1)\mathrm{T}}\boldsymbol{C}$$
$$\boldsymbol{K}_{32} = -\upsilon_s\bar{\boldsymbol{B}}^{(1)\mathrm{T}}\boldsymbol{C}$$

where $\upsilon_1 = (1-\upsilon)/2$, $\upsilon_s = 6\kappa(1-\upsilon)/h^2$, $J = h^3/12$, $\boldsymbol{m}_x$ and $\boldsymbol{m}_y$ are the nodal bending moment vectors with respect to x- and y-directions, and $\boldsymbol{q}_w$ is the nodal force vector with respect to z-direction. The shear rigidity of Mindlin plate is $C = \kappa Gh = \upsilon_s D$ where κ is the shear correction factor and G is the shear modulus.

2.4.5 *3D element*

For 3D problems, the translational displacements in DQFEM are given by

$$[u, v, w] = \sum_{i=1}^{M}\sum_{j=1}^{N}\sum_{k=1}^{L} l_i(x)l_j(y)l_k(z)\,[u_{ijk}, v_{ijk}, w_{ijk}] \tag{2.61}$$

Define the nodal displacement vector as $[\boldsymbol{u}^{\mathrm{T}}\ \boldsymbol{v}^{\mathrm{T}}\ \boldsymbol{w}^{\mathrm{T}}]$ whose elements are arranged as in Eqs. (2.16) and (2.17); similar to in-plane and

Mindlin plate cases, one can determine the 3D element matrices as

$$\boldsymbol{K}=\frac{G}{\upsilon_2}\begin{bmatrix}\boldsymbol{K}_{11} & & \text{sym}\\ \boldsymbol{K}_{21} & \boldsymbol{K}_{22} & \\ \boldsymbol{K}_{31} & \boldsymbol{K}_{32} & \boldsymbol{K}_{33}\end{bmatrix},\quad \boldsymbol{M}=\rho\begin{bmatrix}\boldsymbol{C} & \mathbf{0} & \mathbf{0}\\ \mathbf{0} & \boldsymbol{C} & \mathbf{0}\\ \mathbf{0} & \mathbf{0} & \boldsymbol{C}\end{bmatrix}$$
$$\boldsymbol{R}=\begin{bmatrix}\boldsymbol{C}\boldsymbol{q}_u\\ \boldsymbol{C}\boldsymbol{q}_v\\ \boldsymbol{C}\boldsymbol{q}_w\end{bmatrix} \tag{2.62}$$

$$\begin{bmatrix}\boldsymbol{K}_{11}\\ \boldsymbol{K}_{22}\\ \boldsymbol{K}_{33}\end{bmatrix}=\begin{bmatrix}\upsilon_1 & \upsilon_2 & \upsilon_2\\ \upsilon_2 & \upsilon_1 & \upsilon_2\\ \upsilon_2 & \upsilon_2 & \upsilon_1\end{bmatrix}\begin{bmatrix}\bar{\boldsymbol{A}}^{(1)\mathrm{T}}\boldsymbol{C}\bar{\boldsymbol{A}}^{(1)}\\ \bar{\boldsymbol{B}}^{(1)\mathrm{T}}\boldsymbol{C}\bar{\boldsymbol{B}}^{(1)}\\ \bar{\boldsymbol{C}}^{(1)\mathrm{T}}\boldsymbol{C}\bar{\boldsymbol{C}}^{(1)}\end{bmatrix}$$
$$\boldsymbol{K}_{21}=\upsilon\bar{\boldsymbol{B}}^{(1)\mathrm{T}}\boldsymbol{C}\bar{\boldsymbol{A}}^{(1)}+\upsilon_2\bar{\boldsymbol{A}}^{(1)\mathrm{T}}\boldsymbol{C}\bar{\boldsymbol{B}}^{(1)} \tag{2.63}$$
$$\boldsymbol{K}_{31}=\upsilon\bar{\boldsymbol{C}}^{(1)\mathrm{T}}\boldsymbol{C}\bar{\boldsymbol{A}}^{(1)}+\upsilon_2\bar{\boldsymbol{A}}^{(1)\mathrm{T}}\boldsymbol{C}\bar{\boldsymbol{C}}^{(1)}$$
$$\boldsymbol{K}_{32}=\upsilon\bar{\boldsymbol{C}}^{(1)\mathrm{T}}\boldsymbol{C}\bar{\boldsymbol{B}}^{(1)}+\upsilon_2\bar{\boldsymbol{B}}^{(1)\mathrm{T}}\boldsymbol{C}\bar{\boldsymbol{C}}^{(1)}$$

where $\bar{\boldsymbol{A}}^{(1)}$, $\bar{\boldsymbol{B}}^{(1)}$, and $\bar{\boldsymbol{C}}^{(1)}$ are the weighting coefficient matrices of Eq. (2.15), $\upsilon_1=1-\upsilon$, $\upsilon_2=0.5-\upsilon$, $\boldsymbol{C}=\text{diag}(J_pC_p)$, $J_p=|\boldsymbol{J}|_{ijk}$ is the determinant of the Jacobian of 3D isoparametric transformation, $C_p=C_i^xC_j^yC_k^z$, the scale p is calculated from Eq. (2.17), and C_i^x, C_j^y and C_k^z are the Gauss–Lobatto weights with respect to x-, y-, and z-directions, respectively.

2.4.6 *Thickness-shear vibration analysis of rectangular quartz plates*

Consider a rectangular crystal plate of length a, width b and uniform thickness h, as shown in Fig. 2.5. The total potential functional of the plate is

$$U=\frac{1}{2}\int_0^1\int_0^1\left[\frac{1}{\theta}D_1\left(\frac{\partial\phi}{\partial\xi}\right)^2+2D_2\left(\frac{\partial\phi}{\partial\xi}\frac{\partial\psi}{\partial\eta}\right)\right.$$
$$\left.+\theta D_3\left(\frac{\partial\psi}{\partial\eta}\right)^2+\theta D_5\left(\frac{\partial\phi}{\partial\eta}+\frac{1}{\theta}\frac{\partial\psi}{\partial\xi}\right)^2\right]\mathrm{d}\xi\mathrm{d}\eta$$

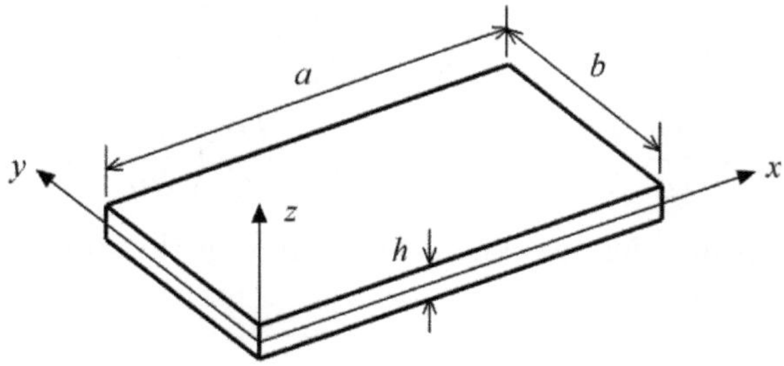

Fig. 2.5. A Mindlin plate and its coordinates.

$$+\frac{1}{2}\int_0^1\int_0^1\left[\frac{1}{\theta}D_6\left(\frac{\partial w}{\partial \xi}-a\phi\right)^2+\theta D_4\left(\frac{\partial w}{\partial \eta}-b\psi\right)^2\right]\mathrm{d}x\mathrm{d}y$$
$$-\frac{ab\rho\omega^2}{2}\int_0^1\int_0^1(hw^2+J\phi^2+J\psi^2)\mathrm{d}\xi\,\mathrm{d}\eta \tag{2.64}$$

where $J = h^3/12$, $\theta = a/b$, w is the deflection, and ϕ and ψ are the angles of rotations of a normal line due to plate bending with respect to x and y coordinates, respectively, and

$$\begin{gathered}
D_1=\frac{h^3}{12}\left(C_{11}-\frac{C_{12}^2}{C_{22}}\right),\quad D_2=\frac{h^3}{12}\left(C_{13}-\frac{C_{12}C_{23}}{C_{22}}\right)\\
D_3=\frac{h^3}{12}\left(C_{33}-\frac{C_{23}^2}{C_{22}}\right)\\
D_4=\frac{\pi^2 h}{12}\left(C_{44}-\frac{C_{24}^2}{C_{22}}\right),\quad D_5=\frac{h^3}{12}C_{55}\\
D_6=\frac{\pi^2 h}{12}C_{66}
\end{gathered}\tag{2.65}$$

in which C_{ij} are elastic constants referred to x-, y-, z-axes fixed in the plate. Note that D_4 and D_6 have involved the shear correction factor. The relations between the internal forces and displacements are

$$M_x=-\frac{1}{a}\left(D_1\frac{\partial \phi}{\partial \xi}+\theta D_2\frac{\partial \psi}{\partial \eta}\right),\quad M_y=-\frac{1}{a}\left(D_2\frac{\partial \phi}{\partial \xi}+\theta D_3\frac{\partial \psi}{\partial \eta}\right)$$

$$M_{xy} = -\frac{D_5}{a}\left(\theta\frac{\partial\phi}{\partial\eta} + \frac{\partial\psi}{\partial\xi}\right) \tag{2.66}$$

$$Q_x = \frac{D_6}{a}\left(\frac{\partial w}{\partial\xi} - a\phi\right),\; Q_y = \frac{D_4}{b}\left(\frac{\partial w}{\partial\eta} - b\psi\right)$$

The fundamental thickness-shear frequency of the magnitude will be solved as follows:

$$\omega_0 = \frac{\pi}{h}\sqrt{\frac{C_{66}}{\rho}} \tag{2.67}$$

The magnitude of the fundamental flexural frequency is $h^2\omega_0$. So the fundamental thickness-shear frequency is very high and it is difficult to get the thickness-shear frequency using a numerical procedure due to the huge computational cost and the difficulty in finding the thickness-shear mode among many frequencies that are close to the thickness-shear frequency.

According to the DQFEM, Eq. (2.64) can be discretized by Eq. (2.9) to obtain the mass and stiffness matrices $\boldsymbol{M}$ and $\boldsymbol{K}$ for the displacement vector $\boldsymbol{U}$ as

$$\boldsymbol{K} = \begin{bmatrix} \boldsymbol{K}_{11} & & \text{sym} \\ \boldsymbol{K}_{21} & \boldsymbol{K}_{22} & \\ \boldsymbol{K}_{31} & \boldsymbol{K}_{32} & \boldsymbol{K}_{33} \end{bmatrix},\; \boldsymbol{M} = ab\rho\begin{bmatrix} J\boldsymbol{C} & \boldsymbol{0} & \boldsymbol{0} \\ \boldsymbol{0} & J\boldsymbol{C} & \boldsymbol{0} \\ \boldsymbol{0} & \boldsymbol{0} & h\boldsymbol{C} \end{bmatrix}$$
$$\boldsymbol{U} = \begin{Bmatrix} \boldsymbol{\phi} \\ \boldsymbol{\psi} \\ \boldsymbol{w} \end{Bmatrix} \tag{2.68}$$

$$\begin{Bmatrix} \boldsymbol{K}_{11} \\ \boldsymbol{K}_{22} \\ \boldsymbol{K}_{33} \end{Bmatrix} = \begin{bmatrix} D_1 & D_5 & D_6 \\ D_5 & D_3 & D_4 \\ D_6 & D_4 & 0 \end{bmatrix}\begin{Bmatrix} \bar{\boldsymbol{A}}^{(1)\mathrm{T}}\boldsymbol{C}\bar{\boldsymbol{A}}^{(1)}/\theta \\ \theta\bar{\boldsymbol{B}}^{(1)\mathrm{T}}\boldsymbol{C}\bar{\boldsymbol{B}}^{(1)} \\ \theta b^2\boldsymbol{C} \end{Bmatrix}$$
$$\boldsymbol{K}_{21} = D_2\bar{\boldsymbol{B}}^{(1)\mathrm{T}}\boldsymbol{C}\bar{\boldsymbol{A}}^{(1)} + D_5\bar{\boldsymbol{A}}^{(1)\mathrm{T}}\boldsymbol{C}\bar{\boldsymbol{B}}^{(1)} \tag{2.69}$$
$$\boldsymbol{K}_{31} = -bD_6\bar{\boldsymbol{A}}^{(1)\mathrm{T}}\boldsymbol{C}$$
$$\boldsymbol{K}_{32} = -aD_4\bar{\boldsymbol{B}}^{(1)\mathrm{T}}\boldsymbol{C}$$

where $\boldsymbol{\phi}$, $\boldsymbol{\psi}$ and $\boldsymbol{w}$ are vectors of the discrete values of ϕ, ψ and w as Eq. (2.54).

2.5 Numerical Comparisons

2.5.1 *Vibration and bending of plates and 3D solids*

The results presented in this section aim to demonstrate the high accuracy and rapid convergence of the DQFEM. This is achieved through 2D and 3D free vibration analyses of plates (Tables 2.1–2.3) and static plate bending analyses (Table 2.4). The performance of this method on plane elasticity problems has been presented in [282]; therefore, such analyses are not given here. For free vibration of thin plates, though analyzed in [281], some new results are presented here. For free vibration analyses, the frequencies are calculated in dimensionless form denoted by Ω which are included in the tables where the results for various boundary conditions are given for a range of the sampling points to show clearly the convergence behavior of the solution method. In all cases, Poisson's ratio is 0.3.

In Table 2.1, comparison and convergence studies are carried out for in-plane free vibration of six types of rectangular plate, namely, two completely free plates, two clapped plates, two simply supported plates, with aspect ratio $a/b = 1$ and 2, respectively. The solutions of the DQFEM are compared with the Rayleigh–Ritz solutions [283]. For the rectangular plates with aspect ratio $a/b = 1$, the results of the completely free, simply supported, and clamped plates converge when the grid sizes equal 7×7, 8×8, and 9×9, respectively. Thus, one can say that a completely free plate converges fastest, while a clamped plate converges slowest. It can be seen that all of the frequencies of the DQFEM are exactly the same as those of Ref. [283].

Table 2.2 presents the comparison studies of flexural free vibration of six triangular thin plates (Fig. 2.6) with three combinations of simply supported, clamped and free edges, namely, CCC, SSS, and SCF which implies the side (1), side (2) and side (3) of the triangle are simply supported, clamped, and free, respectively. For analyses, the triangular plates are divided into three quadrilateral elements. It may be seen the DQFEM is capable of producing accurate enough results when the grid size of each element is 10×10. The DQFEM solutions are in excellent agreement, at least to three significant digits, with the Rayleigh–Ritz solutions [284] and agree with the superposition solutions [285–287] to two to three significant digits.

Table 2.1. Convergence validation of the natural frequency parameters $\Omega = \omega a\sqrt{\rho(1-\mu^2)/E}$ for in-plane free vibration of isotropic rectangular plates.

	Grid points	Mode sequence number					
a/b	$M \times N$	1	2	3	4	5	6
		Completely free plates					
1.0	5×5	2.332	2.464	2.464	2.630	2.991	3.457
	6×6	2.321	2.473	2.473	2.628	2.988	3.453
	7×7	2.321	2.472	2.472	2.628	2.987	3.452
	8×8	2.321	2.472	2.472	2.628	2.987	3.452
	[283]	2.321	2.472	2.472	2.628	2.987	3.452
2.0	8×5	1.954	2.961	3.267	4.731	4.795	5.201
	9×6	1.954	2.961	3.267	4.728	4.784	5.206
	10×7	1.954	2.961	3.267	4.726	4.784	5.205
	11×8	1.954	2.961	3.267	4.726	4.784	5.205
	[283]	1.954	2.961	3.267	4.726	4.784	5.205
		Clamped plates					
1.0	7×7	3.555	3.555	4.236	5.191	5.863	5.863
	8×8	3.555	3.555	4.235	5.186	5.859	5.901
	9×9	3.555	3.555	4.235	5.186	5.859	5.894
	10×10	3.555	3.555	4.235	5.186	5.859	5.895
	[283]	3.555	3.555	4.235	5.186	5.859	5.895
2.0	9×6	4.789	6.379	6.711	7.049	7.609	8.116
	10×7	4.789	6.379	6.712	7.049	7.609	8.142
	11×8	4.789	6.379	6.712	7.049	7.608	8.140
	12×9	4.789	6.379	6.712	7.049	7.608	8.140
	[283]	4.789	6.379	6.712	7.049	7.608	8.140
		Simply supported plates					
1.0	6×6	1.859	1.859	2.628	3.699	3.699	4.157
	7×7	1.859	1.859	2.628	3.718	3.718	4.157
	8×8	1.859	1.859	2.628	3.717	3.717	4.156
	9×9	1.859	1.859	2.628	3.717	3.717	4.156
	[283]	1.859	1.859	2.628	3.717	3.717	4.156
2.0	8×5	1.859	3.716	3.717	4.156	5.258	5.587
	9×6	1.859	3.717	3.717	4.156	5.257	5.574
	10×7	1.859	3.717	3.717	4.156	5.257	5.576
	11×8	1.859	3.717	3.717	4.156	5.257	5.576
	[283]	1.859	3.717	3.717	4.156	5.257	5.576

Table 2.2. DQFEM solutions for the first four flexural free vibration frequency parameters $\Omega = \omega(a^2/\pi^2)\sqrt{\rho h/D}$ of triangular thin plates (Fig. 2.6).

	Mode sequences							
$N_\xi = N_\eta$	1	2	3	4	1	2	3	4
	CCC Plate ($b/a = 1$)				CCC Plate ($b/a = 2$)			
	The Rayleigh–Ritz method based on Mindlin plate theory ($h/a = 0.001$) [284]							
	9.503	15.988	19.741	24.655	5.415	8.355	11.518	12.357
	The superposition method [286]							
	9.510	15.978	19.737	24.601	5.416	8.351	11.500	12.351
	The differential quadrature finite element method							
10	9.489	15.978	19.751	24.589	5.407	8.332	11.519	12.347
12	9.496	15.984	19.742	24.595	5.411	8.342	11.508	12.346
14	9.500	15.986	19.738	24.598	5.413	8.347	11.504	12.346
16	9.501	15.987	19.736	24.600	5.414	8.349	11.501	12.345
18	9.502	15.987	19.735	24.600	5.414	8.350	11.500	12.345
20	9.502	15.987	19.735	24.600	5.415	8.351	11.500	12.345
	SSS Plate ($b/a = 1$)				SSS Plate ($b/a = 2$)			
	The Rayleigh–Ritz method based on Mindlin plate theory ($h/a = 0.001$) [284]							
	5.000	9.999	13.000	17.005	2.813	5.054	7.569	8.241
	The superposition method [285]							
	5.000	10.000	13.000	17.002	2.813	5.054	7.566	8.239
	The differential quadrature finite element method							
10	4.988	9.999	12.999	16.975	2.806	5.047	7.560	8.237
12	4.994	10.000	13.000	16.988	2.809	5.051	7.563	8.238
14	4.997	10.000	13.000	16.994	2.811	5.052	7.565	8.239
16	4.998	10.000	13.000	16.996	2.812	5.053	7.565	8.239
18	4.999	10.000	13.000	16.998	2.812	5.054	7.565	8.239
20	4.999	10.000	13.000	16.999	2.812	5.054	7.566	8.239
	SCF Plate ($b/a = 0.5$)				SCF Plate ($b/a = 2$)			
	The Rayleigh–Ritz method based on Mindlin plate theory ($h/a = 0.001$) [284]							
	9.214	18.156	26.491	29.184	1.465	3.009	4.989	5.435
	The superposition method [287]							
	9.139	18.108	26.319	29.083	1.450	2.984	4.955	5.408
	The differential quadrature finite element method							
10	9.186	18.070	26.291	29.150	1.466	3.009	4.989	5.424
12	9.205	18.122	26.425	29.163	1.465	3.009	4.989	5.429
14	9.211	18.141	26.466	29.164	1.465	3.009	4.989	5.432
16	9.213	18.149	26.479	29.163	1.465	3.009	4.989	5.433
18	9.214	18.153	26.485	29.162	1.465	3.009	4.989	5.434
20	9.214	18.155	26.487	29.161	1.465	3.009	4.989	5.434

Table 2.3. Frequency parameters $\Omega = \omega R\sqrt{\rho/G}$ for 3D free vibration of clamped and free circular plates.

	Grid points	Mode sequence number					
h/R	$N_\xi \times N_\eta \times N_z$	1	2	3	4	5	6
		Completely free circular plates					
0.1	$9 \times 9 \times 4$	0.2576	0.4329	0.5896	0.9655	1.017	1.544
	$11 \times 11 \times 5$	0.2576	0.4329	0.5891	0.9631	1.016	1.530
	$13 \times 13 \times 6$	0.2576	0.4329	0.5891	0.9631	1.016	1.529
	$14 \times 14 \times 7$	0.2576	0.4329	0.5891	0.9631	1.016	1.529
	[288]	0.2576	0.4329	0.5892	0.9633	1.017	1.529
	[289]	0.2576	0.4329	0.5891	0.9631	1.016	1.529
0.2	$9 \times 9 \times 4$	0.4996	0.8315	1.1069	1.765	1.844	2.689
	$11 \times 11 \times 5$	0.4995	0.8314	1.106	1.762	1.843	2.674
	$13 \times 13 \times 6$	0.4995	0.8314	1.106	1.762	1.843	2.673
	$14 \times 14 \times 7$	0.4995	0.8314	1.106	1.762	1.843	2.673
	[288]	0.4997	0.8316	1.107	1.763	1.844	2.677
	[289]	0.4995	0.8314	1.106	1.762	1.843	2.673
		Clamped circular plates					
0.01	$11 \times 11 \times 3$	0.05003	0.1041	0.1707	0.1948	0.2510	0.3038
	$13 \times 13 \times 4$	0.04997	0.1040	0.1703	0.1944	0.2495	0.2979
	$15 \times 15 \times 5$	0.04993	0.1039	0.1703	0.1943	0.2492	0.2970
	$17 \times 17 \times 6$	0.04991	0.1038	0.1702	0.1942	0.2491	0.2969
	[290]	0.04990	0.1038	0.1703	0.1941	0.2490	0.2968
	[291]	0.04985	0.1037	0.1702	0.1941	0.2490	0.2968

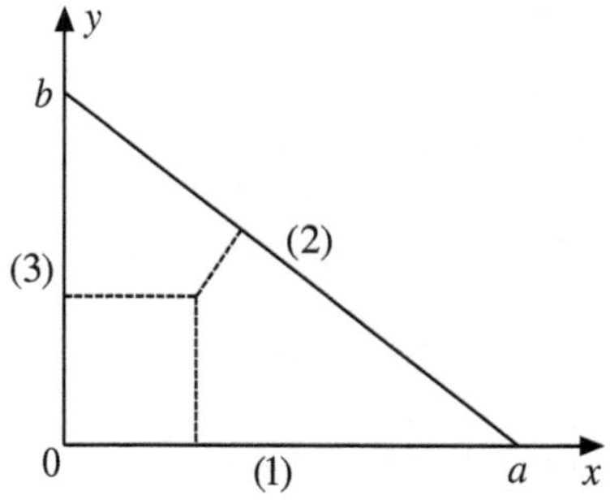

Fig. 2.6. A triangular plate.

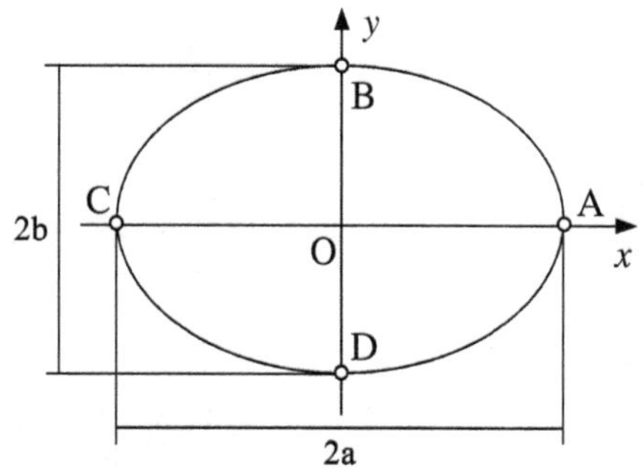

Fig. 2.7. An elliptic plate.

In Table 2.3, a comparison study has been given for 3D free vibration of circular plates with clamped and free boundary conditions. The DQFEM solutions are given for relative thickness $h/R = 0.1$ and 0.2 for two free circular plates, and given for relative thickness $h/R = 0.01$ for a clamped plate. Convergent DQFEM solutions are obtained when the grid sizes equal $13 \times 13 \times 6$ and $15 \times 15 \times 5$ for free and clamped circular plates, respectively. The DQFEM results agree with all the results used for comparisons [288–291] to at least three significant digits.

Table 2.4 presents the comparison studies of bending moments in an elliptical plate (Fig. 2.6) with built-in and simply supported edges subjected to uniformly distributed loads. Values of the geometric and material parameters used in the calculation are as follows: $a = 0.50$ (m), $b = 0.33333$ (m), $h = 0.01$ (m), $E = 1$ (MPa), $q = 1.0$ (Pa). Numerical values at points O, A and B as shown in Fig. 2.4 are presented, for which both exact results [292] and p-type FEM results [293] are available. The DQFEM solutions based on both the thin plate theory and Mindlin plate theory are given in the table. Excellent agreements among the three sets of results are found for both clamped and simply supported conditions.

Thus, one can definitely conclude that the DQFEM is capable of producing accurate and rapid convergent solutions for free vibration analyses of 2D and 3D plates and bending analyses of thin and Mindlin plates with arbitrary shapes.

Table 2.4. Bending moments in an elliptical plate with built-in and simply supported edges subjected to uniformly distributed loads (Fig. 2.7).

$N_\xi = N_\eta$	$100(w_0)$	$100(M_x)_0$	$100(M_y)_0$	$100(M_x)_a$	$100(M_y)_b$	$100(w_0)$	$100(M_x)_0$	$100(M_y)_0$
	Clamped boundary conditions					Simply supported boundary conditions		
	Exact solution							
	−0.3759	−0.9227	−1.4048	1.1016	2.4791	−1.5549	−2.4662	−3.5660
	p-type FEM solution based Mindlin plate theory [293]							
	−0.3776	−0.9238	−1.4064	1.1012	2.5005	−1.5546	−2.4100	−3.5065
	DQFEM solution based Mindlin plate theory							
7	−0.3765	−0.9210	−1.4014	1.0936	2.4575	−1.5604	−2.4332	−3.5919
9	−0.3773	−0.9231	−1.4047	1.1029	2.4772	−1.5510	−2.4239	−3.5698
11	−0.3773	−0.9231	−1.4048	1.1030	2.4781	−1.5481	−2.4213	−3.5625
13	−0.3773	−0.9232	−1.4048	1.1034	2.4781	−1.5470	−2.4203	−3.5598
15	−0.3773	−0.9232	−1.4048	1.1031	2.4780	−1.5465	−2.4197	−3.5587
17	−0.3773	−0.9232	−1.4048	1.1034	2.4779	−1.5462	−2.4195	−3.5583
	DQFEM solution based thin plate theory							
7	−0.3753	−0.9207	−1.4016	1.0919	2.4575	−1.5437	−2.4189	−3.5552
9	−0.3760	−0.9226	−1.4048	1.1010	2.4774	−1.5446	−2.4195	−3.5576
11	−0.3760	−0.9227	−1.4048	1.1013	2.4792	−1.5447	−2.4192	−3.5572
13	−0.3760	−0.9227	−1.4049	1.1012	2.4784	−1.5447	−2.4187	−3.5564
15	−0.3760	−0.9228	−1.4049	1.1013	2.4788	−1.5446	−2.4180	−3.5554

2.5.2 *Thickness-shear vibration analysis of rectangular quartz plates*

This section presents the procedure of the NEKM to solve thickness-shear vibration of a rectangular crystal plate. For ease of comparison, a rectangular quartz crystal plate with the same dimensions and material constants as the one used by Liu *et al.* [294] are considered. The dimensions are $a = 2.5$ m, $b = 1.25$ m, and $h = 0.1$ m. The material constants are

$$C_{ij} = 10^9 \times \left\{ \begin{array}{cccccc} 86.74 & -8.254 & 27.15 & -3.66 & 0 & 0 \\ -8.254 & 129.77 & -7.418 & 5.7 & 0 & 0 \\ 27.15 & -7.418 & 102.83 & 9.92 & 0 & 0 \\ -3.66 & 5.7 & 9.92 & 38.61 & 0 & 0 \\ 0 & 0 & 0 & 0 & 68.81 & 2.534 \\ 0 & 0 & 0 & 0 & 2.534 & 29.01 \end{array} \right\} \mathrm{N/m^2} \tag{2.70}$$

The natural frequency is expressed by frequency parameter Ω used by Liu *et al.* [294]. Its expression is

$$\Omega = \omega\sqrt{\frac{\rho h}{D_1}} \tag{2.71}$$

This frequency parameter is further normalized by the magnitude of fundamental thickness-shear frequency parameter of Eq. (2.72) that is obtained by the substitution of Eq. (2.67) into Eq. (2.71):

$$\Omega_0 = \omega_0\sqrt{\frac{\rho h}{D_1}} = \frac{1}{h}\sqrt{\frac{12D_6}{D_1}} \tag{2.72}$$

The normalization is done for ease of finding the fundamental thickness-shear mode.

The fundamental thickness-shear mode and frequency of a completely free Mindlin plate with the above material constants will be solved. The quartz plate is discretized as 5 × 5 DQFEM elements. The frequency parameters Ω normalized by Ω_0 for different grid points in x- and y-directions in each elements are shown in Table 2.5. The results of this paper were compared with the result by 3D FEM [295] and by a numerical extended Kantorovich method (NEKM) [294]. The

Table 2.5. The frequency parameters Ω normalized by Ω_0 for different number of grid points in each elements.

The Methods Used	Grid Points in x- and y-directions	Frequency Parameters (Ω/Ω_0)
DQFEM	12×12	1.0029467549
	15×15	1.0028498128
	18×18	1.0028496274
	25×18	1.0028496256
3D FEM [295]	$100 \times 50 \times 6^{a}$	1.0030015187
NEKM [294]	25×25	1.0028384677

Note: [a]Here, 6 is the number of grid points in z-direction.

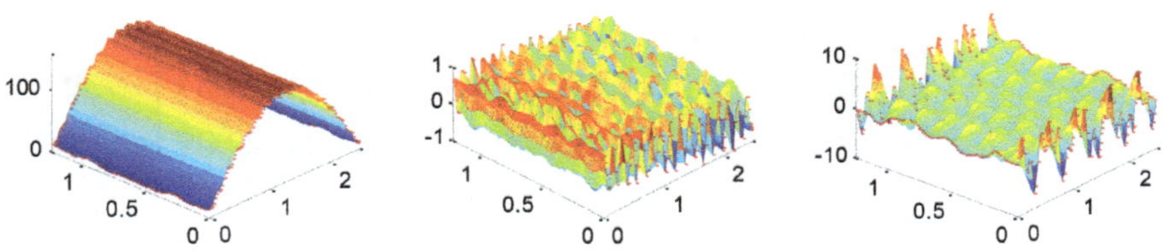

(a) (b) (c)

Fig. 2.8. Mode shapes ϕ (a), w (b) and ψ (c) of the fundamental thickness-shear mode for a completely free quartz plate.

DEFEM result agrees with the 3D FEM for four significant digits and agrees with the NRKM for five significant digits. The three sets of solutions used different equations or reduction techniques, so the difference among them are understandable. The 3D FEM result of Wang *et al.* [295] only have first five significant digits converged. The simulations in [295] were not furthered, which should be limited by the huge computational cost in 3D FEM simulations. With the 2D equations used in this work, the results could converge up to nine significant digits. However, if the nodal points in each element is taken as 25×25, the workstation of the authors could not finish the simulation within 12 hours. The NEKM result is obtained through iterative reduction of the 2D equations to 1D equations, so the boundary conditions could not be satisfied very well [294]. The fundamental thickness-shear modes of DQFEM for ϕ, w, and ψ are shown in Fig. 2.8. The corresponding modes for moments M_x, M_y and M_{xy}, and shear forces Q_x and Q_y are shown in Figs. 2.9 and 2.10,

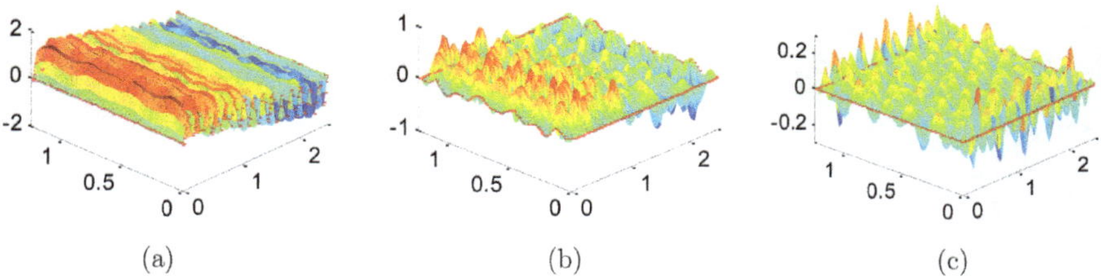

Fig. 2.9. Mode shapes M_x (a), M_y (b) and M_{xy} (c) of the fundamental thickness-shear mode for a completely free quartz plate.

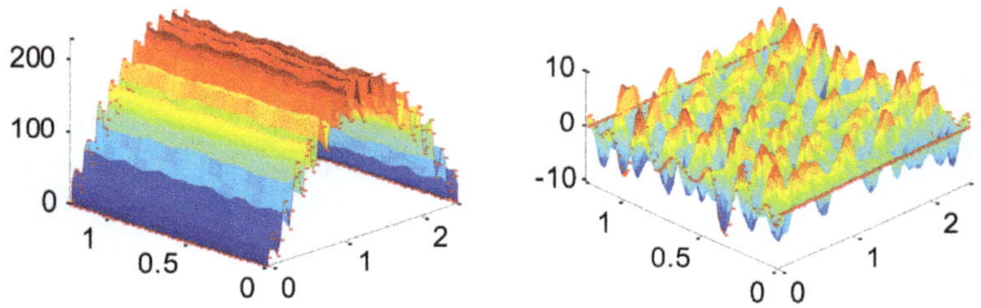

Fig. 2.10. Mode shapes Q_x (a) and Q_y (b) of the fundamental thickness-shear mode for a completely free quartz plate.

respectively. As can be seen from Figs. 2.9 and 2.10, the DQFEM could satisfy the completely free boundary conditions very well.

2.6 Conclusions

A differential quadrature finite element method (DQFEM) has been applied to 1D to 3D elasticity problems. The reformulated differential quadrature rules and the Gauss–Lobatto rule are used to discretize the variational formulation to derive the finite element formulation directly. The method for the construction of the DQFEM formulation satisfying both C^0 and C^1 continuity requirements is presented. The DQFEM matrices are compact and well conditioned. Especially, the mass matrices for C^0 continuity problems are diagonal, which can reduce the computational cost of dynamic problems. Numerical comparison studies with results available in the literature are carried out for free vibration of 2D and 3D plates and bending of thin and Mindlin plates with arbitrary shapes, which validate the high accuracy and rapid convergence of the DQFEM.

Chapter 3

The Differential Quadrature Hierarchical Finite Element Method for Mindlin Plates

A differential quadrature hierarchical finite element method (DQHFEM) is presented by expressing the hierarchical finite element method (HFEM) matrices in similar form as in the differential quadrature finite element method (DQFEM) and introducing interpolation basis on the boundary of HFEM elements. The DQHFEM is similar to the fixed interface mode synthesis method, but the DQHFEM does not need modal analysis. The DQHFEM with NURBS (non-uniform rational B-splines) elements were shown to accomplish the same destination as the isogeometric analysis (IGA). Three key points that determine the accuracy, efficiency and convergence of DQHFEM were addressed: (1) the Gauss–Lobatto–Legendre (GLL) points should be used as nodes, (2) the recursion formula should be used to compute high-order orthogonal polynomials, and (3) the separation variable feature of the basis should be used to save computational cost. Numerical comparison and convergence studies of the DQHFEM were carried out by comparing the DQHFEM results for vibration and bending of Mindlin plates with available exact or highly accurate approximate results in the literatures. The DQHFEM can present highly accurate results using only a few sampling points. Meanwhile, the order of the DQHFEM can be as high-as needed for high-frequency vibration analysis.

3.1 The Reformulated Differential Quadrature Rule

The well-known differential quadrature (DQ) rules approximate the derivate of a function at a point by a weighted linear summation of field variables along a line that passes through the point. Besides Lagrange functions, any other complete basis can be used as basis to formulate DQ rules. Thus, for a field variable $f(x)$, its nth order derivative at a discrete point x_i can be expressed as

$$f_i^{(n)} = \sum_{j=1}^{M} A_{ij}^{(n)} a_j \quad (i = 1, \ldots, N_x) \tag{3.1}$$

where

$$A_{ij}^{(n)} = P_j^{(n)}(x_i) \tag{3.2}$$

$P_j(x), j = 1, \ldots, M$ can be any basis that is complete and a_j may not have clear physical meaning if the basis is not an interpolator. N_x is the number of sampling points and M is the number of basis.

For a (2D) function $f(x, y)$ that is an implicit function of ξ and η as $f = f[x(\xi, \eta), y(\xi, \eta)]$ (Fig. 3.1), the first-order derivatives of $f = f[x(\xi, \eta), y(\xi, \eta)]$ with respect to x and y obtained through the chain rule of differentiation are as follows:

$$\frac{\partial f}{\partial x} = \frac{1}{|\boldsymbol{J}|}\left(\frac{\partial y}{\partial \eta}\frac{\partial f}{\partial \xi} - \frac{\partial y}{\partial \xi}\frac{\partial f}{\partial \eta}\right), \frac{\partial f}{\partial y} = \frac{1}{|\boldsymbol{J}|}\left(\frac{\partial x}{\partial \xi}\frac{\partial f}{\partial \eta} - \frac{\partial x}{\partial \eta}\frac{\partial f}{\partial \xi}\right) \tag{3.3}$$

where the determinant $|\boldsymbol{J}|$ of the Jacobian $\boldsymbol{J} = \partial(x, y)/\partial(\xi, \eta)$ is

$$|\boldsymbol{J}| = \frac{\partial x}{\partial \xi}\frac{\partial y}{\partial \eta} - \frac{\partial y}{\partial \xi}\frac{\partial x}{\partial \eta} \tag{3.4}$$

Thus, the DQ rule for the first-order derivatives of $f(x, y)$ with respect to x and y at a discrete point $x_{ij} = x(\xi_i, \eta_j), y_{ij} = y(\xi_i, \eta_j)(i = 1, 2, \ldots, N_x; j = 1, 2, \ldots, N_y)$ can be given as

$$\left(\frac{\partial f}{\partial x}\right)_{ij} = \frac{1}{|\boldsymbol{J}|_{ij}}\left[\left(\frac{\partial y}{\partial \eta}\right)_{ij}\left(\sum_{m=1}^{M} A_{im}^{(1)} Q_{mj}\right) - \left(\frac{\partial y}{\partial \xi}\right)_{ij}\left(\sum_{n=1}^{N} B_{jn}^{(1)} Q_{in}\right)\right]$$

$$\left(\frac{\partial f}{\partial y}\right)_{ij} = \frac{1}{|\boldsymbol{J}|_{ij}}\left[\left(\frac{\partial x}{\partial \xi}\right)_{ij}\left(\sum_{n=1}^{N} B_{jn}^{(1)} Q_{in}\right)\right.$$
$$\left. - \left(\frac{\partial x}{\partial \eta}\right)_{ij}\left(\sum_{m=1}^{M} A_{im}^{(1)} Q_{mj}\right)\right] \tag{3.5}$$

where

$$Q_{mj} = \sum_{n=1}^{N} P_n(y_j) a_{m,n}, \quad Q_{in} = \sum_{m=1}^{M} P_m(x_i) a_{m,n} \tag{3.6}$$

where M and N denote the number of basis functions in x and y directions, respectively. $A_{im}^{(1)}$ and $B_{jn}^{(1)}$ are the weighting coefficients associated with the first-order derivatives with respect to ξ and η, respectively. The generalized DQ rule defined in Eq. (3.5) can also be written in a single index notation form as

$$\left.\frac{\partial f}{\partial x}\right|_q = \sum_{p=1}^{M\times N} \bar{A}_{qp}^{(1)} \bar{a}_p, \quad \left.\frac{\partial f}{\partial y}\right|_q = \sum_{p=1}^{M\times N} \bar{B}_{qp}^{(1)} \bar{a}_p \tag{3.7}$$

where $\bar{A}_{qp}^{(1)}$ and $\bar{B}_{qp}^{(1)}$ are the weighting coefficients for the first-order derivatives with respect to x and y coordinates, respectively, of the irregular domain, and

$$\bar{a}_p = a_{m,n}, \quad \text{where } p = (n-1)M + m,$$
$$m = 1, 2, \ldots, M \text{ and } n = 1, 2, \ldots, N \tag{3.8}$$

$$q = (j-1)N_x + i \text{ where } i = 1, 2, \ldots, N_x \text{ and } j = 1, 2, \ldots, N_y \tag{3.9}$$

The DQ rule as in Eq. (3.7) will be used to discretize the differentiations in the potential functional for the vibration of Mindlin plates in the DQHFEM.

3.2 The DQHFEM on Quadrilateral Domain

3.2.1 *The method of geometric mapping*

DQM has been applied to irregular domains with the help of the natural-to-Cartesian geometric mapping with the blending functions

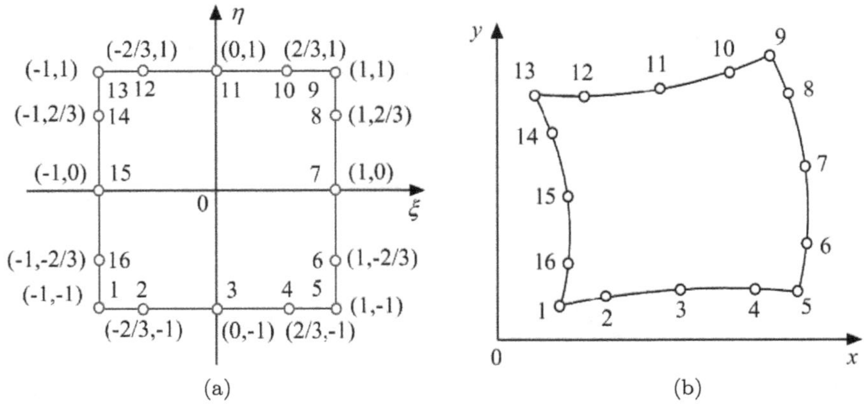

Fig. 3.1. Geometric mapping: (a) a square parent domain in natural $\xi-\eta$ plane; (b) a curvilinear quadrilateral region in Cartesian x–y plane.

which permit exact mapping [280] or the serendipity-family interpolation functions [279]. The blending function method [296, 297] has been used to construct curvilinear coordinate systems for mesh generation and interpolation over arbitrarily curved domains. For an arbitrary quadrilateral domain, see Fig. 3.1, the geometric mapping has the form

$$\begin{aligned}
V(\xi,\eta) &= \frac{1-\xi}{2}M(-1,\eta) + \frac{1+\xi}{2}M(1,\eta) \\
&\quad + \frac{1-\eta}{2}M(\xi,-1) + \frac{1+\eta}{2}M(\xi,1) \\
&\quad - \frac{(1-\xi)(1-\eta)}{4}M(-1,-1) - \frac{(1-\xi)(1+\eta)}{4}M(-1,1) \\
&\quad - \frac{(1+\xi)(1-\eta)}{4}M(1,-1) - \frac{(1+\xi)(1+\eta)}{4}M(1,1)
\end{aligned} \tag{3.10}$$

where $-1 \leq \xi,\eta \leq 1$. The functions $M(\xi_i,\eta)$ and $M(\xi,\eta_j)$ represent the four parametric curves that constitute the boundary of the quadrilateral domain. The function $M(\xi_i,\eta_j)$ represents the x–y coordinates of the points corresponding to the non-dimensional coordinates (ξ_i,η_j). The blending function method can provide exact geometric mapping. Therefore, it is more accurate than the serendipity-family interpolation functions.

Although it provides exact geometric mapping for plane quadrilateral domain, the blending function using analytical function is not

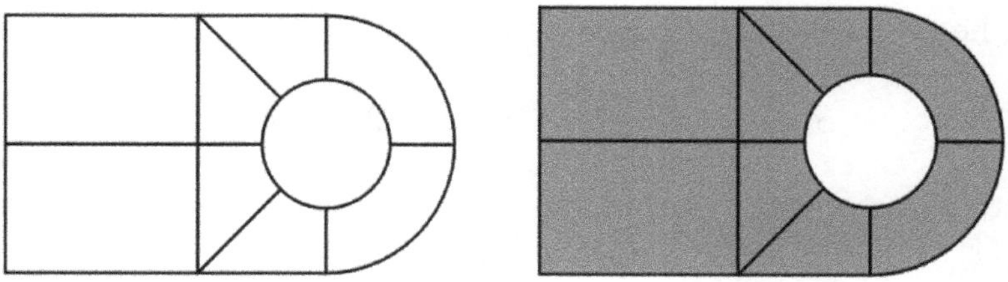

(a) (b)

Fig. 3.2. The construction of Coons surfaces: (a) NURBS curves; (b) NURBS surfaces constructed by NURBS curves.

flexible for complex geometries in practical applications. The surfaces constructed by blending function in NURBS (non-uniform rational B-splines) are known as Coons surfaces [51, 220]. In this work, quadrilateral elements constructed by both NURBS and analytical blending function are used. The geometries were first divided into quadrilateral domains (Fig. 3.2(a)) and then each domain is formed as a blending surface or NURBS surface (Fig. 3.2(b)). The construction of Coons surfaces needs curve cutting and many other NURBS and geometric computations, but it can be used for complex geometries. The open source routines by Mark *et al.* [298] and a package of routines developed by the authors of this paper are used for the modeling in this work.

3.2.2 *The construction of shape functions*

The hierarchical p-method needs shape functions satisfying C^0 or C^1 continuity along element interfaces. For C^0 elements, assume $\xi_i (i = 1, 2, \ldots, M)$ and $\eta_j (j = 1, 2, \ldots, N)$ are the sampling points along the edge of a square domain in the natural ξ–η plane, see Fig. 3.1(a). The preferable nodes are non-uniformly distributed nodes such as Gauss–Lobatto–Legendre nodes, which lead to much lower condition numbers than uniformly distributed nodes in the p-method [53]. Using the blending function as Eq. (3.10), the nodal shape function at $\xi_1 = \eta_1 = -1$ can be constructed from the Lagrange functions as

$$S_1(\xi, \eta) = \frac{1-\eta}{2} L_1^M(\xi) + \frac{1-\xi}{2} L_1^N(\eta) - \frac{(1-\xi)(1-\eta)}{4} \qquad (3.11)$$

The nodal shape functions on edge $\eta_1 = -1$ are

$$S_i(\xi, \eta) = \frac{1-\eta}{2} L_i^M(\xi), i = 2, \ldots, M-1 \tag{3.12}$$

where

$$L_i^M(\xi) = \prod_{j=1, j\neq i}^{M} \frac{\xi - \xi_j}{\xi_i - \xi_j} \tag{3.13}$$

is the Lagrange function. The serendipity shape functions at other corners and edges are of similar form as Eqs. (3.11) and (3.12). The subscripts $k = 1, 2, \ldots, K$ of shape functions are numbered along the edge of a domain, as shown in Fig. 3.1. The shape functions on the boundary of elements constructed herein make the imposition of boundary conditions of HFEM more convenient than that of Refs. [46, 49]. Moreover, besides two neighboring edges, the number of nodes on the two opposite edges of an element can also be different.

The above shape functions are defined on the boundaries of an element. In order to improve the interpolation accuracy, the element should be enriched by hierarchical shape functions that have no influence on inter-element compatibility. The commonly used hierarchical shape functions are the integrated Legendre polynomials that were initially presented by Zhu [299]. For C^0 continuity problems, the integrated Legendre polynomials are

$$\varphi_n(\xi) = \int_{-1}^{\xi} P_n(\xi) d\xi = \frac{(\xi^2 - 1)}{n(n+1)} \frac{dP_n(\xi)}{d\xi}, \xi \in [-1, 1] \tag{3.14}$$

where $n = 1, 2, \ldots$ and $P_n(\xi)$ is the Legendre polynomial of degree n. It can be easily seen that Eq. (3.14) has zero function values at the two endpoints. For C^1 continuous problems, the integrated Legendre polynomials are

$$\psi_n(\xi) = \int_{-1}^{\xi} \int_{-1}^{\xi} P_{n+1}(\xi) d\xi = \frac{(\xi^2 - 1)^2}{n(n+1)(n+2)(n+3)} \times \frac{d^2 P_{n+1}(\xi)}{d\xi^2}, \xi \in [-1, 1] \tag{3.15}$$

It is obvious that Eq. (3.15) has zero function values and first-order derivatives at the two endpoints. Low-order Legendre polynomial can be computed from the Tylor series expression

$$P_n(\xi) = \sum_{k=0}^{[n/2]} \frac{(-1)^k(2n-2k)!}{2^n k!(n-k)!(n-2k)!}\xi^{n-2k} \tag{3.16}$$

This expression was used in early publications on HFEM [8, 47, 57, 299]. Equation (3.16) may have computational problem when n is greater than about 40. By using the recursion formula as follows, the order n can reach several thousand as can be seen from the example of vibration of Euler–Bernoulli beam in the results section of this work:

$$P_{n+1}(\xi) = \frac{2n+1}{n+1}\xi P_n(\xi) - \frac{n}{n+1}P_{n-1}(\xi) \tag{3.17}$$

where $P_0(\xi) = 1$ and $P_1(\xi) = \xi$. The mth-order derivatives of Legendre polynomials can be obtained from the following recursion formula:

$$P_{n+1}^{(m)}(\xi) = \xi P_n^{(m)}(\xi) + (n+m)P_n^{(m-1)}(\xi) \tag{3.18}$$

Orthogonal polynomials constructed from trigonometric functions theoretically also have no computational stability problem and can be tried. Since Gauss–Lobatto or Gauss quadratures are needed in HFEM and the Legendre polynomials can also be of very high order, this work uses integrated Legendre polynomials.

3.2.3 *Gauss–Lobatto quadrature rule*

The well-known Gauss–Lobatto quadrature is the Gauss integration with two endpoints fixed. The Gauss quadrature is not used here mainly because it does not have endpoints that are needed in pre- and post-processing. Theory of the Gauss–Lobatto quadrature rule can be found in mathematics handbooks; it is introduced here to make the paper self-contained. The Gauss–Lobatto quadrature rule with precision degree $(2n-3)$ for function $f(x)$ defined at $[-1, 1]$ is

$$\int_{-1}^{1} f(\xi)\mathrm{d}x = \sum_{j=1}^{n} C_j f(\xi_j) \tag{3.19}$$

where the weights C_j of Gauss–Lobatto integration are given by

$$C_1 = C_n = \frac{2}{n(n-1)}, C_j = \frac{2}{n(n-1)[P_{n-1}(\xi_j)]^2}(j \neq 1, n) \tag{3.20}$$

where ξ_j is the $(j-1)$th zero of $P'_{n-1}(\xi)$. Using Eq. (3.16) to solve the roots of Legendre polynomials, one can only obtain about 40 roots. By using the recursion formula as in Eqs. (3.17) and (3.18), it is easy to get thousands of roots. Assume the span of root ξ_j is $[a_j, b_j]$, which can be easily obtained by searching in points about several times of the number of roots. For example, the Chebyshev–Lobatto (CL) nodes have similar distribution as Gauss–Lobatto nodes

$$\xi'_j = -\cos\left(\frac{j-1}{n-1}\pi\right) \tag{3.21}$$

If the number of CL nodes are five times of that of Gauss–Lobatto nodes, there are about 4–6 CL nodes between $[\xi_j, \xi_{j+1}]$ in general. Therefore, it is easy to locate the span $[a_j, b_j]$ for each root ξ_j. Then we can solve ξ_j using the Newton–Raphson iteration method as follows:

$$\xi_j^{k+1} = \xi_j^k - \frac{P'_n(\xi_j^k)}{P''_n(\xi_j^k)}, \quad k = 1, 2, \ldots \tag{3.22}$$

where $\xi_j^1 = (a_j + b_j)/2$. About five iterations are sufficient to obtain the converged results. Therefore, all roots can be easily obtained. Due to the symmetric property of the roots, one only needs to solve one half of the roots.

The Gauss–Lobatto–Legendre (GLL) nodes can also be solved according to their characteristics. The jth Lagrange polynomial using GLL points has maximum value at jth GLL point and the maximum value is 1, namely,

$$\max L_j^n(\xi) = L_j^n(\xi_j) = 1 \qquad j = 2, 3, \ldots, n-1 \tag{3.23}$$

where the Lagrange polynomial $L_j^n(\xi)$ is defined by Eq. (3.13). These requirements are equivalent to

$$f(\xi_j) = \frac{\partial L_j^n(\xi_j)}{\partial \xi} = \sum_{k=1, k\neq j}^{n} \frac{1}{x_j - x_k} = 0 \qquad j = 2, 3, \ldots, n-1 \tag{3.24}$$

The Newton–Raphson method can be used to solve the system of nonlinear equations given by Eq. (3.24), which requires a Jacobian as follows:

$$\boldsymbol{F}'(\boldsymbol{\xi}) = \left[\frac{\partial f(\xi_j)}{\partial \xi_i}\right]_{(n-2)\times(n-2)} \tag{3.25}$$

where

$$\frac{\partial f(\xi_j)}{\partial \xi_i} = \begin{cases} -\sum_{k=1,k\neq j}^{n} \frac{1}{(x_j - x_k)^2} (i = j) \\ \frac{1}{(x_j - x_i)^2} \quad (i \neq j) \end{cases} \tag{3.26}$$

and $\boldsymbol{\xi} = [\xi_2, \xi_3, \ldots, \xi_{n-1}]^T$. Let $\boldsymbol{F}(\boldsymbol{\xi}) = [f(\xi_2), f(\xi_3), \ldots, f(\xi_{n-1})]^T$, then the Newton–Raphson method can be written as

$$\boldsymbol{\xi}^{k+1} = \boldsymbol{\xi}^k - \boldsymbol{F}'(\boldsymbol{\xi}^k)^{-1}\boldsymbol{F}(\boldsymbol{\xi}^k), \quad k = 0, 1, \ldots \tag{3.27}$$

where the superscript k denotes the value of $\boldsymbol{\xi}$ at kth step of iteration. This method does not need Legendre polynomials to compute the roots and is less sensitive with the initial value than the above method. The values given by Eq. (3.21) can be used as initial values for this method.

The vector quadrature method [300] was shown to be significantly faster than Gaussian quadrature for large matrices whose coefficients are of high order and may also be tried. However, the vector quadrature method used finite difference approximation of derivatives and may not be suitable for high accuracy analysis.

3.3 The DQHFEM on Triangular Domain

3.3.1 *The construction of DQHFEM basis on triangles*

The hierarchical shape functions are usually constructed by blending of the basis of h-version finite element method and orthogonal polynomials [46, 48]. The orthogonal polynomials are not interpolation functions, so the imposition of boundary conditions and inter-element compatibility conditions are not so straightforward. Liu *et al.* recently [34, 301] proposed a differential quadrature hierarchical finite element method (DQHFEM) that uses basis constructed from blending

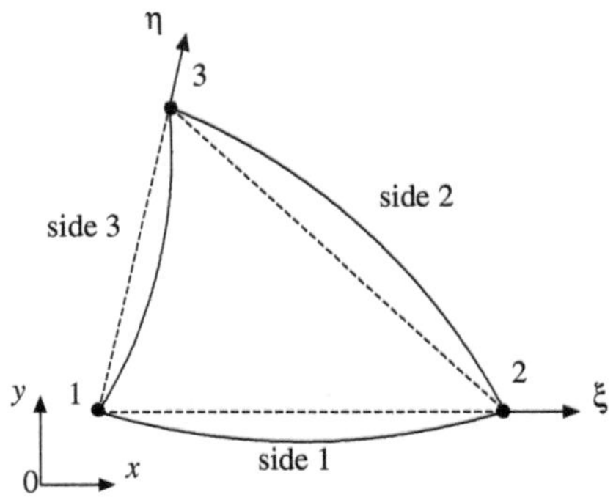

Fig. 3.3. The curved triangular element.

function and Lagrange polynomials on the boundary of elements and orthogonal polynomials inside the element. This work will construct DQHFEM basis for triangles. Similar to the rectangular case, the blending function method of interpolating geometry will be used to construct the basis on the boundary of an element. A curved triangular p-element is shown in Fig. 3.3. The sides 1–2, 2–3 and 1–3 are numbered 1, 2 and 3, respectively. Using the blending function method, the curved sides can be expressed in the following parametric forms [48]:

$$x = x_1(\xi), \quad y = y_1(\xi), \quad 0 \leq \xi \leq 1 \tag{3.28}$$

$$x = x_i(\eta), \quad y = y_i(\eta), \quad 0 \leq \eta \leq 1 \quad (i = 2, 3) \tag{3.29}$$

where the subscripts denote the side numbers. The element mapping functions are

$$\begin{aligned} x(\xi, \eta) = {} & \frac{1-\xi-\eta}{1-\xi} x_1(\xi) + \frac{\xi}{1-\eta} x_2(\eta) + \frac{1-\xi-\eta}{1-\eta} x_3(\eta) \\ & - (1-\xi-\eta) X_1 - \frac{\xi(1-\xi-\eta)}{1-\xi} X_2 \end{aligned} \tag{3.30}$$

$$\begin{aligned} y(\xi, \eta) = {} & \frac{1-\xi-\eta}{1-\xi} y_1(\xi) + \frac{\xi}{1-\eta} y_2(\eta) + \frac{1-\xi-\eta}{1-\eta} y_3(\eta) \\ & - (1-\xi-\eta) Y_1 - \frac{\xi(1-\xi-\eta)}{1-\xi} Y_2 \end{aligned} \tag{3.31}$$

where X_1, X_2 and Y_1, Y_2 are respectively the x and y coordinates of corner points 1 and 2, as shown in Fig. 3.3. The first derivatives of

Eq. (3.30) with respect to ξ and η are

$$\frac{\partial x(\xi,\eta)}{\partial \xi} = \frac{1-\xi-\eta}{1-\xi}\frac{\mathrm{d}x_1(\xi)}{\mathrm{d}\xi} + \frac{x_2(\eta)-x_3(\eta)}{1-\eta} + \frac{\eta[X_2 - x_1(\xi)]}{(1-\xi)^2} + X_1 - X_2$$

$$\frac{\partial x(\xi,\eta)}{\partial \eta} = \frac{\xi}{1-\eta}\frac{\partial x_2(\eta)}{\partial \eta} + \frac{1-\xi-\eta}{1-\eta}\frac{\partial x_3(\eta)}{\partial \eta} + \frac{\xi X_2 - x_1(\xi)}{1-\xi} + \frac{\xi[x_2(\eta)-x_3(\eta)]}{(1-\eta)^2} + X_1 \tag{3.32}$$

The derivatives of Eq. (3.31) with respect to ξ and η are similar. Note that the derivatives given by [48] for Eq. (3.32) have typing errors. Similar to the as rectangular DQHFEM elements [34, 301], the interpolation basis on the boundary of triangles can be constructed from Eq. (3.30) or (3.31). Let $\xi_i (i = 1, 2, \ldots, M)$ and $\eta_j (j = 1, 2, \ldots, N$ or $P)$ be the sampling points along the edge of a unit domain in the natural ξ–η plane (Fig. 3.4). The preferable nodes are non-uniformly distributed nodes such as Gauss–Lobatto–Legendre nodes, which lead to much lower condition numbers than uniformly distributed nodes in the p-method. The serendipity shape

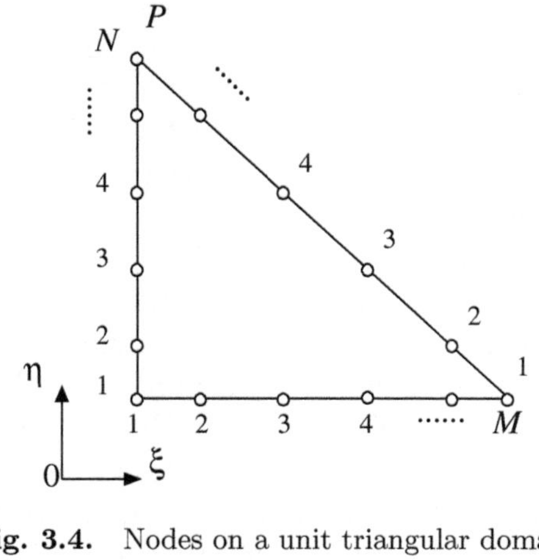

Fig. 3.4. Nodes on a unit triangular domain.

function at the three vertexes can be written as

$$
\begin{aligned}
S_1(\xi,\eta) &= \frac{1-\xi-\eta}{1-\xi}L_1^M(\xi) + \frac{1-\xi-\eta}{1-\eta}L_1^N(\eta) - (1-\xi-\eta) \\
S_2(\xi,\eta) &= \frac{1-\xi-\eta}{1-\xi}L_M^M(\xi) + \frac{\xi}{1-\eta}L_1^P(\eta) - \frac{\xi(1-\xi-\eta)}{1-\xi} \\
S_3(\xi,\eta) &= \frac{\xi}{1-\eta}L_P^P(\eta) + \frac{1-\xi-\eta}{1-\eta}L_N^N(\eta)
\end{aligned} \tag{3.33}
$$

The serendipity shape functions at nodes on sides 1, 2 and 3 are respectively

$$S_{1i}(\xi,\eta) = \frac{1-\xi-\eta}{1-\xi}L_i^M(\xi), \quad i=2,\ldots,M-1 \tag{3.34}$$

$$S_{2j}(\xi,\eta) = \frac{\xi}{1-\eta}L_j^P(\eta), \quad j=2,\ldots,P-1 \tag{3.35}$$

$$S_{3j}(\xi,\eta) = \frac{1-\xi-\eta}{1-\eta}L_j^N(\eta), \quad j=2,\ldots,N-1 \tag{3.36}$$

where

$$L_i^M(\xi) = \prod_{j=1,j\neq i}^{M} \frac{\xi-\xi_j}{\xi_i-\xi_j} \tag{3.37}$$

is the Lagrange functions. The derivatives of Eqs. (3.33)–(3.36) with respect to ξ and η can be obtained from Eq. (3.32) according to their relationship with Eq. (3.30) or (3.31). The nodes on triangular element are usually sampled as the same number on all edges. In this work, they are sampled differently which can bring convenience in applications. Figure 3.5 plots the two basis functions on the boundary of a triangular element.

The above shape functions are defined on the boundaries of a domain. In order to improve the interpolation accuracy, the element should be enriched by hierarchical shape functions that have no influence on inter-element compatibility. The hierarchical basis constructed from Jacobi polynomials by Webb and Abouchacra [54] were shown to be orthogonal on triangles and will be used in this work. The procedures of deriving the orthogonal basis are presented herein so that they can be easily generated to tetrahedrons and even

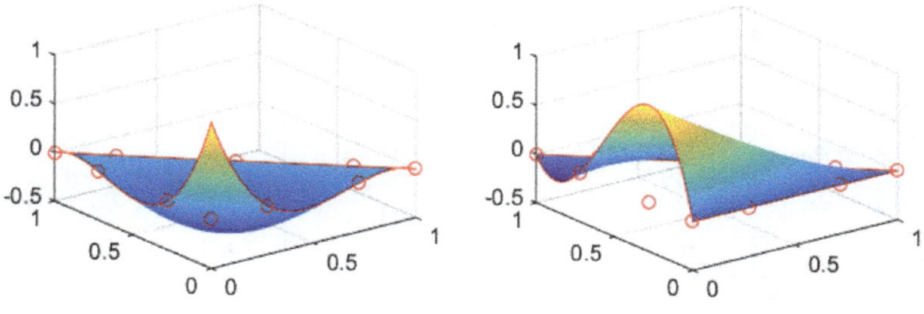

Fig. 3.5. Basis functions on the boundary of a triangular element.

C^1 problems, since they were not presented in [54] and have not been generalized to tetrahedrons and C^1 problems so far. The recursion formula of Jacobi polynomials is as follows:

$$a_{1n}P_{n+1}^{(\alpha,\beta)}(\xi) = (a_{2n} + a_{3n}\xi)P_n^{(\alpha,\beta)}(\xi) - a_{4n}P_{n-1}^{(\alpha,\beta)}(\xi) \tag{3.38}$$

where n is a non-negative integer, $\alpha > -1$, $\beta > -1$ and

$$P_0^{(\alpha,\beta)}(\xi) = 1, \quad P_1^{(\alpha,\beta)}(\xi) = \frac{1}{2}[\alpha - \beta + (\alpha + \beta + 2)\xi] \tag{3.39}$$

and

$$\begin{aligned} a_{1n} &= 2(n+1)(n+\alpha+\beta+1)(2n+\alpha+\beta) \\ a_{2n} &= (2n+\alpha+\beta+1)(\alpha^2-\beta^2) \\ a_{3n} &= (2n+\alpha+\beta)(2n+\alpha+\beta+1)(2n+\alpha+\beta+2) \\ a_{4n} &= 2(n+\alpha)(n+\beta)(2n+\alpha+\beta+2) \end{aligned} \tag{3.40}$$

Note that the Jacobi polynomials are defined on $[-1, 1]$. The Jacobi polynomials have the following orthogonal property:

$$\begin{aligned} &\int_{-1}^{1} (1-\xi)^\alpha (1+\xi)^\beta P_m^{(\alpha,\beta)}(\xi) P_n^{(\alpha,\beta)}(\xi) \\ &\quad = \delta_{mn} \frac{2^{\alpha+\beta+1}}{2n+\alpha+\beta+1} \frac{\Gamma(n+\alpha+1)\Gamma(n+\beta+1)}{n!\Gamma(n+\alpha+\beta+1)} \end{aligned} \tag{3.41}$$

where δ_{mn} is the Kronecker delta. In order to apply the Jacobi polynomials to triangles, coordinate transformation is necessary.

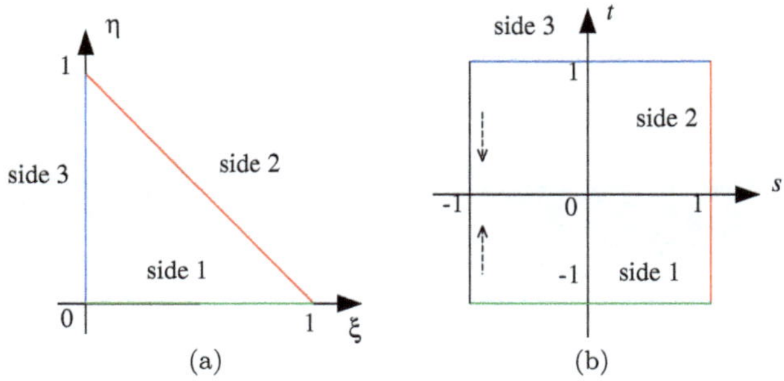

Fig. 3.6. A unit triangle (a) and a unit square domain (b).

The coordinate transformation from a unit square domain and a unit triangular domain as shown in Fig. 3.6 can be given as

$$\xi = \frac{(1+s)(1-t)}{4}, \quad \eta = \frac{(1+s)(1+t)}{4} \tag{3.42}$$

or the inverse transformation as

$$s = 2(\xi + \eta) - 1, \quad t = \frac{\eta - \xi}{\eta + \xi} \tag{3.43}$$

Figure 3.6 also shows the relationship between the edges of the unit triangle and the unit square. The edges with the same color or number correspond to each other. The left edge of the square domain is degenerated into a point. Note that the transformation method is not unique. One can degenerate some other edge to get a different method of transformation. The hierarchical shape function inside the element can be assumed as

$$F_{pi}(\xi, \eta) = (1 - \xi - \eta)\xi\eta H_{pi}(s, t) \tag{3.44}$$

where F_{pi} is defined as face function of the triangle and H_{pi} needs to be determined using the Jacobi polynomials and the following orthogonal condition:

$$\int_S F_{pi} F_{qj} \mathrm{d}S = \int_0^1 \int_0^{1-\xi} F_{pi} F_{qj} \mathrm{d}\xi \mathrm{d}\eta = \delta_{pq}\delta_{ij} C_{pi} \tag{3.45}$$

where C_{pi} is a constant that can be found in [54]. The substitution of Eqs. (3.42) and (3.44) into Eq. (3.45) leads to

$$\int_S F_{pi} F_{qj} \mathrm{d}S = \int_{-1}^1 \int_{-1}^1 \frac{(1-s)^2(1+s)^5(1-t)^2(1+t)^2}{2^{13}} \times H_{pi}(s,t) H_{qj}(s,t) \mathrm{d}s\mathrm{d}t \tag{3.46}$$

One can see from Eqs. (3.41) and (3.46) that, in order to satisfy the orthogonal condition of Eq. (3.45), H_{pi} should be of the following form:

$$H_{pi}(s,t) = [(1-s)^a(1+s)^b P_i^{(2+2a,2b+5)}(s)] \times [(1-t)^c(1+t)^d P_p^{(2+2c,2+2d)}(t)] \quad (3.47)$$

where a, b, c and d should be determined according to convenience in applications and requirements of the Jacobi polynomials. In order to eliminate $\xi+\eta = 1+s$ in the denominator of t in Eq. (3.43), one needs to take $b = p$. For the C^0 problem considered in this work, we can take $a = c = d = 0$. So, we have

$$H_{pi}(s,t) = (1+s)^p P_i^{(2,2p+5)}(s) P_p^{(2,2)}(t) \quad (3.48)$$

which is the same as the equation presented in [54] except the different definition of the index p. The face function in [54] is given as

$$H_{pi}(s,t) = (1+s)^k P_i^{(2,2k+5)}(s) P_k^{(2,2)}(t) \quad (3.49)$$

where $k = p-3-i$, which restricts the total order of the face function to be $k+i+3 = p$. The substitution of Eq. (3.43) into Eq. (3.48) leads to

$$H_{pi}(\xi,\eta) = 2^p P_i^{(2,2p+5)}[2(\eta+\xi)-1]\left[(\eta+\xi)^p P_p^{(2,2)}\left(\frac{\eta-\xi}{\eta+\xi}\right)\right] \quad (3.50)$$

Note that 2^p in Eq. (3.50) was ignored in [54] and will also be ignored in this work. In order to eliminate the computation of $\xi+\eta = 0$ that is in the denominator, make the following variable substitution:

$$\tilde{P}_p^{(\alpha,\beta)}(\tilde{s},\tilde{t}) = \tilde{t}^p P_p^{(\alpha,\beta)}(\tilde{s}/\tilde{t}), \quad \tilde{s} = \eta-\xi, \quad \tilde{t} = \eta+\xi \quad (3.51)$$

where $\alpha = 2, \beta = 2$. The substitution of Eq. (3.51) into Eq. (3.38) leads to

$$a_{1p}\tilde{P}_{p+1}^{(\alpha,\beta)}(\tilde{s},\tilde{t}) = (a_{2p}\tilde{t} + a_{3p}\tilde{s})\tilde{P}_p^{(\alpha,\beta)}(\tilde{s},\tilde{t}) - a_{4p}\tilde{t}^2\tilde{P}_{p-1}^{(\alpha,\beta)}(\tilde{s},\tilde{t}) \quad (3.52)$$

where

$$\tilde{P}_0^{(\alpha,\beta)}(\tilde{s},\tilde{t}) = 1, \quad \tilde{P}_1^{(\alpha,\beta)}(\tilde{s},\tilde{t}) = \frac{1}{2}[(\alpha-\beta)\tilde{t} + (\alpha+\beta+2)\tilde{s}] \quad (3.53)$$

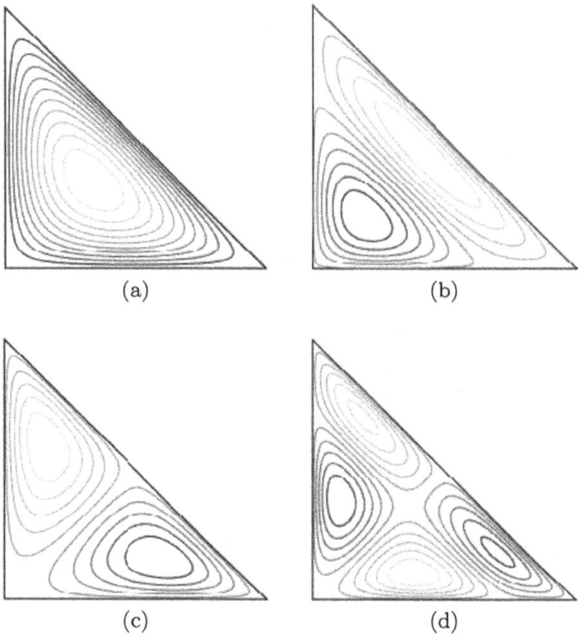

Fig. 3.7. Face functions inside a triangular element: (a) F_{00}; (b) F_{01}; (c) F_{10}; (d) F_{11}.

It is obvious that Eq. (3.44) has zero values on the boundary of the triangular element. Figure 3.7 shows some face functions inside a triangular element computed from Eq. (3.48). Using the general formula of Eq. (3.47), one can get face functions that have zero values and first derivatives on the boundary of the triangular element, which can be used for C^1 problem. Using suitable hexahedron-to-tetrahedron transformation as given in [71], one can easily get orthogonal face functions for tetrahedral regions through the same procedure as above.

3.3.2 *Geometric mapping of triangular elements*

It is obvious that Eqs. (3.30) and (3.31) can be used to describe a plane triangular element and they are exact. However, in computer-aided design (CAD) the NURBS use tensor product of one-dimensional (1D) basis to describe surfaces [220]. Therefore, it is desirable to directly transform such a tensor product surface into

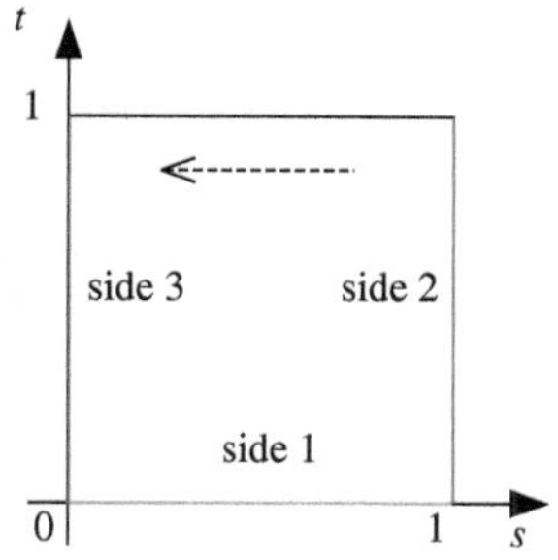

Fig. 3.8. A unit square domain defined at [0, 1].

area coordinates used in triangular element. The coordinate transformation in Fig. 3.6 shows the feasibility of such transformation. For simplicity, a different transformation will be used herein. The parametric coordinates in NURBS are usually defined in [0, 1], as shown in Fig. 3.8. The geometric transformation with the upper edge shown in Fig. 3.8 degenerates to a point as

$$\xi = (1-t)s, \quad \eta = t \tag{3.54}$$

or the inverse transformation as

$$s = \frac{\xi}{1-\eta}, \quad t = \eta \tag{3.55}$$

The edges of Figs. 3.6(a) and 3.8 with the same color or number correspond to each other. With the relations in Eq. (3.55), the derivatives of a points on a surface $\boldsymbol{r}(s,t) = \{x(s,t), y(s,t), z(s,t)\}^T$ with respect to area coordinates are

$$\begin{aligned} \frac{\partial \boldsymbol{r}(s,t)}{\partial \xi} &= \frac{\partial \boldsymbol{r}}{\partial s}\frac{\partial s}{\partial \xi} = \frac{1}{1-\eta}\frac{\partial \boldsymbol{r}}{\partial s} \\ \frac{\partial \boldsymbol{r}(s,t)}{\partial \eta} &= \frac{\partial \boldsymbol{r}}{\partial s}\frac{\partial s}{\partial \eta} + \frac{\partial \boldsymbol{r}}{\partial t} = \frac{\xi}{(1-\eta)^2}\frac{\partial \boldsymbol{r}}{\partial s} + \frac{\partial \boldsymbol{r}}{\partial t} \end{aligned} \tag{3.56}$$

Let $\boldsymbol{r}_s = \partial \boldsymbol{r}/\partial s, \boldsymbol{r}_t = \partial \boldsymbol{r}/\partial t$, and so on. It is obvious that $\boldsymbol{r}_s(s,1) = 0$ and $\boldsymbol{r}_t(s,1)$ changes with the variation of s. Therefore, $\boldsymbol{r}_\xi = 0$ and $\boldsymbol{r}_\eta$ does not have a unique value at parametric point $(0,1)$, which is not reasonable. Therefore, $\boldsymbol{r}_\xi$ and $\boldsymbol{r}_\eta$ at the point should be determined through the other way. One can see from Eq. (3.44) that the basis inside the triangle has zero function value and first derivative at point (0, 1). Therefore, the first derivatives at point (0, 1) can be uniquely determined through the two neighboring edges. One can see from Eqs. (3.30) or (3.31) that the point on the triangle at (0, 1) can be interpolated as

$$\tilde{\boldsymbol{r}}(\xi,\eta) = \frac{1-\xi-\eta}{1-\eta}\boldsymbol{r}(0,t) + \frac{\xi}{1-\eta}\boldsymbol{r}(1,t) \tag{3.57}$$

Therefore, the first derivatives with respect to ξ and η can be given as

$$\begin{aligned}\frac{\partial\tilde{\boldsymbol{r}}(\xi,\eta)}{\partial\xi} &= \frac{\boldsymbol{r}(1,t)-\boldsymbol{r}(0,t)}{1-\eta}\\ \frac{\partial\tilde{\boldsymbol{r}}(\xi,\eta)}{\partial\eta} &= \frac{\xi}{1-\eta}\frac{\partial\boldsymbol{r}(1,t)}{\partial\eta} + \frac{1-\xi-\eta}{1-\eta}\frac{\partial\boldsymbol{r}(0,t)}{\partial\eta}\\ &\quad + \frac{\xi[\boldsymbol{r}(1,t)-\boldsymbol{r}(0,t)]}{(1-\eta)^2}\end{aligned} \tag{3.58}$$

Let $(\xi,\eta)\to(0,1)$, then one obtains

$$\begin{aligned}\frac{\partial\tilde{\boldsymbol{r}}(0,1)}{\partial\xi} &= \boldsymbol{r}_t(0,1)-\boldsymbol{r}_t(1,1)\\ \frac{\partial\tilde{\boldsymbol{r}}(0,1)}{\partial\eta} &= \boldsymbol{r}_t(0,1)\end{aligned} \tag{3.59}$$

Using Eqs. (3.56) and (3.59), one can directly use tensor product surface as shown in Fig. 3.7 and the following equation for NURBS surface [220]:

$$\boldsymbol{S}(s,t) = \sum_{i=0}^{m}\sum_{j=0}^{n} R_{i,j}(s,t)\boldsymbol{P}_{i,j} \quad 0 \le s,t \le 1 \tag{3.60}$$

where $\{\boldsymbol{P}_{i,j}\}$ form a bidirectional control net. The piecewise rational basis functions are

$$R_{i,j}(s,t) = \frac{N_{i,p}(s)N_{j,q}(t)w_{i,j}}{\sum_{k=0}^{m}\sum_{l=0}^{n} N_{k,p}(s)N_{l,q}(t)w_{k,l}} \tag{3.61}$$

where $\{w_{i,j}\}$ are the weights, $\{N_{i,p}(s)\}$ and $\{N_{j,q}(t)\}$ are the non-rational B-spline basis functions, and p, q are the orders of the B-spline basis. One can define some other method of transformation. The procedure is similar to the above. Equation (3.59) is usually

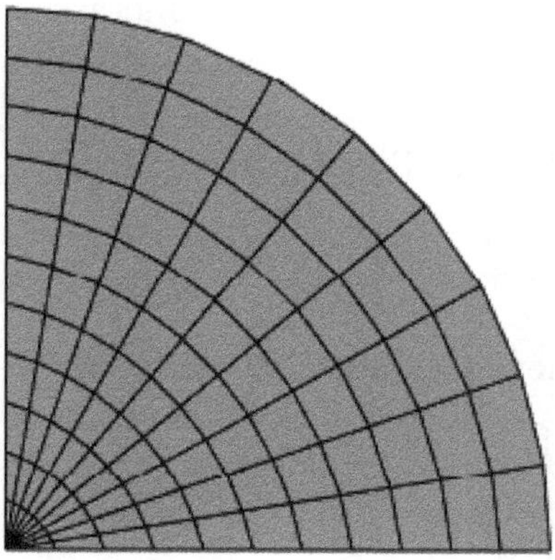

Fig. 3.9. A tensor product plane surface of sector.

sufficient for application. One only needs to rotate the coordinates of the quadrilateral element to let the edge $t = 1$ to be the edge that degenerates to a point (Fig. 3.9).

Note that Eqs. (3.56) and (3.59) can also be used to transform quadrilateral elements given in [34, 35, 301] to triangular elements. One only needs to replace the vector $\boldsymbol{r}$ and their derivatives in Eqs. [34, 35, 301] by the weighting coefficient matrix $\boldsymbol{G}$ and their derivatives on the rectangular element [34, 35, 301]. What needs to be taken care of during the transformation is that the basis and their derivatives on the edge that degenerates to a point should be first added together, and then Eqs. (3.56) and (3.59) should be used for the transformation.

3.4 The DQHFEM for Euler–Bernoulli Beam

This section aims to illustrate how to use the DQHFEM through a simple example, although explicit stiffness matrix has been obtained [302]. For a uniform Euler–Bernoulli beam with length L, the displacement field can be written as

$$w[x(\xi)] = H_1(\xi)w_1 + \frac{L}{2}H_2(\xi)\frac{dw_1}{dx} + H_3(\xi)w_2 + \frac{L}{2}H_4(\xi)\frac{dw_2}{dx} + \sum_{n=1}^{N}\psi_n(\xi)a_n \tag{3.62}$$

where $x = L(\xi + 1)/2$, $\psi_n(\xi)$ are the hierarchical shape functions defined by Eq. (3.15), and

$$H_1(\xi) = \frac{1}{4}(1-\xi)^2(2+\xi), H_2(\xi) = \frac{1}{4}(1-\xi)^2(\xi+1)$$
$$H_3(\xi) = \frac{1}{4}(1+\xi)^2(2-\xi), H_4(\xi) = \frac{1}{4}(1+\xi)^2(\xi-1) \quad (3.63)$$

Denote the element displacement vector as

$$\boldsymbol{w}^T = \begin{Bmatrix} w_1 & w_1' & w_2 & w_2' & a_1 & \cdots & a_N \end{Bmatrix} \quad (3.64)$$

So, Eq. (3.62) becomes

$$w(x) = \boldsymbol{N}^T \boldsymbol{w} \quad (3.65)$$

where

$$\boldsymbol{N}^{\mathrm{T}} = [H_1(\xi) \cdots (L/2)H_4(\xi) \quad \psi_1(\xi) \quad \psi_2(\xi) \quad \cdots \quad \psi_N(\xi)] \quad (3.66)$$

Calculate the Gauss–Lobatto–Legendre nodes $\xi_j, j = 1, 2, \ldots, N_x$, where $N_x = N + 4$. Define the following displacement vector:

$$\bar{\boldsymbol{w}}^T = \begin{Bmatrix} w(x_1) & w(x_2) & \cdots & w(x_{N_x}) \end{Bmatrix} \quad (3.67)$$

According to Eq. (3.65), the relationship between Eqs. (3.64) and (3.67) is

$$\bar{\boldsymbol{w}} = \boldsymbol{G}\boldsymbol{w} \quad (3.68)$$

where

$$\boldsymbol{G} = \begin{bmatrix} \boldsymbol{N}(\xi_1) & \boldsymbol{N}(\xi_2) & \cdots & \boldsymbol{N}(\xi_{N_x}) \end{bmatrix}^{\mathrm{T}} \quad (3.69)$$

The second-order derivatives of $w[x(\xi)]$ with respect to x at point $\xi_j, j = 1, 2, \ldots, N_x$ can be written as

$$\left(\frac{\mathrm{d}^2 w}{\mathrm{d}x^2}\right)_j = \boldsymbol{A}_j^{(2)} \boldsymbol{w} \quad (3.70)$$

where $\boldsymbol{A}_j^{(2)}$ are the jth row of $\boldsymbol{A}^{(2)}$ that corresponds to the second-order partial derivative of $w[x(\xi)]$ with respect to x at the discrete point ξ_j. It is clear that $\psi_n''(\xi) = P_{n+1}(\xi)$. For this simple example, the stiffness and mass matrices can be explicitly provided [302]. For the Mindlin plate problem in rectangular domains in the following section, a similar property can be used to simplify computations.

Now, the computation of element matrices will be discussed. The strain energy and work potential of an Euler–Bernoulli beam is given by

$$U = \frac{1}{2}\int_0^L EI\left(\frac{\mathrm{d}^2 w}{\mathrm{d}x^2}\right)^2 \mathrm{d}x - \int_0^L qw\,\mathrm{d}x \tag{3.71}$$

where E is the Young's modulus, I is the geometric moment of inertia of the cross-section, and q is the transverse distributed force. The kinetic energy functional is given by

$$T = -\int_0^L w\left(-\rho S\frac{\partial^2 w}{\partial t^2}\right)\mathrm{d}x \tag{3.72}$$

where ρ is the density and S is the area of the cross-section. Substitution of Eqs. (3.34) and (3.36) into Eqs. (3.37) and (3.38) leads to

$$\begin{aligned} U &= \frac{EI}{2}\boldsymbol{w}^T(\boldsymbol{A}^{(2)^T}\boldsymbol{I}\boldsymbol{A}^{(2)})\boldsymbol{w} - \boldsymbol{w}^T(\boldsymbol{G}^T\boldsymbol{I}\boldsymbol{q}) \\ T &= \rho S\boldsymbol{w}^T\boldsymbol{G}^T\boldsymbol{I}\boldsymbol{G}\ddot{\boldsymbol{w}} \end{aligned} \tag{3.73}$$

where $\boldsymbol{I} = \mathrm{diag}(C_j)$, C_j $(j = 1, 2, \ldots, N_x)$ are the Gauss–Lobatto weights for integration, and $\boldsymbol{q}$ is the transverse distributed force vector. Then using the variational principle, we have

$$\boldsymbol{K}\boldsymbol{w} + \boldsymbol{M}\ddot{\boldsymbol{w}} = \boldsymbol{R} \tag{3.74}$$

where

$$\boldsymbol{K} = EI(\boldsymbol{A}^{(2)^T}\boldsymbol{I}\boldsymbol{A}^{(2)}), \boldsymbol{M} = \rho S(\boldsymbol{G}^T\boldsymbol{I}\boldsymbol{G}), \boldsymbol{R} = (\boldsymbol{G}^T\boldsymbol{I}\boldsymbol{q}) \tag{3.75}$$

are the stiffness matrix, the mass matrix, and the force vector, respectively. Obviously, Eq. (3.75) is evident to be obtained from Eq. (3.73).

Equation (3.75) does not show the embedding property of hierarchical finite element method that retains the stiffness and mass coefficients as high-order basis is enriched. In order to show this property more clearly, write the weighting coefficient matrices as

$$\boldsymbol{A}^{(2)} = \begin{bmatrix}\boldsymbol{A}_1^{(2)} & \boldsymbol{A}_2^{(2)}\end{bmatrix}, \boldsymbol{G} = \begin{bmatrix}\boldsymbol{G}_1 & \boldsymbol{G}_2\end{bmatrix} \tag{3.76}$$

where $\boldsymbol{A}_1^{(2)}$ and $\boldsymbol{G}_1$ are the original weighting coefficient matrices and $\boldsymbol{A}_2^{(2)}$ and $\boldsymbol{G}_2$ are those of the enriched hierarchical basis. Then Eq. (3.75) can be written as

$$\boldsymbol{K} = EI \begin{bmatrix} \boldsymbol{A}_1^{(2)^T} \boldsymbol{I} \boldsymbol{A}_1^{(2)} & \boldsymbol{A}_1^{(2)^T} \boldsymbol{I} \boldsymbol{A}_2^{(2)} \\ \boldsymbol{A}_2^{(2)^T} \boldsymbol{I} \boldsymbol{A}_1^{(2)} & \boldsymbol{A}_2^{(2)^T} \boldsymbol{I} \boldsymbol{A}_2^{(2)} \end{bmatrix}$$

$$\boldsymbol{M} = \rho S \begin{bmatrix} \boldsymbol{G}_1^T \boldsymbol{I} \boldsymbol{G}_1 & \boldsymbol{G}_1^T \boldsymbol{I} \boldsymbol{G}_2 \\ \boldsymbol{G}_2^T \boldsymbol{I} \boldsymbol{G}_1 & \boldsymbol{G}_2^T \boldsymbol{I} \boldsymbol{G}_2 \end{bmatrix}, \quad \boldsymbol{R} = \begin{bmatrix} \boldsymbol{G}_1^T \boldsymbol{I} \boldsymbol{q} \\ \boldsymbol{G}_2^T \boldsymbol{I} \boldsymbol{q} \end{bmatrix} \tag{3.77}$$

If the stiffness and mass coefficients corresponding to $\boldsymbol{A}_1^{(2)}$ and $\boldsymbol{G}_1$ are obtained already, one only needs to compute the stiffness and mass coefficients related with $\boldsymbol{A}_2^{(2)}$ and $\boldsymbol{G}_2$.

3.5 The DQHFEM for Mindlin Plate

Based on the knowledge discussed above, this section provides the DQHFEM formulations for Mindlin plate. The displacement field of a DQHFEM Mindlin plate element can be given as

$$w[x(\xi,\eta), y(\xi,\eta)] = \sum_{k=1}^{K} S_k(\xi,\eta) w_k + \sum_{m=1}^{N_\xi} \sum_{n=1}^{N_\eta} \varphi_m(\xi)\varphi_n(\eta) a_{mn} \tag{3.78}$$

where $S_k(\xi,\eta)$ are the shape functions defined by Eqs. (3.11) and (3.12), and $\varphi_m(\xi)$ and $\varphi_n(\eta)$ are the hierarchical shape functions defined by Eq. (3.14). Define the following displacement vector:

$$\boldsymbol{w}^T = \{w_1 \quad w_2 \quad \cdots \quad w_K \quad a_{11} \quad \cdots \quad a_{N_\xi 1} \quad \cdots \quad a_{N_\xi N_\eta}\} \tag{3.79}$$

So, Eq. (3.78) can also be written as

$$w[x(\xi,\eta), y(\xi,\eta)] = \boldsymbol{N}^T \boldsymbol{w} \tag{3.80}$$

where

$$\boldsymbol{N}^{\mathrm{T}} = [S_1(\xi,\eta) \ \cdots \ S_K(\xi,\eta) \ \varphi_1(\xi)\varphi_1(\eta) \ \varphi_2(\xi)\varphi_1(\eta) \ \cdots \ \varphi_{N_\xi}(\xi)\varphi_{N_\eta}(\eta)] \tag{3.81}$$

Calculate the Gauss–Lobatto–Legendre nodes $\xi_i, i = 1, 2, \ldots, N_x$ and η_j, $j = 1, 2, \ldots, N_y$, where $N_x \geq \max[M-1, N_\xi + 1] + 2$ and $N_y \geq \max[N-1,\ N_\eta + 1] + 2$. Here, N_x and N_y are the minimum number of nodes required by Gauss–Lobatto integration for precise integration along x- and y-directions, respectively. Define the following displacement vector:

$$\bar{\boldsymbol{w}}^T = \begin{Bmatrix} w_{11} & \cdots & w_{N_x 1} & w_{12} & \cdots & w_{N_x 2} \\ & \cdots & w_{1N_y} & \cdots & w_{N_x N_y} & \end{Bmatrix} \tag{3.82}$$

According to Eq. (3.80), the relationship between Eqs. (3.79) and (3.82) is

$$\bar{\boldsymbol{w}} = \boldsymbol{G}\boldsymbol{w} \tag{3.83}$$

where

$$\boldsymbol{G} = \begin{bmatrix} \boldsymbol{N}(\xi_1, \eta_1) & \boldsymbol{N}(\xi_2, \eta_1) & \cdots & \boldsymbol{N}(\xi_{N_x}, \eta_{N_y}) \end{bmatrix}^{\mathrm{T}} \tag{3.84}$$

The Mindlin plate element matrices will be calculated according to Eq. (3.49), the DQ rules and the variational principle. The first-order derivatives of $w(\xi, \eta)$ with respect to ξ and η at point $(\xi_i, \eta_j)(i = 1, 2, \ldots, N_x; j = 1, 2, \ldots, N_y)$ can be written as

$$\left(\frac{\partial w}{\partial \xi}\right)_{ij} = \boldsymbol{A}_{ij}^{(1)}\boldsymbol{w}, \quad \left(\frac{\partial w}{\partial \eta}\right)_{ij} = \boldsymbol{B}_{ij}^{(1)}\boldsymbol{w} \tag{3.85}$$

where $\boldsymbol{A}_{ij}^{(1)}$ and $\boldsymbol{B}_{ij}^{(1)}$ are the weighting coefficient vectors at the discrete point (ξ_i, η_j) associated with the first-order partial derivative of $w(\xi, \eta)$ with respect to ξ and η, respectively. The substitution of Eq. (3.85) into Eq. (3.5) leads to derivatives of $w[x(\xi, \eta), y(\xi, \eta)]$ with respect to x and y at point $(x(\xi_i, \eta_j), y(\xi_i, \eta_j))$, which are respectively denoted as $\bar{\boldsymbol{A}}^{(1)}$ and $\bar{\boldsymbol{B}}^{(1)}$.

There are two methods of computing the weighting coefficient matrices $\bar{\boldsymbol{A}}^{(1)}$ and $\bar{\boldsymbol{B}}^{(1)}$. One way is the usage of the DQ rules as used in [16, 32]. Let $\tilde{\boldsymbol{A}}^{(1)}$ and $\tilde{\boldsymbol{B}}^{(1)}$ be the weighting coefficient matrices of DQ rules for the first-order partial derivative of $w(\xi, \eta)$ with respect to ξ and η, respectively. Then

$$\bar{\boldsymbol{A}}^{(1)} = \tilde{\boldsymbol{A}}^{(1)}\boldsymbol{G}, \quad \bar{\boldsymbol{B}}^{(1)} = \tilde{\boldsymbol{B}}^{(1)}\boldsymbol{G} \tag{3.86}$$

This method is much more accurate than the finite difference method in HFEM [46] because DQM is very accurate. Another way is to fully use the separation variable characteristics of the shape functions shown in Eq. (3.78) and the definition of the hierarchical basis in Eq. (3.14) that indicates $\varphi'_n(\xi) = P_n(\xi)$. So, we have

$$\frac{\partial \boldsymbol{N}^{\mathrm{T}}}{\partial \xi} = \left[\frac{\partial S_1(\xi,\eta)}{\partial \xi} \quad \cdots \quad \frac{\partial S_K(\xi,\eta)}{\partial \xi} \quad P_1(\xi)\varphi_1(\eta) \right.$$
$$\left. P_2(\xi)\varphi_1(\eta) \quad \cdots \quad P_{N_\xi}(\xi)\varphi_{N_\eta}(\eta)\right]$$
$$\frac{\partial \boldsymbol{N}^{\mathrm{T}}}{\partial \eta} = \left[\frac{\partial S_1(\xi,\eta)}{\partial \eta} \quad \cdots \quad \frac{\partial S_K(\xi,\eta)}{\partial \eta} \quad \varphi_1(\xi)P_1(\eta) \right.$$
$$\left. \varphi_2(\xi)P_1(\eta) \quad \cdots \quad \varphi_{N_\xi}(\xi)P_{N_\eta}(\eta)\right] \tag{3.87}$$

where

$$\frac{\partial S_1(\xi,\eta)}{\partial \xi} = \frac{1-\eta}{2}\frac{\mathrm{d}L_1^M(\xi)}{\mathrm{d}\xi} - \frac{1}{2}L_1^N(\eta) + \frac{1-\eta}{4}$$
$$\frac{\partial S_1(\xi,\eta)}{\partial \eta} = \frac{1-\xi}{2}\frac{\mathrm{d}L_1^N(\eta)}{\mathrm{d}\eta} - \frac{1}{2}L_1^M(\xi) + \frac{1-\xi}{4} \tag{3.88}$$

The derivatives of the basis $S_k(\xi,\eta), k = 2,3,\ldots,K$ can be expressed in the same way as Eq. (3.88) using the definition by Eqs. (3.11) and (3.12). By using this way, the computational cost of $\bar{\boldsymbol{A}}^{(1)}$ and $\bar{\boldsymbol{B}}^{(1)}$ is similar to that of $\boldsymbol{G}$.

The DQHFEM, like the DQFEM, is based on the weak form of mechanics. There are three generalized displacements in Mindlin plate theory, namely, the deflection w and the two rotations θ_x and θ_y of a normal line with respect to the middle surface. The stress–strain relations and strain–displacement relations of Mindlin plate are

$$\begin{Bmatrix} \sigma_x \\ \sigma_y \\ \tau_{xy} \\ \tau_{yz} \\ \tau_{zx} \end{Bmatrix} = \frac{E}{1-\mu^2}\begin{bmatrix} 1 & \mu & 0 & 0 & 0 \\ \mu & 1 & 0 & 0 & 0 \\ 0 & 0 & \mu_1 & 0 & 0 \\ 0 & 0 & 0 & \kappa\mu_1 & 0 \\ 0 & 0 & 0 & 0 & \kappa\mu_1 \end{bmatrix}\begin{Bmatrix} \varepsilon_x \\ \varepsilon_y \\ \gamma_{xy} \\ \gamma_{yz} \\ \gamma_{zx} \end{Bmatrix} \tag{3.89}$$

$$\begin{Bmatrix} \varepsilon_x \\ \varepsilon_y \\ \gamma_{xy} \\ \gamma_{yz} \\ \gamma_{zx} \end{Bmatrix} = -\begin{bmatrix} z\partial/\partial x & 0 & 0 \\ 0 & z\partial/\partial y & 0 \\ z\partial/\partial y & z\partial/\partial x & 0 \\ 0 & 1 & -\partial/\partial y \\ 1 & 0 & -\partial/\partial y \end{bmatrix} \begin{Bmatrix} \theta_x \\ \theta_y \\ w \end{Bmatrix} \tag{3.90}$$

where $\mu_1 = (1-\mu)/2$ and κ is the shear correction factor. The strain energy and work potential of a Mindlin plate is given by

$$\Pi = \frac{1}{2}\iiint_V \boldsymbol{\varepsilon}^T \boldsymbol{\sigma} dv - \iint_S \mathbf{u}^T \boldsymbol{P} dS \tag{3.91}$$

where $\boldsymbol{u} = [\theta_x, \theta_y, w]^T, \boldsymbol{P} = [m_x, m_y, q]^T$, where q is the transverse distributed force, and m_x, m_y are respectively the distributed bending moment with respect to x- and y-directions. The kinetic energy functional is given by

$$T = \frac{1}{2}\iiint_V \rho \dot{\boldsymbol{u}}^T \dot{\boldsymbol{u}} dv \tag{3.92}$$

where $\dot{\boldsymbol{u}}$ is the velocity field vector. For a Mindlin plate element with uniform thickness, the integration with respect to z in the strain energy can be done in advance, which leads to

$$\Pi = \frac{1}{2}\iint_S \bar{\boldsymbol{\varepsilon}}^T \mathbf{D} \bar{\boldsymbol{\varepsilon}} dS - \iint_S \mathbf{u}^T \boldsymbol{P} dS \tag{3.93}$$

where

$$\bar{\boldsymbol{\varepsilon}} = \begin{Bmatrix} \bar{\varepsilon}_x \\ \bar{\varepsilon}_y \\ \bar{\gamma}_{xy} \\ \bar{\gamma}_{yz} \\ \bar{\gamma}_{zx} \end{Bmatrix} = -\begin{bmatrix} \partial/\partial x & 0 & 0 \\ 0 & \partial/\partial y & 0 \\ \partial/\partial y & \partial/\partial x & 0 \\ 0 & 1 & -\partial/\partial y \\ 1 & 0 & -\partial/\partial y \end{bmatrix} \begin{Bmatrix} \theta_x \\ \theta_y \\ w \end{Bmatrix} \tag{3.94}$$

$$\boldsymbol{D} = \begin{bmatrix} D_{11} & D_{12} & 0 & 0 & 0 \\ D_{12} & D_{22} & 0 & 0 & 0 \\ 0 & 0 & D_{66} & 0 & 0 \\ 0 & 0 & 0 & C_{44} & 0 \\ 0 & 0 & 0 & 0 & C_{55} \end{bmatrix} \tag{3.95}$$

where

$$D_{11} = D_{22} = \frac{Eh^3}{12(1-\mu^2)}, \quad D_{12} = \mu D_{11}, \quad D_{66} = \frac{Gh^3}{12}$$
$$C_{44} = C_{55} = \kappa Gh, \quad G = \frac{E}{2(1+\mu)} \tag{3.96}$$

Then Eqs. (3.89) and (3.90) need to be discretized by the DQHFEM. The displacement vectors for θ_x and θ_y can be given in the same form as Eq. (3.82) and are denoted as $\boldsymbol{\theta}_x$ and $\boldsymbol{\theta}_y$. Then Eq. (3.94) can be expressed by the DQ rules as

$$\begin{Bmatrix} \bar{\varepsilon}_x \\ \bar{\varepsilon}_y \\ \bar{\gamma}_{xy} \\ \bar{\gamma}_{xz} \\ \bar{\gamma}_{yz} \end{Bmatrix} = \begin{bmatrix} -\bar{\boldsymbol{A}}^{(1)} & \mathbf{0} & \mathbf{0} \\ \mathbf{0} & -\bar{\boldsymbol{B}}^{(1)} & \mathbf{0} \\ -\bar{\boldsymbol{B}}^{(1)} & -\bar{\boldsymbol{A}}^{(1)} & \mathbf{0} \\ -\boldsymbol{G} & \mathbf{0} & \bar{\boldsymbol{A}}^{(1)} \\ \mathbf{0} & -\boldsymbol{G} & \bar{\boldsymbol{B}}^{(1)} \end{bmatrix} \begin{Bmatrix} \boldsymbol{\theta}_x \\ \boldsymbol{\theta}_y \\ \boldsymbol{w} \end{Bmatrix} \tag{3.97}$$

or

$$\bar{\boldsymbol{\varepsilon}} = \boldsymbol{HU} \tag{3.98}$$

where $\boldsymbol{U} = [\boldsymbol{\theta}_x; \boldsymbol{\theta}_y; \boldsymbol{w}]$ and $\bar{\boldsymbol{A}}^{(1)}$ and $\bar{\boldsymbol{B}}^{(1)}$ are obtained from Eq. (3.7) through Eqs. (3.5), (3.90), (3.92) and (3.93). Substitute Eq. (3.98) into Eq. (3.93) and then discretize Eqs. (3.92) and (3.93) together with Eq. (3.83) and the Gauss–Lobatto quadrature rule, one gets

$$T = \frac{1}{2}\rho\dot{\boldsymbol{U}}^{\mathrm{T}} \begin{bmatrix} J\boldsymbol{G}^T\boldsymbol{IG} & \mathbf{0} & \mathbf{0} \\ \mathbf{0} & J\boldsymbol{G}^T\boldsymbol{IG} & \mathbf{0} \\ \mathbf{0} & \mathbf{0} & h\boldsymbol{G}^T\boldsymbol{IG} \end{bmatrix} \dot{\boldsymbol{U}}$$
$$\Pi = \frac{1}{2}\boldsymbol{U}^{\mathrm{T}}\boldsymbol{H}^{\mathrm{T}}\bar{\boldsymbol{D}}\boldsymbol{HU} - \boldsymbol{U}^{\mathrm{T}} \begin{Bmatrix} \boldsymbol{G}^T\boldsymbol{I}\boldsymbol{m}_x \\ \boldsymbol{G}^T\boldsymbol{I}\boldsymbol{m}_y \\ \boldsymbol{G}^T\boldsymbol{I}\boldsymbol{q}_w \end{Bmatrix} \tag{3.99}$$

where

$$\bar{\boldsymbol{D}} = \begin{bmatrix} D_{11}\boldsymbol{I} & D_{12}\boldsymbol{I} & \boldsymbol{0} & \boldsymbol{0} & \boldsymbol{0} \\ D_{12}\boldsymbol{I} & D_{22}\boldsymbol{I} & \boldsymbol{0} & \boldsymbol{0} & \boldsymbol{0} \\ \boldsymbol{0} & \boldsymbol{0} & D_{66}\boldsymbol{I} & \boldsymbol{0} & \boldsymbol{0} \\ \boldsymbol{0} & \boldsymbol{0} & \boldsymbol{0} & C_{44}\boldsymbol{I} & \boldsymbol{0} \\ \boldsymbol{0} & \boldsymbol{0} & \boldsymbol{0} & \boldsymbol{0} & C_{55}\boldsymbol{I} \end{bmatrix} \tag{3.100}$$

$J = h^3/12$, $\boldsymbol{m}_x$ and $\boldsymbol{m}_y$ are the nodal bending moment vectors with respect to x- and y-directions, and $\boldsymbol{q}_w$ is the nodal force vector with respect to z-direction. The matrix $\boldsymbol{I} = \text{diag}(J_q C_q)$, $C_q = C_i^x C_j^y$, and $J_q = |\boldsymbol{J}|_{ij}$, where the subscript q is defined by Eq. (3.9). Here, C_i^x and $C_j^y (i = 1, 2, \ldots, N_x; j = 1, 2, \ldots, N_y)$ are respectively the Gauss–Lobatto weights for integration with respect to x- and y-directions. Then using the Hamilton principle, we have

$$\boldsymbol{M}\ddot{\boldsymbol{U}} + \boldsymbol{K}\boldsymbol{U} = \boldsymbol{R} \tag{3.101}$$

where

$$\boldsymbol{M} = \rho \begin{bmatrix} J\boldsymbol{G}^T\boldsymbol{I}\boldsymbol{G} & \boldsymbol{0} & \boldsymbol{0} \\ \boldsymbol{0} & J\boldsymbol{G}^T\boldsymbol{I}\boldsymbol{G} & \boldsymbol{0} \\ \boldsymbol{0} & \boldsymbol{0} & h\boldsymbol{G}^T\boldsymbol{I}\boldsymbol{G} \end{bmatrix}, \quad \boldsymbol{R} = \begin{Bmatrix} \boldsymbol{G}^T\boldsymbol{I}\boldsymbol{m}_x \\ \boldsymbol{G}^T\boldsymbol{I}\boldsymbol{m}_y \\ \boldsymbol{G}^T\boldsymbol{I}\boldsymbol{q}_w \end{Bmatrix}$$

$$\boldsymbol{K} = \boldsymbol{H}^{\mathrm{T}}\bar{\boldsymbol{D}}\boldsymbol{H} \tag{3.102}$$

It is clear that the stiffness and mass matrices for Mindlin plate are derived in a simple formula similar to as Eq. (3.41) for beam.

In order to show the embedding property of hierarchical finite element method as Eq. (3.76), write the weighting coefficient matrices as

$$\bar{\boldsymbol{A}}^{(1)} = \begin{bmatrix} \bar{\boldsymbol{A}}_1^{(1)} & \bar{\boldsymbol{A}}_2^{(1)} \end{bmatrix}, \quad \bar{\boldsymbol{B}}^{(1)} = \begin{bmatrix} \bar{\boldsymbol{B}}_1^{(1)} & \bar{\boldsymbol{B}}_2^{(1)} \end{bmatrix},$$
$$\boldsymbol{G} = \begin{bmatrix} \boldsymbol{G}_1 & \boldsymbol{G}_2 \end{bmatrix} \tag{3.103}$$

where $\bar{\boldsymbol{A}}_1^{(1)}$, $\bar{\boldsymbol{B}}_1^{(1)}$ and $\boldsymbol{G}_1$ are the original weighting coefficient matrices and $\bar{\boldsymbol{A}}_2^{(1)}$, $\bar{\boldsymbol{B}}_2^{(1)}$ and $\boldsymbol{G}_2$ are those of the enriched hierarchical basis.

Then the weighting matrix $\boldsymbol{H}$ in Eq. (3.63) can be written as

$$\boldsymbol{H} = \begin{bmatrix} \boldsymbol{H}_1 & \boldsymbol{H}_2 \end{bmatrix} \tag{3.104}$$

where

$$\boldsymbol{H}_i = \begin{bmatrix} -\bar{\boldsymbol{A}}_i^{(1)} & \mathbf{0} & \mathbf{0} \\ \mathbf{0} & -\bar{\boldsymbol{B}}_i^{(1)} & \mathbf{0} \\ -\bar{\boldsymbol{B}}_i^{(1)} & -\bar{\boldsymbol{A}}_i^{(1)} & \mathbf{0} \\ -\boldsymbol{G}_i & \mathbf{0} & \bar{\boldsymbol{A}}_i^{(1)} \\ \mathbf{0} & -\boldsymbol{G}_i & \bar{\boldsymbol{B}}_i^{(1)} \end{bmatrix} \tag{3.105}$$

and $i = 1, 2$. Note that somehow like node numbering, the displacement vector corresponding to Eq. (3.103) is $\boldsymbol{U} = [\boldsymbol{U}_1;\ \boldsymbol{U}_2]$ where $\boldsymbol{U}_i = [\theta_{xi};\ \theta_{yi};\ \boldsymbol{w}_i]$. So, the stiffness matrix in Eq. (3.102) can be written as

$$\boldsymbol{K} = \begin{bmatrix} \boldsymbol{H}_1^T \bar{\boldsymbol{D}} \boldsymbol{H}_1 & \boldsymbol{H}_1^T \bar{\boldsymbol{D}} \boldsymbol{H}_2 \\ \boldsymbol{H}_2^T \bar{\boldsymbol{D}} \boldsymbol{H}_1 & \boldsymbol{H}_2^T \bar{\boldsymbol{D}} \boldsymbol{H}_2 \end{bmatrix} \tag{3.106}$$

The derivation of the embedded mass matrix and force vector are similar to Eqs. (3.77) and so they are not presented here. Note that one should rearrange the coefficients of the embedded mass matrix and force vector to correspond to the displacement vector $\boldsymbol{U} = [\boldsymbol{U}_1;\ \boldsymbol{U}_2]$.

3.6 The DQHFEM for Quartz Crystal Plate

Consider a quartz crystal plate in a curvilinear quadrilateral region with uniform thickness h, as shown in Fig. 3.10. The total potential functional of the plate is

$$\Pi = \frac{1}{2}\int_{-1}^{1}\int_{-1}^{1}\left[D_1\left(\frac{\partial\psi_x}{\partial x}\right)^2 + 2D_2\left(\frac{\partial\psi_x}{\partial x}\frac{\partial\psi_y}{\partial y}\right) + D_3\left(\frac{\partial\psi_y}{\partial y}\right)^2 \right.$$
$$\left. + D_5\left(\frac{\partial\psi_x}{\partial y} + \frac{\partial\psi_y}{\partial x}\right)^2\right]|\boldsymbol{J}|\mathrm{d}\xi\,\mathrm{d}\eta$$

$$+\frac{1}{2}\int_{-1}^{1}\int_{-1}^{1}\left[D_6\left(\frac{\partial w}{\partial x}-\psi_x\right)^2+D_4\left(\frac{\partial w}{\partial y}-\psi_y\right)^2\right]|\boldsymbol{J}|\mathrm{d}\xi\,\mathrm{d}\eta$$
$$-\frac{\omega^2\rho}{2}\int_{-1}^{1}\int_{-1}^{1}(hw^2+J\psi_x^2+J\psi_y^2)|\boldsymbol{J}|\mathrm{d}\xi\,\mathrm{d}\eta \tag{3.107}$$

where $J = h^3/12$, $|\boldsymbol{J}|$ is the Jacobian defined in Eq. (3.4), w is the deflection, and ψ_x and ψ_y are the angles of rotations of a normal line due to plate bending with respect to x and y coordinates, respectively, and

$$D_1 = \frac{h^3}{12}\left(C_{11}-\frac{C_{12}^2}{C_{22}}\right),\quad D_2 = \frac{h^3}{12}\left(C_{13}-\frac{C_{12}C_{23}}{C_{22}}\right)$$
$$D_3 = \frac{h^3}{12}\left(C_{33}-\frac{C_{23}^2}{C_{22}}\right)$$
$$D_4 = \frac{\pi^2 h}{12}\left(C_{44}-\frac{C_{24}^2}{C_{22}}\right),\quad D_5 = \frac{h^3}{12}C_{55},\quad D_6 = \frac{\pi^2 h}{12}C_{66} \tag{3.108}$$

in which C_{ij} are elastic constants referred to x-, y-, z-axes fixed in the plate. Note that D_4 and D_6 have involved the shear correction factor.

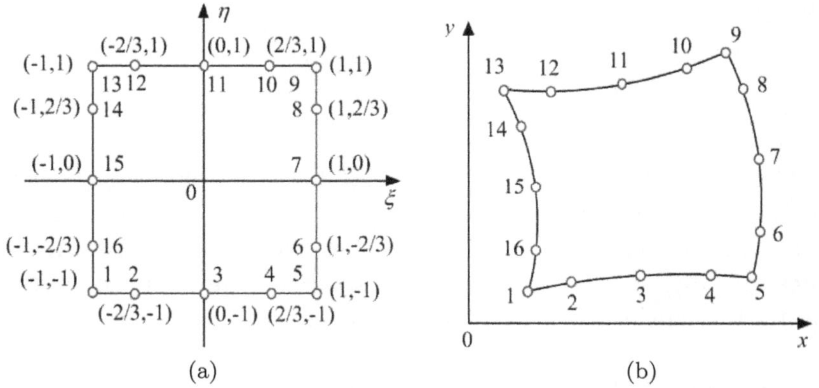

Fig. 3.10. (a) A square parent domain in natural $\xi-\eta$ plane; (b) a curvilinear quadrilateral region in Cartesian x–y plane.

The relations between the internal forces and displacements are

$$M_x = -\left(D_1\frac{\partial\psi_x}{\partial x} + D_2\frac{\partial\psi_y}{\partial y}\right), \quad M_y = -\left(D_2\frac{\partial\psi_x}{\partial x} + D_3\frac{\partial\psi_y}{\partial y}\right)$$

$$M_{xy} = -D_5\left(\frac{\partial\psi_x}{\partial y} + \frac{\partial\psi_y}{\partial x}\right)$$

$$Q_x = D_6\left(\frac{\partial w}{\partial x} - \psi_x\right), \quad Q_y = D_4\left(\frac{\partial w}{\partial y} - \psi_y\right) \tag{3.109}$$

The relationship between Cartesian coordinates and polar coordinates are

$$r = \sqrt{x^2 + y^2}, \quad \cos\theta = \frac{x}{r}, \quad \sin\theta = \frac{y}{r} \tag{3.110}$$

Therefore, the displacements and force components in polar coordinates are

$$\begin{aligned}
\psi_r &= \psi_x\cos\theta + \psi_y\sin\theta, \quad \psi_\theta = \psi_y\cos\theta - \psi_x\sin\theta \\
M_r &= M_x\cos^2\theta + M_y\sin^2\theta + 2M_{xy}\sin\theta\cos\theta \\
M_\theta &= M_x\sin^2\theta + M_y\cos^2\theta - 2M_{xy}\sin\theta\cos\theta \\
M_{r\theta} &= (M_y - M_x)\sin\theta\cos\theta + M_{xy}(\cos^2\theta - \sin^2\theta) \\
Q_r &= Q_x\cos\theta + Q_y\sin\theta, \quad Q_\theta = Q_y\cos\theta - Q_x\sin\theta
\end{aligned} \tag{3.111}$$

The following fundamental thickness-shear frequency of the magnitude [303] will be solved:

$$\omega_0 = \frac{\pi}{h}\sqrt{\frac{C_{66}}{\rho}} \tag{3.112}$$

The magnitude of the fundamental flexural frequency is $h^2\omega_0$. So, the fundamental thickness-shear frequency is very high and it is difficult to get the thickness-shear frequency using a numerical procedure due to the huge computational cost and the difficulty in finding the thickness-shear mode among many frequencies that are close to the thickness-shear frequency.

The Gauss–Lobatto quadrature rule is necessary for discretizing the integration in Eq. (3.107). Details of the integration method have been given above. The Gauss–Lobatto quadrature rule for function

$f(x, y)$ defined in $[(x, y)| -1 \leq x \leq 1, -1 \leq y \leq 1]$ can be expressed as

$$\int_{-1}^{1}\int_{-1}^{1} f(x, y)\mathrm{d}x\mathrm{d}y = \sum_{i=1}^{N_x}\sum_{j=1}^{N_y} C_i^x C_j^y f(x_i, y_j) = \boldsymbol{C}\boldsymbol{f} \tag{3.113}$$

where

$$\boldsymbol{C} = \mathrm{diag}(\boldsymbol{C}_1, \quad \boldsymbol{C}_2, \quad \ldots, \quad \boldsymbol{C}_N) \tag{3.114}$$

is a $(N_x \times N_y) \times (N_x \times N_y)$ matrix, $\boldsymbol{C}_j = C_j^y \mathrm{diag}(C_1^x, \ldots, C_{N_x}^x)_{N_x \times N_x}$, C_i^x and C_j^y $(i = 1, 2, \ldots, N_x; j = 1, 2, \ldots, N_y)$ are the Gauss–Lobatto weights with respect to x- and y-directions, respectively.

According to the DQHFEM, Eq. (3.107) can be discretized by Eqs. (3.86) and (3.113) to obtain the mass and stiffness matrices $\boldsymbol{M}$ and $\boldsymbol{K}$ with the displacement vector $\boldsymbol{U}$ as

$$\boldsymbol{K} = \begin{bmatrix} \boldsymbol{K}_{11} & & \mathrm{sym} \\ \boldsymbol{K}_{21} & \boldsymbol{K}_{22} & \\ \boldsymbol{K}_{31} & \boldsymbol{K}_{32} & \boldsymbol{K}_{33} \end{bmatrix},$$

$$\boldsymbol{M} = \rho \begin{bmatrix} J\boldsymbol{G}^T\boldsymbol{C}\boldsymbol{G} & \boldsymbol{0} & \boldsymbol{0} \\ \boldsymbol{0} & J\boldsymbol{G}^T\boldsymbol{C}\boldsymbol{G} & \boldsymbol{0} \\ \boldsymbol{0} & \boldsymbol{0} & h\boldsymbol{G}^T\boldsymbol{C}\boldsymbol{G} \end{bmatrix}, \boldsymbol{U} = \begin{Bmatrix} \boldsymbol{\psi}_x \\ \boldsymbol{\psi}_y \\ \boldsymbol{w} \end{Bmatrix} \tag{3.115}$$

$$\begin{Bmatrix} \boldsymbol{K}_{11} \\ \boldsymbol{K}_{22} \\ \boldsymbol{K}_{33} \end{Bmatrix} = \begin{bmatrix} D_1 & D_5 & D_6 \\ D_5 & D_3 & D_4 \\ D_6 & D_4 & 0 \end{bmatrix} \begin{Bmatrix} \boldsymbol{A}^{(1)^{\mathrm{T}}}\boldsymbol{C}\boldsymbol{A}^{(1)} \\ \boldsymbol{B}^{(1)^{\mathrm{T}}}\boldsymbol{C}\boldsymbol{B}^{(1)} \\ \boldsymbol{G}^T\boldsymbol{C}\boldsymbol{G} \end{Bmatrix}$$

$$\begin{aligned} \boldsymbol{K}_{21} &= D_2\boldsymbol{B}^{(1)\mathrm{T}}\boldsymbol{C}\boldsymbol{A}^{(1)} + D_5\boldsymbol{A}^{(1)\mathrm{T}}\boldsymbol{C}\boldsymbol{B}^{(1)} \\ \boldsymbol{K}_{31} &= -D_6\boldsymbol{A}^{(1)^T}\boldsymbol{C}\boldsymbol{G} \\ \boldsymbol{K}_{32} &= -D_4\boldsymbol{B}^{(1)^T}\boldsymbol{C}\boldsymbol{G} \end{aligned} \tag{3.116}$$

where $\boldsymbol{\psi}_x$, $\boldsymbol{\psi}_y$ and $\boldsymbol{w}$ are vectors of the discrete values of ψ_x, ψ_y and w.

The derivation of Eq. (3.115) needs matrix differentiation; this could be found in matrix theory textbooks. For completeness, a simple introduction is included. Let $z_1 = \boldsymbol{y}^{\mathrm{T}}\boldsymbol{A}\boldsymbol{x}$ and $z_2 = \boldsymbol{x}^{\mathrm{T}}\boldsymbol{B}\boldsymbol{x}$ where

$\boldsymbol{y}$ is $m \times 1$, $\boldsymbol{x}$ is $n \times 1$, $\boldsymbol{A}$ is $m \times n$, $\boldsymbol{B}$ is $n \times n$, and $\boldsymbol{A}$ and $\boldsymbol{B}$ do not depend on $\boldsymbol{x}$ or $\boldsymbol{y}$, then

$$\frac{\partial z_1}{\partial \boldsymbol{x}} = \boldsymbol{y}^{\mathrm{T}} \boldsymbol{A}, \quad \frac{\partial z_1}{\partial \boldsymbol{y}} = \boldsymbol{x}^{\mathrm{T}} \boldsymbol{A}^{\mathrm{T}}, \quad \frac{\partial z_2}{\partial \boldsymbol{x}} = \boldsymbol{x}^{\mathrm{T}} \left(\boldsymbol{B}^{\mathrm{T}} + \boldsymbol{B} \right) \tag{3.117}$$

if $\boldsymbol{B}$ is a symmetric matrix and then

$$\frac{\partial z_2}{\partial \boldsymbol{x}} = 2\boldsymbol{x}^{\mathrm{T}} \boldsymbol{B} \tag{3.118}$$

These equations were not presented in [281, 304], but this should be necessary for those who are not familiar with matrix differentiation. So, the process is included here. Using Eqs. (3.117) and (3.118), anyone who has some knowledge of linear algebra is able to derive the stiffness and mass matrices in Eq. (3.115).

3.7 Results and Discussion

3.7.1 *The DQHFEM for isotropic Mindlin plates*

This section presents the DQHFEM results for vibration and bending of Mindlin plates to show the accuracy, convergence and procedure of simulation of the DQHFEM. For vibration analyses, the frequencies are calculated in dimensionless form denoted by Ω which are included in the table captions. The results for various planforms and boundary conditions are given for a range of the sampling points to clearly show the convergence behavior of the solution method. In all cases, Poisson's ratio is 0.3. For numerical convenience, the number of sampling points on the edges in the ξ- and η-directions is taken as the same and is denoted by N_e, and the number of hierarchical shape functions in the two directions is also taken as the same and is denoted by N_h.

Figure 3.11 shows the flowchart of the simulation process. The creation of geometries and discretization of elements were discussed above. In this work, both blending function surfaces (Tables 3.1–3.4) and NURBS elements (Tables 3.5 and 3.7) are used. The simulation steps are explained as follows:

(1) **Step 1 (Assign material constants):** In this work, only isotropic materials are considered. However, it is easy to change

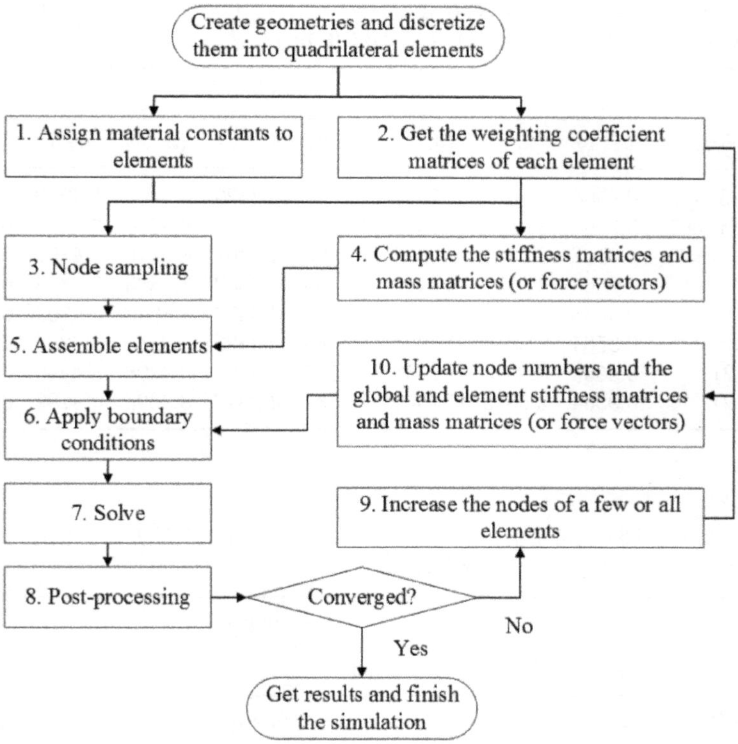

Fig. 3.11. Flowchart of the simulation process by the DQHFEM.

Table 3.1. The dimensionless frequencies $\Omega = (\omega a^2/\pi^2)\sqrt{\rho h/D}$ for SCSF rectangular Mindlin plate ($b/a = 0.5, h/a = 0.1, \kappa = 0.86667$).

	Mode sequences							
(N_e, N_h)	1	2	3	4	5	6	7	8
Exact [305]	2.1467	4.6175	8.2106	8.5189	10.496	13.405	13.935	18.105
(6, 6)	2.1446	4.6136	8.2058	9.1209	10.487	14.265	14.834	18.102
(6, 8)	2.1441	4.6120	8.2050	9.0923	10.478	14.207	14.629	18.099
(6, 10)	2.1440	4.6117	8.2049	9.0894	10.478	14.204	14.621	18.099
(8, 10)	2.1460	4.6155	8.2091	8.5250	10.491	13.472	13.931	18.103
(10, 12)	2.1464	4.6167	8.2101	8.5177	10.494	13.404	13.932	18.104
(12, 12)	2.1466	4.6172	8.2104	8.5183	10.495	13.404	13.934	18.105

Table 3.2. The dimensionless frequencies $\Omega = (\omega a^2/\pi^2)\sqrt{\rho h/D}$ for SFSF square Mindlin plate ($h/a = 0.1, \kappa = 0.82305$).

	Mode sequences							
(N_e, N_h)	(1, 1)	(2, 1)	(3, 1)	(1, 2)	(2, 2)	(3, 2)	(1, 3)	(2, 3)
Liew [306]	0.9564	3.6815	7.7558	1.5588	4.3329	—	3.4290	6.2910
Shufrin [307]	0.9564	3.6815	7.7557	1.5587	4.3327	8.3396	3.4289	6.2908
(6, 6)	0.9567	3.6857	8.0805	1.5603	4.3402	8.9216	3.4367	6.3089
(6, 8)	0.9565	3.6847	8.0440	1.5595	4.3382	8.8883	3.4318	6.2990
(6, 10)	0.9565	3.6845	8.0391	1.5594	4.3379	8.8849	3.4309	6.2973
(8, 10)	0.9564	3.6816	7.7638	1.5588	4.3329	8.3535	3.4295	6.2922
(10, 12)	0.9564	3.6815	7.7558	1.5587	4.3327	8.3398	3.4290	6.2909
(12, 12)	0.9564	3.6815	7.7557	1.5587	4.3327	8.3396	3.4289	6.2908

Table 3.3. The dimensionless frequencies $\Omega = \omega a^2\sqrt{\rho h/D}$ for free circular Mindlin plate ($h/a = 0.1, \kappa = \pi^2/12$).

	Mode sequences							
(N_e, N_h)	1	2	3	4	5	6	7	8
Exact [308]	5.278	8.868	12.064	19.711	20.801	31.270	33.033	36.041
(6, 6)	5.281	8.873	12.110	19.779	20.935	33.374	33.803	35.949
(6, 8)	5.278	8.868	12.086	19.714	20.848	32.630	33.049	36.256
(6, 10)	5.278	8.868	12.080	19.711	20.837	32.523	33.035	36.049
(8, 10)	5.278	8.868	12.065	19.711	20.802	31.292	33.034	36.049
(10, 12)	5.278	8.868	12.064	19.711	20.801	31.270	33.033	36.041
(12, 12)	5.278	8.868	12.064	19.711	20.801	31.270	33.033	36.041

to orthotropic materials as can be seen from Eqs. (3.95) and (3.96).

(2) **Step 2 (Compute the weighting coefficient matrices):** This process is discussed in detail in Eqs. (3.5)–3.88).

(3) **Step 3 (Node sampling):** the automatic node sampling was done according to the distance among nodes. Nodes of the same coordinate are assigned the same number. In order to save computational cost, one only needs to compute the distances among nodes on the boundary of elements.

Table 3.4. Bending moments in an elliptical plate with clamped and simply supported edges subjected to uniformly distributed loads (Fig. 3.12).

(N_e, N_h)	$100(w)_0$	$100(M_x)_0$	$100(M_y)_0$	$100(M_x)_a$	$100(M_y)_b$	$100(w)_0$	$100(M_x)_0$	$100(M_y)_0$
	Clamped boundary conditions					Simply supported boundary conditions		
Exact [292]	−0.3759	−0.9227	−1.4048	1.1016	2.4791	−1.5549	−2.4662	−3.5660
FEM [293]	−0.3776	−0.9238	−1.4064	1.1012	2.5005	−1.5546	−2.4100	−3.5065
DQFEM [32]	−0.3773	−0.9232	−1.4048	1.1034	2.4779	−1.5462	−2.4195	−3.5583
	DQHFEM solution based on Mindlin plate theory($h = 0.01$m)							
(6, 6)	−0.3763	−0.9119	−1.3872	1.0675	2.4052	−1.5447	−2.4057	−3.5397
(6, 8)	−0.3773	−0.9228	−1.4042	1.1045	2.4814	−1.5459	−2.4193	−3.5578
(6, 10)	−0.3773	−0.9232	−1.4048	1.1036	2.4786	−1.5459	−2.4192	−3.5585
(8, 10)	−0.3773	−0.9232	−1.4048	1.1036	2.4786	−1.5459	−2.4198	−3.5582
(10, 10)	−0.3773	−0.9232	−1.4048	1.1036	2.4786	−1.5459	−2.4198	−3.5581
(12, 12)	−0.3773	−0.9232	−1.4049	1.1037	2.4788	−1.5459	−2.4198	−3.5582

(4) **Step 4 (Compute the stiffness matrices and mass matrices or force vectors):** This process is discussed in Eqs. (3.97)–(3.106).
(5) **Steps 5–7:** These processes are the common procedures of finite element analysis that can be found in many finite element analysis books.
(6) **Step 8:** In this work, the mode shapes or deformation shapes are plotted using the integration nodes. So, the Gauss–Lobatto quadrature instead of the Gauss quadrature is used.
(7) **Steps 9 and 10:** The formulations of p-adaptive updating stiffness, mass matrices and force vector are presented in Eqs. (3.103)–(3.106).

It is clear that the process presented in Fig. 3.11 together with the construction of Coons surfaces described in Sections 3.2.1 and 3.3.2 composes the basic functionalities of a package of linear simulation for vibration and bending of Mindlin plates.

In Tables 3.1 and 3.2, the DQHFEM results for free vibration of rectangular Mindlin plates were compared with those of the exact results [305], p-version Ritz results [306] and semi-analytical solutions [307]. The length, width and thickness of the rectangular Mindlin plates are denoted by a, b and h, respectively. In Table 3.3, the DQHFEM results for free vibration of a free circular Mindlin plate were compared with those of the exact results [308]. The radius and thickness of the circular plate were denoted by a and h, respectively. The dimension of the plate and the value of the shear correction factor were included in the table captions. The formulation of Ω used in the tables are the same as those of the literatures whose results were used for comparisons. The geometries in Tables 3.1–3.3 were all mapped by the blending functions with analytical boundary curves. One can see from all tables that the DQHFEM results using only a few grid points agree with the exact results and highly accurate approximate results almost for all digits compared.

Table 3.4 presents the comparisons of bending moments in an elliptical plate (Fig. 3.12) with clamped and simply supported edges subjected to uniformly distributed loads. Values of the geometric and material parameters used in the calculation are $a = 0.50\,\text{m}$, $b = 0.33333\,\text{m}$, $h = 0.01\,\text{m}$, $E = 1\,\text{MPa}$, $q = 1.0\,\text{Pa}$, and $\kappa = \pi^2/12$. Numerical values of DQHFEM at points O, A and B as shown in

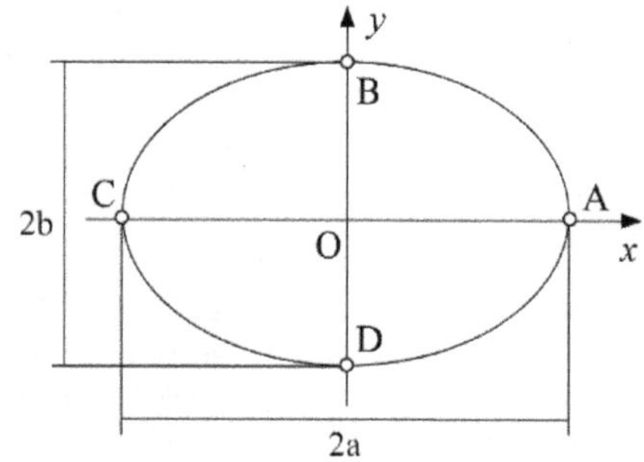

Fig. 3.12. An elliptic plate.

Fig. 3.12 are presented, for which the exact results [292], results of a p-type FEM [293] and the DQFEM results [32] based on Mindlin plate theory are available. The geometries in Tables 3.4 are mapped by blending functions. One can see that the DQHFEM results converge very fast with about 10 sampling points in each direction, which is much better than the DQFEM as can be seen in [32], where about 15 sampling points were used to obtain convergent results. Excellent agreements among the four sets of results are found for both clamped and simply supported conditions.

In Table 3.5, free vibration analysis of the circular Mindlin plate analyzed in Table 3.3 was carried out by discretizing it into five elements as shown in Fig. 3.13 and using the DQHFEM. Both blending function and NURBS surfaces were used to map the geometry. The DQHFEM results were compared with those of the exact results [308]. The material constants are the same as those in Table 3.3. One can see from Table 3.5 that the accuracy of the DQHFEM results using blending function is slightly better than that using NURBS, since the former use analytical expressions for the geometry. Comparing Tables 3.3 and 3.5, it is clear that a global method can provide results with better accuracy using less number of grid points than a domain decomposition method.

One may have noted from Tables 3.1–3.5 that for the results presented in this work, some of the frequencies the convergence is from above and for others it is from the bottom. The phenomenon is caused by geometric singularity and boundary conditions of corner points. Otherwise, the DQHFEM converges from above. For a rectangular plate, a corner point belongs to both neighboring edges. In Table 3.1,

Table 3.5. The dimensionless frequencies $\Omega = \omega a^2\sqrt{\rho h/D}$ for free Mindlin circular plate using five DQHFEM elements (Fig. 3.13, $h/a = 0.1, \kappa = \pi^2/12$).

(N_e, N_h)	1	2	3	4	5	6	7	8
				Blending function				
(6, 4)	5.280	8.868	12.091	19.715	20.879	32.753	33.057	36.051
(8, 4)	5.278	8.868	12.071	19.712	20.823	31.487	33.044	36.042
(8, 8)	5.278	8.868	12.064	19.711	20.801	31.280	33.033	36.041
(10, 10)	5.278	8.868	12.064	19.711	20.801	31.270	33.033	36.041
				NURBS surface				
(6, 4)	5.282	8.868	12.136	19.718	21.000	33.107	33.547	36.059
(8, 4)	5.280	8.868	12.083	19.714	20.853	31.782	33.051	36.044
(8, 8)	5.278	8.868	12.064	19.711	20.803	31.316	33.033	36.041
(10, 10)	5.278	8.868	12.064	19.711	20.801	31.270	33.033	36.041
Exact [308]	5.278	8.868	12.064	19.711	20.801	31.270	33.033	36.041

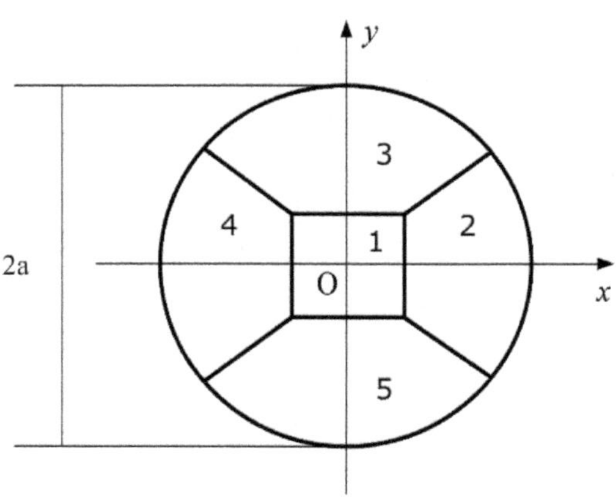

Fig. 3.13. A circular plate discretized as 5 elements.

the corner points of the free edge is taken as free. But the two corner points also belong to the two neighboring simply supported edges, so the results converge from the bottom. The influence of the corner points is small, as can be seen from Table 3.1, which indicates the high accuracy of the DQHFEM. If a circular plate is modeled as a quadrilateral element, the determinant of the Jacobian given in Eq. (3.4) is singular at the four corner points. Such an element cannot be used in FEM in general. But it can be used in the DQHFEM

Table 3.6. DQHFEM solutions for the first four natural frequencies of a square plate ($\Omega_{ij} = \omega_{ij}a^2\sqrt{\rho h/D}$).

	Mode shape							
(N_e, N_h)	(1, 1)	(2, 1)	(1, 2)	(2, 2)	(1, 1)	(2, 1)	(1, 2)	(2, 2)
	CSCS plate				*FSFS* plate			
Kantorovich [309]	28.951	69.327	54.743	94.585	9.631	16.135	38.945	46.738
DQFEM [16]	28.951	69.327	54.743	94.585	9.631	16.135	38.945	46.738
	DQHFEM solutions based on Mindlin plate theory ($h/a = 0.0001$)							
(6, 6)	28.951	69.364	54.765	94.620	9.633	16.136	39.006	46.828
(6, 8)	28.951	69.348	54.759	94.607	9.633	16.136	39.003	46.826
(6, 10)	28.951	69.348	54.759	94.607	9.633	16.136	39.002	46.826
(8, 10)	28.951	69.327	54.743	94.586	9.631	16.135	38.945	46.738
(10, 10)	28.951	69.327	54.743	94.585	9.631	16.135	38.945	46.738
	CCCC plate				*FFFF* plate			
Kantorovich [309]	35.999	73.405	73.405	108.24	—	—	—	—
DQFEM [16]	35.985	73.394	73.394	108.22	13.468	19.596	24.270	34.801
	DQHFEM solutions based on Mindlin plate theory ($h/a = 0.0001$)							
(6, 6)	35.990	73.420	73.420	108.26	13.469	19.641	24.374	35.111
(6, 8)	35.986	73.395	73.395	108.22	13.469	19.641	24.369	35.105
(6, 10)	35.985	73.394	73.394	108.22	13.469	19.641	24.368	35.105
(8, 10)	35.985	73.394	73.394	108.22	13.468	19.596	24.270	34.801
(10, 10)	35.985	73.394	73.394	108.22	13.468	19.596	24.270	34.801

by obtaining the derivatives in Eq. (3.3) from the boundary curve of the circular plate instead of the circular domain. As can be seen from Tables 3.3 and 3.4, the DQHFEM results are still very stable and have high accuracy, which indicates that the DQHFEM is not sensitive with geometric singularity, although the results did not converge from above. If the circular plate is discretized as five elements (Fig. 3.5), the DQHFEM results converge from above, as can be seen from Table 3.5.

In Table 3.6, numerical tests of DQHFEM are carried out through vibration analyses of square thin plates with four combinations of simply supported, clamped and free edges. Because the results of Refs. [309] and [16] are based on classical thin plate theory, this paper takes the relative thickness h/a as 0.0001. It is shown that DQHFEM solutions are very close to the Kantorovich solutions [309]

Table 3.7. Natural frequencies of an isotropic straight lug (Fig. 3.14) with $E = 70\,\text{GPa}$, $\rho = 2700\ \text{kg/m}^3$ and $\upsilon = 0.3$.

	Mode sequence number							
(N_e, N_h)	1	2	3	4	5	6	7	8
(6, 2)	8.6397	29.561	42.474	89.897	111.34	152.58	182.64	213.60
(8, 2)	8.6352	29.511	42.436	89.740	111.29	152.49	182.19	213.14
(8, 4)	8.6213	29.431	42.385	89.524	111.12	152.11	181.61	212.62
(10, 8)	8.6192	29.399	42.369	89.449	111.09	151.99	181.44	212.54
(15, 15)	8.6190	29.397	42.368	89.445	111.09	151.99	181.43	212.53
FEM ($l = 0.1\,\text{m}$)	8.6259	29.576	42.587	90.092	112.01	152.96	183.30	215.00
FEM ($l = 0.05\,\text{m}$)	8.6212	29.461	42.430	89.653	111.35	152.29	181.98	213.27
FEM ($l = 0.1/3\,\text{m}$)	8.6203	29.434	42.401	89.551	111.21	152.12	181.71	212.85
FEM ($l = 0.025\,\text{m}$)	8.6198	29.419	42.387	89.512	111.17	152.07	181.61	212.74
FEM ($l = 0.02\,\text{m}$)	8.6196	29.413	42.382	89.494	111.15	152.05	181.56	212.69

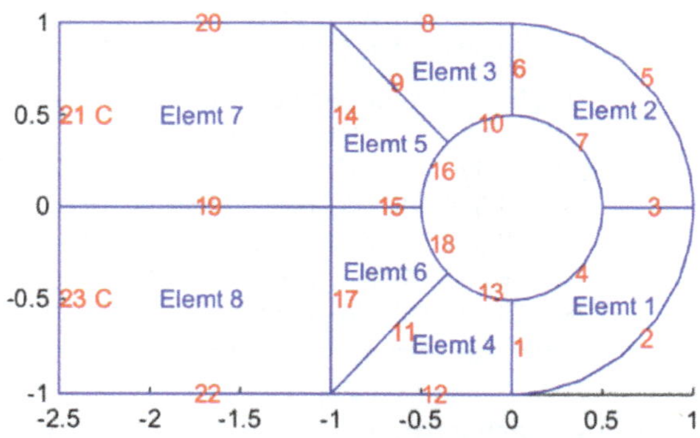

Fig. 3.14. A straight lug discretized into eight elements.

and the same as the DQFEM solutions [16] for all available significant digits used for comparisons. This validates the high accuracy of the DQHFEM for plates with various boundary conditions. Moreover, it is clear that the DQHFEM is free of shear locking. Again, one may find that the DQHFEM is capable of presenting very accurate results with only a few sampling points.

In Table 3.7, the DQHFEM results for natural vibration frequencies of a straight lug (Fig. 3.14) with the left straight edge clamped is compared with FEM results obtained from commercial software package ANSYS. The red numbers in Fig. 3.14 denote the edge numbers of NURBS elements and C denotes that the edge is clamped.

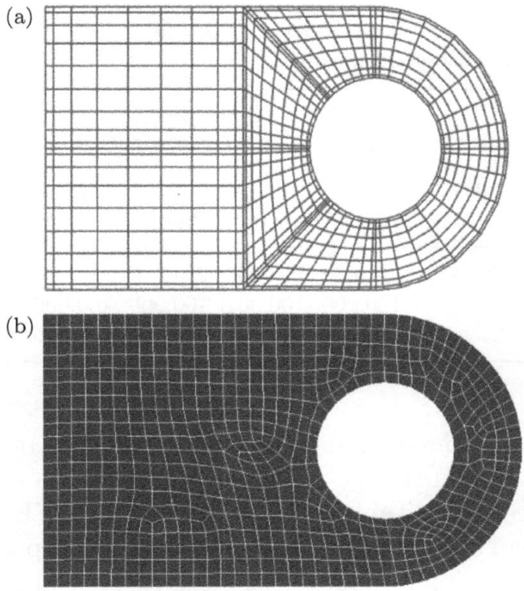

Fig. 3.15. Meshes of two methods: (a) DQHFEM with $(N_e, N_h) = (10, 8)$; (b) FEM with $l = 0.1$ m.

The element type used in FEM analysis of ANSYS is SHELL181 element that is suitable for analyzing thin to moderately thick shell structures. The radius of the circular hole is 0.5 m. The thickness h is 0.1 m. The material is isotropic with $E = 70$ GPa, $\rho = 2700\,\text{kg/m}^3$, $\upsilon = 0.3$ and $\kappa = \pi^2/12$. The dimension of the lug in Fig. 3.6 is meter. The mesh of the DQHFEM with sampling points $(N_e, N_h) = (10, 8)$ is similar with the mesh of the FEM with element edge length $l = 0.1$ m, as can be seen from Fig. 3.15. Note that the DQHFEM mesh in Fig. 3.15 is an equivalent of basis numbers with node numbers, except the nodes on element boundaries are physical nodes. The DQHFEM results with $(N_e, N_h) = (10, 8)$ already have four significant digits converged and the accuracy seems to be better than the FEM results with $l = 0.02$ m. DQHFEM needs less than one second to get the results but ANSYS needs about one minute to get the results, with $l = 0.02$ m on the same computer. So, DQHFEM is very efficient for high-accuracy analysis.

From Tables 3.5 and 3.7, one can see that the DQHFEM with a certain number of nodes but only a few nodes inside elements can present results with very good accuracy. This property is similar to

Table 3.8. The dimensionless frequencies $\Omega = \omega(L/100\pi)^2\sqrt{\rho S/EI}$ for a simply supported Euler–Bernoulli beam.

	Mesh size							
	1001		2001		3001		4001	
Mode number	Present	DSC [311]	Present	DSC [311]	Present	DSC [311]	Present	DSC [311]
500	25.0000	25.0002	25.0000	25.0000	25.0000	25.0000	25.0000	25.0000
1000	—	—	100.000	100.001	100.000	100.000	100.000	100.000
2000	—	—	—	—	410.976	401.206	400.000	400.004

the fixed interface mode synthesis method instead of the p-version FEM in the common sense.

It can be seen from Tables 3.1–3.7 that the DQHFEM can provide results with high accuracy using only a few sampling points. However, the number of sampling points may be very high in some problems [294, 310]. The results in Table 3.8 for a simply supported Euler–Bernoulli beam are aimed at showing the reliability and accuracy of the integration method and the orthogonal polynomials using the recursion formula in this work. Therefore, although the stiffness and mass matrices for this problem can be explicitly obtained, numerical method is used herein. The DQHFEM results were compared with the available discrete singular convolution (DSC) results [311]. The dimensionless frequency parameter Ω is shown in the table caption. The exact values of the frequency parameters are $(n/100)^2$, where n is the mode number. It is clear that the DQHFEM is more accurate than the DSC for the first 60% of the total number of frequencies. The accuracy of DQHFEM frequencies will reduce after 60% because the number of base is not enough to describe the mode shapes. The DQHFEM results for the problem is also compared with the results of a cubic Euler–Bernoulli beam element of FEM computed by the authors. Totally 1000 (N) frequencies were computed. The relative error of the two methods with respect to frequency numbers (n) is plotted in Fig. 3.16. As can be seen from Fig. 3.16 the relative error of the FEM begin to increase after the first 20% and has a sudden increase in the first 50%. Similar phenomenon has also been reported in [312] for a fixed–fixed bar. The relative error of the DQHFEM only begins to increase after the first 60%. The first 60% frequencies are

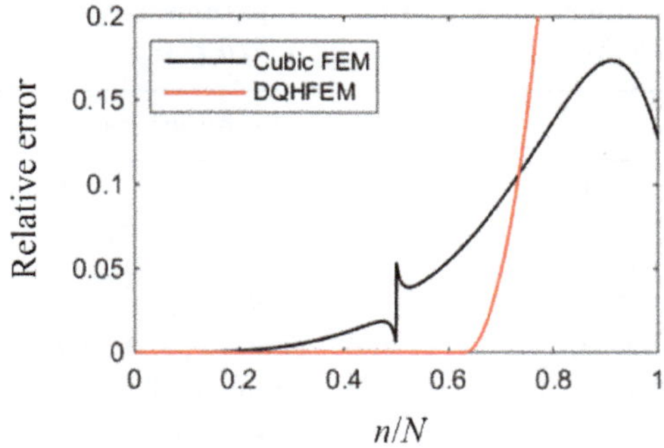

Fig. 3.16. Simply supported beam. Normalized discrete spectra using cubic finite elements and DQHFEM.

very accurate. The DSC can give frequencies with good accuracy higher than 60% may be due to the use of singular basis. But the DSC seems to be mainly used in regular domains so far, and like the DQM or the DQHFEM, the DSC does not have the hierarchical property.

For Mindlin plates, a simply supported (S) edge has deflection and tangent angle of rotation being zeros. First, a square Mindlin plate with SFSF boundary conditions is analyzed in Table 3.9. Since all edges are straight, linear elements for geometries are sufficient. For this plate, only two elements are used, as shown in Fig. 3.17. The DQHFEM results of this work using triangular elements are compared with the p-version Ritz results [306], semi-analytical solutions [307] and the DQHFEM results of [35] using quadrilateral elements. One can see from Table 3.9 the fast convergence rate and high accuracy of the present method using only a few nodes. In Table 3.10, a rectangular Mindlin plate with an aspect ratio of being 0.5 and SCSF boundary conditions (Fig. 3.18) is analyzed by four triangular DQHFEM elements. The present results are compared with the exact results [305] and the DQHFEM results of [35] using rectangular elements. One can again find the fast convergence rate and high accuracy of the present method using only a few nodes.

In Table 3.11, a sectorial Mindlin plate with simply supported radial edges and clamped circular edge is analyzed. The mesh of the plate is the same as Fig. 3.19 and cubic elements as shown in Fig. 3.19(b) are used. The DQM results of [313] and exact results of [314] are used for comparison. One can see from Table 3.11 that

Table 3.9. The dimensionless frequency parameters $\Omega = (\omega a^2/\pi^2)\sqrt{\rho h/D}$ for an SFSF square Mindlin plate ($v = 0.3, h/a = 0.1, \kappa = 0.82305$, Fig. 3.17).

	Mode sequences							
(P, N_e)	(1, 1)	(2, 1)	(3, 1)	(1, 2)	(2, 2)	(3, 2)	(1, 3)	(2, 3)
(2,5)	0.9587	4.0651	—	1.5916	—	—	3.5913	—
(3, 6)	0.9572	3.7029	—	1.5683	4.4869	—	3.5178	6.6371
(4, 7)	0.9568	3.6885	7.8620	1.5628	4.3529	8.7432	3.4429	6.4870
(5, 8)	0.9565	3.6831	7.7926	1.5607	4.3418	8.4067	3.4344	6.3200
(6, 9)	0.9564	3.6821	7.7601	1.5594	4.3354	8.3646	3.4308	6.3057
(7, 10)	0.9564	3.6817	7.7572	1.5590	4.3336	8.3441	3.4294	6.2935
(8, 11)	0.9564	3.6816	7.7561	1.5588	4.3330	8.3411	3.4291	6.2915
(9, 12)	0.9564	3.6815	7.7558	1.5588	4.3328	8.3400	3.4289	6.2910
(10, 13)	0.9564	3.6815	7.7557	1.5587	4.3327	8.3397	3.4289	6.2908
(12, 15)	0.9564	3.6815	7.7557	1.5587	4.3327	8.3396	3.4289	6.2908
(15, 18)	0.9564	3.6815	7.7557	1.5587	4.3327	8.3396	3.4289	6.2908
Shufrin [307]	0.9564	3.6815	7.7557	1.5587	4.3327	8.3396	3.4289	6.2908
DQHFEM [35]	0.9564	3.6815	7.7557	1.5587	4.3327	8.3396	3.4289	6.2908
Liew [306]	0.9564	3.6815	7.7558	1.5588	4.3329	—	3.4290	6.2910

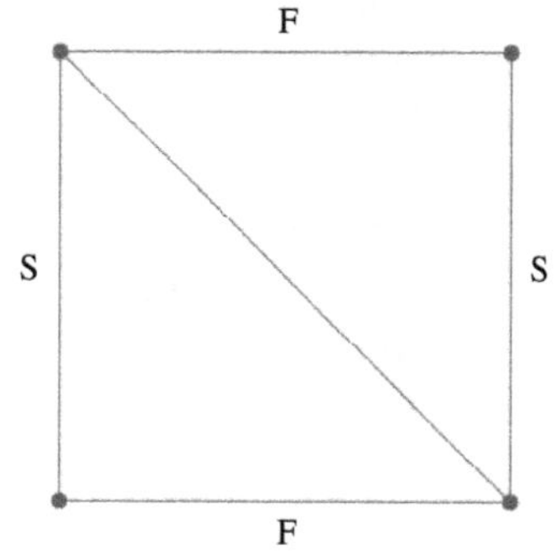

Fig. 3.17. A square plate with two linear elements.

the present results agree with the DQM results of [313] for about four digits. One can find from [313] that the DQM results only have about four digits converged, but the present DQHFEM results have all five digits used for comparison converged. The present results also have an excellent agreement with the exact results of [314]. In Table 3.12, a clamped annular sectorial plate (Fig. 3.20) with the ratio of inner and outer radii $b/a = 0.5$ is analyzed. The mesh seed

Table 3.10. The dimensionless frequency parameters $\Omega = (\omega a^2/\pi^2)\sqrt{\rho h/D}$ for an SCSF rectangular Mindlin plate ($\upsilon = 0.3$, $b/a = 0.5$, $h/a = 0.1$, $\kappa = 0.86667$, Fig. 3.18).

	Mode sequences							
(P, N_e)	1	2	3	4	5	6	7	8
(2, 5)	2.1539	4.6693	8.3562	8.7081	10.869	14.283	15.282	20.465
(3, 6)	2.1487	4.6244	8.2282	8.5465	10.617	13.534	14.257	18.473
(4, 7)	2.1471	4.6194	8.2126	8.5240	10.505	13.426	13.990	18.188
(5, 8)	2.1468	4.6178	8.2109	8.5199	10.498	13.409	13.944	18.112
(6, 9)	2.1467	4.6175	8.2107	8.5191	10.496	13.405	13.936	18.106
(7, 10)	2.1467	4.6175	8.2106	8.5190	10.496	13.405	13.935	18.105
(8, 11)	2.1467	4.6175	8.2106	8.5189	10.496	13.405	13.935	18.105
(15, 18)	2.1467	4.6175	8.2106	8.5189	10.496	13.405	13.935	18.105
Exact [305]	2.1467	4.6175	8.2106	8.5189	10.496	13.405	13.935	18.105
DQHFEM [35]	2.1466	4.6172	8.2104	8.5183	10.495	13.404	13.934	18.105

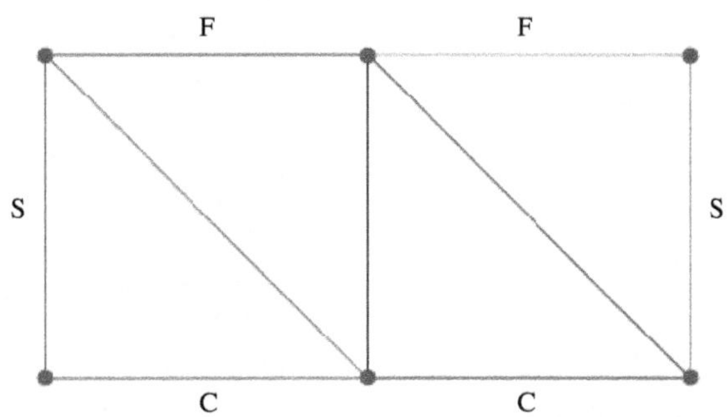

Fig. 3.18. A rectangular plate discretized as four linear elements.

length is $h_0 = 0.5(a - b)$. The order of the mesh is cubic. Note that since only mode shapes without radial node lines were considered, the fourth mode in Table 3.12 is higher than 100th total mode. This requires more elements and more nodes in each element. One can see from Table 3.12 that the cubic elements with node numbers $(P, N_e) = (7, 10)$ already converged and have five digits that agree with the exact results. If quartic elements are used, all digits used for comparison agree with the exact results.

Table 3.11. The dimensionless frequency parameters $\Omega = \omega a^2\sqrt{\rho h/D}$ for a sectorial Mindlin plate with simply supported radial edges and clamped circular edge (Fig. 3.19, mode shapes have no radial node lines, $\upsilon = 0.3, \kappa = \pi^2/12$).

	Mode sequences									
(P, N_e)	1	2	3	4	5	1	2	3	4	5
	$h/a = 0.1$					$h/a = 0.2$				
(2,5)	32.226	72.485	122.98	—	—	26.999	54.550	84.880	—	—
(3,6)	32.206	72.243	120.77	175.75	—	26.994	54.472	83.993	114.88	—
(4,7)	32.205	72.222	120.47	174.30	232.45	26.994	54.466	83.910	114.32	145.32
(5,8)	32.205	72.221	120.45	174.00	231.12	26.994	54.466	83.904	114.23	144.92
(6,9)	32.205	72.221	120.45	173.99	230.90	26.994	54.466	83.903	114.23	144.84
(7, 10)	32.205	72.221	120.45	173.98	230.88	26.994	54.466	83.903	114.23	144.84
(15, 18)	32.205	72.221	120.45	173.98	230.88	26.994	54.466	83.903	114.23	144.84
DQM [313]	32.205	72.221	120.45	173.98	230.89	26.993	54.466	83.903	114.23	144.84
Exact [314]	32.262	72.250	120.56	174.24	231.04	27.040	54.612	83.906	114.28	144.96

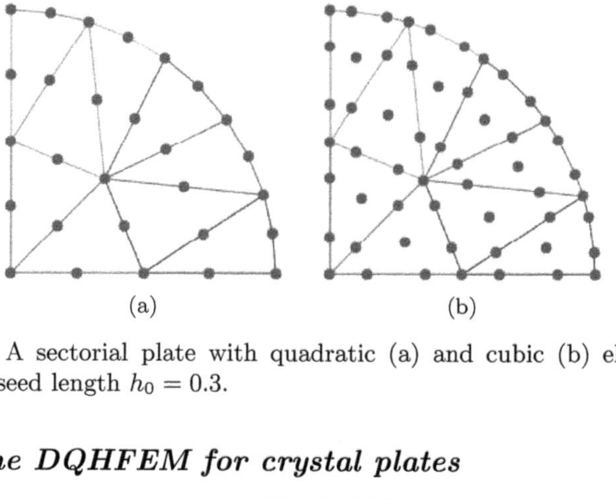

(a) (b)

Fig. 3.19. A sectorial plate with quadratic (a) and cubic (b) elements and global mesh seed length $h_0 = 0.3$.

3.7.2 *The DQHFEM for crystal plates*

This section aims to present the DQHFEM results for thickness-shear vibration of quartz crystal plates and validate the high accuracy of the DQHFEM. The mappings of a square parent domain to the interest domains are carried out via the blending function. For simplicity of denotation, the number of sampling points on the edge in the ξ- and η-directions is taken as the same and is denoted by N_e, and the number of hierarchical shape functions in the two directions is also taken as the same and is denoted by N_h. The natural frequency

Table 3.12. The dimensionless frequency parameters $\Omega = \omega a^2\sqrt{\rho h/D}$ for an annular sectorial Mindlin plate with clamped circular edges (Fig. 3.20, mode shapes have no radial node lines, $\upsilon = 0.3, b/a = 0.5, \kappa = \pi^2/12$).

(P, N_e)	1	2	3	4	1	2	3
		$h/a = 0.1$				$h/a = 0.2$	
(2, 5)	70.2846	159.871	267.145	—	48.3114	97.4053	156.010
(3, 6)	70.2770	159.795	265.622	379.259	48.3103	97.3903	155.509
(4, 7)	70.2769	159.784	265.449	378.610	48.3103	97.3886	155.470
(5, 8)	70.2769	159.784	265.443	378.426	48.3103	97.3886	155.468
(6, 9)	70.2769	159.784	265.442	378.422	48.3103	97.3885	155.468
(7, 10)	70.2769	159.784	265.442	378.421	48.3103	97.3885	155.468
(10, 13)	70.2769	159.784	265.442	378.421	48.3103	97.3885	155.468
(7, 10)*	70.2770	159.784	265.442	378.421	48.3103	97.3886	155.468
Exact [315]	70.2770	159.784	265.442	378.421	48.3103	97.3886	155.468

Note: *Quartic mesh.

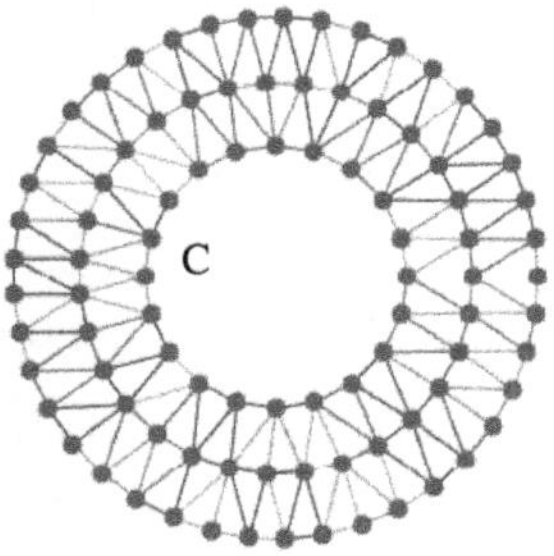

Fig. 3.20. An annular sectorial plate and the mesh.

is expressed by frequency parameter Ω used by Liu *et al.* [316, 317]. Its expression is

$$\Omega = \omega\sqrt{\frac{\rho h}{D_1}} \tag{3.119}$$

This frequency parameter is further normalized by the magnitude of fundamental thickness-shear frequency parameter of Eq: (3.112), which is done by substituting Eq. (3.112) into Eq. (3.119)

$$\Omega_0 = \omega_0\sqrt{\frac{\rho h}{D_1}} = \frac{1}{h}\sqrt{\frac{12D_6}{D_1}} \tag{3.120}$$

The normalization is done for ease of finding the fundamental thickness-shear mode.

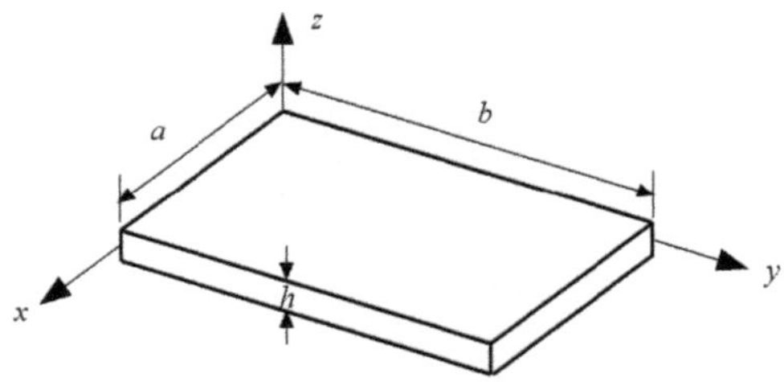

Fig. 3.21. A rectangular plate and its coordinates.

Vibration of a rectangular quartz crystal plate (Fig. 3.21) is analyzed first for validating the accuracy of the DQHFEM for thickness-shear vibrations, since there are plenty of such results in the literatures. For ease of comparison, the dimensions and material constants of the rectangular quartz crystal plate are taken as the same as the one used by Liu *et al.* [316, 317]. The dimensions are $a = 2.5\,\text{mm}$, $b = 1.25\,\text{mm}$, and $h = 0.1\,\text{mm}$. The material constants are

$$\mathrm{C}_{ij} = 10^9 \times \begin{Bmatrix} 86.74 & -8.254 & 27.15 & -3.66 & 0 & 0 \\ -8.254 & 129.77 & -7.418 & 5.7 & 0 & 0 \\ 27.15 & -7.418 & 102.83 & 9.92 & 0 & 0 \\ -3.66 & 5.7 & 9.92 & 38.61 & 0 & 0 \\ 0 & 0 & 0 & 0 & 68.81 & 2.534 \\ 0 & 0 & 0 & 0 & 2.534 & 29.01 \end{Bmatrix} N/m^2 \tag{3.121}$$

In Table 3.13, the DQHFEM results are compared with the DQFEM results [317], 3D FEM results [318] and the results obtained by a numerical extended Kantorovich method (NEKM) [316]. Both DQFEM and NEKM used 5×5 elements. The NEKM [316] used the separation variable technique and is very efficient but may lose accuracy. Both DQHFEM and DQFEM are 2D methods of high accuracy. The number of grid points (or number of basis functions) of the DQHFEM is taken with the same degree of freedom as DQFEM for comparison of the two methods. Due to the limitation of the number of grid points in one element, the DQFEM has to use 5×5 elements to analyze the problem, which increases the complexity. The DQHFEM can use more than 2000 grid points in one element, so the

Table 3.13. Comparison of the frequency parameters Ω normalized by Ω_0 for a rectangular quartz crystal plate (Fig. 3.21).

The methods used	Elements	Grid points in one element	Frequency parameters (Ω/Ω_0)
DQHFEM	1×1	60×60	1.0028496408
		75×75	1.0028496274
		90×90	1.0028496274
	5×5	(15, 5)[a]	1.0028605457
		(18, 10)[a]	1.0028501128
		(18, 15)[a]	1.0028496292
		(20, 18)[a]	1.0028496257
DQFEM [317]	5×5	12×12	1.0029467549
		15×15	1.0028498128
		18×18	1.0028496274
		25×18	1.0028496256
3D FEM [318]	$100 \times 50 \times 6$[b]	20	1.0030015187
NEKM [316]	5×5	25×25	1.0028384677

Notes: [a]These numbers are (N_e, N_h).
[b]Here, 6 is the number of grid points in z-direction.

problem can be analyzed by one element using the DQHFEM. As can be seen from Table 3.13 that the DQHFEM using one element with 75×75 grid points in each element has converged and is as accurate as the DQFEM using 5×5 elements with 18×18 grid points in each element. The degrees of freedom or the p-order of the hierarchical basis functions inside DQHFEM elements can be increased without mesh refinement. Table 3.13 also includes the results for the plate with 5×5 elements and different degrees of freedom on the edge and inside a DQHFEM element. It can be seen that DQHFEM converges similar or faster than the DQFEM. The relative error of the DQHFEM results for rectangular quartz crystal plate compared with the corresponding 3D FEM result is 0.016%. The 3D FEM results of Wang *et al.* [318] only have first five significant digits converged. The simulations in [318] were not furthered, which should be limited by the huge computational cost in 3D FEM simulations.

Then thickness-shear vibration of a circular quartz crystal plate (Fig. 3.22) is analyzed by the DQHFEM. In Table 3.14, the DQHFEM results is compared with the 3D FEM results [318]. The material constants are taken as the same as Ref. [318] that rounded the

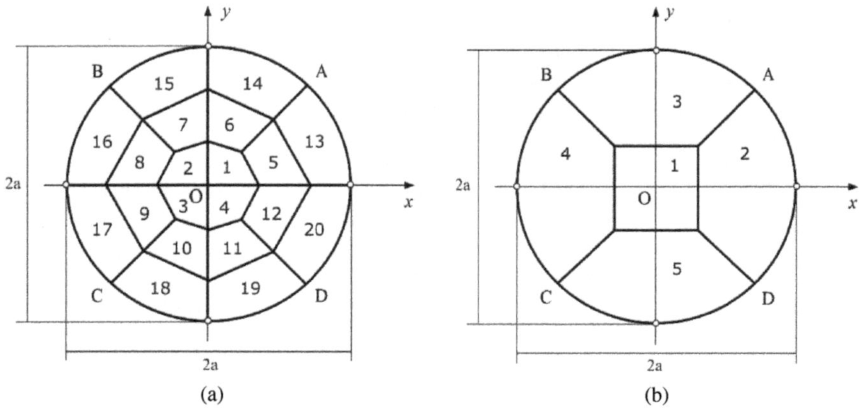

Fig. 3.22. A circular plate and its coordinates: (a) 20 elements; (b) 5 elements.

Table 3.14. Comparison of the frequency parameters Ω normalized by Ω_0 for a circular quartz crystal plate (Fig. 3.22).

The methods used	Elements	Grid points in one element (N_e, N_h)	Frequency parameters (Ω/Ω_0)
DQHFEM	20	(20, 5)	1.0051067135
		(20, 10)	1.0052688533
		(20, 15)	1.0056913595
		(25, 15)	1.0056911214
		(25, 20)	1.0056865113
	5	(40, 10)	1.0050782991
		(40, 20)	1.0051659933
		(40, 30)	1.0057205417
		(50, 30)	1.0057200725
		(50, 40)	1.0056865105
3D FEM [318]	—	—	1.005462

material constants of Eq. (3.50) to have two digits after decimal point. The radius of the circular quartz crystal plate is $a = 1.2\,\text{mm}$ and the thickness is $h = 0.1\,\text{mm}$. The circular plate is first discretized into 20 elements (Fig. 3.22(a)) and then discretized into five elements (Fig. 3.22(b)). The converged results of the two sets of DQHFEM results agree with each other for nine significant digits, as can be seen

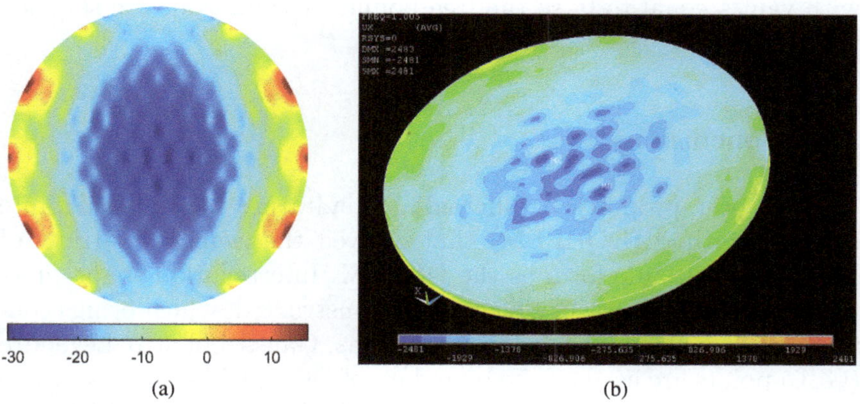

Fig. 3.23. Thickness-shear vibration mode shape ψ_x of the circular plate: (a) DQHFEM; (b) 3D FEM.

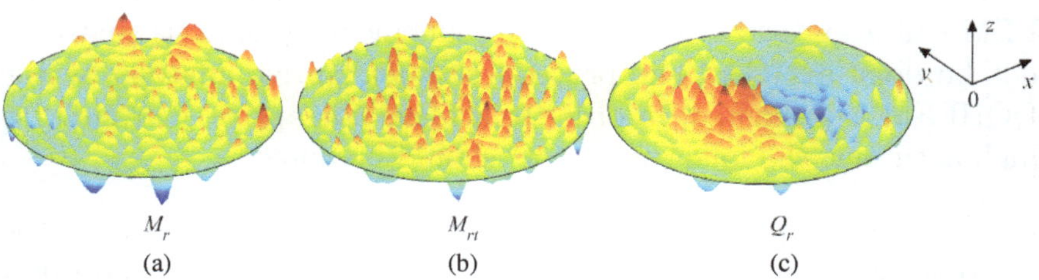

Fig. 3.24. Thickness-shear vibration mode shapes M_r, M_{rt} and Q_r of the circular plate.

from Table 3.14. The relative error of the DQHFEM results for circular quartz crystal plate compared with the corresponding 3D FEM result is 0.022%. So, the accuracy of the 2D result of DQHFEM for rectangular and circular quartz crystal plates is similar compared to the 3D FEM results [318]. Figure. 3.23(a) shows the thickness-shear vibration mode shape ψ_x of the circular plate obtained by the DQHFEM. The corresponding mode shape obtained by 3D FEM [318] is given in Fig. 3.23(b). One can see that the mode shapes obtained by the two methods agree with each other. Figure 3.24 shows the thickness-shear vibration mode shapes M_r, M_{rt} and Q_r of the circular plate. The boundaries of the mode shapes are shown in black lines to clearly show how well the free boundary conditions are satisfied. One can see that the boundaries are smooth circular curves with values equal to 0, so the free boundary conditions are satisfied very well.

3.8 Conclusions

This chapter proposes a differential quadrature hierarchical finite element method (DQHFEM) that derived the element matrices of HFEM in the same form as the DQFEM. Interpolation shape functions on the boundary of elements are constructed, which brings convenience in imposing boundary conditions. Gauss–Lobatto–Legendre (GLL) points are used on the boundary of DQHFEM elements, which guarantees the high accuracy of the DQHFEM. The DQHFEM can also directly use the NURBS elements that match the exact CAD geometry and further mesh refinement is not necessary. Matching exact CAD geometry is the primary objective of IGA. The DQHFEM can also carry out adaptive p-refinement in analysis that cannot be done by IGA. The procedure of p-adaptive analysis by the DQHFEM is presented. In order to achieve p-adaptive analysis with high accuracy, the following points should be noted:

(1) If sampling points are needed, they should be the roots of the orthogonal polynomials that satisfy the boundary conditions. Lagrange polynomials using GLL points provide the best accuracy for C^0 problem because the jth Lagrange polynomial has maximum value at jth GLL point and the maximum value is 1.
(2) If high-order orthogonal polynomials are needed, they should be computed from the recursion formula that has the minimum round-off errors. Many publications still use symbolic computation to overcome the computational stability problem of HFEM just because they were not aware of this fact.
(3) The derivatives in HFEM should be computed analytically whenever possible to save the computational cost. For 2D or 3D problem, the separation variable feature of the basis should be used so that 2D or 3D differentiation can be computed from 1D cases and computational cost can be saved greatly.

The three points can be found in the literatures, but they were usually discussed separately in different papers. Successful application of the DQHFEM needs to consider the three points simultaneously. The DQHFEM was applied to vibration and bending analysis of Mindlin plates, which showed that the DQHFEM can provide highly accurate results by only a few nodes in each element. Meanwhile, high-frequency analysis of Euler–Bernoulli beam showed that the maximum order of the DQHFEM elements can be as high as a workstation can compute.

The DQHFEM with interpolation basis on the boundary of elements were found to be equivalent to the fixed interface mode synthesis method(MSM) to some extent, but the DQHFEM does not need mode analysis needed in MSM. The DQHFEM can analyze a structure by a few elements that have a certain number of nodes on the boundary of the elements but several clamped modes inside the elements. These property may open a door to powerful applications. Therefore, triangular, tetrahedral and hexahedral DQHFEM elements should be developed. This is obviously not a trivial task and one that deserves much study. With the three points or three principles above in mind, hexahedral C^0 DQHFEM elements are straightforward. Triangular and tetrahedral elements and C^1 elements are also possible.

The differential quadrature hierarchical finite element method (DQHFEM) was also applied to thickness-shear vibration of quartz circular crystal plate. Thickness-shear vibration of circular crystal plates is recognized as difficult for analytical methods because the scalar differential equation for the problem is not separable in polar coordinates due to material anisotropy. For numerical methods, thickness-shear vibration analysis needs huge computational cost and requires high accuracy at the same time. The DQHFEM overcomes the numerical stability problem in the hierarchical finite element method (HFEM) and is as accurate as the DQFEM. The efficiency of the DQHFEM is improved by using the separation variable feature of the hierarchical basis. Therefore, the DQHFEM is suitable for thickness-shear vibration of quartz circular crystal plate. It was shown that the DQHFEM can analyze the problem using only one element or a few elements, which greatly simplifies the pre- and post-processing. Numerical results of the DQHFEM were compared with the results of other methods, which validated the high accuracy of the DQHFEM. The DQHFEM, as an improved version of the HFEM, should have great potential in developing finite element software in future.

Chapter 4

The Differential Quadrature Hierarchical Finite Element Method for Plane Problems

Free in-plane vibration analysis of plates is carried out by a differential quadrature hierarchical finite element method (DQHFEM). The NURBS (non-uniform rational B-splines) patches of geometries were first transformed into differential quadrature hierarchical (DQH) patches, and then the elastic field was discretized by the same DQH basis. The DQHFEM solved the compatibility problem caused by different parametrization of neighboring patches of isogeometric analysis (IGA) using NURBS, and mesh refinement in DQHFEM does not propagate from patch to patch. The DQHFEM matrices also have the same embedding property as the hierarchical finite element method (HFEM). In-plane vibration analyses of plates of several planforms showed that the DQHFEM is similar to the fixed interface mode synthesis method that can analyze a structure using a few nodes on the boundary of substructure elements but several clamped modes inside each substructure element but the DQHFEM does not need modal analysis and is of high accuracy. The accuracy and convergence of the DQHFEM were validated through comparison with exact and approximate results in the literatures and with those computed by the authors.

4.1 The Differential Quadrature Hierarchical Method

4.1.1 *Rectangular elements*

The differential quadrature hierarchical method (DQHM) is based on the approximation of a field variable $f[x(\xi)]$, where $\xi \in [-1, 1]$, by a linear summation of interpolation basis on the two ends and hierarchical basis inside the domain, namely,

$$f[x(\xi)] = \frac{1-\xi}{2} f[x(\xi_1)] + \frac{1+\xi}{2} f[x(\xi_2)] + \sum_{m=1}^{N_\xi} \varphi_m(\xi) a_m \qquad (4.1)$$

where $\xi_1 = -1$ and $\xi_2 = 1$, N_ξ is the number of hierarchical basis in the ξ-direction, a_m, $m = 1, 2, \ldots, N_\xi$ are coefficients of the hierarchical basis, and $\phi_m(\xi)$ are the integrated Legendre polynomials as follows:

$$\varphi_n(\xi) = \int_{-1}^{\xi} P_n(\xi) d\xi = \frac{(\xi^2 - 1)}{n(n+1)} \frac{dP_n(\xi)}{d\xi}, \quad \xi \in [-1, 1] \qquad (4.2)$$

where $n = 1, 2, \ldots$, $P_n(\xi)$ is the Legendre polynomial of degree n. One can see from Eq. (4.2) that the hierarchical shape functions have zero function values at the two endpoints, so they have no influence on inter-element compatibility. The interpolation basis and the hierarchical basis are respectively shown in Figs. 4.1(a) and 4.1(b). In order to overcome the numerical stability problem, the Legendre polynomials should be computed from recursion formula if the order n is high [34].

A two-dimensional (2D) function $f(x, y)$ considered to be an implicit function of ξ and η as $f = f[x(\xi, \eta), y(\xi, \eta)]$ is approximated

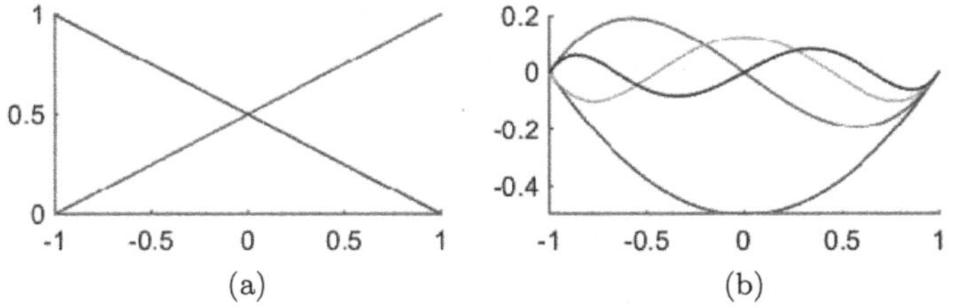

Fig. 4.1. One-dimensional (1D) basis of DQHM: (a) interpolation basis on the two ends; (b) hierarchical basis inside the domain.

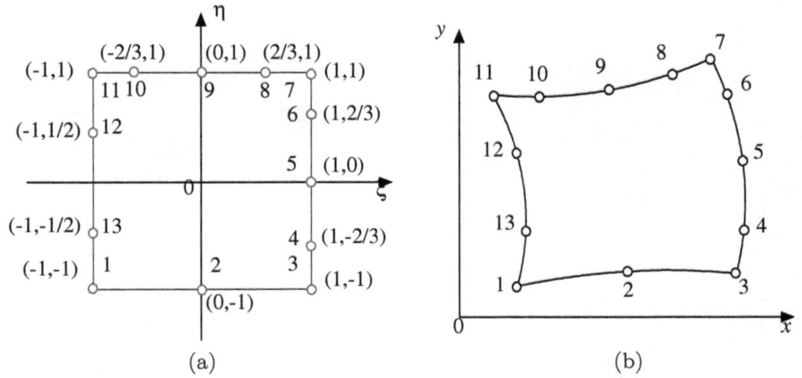

Fig. 4.2. Geometric mapping: (a) a square parent domain in natural ξ–η plane; (b) a curvilinear quadrilateral region in Cartesian x–y plane.

by interpolation shape functions constructed through blending functions and Lagrange functions on the boundary of the domain and tensor product of the hierarchical basis inside the domain [34] as follows:

$$f[x(\xi,\eta),y(\xi,\eta)] = \sum_{k=1}^{K} S_k(\xi,\eta) f_k + \sum_{m=1}^{N_\xi} \sum_{n=1}^{N_\eta} \varphi_m(\xi)\varphi_n(\eta) a_{mn} \quad (4.3)$$

where $S_k(\xi,\eta)$ denote the interpolation basis on the boundary of the domain (Fig. 4.2), and $\varphi_m(\xi)$ and $\varphi_n(\eta)$ are the hierarchical shape functions defined by Eq. (4.2). Let $\xi_i, (i = 1,2,\ldots,M)$ and $\eta_j(j = 1,2,\ldots,N)$ be the sampling points along the edge of a square domain in the natural ξ–η plane (Fig. 4.2(a)). The preferable nodes are non-uniformly distributed nodes such as Gauss–Lobatto–Legendre nodes, which lead to much lower condition numbers than uniformly distributed nodes in the p-method [53]. The interpolation shape function at node $\xi_1 = \eta_1 = -1$ can be written as

$$S_1(\xi,\eta) = \frac{1-\eta}{2} L_1^M(\xi) + \frac{1-\xi}{2} L_1^N(\eta) - \frac{(1-\xi)(1-\eta)}{4} \quad (4.4)$$

The interpolation shape functions at nodes on the edge $\eta_1 = -1$ are

$$S_i(\xi,\eta) = \frac{1-\eta}{2} L_i^M(\xi), \quad i = 2,\ldots,M-1 \quad (4.5)$$

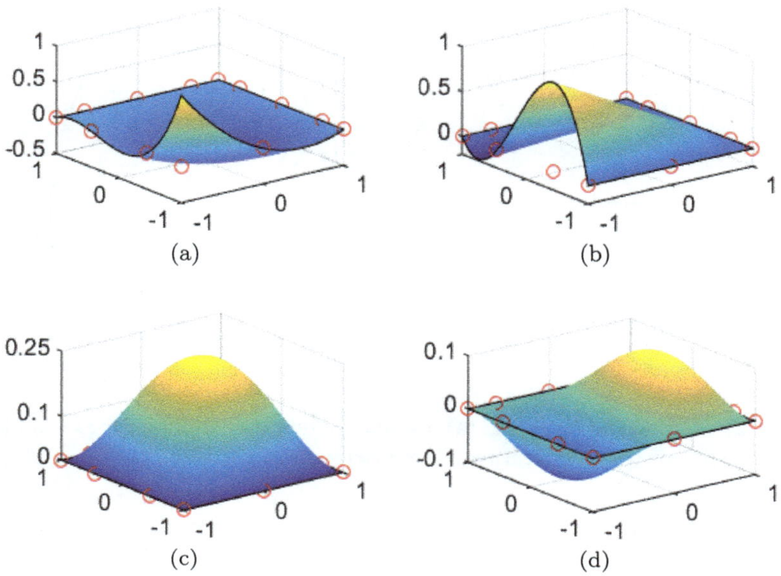

Fig. 4.3. Basis functions: (a) $S_1(\xi,\eta)$; (b) $S_{13}(\xi,\eta)$; (c) $\varphi_1(\xi)\varphi_1(\eta)$; (d) $\varphi_2(\xi)\varphi_1(\eta)$.

where

$$L_i^M(\xi) = \prod_{j=1,j\neq i}^{M} \frac{\xi-\xi_j}{\xi_i-\xi_j} \tag{4.6}$$

is the Lagrange function. The interpolation shape functions at other corners and edges are of the same form as Eqs. (4.4) and (4.5). The subscripts $k = 1, 2, \ldots, K$ of shape functions are numbered along the edge of a domain, as shown in Fig. 4.2. Some basis functions corresponding to the natural ξ–η plane in Fig. 4.2 are shown in Fig. 4.3. The interpolation basis on the boundary brings convenience of imposing boundary conditions and geometric compatibility conditions.

The DQHM is an analogue of the differential quadrature method (DQM) [319, 320] but uses the approximation expression as shown in Eq. (4.1) or Eq. (4.3) to approximate the derivatives. Thus, the first-order partial derivatives of $f(\xi,\eta)$ with respect to ξ and η at point $(\xi_i, \eta_j)(i = 1, 2, \ldots, N_x; j = 1, 2, \ldots, N_y)$ can be expressed as

$$\left(\frac{\partial f}{\partial \xi}\right)_{ij} = \sum_{k=1}^{K} \frac{\partial S_k(\xi_i, \eta_j)}{\partial \xi} f_k + \sum_{m=1}^{N_\xi} \sum_{n=1}^{N_\eta} P_m(\xi_i)\varphi_n(\eta_j)a_{mn}$$
$$\left(\frac{\partial f}{\partial \eta}\right)_{ij} = \sum_{k=1}^{K} \frac{\partial S_k(\xi_i, \eta_j)}{\partial \eta} f_k + \sum_{m=1}^{N_\xi} \sum_{n=1}^{N_\eta} \varphi_m(\xi_i)P_n(\eta_j)a_{mn} \tag{4.7}$$

Using the chain rule of differentiation, the partial derivatives $\partial f/\partial x$ and $\partial f/\partial y$ at the grid point $x_{ij} = x(\xi_i, \eta_j), y_{ij} = y(\xi_i, \eta_j)$ in a curvilinear quadrilateral domain can be computed as

$$\left(\frac{\partial f}{\partial x}\right)_{ij} = \frac{1}{|\boldsymbol{J}|_{ij}} \left[\left(\frac{\partial y}{\partial \eta}\right)_{ij}\left(\frac{\partial f}{\partial \xi}\right)_{ij} - \left(\frac{\partial y}{\partial \xi}\right)_{ij}\left(\frac{\partial f}{\partial \eta}\right)_{ij}\right]$$
$$\left(\frac{\partial f}{\partial y}\right)_{ij} = \frac{1}{|\boldsymbol{J}|_{ij}} \left[\left(\frac{\partial x}{\partial \xi}\right)_{ij}\left(\frac{\partial f}{\partial \eta}\right)_{ij} - \left(\frac{\partial x}{\partial \eta}\right)_{ij}\left(\frac{\partial f}{\partial \xi}\right)_{ij}\right] \tag{4.8}$$

where the determinant $|\boldsymbol{J}|$ of the Jacobian $\boldsymbol{J} = \partial(x, y)/\partial(\xi, \eta)$ is

$$|\boldsymbol{J}| = \frac{\partial x}{\partial \xi}\frac{\partial y}{\partial \eta} - \frac{\partial y}{\partial \xi}\frac{\partial x}{\partial \eta} \tag{4.9}$$

and the partial derivatives $\partial f/\partial \xi$ and $\partial f/\partial \eta$ at the grid point $x_{ij} = x(\xi_i, \eta_j)$, $y_{ij} = y(\xi_i, \eta_j)$ are given by Eq. (4.7). Equations (4.8) define the DQH rules of the first-order partial derivatives with respect to the Cartesian x, y coordinates for the irregular domain. In order to make the notification simpler, the following vectors of variables are defined:

$$\boldsymbol{f}^T = \{f_1 \quad f_2 \quad \cdots \quad f_K \quad a_{11} \quad \cdots \quad a_{N_\xi 1} \quad \cdots \quad a_{N_\xi N_\eta}\}$$
$$\bar{\boldsymbol{f}}^T = \{f_{11} \quad \cdots \quad f_{N_x 1} \; f_{12} \quad \cdots \quad f_{N_x 2} \quad \cdots \quad f_{1N_y} \quad \cdots \quad f_{N_x N_y}\} \tag{4.10}$$

where $f_{ij} = f[x(\xi_i, \eta_j), y(\xi_i, \eta_j)]$. According to Eq. (4.10), Eq. (4.3) can be expressed as

$$f(x, y) = \boldsymbol{N}^T \boldsymbol{f} \tag{4.11}$$

where

$$\begin{aligned}\boldsymbol{N}^{\mathrm{T}} = [S_1(\xi,\eta) \quad \cdots \quad S_K(\xi,\eta) \quad \varphi_1(\xi)\varphi_1(\eta) \\ \varphi_2(\xi)\varphi_1(\eta) \quad \cdots \quad \varphi_{N_\xi}(\xi)\varphi_{N_\eta}(\eta)]\end{aligned} \tag{4.12}$$

According to Eqs. (4.11) and (4.12), the relationship between the two vectors in Eqs. (4.10) is

$$\bar{\boldsymbol{f}} = \boldsymbol{G}\boldsymbol{f} \tag{4.13}$$

where

$$\boldsymbol{G} = \left[\boldsymbol{N}(\xi_1,\eta_1) \quad \boldsymbol{N}(\xi_2,\eta_1) \quad \cdots \quad \boldsymbol{N}(\xi_{N_x},\eta_{N_y})\right]^{\mathrm{T}} \tag{4.14}$$

The DQH rules defined by Eq. (4.8) can be written in a compact form using a single index notation for grid points as

$$\left.\frac{\partial f}{\partial x}\right|_q = \sum_{p=1}^{M\times N} A_{qp}^{(1)} f_p, \quad \left.\frac{\partial f}{\partial y}\right|_q = \sum_{p=1}^{M\times N} B_{qp}^{(1)} f_p \tag{4.15}$$

or

$$\bar{\boldsymbol{f}}_x = \boldsymbol{A}\boldsymbol{f}, \quad \bar{\boldsymbol{f}}_y = \boldsymbol{B}\boldsymbol{f} \tag{4.16}$$

where $A_{qp}^{(1)}$ and $B_{qp}^{(1)}$ are respectively the weighting coefficients for the first-order derivatives with respect to the x and y coordinates for the irregular domain; f_p corresponds to the variables in Eq. $(4.10)_1$; the indexes p and q are respectively the indexes of the vectors of Eq. (4.8). The reformulated DQH rules as in Eq. (4.16) will be used to discretize the differentiations in the potential functional for in-plane vibrations.

4.1.2 *Triangular elements*

This section derives orthogonal polynomials on triangles to formulate the displacement field of hierarchical finite element method (HFEM) on triangles. The method of computing the orthogonal polynomials is also discussed. The displacement field u of a HFEM element on

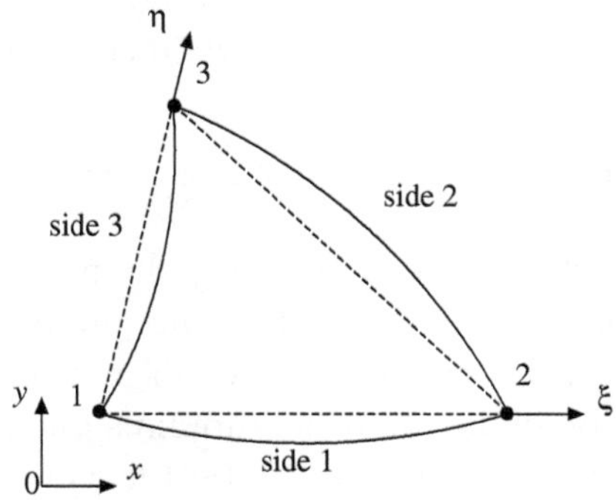

Fig. 4.4. A curved triangular element.

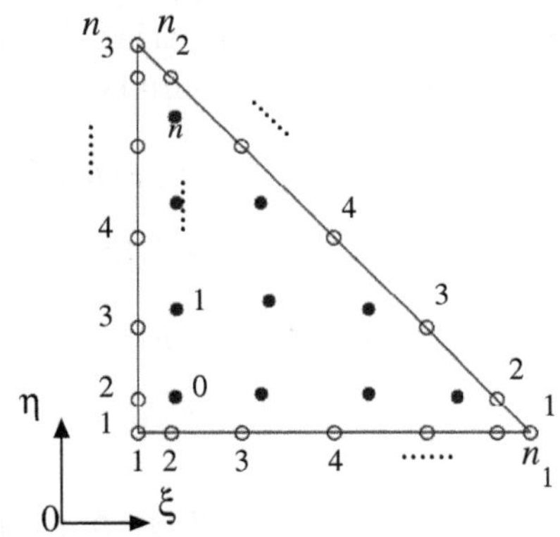

Fig. 4.5. Nodes on a unit triangular domain.

triangles can be given as

$$u[x(\xi,\eta),y(\xi,\eta),z(\xi,\eta)] = \sum_{v=1}^{3} V^v(\xi,\eta)u_v$$
$$+\sum_{e=1}^{3}\sum_{i=0}^{n_e-1} E_i^e(\xi,\eta)a_i^e + \sum_{p=0}^{n}\sum_{i=0}^{n-p} F_{pi}(\xi,\eta)c_{pi}^f \tag{4.17}$$

where $v = 1,2,3$ denotes the vertexes and $e = 1,2,3$ denotes the edges (Fig. 4.4), n_e $(e = 1,2,3)$ denotes the number of functions on each edge (Fig. 4.5), and n denotes the highest order of the polynomials on the triangle. The coefficient a_i^e denotes the hierarchical bases on the edges of the element and c_{pi}^f denotes the hierarchical bases on the face of the triangle. These coefficients do not have a clear physical meaning. The bases V^v defined on the three vertexes of the triangle can be given as

$$V^1 = 1 - \xi - \eta, \quad V^2 = \xi, \quad V^3 = \eta \tag{4.18}$$

The bases E_i^e and F_{pi} are respectively defined on the three edges and on the face of the triangle. There are many ways of defining hierarchical bases on the edges and face of a triangle. However, the best choice should be orthogonal polynomials on triangles, by which the resulting element matrices can be better conditioned [54].

The Jacobi polynomials are very general since they include the Legendre polynomials and the Chebyshev polynomials as special cases. Therefore, the usage of the Jacobi polynomials is a good choice of deriving orthogonal polynomials on triangles and tetrahedrons. The recursion formula of Jacobi polynomials is as follows:

$$a_{1n}P_{n+1}^{(\alpha,\beta)}(\xi) = (a_{2n} + a_{3n}\xi)P_n^{(\alpha,\beta)}(\xi) - a_{4n}P_{n-1}^{(\alpha,\beta)}(\xi) \tag{4.19}$$

where n is a non-negative integer, $\alpha > -1$, $\beta > -1$, and

$$P_0^{(\alpha,\beta)}(\xi) = 1, \quad P_1^{(\alpha,\beta)}(\xi) = \frac{1}{2}[\alpha - \beta + (\alpha + \beta + 2)\xi] \tag{4.20}$$

and

$$\begin{aligned} a_{1n} &= 2(n+1)(n+\alpha+\beta+1)(2n+\alpha+\beta) \\ a_{2n} &= (2n+\alpha+\beta+1)(\alpha^2-\beta^2) \\ a_{3n} &= (2n+\alpha+\beta)(2n+\alpha+\beta+1)(2n+\alpha+\beta+2) \\ a_{4n} &= 2(n+\alpha)(n+\beta)(2n+\alpha+\beta+2) \end{aligned} \tag{4.21}$$

Note that the Jacobi polynomials are defined on $[-1, 1]$. The Jacobi polynomials have the following orthogonal property:

$$\begin{aligned} &\int_{-1}^{1} (1-\xi)^{\alpha}(1+\xi)^{\beta} P_m^{(\alpha,\beta)}(\xi) P_n^{(\alpha,\beta)}(\xi) \\ &\quad = \delta_{mn} \frac{2^{\alpha+\beta+1}}{2n+\alpha+\beta+1} \frac{\Gamma(n+\alpha+1)\Gamma(n+\beta+1)}{n!\Gamma(n+\alpha+\beta+1)} \end{aligned} \tag{4.22}$$

where δ_{mn} is the Kronecker delta. In order to apply the Jacobi polynomials to triangles, coordinate transformation is necessary.

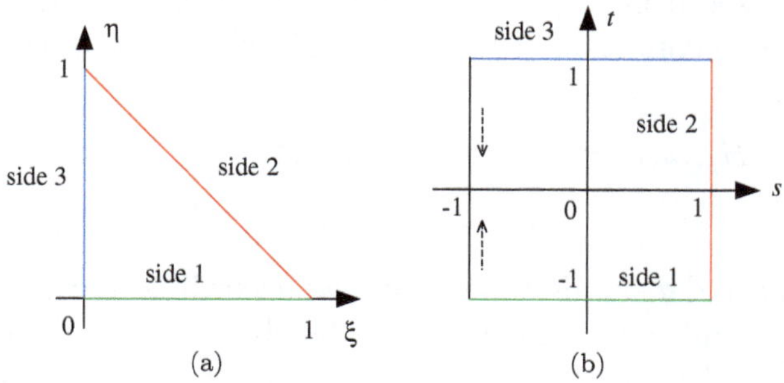

Fig. 4.6. A unit triangle and a unit square domain.

The coordinate transformation from a unit square domain to a unit triangular domain as shown in Fig. 4.6 can be given as

$$\xi = \frac{(1+s)(1-t)}{4}, \quad \eta = \frac{(1+s)(1+t)}{4} \tag{4.23}$$

The corresponding inverse transformation is

$$s = 2(\xi+\eta) - 1, \quad t = \frac{\eta-\xi}{\eta+\xi} \tag{4.24}$$

Figure 4.6 also shows the relationship between the edges of a unit triangle and a unit square. The edges with the same color or number correspond to each other. The left edge of the square domain is degenerated into a point. Note that the transformation method is not unique. One can degenerate some other edge to obtain a different method of transformation. The hierarchical shape function on the face of the triangle can be expressed as

$$F_{pi}(\xi,\eta) = (1-\xi-\eta)\xi\eta H_{pi}(s,t) \tag{4.25}$$

where $F_{pi}(i, p = 0, 1, 2, \ldots)$ are defined as face functions of the triangle and H_{pi} need to be determined using the Jacobi polynomials and the following orthogonal condition:

$$\int_S F_{pi}F_{qj}\mathrm{d}S = \int_0^1 \int_0^{1-\xi} F_{pi}F_{qj}\mathrm{d}\xi\mathrm{d}\eta = \delta_{pq}\delta_{ij}C_{pi} \tag{4.26}$$

where C_{pi} is a constant. The substitution of Eqs. (4.23) and (4.25) into Eq. (4.26) leads to

$$\int_S F_{pi}F_{qj}\mathrm{d}S = \int_{-1}^1 \int_{-1}^1 \frac{(1-s)^2(1+s)^5(1-t)^2(1+t)^2}{2^{13}} \times H_{pi}(s,t)H_{qj}(s,t)\mathrm{d}s\,\mathrm{d}t \tag{4.27}$$

One can see from Eqs. (4.22) and (4.27) that, in order to satisfy the orthogonal condition of Eq. (4.26), H_{pi} should be of the following form:

$$H_{pi}(s,t) = [(1-s)^a(1+s)^b P_i^{(2+2a,2b+5)}(s)] \times [(1-t)^c(1+t)^d P_p^{(2+2c,2+2d)}(t)] \tag{4.28}$$

where a, b, c and d should be determined according to conveniences in applications and requirements of the Jacobi polynomials; the indices p and i should have the relation $p+i \leq \gamma-3$, where $\gamma \geq 3$ is the expected maximum order of the bases if one wants to keep the bases below a certain maximum order. Using Eqs. (4.22), (4.26)–(4.28), the constant C_{pi} can be determined as

$$C_{pi} = \frac{\begin{array}{c}2^{2(a+b+c+d)}\Gamma(i+2a+3)\Gamma(i+2b+6)\\ \Gamma(p+2c+3)\Gamma(p+2d+3)\end{array}}{\begin{array}{c}2(i+a+b+4)(2p+2c+2d+5)i!p!\Gamma(i+2a+2b+8)\\ \Gamma(p+2c+2d+5)\end{array}} \tag{4.29}$$

For the C^0 problem considered in this work, we can take $a = c = d = 0$. In order to eliminate $\xi + \eta = (1+s)/2$ in the denominator of t in Eq. (4.24), one needs to take $b = p$. So, we have

$$H_{pi}(s,t) = (1+s)^p P_i^{(2,2p+5)}(s) P_p^{(2,2)}(t) \tag{4.30}$$

and

$$C_{pi} = \frac{2^{2p}(p+1)(p+2)(i+1)(i+2)}{2(2p+5)(p+4)(p+3)(i+p+4)(2p+i+6)(2p+i+7)} \tag{4.31}$$

The relation $\Gamma(n+1) = n!$ was used to derive Eq. (4.31) from Eq. (4.29). Equations (4.30) and (4.31) are the same as the corresponding equations presented in [54] except the different definition of the index p. The substitution of Eq. (4.24) into Eq. (4.30) leads to

$$H_{pi}(\xi,\eta) = 2^p P_i^{(2,2p+5)}[2(\eta+\xi)-1]\left[(\eta+\xi)^p P_p^{(2,2)}\left(\frac{\eta-\xi}{\eta+\xi}\right)\right] \tag{4.32}$$

Note that the coefficient 2^p in Eq. (4.32) may not be necessary. In this case, the coefficient 2^{2p} in Eq. (4.31) should be eliminated.

Orthogonal polynomials on the edges of a triangle can be derived in the same way as Eqs. (4.25)–(4.32). The edge functions on edges 1–3 can be respectively written as

$$E_i^1(\xi,\eta) = (1-\xi-\eta)\xi P_i^{(2,3)}[2(\xi+\eta)-1] \tag{4.33}$$

$$E_p^2(\xi,\eta) = \xi\eta(\eta+\xi)^p P_p^{(2,2)}\left(\frac{\eta-\xi}{\eta+\xi}\right) \tag{4.34}$$

$$E_i^3(\xi,\eta) = (1-\xi-\eta)\eta P_i^{(2,3)}[2(\xi+\eta)-1] \tag{4.35}$$

In order to eliminate the computation of $\xi+\eta=0$ that is in the denominator, make the following variable substitution:

$$\tilde{P}_n^{(\alpha,\beta)}(\tilde{s},\tilde{t}) = \tilde{t}^p P_n^{(\alpha,\beta)}(\tilde{s}/\tilde{t}), \quad \tilde{s}=\eta-\xi, \quad \tilde{t}=\eta+\xi \tag{4.36}$$

The substitution of Eq. (4.36) into Eq. (4.19) leads to

$$a_{1n}\tilde{P}_{n+1}^{(\alpha,\beta)}(\tilde{s},\tilde{t}) = (a_{2n}\tilde{t}+a_{3n}\tilde{s})\tilde{P}_n^{(\alpha,\beta)}(\tilde{s},\tilde{t}) - a_{4n}\tilde{t}^2\tilde{P}_{n-1}^{(\alpha,\beta)}(\tilde{s},\tilde{t}) \tag{4.37}$$

where

$$\tilde{P}_0^{(\alpha,\beta)}(\tilde{s},\tilde{t}) = 1, \quad \tilde{P}_1^{(\alpha,\beta)}(\tilde{s},\tilde{t}) = \frac{1}{2}[(\alpha-\beta)\tilde{t}+(\alpha+\beta+2)\tilde{s}] \tag{4.38}$$

It is obvious that Eq. (4.25) has zero values on the boundary of a triangular element. Figure 4.7 shows some face functions of a triangular element computed from Eq. (4.32). Some edge functions are plotted in Fig. 4.8. One can see from the above that the singular point $(\xi,\eta) = (0,0)$ of the collapsed coordinate system has been eliminated in computation. This should be a major difference between the present method and the spectral *hp* element method of Ref. [321] where such direct elimination does not seem to have been used and the Gauss–Radau quadrature method may be needed to avoid the calculation on this singular point.

4.1.3 *The Fekete points and the DQ rules on triangles*

In this section, a simple formula of direct map 1D non-uniform interpolation nodes to triangles that are close to the Fekete points

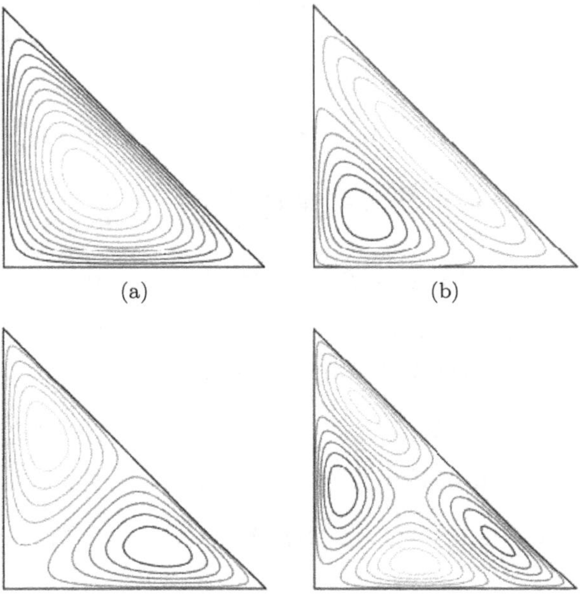

Fig. 4.7. Face functions inside a triangular element: (a) F_{00}; (b) F_{01}; (c) F_{10}; (d) F_{11}.

is presented. The non-uniform interpolation nodes together with the orthogonal polynomials on triangles presented above are used to solve the Fekete points through the Newton–Raphson method. Using the Fekete points and the orthogonal polynomials, interpolation bases and DQ rules on triangles are formulated. Similar to the one-dimensional (1D) Gauss–Lobatto–Chebyshev nodes that are transformed from 1D uniformly distributed nodes, this is also possible for triangles. Let the transformation formula from uniform nodes $(\xi, \eta, 1-\xi-\eta)$ to non-uniform nodes $(\lambda_1, \lambda_2, \lambda_3)$ be

$$\lambda_1 = f(\xi), \quad \lambda_2 = f(\eta), \quad \lambda_3 = f(1-\xi-\eta) \tag{4.39}$$

If $(\lambda_1, \lambda_2, \lambda_3)$ are also area coordinates, then $\lambda_i (i = 1, 2, 3)$ is symmetric with respect to each other and we can let

$$s(\lambda_1, \lambda_2, \lambda_3) = \frac{\lambda_1 + \alpha\lambda_1\lambda_2\lambda_3}{\lambda_1 + \lambda_2 + \lambda_3 + 3\alpha\lambda_1\lambda_2\lambda_3},$$

$$t(\lambda_1, \lambda_2, \lambda_3) = \frac{\lambda_2 + \alpha\lambda_1\lambda_2\lambda_3}{\lambda_1 + \lambda_2 + \lambda_3 + 3\alpha\lambda_1\lambda_2\lambda_3} \tag{4.40}$$

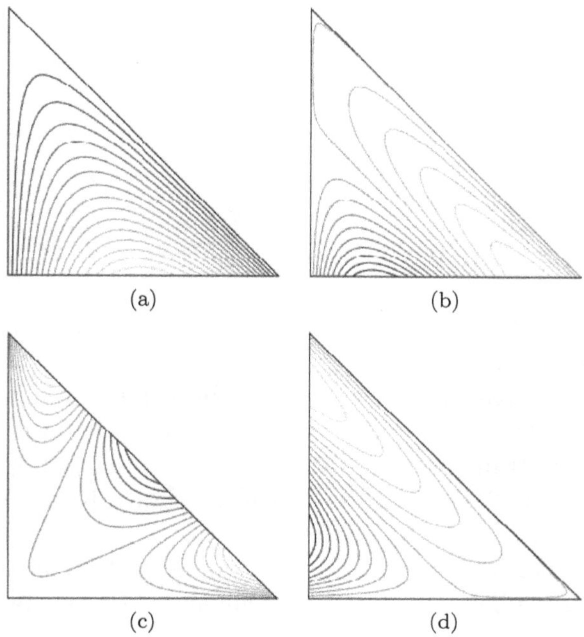

Fig. 4.8. Edge functions on a triangular element: (a) E_0^1; (b) E_1^1; (c) E_2^2; (d) E_1^3.

where α is an arbitrary constant. Note that λ_3 is not equal to $1 - \lambda_1 - \lambda_2$. It is equal to $f(1-\xi-\eta)$, where ξ and η are uniform nodes. If 1D Chebyshev–Lobatto nodes are used, $\lambda_1 = \sin(\pi\xi/2)$, $\lambda_2 = \sin(\pi\eta/2)$ and $\lambda_3 = \sin[\pi(1-\xi-\eta)/2]$. If 1D Gauss–Lobatto nodes are used, for example, let $\boldsymbol{\alpha} = [\alpha_1, \alpha_2, \ldots, \alpha_n]$ be the 1D Gauss–Lobatto nodes defined on [0, 1]. Let $\lambda_1 = \alpha_i$ and $\lambda_2 = \alpha_j$, then we have $\lambda_3 = \alpha_{n+2-i-j}$ instead of $\lambda_3 = 1 - \lambda_1 - \lambda_2$. It is empirically found that $\alpha = 3$ is a good choice. One can easily verify that s and t can also be used as coordinates of a unit triangle (Fig. 4.6(a)) as ξ and η. Especially,

$$s(\lambda_1, 0, \lambda_3) = \lambda_1, \quad t(0, \lambda_2, \lambda_3) = \lambda_2 \tag{4.41}$$

which means if Gauss–Lobatto nodes are used to generate the non-uniform nodes in a unit triangle, the generated nodes along the edges of the triangle are still the Gauss–Lobatto quadrature points. This property is important, since the Fekete points on the edge of a triangle have been proved to be the Gauss–Lobatto quadrature points [90]. Using 1D Gauss–Lobatto nodes and indexes of uniform nodes shown

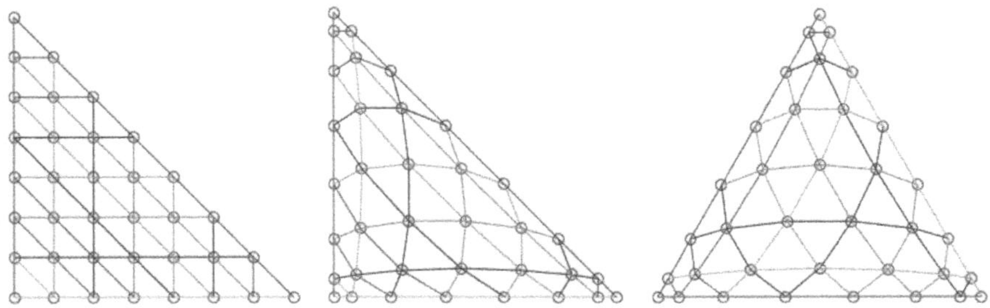

Fig. 4.9. A unit triangle with uniform (a) and non-uniform (b, c) nodes.

in Fig. 4.9(a), non-uniform nodes can be generated by Eq. (4.40) as shown in Fig. 4.9(b). The nodes on an equilateral triangle of Fig. 4.9(c) are transformed from the nodes in Fig. 4.9(b), by taking $\tilde{\eta} = \sqrt{3}\eta$, $\tilde{\xi} = 2\xi + \eta - 1$. They will be shown to be very close to the Fekete points in the following.

For convenience of optimizing the nodes and for convenience of analysis in the following, the displacement field given by Eq. (4.17) is written as

$$u[x(\xi,\eta), y(\xi,\eta), z(\xi,\eta)] = \boldsymbol{\psi}_{N_t}^{\mathrm{T}}(\xi,\eta)\tilde{\boldsymbol{u}} \tag{4.42}$$

where N_t is the total number of bases. Let the number of bases on all edges be $m+2$, i.e., let $n_1 = n_2 = n_3 = m+2$, where $m = n$ if face functions are included, otherwise, $m = 0$. Use $m+2$ 1D Gauss–Lobatto quadrature points to generate non-uniform nodes on the triangle through Eq. (4.40) and denote the resulted node vector as $(\boldsymbol{\xi}, \boldsymbol{\eta})$. So, we can construct a generalized Vandermonde matrix as

$$\boldsymbol{V} = \boldsymbol{\psi}_{N_t}^{\mathrm{T}}(\boldsymbol{\xi}, \boldsymbol{\eta}) \tag{4.43}$$

So, we have

$$\begin{aligned} u[x(\xi,\eta), y(\xi,\eta), z(\xi,\eta)] &= \boldsymbol{\psi}_{N_t}^{\mathrm{T}}(\xi,\eta)\boldsymbol{V}^{-1}u(\boldsymbol{\xi}, \boldsymbol{\eta}) \\ &= \sum_{i=1}^{N_t} L_i(\xi,\eta)u(\xi_i,\eta_i) \end{aligned} \tag{4.44}$$

where $L_i(\xi,\eta)$ is defined as a Lagrange function on the triangle at point (ξ_i,η_i). The condition numbers of the generalized Vandermonde matrix in Eq. (4.43) can be approximately given by $e^{0.45m}$, where

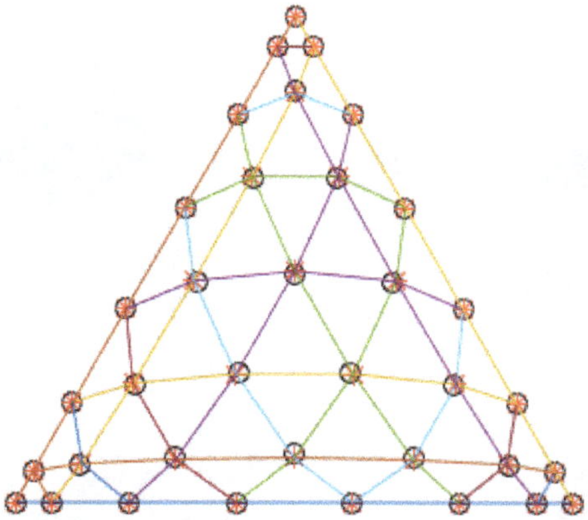

Fig. 4.10. Comparison of the solved Fekete points (red *) and the explicitly generated nodes of this work (black ◯).

$(m+1)$ is the order of the element. The Fekete points $(\boldsymbol{\xi}, \boldsymbol{\eta})$ should make

$$|L_i(\xi, \eta)| \leq 1 \text{ and } L_i(\xi_i, \eta_i) = 1 \tag{4.45}$$

This relation usually can be replaced by

$$\frac{\partial L_i(\boldsymbol{\xi}, \boldsymbol{\eta})}{\partial \xi_i} = 0, \quad \frac{\partial L_i(\boldsymbol{\xi}, \boldsymbol{\eta})}{\partial \eta_i} = 0 \tag{4.46}$$

Since the nodes on the edges of the triangle have already been chosen as Gauss–Lobatto nodes and the Fekete points on the edges of a triangle have been proved to be Gauss–Lobatto quadrature points, only the nodes inside the triangle need to be optimized. Since the initial guess of the Fekete points is very close to the real Fekete points, the Newton–Raphson method that is more efficient than the steepest ascent algorithm can be used to solve the system of nonlinear equations. In this work, to get Fekete points with degree 30 only needs a few minutes through MATLAB codes on a DELL Precision M6800 mobile workstation. Figure 4.10 compares the solved Fekete points with the explicitly generated nodes through Eq. (4.40). One can see that they are very close to each other. Figure 4.11 plots four Lagrange bases obtained through the Fekete points and the orthogonal polynomials of this work.

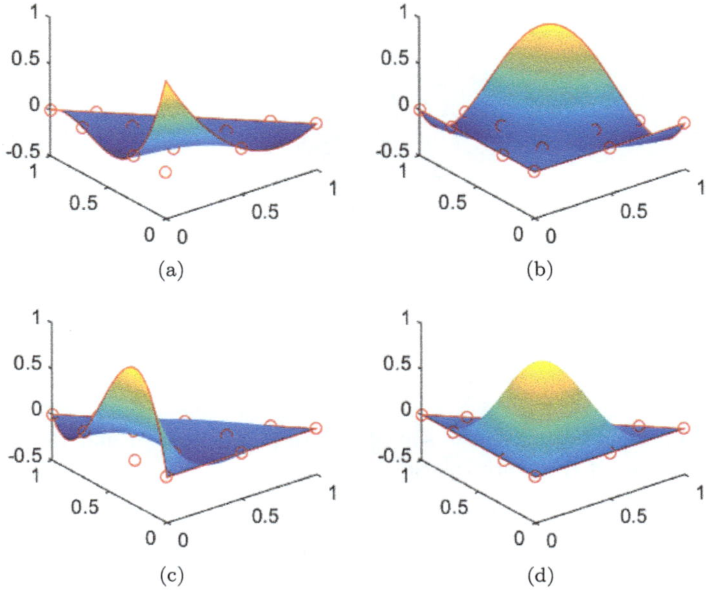

Fig. 4.11. Four interpolation bases on a unit triangle.

4.1.4 *The differential quadrature hierarchical rules*

Using the Fekete points presented above, one can also construct triangular element with node distribution as shown in Fig. 4.5, namely, the numbers of nodes on different edges of a triangle can be different and do not relate with (this word only refers to the node numbers) the nodes inside the triangle. The bases shown in Fig. 4.11 used the node distribution shown in Fig. 4.5. Take the derivatives of Eq. (4.44) with respect to the parametric coordinates and then substitute the Fekete points into it, one gets the differential quadrature (DQ) rules on a unit triangle with non-uniform node distribution:

$$\frac{\partial u(\bar{\boldsymbol{\xi}}, \bar{\boldsymbol{\eta}})}{\partial \xi} = \boldsymbol{A}^{(1)} u(\boldsymbol{\xi}, \boldsymbol{\eta}), \quad \frac{\partial u(\bar{\boldsymbol{\xi}}, \bar{\boldsymbol{\eta}})}{\partial \eta} = \boldsymbol{B}^{(1)} u(\boldsymbol{\xi}, \boldsymbol{\eta}) \tag{4.47}$$

The high-order derivatives can be obtained as

$$\frac{\partial^r u(\bar{\boldsymbol{\xi}}, \bar{\boldsymbol{\eta}})}{\partial \xi^r} = \boldsymbol{A}^{(r)} u(\boldsymbol{\xi}, \boldsymbol{\eta}), \quad \frac{\partial^s u(\bar{\boldsymbol{\xi}}, \bar{\boldsymbol{\eta}})}{\partial \eta^s} = \boldsymbol{B}^{(s)} u(\boldsymbol{\xi}, \boldsymbol{\eta}),$$

$$\frac{\partial^{r+s} u(\bar{\boldsymbol{\xi}}, \bar{\boldsymbol{\eta}})}{\partial \xi^r \partial \eta^s} = \boldsymbol{F}^{(r+s)} u(\boldsymbol{\xi}, \boldsymbol{\eta}) \tag{4.48}$$

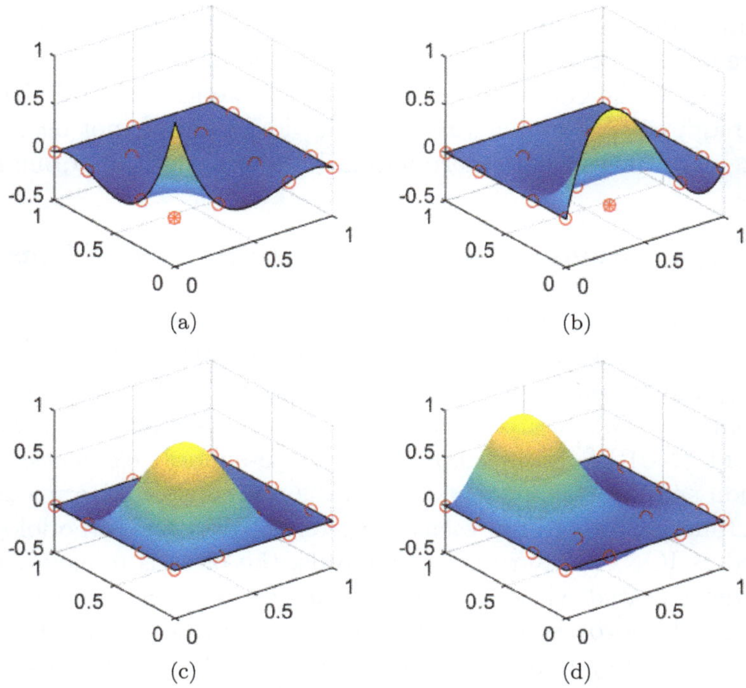

Fig. 4.12. Four interpolation bases on a unit quadrangle.

Note that the evaluation points $(\bar{\boldsymbol{\xi}}, \bar{\boldsymbol{\eta}})$ and the nodes $(\boldsymbol{\xi}, \boldsymbol{\eta})$ were taken as different. In this case, the following interpolation relation may be needed:

$$\boldsymbol{u}(\bar{\boldsymbol{\xi}}, \bar{\boldsymbol{\eta}}) = \boldsymbol{G}\boldsymbol{u}(\boldsymbol{\xi}, \boldsymbol{\eta}) \quad \text{or} \quad \bar{\boldsymbol{u}} = \boldsymbol{G}\boldsymbol{u} \tag{4.49}$$

If the evaluation points and the nodes are taken as the same, the following recurrence formulations hold:

$$\boldsymbol{A}^{(r)} = \boldsymbol{A}^{(1)}\boldsymbol{A}^{(r-1)}, \quad \boldsymbol{B}^{(s)} = \boldsymbol{B}^{(1)}\boldsymbol{B}^{(s-1)}, \quad \boldsymbol{F}^{(r+s)} = \boldsymbol{A}^{(r)}\boldsymbol{B}^{(s)} \tag{4.50}$$

The direct usage of Eq. (4.48) is usually more efficient and accurate than Eq. (4.50). However, it is obvious that the usage of Eq. (4.50) is simpler than Eq. (4.47). Using the orthogonal polynomials on a quadrilateral domain [35], one can define DQ rules as Eqs. (4.47)–(4.50) on the domain with node distribution on the boundary of the element similar to Fig. 4.5 through the same procedure as Eqs. (4.42)–(4.44). Since the Fekete points on a quadrilateral region is the Gauss–Lobatto integration nodes, the optimization with requirements similar to Eqs. (4.45) and (4.46) is not necessary. Figure 4.12 shows some interpolation bases on a unit quadrangle.

4.2 Applications of the DQHFEM to Plane Problems

This section aims to solve the problem of trimmed NURBS elements not being watertight [322, 323] by using the DQHM and then analyze the in-plane vibration of plates using the DQHFEM. The isogeometric analysis (IGA) matched using exact CAD geometry by a coarse mesh of "NURBS (non-uniform rational B-splines) elements." The solution space for dependent variables is represented in terms of the same functions which represent the geometry, so the methodology is named as IGA. For purposes of analysis, the basis is refined and/or its order elevated without changing the geometry or its parameterization. This process does not require any further communication with the CAD system and therefore IGA was expected to facilitate more widespread adoption in industry. However, trimmed NURBS are not watertight [322, 323], even NURBS elements with the same number of nodes may not be compatible with each other, as shown in Fig. 4.13(a). As can be seen from Eq. (4.3), the DQHM element is not based on tensor product of 1D basis, and the number of nodes on edges of a DQHM element and inside the DQHM element has no influence with each other and therefore can be used to solve the problem of trimmed NURBS elements.

The basis of an NURBS surface is an interpolator on the four corners of the surface, so NURBS elements can be compatible on

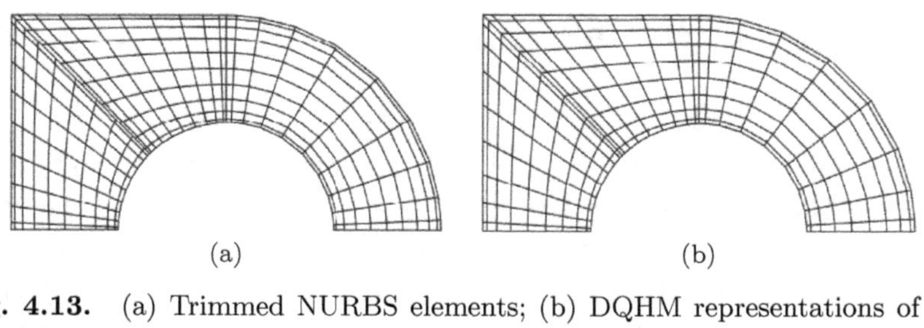

Fig. 4.13. (a) Trimmed NURBS elements; (b) DQHM representations of the elements.

the four corner points. This information can be used to transform NURBS elements into DQH elements, since two neighboring NURBS elements only have two common corners shared by their common edges. For plane surfaces, the hierarchical basis in Eq. (4.3) is not necessary, so Eq. (4.3) can be written as

$$f[x(\xi,\eta),y(\xi,\eta)]=\sum_{k=1}^{K}S_k(\xi,\eta)f_k \tag{4.51}$$

which means all bases are interpolators and only boundary points are needed to approximate a domain. Equation (4.51) can be regarded as a blending function surface expressed by Lagrange functions. The NURBS elements shown in Fig. 4.13(a) and transformed into DQH elements by Eq. (4.51) are shown in Fig. 4.13(b). One can see that the DQH elements are compatible with each other. During the transformation, two neighboring elements only used points from one of their common edges.

The solution space of IGA for dependent variables is the same as the functions which represent the geometry. Therefore, analysis similar to IGA can be carried out by first transforming the NURBS elements into DQH elements and then analyzing the elastic field by the DQHM. The DQHM in weak forms was named as a differential quadrature hierarchical finite element method (DQHFEM) [34]. The strain energy (U) and work potential (W) of an elastic body can be given by

$$\begin{aligned}U&=\frac{1}{2}\iiint_V\boldsymbol{\varepsilon}^T\boldsymbol{\sigma}dV=\frac{1}{2}\iiint_V\boldsymbol{\varepsilon}^T\boldsymbol{D}\boldsymbol{\varepsilon}dV\\W&=-\iiint_V\boldsymbol{u}^T(-\rho\frac{\partial^2\boldsymbol{u}}{\partial t^2})dV-\iiint_V\boldsymbol{u}^T\boldsymbol{q}dV\end{aligned} \tag{4.52}$$

where ε is the strain field vector, σ the stress field vector, $\boldsymbol{D}$ the material matrix, $\boldsymbol{u}$ the displacement field vector, $\boldsymbol{q}$ the distributed force vector, ρ the volume density and t the time. The stress–strain relations and strain–displacement relations of plane stress problems are

$$\begin{Bmatrix}\sigma_x\\\sigma_y\\\tau_{xy}\end{Bmatrix}=\frac{E}{1-\upsilon^2}\begin{bmatrix}1&\upsilon&0\\\upsilon&1&0\\0&0&(1-\upsilon)/2\end{bmatrix}\begin{Bmatrix}\varepsilon_x\\\varepsilon_y\\\gamma_{xy}\end{Bmatrix} \tag{4.53}$$

$$\begin{bmatrix} \varepsilon_x \\ \varepsilon_y \\ \gamma_{xy} \end{bmatrix} = \begin{bmatrix} \partial/\partial x & 0 \\ 0 & \partial/\partial y \\ \partial/\partial y & \partial/\partial x \end{bmatrix} \begin{bmatrix} u \\ v \end{bmatrix} \tag{4.54}$$

where E is the Young modulus and υ is the Poisson ratio. For plane strain problem, they should be replaced by $E = E/(1-\upsilon^2)$ and $\upsilon = \upsilon/(1-\upsilon)$. The stress–strain relations for orthotropic materials are

$$\begin{Bmatrix} \sigma_x \\ \sigma_y \\ \tau_{xy} \end{Bmatrix} = \begin{bmatrix} \bar{Q}_{11} & \bar{Q}_{12} & \bar{Q}_{16} \\ \bar{Q}_{12} & \bar{Q}_{22} & \bar{Q}_{26} \\ \bar{Q}_{16} & \bar{Q}_{26} & \bar{Q}_{66} \end{bmatrix} \begin{Bmatrix} \varepsilon_x \\ \varepsilon_y \\ \gamma_{xy} \end{Bmatrix} = \boldsymbol{D} \begin{Bmatrix} \varepsilon_x \\ \varepsilon_y \\ \gamma_{xy} \end{Bmatrix} \tag{4.55}$$

where

$$\boldsymbol{D} = \boldsymbol{T} \begin{bmatrix} Q_{11} & Q_{12} & 0 \\ Q_{12} & Q_{22} & 0 \\ 0 & 0 & Q_{66} \end{bmatrix} \boldsymbol{T}^{\mathrm{T}} \tag{4.56}$$

$$\boldsymbol{T} = \begin{bmatrix} \cos^2\theta & \sin^2\theta & -2\sin\theta\cos\theta \\ \sin^2\theta & \cos^2\theta & 2\sin\theta\cos\theta \\ \sin\theta\cos\theta & -\sin\theta\cos\theta & \cos^2\theta - \sin^2\theta \end{bmatrix} \tag{4.57}$$

where θ is the angle between x-axis and the first material direction, and Q_{ij}, called plane stress-induced stiffness [324], are given by

$$Q_{11} = \frac{E_1}{1-\upsilon_{12}\upsilon_{21}}, \quad Q_{12} = \frac{\upsilon_{12}E_2}{1-\upsilon_{12}\upsilon_{21}},$$
$$Q_{22} = \frac{E_2}{1-\upsilon_{12}\upsilon_{21}}, \quad Q_{66} = G_{12} \tag{4.58}$$

where E_1 and E_2 are Young's moduli in 1 and 2 material directions, respectively, υ_{ij} are Poisson's ratio with relation $\upsilon_{12}E_2 = \upsilon_{21}E_1$; and G_{12} is shear moduli in 1–2 plane. By inserting the DQH rules for the first-order partial derivatives defined by Eq. (4.17) into the strain–displacement relations Eq. (4.54), one can obtain the corresponding nodal strain vector

$$\begin{bmatrix} \varepsilon_x \\ \varepsilon_y \\ \gamma_{xy} \end{bmatrix} = \begin{bmatrix} \boldsymbol{A}^{(1)} & \mathbf{0} \\ \mathbf{0} & \boldsymbol{B}^{(1)} \\ \boldsymbol{B}^{(1)} & \boldsymbol{A}^{(1)} \end{bmatrix} \begin{bmatrix} \boldsymbol{u} \\ \boldsymbol{v} \end{bmatrix} \tag{4.59}$$

or

$$\boldsymbol{\varepsilon} = \boldsymbol{HU} \tag{4.60}$$

where $\boldsymbol{U} = [\boldsymbol{u}; \boldsymbol{v}]$. The grid points used to compute the weighting coefficient matrices $\boldsymbol{A}^{(1)}$ and $\boldsymbol{B}^{(1)}$ are the Gauss–Lobatto–Legendre (GLL) nodes [34], so that further interpolations are not necessary. The GLL quadrature rule for a function $f(x, y)$ can be expressed as

$$\iint_S f(x,y)\mathrm{d}x\mathrm{d}y = \sum_{i=1}^{N_x}\sum_{j=1}^{N_y} |\boldsymbol{J}|_{ij} C_i^x C_j^y f(x_i, y_j) = \boldsymbol{Cf} \tag{4.61}$$

where

$$\boldsymbol{C} = \mathrm{diag}(\boldsymbol{C}_1,\ \boldsymbol{C}_2, \ldots, \boldsymbol{C}_{N_y}) \tag{4.62}$$

is a $(N_x \times N_y) \times (N_x \times N_y)$ matrix, $\boldsymbol{C}_j = C_j^y \mathrm{diag}(|\boldsymbol{J}|_{1j}C_1^x, \ldots, |\boldsymbol{J}|_{N_x j}C_{N_x}^x)_{N_x \times N_x}$, C_i^x and $C_j^y (i = 1, 2, \ldots, N_x;\ j = 1, 2, \ldots, N_y)$ are respectively the GLL weights with respect to x and y directions and $|\boldsymbol{J}|_{ij}$ are discrete values of the determinant $|\boldsymbol{J}|$ of the Jacobian given by Eq. (4.9). See Ref. [34] for the method of computing GLL points. Discretizing Eq. (4.52) by the DQH rules and GLL integration rules leads to

$$\begin{aligned} U &= \frac{1}{2}\boldsymbol{U}^T\boldsymbol{H}^T\bar{\boldsymbol{D}}\boldsymbol{HU} \\ W &= \rho h \boldsymbol{U}^T \begin{bmatrix} \boldsymbol{G}^T\boldsymbol{CG} & \mathbf{0} \\ \mathbf{0} & \boldsymbol{G}^T\boldsymbol{CG} \end{bmatrix} \ddot{\boldsymbol{U}} - \boldsymbol{U}^T\boldsymbol{R} \end{aligned} \tag{4.63}$$

where h is the thickness and

$$\bar{\boldsymbol{D}} = C \begin{bmatrix} \boldsymbol{C} & \upsilon\boldsymbol{C} & \mathbf{0} \\ \upsilon\boldsymbol{C} & \boldsymbol{C} & \mathbf{0} \\ \mathbf{0} & \mathbf{0} & \boldsymbol{C}(1-\upsilon)/2 \end{bmatrix}$$

$$\text{or} \quad \bar{\boldsymbol{D}} = h \begin{bmatrix} \bar{Q}_{11}\boldsymbol{C} & \bar{Q}_{12}\boldsymbol{C} & \bar{Q}_{16}\boldsymbol{C} \\ \bar{Q}_{12}\boldsymbol{C} & \bar{Q}_{22}\boldsymbol{C} & \bar{Q}_{26}\boldsymbol{C} \\ \bar{Q}_{16}\boldsymbol{C} & \bar{Q}_{26}\boldsymbol{C} & \bar{Q}_{66}\boldsymbol{C} \end{bmatrix} \tag{4.64}$$

where $C = Eh/(1 - \upsilon^2)$ is the stretch rigidity. Using the variational principle, one obtains the equilibrium equation as

$$\boldsymbol{KU} + \boldsymbol{M}\ddot{\boldsymbol{U}} = \boldsymbol{R} \tag{4.65}$$

where the stiffness matrix $\boldsymbol{K}$, mass matrix $\boldsymbol{M}$ and force vector $\boldsymbol{R}$ are

$$\begin{aligned} \boldsymbol{K} &= \boldsymbol{H}^T \bar{\boldsymbol{D}} \boldsymbol{H} \\ \boldsymbol{M} &= \rho h \begin{bmatrix} \boldsymbol{G}^{\mathrm{T}}\boldsymbol{C}\boldsymbol{G} & \boldsymbol{0} \\ \boldsymbol{0} & \boldsymbol{G}^{\mathrm{T}}\boldsymbol{C}\boldsymbol{G} \end{bmatrix} \\ \boldsymbol{R} &= h \begin{bmatrix} \boldsymbol{G}_\Gamma^T \boldsymbol{C}_\Gamma \boldsymbol{q}_u \\ \boldsymbol{G}_\Gamma^T \boldsymbol{C}_\Gamma \boldsymbol{q}_v \end{bmatrix} \end{aligned} \tag{4.66}$$

where $\boldsymbol{q}_u$ and $\boldsymbol{q}_v$ are the nodal load vectors with respect to x and y directions, respectively. $\boldsymbol{G}_\Gamma^T$ and $\boldsymbol{C}_\Gamma$ in Eq. (4.66) are the interpolation and integration matrices on the force boundaries. Note that only the interpolation part of Eq. (4.3) is used to approximate the plane surface as shown in Eq. (4.51), so the approximation relations given by Eq. (4.13) and the DQH rules given by Eq. (4.16) can be written in two parts as

$$\begin{aligned} \bar{\boldsymbol{f}} &= \begin{bmatrix} \boldsymbol{G}_1 & \boldsymbol{G}_2 \end{bmatrix} \boldsymbol{f} \\ \bar{\boldsymbol{f}}_x &= [\boldsymbol{A}_1 \quad \boldsymbol{A}_2]\boldsymbol{f}, \quad \bar{\boldsymbol{f}}_y = [\boldsymbol{B}_1 \quad \boldsymbol{B}_2]\boldsymbol{f} \end{aligned} \tag{4.67}$$

where $\boldsymbol{A}_1^{(1)}$, $\boldsymbol{B}_1^{(1)}$ and $\boldsymbol{G}_1$ correspond to the interpolation part as shown in Eq. (4.51) and $\boldsymbol{A}_2^{(1)}$, $\boldsymbol{B}_2^{(1)}$ and $\boldsymbol{G}_2$ are those of the enriched hierarchical part. Thus, the weighting matrix $\boldsymbol{H}$ in Eq. $(4.63)_1$ can be written as

$$\boldsymbol{H} = [\boldsymbol{H}_1 \quad \boldsymbol{H}_2] \tag{4.68}$$

where

$$\boldsymbol{H}_i = \begin{bmatrix} \boldsymbol{A}_i^{(1)} & \mathbf{0} \\ \mathbf{0} & \boldsymbol{B}_i^{(1)} \\ \boldsymbol{B}_i^{(1)} & \boldsymbol{A}_i^{(1)} \end{bmatrix} \tag{4.69}$$

and $i = 1, 2$. Just like node numbering, the displacement vector corresponding to Eq. (4.68) is $\boldsymbol{U} = [\boldsymbol{U}_1; \boldsymbol{U}_2]$ where $\boldsymbol{U} = [\boldsymbol{u}_i; \boldsymbol{v}_i]$. So, the stiffness matrix in Eq. $(4.66)_1$ can be written as

$$\boldsymbol{K} = \begin{bmatrix} \boldsymbol{H}_1^T \bar{\boldsymbol{D}} \boldsymbol{H}_1 & \boldsymbol{H}_1^T \bar{\boldsymbol{D}} \boldsymbol{H}_2 \\ \boldsymbol{H}_2^T \bar{\boldsymbol{D}} \boldsymbol{H}_1 & \boldsymbol{H}_2^T \bar{\boldsymbol{D}} \boldsymbol{H}_2 \end{bmatrix} \tag{4.70}$$

The derivation of the embedded mass matrix is similar to Eq. (4.70) and so they are not presented here. Note that the coefficients of the embedded mass matrix should be rearranged to correspond to the displacement vector $\boldsymbol{U} = [\boldsymbol{U}_1; \boldsymbol{U}_2]$.

In order to carry out an analysis similar to IGA by the DQHFEM, the NURBS elements are first discretized by the sampling points on the boundary of elements, namely, only $\boldsymbol{G}_1$ of Eq. $(4.67)_1$ is used in this process. Then the NURBS elements are transformed into DQH elements so that the mismatch problem of trimmed NURBS elements can be overcome. The interpolation part $\boldsymbol{G}_1$ of Eq. $(4.66)_1$ will also be used to analyze the elastic field, so $\boldsymbol{A}_1^{(1)}$ and $\boldsymbol{B}_1^{(1)}$ should be computed together with $\boldsymbol{G}_1$. In order to carry out elastic analysis, $\boldsymbol{C}$, $\boldsymbol{A}_2^{(1)}$, $\boldsymbol{B}_2^{(1)}$ and $\boldsymbol{G}_2$ should be computed and substituted into Eq. (4.70). Flowchart of the procedure of analysis by the DQHFEM is shown in Fig. 4.14. The entire stiffness and mass matrices need not to be recalculated by using the present method if one wants to increase the polynomial order. The same process as Eq. (4.70) can be used to update the stiffness and mass matrices.

4.3 Results for Free Vibrations

4.3.1 *Rectangular elements*

This section aims to present DQHFEM results for in-plane vibration of isotropic plates in curvilinear domains and validate the high accuracy of the DQHFEM. The mapping of a square parent domain to the

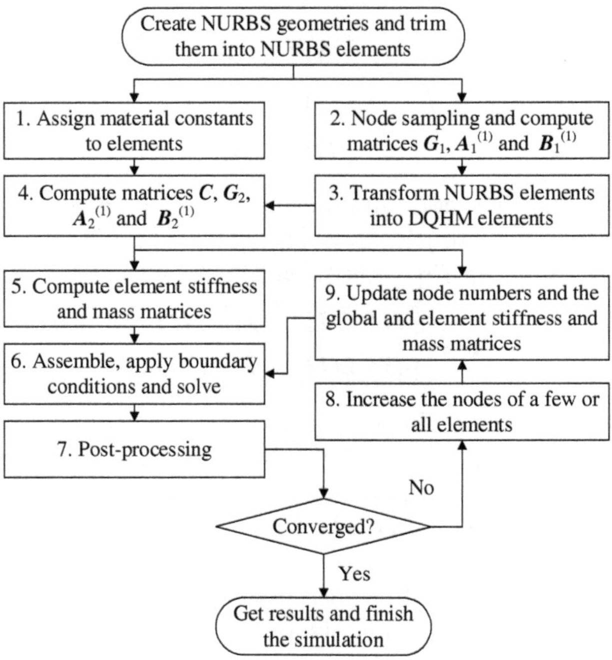

Fig. 4.14. Flowchart of procedure of analysis by the DQHFEM.

interest domains are carried out via either the blending functions constructed by analytical expressions as used in [34] or those constructed by Lagrange functions as shown in Eq. (4.51). The two methods do not have any difference for all the digits used for comparison, so they are not compared with each other. For numerical convenience, in Tables 4.1–4.3 and 4.5–4.7, the number of sampling points on the edges in the ξ- and η-directions is taken as the same and is denoted by N_e, and the number of hierarchical shape functions in the two directions is also taken as the same and is denoted by N_h. The natural frequencies are expressed by the same frequency parameters as used in references, which is done not to compare with round-off errors that are unrealistic. Being different from out-plane vibrations, in-plane vibrations have two distinct simply supported boundary conditions, where the normal stress and tangential displacement are zeros (S1) or normal displacement and tangential stress are zeros (S2).

In-plane vibration of a square plate is analyzed first in Table 4.1 for the plate with two combinations of simply supported, clamped

Table 4.1. Comparison of the dimensionless frequency parameters $\gamma = (\omega a/\pi)\sqrt{\rho/G}$ for a square plate.

B.C.	(N_e, N_h)	Mode sequence number							
		1	2	3	4	5	6	7	8
S1–S1–S1–S1	(10,8)	0.9995	0.9995	1.4142	1.9980	2.0000	2.2353	2.2353	2.3897
	(13, 10)	0.9999	0.9999	1.4142	1.9996	2.0000	2.2358	2.2360	2.3902
	(10, 8)[a]	1.0000	1.0000	1.4142	2.0000	2.0000	2.2361	2.2361	2.3905
	(25, 20)	1.0000	1.0000	1.4142	2.0000	2.0000	2.2361	2.2361	2.3905
	Exact [325]	1	1	1.4142	2	2	2.2361	2.2361	2.3905
S1–C–S2–F	(5, 3)	0.8600	0.9468	1.4309	1.7482	2.2306	2.4229	2.4897	2.7977
	(8, 3)	0.8600	0.9468	1.4276	1.7421	2.2273	2.3529	2.3627	2.7935
	(10, 3)	0.8600	0.9468	1.4276	1.7421	2.2273	2.3529	2.3626	2.7935
	(10, 5)	0.8600	0.9468	1.4258	1.7420	2.2193	2.2978	2.3610	2.7525
	(10, 8)	0.8600	0.9468	1.4258	1.7420	2.2193	2.2967	2.3610	2.7517
	(25, 20)	0.8600	0.9468	1.4258	1.7420	2.2193	2.2967	2.3610	2.7517
	Exact [325]	0.8600	0.9468	1.4258	1.7420	2.2193	2.2967	2.3610	2.7517

[a]The influence of corner boundary conditions are taken into account.

Table 4.2. Comparison of the natural frequencies of a clamped circular plate.

(N_e, N_h)	Mode sequence number								
	1	2	3	4	5	6	7	8	9
(10, 8)	3361.7	3834.9	5219.5	5384.1	6625.6	6774.6	6946.4	7035.9	8181.60
(13, 10)	3361.7	3834.9	5219.4	5382.9	6625.6	6764.2	6938.8	7021.8	8132.94
(16, 14)	3361.7	3834.9	5219.4	5382.9	6625.6	6763.8	6938.5	7021.3	8130.48
(25, 20)	3361.7	3834.9	5219.4	5382.9	6625.6	6763.8	6938.5	7021.3	8130.48
Exact [331]	3362	3835	5219	5383	6626	6764	6939	7021	8130
FEM [331]	3363.6	3836.4	5217.5	5380.5	6624	6749.3	6929	7019.3	8093

(C) and free (F) boundaries, since there are plenty of such results in the literatures [325, 326]. A plate with S1–C–S2–F boundary condition means that the plate has S1–S2 at x-direction and C–F at y-direction. The natural frequencies in Table 4.1 are expressed by the same frequency parameter γ used in [325, 326]. As can be seen from Table 4.1, for the S1-S1-S1-S1 plate, the DQHFEM results are very close to the exact solution of [325] with sampling points

Table 4.3. Comparison of frequency parameters $\Omega = \omega a\sqrt{\rho/E}$ for the in-plane vibrating rhombic plate with the Poisson's ratio $\upsilon = 0.3$ (Fig. 4.17).

		$\alpha = 0.0°$		$\alpha = 30.0°$		$\alpha = 60.0°$	
B.C.	Mode	DQHFEM	Spectral [332]	DQHFEM	Spectral [332]	DQHFEM	Spectral [332]
CCCC	1	3.7269	3.7269	3.8498	3.8498	5.5368	5.5368
	2	3.7269	3.7269	4.4856	4.4856	7.5135	7.5134
	3	4.4395	4.4395	5.0125	5.0125	7.5420	7.5420
	4	5.4361	5.4361	5.6950	5.6950	8.0486	8.0486
	5	6.1415	6.1415	6.5249	6.5249	9.3357	9.3357
	6	6.1790	6.1790	6.7838	6.7838	9.5452	9.5452
CFCF	1	1.7748	1.7747	1.9968	1.9965	3.1393	3.1361
	2	3.1635	3.1635	3.4586	3.4586	4.9110	4.9110
	3	3.2711	3.2711	3.5192	3.5191	4.9655	4.9654
	4	3.5125	3.5125	3.8485	3.8485	5.1433	5.1440
	5	3.9238	3.9238	4.3172	4.3170	5.9058	5.9043
	6	4.0908	4.0908	4.3554	4.3553	6.4792	6.4774

$(N_e, N_h) = (13, 10)$ but not the same as the exact results for the digits used for comparison. This is caused by the fact that the corner point belonging to both neighboring S1 edges is clamped. This fact should be noted in all numerical methods. When the number of sampling points is big enough, this influence becomes negligible, as can be seen from Table 4.1. If the corner points for the S1–S1–S1–S1 plate are taken as clamped, the DQHFEM results with $(N_e, N_h) = (10, 8)$ agree with the exact results for all digits used for comparisons, as shown in Table 4.1. For the S1-C–S2-F plate that considered the influence of boundary conditions on corner points, the DQHFEM results converge very fast. One can see that the DQHFEM is similar to the fixed interface mode synthesis method that can analyze a structure using a few nodes on the boundary of each element but several clamped modes inside the elements, however the DQHFEM does not need modal analysis and can provide results with high accuracy.

In order to show the convergence property of the entire spectrum and to the influence of boundary conditions on corner points, the S1S1S1S1 plate is further analyzed. The DQHFEM results with mesh size $(N_e, N_h) = (30, 28)$ are compared with those of FEM obtained by the authors using commercial software ANSYS. The material constants of the plate are Young's modulus of 71 GPa, a density of $2700\,\mathrm{kg/m^3}$ and Poisson's ratio of 0.3. The elements used in ANSYS are SHELL181 quadrilateral elements. The element length of FEM is $l = 1/30$ m which is the same as DQHFEM. The exact results are obtained from the method of [325–327]. Relative error of FEM and DQHFEM are plotted in Fig. 4.15(a). One can see that the DQHFEM results are better than the FEM results for the first 80% frequencies. Especially, the first 20% frequencies of the DQHFEM are of high accuracy. In Fig. 4.15(a), DQHFEM* indicates the results that the influence of corner boundary conditions is not taken into account. The curve almost coincides with the one considering corner boundary conditions. So, the corner boundary conditions have little influence over the accuracy of the entire spectra.

In order to compare the DQHFEM results with the excellent results of IGA using NURBS basis, both the S1S1S1S1 square plate analyzed above and a simple fixed rod are analyzed and the relative errors for the entire spectra are respectively plotted in Figs. 4.15 and 4.16. As can be seen from Figs. 4.15(b) and 4.16, the DQHFEM results are very accurate compared with the IGA results for the first 20% natural frequencies of the S1S1S1S1 square plate and the first 60% natural frequencies of the fixed rod. The FEM and quadratic NURBS results in Figs. 4.15 and 4.16 are obtained by the authors of this work. The FEM and NURBS results for the fixed rod have the same distribution as in [328–330]. The NURBS results for in-plane vibration cannot be found in the literatures and are presented by the authors of this work for the first time. The computation of NURBS elements should take care of two points: (1) the integration should be done in each knot subdivision instead of the entire interval like the DQHFEM; (2) the knots should be equally spaced instead of non-equally spaced as in the DQHFEM. Otherwise, the IGA using NURBS cannot provide good results for high-order frequencies, which is meaningful for explicit transient analysis [328]. The relative error of the high-order frequencies of the DQHFEM can also be reduced through mesh refinement, as can be seen from Fig. 4.6(c), where the grid size of each element is $(N_e, N_h) = (6, 4)$. The IGA using NURBS like the DQM or DQFEM does not have the hierarchical property of the DQHFEM.

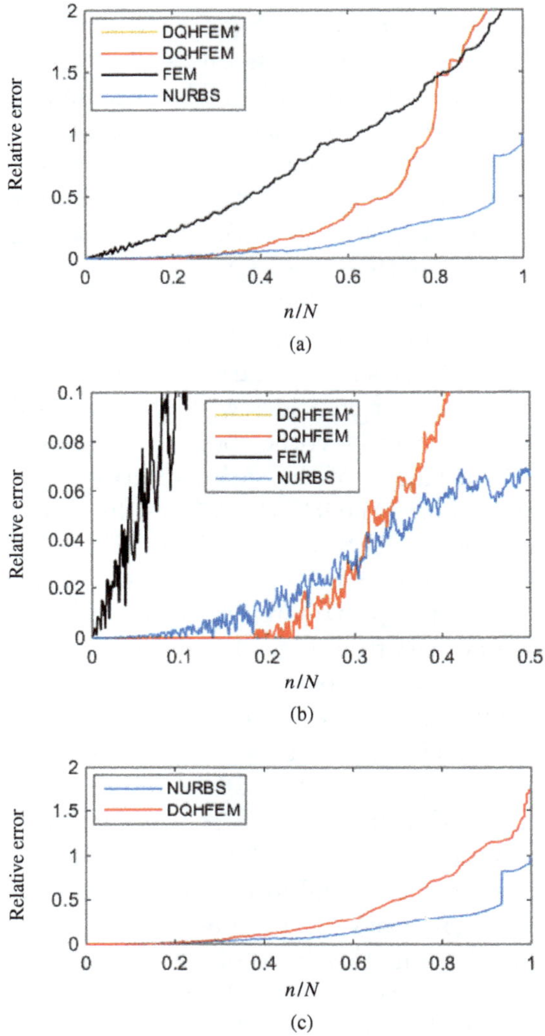

Fig. 4.15. Simply supported (S1) plate: (a) Relative error of discrete spectra using FEM, NURBS and DQHFEM; (b) partial enlargement of (a); (c) the square plate discretized as 6×6 elements and analyzed by the DQHFEM.

In Table 4.2, the DQHFEM results of the natural frequencies of a clamped circular plate are compared with the exact and FEM results of Ref. [331]. The radius and thickness of the circular plate are 0.5 m and 5 mm, respectively. The material of the plate is aluminum, which has Young's modulus of 71 GPa, a density of 2700 kg/m^3, and

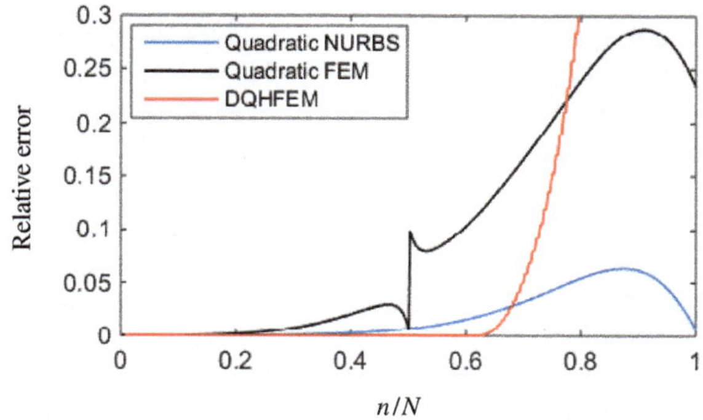

Fig. 4.16. Fixed rod. Relative error of discrete spectra using FEM, NURBS and DQHFEM.

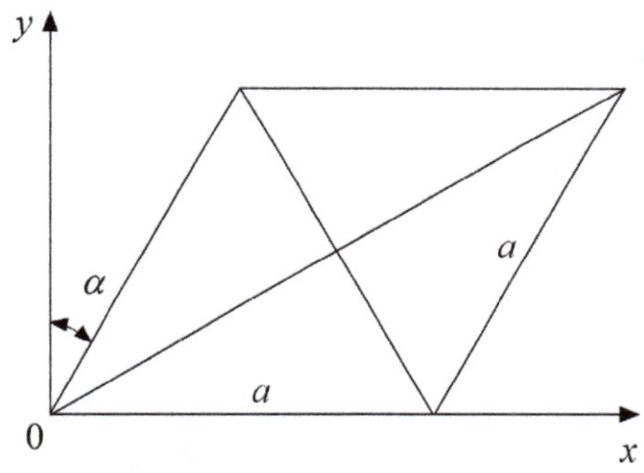

Fig. 4.17. A rhombic plate.

Poisson's ratio of 0.33. It can be seen that the DQHFEM results converge when $(N_e, N_h) = (16, 14)$ and agree with the exact results in [331] for all digits used for comparisons.

In Table 4.3, the DQHFEM results for in-plane vibration of a rhombic plate (Fig. 4.17) are compared with those obtained by a spectral collocation method [332]. The Poisson's ratio is taken as $\upsilon = 0.3$. The DQHFEM results agree with the spectral collocation solution for almost all digits used for comparison for clamped rhombic plate, as can be seen from Table 4.3. For the rhombic plate with CFCF boundaries and $\alpha = 60.0°$, the DQHFEM results agree with the spectral collocation solution for about three significant digits. In other cases, the agreement is in 4–5 significant digits. The number of sampling points in Table 4.3 is taken as $(N_e, N_h) = (25, 20)$.

Table 4.4. The frequency parameters $\gamma = \omega\sqrt{\rho/G}$ for the 10 lowest modes of the open S2–C–S2–F sectorial plate with $\upsilon = 0.3$, $a = 1$, $b = 0.5$ (Fig. 4.18).

	Mode sequence number									
(N_ξ, N_η)	1	2	3	4	5	6	7	8	9	10
(3, 6)	3.809	5.381	5.406	5.513	7.206	7.757	10.561	10.9722	11.283	13.454
(4, 8)	3.806	5.372	5.381	5.512	6.899	7.751	8.913	10.5345	10.812	11.665
(5, 10)	3.806	5.371	5.381	5.512	6.886	7.749	8.674	10.5279	10.650	10.807
(6, 12)	3.806	5.371	5.381	5.512	6.885	7.749	8.664	10.5267	10.528	10.807
(7, 14)	3.806	5.371	5.381	5.512	6.885	7.749	8.664	10.5205	10.528	10.807
(15, 30)	3.806	5.371	5.381	5.512	6.885	7.749	8.664	10.5205	10.528	10.807
Exact [48]	3.806	5.371	5.381	5.512	6.885	7.749	8.664	10.525	10.528	10.807
p-FEM [48]	3.806	5.372	5.381	5.512	6.887	7.750	8.667	10.525	10.529	10.808

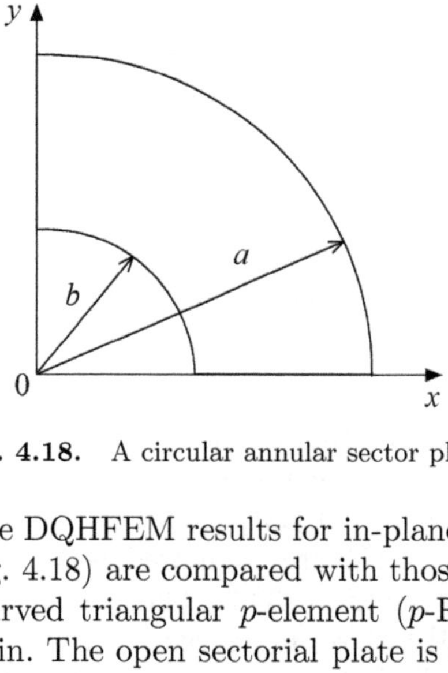

Fig. 4.18. A circular annular sector plate.

In Table 4.4, the DQHFEM results for in-plane vibration of open sectorial plate (Fig. 4.18) are compared with those by a trigonometrically enriched curved triangular *p*-element (*p*-FEM) [48] and the exact results therein. The open sectorial plate is clamped along the inner curved edge, free along the outer curved edge, and guided along the straight edges. The number of sampling points on the edge and the number of hierarchical shape functions in ξ-directions are taken as the same and denoted by N_ξ, and those in η-directions are also taken as the same and denoted by N_η. It can be seen that the DQHFEM results with $(N_\xi, N_\eta) = (6, 12)$ already agree with the exact

Table 4.5. The frequency parameters $\gamma = \omega\sqrt{\rho/G}$ for the 10 lowest symmetric–symmetric and symmetric–antisymmetric modes of the square plate with a central circular opening and $\upsilon = 0.3$ (Fig. 4.19).

	Mode sequence number									
(N_e, N_h)	1	2	3	4	5	6	7	8	9	10
				Symmetric–symmetric cases						
(5, 6)	0.0000	0.9220	3.2337	4.4344	4.6215	5.6997	6.2583	7.0118	7.0829	8.5128
(8, 6)	0.0000	0.9215	3.2312	4.4262	4.6201	5.6806	6.1842	6.9940	7.0093	8.4452
(10, 8)	0.0000	0.9215	3.2311	4.4261	4.6201	5.6805	6.1841	6.9935	7.0093	8.4450
(12, 10)	0.0000	0.9215	3.2311	4.4261	4.6201	5.6805	6.1841	6.9934	7.0093	8.4450
(18, 16)	0.0000	0.9215	3.2311	4.4261	4.6201	5.6805	6.1841	6.9934	7.0093	8.4450
p-FEM [48]	0.003	0.922	3.233	4.427	4.620	5.682	6.187	7.000	7.010	8.448
				Symmetric–antisymmetric cases						
(5, 6)	2.0483	2.2723	4.2939	5.0010	5.4476	6.5981	7.2555	7.8318	8.3722	8.9076
(8, 6)	2.0476	2.2711	4.2898	4.9964	5.4196	6.5160	7.2213	7.8087	8.1745	8.7333
(10, 8)	2.0476	2.2711	4.2897	4.9963	5.4196	6.5158	7.2211	7.8085	8.1729	8.7317
(18, 16)	2.0476	2.2711	4.2897	4.9963	5.4196	6.5158	7.2211	7.8085	8.1729	8.7317
p-FEM [48]	2.049	2.272	4.292	4.998	5.421	6.521	7.223	7.810	8.180	8.735

solutions for at least four significant digits. The converged results of this work agree with the exact solutions for all digits used for comparison in general. The eighth frequency in [48] did not completely converge and the paper did not indicate where the exact results come, so it is hard to tell whether there are typing errors. The converged result for this frequency in this work is 10.5205 instead of 10.525 as in [48].

In Table 4.5, the DQHFEM results for in-plane vibration of square plate with a central circular opening (Fig. 4.19) are presented to show the applicability of the DQHFEM to plates of more complex planforms. The plate has free boundary conditions. Because of symmetry in geometry and in boundary conditions, it is necessary to consider only one quarter of the plate (Fig. 4.20). The modes for symmetric–symmetric and symmetric–antisymmetric cases are presented in Table 4.5 and compared with the p-FEM [48] results. The first four mode shapes for symmetric–symmetric cases are presented in Fig. 4.21. In the symmetric–symmetric modes, the displacements u and v are respectively even functions of the x coordinate and the y coordinate. In the symmetric–antisymmetric modes, u and v are

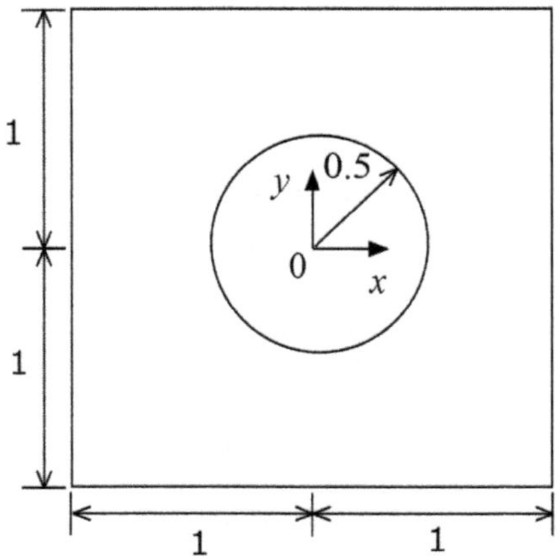

Fig. 4.19. A square plate with a central circular opening.

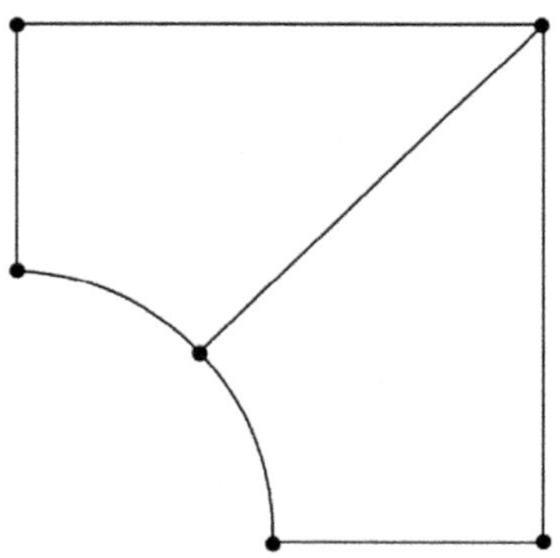

Fig. 4.20. DQHFEM mesh for solutions in Table 4.5.

respectively an even function of the x coordinate and an odd function of the y coordinate. One can see from Table 4.5 that all the five digits of the results of this work converge with $(N_e, N_h) = (12, 10)$, while only three significant digits of the results of [48] converged. The results of this work agree with the first three converged digits of [48] in general. The first mode for the symmetric–symmetric case is a rigid mode as can be seen from Fig. 4.21(a), but reference [48] still regards it as a normal one.

In Table 4.6, the DQHFEM results of the natural frequencies of a titanium straight lug (Fig. 4.22) are compared with the FEM results computed by the authors through the commercial software ANSYS.

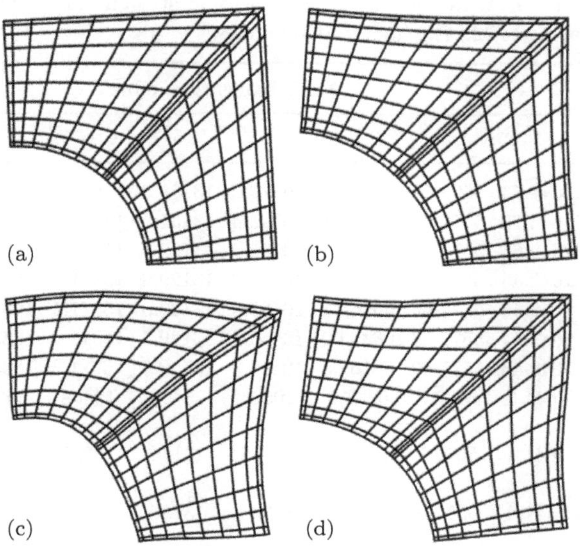

Fig. 4.21. Four modes for the symmetric–symmetric cases in Table 4.5: (a) mode 1; (b) mode 2; (c) mode 3; (d) mode 4.

Due to the symmetry of the lug, only one half is considered. The boundary conditions shown in Fig. 4.22 indicates that the edges with C and S1 are respectively clamped and simply supported (the first type). All the other edges are free. The length a shown in Fig. 4.22 is $a = 70\,\text{mm}$. The thickness of the lug is 6 mm. The material of the plate is titanium alloy, which has Young's modulus of 195 GPa, a density of $7722.7\,\text{kg/m}^3$, and Poisson's ratio of 0.28. The elements used in ANSYS are SHELL181 quadrilateral elements. The element length is denoted by l, as shown in Table 4.6. It can be seen that the first five digits of the DQHFEM results converge when $(N_e, N_h) = (12, 10)$. This mesh is similar to the element length $l = 6\,\text{mm}$ in ANSYS. The FEM results by ANSYS have four digits converged with $l = 2\,\text{mm}$, which is very fine mesh similar to $(N_e, N_h) = (35, 30)$ of DQHFEM. So, the DQHFEM is more efficient than the low-order schemes used in commercial software to obtain results with the same accuracy.

In Table 4.7, the DQHFEM results of the natural frequencies of a titanium torque arm (Fig. 4.23) are compared with the FEM results computed by the authors through commercial software ANSYS. The DQHFEM mesh is shown in Fig. 4.24. The right inner circle is clamped. All the other edges are free. The length and radius in

Table 4.6. Natural frequencies of an isotropic straight lug (Fig. 4.22) with $E = 195\,\text{GPa}$, $\rho = 7722.7\text{kg/m}^3$, and $v = 0.28$.

	Mode sequence number							
(N_e, N_h)	1	2	3	4	5	6	7	8
(6, 2)	1647.53	5598.25	6644.62	10770.0	13155.3	15827.3	18742.3	20778.3
(7, 5)	1642.70	5579.00	6617.81	10596.5	13131.9	15375.1	18652.2	20105.7
(9, 5)	1642.61	5577.88	6616.74	10585.2	13129.8	15315.9	18633.1	19917.3
(10, 8)	1642.48	5577.61	6616.24	10583.8	13129.6	15311.9	18631.2	19900.3
(12, 10)	1642.46	5577.56	6616.15	10583.5	13129.5	15311.5	18631.0	19898.9
(15, 15)	1642.44	5577.53	6616.09	10583.4	13129.5	15311.4	18631.0	19898.8
(35, 30)	1642.43	5577.50	6616.05	10583.3	13129.5	15311.3	18631.0	19898.8
FEM ($l = 10\,\text{mm}$)	1641.6	5577.9	6631.1	10626	13166	15408	18716	20062
FEM ($l = 6\,\text{mm}$)	1641.4	5572.7	6623.8	10598	13144	15339	18663	19947
FEM ($l = 4\,\text{mm}$)	1641.4	5571.2	6621.8	10590	13138	15318	18648	19913
FEM ($l = 2\,\text{mm}$)	1641.4	5570.3	6620.8	10585	13134	15306	18639	19893
FEM ($l = 1\,\text{mm}$)	1641.4	5570.0	6621.0	10585	13133	15304	18637	19888

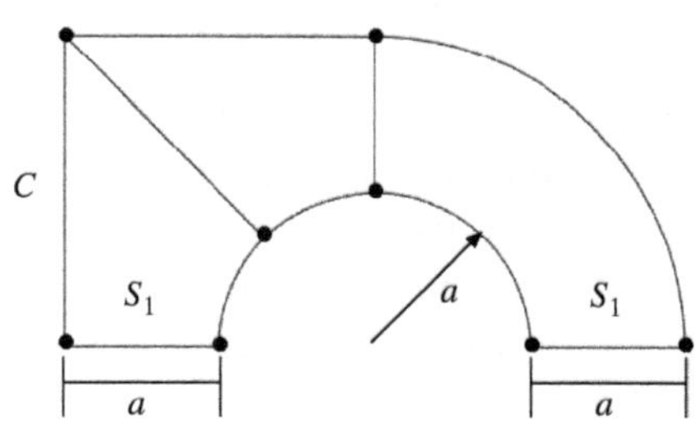

Fig. 4.22. DQHFEM mesh for a straight lug.

Fig. 4.23 are $r_1 = 0.1$ m, $r_2 = r_3 = 0.2$ m, $r_4 = 0.3$ m and $a = 1$ m. The thickness of the lug is 10 mm. The material of the plate is the same as the one in Table 4.6. The elements used in ANSYS are SHELL181 quadrilateral elements. The element length is denoted by l, as shown in Table 4.7. The mesh of the DQHFEM with $(N_e, N_h) = (10, 8)$ is similar to the FEM results with element length $l = 10\,\text{mm}$. The DQHFEM results with the mesh have 5–6 digits

Table 4.7. Natural frequencies of an isotropic torque arm (Fig. 4.23) with $E = 195\,\text{GPa}$, $\rho = 7722.7\,\text{kg/m}^3$, and $\upsilon = 0.28$.

	Mode sequence number							
(N_e, N_h)	1	2	3	4	5	6	7	8
(6, 2)	403.365	1280.69	1399.45	2427.55	2641.34	3130.89	3983.26	4301.92
(7, 5)	402.986	1275.27	1398.02	2408.11	2618.54	3118.82	3972.47	4255.70
(9, 5)	402.976	1275.07	1397.99	2407.45	2617.96	3118.52	3972.29	4255.25
(10, 8)	402.962	1274.87	1397.95	2407.01	2617.31	3118.33	3972.11	4254.41
(12, 10)	402.961	1274.87	1397.95	2407.00	2617.30	3118.32	3972.10	4254.39
(15, 15)	402.961	1274.86	1397.95	2407.00	2617.29	3118.32	3972.10	4254.38
(35, 30)	402.961	1274.86	1397.95	2407.00	2617.29	3118.32	3972.10	4254.38
FEM ($l = 15\,\text{mm}$)	403.17	1276.1	1398.8	2410.3	2617.8	3122.6	3975.0	4265.1
FEM ($l = 10\,\text{mm}$)	403.26	1275.8	1398.8	2409.4	2615.1	3120.8	3973.0	4261.4
FEM ($l = 5\,\text{mm}$)	403.29	1275.6	1398.8	2408.7	2615.1	3119.4	3972.2	4258.5

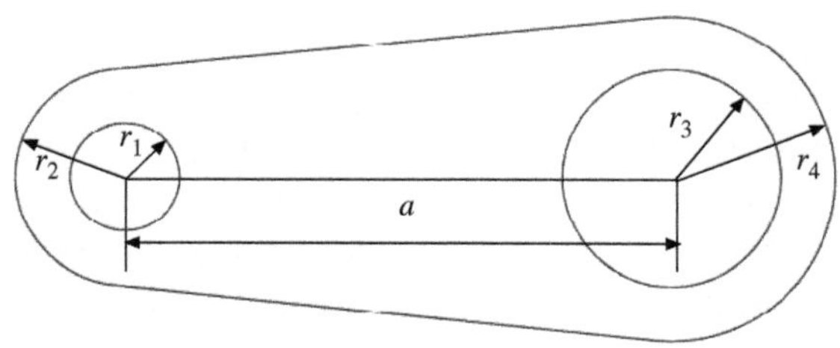

Fig. 4.23. A torque arm and its dimensions.

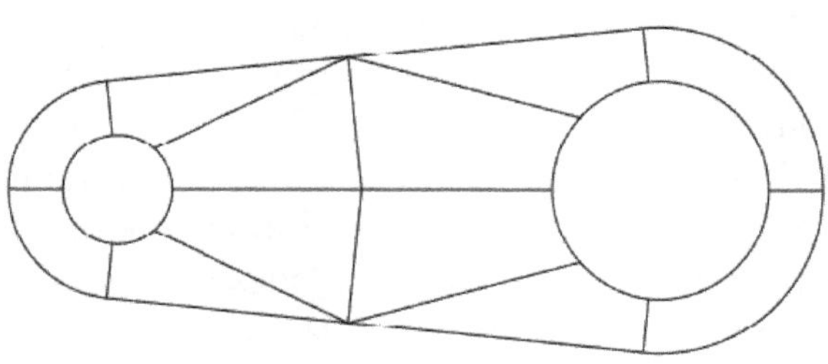

Fig. 4.24. DQHFEM mesh for a torque arm.

converged, while the FEM results only have 3–4 digits converged for most frequencies. One can see that the DQHFEM can provide results with good accuracy using a few nodes on the boundary of elements but several nodes inside the elements.

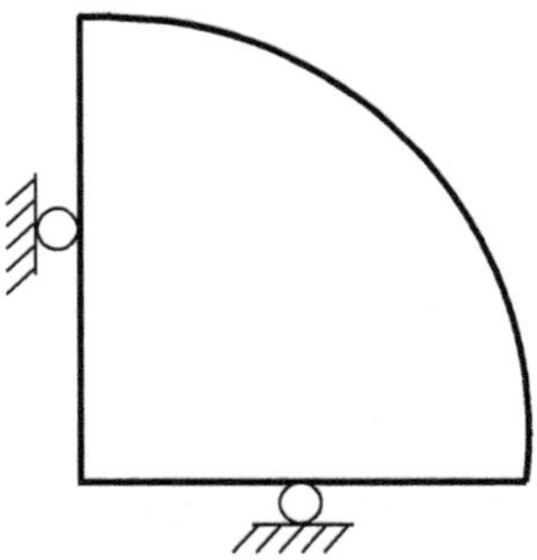

Fig. 4.25. A sectorial plate with S2 boundary conditions on the two straight edges.

4.3.2 *Triangular elements*

First, a sector plate shown in Fig. 4.25 with radius $R = 1\,\text{m}$ and S2 boundary conditions on the two straight edges is considered. The circumferential edge of the sector is free. The geometric mapping method by the transformation method given in Section 3.3.2 for NURL patches is used. The results are presented in Table 4.8. The exact and p-FEM results of [48] are used for comparison. The number of inside hierarchical bases is taken as P. The number of nodes on the three edges is taken as the same and denoted as N_e, namely, $N_1 = N_2 = N_3 = N_e$. First, take $N_e = P + 3$. One can see from Table 4.8 that with the increase of the number of nodes, the DQHFEM results monotonously converge from the above and the convergence rate is very fast. The first four digits of the first 10 frequencies completely converged when $(P, N_e) = (10, 13)$. Moreover, one can see that the DQHFEM is very stable even with a great number of nodes. Then take $(P, N_e) = (6, 13)$, see the bold italic line in Table 4.8. One can see that with a certain number of nodes on the boundary of the element and only a few nodes inside the element, the accuracy of the DQHFEM can still be very good. This property is similar to the fixed interface modal synthesis method. But the DQHFEM does not need modal analysis and is of high accuracy. Figure 4.26 presents a convergence study of the first, fourth and ninth frequency parameters in Table 4.8. One can see that the DQHFEM converges very fast using only a few DOFs. Note that the minimum relative error of the DQHFEM in Fig. 4.26 and in other figures is not e^{-8} or the following other values shown in the figures, since the given digits of

Table 4.8. The frequency parameters $\Omega = \omega\sqrt{\rho/G}$ for the 10 lowest modes of an open S2–S2–F sectorial plate with $\upsilon = 0.3$, $R = 1\,\text{m}$ (Fig. 4.25).

		Mode sequence number									
(P, N_e)	DOFs	1	2	3	4	5	6	7	8	9	10
(2, 5)	20	2.350	3.464	4.277	4.905	—	—	—	—	—	—
(3, 6)	30	2.347	3.464	4.249	4.742	7.310	7.570	7.683	8.899	—	—
(4, 7)	42	2.346	3.463	4.246	4.702	6.905	7.468	7.652	8.814	—	—
(5, 8)	56	2.346	3.463	4.245	4.691	6.760	7.446	7.645	8.797	9.110	9.147
(6, 9)	72	2.346	3.463	4.245	4.689	6.707	7.440	7.643	8.795	8.804	9.110
(7, 10)	90	2.346	3.463	4.245	4.689	6.698	7.439	7.642	8.666	8.795	9.110
(8, 11)	110	2.346	3.463	4.245	4.689	6.696	7.439	7.642	8.634	8.795	9.110
(9, 12)	132	2.346	3.463	4.245	4.689	6.696	7.439	7.642	8.622	8.795	9.110
(10, 13)	156	2.346	3.463	4.245	4.689	6.695	7.439	7.642	8.620	8.795	9.110
(15, 18)	306	2.346	3.463	4.245	4.689	6.695	7.439	7.642	8.620	8.795	9.110
(25, 28)	756	2.346	3.463	4.245	4.689	6.695	7.439	7.642	8.620	8.795	9.110
(6,13)	***88***	***2.346***	***3.463***	***4.245***	***4.689***	***6.704***	***7.439***	***7.642***	***8.721***	***8.795***	***9.110***
Exact [48]	—	2.346	3.463	4.245	4.689	6.695	7.439	7.642	8.620	8.795	9.110
p-FEM [48]	—	2.346	3.463	4.245	4.689	6.695	7.439	7.642	8.620	8.795	9.110

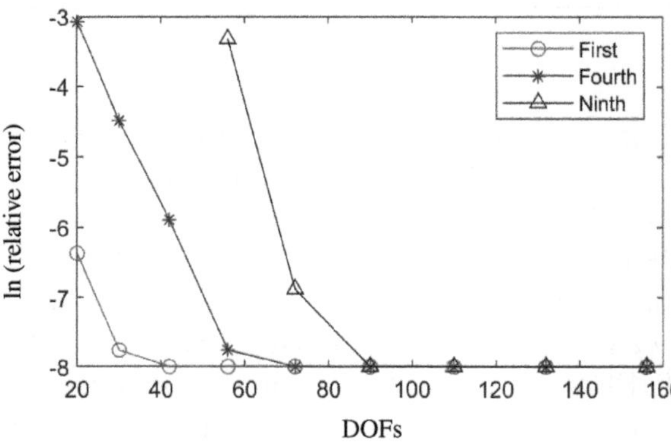

Fig. 4.26. Convergence rate of the first, forth and ninth natural frequency parameters of Table 4.8, where ln(relative error) gives the natural logarithm of the relative error.

the exact (or other reference) results are limited. One can find from [333] that the DQHFEM can almost reach computer accuracy.

In Table 4.9, the DQHFEM results of the natural frequencies of a clamped circular plate using four triangular elements are compared with the exact results of [331] and the DQHFEM results of [301] using quadrilateral element. The radius of the circular plate is 0.5 m.

Table 4.9. Comparison of the natural frequencies of a clamped circular plate with $E = 71\,\text{GPa}$, $\rho = 2700\,\text{kg/m}^3$, $\upsilon = 0.33$, and $R = 0.5\,\text{m}$.

	Mode sequence number								
(P, N_e)	1	2	3	4	5	6	7	8	9
(2, 5)	3362.2	3841.3	5241.0	5436.2	6628.6	6970.8	—	—	—
(3, 6)	3361.8	3837.6	5223.6	5387.8	6626.8	6814.2	6975.3	7033.7	8327.26
(4, 7)	3361.7	3836.3	5221.1	5383.5	6626.2	6772.8	6945.2	7027.3	8192.05
(5, 8)	3361.7	3835.6	5220.1	5383.0	6625.9	6765.3	6940.0	7024.4	8141.72
(6, 9)	3361.7	3835.3	5219.8	5382.9	6625.8	6764.0	6938.9	7022.8	8132.73
(7, 10)	3361.7	3835.1	5219.6	5382.9	6625.7	6763.8	6938.7	7022.3	8130.88
(8, 11)	3361.7	3835.0	5219.5	5382.9	6625.7	6763.8	6938.6	7022.0	8130.54
(9, 12)	3361.7	3835.0	5219.5	5382.9	6625.7	6763.8	6938.6	7021.8	8130.49
(10, 13)	3361.7	3834.9	5219.5	5382.9	6625.7	6763.8	6938.6	7021.7	8130.48
(15, 18)	3361.7	3834.9	5219.4	5382.9	6625.6	6763.8	6938.5	7021.4	8130.48
(25, 28)	3361.7	3834.9	5219.4	5382.9	6625.6	6763.8	6938.5	7021.4	8130.48
DQHFEM [301]	3361.7	3834.9	5219.4	5382.9	6625.6	6763.8	6938.5	7021.3	8130.48
Exact [331]	3362	3835	5219	5383	6626	6764	6939	7021	8130

The material of the plate is aluminum, which has Young's modulus of 71 GPa, a density of $2700\,\text{kg/m}^3$, and Poisson's ratio of 0.33. The geometries are still mapped by the NURL as above. It can be seen that the DQHFEM results converge when $(P, N_e) = (15, 18)$ and agree with the exact results in [331] for all digits used for comparisons. The DQHFEM of this work using triangular elements agree with the DQHFEM results of [301] using quadrilateral element also for all digits used for comparison except for the last digits of the eighth frequency. Since Ref. [301] used one quadrilateral element to analyze a circular plate that has singularity at the four corner points, the results of this work should be more accurate. Figure 4.27 plots the first three mode shapes of the clamped circular plate.

In-plane vibration of a square plate using two triangular elements (Fig. 4.28) is analyzed in Tables 4.10 and 4.11 for the plate with two combinations of simply supported, clamped (C) and free (F) boundaries, since there are plenty of such results in the literatures [334, 335]. A plate with S1–C–S2–F boundary condition indicates that the plate has S1–S2 at x-direction and C–F at y-direction. The

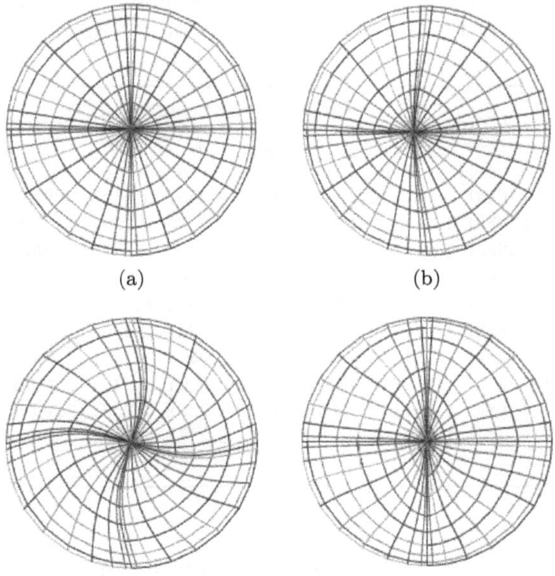

(a) (b) (c) (d)

Fig. 4.27. First three mode shapes of the clamped circular plate analyzed in Table 4.9: (a) the plate before deformation; (b) mode 1; (c) mode 2; (d) mode 3.

natural frequencies in Tables 4.10 and 4.11 are expressed by the same frequency parameter Ω as used in [334, 335]. One can see that the convergence rate of the DQHFEM on straight-sided domain is slightly faster than on curvilinear domain compared with Tables 4.8 and 4.9. Irrespective of whether it is in curvilinear or straight-sided domain, the DQHFEM is very stable. Figure 4.29 plots a few mode shapes for the S1–S1–S1–S1 plate.

In Tables 4.12 and 4.13, the sectorial plate analyzed in Table 4.8 using NURL (non-uniform rational Lagrange functions) geometry is analyzed again using linear triangular elements (order $= 1$). In Table 4.12, the node number is taken as $(P, N_e) = (2, 5)$ for each element, while the mesh seed length h_0 varies from 0.2 to 0.03. Fig. 4.30 plots the meshes for $h_0 = 0.2$ and $h_0 = 0.1$. In Table 4.13, the plate with mesh seed length $h_0 = 0.1$ is analyzed again by varying the node number (P, N_e). One can see from Table 4.13 that the plate with $(P, N_e) = (2, 5)$ and $h_0 = 0.1$ already converged but had not converged to the exact results due to geometric errors. One can see from Table 4.12 that the convergence rate of an h-version method is

Table 4.10. Comparison of the dimensionless frequency parameters $\Omega = (\omega a/\pi)\sqrt{\rho/G}$ for in-plane vibration of an S1–S1–S1–S1 square plate (Fig. 4.29(a)).

	Mode sequence number									
(P, N_e)	1	2	3	4	5	6	7	8	9	10
(2, 5)	1.0000	1.0000	1.4275	2.0479	2.0495	2.2837	2.3153	2.4020	3.2168	3.2168
(3, 6)	1.0000	1.0000	1.4144	2.0006	2.0006	2.2576	2.2922	2.3906	2.9444	3.1524
(4, 7)	1.0000	1.0000	1.4143	2.0004	2.0004	2.2370	2.2376	2.3905	2.8962	3.0059
(5, 8)	1.0000	1.0000	1.4142	2.0000	2.0000	2.2365	2.2372	2.3905	2.8334	3.0040
(6, 9)	1.0000	1.0000	1.4142	2.0000	2.0000	2.2361	2.2361	2.3905	2.8317	3.0001
(7, 10)	1.0000	1.0000	1.4142	2.0000	2.0000	2.2361	2.2361	2.3905	2.8285	3.0000
(8, 11)	1.0000	1.0000	1.4142	2.0000	2.0000	2.2361	2.2361	2.3905	2.8285	3.0000
(9, 12)	1.0000	1.0000	1.4142	2.0000	2.0000	2.2361	2.2361	2.3905	2.8284	3.0000
(20, 23)	1.0000	1.0000	1.4142	2.0000	2.0000	2.2361	2.2361	2.3905	2.8284	3.0000
Exact [334]	1	1	1.4142	2	2	2.2361	2.2361	2.3905	2.8284	3

Table 4.11. Comparison of the dimensionless frequency parameters $\Omega = (\omega a/\pi)\sqrt{\rho/G}$ for in-plane vibration of a S1–C–S2–F square plate (Fig. 4.28(b)).

	Mode sequence number									
(P, N_e)	1	2	3	4	5	6	7	8	9	10
(2, 5)	0.8603	0.9468	1.4484	1.7593	2.3348	2.3939	2.7036	2.8768	2.9273	2.9410
(3, 6)	0.8600	0.9468	1.4287	1.7423	2.2331	2.3714	2.3824	2.7820	2.8379	2.8874
(4, 7)	0.8600	0.9468	1.4260	1.7422	2.2258	2.3149	2.3615	2.7694	2.8019	2.8226
(5, 8)	0.8600	0.9468	1.4258	1.7420	2.2195	2.2995	2.3613	2.7528	2.7987	2.8220
(6, 9)	0.8600	0.9468	1.4258	1.7420	2.2194	2.2971	2.3610	2.7524	2.7934	2.8207
(7, 10)	0.8600	0.9468	1.4258	1.7420	2.2193	2.2968	2.3610	2.7517	2.7933	2.8207
(8, 11)	0.8600	0.9468	1.4258	1.7420	2.2193	2.2967	2.3610	2.7517	2.7932	2.8207
(9, 12)	0.8600	0.9468	1.4258	1.7420	2.2193	2.2967	2.3610	2.7517	2.7932	2.8207
(20, 23)	0.8600	0.9468	1.4258	1.7420	2.2193	2.2967	2.3610	2.7517	2.7932	2.8207
Exact [334]	0.8600	0.9468	1.4258	1.7420	2.2193	2.2967	2.3610	2.7517	2.7932	2.8207

very slow. When $h_0 = 0.03$, the number of nodes on the edge of the plate reaches about 166, which is very fine mesh. However, the last digit used for comparison of the fifth frequency in Table 4.12 still did not converge to the exact result.

In order to further study the influence of geometric error on the simulation accuracy, the sectorial plate analyzed in Tables 4.8, 4.12

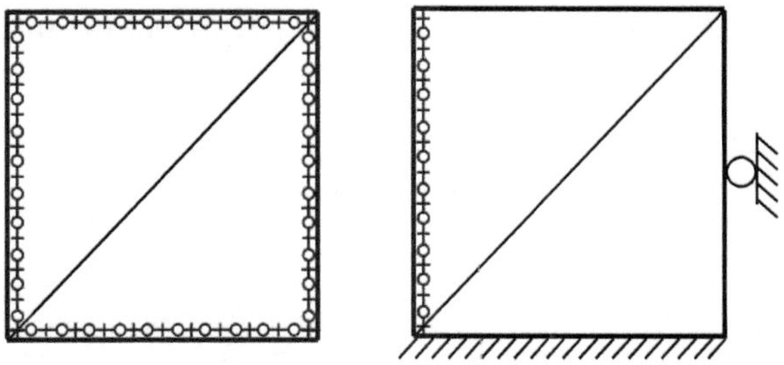

(a) (b)

Fig. 4.28. A square plate with two combinations of simply supported, clamped (C) and free (F) boundaries: (a) S1–S1–S1–S1; (b) S1–C–S2–F.

Table 4.12. The frequency parameters $\Omega = \omega\sqrt{\rho/G}$ for the 10 lowest modes of an open S2–S2–F sectorial plate with $(P, N_e) = (2, 5)$, $\upsilon = 0.3$, $R = 1\,\text{m}$ (Fig. 4.25).

	Mode sequence number									
h_0	1	2	3	4	5	6	7	8	9	10
0.20 m	2.355	3.478	4.266	4.708	6.723	7.469	7.674	8.655	8.831	9.148
0.10 m	2.348	3.466	4.249	4.693	6.701	7.446	7.649	8.627	8.802	9.118
0.08 m	2.347	3.465	4.248	4.692	6.699	7.443	7.646	8.625	8.799	9.115
0.05m	2.346	3.464	4.246	4.690	6.697	7.440	7.643	8.622	8.796	9.112
0.04 m	2.346	3.464	4.245	4.690	6.696	7.440	7.643	8.621	8.796	9.111
0.03 m	2.346	3.463	4.245	4.689	6.696	7.439	7.642	8.620	8.795	9.110
Exact [48]	2.346	3.463	4.245	4.689	6.695	7.439	7.642	8.620	8.795	9.110
p-FEM [48]	2.346	3.463	4.245	4.689	6.695	7.439	7.642	8.620	8.795	9.110

and 4.13 is analyzed again in Table 4.14 by elevating the order from linear to cubic. The mesh seed length is taken as $h_0 = 0.3$, which is very coarse. When linear elements (order = 1) are used, the DQHFEM results converge with node number $(P, N_e) = (4, 7)$, but only first two digits agree with the exact results. When quadratic elements (order = 2) are used, one can see from Table 4.14 that the results changed dramatically. Almost all digits used for comparison, except for the last digit of the fifth frequency, converge to exact results with $(P, N_e) = (4, 7)$. When cubic elements (order = 3) are used, all digits used for comparison converge to exact results with $(P, N_e) = (4, 7)$. Therefore, for high-accuracy analyses, order elevation (*p*-version

Table 4.13. The frequency parameters $\Omega = \omega\sqrt{\rho/G}$ for the 10 lowest modes of an open S2–S2–F sectorial plate with order= 1, $h_0 = 0.1$, $\upsilon = 0.3$, $R = 1\,\text{m}$ (Fig. 4.25).

	Mode sequence number									
(P, N_e)	1	2	3	4	5	6	7	8	9	10
(0, 2)	2.362	3.472	4.293	4.831	7.095	7.596	7.902	9.065	9.247	9.422
(0, 3)	2.348	3.466	4.250	4.694	6.705	7.448	7.651	8.640	8.806	9.119
(1, 4)	2.348	3.466	4.250	4.693	6.702	7.446	7.649	8.628	8.802	9.118
(2, 5)	2.348	3.466	4.249	4.693	6.701	7.446	7.649	8.627	8.802	9.118
(3, 6)	2.348	3.466	4.249	4.693	6.701	7.446	7.649	8.627	8.802	9.118
(4, 7)	2.348	3.466	4.249	4.693	6.701	7.446	7.649	8.627	8.802	9.118
Exact [48]	2.346	3.463	4.245	4.689	6.695	7.439	7.642	8.620	8.795	9.110
p-FEM [48]	2.346	3.463	4.245	4.689	6.695	7.439	7.642	8.620	8.795	9.110

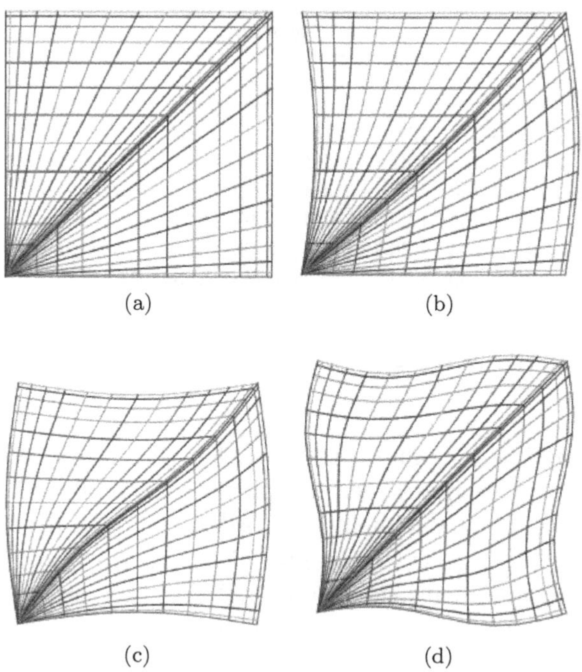

(a) (b) (c) (d)

Fig. 4.29. Three mode shapes of the S1–S1–S1–S1 square plate (Fig. 4.28(a)) analyzed in Table 4.10: (a) the plate before deformation; (b) mode 1; (c) mode 3; (d) mode 6.

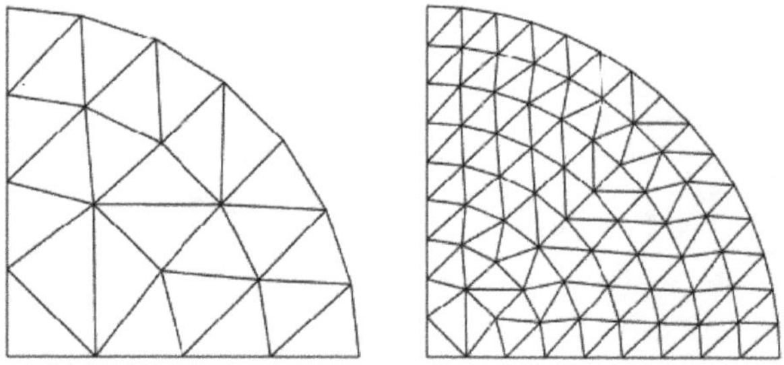

(a) (b)

Fig. 4.30. H-version meshes for a sectorial plate: (a) $h_0 = 0.2$; (b) $h_0 = 0.1$.

Table 4.14. The frequency parameters $\Omega = \omega\sqrt{\rho/G}$ for the 10 lowest modes of an open S2–S2–F sectorial plate with $h_0 = 0.3$, $v = 0.3$, $R = 1\,\text{m}$ (Fig. 4.25).

	Mode sequence number									
(P, N_e)	1	2	3	4	5	6	7	8	9	10
					Linear mesh					
(2, 5)	2.364	3.492	4.286	4.727	6.751	7.498	7.707	8.698	8.871	9.186
(3, 6)	2.364	3.492	4.285	4.727	6.748	7.495	7.706	8.687	8.866	9.185
(4, 7)	2.364	3.492	4.285	4.727	6.748	7.494	7.705	8.685	8.866	9.185
(8, 11)	2.364	3.492	4.285	4.726	6.748	7.493	7.705	8.684	8.866	9.185
					Quadratic mesh					
(2, 5)	2.346	3.463	4.245	4.689	6.697	7.440	7.643	8.628	8.799	9.110
(3, 6)	2.346	3.463	4.245	4.689	6.696	7.439	7.642	8.621	8.795	9.110
(4, 7)	2.346	3.463	4.245	4.689	6.696	7.439	7.642	8.620	8.795	9.110
(8, 11)	2.346	3.463	4.245	4.689	6.696	7.439	7.642	8.620	8.795	9.110
					Cubic mesh					
(2, 5)	2.346	3.463	4.245	4.689	6.697	7.440	7.643	8.631	8.798	9.110
(3, 6)	2.346	3.463	4.245	4.689	6.696	7.439	7.642	8.620	8.795	9.110
(4, 7)	2.346	3.463	4.245	4.689	6.695	7.439	7.642	8.620	8.795	9.110
(8, 11)	2.346	3.463	4.245	4.689	6.695	7.439	7.642	8.620	8.795	9.110
Exact [48]	2.346	3.463	4.245	4.689	6.695	7.439	7.642	8.620	8.795	9.110
p-FEM [48]	2.346	3.463	4.245	4.689	6.695	7.439	7.642	8.620	8.795	9.110

refinements) is much more efficient than h-version mesh refinements. Figure 4.31 plots the quadratic and cubic elements with global mesh seed length $h_0 = 0.3$. Figure 4.32 analyzes the influence of geometric errors on simulation accuracy through the fundamental frequency, where mesh seed length $h_0 = 0.1, 0.2, 0.3, 0.4, 0.5, 0.7$ and

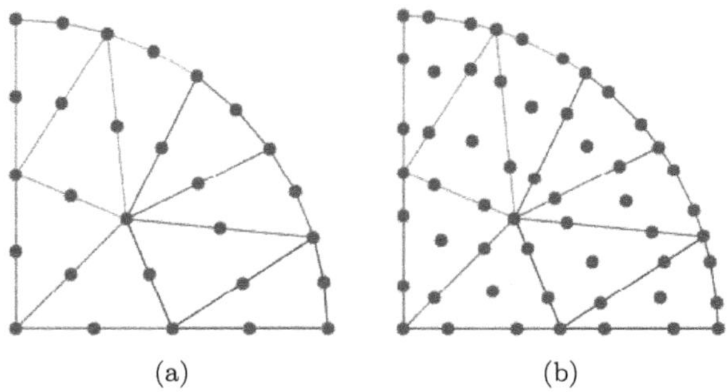

Fig. 4.31. A sectorial plate with quadratic (a) and cubic (b) elements and global mesh seed length $h_0 = 0.3$.

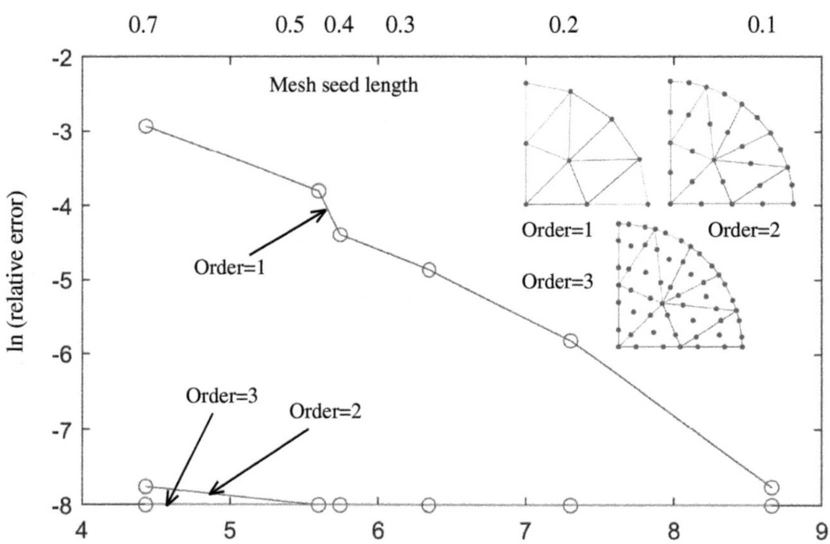

Fig. 4.32. An analysis of the influence of geometric errors (order of mesh) on simulation accuracy through fundamental frequency, where ln (•) gives the natural logarithm of the data "•" and node number is $(P, N_e) = (4, 7)$.

node number $(P, N_e) = (4, 7)$. One can clearly see that slight order elevation of geometry can greatly increase the simulation accuracy using much less DOFs.

In Table 4.15, the circular plate analyzed in Table 4.9 is analyzed again by elevating the order from linear to cubic and conclusions similar to the sectorial plate above are made. The mesh for the circular

Table 4.15. Comparison of the natural frequencies of a clamped circular plate with $E = 71\,\text{GPa}$, $\rho = 2700\,\text{kg/m}^3$, $\upsilon = 0.33$, and $R = 0.5\,\text{m}$.

	Mode sequence number								
(P, N_e)	1	2	3	4	5	6	7	8	9
					Linear mesh				
(2, 5)	3390.5	3869.5	5261.3	5431.4	6682.0	6822.5	6998.2	7085.1	8203.7
(3, 6)	3390.4	3869.0	5261.1	5430.6	6681.9	6821.4	6997.1	7083.6	8199.9
(4, 7)	3390.4	3868.7	5261.1	5430.3	6681.8	6821.3	6996.8	7083.2	8199.6
(8, 11)	3390.3	3868.5	5261.1	5430.1	6681.8	6821.3	6996.5	7083.0	8199.5
					Quadratic mesh				
(2, 5)	3361.8	3834.9	5219.5	5383.0	6625.7	6764.6	6938.9	7022.0	8133.4
(3, 6)	3361.8	3834.9	5219.4	5382.9	6625.7	6763.8	6938.6	7021.5	8130.7
(4, 7)	3361.8	3834.9	5219.4	5382.9	6625.7	6763.8	6938.6	7021.4	8130.6
(8, 11)	3361.8	3834.9	5219.4	5382.9	6625.7	6763.8	6938.6	7021.4	8130.6
					Cubic mesh				
(2, 5)	3361.7	3835.0	5219.5	5383.2	6625.7	6765.1	6939.0	7022.1	8133.4
(3, 6)	3361.7	3834.9	5219.4	5382.9	6625.6	6763.8	6938.6	7021.6	8130.7
(4, 7)	3361.7	3834.9	5219.4	5382.9	6625.6	6763.8	6938.5	7021.4	8130.48
(8, 11)	3361.7	3834.9	5219.4	5382.9	6625.6	6763.8	6938.5	7021.4	8130.48
DQHFEM [301]	3361.7	3834.9	5219.4	5382.9	6625.6	6763.8	6938.5	7021.3	8130.48
Exact [331]	3362	3835	5219	5383	6626	6764	6939	7021	8130

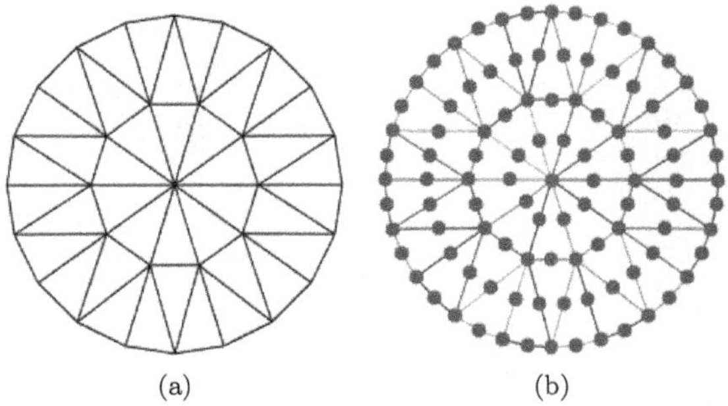

(a) (b)

Fig. 4.33. Triangulation of circular domain (a) and quadratic approximation of the circular domain (b).

plate is in Fig. 4.33. Note that for the eighth frequency, the DQHFEM of [301] used one quadrilateral element to analyze a circular plate that has singularity at the four corner points, so the results of this work should be more accurate as discussed for Table 4.9.

4.4 Results for Static Analysis

This section aims to validate the performance of the DQHFEM for static plane problems. Analytical results used to validate the DQHFEM results include a thick-walled cylinder under uniform boundary pressure and a stress-free hole in an infinite medium under uniform far-field tension loading. For simplicity, the number of sampling points on the edges of elements is taken as the same and denoted as N_e, and the number of hierarchical basis on the two parametric directions (ξ and η) is also taken as the same and denoted as N_h. Then the DQHFEM will be applied to the interaction of grains and grain boundaries of metals, and the interface between nanoparticles and the matrix of nanoparticle composites.

4.4.1 *Thick-walled cylinder under uniform boundary pressure*

The stress components for a thick-walled cylinder under the action of uniform internal and external pressure loadings, as shown in Fig. 4.34, are

$$\begin{aligned}
\sigma_r &= \frac{a^2b^2(p_2-p_1)}{b^2-a^2}\frac{1}{r^2}+\frac{a^2p_1-b^2p_2}{b^2-a^2} \\
\sigma_\theta &= -\frac{a^2b^2(p_2-p_1)}{b^2-a^2}\frac{1}{r^2}+\frac{a^2p_1-b^2p_2}{b^2-a^2}
\end{aligned} \tag{4.71}$$

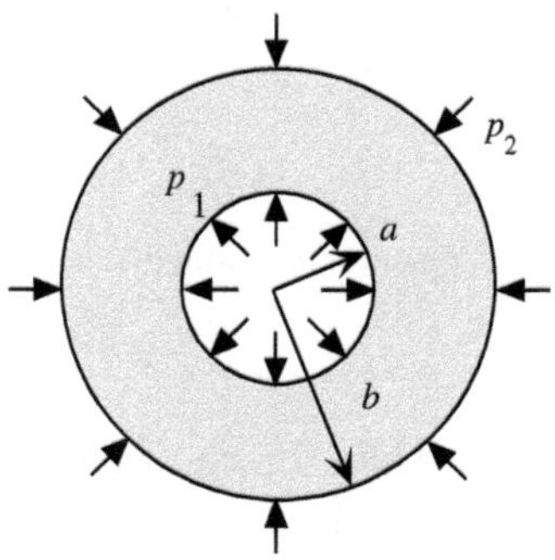

Fig. 4.34. Thick-walled cylinder problem.

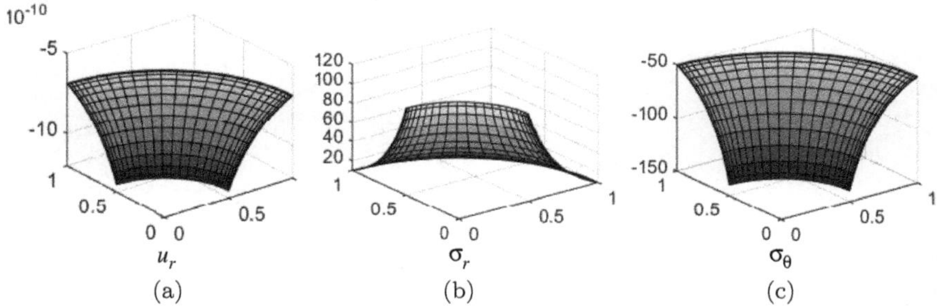

Fig. 4.35. Exact results for a thick-walled cylinder under boundary pressure $p_1 = -100\,\text{N}$ and $p_2 = 10$.

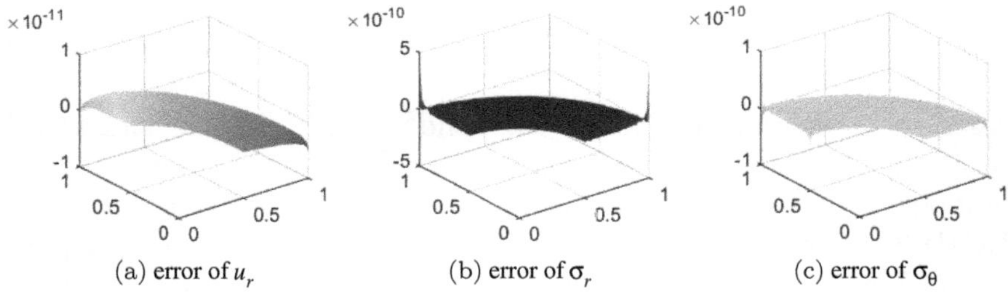

Fig. 4.36. Relative error of the DQHFEM results for the thick-walled cylinder.

Using the strain–displacement relations and Hooke's law, the radial displacement is easily determined as

$$u_r = \frac{1+v}{E}\left[-\frac{a^2b^2(p_2-p_1)}{b^2-a^2}\frac{1}{r} + (1-2v)\frac{a^2p_1 - b^2p_2}{b^2-a^2}r\right] \tag{4.72}$$

where v is Poisson's ratio and E is Young's modulus. The above solutions are exact. Take $a = 0.5$, $b = 1.0$, $p_1 = -100\,\text{Pa}$, $p_2 = 10\,\text{Pa}$, $E = 72\,\text{GPa}$ and $v = 0.3$, the exact solutions are shown in Fig. 4.35. Note that this is a plane strain problem.

The DQHFEM modeled the cylinder by one element using the symmetry of the geometry and the loads. The number of basis on the boundary of the element and inside the element in both directions are 25. Relative errors of the DQHFEM for the displacement u_r and stresses σ_r and σ_θ are plotted in Fig. 4.36. One can see from Fig. 4.37 that the DQHFEM almost reaches computer accuracy. In order to check the convergence of the DQHFEM, the following error index for

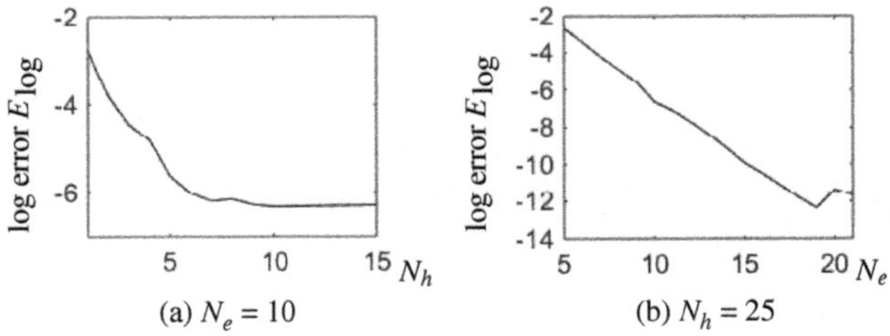

Fig. 4.37. Convergence study of the DQHFEM for displacement u_r: (a) Fix N_e; (b) Fix N_h.

a variable u_r is used:

$$E_{\log} = \log_{10} \frac{\bar{u}_r - u_r}{u_r} \tag{4.73}$$

where u_r indicates the exact value and $\bar{u}_r$ indicates the numerical results of this work. As can be seen from Fig. 4.37(a) with the node number $N_e = 10$ and N_h varying from 1 to 15, the accuracy of DQHFEM improves fast but reaches the maximum when N_h equals N_e–2. The accuracy of the DQHFEM is very good even when N_h equals 1. On this aspect, the DQHFEM is similar to the substructure method. In Fig. 4.37(b), the node number $N_h = 25$ and N_e varies from 5 to 21. One can see that DQHFEM can almost reach computer accuracy with the increase of the number of sampling points on the edges.

4.4.2 *Stress concentration problem of circular hole in plate*

The stress field of this problem is obtained from a stress-free hole in an infinite medium under uniform far-field tension loading, as shown in Fig. 4.38. The stress components are

$$\begin{aligned}
\sigma_r &= \frac{\sigma}{2}\left(1 - \frac{a^2}{r^2}\right) + \frac{\sigma}{2}\left(1 - \frac{4a^2}{r^2} + \frac{3a^4}{r^4}\right)\cos 2\theta \\
\sigma_\theta &= \frac{\sigma}{2}\left(1 + \frac{a^2}{r^2}\right) - \frac{\sigma}{2}\left(1 + \frac{3a^4}{r^4}\right)\cos 2\theta \\
\sigma_{r\theta} &= -\frac{\sigma}{2}\left(1 + \frac{2a^2}{r^2} - \frac{3a^4}{r^4}\right)\sin 2\theta
\end{aligned} \tag{4.74}$$

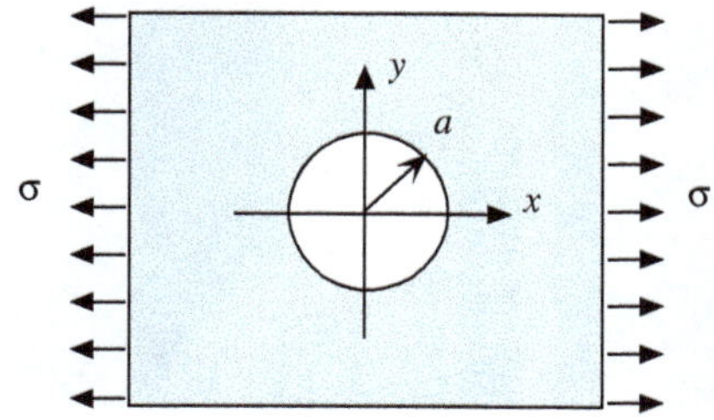

Fig. 4.38. Stress-free hole in an infinite medium under uniform far-field tension loading.

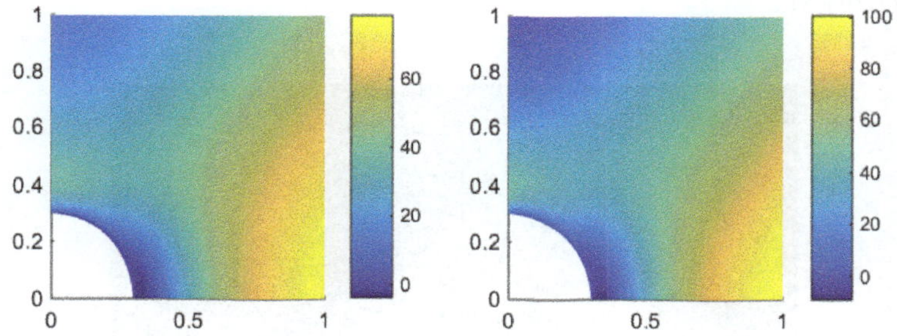

Fig. 4.39. Results of σ_r for the stress concentration problem: (a) analytical; (b) DQHFEM.

The above solutions are obtained under the assumption of "far-field tension loading". So, they are not exact. Take $a = 0.3$, $\sigma_1 = 100\,\text{Pa}$, and Poisson's ratio $\upsilon = 0.3$. The size of the plate is taken as square and with length 2. Analytical and DQHFEM results for σ_r and Mises stress are respectively given in Figs. 4.39 and 4.40. The DQHFEM used two elements. One can see that the analytical results under "far-field tension loading" assumption are smaller than the DQHFEM results. The DQHFEM results satisfy the loading condition very well as can be seen from Fig. 4.39(b), where the maximum stress of σ_r equals the loading stress.

4.4.3 *The interaction of grains and grain boundaries of metals*

With the accuracy of the DQHFEM verified above, this section analyzes the interaction of grains and grain boundaries of metals through

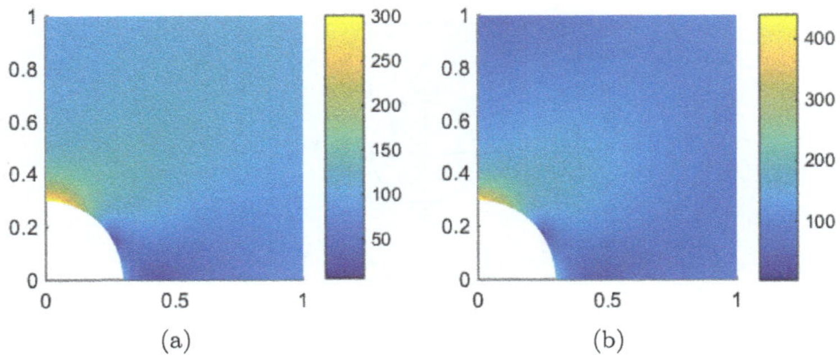

Fig. 4.40. Mises stress for the stress concentration problem: (a) analytical; (b) DQHFEM.

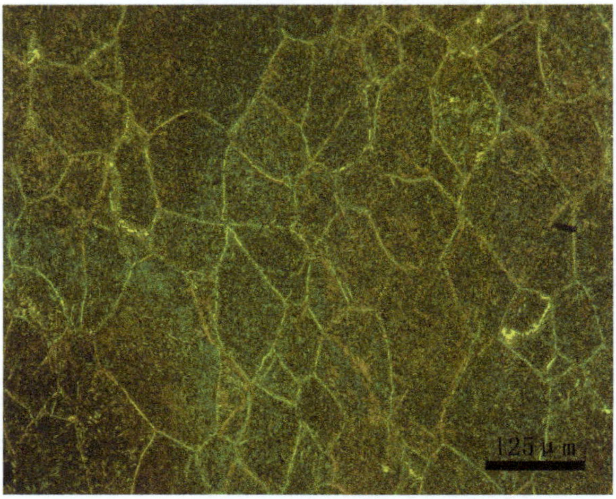

Fig. 4.41. Metallographic structure of TC18 Titanium alloy.

the DQHFEM. Figure 4.41 presents the metallographic structure of TC18 Titanium alloy. The moduli of the grain boundaries is usually taken as 70–80% of that of the grains [336, 337]. Therefore, the metallographic structure of metals can be regarded as a type of composite structure. Take $x = kx_0$ and $y = ky_0$, where k is the length scale of the grain. One can easily see from Eqs. (4.54) and (4.65) that the scale k does not influence the computation of the modulus of the metallographic structure. Therefore, a unit square representative volume element (RVE) will be considered. Take m and n respectively to be the number of grains on each direction, and take d to be the ration

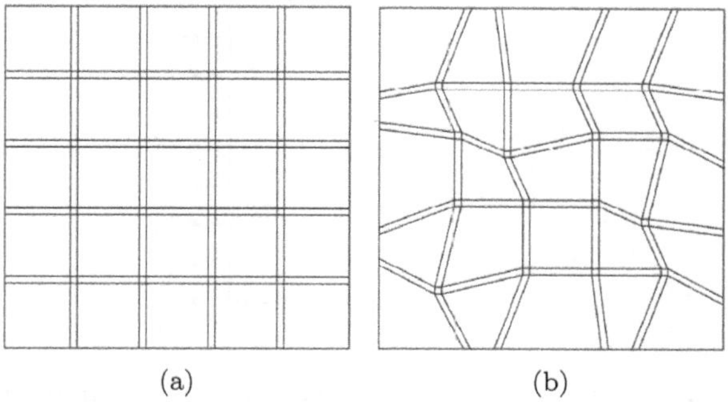

Fig. 4.42. The method of generating a metallographic structure model: (a) a regular mash; (b) metallographic structure model generated by adding a random number to the x and y coordinates of each intersection of grain boundaries.

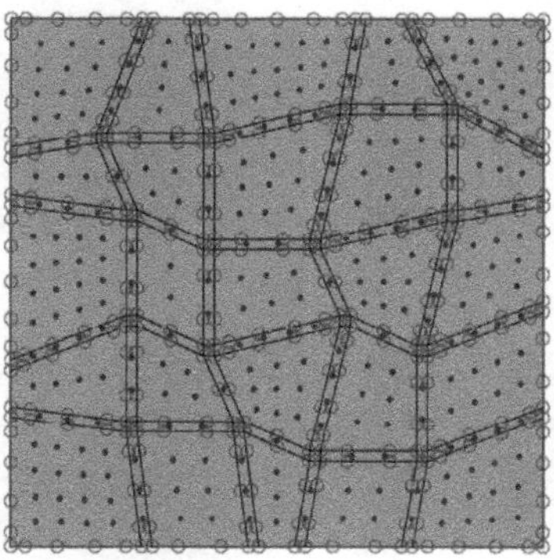

Fig. 4.43. DQHFEM mesh of a metallographic structure model, where o denotes a node on the boundary of DQHFEM elements and · illustrates the hierarchical basis inside a DQHFEM element.

of the width of grain boundary over the average length of the grains. First, a regular mesh as in Fig. 4.42(a) is generated and then a random number is added to the x and y coordinates of each intersection of grain boundaries to generate a metallographic structure model, as shown in Fig. 4.42(b). The finite element mesh generation by common FEM may not be an easy task. By DQHFEM, each grain, grain boundary and intersection of grain boundaries can be modeled as a DQHFEM element, as shown in Fig. 4.43.

With the method of generating metallographic structure model and DQHFEM mesh above, a unit square plate under uniaxial

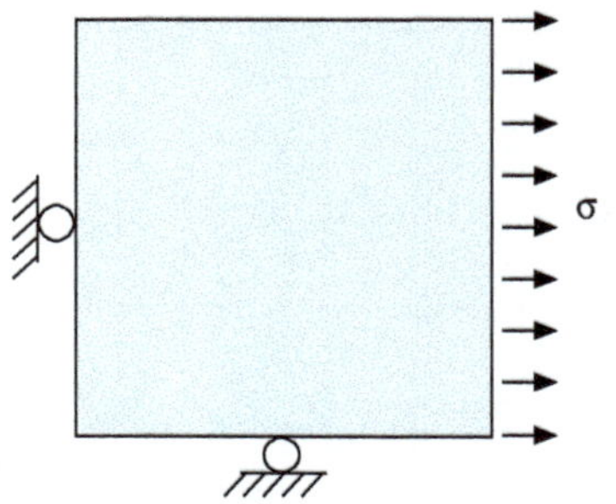

Fig. 4.44. Uniaxial tension problem of a unit square plate.

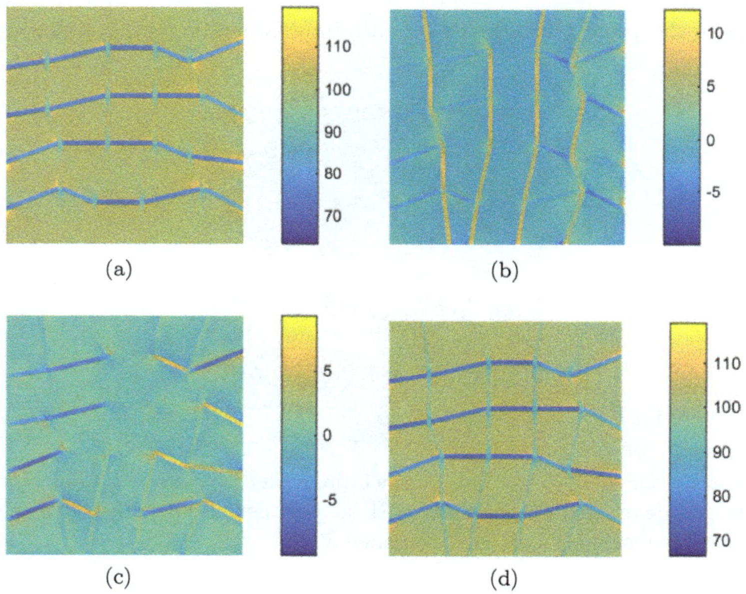

Fig. 4.45. Stresses of a unit square plate with isotropic grains and grain boundaries under uniaxial tension: (a) σ_x; (b) σ_y; (c) σ_{xy}; (d) Mises stress.

tension as shown in Fig. 4.44 will be considered herein. It is obvious that the average displacement of the loaded edge divided by the stress is the Young's modulus of the structure. First, isotropic material with Young's modulus $E = 260\,\text{GPa}$ for grains and 70% of that value for grain boundaries are considered. The Poisson's ratio is taken as $\upsilon = 0.28$. The loading stress σ is taken as 100 Pa. The corresponding stress components is given in Fig. 4.45. The approximated global Young's modulus obtained by this method is $E_a = 245.4\,\text{GPa} = 0.9438E$. According to the role of mixture, the approximated global Young's modulus can be obtained as

$$E_a = \frac{A_G}{A_G + A_{GB}} E_G + \frac{A_{GB}}{A_G + A_{GB}} E_{GB} \tag{4.75}$$

where E_G and E_{GB} are respectively the Young's modulus of grains and grain boundaries; and A_G and A_{GB} are respectively the area of grains and grain boundaries. The Young's modulus obtained by Eq. (4.75) is $E_a = 246.6\,\text{GPa} = 0.9484E$, which verifies the correctness of the result obtained by this work.

The grains are taken as orthotropic and Young's moduli in 1 and 2 material directions are respectively taken as $E_1 = 260\,\text{GPa}$ and $E_2 = 0.7E_1$. The Poisson's ratio υ_{12} is taken as 0.28. The angles between x-axis and the first material directions are randomly given, as shown in Fig. 4.46. The corresponding stress components with loading stress σ being 100 MPa is given in Fig. 4.47. The Young's modulus for this case obtained by this method is $E_a = 198.3\,\text{GPa} = 0.7625E_1$. It is obvious that if the grains are orthotropic, the global Young's modulus will be close to the Young's moduli of the weakest material direction. The approximated global Young's modulus for this case can be given as

$$\tilde{E}_a = \frac{A_G}{A_G + A_{GB}} E_2 + \frac{A_{GB}}{A_G + A_{GB}} E_1 \tag{4.76}$$

The Young's modulus obtained by Eq. (4.76) is $\tilde{E}_a = 195.4\,\text{GPa} = 0.7516E_1$. The relative error of this value compared with the FEM results is within 2%, which verifies the effectiveness of Eq. (4.76).

4.4.4 *The interface between nanoparticles and the matrix of nanoparticle composites*

Polymer–nanoparticle composites (PNPCs) were extensively studied in recent years due to their lower cost and notably better mechanical properties compared to than conventional polymer–particle composites. Figure 4.48 presents a periodic PNPC molecular model. The 2D continuous model is given in Fig. 4.49(a), where the area inside the inner circle represents the particle, the area outside the outer circle represents the polymer matrix, and the area between the two

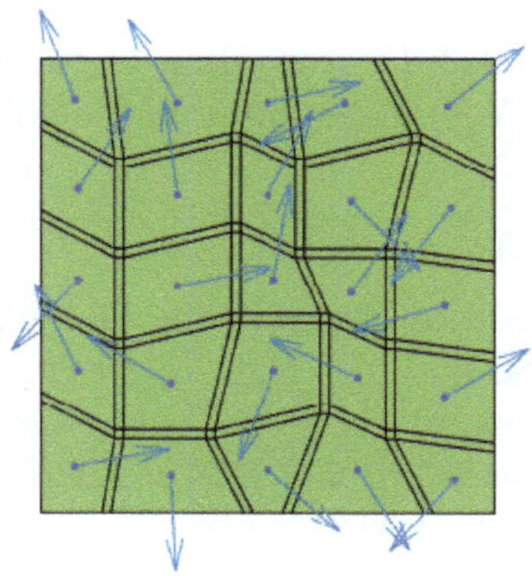

Fig. 4.46. The first material direction of grains.

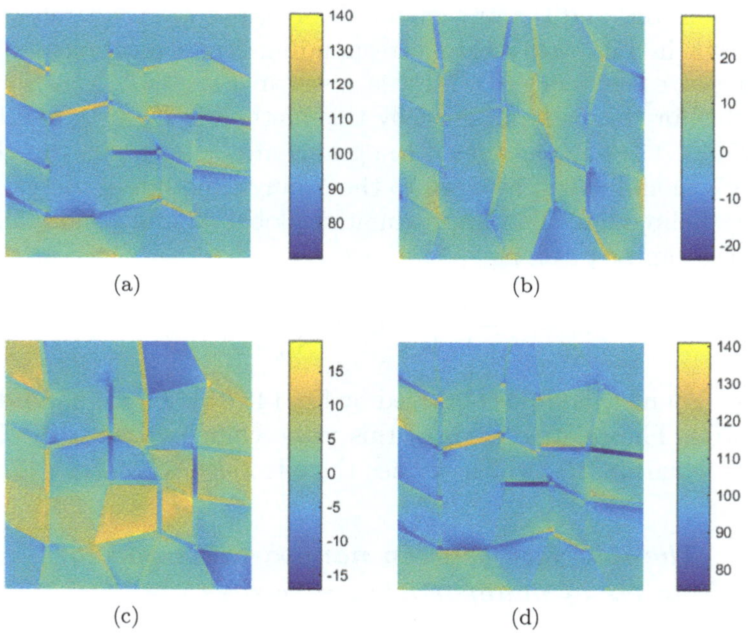

Fig. 4.47. Stresses of a unit square plate with orthotropic grains and grain boundaries under uniaxial tension: (a) σ_x; (b) σ_y; (c) σ_{xy}; (d) sises stress.

circles represents the interfaces between the particle and the matrix. DQHFEM mesh of the model is given in Fig. 4.49(b). The Young's modulus of the polymer matrix obtained by a cubic molecular model with edge length 39.21 Å using Materials Studio 7.0 is 2.20 GPa. The

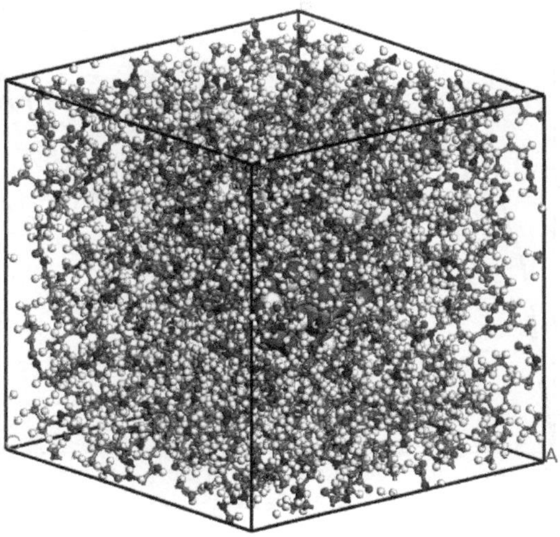

Fig. 4.48. A periodic PNPC molecular model.

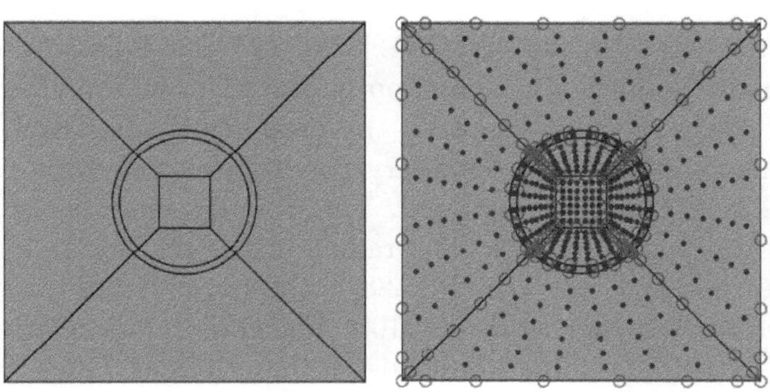

(a) (b)

Fig. 4.49. 2D continuous PNPC model (a) and its DQHFEM mesh (b), where o denotes a node on the boundary of DQHFEM elements and · illustrates the hierarchical basis inside a DQHFEM element.

Young's modulus of PNPC obtained by a cubic molecular model with edge length 37.88 Å particle radius 7.2 Å using Materials Studio 7.0 is 3.12 GPa. The stress-strain curves of Nylon6 and Nylon6/SiO_2 are shown in Fig. 4.50.

With the continuous model developed above and the Young's modulus of polymer matrix obtained above, DQHFEM simulation of the PNPC will be carried out herein. The Young's modulus of

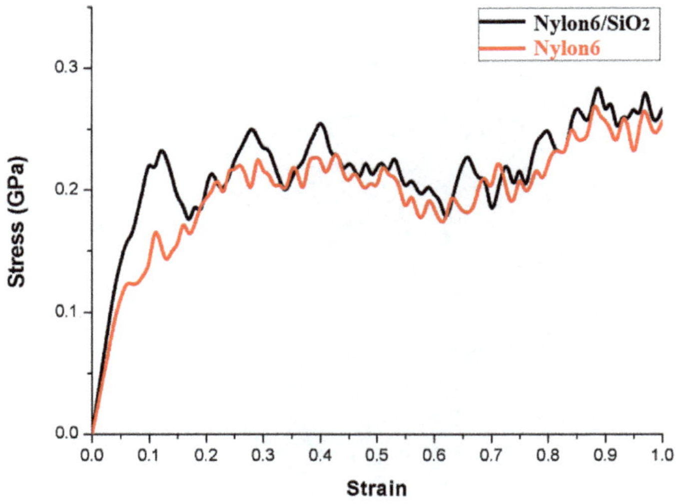

Fig. 4.50. Stress–strain curves of pure Nylon6 and Nylon6/SiO_2 PNPC.

the particle is taken as 104 GPa [338] and the Poisson's ratio is approximately taken as −0.4 (approximated from the shear modulus of [338]). In order to verify the property of the DQHFEM, a homogeneous plate with boundary conditions as in Fig. 4.51 and mesh as in Fig. 4.49 is analyzed. The Young's modulus of all elements is taken as 104 GPa. The radius of the particle (include the interface) is taken as $r = 7.2/37.88 = 0.1901$, and the thickness of the interface is taken as $d = r/100$ (Fig. 4.51). The relative error of the Young's modulus obtained by this method is 9.5×10^{-12}, which shows the high accuracy of the DQHFEM and its less sensitivity with mesh distortion (the interface elements are almost a line). Then the Young's modulus of the polymer matrix and the interface are all taken as 2.20 GPa. The Young's modulus obtained by the PNPC model is 2.799 GPa. If the Young's modulus of the interface is taken as 0.22 GPa, the Young's modulus obtained by the PNPC model is 2.700 GPa, which shows the necessity of surface modification of nanoparticle. Figure 4.52 shows the corresponding stress components with loading stress σ being 100 MPa. Since this work uses a 2D model and the local elastic theory, the difference between the DQHFEM results and the molecular dynamic (MD) results is reasonable.

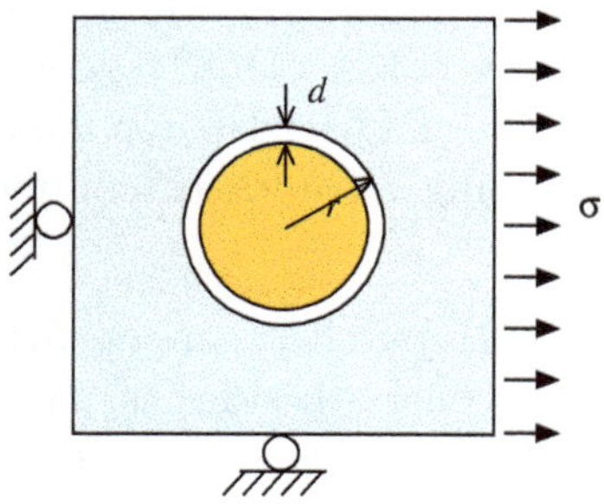

Fig. 4.51. Uniaxial tension problem of a unit square PNPC model.

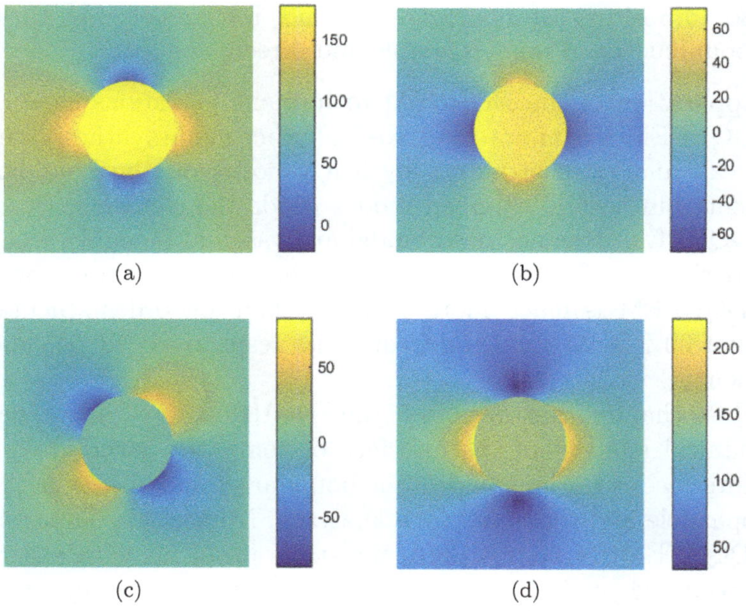

Fig. 4.52. Stresses of a unit square PNPC model under uniaxial tension: (a) σ_x; (b) σ_y; (c) σ_{xy}; (d) Mises stress.

4.5 Conclusions

In-plane vibration of plates in curvilinear domains are carried out by a differential quadrature hierarchical finite element method (DQH-FEM). A procedure of analysis similar to IGA is presented for the DQHFEM. Both curvilinear quadrilateral plates in several planforms and structures composed of several curvilinear quadrilateral elements are analyzed. The NURBS geometries are first transformed into differential quadrature hierarchical (DQH) geometries and then the solutions field also uses the DQH basis for analysis. The DQHFEM is seen to have the following advantages compared with IGA using NURBS:

(1) Compatibility of DQH patches can be attained by employing the same DQH edges on both sides of patch interfaces without mesh refinement propagation from patch to patch. Therefore, elements of different polynomial degrees can be easily joined and additional degrees of freedom where needed can be included.
(2) The DQHM uses interpolation basis on the boundary of elements, so imposition of homogeneous and inhomogeneous Dirichlet boundary conditions is simple and direct.

Application of the DQHFEM to in-plane vibrations shows that the DQHFEM is similar to the fixed interface mode synthesis method that can analyze a structure by a few nodes on the boundary of elements but several clamped modes inside the elements. However, the DQHFEM does not need modal analysis and is highly accurate. Comparisons with other methods for the entire spectra show that the DQHFEM is much more accurate than low-order schemes for the first 60% or 20% natural frequencies respectively for 1D and 2D problems.

This chapter also applies the DQHFEM to micro/macro-mechanical analysis of the interface of composite structures. Both the interfaces of grains and grain boundaries and the interfaces of nanoparticle and the matrix are analyzed. Numerical comparison of the DQHFEM results for the plane static problems with exact and approximated analytical results shows the high accuracy and efficiency of the DQHFEM. Young's modulus analysis of the interfaces of the two types of composite structures shows that the strength of the interfaces has obvious influence over the global strength of composite structures and the DQHFEM is suitable for two-scale analysis.

Chapter 5

The Hierarchical Quadrature Element Method for 3D Solids

A three-dimensional (3D) solid element can be considered to be the most general form of all solid finite elements because all the field variables are dependent on x, y and z. A 3D solid can have any arbitrary shape, material properties and boundary conditions in space. As such, there are altogether six possible stress components, three normal and three shear components that need to be taken into consideration. Typically, a 3D solid element can be a tetrahedron, pyramid, prism or hexahedron in shape with either flat or curved surfaces. Each node of the element will have three translational degrees of freedom. The element can thus deform in all three directions in space. The formulation of 3D solids elements is straightforward because it is basically an extension of two dimensional (2D) solids elements. All the techniques used in 2D solids can be utilized except that all the variables are now functions of x, y and z.

5.1 Element Energy Functions

Consider a three-dimensional (3D) solid of volume V which is enclosed by a surface S. The state of stress in a three-dimensional elastic body is defined by the stress components (which are referred to Cartesian axes x, y, z).

$$\boldsymbol{\sigma} = \begin{bmatrix} \sigma_x & \tau_{xy} & \tau_{xz} \\ \tau_{yx} & \sigma_y & \tau_{yz} \\ \tau_{zx} & \tau_{zy} & \sigma_z \end{bmatrix} \tag{5.1}$$

with

$$\tau_{yx} = \tau_{xy}, \quad \tau_{zx} = \tau_{xz}, \quad \tau_{zy} = \tau_{yz} \tag{5.2}$$

In relating the stresses to the strains, anisotropic, orthotropic and isotropic materials will be considered except in the case of one-dimensional (1D) elements (axial, torque and beam) when only the isotropic case is treated.

The state of strain in an elastic body is defined by the strain components

$$\boldsymbol{\varepsilon} = \begin{bmatrix} \varepsilon_x & \gamma_{xy} & \gamma_{xz} \\ \gamma_{xy} & \varepsilon_y & \gamma_{yz} \\ \gamma_{xz} & \gamma_{yz} & \varepsilon_z \end{bmatrix} \tag{5.3}$$

If the displacement components in the directions of the axes are denoted by (u, v, w), then the strain–displacement relations are

$$\begin{gathered} \varepsilon_x = \frac{\partial u}{\partial x}, \quad \varepsilon_y = \frac{\partial v}{\partial y}, \quad \varepsilon_z = \frac{\partial w}{\partial z} \\ \gamma_{xy} = \frac{\partial u}{\partial y} + \frac{\partial v}{\partial x}, \quad \gamma_{xz} = \frac{\partial u}{\partial z} + \frac{\partial w}{\partial x}, \quad \gamma_{yz} = \frac{\partial v}{\partial z} + \frac{\partial w}{\partial y} \end{gathered} \tag{5.4}$$

It should be noted that all displacement components are time- dependent. The dynamic equilibrium equations are

$$\begin{aligned} \frac{\partial \sigma_x}{\partial x} + \frac{\partial \tau_{yx}}{\partial y} + \frac{\partial \tau_{zx}}{\partial z} + f_x &= \rho \ddot{u} \\ \frac{\partial \tau_{xy}}{\partial x} + \frac{\partial \sigma_y}{\partial y} + \frac{\partial \tau_{zy}}{\partial z} + f_y &= \rho \ddot{v} \\ \frac{\partial \tau_{xz}}{\partial x} + \frac{\partial \tau_{yz}}{\partial y} + \frac{\partial \sigma_z}{\partial z} + f_z &= \rho \ddot{w} \end{aligned} \tag{5.5}$$

In considering the spatial variation of these quantities, explicit mention of time dependence is omitted.

The strain energy is therefore given by

$$U = \frac{1}{2}\int_V \{\sigma_x\varepsilon_x + \sigma_y\varepsilon_y + \sigma_z\varepsilon_z + \tau_{xy}\gamma_{xy} + \tau_{xz}\gamma_{xz} + \tau_{yz}\gamma_{yz}\}\mathrm{d}V \quad (5.6)$$

which can be expressed in matrix form as

$$U = \frac{1}{2}\int_V \boldsymbol{\sigma}^{\mathrm{T}}\boldsymbol{\varepsilon}\mathrm{d}V \quad (5.7)$$

where

$$\begin{aligned} \boldsymbol{\sigma}^{\mathrm{T}} &= \lfloor \sigma_x \quad \sigma_y \quad \sigma_z \quad \tau_{xy} \quad \tau_{xz} \quad \tau_{yz} \rfloor \\ \boldsymbol{\varepsilon}^{\mathrm{T}} &= \lfloor \varepsilon_x \quad \varepsilon_y \quad \varepsilon_z \quad \gamma_{xy} \quad \gamma_{xz} \quad \gamma_{yz} \rfloor \end{aligned} \quad (5.8)$$

The stress–strain relationships take the form

$$\boldsymbol{\sigma} = \boldsymbol{D}\boldsymbol{\varepsilon} \quad (5.9)$$

where **D** is a symmetric matrix. For an anisotropic material, it contains 21 independent material constants. In the case of an isotropic material, it is

$$\boldsymbol{D} = \frac{E}{(1+v)(1-2v)} \times \begin{bmatrix} 1-v & v & v & 0 & 0 & 0 \\ & 1-v & v & 0 & 0 & 0 \\ & & 1-v & 0 & 0 & 0 \\ & & & \frac{1}{2}(1-2v) & 0 & 0 \\ & & & & \frac{1}{2}(1-2v) & 0 \\ \text{Sym} & & & & & \frac{1}{2}(1-2v) \end{bmatrix} \quad (5.10)$$

where E = Young's modulus, and v = Poisson's ratio.

Substituting Eq. (5.9) into Eq. (5.7) gives

$$U = \frac{1}{2}\int_V \boldsymbol{\varepsilon}^{\mathrm{T}} \boldsymbol{D}\, \boldsymbol{\varepsilon}\, \mathrm{d}V \tag{5.11}$$

The strain–displacement relations to be used in Eq. (5.11) are given by (5.4):

$$\boldsymbol{\varepsilon} = \begin{bmatrix} \dfrac{\partial u}{\partial x} \\ \dfrac{\partial v}{\partial y} \\ \dfrac{\partial w}{\partial z} \\ \dfrac{\partial u}{\partial y} + \dfrac{\partial v}{\partial x} \\ \dfrac{\partial u}{\partial z} + \dfrac{\partial w}{\partial x} \\ \dfrac{\partial v}{\partial z} + \dfrac{\partial w}{\partial y} \end{bmatrix} \tag{5.12}$$

The kinetic energy is given by

$$T = \frac{1}{2}\int_V \rho(\dot{u}^2 + \dot{v}^2 + \dot{w}^2)\mathrm{d}V \tag{5.13}$$

If p_x, p_y, p_z are the components of the applied surface forces per unit area, then the virtual work is

$$\delta W = \int_S (p_x \delta u + p_y \delta v + p_z \delta w)\mathrm{d}S \tag{5.14}$$

5.2 Shape Function Construction

Through FEM for two-dimensional (2D) solids, we can find that the most fundamental and important procedure in FEM is the construction of shape functions. The procedures left for elements of various shapes are basically similar. Therefore, the shape functions for various 3D elements will be presented in this section. Since HQEM is a general method, namely, other types of elements can be a special case of HQEM, this section mainly introduces 3D HQEM elements. The element types considered include tetrahedral, wedge, pyramid and hexahedral elements. Hexahedral elements, like quadrilateral elements in 2D cases, are the simplest and most accurate 3D elements. However, automatic generation of hexahedral meshes for complex engineering problems is not possible so far. Most commercial codes support automatic generation of tetrahedral meshes. Therefore, wedge and pyramid elements such as transition elements are necessary for a hexa-dominated mesh.

The displacement field for tetrahedral, wedge, pyramid and hexahedral elements can be written in a general form as follows:

$$\begin{aligned} & u[x(\xi,\eta,\zeta), y(\xi,\eta,\zeta), z(\xi,\eta,\zeta)] \\ & \quad = \sum_{i=1}^{N_{\mathrm{v}}} S_i^{\mathrm{v}}(\xi,\eta,\zeta) u_i + \sum_{i=1}^{L_k}\sum_{k=1}^{N_{\mathrm{e}}} S_i^{\mathrm{e},k}(\xi,\eta,\zeta) a_i^{\mathrm{e},k} \\ & \qquad + \sum_{i=1}^{M_k}\sum_{j=1}^{N_k}\sum_{k=1}^{N_{\mathrm{f}}} S_{ij}^{\mathrm{f},k}(\xi,\eta,\zeta) a_{ij}^{\mathrm{f},k} + \sum_{i=1}^{L}\sum_{j=1}^{M}\sum_{k=1}^{N} S_{ijk}^{\mathrm{b}}(\xi,\eta,\zeta) a_{ijk}^{\mathrm{b}} \end{aligned} \tag{5.15}$$

where v, e, f and b respectively denote vertex, edge, face and body, N_v, N_e and N_f are respectively the number of vertices, edges and faces, i, j and k are integer indices; other italic variables are bases numbers, $S_i^v(\xi,\eta,\zeta)$ is a vertex basis, $S_i^{e,k}(\xi,\eta,\zeta)$ is an edge basis, $S_{ij}^{f,k}(\xi,\eta,\zeta)$ is a face basis, and $S_{ijk^b}(\xi,\eta,\zeta)$ is a body basis.

5.2.1 *Tetrahedral elements*

A tetrahedral element has four vertices, six edges and four faces (see Fig. 5.1). The vertex bases of a tetrahedral element are

$$\begin{aligned} & S_1^{\mathrm{v}}(\xi,\eta,\zeta) = 1-\xi-\eta-\zeta, \quad S_2^{\mathrm{v}}(\xi,\eta,\zeta) = \xi, \\ & S_3^{\mathrm{v}}(\xi,\eta,\zeta) = \zeta, \quad S_4^{\mathrm{v}}(\xi,\eta,\zeta) = \eta \end{aligned} \tag{5.16}$$

According to the derivations for triangular HQEM elements, the general form of the modified Dubiner's tetrahedral bases can be

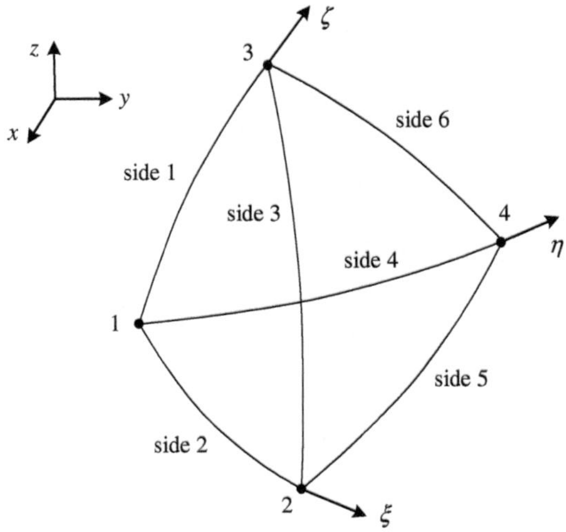

Fig. 5.1. A curved tetrahedral element.

given as

$$N_{ijk}(\xi,\eta,\zeta) = (1-\xi-\eta-\zeta)^h \xi^l \eta^m \zeta^n H_{ijk}(\xi,\eta,\zeta) \tag{5.17}$$

where h, l, m and n are non-negative integers. In order to apply Jacobi polynomials to the tetrahedron element, coordinate transformation is necessary. The Duffy transformation from a unit hexahedral domain to a unit tetrahedral domain as shown in Fig. 5.2 can be given as

$$\xi = \frac{(1+r)(1+t)}{4}, \quad \eta = \frac{(1-r)(1+s)(1+t)}{8}$$
$$\zeta = \frac{(1-r)(1-s)(1+t)}{8} \tag{5.18}$$

The corresponding inverse transformation is

$$r = 1 - \frac{2(\eta+\zeta)}{(\xi+\eta+\zeta)}, \quad s = \frac{\eta-\zeta}{\eta+\zeta}, \quad t = 2(\xi+\eta+\zeta)-1 \tag{5.19}$$

By the transformation given in Eq. (5.18), the face $r = -1$ of the unit hexahedron corresponds to the face $\xi = 0$ of the unit tetrahedron, $s = -1$ corresponds to $\eta = 0$, $s = 1$ corresponds to $\zeta = 0$, $t = 1$ corresponds to $\xi + \eta + \zeta = 1$, the face $t = -1$ is degenerated to be a point and the face $r = 1$ is degenerated to be a line. According to

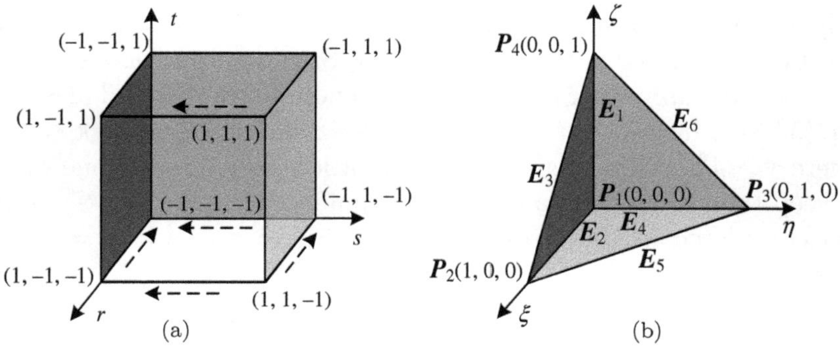

Fig. 5.2. A unit hexahedron (a) and tetrahedron (b).

the orthogonal requirements, the following formulation should hold for the shape functions on the unit tetrahedron:

$$\int_V N_{ijk}N_{pqw}\mathrm{d}V = \int_0^1\int_0^{1-\xi}\int_0^{1-\xi-\eta} N_{ijk}N_{pqw}\mathrm{d}\zeta\mathrm{d}\eta\mathrm{d}\xi = \delta_i^p\delta_j^q\delta_k^w C_{ijk} \tag{5.20}$$

The substitution of Eq. (5.17) and (5.18) into Eq. (5.20) leads to

$$\int_0^1\int_0^{1-\xi-\eta} N_{ijk}N_{pqw}d\zeta d\eta d\xi = \int_{-1}^1\int_{-1}^1\int_{-1}^1 \tilde{J}H_{ijk}H_{pqw}\mathrm{d}r\mathrm{d}s\mathrm{d}t \tag{5.21}$$

where

$$\tilde{J} = \frac{(1+r)^{2l}(1-r)^{2m+2n+1}(1+s)^{2m}(1-s)^{2n}(1+t)^{2l+2m+2n+2}(1-t)^{2h}}{2^{2h+4l+6m+6n+6}} \tag{5.22}$$

Using the orthogonal property of Jacobi polynomials as shown in Section 3.3.1, $H_{ijk}(r,s,t)$ can be determined as

$$\begin{aligned} H_{ijk}(r,s,t) = {} & [(1-r)^a(1+r)^b P_i^{(2m+2n+1+2a,2l+2b)}(r)] \\ & \times [(1-s)^c(1+s)^d P_j^{(2n+2c,2m+2d)}(s)] \\ & \times [(1-t)^e(1+t)^f P_k^{(2h+2e,2l+2m+2n+2+2f)}(t)] \end{aligned} \tag{5.23}$$

where a, b, c, d, e and f should be determined according to convenience of applications and the requirements of the Jacobi polynomials. In order to eliminate $(\eta+\zeta)$ in the denominator of s in Eq. (5.23), one needs to take $a=j$. Similarly, one needs to take $f=i+j$ to eliminate $(\xi+\eta+\zeta)$ in the denominator of r and $(1-r)$ in Eq. (5.23). The indices i, j, k should have the relation $i+j+k \le \gamma-4$, where $\gamma \ge 4$ is the expected maximum order of the bases if one wants to keep the bases below a certain maximum order. For the C^0 problem considered in this work, we can take $b=c=d=e=0$. We have

$$H_{ijk}(r,s,t) = [(1-r)^j P_i^{(2m+2n+1+2j,2l)}(r)]P_j^{(2n,2m)}(s) \times [(1+t)^{i+j} P_k^{(2h,2(l+m+n+1)+2(i+j))}(t)] \tag{5.24}$$

The substitution of Eq. (5.19) into Eq. (5.24) leads to

$$H_{ijk}(\xi,\eta,\zeta) = \left[2^{i+2j}(\xi+\eta+\zeta)^i P_i^{(2m+2n+1+2j,2l)}\left(\frac{\xi-\eta-\zeta}{\xi+\eta+\zeta}\right)\right] \times \left[(\eta+\zeta)^j P_j^{(2n,2m)}\left(\frac{\eta-\zeta}{\eta+\zeta}\right)\right] \times P_k^{(2h,2(l+m+n+1)+2(i+j))}[2(\xi+\eta+\zeta)-1] \tag{5.25}$$

The constant 2^{i+2j} in Eq. (5.25) will not be considered in computation. The computation of the rational terms in the denominator of Eq. (5.25) can be avoided by using the procedure shown in Section 3.3.1. The constants C_{ijk} in Eq. (5.20) is given by

$$C_{ijk} = \frac{C_i^{(2m+2n+1+2j,2l)} C_j^{(2n,2m)} C_k^{(2h,2(l+m+n+1)+2(i+j))}}{2^{2h+4l+6m+6n+6}} \tag{5.26}$$

Dubiner's warped orthogonal bases can be obtained by taking $h=l=m=n=0$. The modified Dubiner's bases presented can be obtained by taking at least one of h, l, m, n as non-zero. Figure 5.3 plots a few bases on a unit tetrahedral domain.

Similar to the triangular element, a simple formula of directly mapping 1D Gauss–Lobatto nodes to tetrahedrons can be presented. Let the transformation formula from uniform nodes $(\xi,\eta,\zeta,1-\xi-\eta-\zeta)$ to non-uniform nodes $(\lambda_1,\lambda_2,\lambda_3,\lambda_4)$ be

$$\lambda_1 = f(\xi), \quad \lambda_2 = f(\eta), \quad \lambda_3 = f(\zeta), \quad \lambda_4 = f(1-\xi-\eta-\zeta) \tag{5.27}$$

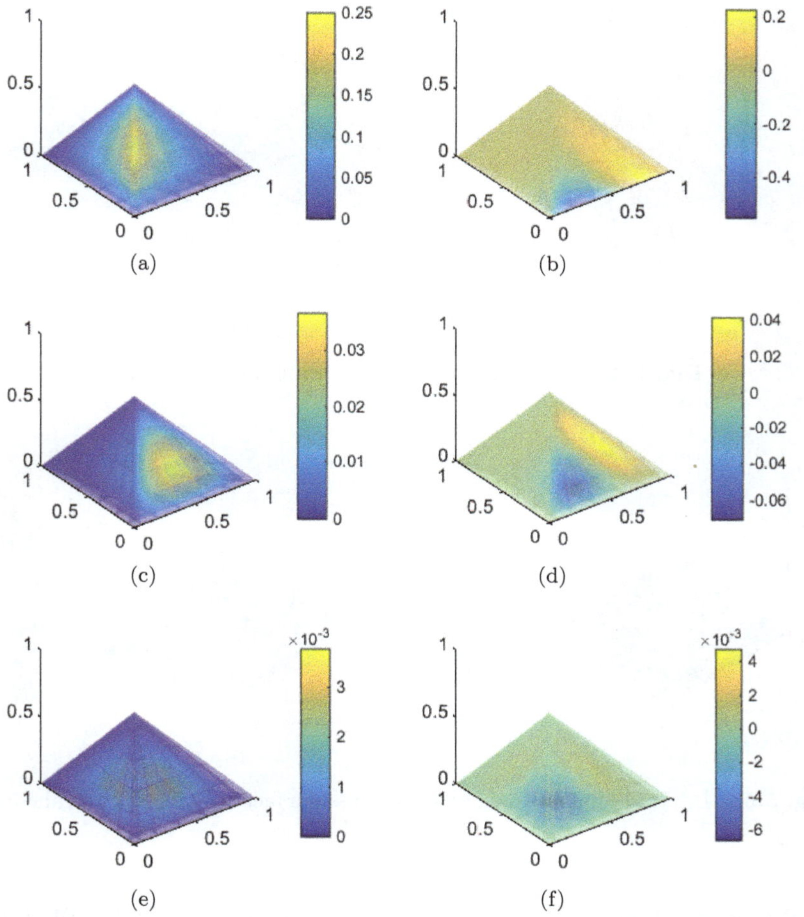

Fig. 5.3. Bases on edges 1 and 2 (a, b), face $\eta = 0$ (c, d) and body (e, f) of a unit tetrahedron.

If the number of nodes on each edge is taken as the same, then $\lambda_i (i = 1, 2, 3, 4)$ is symmetric with respect to each other. Therefore, we can let

$$\begin{aligned} r(\lambda_1, \lambda_2, \lambda_3, \lambda_4) &= \frac{\lambda_1 + \alpha\lambda_1\lambda_2\lambda_3\lambda_4}{\lambda_1 + \lambda_2 + \lambda_3 + 4\alpha\lambda_1\lambda_2\lambda_3\lambda_4} \\ s(\lambda_1, \lambda_2, \lambda_3, \lambda_4) &= \frac{\lambda_2 + \alpha\lambda_1\lambda_2\lambda_3\lambda_4}{\lambda_1 + \lambda_2 + \lambda_3 + 4\alpha\lambda_1\lambda_2\lambda_3\lambda_4} \\ t(\lambda_1, \lambda_2, \lambda_3, \lambda_4) &= \frac{\lambda_3 + \alpha\lambda_1\lambda_2\lambda_3\lambda_4}{\lambda_1 + \lambda_2 + \lambda_3 + 4\alpha\lambda_1\lambda_2\lambda_3\lambda_4} \end{aligned} \tag{5.28}$$

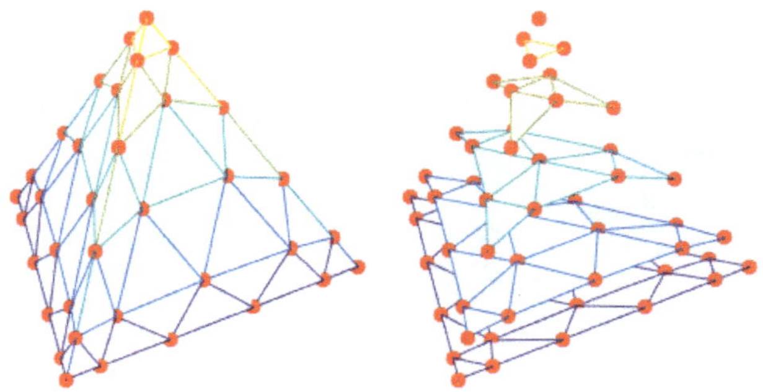

(a) (b)

Fig. 5.4. A unit tetrahedron with non-uniform nodes.

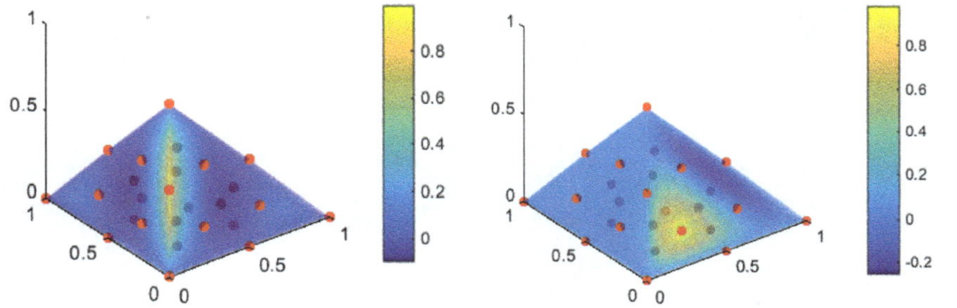

(a) (b)

Fig. 5.5. Interpolation bases on an edge (a) and a face (b) of a unit tetrahedron.

where α is a constant to be determined. Note that λ_4 is not equal to $1 - \lambda_1 - \lambda_2 - \lambda_3$ as has been discussed for triangles. If 1D Gauss–Lobatto nodes are used, for example, let $\boldsymbol{\alpha} = [\alpha_1, \alpha_2, \ldots, \alpha_n]$ be the 1D Gauss–Lobatto nodes defined on [0, 1]. If $\lambda_1 = \alpha_i, \lambda_2 = \alpha_j$ and $\lambda_3 = \alpha_k$, then we have $\lambda_4 = \alpha_{n+3-i-j-k}$. It was empirically found that the best choice of α on a unit tetrahedron is 64. The approximated nodes on a unit tetrahedron (Fig. 5.4) can be optimized by the bases presented above through the same procedure as for triangles, as shown in Section 4.1.3. With the optimized nodes, interpolation bases can be obtained as shown in Fig. 5.5 and DQ rules can be defined. The condition numbers of the generalized Vandermonde matrix for a unit tetrahedron using the Fekete points can be approximately given by $e^{0.9(n-2)}$, where $(n - 1)$ is the order of the element.

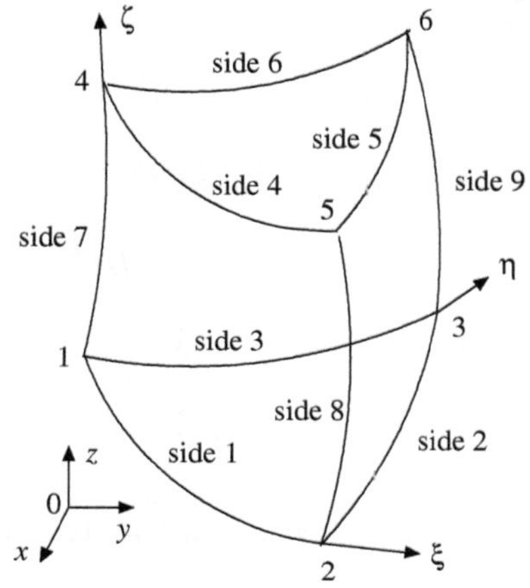

Fig. 5.6. A triangular prism (wedge) element.

5.2.2 *Wedge elements*

A wedge element (Fig. 5.6) is the product of a triangular element and a 1D line element. The vertex bases for a wedge element are given by

$$\begin{aligned} &S_1^{\mathrm{v}}(\xi,\eta,\zeta) = (1-\xi-\eta)(1-\zeta), \quad S_2^{\mathrm{v}}(\xi,\eta,\zeta) = \xi(1-\zeta) \\ &S_3^{\mathrm{v}}(\xi,\eta,\zeta) = \eta(1-\zeta), \quad S_4^{\mathrm{v}}(\xi,\eta,\zeta) = (1-\xi-\eta)\zeta \\ &S_5^{\mathrm{v}}(\xi,\eta,\zeta) = \xi\zeta, \quad S_6^{\mathrm{v}}(\xi,\eta,\zeta) = \eta\zeta \end{aligned} \tag{5.29}$$

The orthogonal polynomials on the thickness direction of the wedge element can be given as

$$H_k(\zeta) = \zeta(1-\zeta)P_k^{(2,2)}(2\zeta-1) \tag{5.30}$$

Thus, the orthogonal bases $S_{ie,}^{k}(\xi,\eta,\zeta)$ on the nine edges of the triangular prism element can be given as

$$\begin{aligned} S_i^{\mathrm{e},1}(\xi,\eta,\zeta) &= (1-\zeta)E_i^1(\xi,\eta) \\ S_i^{\mathrm{e},2}(\xi,\eta,\zeta) &= (1-\zeta)E_i^2(\xi,\eta) \\ S_i^{\mathrm{e},3}(\xi,\eta,\zeta) &= (1-\zeta)E_i^3(\xi,\eta) \end{aligned}$$

$$
\begin{aligned}
S_i^{e,4}(\xi,\eta,\zeta) &= \zeta E_i^1(\xi,\eta)\\
S_i^{e,5}(\xi,\eta,\zeta) &= \zeta E_i^2(\xi,\eta)\\
S_i^{e,6}(\xi,\eta,\zeta) &= \zeta E_i^3(\xi,\eta)\\
S_i^{e,7}(\xi,\eta,\zeta) &= (1-\xi-\eta)H_i(\zeta)\\
S_i^{e,8}(\xi,\eta,\zeta) &= \xi H_i(\zeta)\\
S_i^{e,9}(\xi,\eta,\zeta) &= \eta H_i(\zeta)
\end{aligned}
\tag{5.31}
$$

The orthogonal bases $S_{ij}^{f,k}(\xi,\eta,\zeta)$ on the five faces can be given as

$$
\begin{aligned}
S_{ij}^{f,1}(\xi,\eta,\zeta) &= (1-\zeta)F_{ij}(\xi,\eta)\\
S_{ij}^{f,2}(\xi,\eta,\zeta) &= \zeta F_{ij}(\xi,\eta)\\
S_{ij}^{f,3}(\xi,\eta,\zeta) &= E_i^1(\xi,\eta)H_j(\zeta)\\
S_{ij}^{f,4}(\xi,\eta,\zeta) &= E_i^2(\xi,\eta)H_j(\zeta)\\
S_{ij}^{f,5}(\xi,\eta,\zeta) &= E_i^3(\xi,\eta)H_j(\zeta)
\end{aligned}
\tag{5.32}
$$

The orthogonal bases $S_{ijk}^{b}(\xi,\eta,\zeta)$ inside the body of the element can be given as

$$S_{ijk}^{b}(\xi,\eta,\zeta) = F_{ij}(\xi,\eta)H_k(\zeta) \tag{5.33}$$

The edge, face and body bases of the wedge element above used the edge and face bases of the triangular elements. One can also present a general formula based on Eq. (5.30) and the orthogonal bases for triangles. This task is left to the reader. Figure 5.7 plots two orthogonal bases and two Lagrange functions on a triangular prism element.

5.2.3 *Hexahedral elements*

The bases for a hexahedral element (Fig. 5.8) can be derived in the same way as the edge element. In this case, only the 1D orthogonal

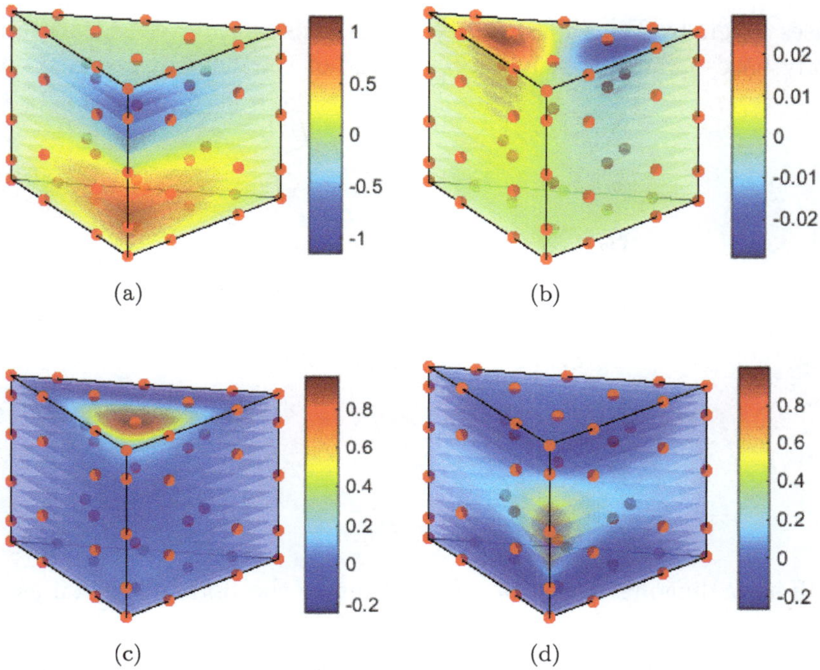

Fig. 5.7. Orthogonal bases (a, b) and interpolation bases (c, d) on a unit wedge element, where • denote nodes.

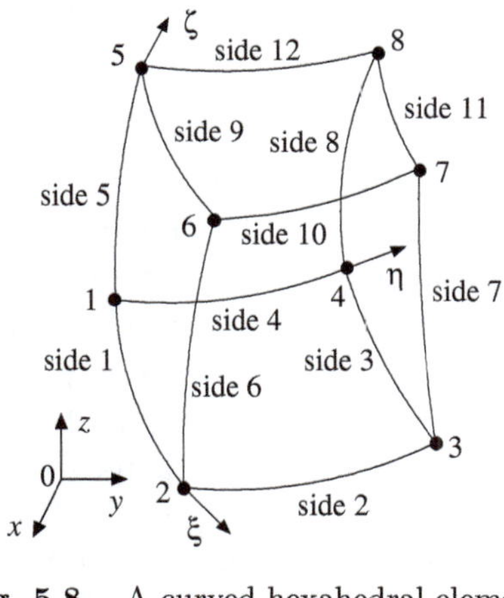

Fig. 5.8. A curved hexahedral element.

bases given by Eq. (5.30) is needed in constructing edge, face and body bases. The vertex bases $S_i^v(\xi, \eta, \zeta)$ of the element are given as

$$\begin{aligned} S_1^{\mathrm{v}}(\xi,\eta,\zeta) &= (1-\xi)(1-\eta)(1-\zeta) \\ S_2^{\mathrm{v}}(\xi,\eta,\zeta) &= \xi(1-\eta)(1-\zeta) \\ S_3^{\mathrm{v}}(\xi,\eta,\zeta) &= \xi\eta(1-\zeta) \\ S_4^{\mathrm{v}}(\xi,\eta,\zeta) &= (1-\xi)\eta(1-\zeta) \\ S_5^{\mathrm{v}}(\xi,\eta,\zeta) &= (1-\xi)(1-\eta)\zeta \\ S_6^{\mathrm{v}}(\xi,\eta,\zeta) &= \xi(1-\eta)\zeta \\ S_7^{\mathrm{v}}(\xi,\eta,\zeta) &= \xi\eta\zeta \\ S_8^{\mathrm{v}}(\xi,\eta,\zeta) &= (1-\xi)\eta\zeta \end{aligned} \tag{5.34}$$

The orthogonal bases $S_{ijk}^b(\xi, \eta, \zeta)$ inside the body are given as

$$\begin{aligned} S_{ijk}^{\mathrm{b}}(\xi,\eta,\zeta) &= \xi(1-\xi)\eta(1-\eta)\zeta(1-\zeta)P_i^{(2,2)} \\ &\quad \times (2\xi-1)P_j^{(2,2)}(2\eta-1)P_k^{(2,2)}(2\zeta-1) \end{aligned} \tag{5.35}$$

The edge and face bases can be obtained by removing some factors in Eq. (5.35). This process is left to the reader.

5.2.4 *Pyramid elements*

The choice of degrees of freedom and the types of basis functions that can be used in the pyramidal elements are constrained by the requirement for the resulting elements to be compatible with neighboring hexahedral and tetrahedral finite elements [123]. Indeed, the need for compatibility between hexahedra-pyramids and between tetrahedra-pyramids dictates the choice of the degrees of freedom on the faces of the pyramid. The problems start when one seeks to select a basis for the pyramidal elements. In 1971, Zienkiewicz [122] first proposed to collapse all of the nodes of one face of a hexahedral element into one node to obtain a pyramidal element (Fig. 5.9). The shape functions for the uncollapsed nodes are the same as before, while the shape function for the collapsed (degenerated) node is the

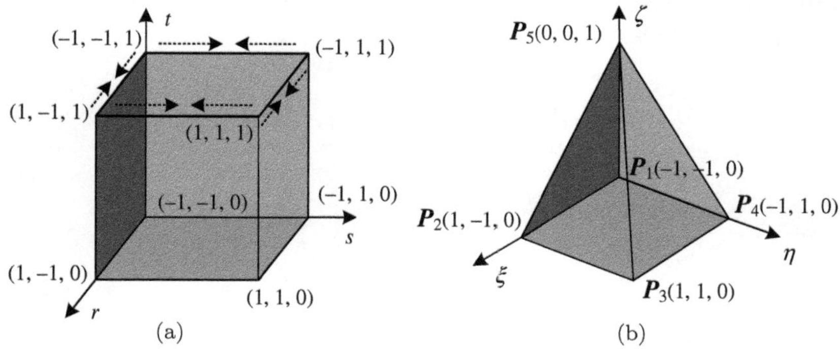

Fig. 5.9. A unit hexahedron (a) and a unit pyramid (b).

sum of the original shape functions for that face, namely, the vertex shape functions are

$$\begin{aligned} S_1^{\mathrm{V}} &= \frac{1}{4}(1-r)(1-s)(1-t) \\ S_2^{\mathrm{V}} &= \frac{1}{4}(1+r)(1-s)(1-t) \\ S_3^{\mathrm{V}} &= \frac{1}{4}(1+r)(1+s)(1-t) \\ S_4^{\mathrm{V}} &= \frac{1}{4}(1-r)(1+s)(1-t) \\ S_5^{\mathrm{V}} &= t \end{aligned} \tag{5.36}$$

The collapsing of the four vertices means that the Jacobian locally tends to zero, making derivatives impossible to compute. The errors introduced seem in practice to be fairly local, so these elements have some applications although their description is accompanied by a warning regarding unreliable local stress results (for example, in the ANSYS code, the degeneration form of the element designated SOLID186) [104]. In 1973, Newton [124] showed that a more extensive modification is required for the shape functions of degenerated brick-type isoparametric elements.

First work about nodal pyramidal elements was released by Bedrosian [108] in 1992 who noted the impossibility of choosing only polynomial shape functions if we want to preserve the conformity with neighboring elements. As a solution, he proposes what he calls "rabbit functions" that are rational basis functions on the pyramids. They are as follows:

$$
\begin{aligned}
S_1^{\mathrm{v}} &= \frac{1}{4}\left(1-\xi-\eta-\zeta+\frac{\xi\eta}{1-\zeta}\right)\\
S_2^{\mathrm{v}} &= \frac{1}{4}\left(1+\xi-\eta-\zeta-\frac{\xi\eta}{1-\zeta}\right)\\
S_3^{\mathrm{v}} &= \frac{1}{4}\left(1+\xi+\eta-\zeta+\frac{\xi\eta}{1-\zeta}\right)\\
S_4^{\mathrm{v}} &= \frac{1}{4}\left(1-\xi+\eta-\zeta-\frac{\xi\eta}{1-\zeta}\right)\\
S_5^{\mathrm{v}} &= \zeta
\end{aligned}
\tag{5.37}
$$

It is verified that these shape functions satisfy the requirements of C^0 continuity at all faces. For any face, it suffices to show that the shape functions reduce to a two-dimensional triangular or quadrilateral shape function, since adjacent tetrahedrons, prisms or hexahedrons have the same property. It is further observed that the Jacobian matrix is constant for any pyramid element of this kind with a planar parallelogram base. In the general case of a base of any shape, not necessarily in a plane, the Jacobian will, of course, vary with position, but far less strongly than for the degenerate hexahedron.

The rational "rabbit functions" given by Bedrosian [108] for first- and second-order pyramidal elements have been demonstrated to be at least as good as classic elements. Coulomb *et al.* discussed the behavior of the "rabbit functions" and their derivatives. At the top of the pyramid, $zeta = 1$ and consequently $\xi = \eta = 0$ nullify both numerators and denominators of the rational terms. These local coordinates verify the following inequalities:

$$
\begin{gathered}
-(1-\zeta) \le \xi \le (1-\zeta)\\
-(1-\zeta) \le \eta \le (1-\zeta)\\
0 \le \zeta \le 1
\end{gathered}
\tag{5.38}
$$

As the rational terms $\xi\eta/(1-\zeta) \to 0$ when $\zeta \to 1$, the nodal functions S_1^v, S_2^v, S_3^v and S_4^v are well defined at the summit. Inequalities

Eq.(5.38) in local coordinates can also be written as follows:

$$-1 \leq \frac{\xi}{1-\zeta} \leq 1$$
$$-1 \leq \frac{\eta}{1-\zeta} \leq 1 \tag{5.39}$$

which shows that derivatives are indeterminate. However, Bravo *et al.* proved that the rational term in the vertex bases itself and its derivatives vanish at the top of the pyramid, so it does not cause any inconvenience in computation.

In 2010, Bergot *et al.* [118] constructed highly efficient nodal pyramidal finite elements of any order using rational "rabbit functions" on the pyramid vertices and Dubiner's orthogonal expansion polynomials inside the pyramid domain. Numerical experiments conducted with these elements (up to order six) exhibit a low phase error, a good stability condition, and a very good behavior in hybrid meshes.

In 2014, Bravo *et al.* [105] defined a set of rational form functions of degree p in a hierarchical way for pyramidal elements. The concept of side of the pyramidal element was extended to include its vertices, edges, faces and interior. The rational terms are limited within their domains and they go to zero at the element's boundary. The edge, face and body bases were constructed through rational "rabbit functions" and the Chebyshev polynomials. One of the main tests consisted in interpolating every monomial of degree 0–7 in three variables, including the rational monomials.

Modified Dubiner's orthogonal polynomials will be derived herein using the rational vertex bases shown in Eq. (5.37). The orthogonal bases are split into interior and boundary modes (Fig. 5.10). A unit hexahedral domain can be transformed into a unit pyramid domain through Duffy's transformation as follows:

$$\xi = r(1-t) \quad \eta = s(1-t) \quad \zeta = t \tag{5.40}$$

Its inverse transformation is

$$r = \frac{\xi}{1-\zeta} \quad s = \frac{\eta}{1-\zeta} \quad t = \zeta \tag{5.41}$$

One can see from Fig. 5.11 that the coordinate transformations given by Eq. (5.41) link the parametric domains of the unit pyramid

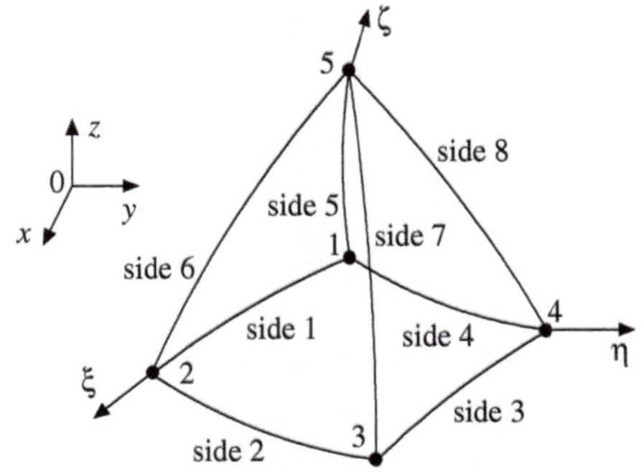

Fig. 5.10. A curved pyramidal element.

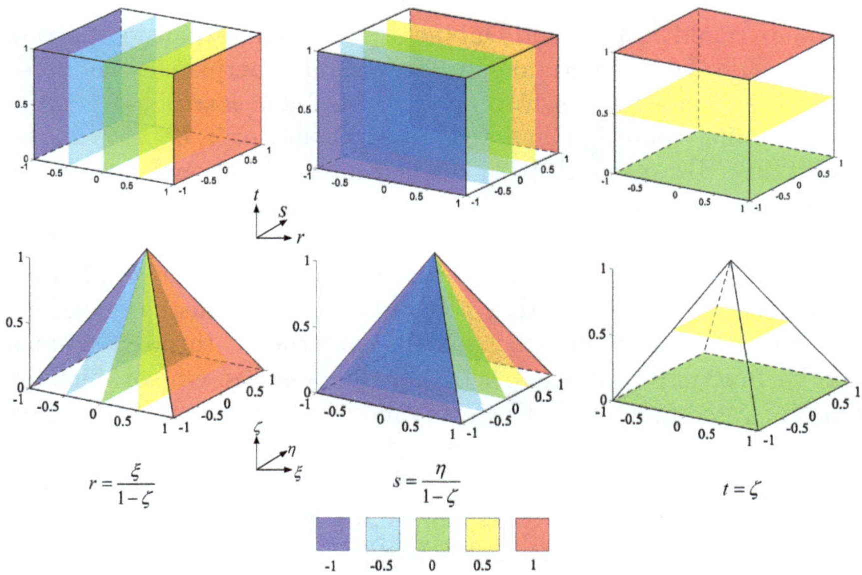

Fig. 5.11. Parametric relations between a unit hexahedron (up) and a unit pyramid (down).

and the unit hexahedron together. For example, a plane $r = -1$ of a unit hexahedron corresponds to a plane $\xi/(1-\zeta) = -1$ of a unit pyramid. One can easily verify that Eq. (5.37) can be obtained by substituting Eq. (5.41) into Eq. (5.36).

With the vertex shape functions, we can derive the edge, face and body shape functions (Fig. 5.12). For this purpose, we write

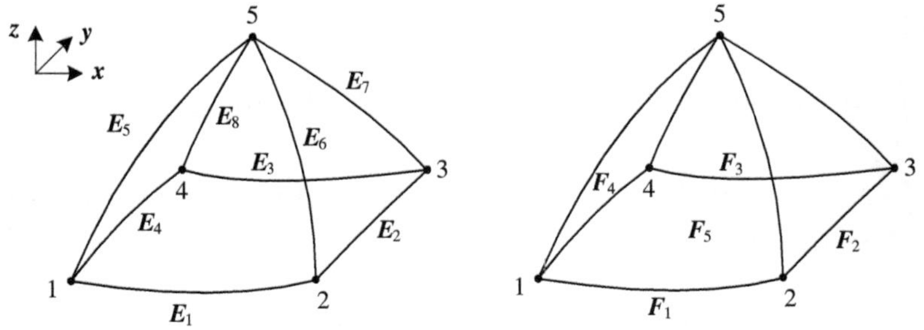

Fig. 5.12. The edge and face numbers of a pyramidal element.

Eq. (5.37) in a different form as

$$\begin{aligned} S_1^{\mathrm{v}} &= \frac{1}{4}\frac{(1-\xi-\zeta)(1-\eta-\zeta)}{1-\zeta} \\ S_2^{\mathrm{v}} &= \frac{1}{4}\frac{(1+\xi-\zeta)(1-\eta-\zeta)}{1-\zeta} \\ S_3^{\mathrm{v}} &= \frac{1}{4}\frac{(1+\xi-\zeta)(1+\eta-\zeta)}{1-\zeta} \\ S_4^{\mathrm{v}} &= \frac{1}{4}\frac{(1-\xi-\zeta)(1+\eta-\zeta)}{1-\zeta} \\ S_5^{\mathrm{v}} &= \zeta \end{aligned} \tag{5.42}$$

One can see that a vertex shape function on the square face vanishes on the two triangular surfaces opposite to the vertex. The top vertex shape function, namely, S_5^v, vanishes on the bottom square face.

With the properties of the vertex shape functions described above, the edge function $S_i^{\mathrm{e},1}$, namely, E_1, as shown in Fig. 5.12, formed by vertices 1 and 2 can be written as

$$S_i^{\mathrm{e},1} = S_1^{\mathrm{v}} S_2^{\mathrm{v}} H_i \tag{5.43}$$

where H_i is an undetermined polynomial and i is an integer index. In order to improve the numerical properties of the bases, we hope they can be orthogonal, i.e., the edge shape functions satisfy the following condition:

$$\int_V S_i^{\mathrm{e},1} S_j^{\mathrm{e},1} \mathrm{d}V = \int_0^1 \int_{\zeta-1}^{1-\zeta} \int_{\zeta-1}^{1-\zeta} S_i^{\mathrm{e},1} S_j^{\mathrm{e},1} \mathrm{d}\xi \mathrm{d}\eta \mathrm{d}\zeta = \delta_i^j C_i \tag{5.44}$$

where δ_{ij} is the Kronecker delta. The substitution of Eq. (5.43) into Eq. (5.44) leads to

$$\begin{aligned} &\int_0^1 \int_{\zeta-1}^{1-\zeta} \int_{\zeta-1}^{1-\zeta} S_i^{\mathrm{e},1} S_j^{\mathrm{e},1} \mathrm{d}\xi \mathrm{d}\eta \mathrm{d}\zeta \\ &\quad = \int_0^1 \int_{\zeta-1}^{1-\zeta} \int_{\zeta-1}^{1-\zeta} (S_1^{\mathrm{v}} S_2^{\mathrm{v}})^2 H_i H_j \mathrm{d}\xi \mathrm{d}\eta \mathrm{d}\zeta \end{aligned} \tag{5.45}$$

Let

$$r = \frac{\xi}{1-\zeta}, \quad s = \frac{\eta}{1-\zeta}, \quad t = 2\zeta - 1 \tag{5.46}$$

Compared to Eq. (5.41), the t coordinate is expanded to $[-1, 1]$ in Eq. (5.46). Thus, we have

$$\xi = \frac{1}{2} r(1-t), \quad \eta = \frac{1}{2} s(1-t), \quad \zeta = \frac{1}{2}(1+t) \tag{5.47}$$

The determinate of the Jacobian of coordinates (ξ, η, ζ) with respect to (r, s, t) is

$$\|\boldsymbol{J}| = \left| \frac{\partial(\xi, \eta\zeta)}{\partial(r, st)} \right| = \begin{vmatrix} \dfrac{1-t}{2} & 0 & -\dfrac{r}{2} \\ 0 & \dfrac{1-t}{2} & -\dfrac{s}{2} \\ 0 & 0 & \dfrac{1}{2} \end{vmatrix} = \frac{1}{2^3}(1-t)^2 \tag{5.48}$$

Thus, the substitution of Eq. (5.47) into Eq. (5.45) leads to

$$\int_0^1 \int_{\zeta-1}^{1-\zeta} \int_{\zeta-1}^{1-\zeta} (S_1^{\mathrm{v}} S_2^{\mathrm{v}})^2 H_i H_j \mathrm{d}\xi \mathrm{d}\eta \mathrm{d}\zeta = \int_{-1}^1 \int_{-1}^1 \int_{-1}^1 \bar{J} H_i H_j \mathrm{d}r \mathrm{d}s \mathrm{d}t \tag{5.49}$$

where

$$\bar{J} = \frac{1}{2^{19}}(1-r)^2(1+r)^2(1-s)^4(1-t)^6 \tag{5.50}$$

The bases $H_i(r, s, t)$ can be determined as

$$H_i = (1-r)^a(1+r)^b(1-s)^c(1+s)^d(1-t)^e(1+t)^f P_i^{(2a+2,2b+2)}(r) \tag{5.51}$$

where a, b, c, d, e and f should be determined according to convenience in applications and the requirements of the Jacobi polynomials. In order to eliminate $(1-\zeta)$ in the denominator of r and s in Eq. (5.51), one needs to take $e = i$. For the C^0 problem considered in this work, we can take $a = b = c = d = f = 0$. So, we have

$$H_i(r, s, t) = (1-t)^i P_i^{(2,2)}(r) \tag{5.52}$$

Thus, the edge shape function $S_i^{\mathrm{e},1}$ can be written as

$$S_i^{\mathrm{e},1} = S_1^{\mathrm{v}} S_2^{\mathrm{v}} 2^i (1-\zeta)^i P_i^{(2,2)}\left(\frac{\xi}{1-\zeta}\right) \tag{5.53}$$

The constant coefficient 2^i in Eq. (5.53) is usually omitted in application. The computation of $(1-\zeta)$ that is in the denominator can be avoided using the procedure shown in Section 3.3.1.

Using similar procedures as above, one can get other edge shape functions. The edge shape function Eq. (5.53), together with other edge shape functions are listed as follows:

$$\begin{aligned}
S_i^{\mathrm{e},1} &= S_1^{\mathrm{v}} S_2^{\mathrm{v}} \tilde{P}_i^{(2,2)}(\xi, 1-\zeta)\\
S_j^{\mathrm{e},2} &= S_2^{\mathrm{v}} S_3^{\mathrm{v}} \tilde{P}_j^{(2,2)}(\eta, 1-\zeta)\\
S_i^{\mathrm{e},3} &= S_3^{\mathrm{v}} S_4^{\mathrm{v}} \tilde{P}_i^{(2,2)}(\xi, 1-\zeta)\\
S_j^{\mathrm{e},4} &= S_4^{\mathrm{v}} S_1^{\mathrm{v}} \tilde{P}_j^{(2,2)}(\eta, 1-\zeta)\\
S_k^{\mathrm{e},5} &= S_1^{\mathrm{v}} S_5^{\mathrm{v}} P_k^{(4,2)}(2\zeta-1)\\
S_k^{\mathrm{e},6} &= S_2^{\mathrm{v}} S_5^{\mathrm{v}} P_k^{(4,2)}(2\zeta-1)\\
S_k^{\mathrm{e},7} &= S_3^{\mathrm{v}} S_5^{\mathrm{v}} P_k^{(4,2)}(2\zeta-1)\\
S_k^{\mathrm{e},8} &= S_4^{\mathrm{v}} S_5^{\mathrm{v}} P_k^{(4,2)}(2\zeta-1)
\end{aligned} \tag{5.54}$$

where the indices i, j, k are respectively the bases sequences in ξ, η, ζ parametric directions. One can see that an edge shape function can be expressed as $S^k_{ie,} = S_{m^v} S_{n^v} H_i$, where the vertex indices m and n define the edge number k, as shown in Fig. 5.12.

The derivation of face and body shape factions is similar to the above for edge shape functions. They are listed as follows:

$$\begin{aligned}
S^{\mathrm{f},1}_{ik} &= S^{\mathrm{v}}_1 S^{\mathrm{v}}_2 S^{\mathrm{v}}_5 \tilde{P}^{(2,2)}_i(\xi, 1-\zeta) P^{(2i+6,2)}_k(2\zeta - 1)\\
S^{\mathrm{f},2}_{jk} &= S^{\mathrm{v}}_2 S^{\mathrm{v}}_3 S^{\mathrm{v}}_5 \tilde{P}^{(2,2)}_j(\eta, 1-\zeta) P^{(2j+6,2)}_k(2\zeta - 1)\\
S^{\mathrm{f},3}_{ik} &= S^{\mathrm{v}}_3 S^{\mathrm{v}}_4 S^{\mathrm{v}}_5 \tilde{P}^{(2,2)}_i(\xi, 1-\zeta) P^{(2i+6,2)}_k(2\zeta - 1)\\
S^{\mathrm{f},4}_{jk} &= S^{\mathrm{v}}_4 S^{\mathrm{v}}_1 S^{\mathrm{v}}_5 \tilde{P}^{(2,2)}_j(\eta, 1-\zeta) P^{(2j+6,2)}_k(2\zeta - 1)\\
S^{\mathrm{f},5}_{ij} &= S^{\mathrm{v}}_1 S^{\mathrm{v}}_3 \tilde{P}^{(2,2)}_i(\xi, 1-\zeta) \tilde{P}^{(2,2)}_j(\eta, 1-\zeta)\\
S^{\mathrm{b}}_{ijk} &= S^{\mathrm{v}}_1 S^{\mathrm{v}}_3 S^{\mathrm{v}}_5 \tilde{P}^{(2,2)}_i(\xi, 1-\zeta) \tilde{P}^{(2,2)}_j(\eta, 1-\zeta) P^{(2i+2j+6,2)}_k(2\zeta - 1)
\end{aligned} \tag{5.55}$$

Figure 5.13 presents the figures of a few edge, face and body shape functions. One can see from Fig. 5.13 that the orthogonal bases provided in this work have some symmetric or anti-symmetric axes.

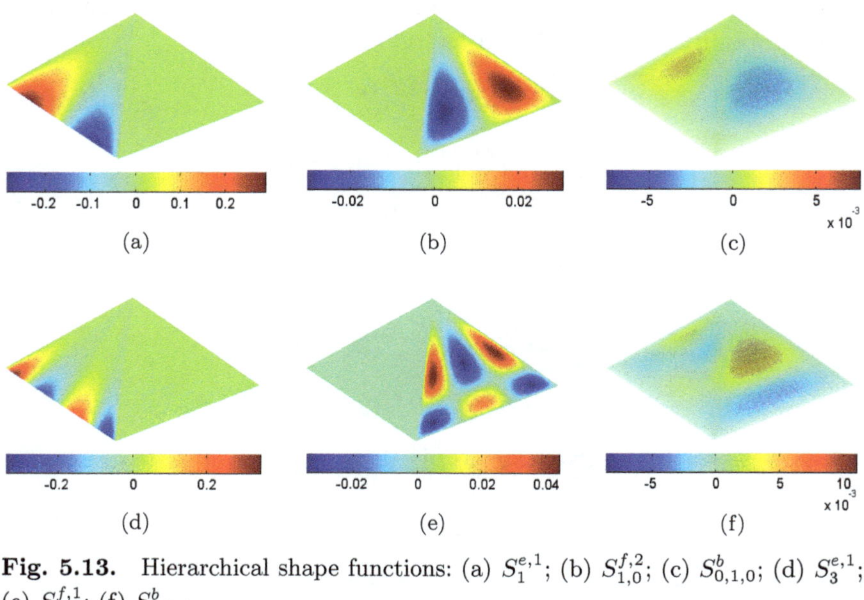

Fig. 5.13. Hierarchical shape functions: (a) $S^{e,1}_1$; (b) $S^{f,2}_{1,0}$; (c) $S^b_{0,1,0}$; (d) $S^{e,1}_3$; (e) $S^{f,1}_{2,1}$; (f) $S^b_{0,2,1}$.

Using the derivation procedures above, the edge, face and body shape functions can be written in a general form as

$$N_{ijk}(\xi,\eta,\zeta) = (S_1^{\mathrm{v}})^g (S_2^{\mathrm{v}})^h (S_3^{\mathrm{v}})^l (S_4^{\mathrm{v}})^m (S_5^{\mathrm{v}})^n H_{ijk}(\xi,\eta,\zeta) \tag{5.56}$$

which should satisfy the following orthogonal condition:

$$\begin{aligned} &\int_0^1 \int_{\zeta-1}^{1-\zeta} \int_{\zeta-1}^{1-\zeta} N_{ijk} N_{pqw} \mathrm{d}\xi \mathrm{d}\eta \mathrm{d}\zeta \\ &\quad = \int_{-1}^1 \int_{-1}^1 \int_{-1}^1 \tilde{J} H_{ijk} H_{pqw} \mathrm{d}r \mathrm{d}s \mathrm{d}t = \delta_{ip}\delta_{jq}\delta_{kw} C_{ijk} \end{aligned} \tag{5.57}$$

where

$$\tilde{J} = \frac{\begin{matrix}(1-r)^{2(g+m)}(1+r)^{2(h+l)}(1-s)^{2(g+h)}(1+s)^{2(l+m)} \\ \times (1-t)^{2(g+h+l+m)+2}(1+t)^{2n}\end{matrix}}{2^{6(g+h+l+m)+2n+3}} \tag{5.58}$$

and H_{ijk} can be determined through the same way as Eqs. (5.46)–(5.53) as

$$\begin{aligned} H_{ijk}(r,s,t) &= [P_i^{(2g+2m,2h+2l)}(r)][P_j^{(2g+2h,2l+2m)}(s)] \\ &\quad \times [(1-t)^{i+j} P_k^{(2g+2h+2l+2m+2+2i+2j,2n)}(t)] \end{aligned} \tag{5.59}$$

or

$$\begin{aligned} H_{ijk}(\xi,\eta,\zeta) &= 2^{i+j} \tilde{P}_i^{(2g+2m,2h+2l)}(\xi, 1-\zeta) \tilde{P}_j^{(2g+2h,2l+2m)}(\eta, 1-\zeta) \\ &\quad \times P_k^{(2g+2h+2l+2m+2+2i+2j,2n)}(2\zeta - 1) \end{aligned} \tag{5.60}$$

where g, h, l, m, n equal 0 or 1. The constant coefficient 2^{i+j} is omitted in this work. The constant C_{ijk} in Eq. (5.57) is

$$C_{ijk} = \frac{C_i^{(2g+2m,2h+2l)} C_j^{(2g+2h,2l+2m)} C_k^{(2g+2h+2l+2m+2+2i+2j,2n)}}{2^{6(g+h+l+m)+2n+3}} \tag{5.61}$$

One can see that Eq. (5.56) becomes the body function S_{ijk}^b in Eq. (5.55) when $g = l = n = 1$ and $h = m = 0$. The other edge and face shape functions above can be obtained in similar ways.

With the hierarchical bases on a pyramid in hand, the procedure of computing element matrices becomes routine and it is similar to that for other types of three-dimensional DQHFEM elements [76]. For completeness, a simple introduction is presented here. First, for simplicity of boundary condition imposition and element assembly, we need to transform the edge and face bases into interpolation ones. Let $(\bar{\xi}_i, \bar{\eta}_i, \bar{\zeta}_i)$ be a new set of nodes whose number on the edges, faces and body are exactly the same as the number of bases there. Substituting the new nodes into Eq. (5.15), we can get

$$\bar{\boldsymbol{u}} = u(\bar{\boldsymbol{\xi}}, \bar{\boldsymbol{\eta}}, \bar{\boldsymbol{\zeta}}) = \boldsymbol{N}^{\mathrm{T}}(\bar{\boldsymbol{\xi}}, \bar{\boldsymbol{\eta}}, \bar{\boldsymbol{\zeta}})\tilde{\boldsymbol{u}} \tag{5.62}$$

Define a generalized Vandermonde matrix as follows:

$$\boldsymbol{V} = \boldsymbol{N}^{\mathrm{T}}(\bar{\boldsymbol{\xi}}, \bar{\boldsymbol{\eta}}, \bar{\boldsymbol{\zeta}}) \tag{5.63}$$

Then the substitution of Eq. (5.62) into Eq. (5.15) leads to

$$\begin{aligned} u[x(\xi,\eta,\zeta), y(\xi,\eta,\zeta), z(\xi,\eta,\zeta)] &= \boldsymbol{N}^{\mathrm{T}}(\xi,\eta,\zeta)\boldsymbol{V}^{-1}u(\bar{\boldsymbol{\xi}}, \bar{\boldsymbol{\eta}}, \bar{\boldsymbol{\zeta}}) \\ &= \sum_{i=1}^{N_{\mathrm{T}}} L_i(\xi,\eta,\zeta)u(\bar{\xi}_i, \bar{\eta}_i, \bar{\zeta}_i) \\ &= \bar{\boldsymbol{N}}^{\mathrm{T}}(\xi,\eta,\zeta)\bar{\boldsymbol{u}} \end{aligned} \tag{5.64}$$

where N_T denotes the total number of nodes, $L_i(\xi, \eta, \zeta)$ is defined as a Lagrange function on the domain at point $(\bar{\xi}_i, \bar{\eta}_i, \bar{\zeta}_i)$. The best choice of the new set of nodes $(\bar{\xi}_i, \bar{\eta}_i, \bar{\zeta}_i)$ is the Fekete points, which should satisfy the following conditions:

$$|L_i(\xi,\eta,\zeta)| \leq 1 \quad \text{and} \quad L_i(\xi_i,\eta_i,\zeta_i) = 1 \tag{5.65}$$

This relation usually can be replaced by

$$\frac{\partial L_i(\bar{\boldsymbol{\xi}}, \bar{\boldsymbol{\eta}}, \bar{\boldsymbol{\zeta}})}{\partial \bar{\xi}_i} = 0, \quad \frac{\partial L_i(\bar{\boldsymbol{\xi}}, \bar{\boldsymbol{\eta}}, \bar{\boldsymbol{\zeta}})}{\partial \bar{\eta}_i} = 0, \quad \frac{\partial L_i(\bar{\boldsymbol{\xi}}, \bar{\boldsymbol{\eta}}, \bar{\boldsymbol{\zeta}})}{\partial \bar{\zeta}_i} = 0 \tag{5.66}$$

In order to preserve the conformity with neighboring elements, the face nodes are taken as those of triangles and quadrangles (Fig. 5.14(a)). The Fekete nodes for triangles and quadrangles can be found in [76]. The interior Fekete nodes on pyramids should be solved from Eq. (4.46), as shown in Fig. 5.14(b). Fig. 5.15 presents a few

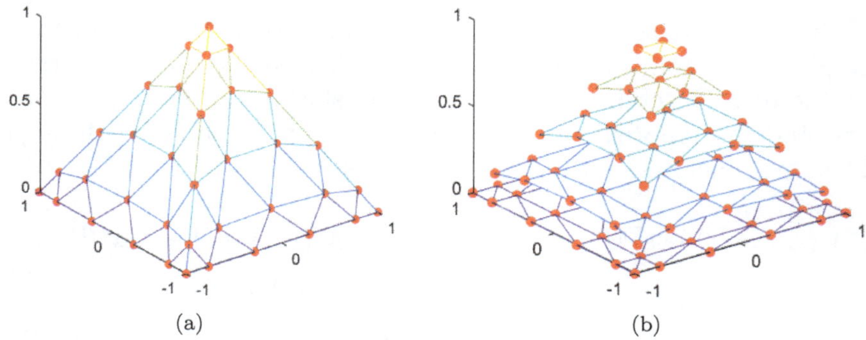

Fig. 5.14. Fekete nodes on triangles and quadrangles (a) and interior Fekete points on pyramids solved by Eq. (4.46).

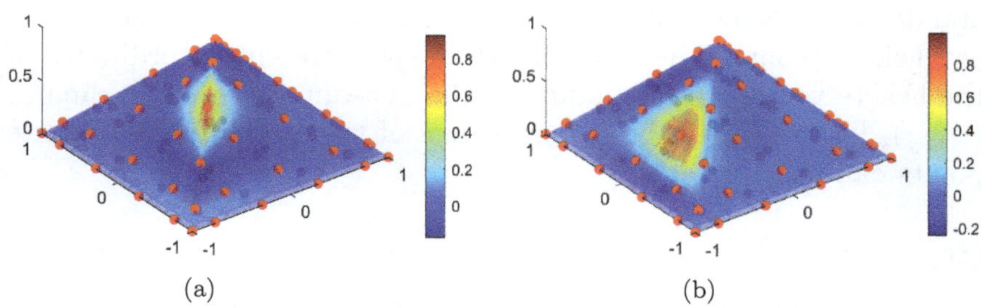

Fig. 5.15. Interpolation bases on an edge (a) and a face (b) of a unit pyramid.

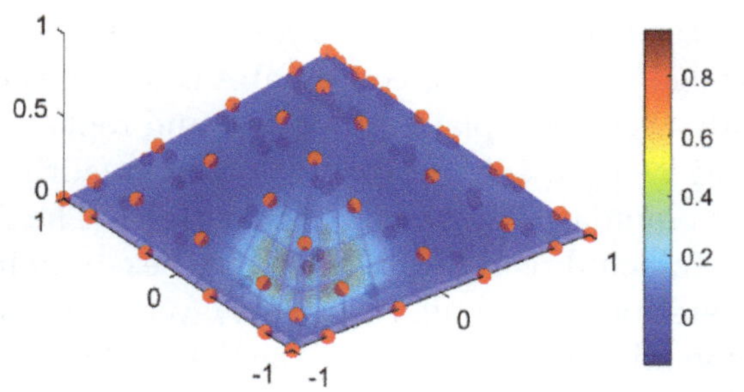

Fig. 5.16. An interior interpolation basis based on a unit pyramid.

interpolation bases on the edges and faces of a unit pyramid. An interior interpolation basis based on Fekete points is shown in Fig. 5.16. One can see that the requirements of Eq. (4.45) are satisfied.

In order to solve Eq. (4.46) efficiently though the Newton–Raphson method, approximated Fekete nodes with good accuracy are needed. Observing the Fekete points shown in Fig. 5.14(b), one can find that approximated Fekete points can be obtained through blending function interpolation as follows:

$$\begin{aligned} V(r,s) = {} & (1-s)M(r,0) + rM(1,s) + sM(r,1) + (1-r)M(0,s) \\ & - (1-r)(1-s)M(0,0) - r(1-s)M(1,0) \\ & - rsM(1,1) - (1-r)sM(0,1) \end{aligned} \tag{5.67}$$

where $0 \le r, s \le 1$. The functions $M(r_i, s)$ and $M(r, s_j)$ represent the four parametric curves that constitute the boundary of a quadrilateral domain (the red solid lines in Fig. 5.17(a)) formed by the boundary Fekete points. The function $M(r_i, s_j)$ represents coordinates of the Fekete points corresponding to the non-dimensional coordinates (r_i, s_j). For example, the ξ coordinates of the approximated Fekete points can be given as

$$\begin{aligned} \xi(r_i,s_j) = {} & (1-s_j)R(r_i,0) + r_iR(1,s_j) + s_jR(r_i,1) + (1-r_i)R(0,s_j) \\ & - (1-r_i)(1-s_j)R(0,0) - r_i(1-s_j)R(1,0) \\ & - r_is_jR(1,1) - (1-r_i)s_jR(0,1) \end{aligned} \tag{5.68}$$

where the indices i, j range from 2 to $n-1$. Here n is the number of nodes on an edge. The η and ζ coordinates can be given in a similar form. One only needs to replace ξ by η or ζ and replace R by S or T. Here R, S and T represent coordinates of the quadrilateral domain formed by the boundary Fekete points (Fig. 5.17). The T coordinates can also be obtained directly from boundary Fekete points by taking the ζ coordinates of each circular ring of a layer to be the same value as the corresponding boundary one. The T coordinates in this work are obtained by averaging values of the two methods. Figure 5.18 presents a comparison of the approximated Fekete points given by Eq. (5.66) and the exact Fekete points solved by Eq. (4.46). One can see that the present approximated Fekete points are very close to the exact ones.

In order to test the performance of the approximated Fekete points, we compute relative interpolation errors in the max norm

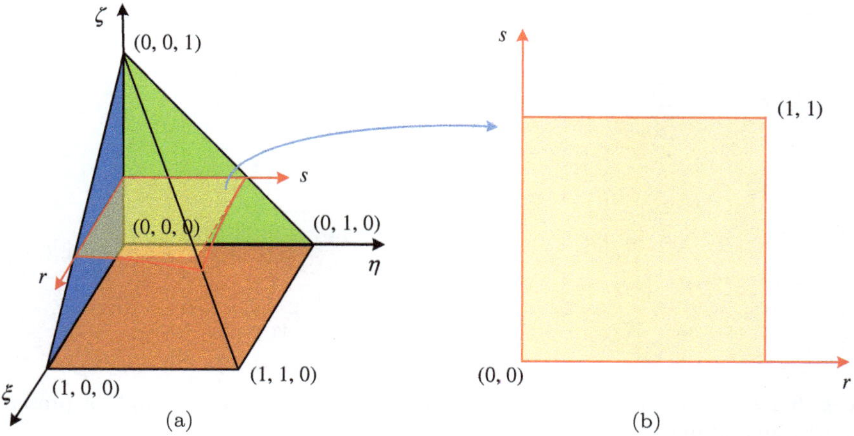

Fig. 5.17. A quarter of a pyramid (a) and its layer (b) used to interpolate Fekete points.

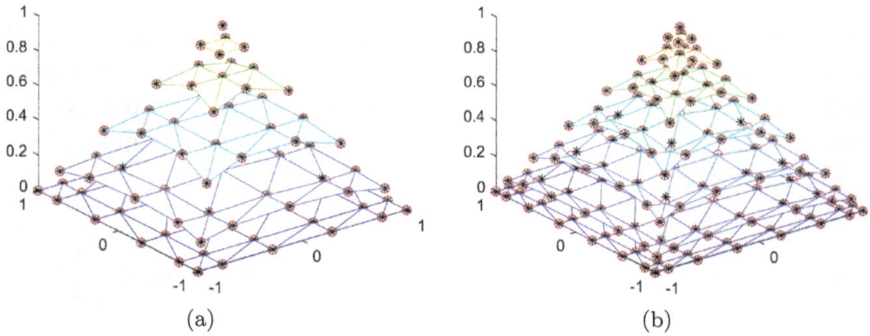

Fig. 5.18. A comparison of approximated Fekete points (*) given by Eq. (5.67) and exact Fekete points (o) solved by Eq. (4.46): (a) $n = 6$; (b) $n = 8$.

for two functions that were used in [100] as follows:

$$\begin{aligned} f_1(r, s, t) &= (r+1)(s+1)(t+1)\cos(r+s+t+1) \\ f_2(r, s, t) &= \frac{1}{1+(r^2+s^2+t^2)/2} \end{aligned} \tag{5.69}$$

The relative interpolation errors are defined as $\log[\|f - f_N\|/\|f\|]$, where N is the number of nodes in one direction of the pyramidal domain. One can see from Fig. 5.19 that the accuracy of the approximated Fekete points obtained through blending function interpolation is very close to that of the exact ones and is stable.

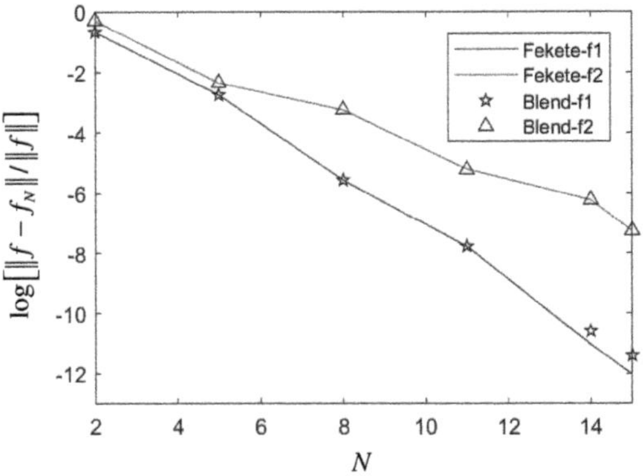

Fig. 5.19. A comparison of interpolation errors by Fekete points and the present approximated ones.

5.3 Strain Matrix

Since the faces of the tetrahedral, wedge, pyramid and hexahedral elements are triangles or quadrangles, to transform the boundary bases into interpolation, ones only needs the method presented in Section 4.1.3. Similar to Section 4.1.3, the displacement field given by Eq. (5.15) can be written as

$$\begin{aligned} u[x(\xi,\eta,\zeta),y(\xi,\eta,\zeta),z(\xi,\eta,\zeta)] &= \left[\boldsymbol{S}_1^{\mathrm{T}}(\xi,\eta,\zeta) \quad \boldsymbol{S}_2^{\mathrm{T}}(\xi,\eta,\zeta)\right]\begin{Bmatrix}\tilde{\boldsymbol{u}}_1\\ \tilde{\boldsymbol{u}}_2\end{Bmatrix} \\ &= \boldsymbol{S}^{\mathrm{T}}(\xi,\eta,\zeta)\tilde{\boldsymbol{u}} \end{aligned} \tag{5.70}$$

where $\boldsymbol{S}_1^T$ and $\boldsymbol{S}_2^T$ are respectively a vector of the boundary and interior bases; and $\tilde{\boldsymbol{u}}_1$ and $\tilde{\boldsymbol{u}}_2$ are the corresponding coefficients of generalized displacements. After transforming the boundary bases into interpolation ones, Eq. (5.70) is written as

$$u[x(\xi,\eta,\zeta),y(\xi,\eta,\zeta),z(\xi,\eta,\zeta)] = \boldsymbol{S}^{\mathrm{T}}(\xi,\eta,\zeta)\boldsymbol{u} \tag{5.71}$$

Substitution of the evaluation points $(\bar{\boldsymbol{\xi}},\bar{\boldsymbol{\eta}},\bar{\zeta})$ into Eq. (5.71) yields the following interpolation relation:

$$u(\bar{\boldsymbol{\xi}},\bar{\boldsymbol{\eta}},\bar{\boldsymbol{\zeta}}) = \boldsymbol{G}\boldsymbol{u} \quad \text{or} \quad \bar{\boldsymbol{u}} = \boldsymbol{G}\boldsymbol{u} \tag{5.72}$$

The first-order partial derivatives with respect to the parametric coordinates are

$$\frac{\partial u(\bar{\boldsymbol{\xi}},\bar{\boldsymbol{\eta}},\bar{\boldsymbol{\zeta}})}{\partial \xi} = \boldsymbol{A}^{(1)}\boldsymbol{u}, \quad \frac{\partial u(\bar{\boldsymbol{\xi}},\bar{\boldsymbol{\eta}},\bar{\boldsymbol{\zeta}})}{\partial \eta} = \boldsymbol{B}^{(1)}\boldsymbol{u}, \quad \frac{\partial u(\bar{\boldsymbol{\xi}},\bar{\boldsymbol{\eta}},\bar{\boldsymbol{\zeta}})}{\partial \zeta} = \boldsymbol{C}^{(1)}\boldsymbol{u} \tag{5.73}$$

Note that the evaluation points $(\bar{\boldsymbol{\xi}},\bar{\boldsymbol{\eta}},\bar{\boldsymbol{\zeta}})$ are usually integration nodes.

For a 3D function $u(x,y,z)$ that is an implicit function of ξ,η and ζ as $u = u[x(\xi,\eta,\zeta), y(\xi,\eta,\zeta), z(\xi,\eta,\zeta)]$, the first-order derivatives of $u = u[x(\xi,\eta,\zeta), y(\xi,\eta,\zeta), z(\xi,\eta,\zeta)]$ with respect to x, y and z can be obtained through the chain rule of differentiations as

$$\begin{bmatrix} \dfrac{\partial u}{\partial x} \\ \dfrac{\partial u}{\partial y} \\ \dfrac{\partial u}{\partial z} \end{bmatrix} = \frac{1}{|\boldsymbol{J}|} \begin{bmatrix} \dfrac{\partial y}{\partial \eta}\dfrac{\partial z}{\partial \zeta} - \dfrac{\partial z}{\partial \eta}\dfrac{\partial y}{\partial \zeta} & \dfrac{\partial y}{\partial \zeta}\dfrac{\partial z}{\partial \xi} - \dfrac{\partial z}{\partial \zeta}\dfrac{\partial y}{\partial \xi} & \dfrac{\partial y}{\partial \xi}\dfrac{\partial z}{\partial \eta} - \dfrac{\partial z}{\partial \xi}\dfrac{\partial y}{\partial \eta} \\ \dfrac{\partial z}{\partial \eta}\dfrac{\partial x}{\partial \zeta} - \dfrac{\partial x}{\partial \eta}\dfrac{\partial z}{\partial \zeta} & \dfrac{\partial z}{\partial \zeta}\dfrac{\partial x}{\partial \xi} - \dfrac{\partial x}{\partial \zeta}\dfrac{\partial z}{\partial \xi} & \dfrac{\partial z}{\partial \xi}\dfrac{\partial x}{\partial \eta} - \dfrac{\partial x}{\partial \xi}\dfrac{\partial z}{\partial \eta} \\ \dfrac{\partial x}{\partial \eta}\dfrac{\partial y}{\partial \zeta} - \dfrac{\partial y}{\partial \eta}\dfrac{\partial x}{\partial \zeta} & \dfrac{\partial x}{\partial \zeta}\dfrac{\partial y}{\partial \xi} - \dfrac{\partial y}{\partial \zeta}\dfrac{\partial x}{\partial \xi} & \dfrac{\partial x}{\partial \xi}\dfrac{\partial y}{\partial \eta} - \dfrac{\partial y}{\partial \xi}\dfrac{\partial x}{\partial \eta} \end{bmatrix} \begin{bmatrix} \dfrac{\partial u}{\partial \xi} \\ \dfrac{\partial u}{\partial \eta} \\ \dfrac{\partial u}{\partial \zeta} \end{bmatrix} \tag{5.74}$$

where the determinant $|\boldsymbol{J}|$ of the Jacobian $\boldsymbol{J} = \partial(x,y,y)/\partial(\xi,\eta,\zeta)$ is

$$\begin{aligned} |\boldsymbol{J}| = & \frac{\partial x}{\partial \xi}\left(\frac{\partial y}{\partial \eta}\frac{\partial z}{\partial \zeta} - \frac{\partial z}{\partial \eta}\frac{\partial y}{\partial \zeta}\right) + \frac{\partial y}{\partial \xi}\left(\frac{\partial z}{\partial \eta}\frac{\partial x}{\partial \zeta} - \frac{\partial x}{\partial \eta}\frac{\partial z}{\partial \zeta}\right) \\ & + \frac{\partial z}{\partial \xi}\left(\frac{\partial x}{\partial \eta}\frac{\partial y}{\partial \zeta} - \frac{\partial y}{\partial \eta}\frac{\partial x}{\partial \zeta}\right) \end{aligned} \tag{5.75}$$

Using the same procedure as in Section 4.1.4, the substitution of Eq. (5.73) into Eq. (5.74) yields

$$\bar{\boldsymbol{u}}_x = \boldsymbol{G}_x\boldsymbol{u}, \quad \bar{\boldsymbol{u}}_y = \boldsymbol{G}_y\boldsymbol{u}, \quad \bar{\boldsymbol{u}}_z = \boldsymbol{G}_z\boldsymbol{u} \tag{5.76}$$

This is the generalized differential quadrature (DQ) rules for 3D solid elements.

The strain–displacement relation given by Eq. (5.12) can be discretized by the generalized DQ rules (5.76) as

$$\boldsymbol{\varepsilon} = \begin{bmatrix} \boldsymbol{G}_x & \boldsymbol{0} & \boldsymbol{0} \\ \boldsymbol{0} & \boldsymbol{G}_y & \boldsymbol{0} \\ \boldsymbol{0} & \boldsymbol{0} & \boldsymbol{G}_z \\ \boldsymbol{G}_y & \boldsymbol{G}_x & \boldsymbol{0} \\ \boldsymbol{G}_z & \boldsymbol{0} & \boldsymbol{G}_x \\ \boldsymbol{0} & \boldsymbol{G}_z & \boldsymbol{G}_y \end{bmatrix} \begin{Bmatrix} \boldsymbol{u} \\ \boldsymbol{v} \\ \boldsymbol{w} \end{Bmatrix} = \boldsymbol{B}\boldsymbol{d}_e \tag{5.77}$$

One can see that it is similar to the low-order case.

5.4 Element Matrices

5.4.1 *Isotropic solids*

The method of deriving element matrices is similar to that in Section 4.2, so only the final results are presented here. The stiffness matrix for 3D isotropic solid is obtained as

$$\boldsymbol{k}_e = \frac{G}{\upsilon_2} \begin{bmatrix} \boldsymbol{K}_{11} & & \text{sym} \\ \boldsymbol{K}_{21} & \boldsymbol{K}_{22} & \\ \boldsymbol{K}_{31} & \boldsymbol{K}_{32} & \boldsymbol{K}_{33} \end{bmatrix} \tag{5.78}$$

where E is the Young's modulus, $G = E/2(1+\upsilon)$ is the shear modulus, υ is the Poisson's ratio, $\upsilon_1 = (1-\upsilon)/2, \upsilon_2 = 0.5 - \upsilon$, and

$$\begin{aligned} \begin{bmatrix} \boldsymbol{K}_{11} \\ \boldsymbol{K}_{22} \\ \boldsymbol{K}_{33} \end{bmatrix} &= \begin{bmatrix} 2\upsilon_1 & \upsilon_2 & \upsilon_2 \\ \upsilon_2 & 2\upsilon_1 & \upsilon_2 \\ \upsilon_2 & \upsilon_2 & 2\upsilon_1 \end{bmatrix} \begin{bmatrix} \boldsymbol{G}_x^T\boldsymbol{C}\boldsymbol{G}_x \\ \boldsymbol{G}_y^T\boldsymbol{C}\boldsymbol{G}_y \\ \boldsymbol{G}_z^T\boldsymbol{C}\boldsymbol{G}_z \end{bmatrix}, \\ \boldsymbol{K}_{21} &= \upsilon\boldsymbol{G}_y^T\boldsymbol{C}\boldsymbol{G}_x + \upsilon_2\boldsymbol{G}_x^T\boldsymbol{C}\boldsymbol{G}_y \\ \boldsymbol{K}_{31} &= \upsilon\boldsymbol{G}_z^T\boldsymbol{C}\boldsymbol{G}_x + \upsilon_2\boldsymbol{G}_x^T\boldsymbol{C}\boldsymbol{G}_z \\ \boldsymbol{K}_{32} &= \upsilon\boldsymbol{G}_z^T\boldsymbol{C}\boldsymbol{G}_y + \upsilon_2\boldsymbol{G}_y^T\boldsymbol{C}\boldsymbol{G}_z \end{aligned} \tag{5.79}$$

The mass matrix and force vector are obtained as

$$\boldsymbol{m}_e = \rho \begin{bmatrix} \boldsymbol{G}^T\boldsymbol{C}\boldsymbol{G} & \boldsymbol{0} & \boldsymbol{0} \\ \boldsymbol{0} & \boldsymbol{G}^T\boldsymbol{C}\boldsymbol{G} & \boldsymbol{0} \\ \boldsymbol{0} & \boldsymbol{0} & \boldsymbol{G}^T\boldsymbol{C}\boldsymbol{G} \end{bmatrix} \tag{5.80}$$

$$\boldsymbol{f}_e = \begin{bmatrix} \boldsymbol{G}^T\boldsymbol{C}\boldsymbol{q}_u \\ \boldsymbol{G}^T\boldsymbol{C}\boldsymbol{q}_v \\ \boldsymbol{G}^T\boldsymbol{C}\boldsymbol{q}_w \end{bmatrix} \tag{5.81}$$

where ρ is the density, $\boldsymbol{q}_u\boldsymbol{q}_v$ and $\boldsymbol{q}_w$ are respectively the distributed nodal load vectors with respect to x-, y- and z-directions.

5.4.2 *Thermo-mechanical cross-ply laminates*

This section presents 3D kinematic relations for cross-ply laminated shells (Fig. 5.20) considering thermal effects. The principal material axes $(1, 2, 3)$ and laminated axes (x, y, z) in the curvilinear coordinate system are defined in Fig. 5.21. The 3D stress–strain relations for k^{th} orthotropic lamina with reference to the principal material axes are defined as follows [173, 339]:

$$\begin{Bmatrix} \sigma_{11} \\ \sigma_{22} \\ \sigma_{33} \\ \tau_{23} \\ \tau_{13} \\ \tau_{12} \end{Bmatrix}^{(k)} = \begin{bmatrix} C_{11} & C_{12} & C_{13} & 0 & 0 & 0 \\ C_{12} & C_{22} & C_{23} & 0 & 0 & 0 \\ C_{13} & C_{23} & C_{33} & 0 & 0 & 0 \\ 0 & 0 & 0 & C_{44} & 0 & 0 \\ 0 & 0 & 0 & 0 & C_{55} & 0 \\ 0 & 0 & 0 & 0 & 0 & C_{66} \end{bmatrix}^{(k)} \times \begin{Bmatrix} \varepsilon_{11} - \alpha_{11}\Delta T \\ \varepsilon_{22} - \alpha_{22}\Delta T \\ \varepsilon_{33} - \alpha_{33}\Delta T \\ \gamma_{23} \\ \gamma_{13} \\ \gamma_{12} \end{Bmatrix}^{(k)} \tag{5.82}$$

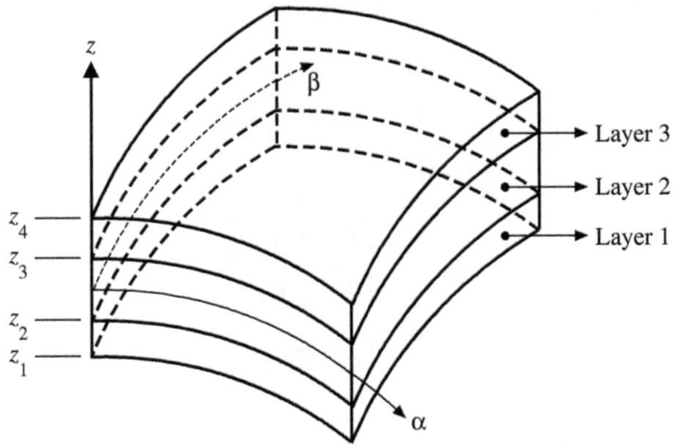

Fig. 5.20. A laminated shell section.

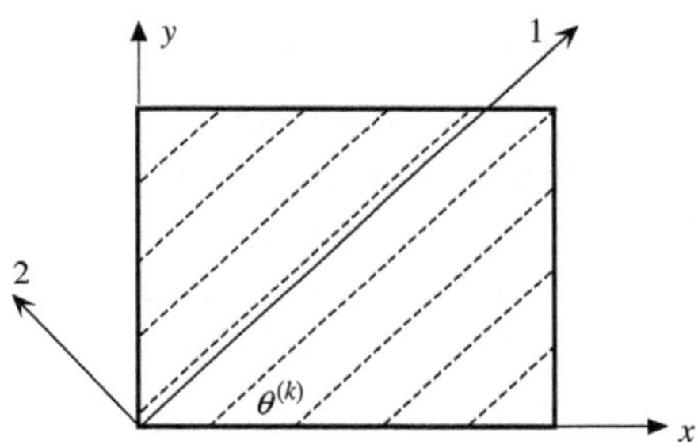

Fig. 5.21. Coordinate systems of the kth layer.

in which

$$
\begin{aligned}
C_{11} &= \frac{S_{22}S_{33} - S_{23}^2}{S}, \quad C_{12} = \frac{S_{13}S_{23} - S_{12}S_{33}}{S} \\
C_{13} &= \frac{S_{12}S_{23} - S_{13}S_{22}}{S}, \quad C_{22} = \frac{S_{33}S_{11} - S_{13}^2}{S} \\
C_{23} &= \frac{S_{12}S_{13} - S_{23}S_{11}}{S}, \quad C_{33} = \frac{S_{11}S_{22} - S_{12}^2}{S} \\
C_{44} &= G_{23}, \quad C_{55} = G_{13}, \quad C_{66} = G_{12} \\
S &= S_{11}S_{22}S_{33} - S_{11}S_{23}^2 - S_{22}S_{13}^2 - S_{33}S_{12}^2 + 2S_{12}S_{23}S_{13}
\end{aligned}
\tag{5.83}
$$

where

$$
S_{ii} = \frac{1}{E_i}, \quad S_{ij} = -\frac{v_{ij}}{E_i} (\text{no sum on } i, j \text{ and } i \neq j) \tag{5.84}
$$

and ΔT indicates the temperature rise and $\alpha_i (i = 1, 2, 3)$ are the linear thermal expansion coefficients in the direction of the principle material axes; and E_{ij} and v_{ij} are respectively the Young's moduli and the Poisson's ratios, which are not independent but relate with each other by the expressions

$$\frac{v_{ij}}{E_i} = \frac{v_{ji}}{E_j} \text{ (no sum on } i, j) \tag{5.85}$$

In conformity with the assumption that the normal stress σ_3 may be assumed to be small and negligible and the corresponding strain ε_3 is equal to zero, Equation (5.82) referred to lamina's principle material axes (1, 2, 3). Its compacted form may be written as

$$\tilde{\boldsymbol{\sigma}}^{(k)} = \boldsymbol{C}^{(k)} \tilde{\boldsymbol{\varepsilon}}^{(k)} \tag{5.86}$$

The principle material axes of lamina may not coincide with the reference axes (x, y, z) of the laminate. This can be conveniently accomplished through the transformation as follows:

$$\begin{Bmatrix} \sigma_{xx} \\ \sigma_{yy} \\ \sigma_{zz} \\ \tau_{yz} \\ \tau_{xz} \\ \tau_{xy} \end{Bmatrix}^{(k)} = \boldsymbol{T}^{(k)} \boldsymbol{C}^{(k)} \boldsymbol{T}^{(k)\mathrm{T}} \begin{Bmatrix} \varepsilon_{xx} - \alpha_{xx}\Delta T \\ \varepsilon_{yy} - \alpha_{yy}\Delta T \\ \varepsilon_{zz} - \alpha_{zz}\Delta T \\ \gamma_{yz} \\ \gamma_{xz} \\ \gamma_{xy} - \alpha_{xy}\Delta T \end{Bmatrix}^{(k)} \tag{5.87}$$

or

$$\boldsymbol{\sigma}^{(k)} = \boldsymbol{Q}^{(k)} \boldsymbol{\varepsilon}^{(k)} \tag{5.88}$$

where

$$\boldsymbol{T}^{(k)} = \begin{bmatrix} \cos^2\theta & \sin^2\theta & 0 & 0 & 0 & -\sin 2\theta \\ \sin^2\theta & \cos^2\theta & 0 & 0 & 0 & \sin 2\theta \\ 0 & 0 & 1 & 0 & 0 & 0 \\ 0 & 0 & 0 & \cos\theta & \sin\theta & 0 \\ 0 & 0 & 0 & -\sin\theta & \cos\theta & 0 \\ \sin\theta\cos\theta & -\sin\theta\cos\theta & 0 & 0 & 0 & \cos^2\theta - \sin^2\theta \end{bmatrix} \tag{5.89}$$

$$
\begin{aligned}
\alpha_{xx} &= \alpha_{11}\cos^2\theta + \alpha_{22}\sin^2\theta \\
\alpha_{yy} &= \alpha_{11}\sin^2\theta + \alpha_{22}\cos^2\theta \\
\alpha_{zz} &= \alpha_{33} \\
\alpha_{xy} &= (\alpha_{11} - \alpha_{22})\sin 2\theta
\end{aligned} \tag{5.90}
$$

For equilibrium, the total potential energy and kinetic energy must be stationary. Using the definitions of stress resultants stated above, one obtains

$$
\begin{aligned}
\delta\Pi = \int_{\Omega} (&\sigma_{xx}\delta\varepsilon_{xx} + \sigma_{yy}\delta\varepsilon_{yy} + \sigma_{zz}\delta\varepsilon_{zz} + \tau_{yz}\delta\gamma_{yz} \\
&+ \tau_{zx}\delta\gamma_{zx} + \tau_{xy}\delta\gamma_{xy})\mathrm{d}\Omega - \int_{\Gamma} q\delta w \mathrm{d}\Gamma
\end{aligned} \tag{5.91}
$$

where Ω denotes the entire domain, Γ denotes boundary of the domain and q is distributed transverse boundary force. The kinetic energy is given by

$$
T = -\iiint_V \boldsymbol{u}^T \left(-\rho\frac{\partial^2 \boldsymbol{u}}{\partial t^2}\right) dV \tag{5.92}
$$

where $\boldsymbol{u} = \{u, v, w\}^T$ is the displacement field vector.

The potential energy and kinetic energy are discretized to get element matrices and vectors. The resultant equilibrium equations of an elastic solid can be written as

$$
\boldsymbol{KU} + \boldsymbol{M\ddot{U}} = \boldsymbol{R} \tag{5.93}
$$

where $\boldsymbol{K}$, $\boldsymbol{M}$, $\boldsymbol{R}$ and $\boldsymbol{U}$ are respectively the stiffness matrix, mass matrix, force vector and displacement vector. For isotropic 3D elasticity problems, the stiffness and mass matrices and force and displacement vectors can be written as

$$
\boldsymbol{K} = \frac{G}{\upsilon_2}\begin{bmatrix} \boldsymbol{K}_{11} & & \text{sym} \\ \boldsymbol{K}_{21} & \boldsymbol{K}_{22} & \\ \boldsymbol{K}_{31} & \boldsymbol{K}_{32} & \boldsymbol{K}_{33} \end{bmatrix}
$$

$$
\boldsymbol{M} = \rho\begin{bmatrix} \boldsymbol{G}^T\boldsymbol{C}\boldsymbol{G} & \boldsymbol{0} & \boldsymbol{0} \\ \boldsymbol{0} & \boldsymbol{G}^T\boldsymbol{C}\boldsymbol{G} & \boldsymbol{0} \\ \boldsymbol{0} & \boldsymbol{0} & \boldsymbol{G}^T\boldsymbol{C}\boldsymbol{G} \end{bmatrix}, \quad \boldsymbol{R} = \begin{bmatrix} \boldsymbol{G}^T\boldsymbol{C}\boldsymbol{q}_u \\ \boldsymbol{G}^T\boldsymbol{C}\boldsymbol{q}_v \\ \boldsymbol{G}^T\boldsymbol{C}\boldsymbol{q}_w \end{bmatrix} \tag{5.94}
$$

$$\begin{bmatrix} \boldsymbol{K}_{11} \\ \boldsymbol{K}_{22} \\ \boldsymbol{K}_{33} \end{bmatrix} = \begin{bmatrix} 2\upsilon_1 & \upsilon_2 & \upsilon_2 \\ \upsilon_2 & 2\upsilon_1 & \upsilon_2 \\ \upsilon_2 & \upsilon_2 & 2\upsilon_1 \end{bmatrix} \begin{bmatrix} \boldsymbol{G}_x^T \boldsymbol{C} \boldsymbol{G}_x \\ \boldsymbol{G}_y^T \boldsymbol{C} \boldsymbol{G}_y \\ \boldsymbol{G}_z^T \boldsymbol{C} \boldsymbol{G}_z \end{bmatrix}$$

$$\begin{aligned} \boldsymbol{K}_{21} &= \upsilon \boldsymbol{G}_y^T \boldsymbol{C} \boldsymbol{G}_x + \upsilon_2 \boldsymbol{G}_x^T \boldsymbol{C} \boldsymbol{G}_y \\ \boldsymbol{K}_{31} &= \upsilon \boldsymbol{G}_z^T \boldsymbol{C} \boldsymbol{G}_x + \upsilon_2 \boldsymbol{G}_x^T \boldsymbol{C} \boldsymbol{G}_z, \quad \mathbf{U} = \begin{bmatrix} \boldsymbol{u} \\ \boldsymbol{v} \\ \boldsymbol{w} \end{bmatrix} \\ \boldsymbol{K}_{32} &= \upsilon \boldsymbol{G}_z^T \boldsymbol{C} \boldsymbol{G}_y + \upsilon_2 \boldsymbol{G}_y^T \boldsymbol{C} \boldsymbol{G}_z \end{aligned} \tag{5.95}$$

where $\upsilon_2 = 0.5 - \upsilon$, $G = E/2(1+\upsilon)$ is the shear modulus, and $\boldsymbol{q}_u$, $\boldsymbol{q}_v$ and $\boldsymbol{q}_w$ are respectively the nodal load vectors with respect to x-, y- and z-directions, and $\boldsymbol{u}$, $\boldsymbol{v}$ and $\boldsymbol{w}$ are the discretized vectors of displacements on the three directions.

The stiffness and mass matrices and force vector of cross-ply laminated shells should be derived from the DQH rules above together with the stress–strain relations given by Eq. (5.87), the total potential energy given by Eq. (5.91) and the strain–displacement relations given by Eq. (5.77). The mass matrix is the same as Eq. (5.94). The stiffness matrix is

$$\boldsymbol{K} = \begin{bmatrix} \boldsymbol{k}_{1x}^1 + \boldsymbol{k}_{5z}^1 + \boldsymbol{k}_{6y}^1 & \boldsymbol{k}_{2y}^1 + \boldsymbol{k}_{4z}^1 + \boldsymbol{k}_{6x}^1 & \boldsymbol{k}_{3z}^1 + \boldsymbol{k}_{4y}^1 + \boldsymbol{k}_{5x}^1 \\ & \boldsymbol{k}_{2y}^2 + \boldsymbol{k}_{4z}^2 + \boldsymbol{k}_{6x}^2 & \boldsymbol{k}_{3z}^2 + \boldsymbol{k}_{4y}^2 + \boldsymbol{k}_{5x}^2 \\ \text{symmetric} & & \boldsymbol{k}_{3z}^3 + \boldsymbol{k}_{4y}^3 + \boldsymbol{k}_{5x}^3 \end{bmatrix} \tag{5.96}$$

where

$$\begin{aligned} \boldsymbol{k}_{ij}^1 &= Q_{i1} \boldsymbol{G}_x \boldsymbol{C} \boldsymbol{G}_j^T + Q_{i5} \boldsymbol{G}_z \boldsymbol{C} \boldsymbol{G}_j^T + Q_{i6} \boldsymbol{G}_y \boldsymbol{C} \boldsymbol{G}_j^T \\ \boldsymbol{k}_{ij}^2 &= Q_{i2} \boldsymbol{G}_y \boldsymbol{C} \boldsymbol{G}_j^T + Q_{i4} \boldsymbol{G}_z \boldsymbol{C} \boldsymbol{G}_j^T + Q_{i6} \boldsymbol{G}_x \boldsymbol{C} \boldsymbol{G}_j^T \\ \boldsymbol{k}_{ij}^3 &= Q_{i3} \boldsymbol{G}_z \boldsymbol{C} \boldsymbol{G}_j^T + Q_{i4} \boldsymbol{G}_y \boldsymbol{C} \boldsymbol{G}_j^T + Q_{i5} \boldsymbol{G}_x \boldsymbol{C} \boldsymbol{G}_j^T \end{aligned} \tag{5.97}$$

where $i = 1, 2, \ldots, 6$ and $j = x, y, z$. The force vector is

$$\boldsymbol{R} = \begin{bmatrix} \sum_{j=1}^{6} (Q_{1j}\alpha_j \boldsymbol{G}_x^{\mathrm{T}} + Q_{5j}\alpha_j \boldsymbol{G}_z^{\mathrm{T}} + Q_{6j}\alpha_j \boldsymbol{G}_y^{\mathrm{T}}) \boldsymbol{C} \Delta \boldsymbol{T} \\ \sum_{j=1}^{6} (Q_{2j}\alpha_j \boldsymbol{G}_y^{\mathrm{T}} + Q_{4j}\alpha_j \boldsymbol{G}_z^{\mathrm{T}} + Q_{6j}\alpha_j \boldsymbol{G}_x^{\mathrm{T}}) \boldsymbol{C} \Delta \boldsymbol{T} \\ \boldsymbol{G}^T \boldsymbol{C} \boldsymbol{q}_w + \sum_{j=1}^{6} (Q_{3j}\alpha_j \boldsymbol{G}_z^{\mathrm{T}} + Q_{4j}\alpha_j \boldsymbol{G}_y^{\mathrm{T}} + Q_{5j}\alpha_j \boldsymbol{G}_x^{\mathrm{T}}) \boldsymbol{C} \Delta \boldsymbol{T} \end{bmatrix} \tag{5.98}$$

where $\alpha_1 = \alpha_{xx}, \alpha_2 = \alpha_{yy}, \alpha_3 = \alpha_{zz}, \alpha_4 = 0, \alpha_5 = 0, \alpha_6 = \alpha_{xy}$, and $\Delta \boldsymbol{T}$ is the vector of the discrete value of ΔT on integration points of the element.

5.5 Results and Discussions

5.5.1 *3D vibration analyses*

In this section, 3D vibration analyses of solid structures are carried out. For the tetrahedral elements, the number of sampling points on all edges is taken as the same and denoted as N_e, the number of sampling points on both directions of all faces is taken as N_f, and the number of sampling points on all directions of the body is taken as N_b. For the wedge elements, the number of nodes on the edges of the two triangular faces is taken as the same and denoted as N_e, the number of nodes on both directions of the faces of the two triangular faces is taken as the same and is denoted as N_f, the number of nodes on the thickness direction of the quadrilateral faces on all edges and faces is taken as the same and denoted as N_t (without two ends). The number of nodes inside the wedge elements correlates with the face and edges nodes by keeping the maximum polynomial degree. The geometries are first constructed from NURBS and then expressed by the DQH bases. On the physical field, one can either use interpolation bases on the boundary surfaces of an element or use interpolation bases on the entire domain. The former method should be more accurate since the variational principle can automatically optimize the results. The inclusion of interpolation bases on the boundary of an element is convenient for element assembly as one does not need to check the parametric direction of an odd basis. First, a few examples of the tetrahedral elements for vibration analyses of thick plates are studied. Then the wedge element of the DQHFEM is analyzed through application to vibration of plates and shells.

First, vibration analyses of a square plate with simply supported (S) and clamped (C) boundary conditions are respectively carried out in Tables 5.1 and 5.2. The plate is discretized as six tetrahedral elements (Fig. 5.22). Poisson's ratio of the plate is 0.3. One can see from Table 5.1 that, for the simply supported square plate, the DQHFEM results converge to the 3D Ritz solutions [340] using only a few nodes. For the clamped thin square plate (relative thickness $h/b = 0.01$) shown in Table 5.2, the convergence rate of the

Table 5.1. Comparison of frequency parameters $\Omega = (\omega b^2/\pi^2)\sqrt{\rho h/D}$ for a square plate with SSSS boundary conditions (Fig. 5.22).

		Mode sequence number					
h/b	(N_e, N_f, N_b)	1	2	3	4	5	6
0.01	(6, 3, 3)	2.1291	6.0211	6.0346	65.235	65.235	92.304
	(7, 4, 4)	1.9999	5.9866	5.9944	9.4030	14.251	14.313
	(8, 5, 5)	1.9999	5.0125	5.0126	9.3670	13.351	13.952
	(9, 6, 6)	1.9993	5.0101	5.0111	8.0337	10.129	10.130
	(10, 7, 7)	1.9993	4.9957	4.9957	8.0322	10.072	10.104
	(11, 8, 8)	1.9993	4.9957	4.9957	7.9895	9.9846	9.9847
	(12, 9, 9)	1.9993	4.9956	4.9956	7.9894	9.9836	9.9839
	(13, 10, 10)	1.9993	4.9956	4.9956	7.9888	9.9826	9.9826
	(14, 11, 11)	1.9993	4.9956	4.9956	7.9888	9.9826	9.9826
	3D Ritz [340]	1.9993	4.9956	4.9956	7.9888	9.9826	9.9826
	Classical theory [342]	2.0	5.0	5.0	8.0	10.0	10.0
0.1	(5, 2, 2)	2.0216	6.5235	6.5235	9.3696	11.143	13.424
	(6, 3, 3)	1.9839	4.9852	5.0092	6.5234	6.5234	9.2270
	(7, 4, 4)	1.9345	4.7414	4.7814	6.5234	6.5234	7.6553
	(8, 5, 5)	1.9344	4.6259	4.6261	6.5234	6.5234	7.4051
	(9, 6, 6)	1.9342	4.6234	4.6238	6.5234	6.5234	7.1167
	(10, 7, 7)	1.9342	4.6222	4.6222	6.5234	6.5234	7.1098
	(11, 8, 8)	1.9342	4.6222	4.6222	6.5234	6.5234	7.1032
	(12, 9, 9)	1.9342	4.6222	4.6222	6.5234	6.5234	7.1031
	(13, 10, 10)	1.9342	4.6222	4.6222	6.5234	6.5234	7.1030
	(14, 11, 11)	1.9342	4.6222	4.6222	6.5234	6.5234	7.1030
	3D Ritz [340]	1.9342	4.6222	4.6222	6.5234	6.5234	7.1030
	3D exact [343]	1.9342	4.6222	4.6222	—	—	7.1030

DQHFEM is a little bit slow. But one can see that the DQM results [341] have the same convergence rate as the present method, i.e., for the same face node number N_f, the DQHFEM results and the DQM results are close to each other. So, the DQHFEM has the same convergence rate as the DQM, which is known to be able to present highly accurate results using only a few DQFs [11]. But the DQM does not have the hierarchical property of the DQHFEM that can easily join elements with different polynomial degrees. For the thick clamped square plate, the DQHFEM still can agree with the 3D Ritz solutions [340] very well using only a few nodes.

In Table 5.3, the DQHFEM results for vibration of a free pentagonal plate (Fig. 5.23) are compared with the FEM results computed

Table 5.2. Comparison of frequency parameters $\Omega = (\omega b^2/\pi^2)\sqrt{\rho h/D}$ for a square plate with CCCC boundary conditions (Fig. 5.22).

		Mode sequence number					
h/b	(N_e, N_f, N_b)	1	2	3	4	5	6
0.01	(7, 4, 4)	7.1779	56.415	57.955	84.976	109.81	120.52
	(8, 5, 5)	4.4206	11.428	12.102	67.031	67.578	71.466
	(10, 7, 7)	3.6646	7.4719	7.4822	12.055	14.579	14.817
	(11, 8, 8)	3.6609	7.4588	7.4612	11.001	13.408	13.454
	(12, 9, 9)	3.6564	7.4490	7.4554	10.983	13.365	13.418
	(13, 10, 10)	3.6546	7.4450	7.4476	10.972	13.331	13.402
	(15, 12, 12)	3.6511	7.4384	7.4404	10.961	13.319	13.387
	(16, 13, 13)	3.6496	7.4359	7.4388	10.956	13.316	13.381
	(20, 14, 14)	**3.6489**	**7.4344**	**7.4360**	**10.954**	**13.312**	**13.378**
	3D Ritz[340]	3.6492	7.4352	7.4352	10.953	13.315	13.379
	DQM [341] (9, 7, 7)	3.6648	7.4730	7.4758	11.021	13.332	13.384
	DQM [341] (10, 8, 8)	3.6671	7.4599	7.4705	10.995	13.367	13.430
	DQM [341] (11, 9, 9)	3.6516	7.4505	7.4664	10.975	13.356	13.419
0.1	(6, 3, 3)	4.2158	8.5973	8.8349	12.656	12.673	15.261
	(7, 4, 4)	3.3716	7.1130	7.2776	11.328	12.545	12.555
	(10, 7, 7)	3.3247	6.3505	6.3538	8.9552	10.527	10.656
	(13, 10, 10)	3.3201	6.3428	6.3437	8.9001	10.494	10.595
	(15, 12, 12)	3.3188	6.3407	6.3413	8.8972	10.491	10.591
	(16, 13, 13)	3.3184	6.3400	6.3406	8.8961	10.490	10.590
	3D Ritz [340]	3.3184	6.3402	6.3402	8.8961	10.490	10.590
	DQM [341]	3.3282	6.3547	6.3547	8.9135	10.493	10.592

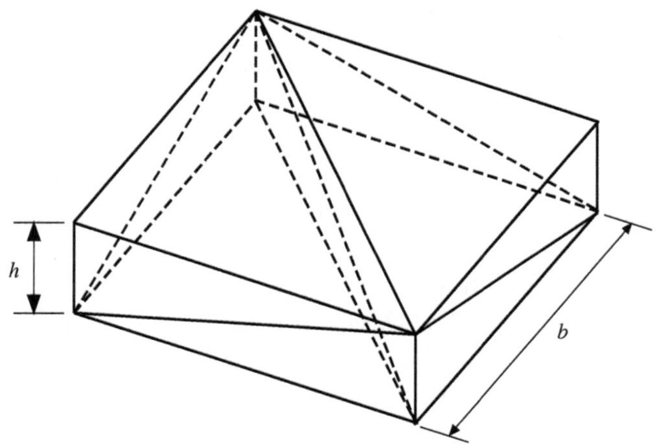

Fig. 5.22. A rectangular thick plate discretized as six tetrahedral elements.

Table 5.3. Comparison of natural frequencies (Hz) for a free pentagonal plate with $E = 195\,\text{GPa}$, $\upsilon = 0.3$, $\rho = 7722.7\text{Kg/m}^3$, $h = 0.1\text{m}$, $L = 1\,\text{m}$, (Fig. 5.23).

Methods	(N_e, N_f, N_b)/ mesh size	DOFs	Mode sequence number 1	2	3	4	5	6
DQHFEM	(3, 0,0)	105	244.2532	244.7002	446.994	1002.058	1798.542	1811.431
	(4, 1, 1)	276	233.9903	237.2928	426.7468	583.213	593.9454	1016.215
	(5, 2, 2)	735	225.1900	225.3030	377.4424	555.7588	558.6160	803.4888
	(6, 3, 3)	1626	223.2692	223.4222	371.6648	519.2974	519.4323	742.1603
	(7, 4, 4)	3093	222.3509	222.3914	370.1725	515.3650	515.5405	727.9085
	(8, 5, 5)	5280	222.0856	222.1346	370.0288	513.9288	513.9543	724.1810
	(9, 6, 6)	8331	222.0096	222.0282	370.0243	513.6351	513.6832	723.5787
	(10, 7, 7)	12390	221.9789	221.9893	370.0242	513.5358	513.5573	723.4164
FEM	1.0	105	244.25	244.7	446.99	1002.1	1798.5	1811.4
	0.5	432	233.49	233.73	404.57	660.99	678.87	962.66
	0.3	774	227.12	227.16	379.95	550.37	550.46	800.48
	0.2	1971	224.33	224.41	373.47	524.04	524.48	745.30
	0.15	3798	223.50	223.52	371.60	519.70	519.81	735.67
	0.1	9189	222.96	223.01	370.76	517.19	517.25	730.30
	0.07	23520	222.42	222.43	370.15	514.95	514.97	725.73
	0.05	61581	222.08	222.08	370.06	513.84	513.85	723.96
	0.03	216966	221.98	221.99	370.03	513.55	513.55	723.47

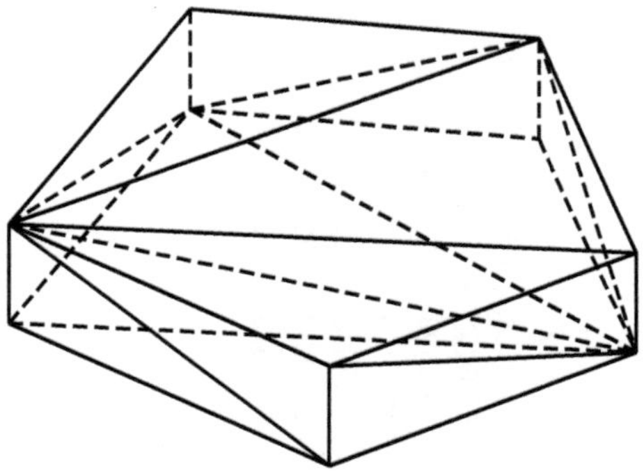

Fig. 5.23. A pentagonal thick plate discretized as eight tetrahedral elements.

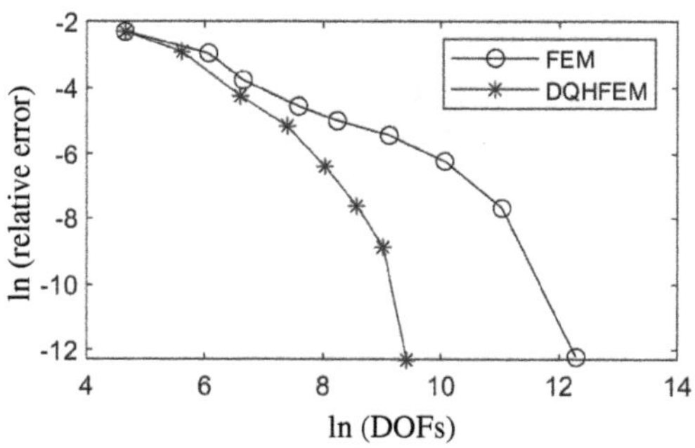

Fig. 5.24. Convergence rate of the fundamental natural frequency parameter in Table 5.15, where ln(•) gives the natural logarithm of the data "•".

by the authors of this work through commercial FEM codes. One can see that the common FEM needs about 20 times of the DOFs of the DQHFEM to obtain results with the same convergence as the present method. Figure 5.24 shows the convergence rate study of the DQHFEM for the pentagonal plate; one can see that the DQHFEM in 3D also has high convergence rate as the 2D cases. Note that the tetrahedral elements for the plate are much distorted. The conventional FEM may have singularity problem in such cases, but the DQHFEM is still very stable.

In Table 5.4, free vibration analyses of a cantilever isosceles triangular thick plate are respectively carried out using one wedge DQHFEM element and three tetrahedral elements (Fig. 5.25). The angle between the two equivalent edges is 90°. The edge opposite to the

Table 5.4. Comparison of frequency parameters $\Omega = (\omega b^2/2\pi)\sqrt{\rho h/D}$ for isosceles triangular plates with CFF boundary condition and Poisson's ratio $\upsilon = 0.3$ (Fig. 5.25).

h/b	Element types-(N_e, N_f, N_b)/ (N_e, N_f, N_t)	Mode sequence number					
		1	2	3	4	5	6
0.1	Tetrahedron-(5, 2, 2)	3.853	10.54	14.03	22.65	28.98	36.93
	Tetrahedron-(6, 3, 3)	3.788	9.912	13.82	21.45	24.51	30.29
	Tetrahedron-(7, 4, 4)	3.772	9.768	13.59	20.73	22.95	28.97
	Tetrahedron-(8, 5, 5)	3.767	9.747	13.55	20.37	22.52	28.46
	Tetrahedron-(9, 6, 6)	3.764	9.740	13.54	20.34	22.44	28.35
	Tetrahedron-(10, 7, 7)	3.763	9.737	13.53	20.33	22.43	28.32
	Tetrahedron-(11, 8, 8)	3.762	9.734	13.53	20.33	22.42	28.31
	Tetrahedron-(12, 9, 9)	3.762	9.733	13.53	20.32	22.42	28.31
	Tetrahedron-(13, 10, 10)	3.761	9.732	13.53	20.32	22.42	28.31
	Tetrahedron-(14, 11, 11)	**3.761**	**9.732**	**13.53**	**20.32**	**22.42**	**28.30**
	Prism-(6, 3, 3)	3.796	9.933	13.84	21.66	24.50	30.41
	Prism-(7, 4, 3)	3.777	9.791	13.64	20.82	23.11	29.12
	Prism-(8, 5, 3)	3.771	9.759	13.57	20.42	22.60	28.60
	Prism-(9, 6, 4)	3.767	9.748	13.55	20.36	22.47	28.41
	Prism-(11, 8, 4)	3.764	9.739	13.54	20.34	22.43	28.33
	Prism-(13, 10, 5)	3.763	9.735	13.53	20.33	22.42	28.31
	Prism-(15, 12, 5)	**3.762**	**9.734**	**13.53**	**20.33**	**22.42**	**28.31**
	Prism-(17, 14, 5)	3.762	9.733	13.53	20.32	22.42	28.31
	Prism-(17, 14, 3)	3.763	9.736	13.53	20.33	22.43	28.32
	3D elastic theory [344]	3.762	9.734	13.53	20.33	22.42	28.31
	Mindlin theory [345]	3.741	9.661	13.41	20.11	22.19	27.96
0.05	Prism-(8, 5, 1)	3.966	10.99	15.87	24.92	27.98	37.51
	Prism-(10, 7, 1)	3.955	10.95	15.79	24.74	27.75	36.85
	Prism-(12, 9, 2)	3.941	10.89	15.66	24.51	27.43	36.29
	Prism-(14, 11, 3)	**3.939**	**10.88**	**15.65**	**24.49**	**27.41**	**36.26**
	Prism-(16, 13, 3)	3.938	10.88	15.64	24.49	27.40	36.25
	Prism-(16, 13, 1)	3.949	10.93	15.76	24.69	27.66	36.71
	3D elastic theory [344]	3.938	10.88	15.65	24.49	27.41	36.26
	Mindlin theory [345]	3.926	10.84	15.58	24.36	27.26	36.03

right angle is clamped (C). The two equivalent edges are free (F). The Poisson's ratio $\upsilon = 0.3$. Relative thickness ratios $h/b = 0.05$ and 0.1 are considered. One can again see the fast convergence of the tetrahedron elements. However, it requires similar number of nodes on almost all edges according to edge length. In this aspect, the wedge element is more efficient, since one can only use a few nodes

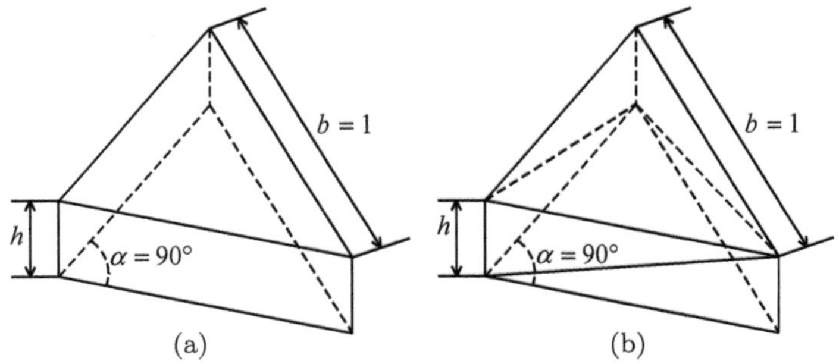

Fig. 5.25. A cantilever isosceles triangular thick plate discretized as one wedge DQHFEM element (a) and three tetrahedral elements(b).

on the thickness direction, as shown in Table 5.4. For the thick plate ($h/b = 0.1$), the results of the wedge elements using about three nodes (without the two ends) on the thickness direction are accurate enough compared with 3D elasticity solutions [344] and Mindlin plate solutions [345]. For the thin plate ($h/b = 0.05$), one node (without the two ends) on the thickness direction is enough. Moreover, the wedge element seems non-sensitive with shear or membrane locking. So, the wedge element provides a good way of solving plate and shell structures.

In Table 5.5, free vibration of a thick circular plate is analyzed by the DQHFEM. The plate is discretized as eight wedge elements, as shown in Fig. 5.26. The wedge mesh is generated from the triangular meshes. The present DQHFEM results are compared with the 3D Chebyshev–Ritz solutions of [290] and 3D polynomials–Ritz solutions of [346]. One can again find that the DQHFEM using wedge elements has great similarities with high-order shear deformation theory of plates and shells, that is, the present DQHFEM using the computational cost of shear deformation theories can reach the accuracy of 3D elasticity theory.

In Tables 5.6 and 5.7, a shallow cylindrical shell and a shallow spherical shell of square planforms are respectively analyzed by the DQHFEM. The DQHFEM meshes of the two types of shell structures are respectively shown in Figs. 5.27 and 5.28. The frequencies are given according to the symmetry (S) and anti-symmetry (A) of the mode shapes against the coordinate directions. One can again find the fast convergence of the present method.

Table 5.5. Comparison of frequency parameters $\Omega = \omega R^2\sqrt{\rho h/D}$ for a clamped circular plate with Poisson's ratio $\upsilon = 0.3$ (Fig. 5.26).

		Mode sequence number							
h/b	(N_e, N_f, N_t)	1	2	3	4	5	6	7	8
0.1	(6, 3, 1)	10.263	20.973	33.482	38.293	47.845	56.542	64.328	68.207
	(7, 4, 1)	10.079	20.552	33.075	37.611	47.228	55.763	62.815	68.104
	(8, 5, 2)	10.024	20.364	32.545	36.878	46.368	54.558	61.553	68.049
	(15, 12, 2)	9.9824	20.282	32.408	36.720	46.114	54.279	61.153	67.956
	(15, 12, 5)	9.9810	20.278	32.400	36.711	46.099	54.261	61.130	67.951
	3D Ritz [290]	9.9755	20.267	32.383	36.692	46.076	54.234	61.113	67.941
	3D *p*-Ritz [346]	9.9909	20.297	32.430	36.744	46.140	54.308	61.186	67.969
0.3	(6, 3, 1)	8.6966	15.976	22.771	23.502	26.162	26.186	31.284	35.460
	(8, 5, 2)	8.4897	15.497	22.739	22.741	25.237	26.176	30.209	34.387
	(15, 12, 2)	8.4804	15.485	22.725	22.727	25.221	26.176	30.185	34.360
	(15, 12, 3)	8.4690	15.458	22.678	22.722	25.163	26.176	30.110	34.261
	(17, 14, 5)	8.4636	15.447	22.661	22.717	25.142	26.176	30.086	34.232
	3D Ritz [290]	8.4606	15.442	22.654	22.715	25.134	26.176	30.078	34.222
	3D *p*-Ritz [346]	8.4676	15.453	22.667	22.721	25.150	—	30.093	34.239
0.5	(6, 3, 2)	6.8497	11.576	13.683	15.712	16.352	17.992	21.229	21.234
	(8, 5, 3)	6.8209	11.520	13.665	15.705	16.261	17.862	21.073	21.201
	(10, 7, 4)	6.8108	11.503	13.661	15.705	16.236	17.833	21.037	21.194
	(12, 9, 5)	6.8080	11.499	13.658	15.705	16.232	17.827	21.032	21.188
	(14, 11, 7)	6.8052	11.495	13.656	15.705	16.227	17.822	21.027	21.185
	3D Ritz [290]	6.8027	11.492	13.654	15.705	16.224	17.819	21.024	21.183
	3D *p*-Ritz [346]	6.8068	11.497	13.658	—	16.229	17.825	21.029	21.188

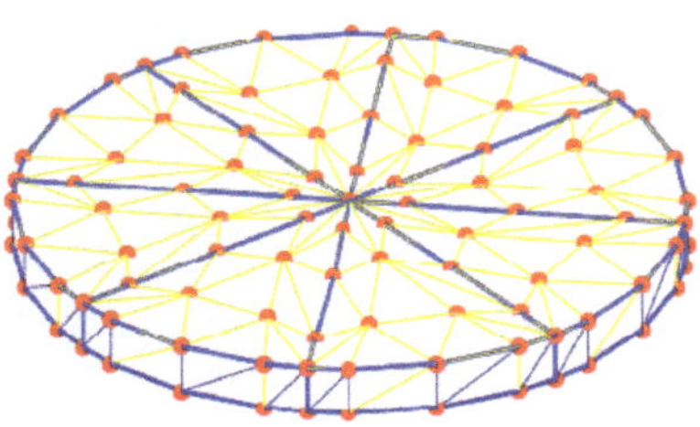

Fig. 5.26. A circular plate discretized as eight wedge elements: the bold blue lines denote element boundaries; the red dots denote nodes of elements; the thin lines are used to draw the geometry.

5.5.2 *Applications of the pyramid elements*

In order to verify the *p*-convergence of the pyramidal element, convergence study is carried out through free vibration analyses using a single pyramidal element (Fig. 5.29) and three pyramidal

Table 5.6. Comparison of frequency parameters $\Omega = \omega a\sqrt{\rho/E}$ for a clamped shallow cylindrical shell of square planform with $\upsilon = 0.3$, $a/b = 1$, $h/b = 0.1$ and $b/R_y = 0.5$ (Fig. 5.27).

	Mode shapes							
(N_e, N_f, N_t)	SS-1	SS-2	SA-1	SA-2	AS-1	AS-2	AA-1	AA-2
(6, 3, 1)	1.0973	3.2408	1.9847	3.7402	1.9323	3.7566	2.7538	4.4477
(8, 5, 2)	1.0841	3.1466	1.9471	3.7252	1.8936	3.7492	2.6722	4.4463
(10, 7, 3)	1.0828	3.1419	1.9446	3.7229	1.8910	3.7476	2.6686	4.4461
(12, 9, 5)	1.0822	3.1395	1.9432	3.7219	1.8897	3.7470	2.6667	4.4460
HOSDT [347]	1.0822	3.1337	1.9376	3.7110	1.8888	3.7355	2.6608	4.4441

Table 5.7. Comparison of frequency parameters $\Omega = \omega a\sqrt{\rho/E}$ for a clamped shallow spherical shell of square planform with $\upsilon = 0.3$, $h/R = 0.5a$ and $a/R = 0.5$ (Fig. 5.28).

	Mode shapes								
(N_e, N_f, N_t)	SS-1	SS-2	SS-3	SA-1	SA-2	SA-3	AA-1	AA-2	AA-3
(6, 3, 2)	2.4057	5.2666	5.3528	3.4758	3.7871	5.5953	4.2764	4.7296	5.2595
(8, 5, 3)	2.3926	5.2331	5.3160	3.4647	3.7759	5.5768	4.2761	4.7005	5.2515
(8, 5, 5)	2.3869	5.2183	5.2993	3.4587	3.7700	5.5735	4.2760	4.6879	5.2470
(10, 7, 7)	2.3856	5.2164	5.2971	3.4575	3.7690	5.5730	4.2759	4.6861	5.2459
3D results [348]	2.3880	5.2207	5.3021	3.4662	3.7772	5.5791	4.2762	4.6901	5.2486

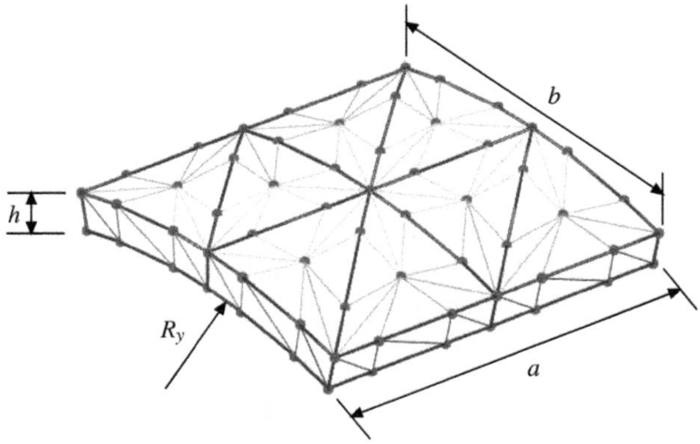

Fig. 5.27. A shallow cylindrical shell discretized as eight wedge elements.

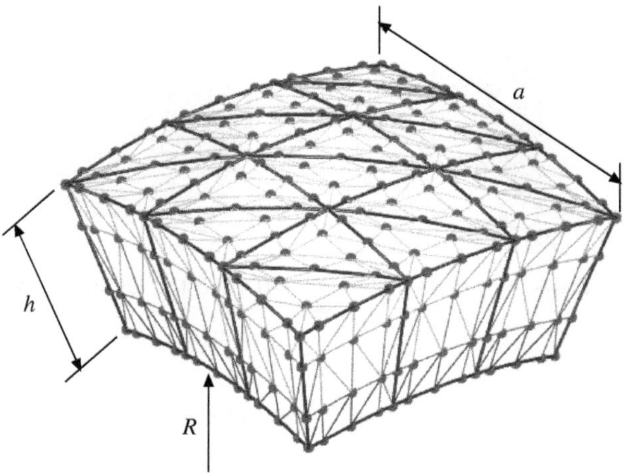

Fig. 5.28. A shallow spherical shell discretized as 18 wedge elements.

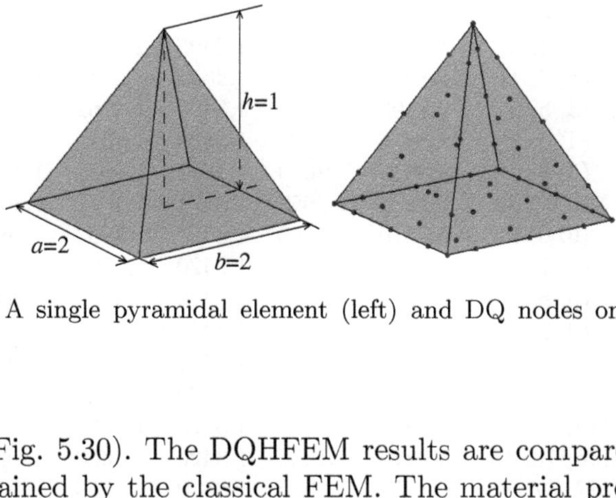

Fig. 5.29. A single pyramidal element (left) and DQ nodes on the surfaces (right).

elements (Fig. 5.30). The DQHFEM results are compared with the results obtained by the classical FEM. The material properties are density $\rho = 7722.7\,\text{kg/m}^3$, elastic modulus $E = 190\,\text{Gpa}$ and Poisson's ratio $v = 0.3$. The dimensions of the model are shown in Figs. 5.29 and 5.30. The boundary conditions of the two cases are completely free. Table 5.8 shows the results obtained by the DQHFEM and the FEM.

As can be seen from Table 5.8, the results of a single pyramidal element have rapid p-convergence. The results in order (6, 6, 5) converge to four significant digits and that in order (8, 8, 7) converge to five significant digits. It can be found that, under the condition

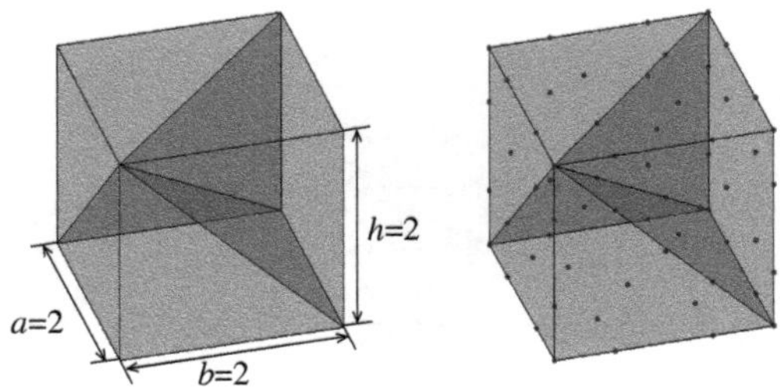

Fig. 5.30. Multiple pyramidal elements (left) and the DQ nodes on the surfaces (right).

Table 5.8. Convergence study of free vibration frequency parameters $\lambda = (\omega a^2/2\pi)\sqrt{\rho h/D_0}$ of a single pyramidal element with free boundary conditions (Fig. 5.29).

Methods	Element order	Frequency sequence					
		1	2	3	4	5	6
DQHFEM	(1, 1, 0)	1.1695	1.6102	1.8496	2.1452	2.1452	2.4953
	(2, 2, 1)	0.8972	1.3080	1.4390	1.6319	1.6319	2.0467
	(3, 3, 2)	0.8893	1.2705	1.3475	1.4762	1.4762	1.9547
	(4, 4, 3)	0.8878	1.2678	1.3433	1.4690	1.4690	1.9520
	(5, 5, 4)	0.8876	1.2660	1.3423	1.4672	1.4672	1.9494
	(6, 6, 5)	0.8876	1.2656	1.3421	1.4667	1.4667	1.9491
	(6, 6, 7)	0.8876	1.2656	1.3421	1.4667	1.4667	1.9490
	(7, 7, 7)	0.8876	1.2655	1.3420	1.4666	1.4666	1.9490
	(8, 8, 7)	0.8876	1.2655	1.3420	1.4666	1.4666	1.9490
	(9, 9, 7)	0.8876	1.2655	1.3420	1.4666	1.4666	1.9490
	(10, 10, 7)	0.8876	1.2655	1.3420	1.4666	1.4666	1.9490
	(13, 13, 7)	0.8876	1.2655	1.3420	1.4666	1.4666	1.9490
	(14, 14, 13)	0.8876	1.2655	1.3420	1.4666	1.4666	1.9490
	(16, 16, 13)	0.8876	1.2655	1.3420	1.4666	1.4666	1.9490
	(17, 17, 13)	0.8876	1.2655	1.3420	1.4666	1.4666	1.9490
	(17, 17, 16)	0.8876	1.2655	1.3420	1.4666	1.4666	1.9490
3D FEM	1	0.888	1.266	1.342	1.467	1.467	1.949

of the same computation accuracy, the DOFs required by 3D DQHFEM and FEM have huge differences. When four significant digits are converged, the FEM needs 13868×3 DOFs, while the DQHFEM only needs 184×3 DOFs.

Table 5.9. Convergence study of frequency parameters $\lambda = (\omega a^2/2\pi)\sqrt{\rho h/D_0}$ of a cube with free boundary conditions using multiple pyramidal elements (Fig. 5.30).

Methods	Element order	Frequency sequence					
		1	2	3	4	5	6
DQHFEM	(1, 1, 0)	0.7024	0.7024	0.8785	0.8785	0.8866	0.9254
	(2, 2, 1)	0.6085	0.6085	0.8155	0.8238	0.8238	0.8238
	(3, 3, 2)	0.5952	0.5952	0.8001	0.8004	0.8004	0.8151
	(4, 4, 3)	0.5936	0.5936	0.7989	0.7990	0.7990	0.8125
	(5, 5, 4)	0.5932	0.5932	0.7987	0.7987	0.7987	0.8122
	(6, 6, 5)	0.5931	0.5931	0.7986	0.7986	0.7986	0.8122
	(7, 7, 6)	0.5931	0.5931	0.7986	0.7986	0.7986	0.8122
	(8, 8, 7)	0.5930	0.5930	0.7986	0.7986	0.7986	0.8122
	(10, 10, 9)	0.5930	0.5930	0.7986	0.7986	0.7986	0.8122
	(12, 12, 11)	0.5930	0.5930	0.7986	0.7986	0.7986	0.8122
3D FEM	1	0.593	0.593	0.799	0.799	0.799	0.812

As can be seen from Table 5.9, when multiple pyramidal elements are used, the results still converge fast. It indicates that the displacement consistency among pyramidal elements is well satisfied. Moreover, there is no obvious difference between the convergence rate of multiple elements and that of a single element, which further demonstrates the effectiveness of the presented pyramidal element. In addition, the results of Table 5.8 are calculated to order (17, 17, 16) without numerical instability, which indicates that DQHFEM is numerically stable.

In the above, only straight sided structures were studied. Now, the vibration of a circular plate is studied to verify the assembly of curved elements. The mesh of the circular plate is given in Fig. 5.31. In order to make the mesh contain pyramidal elements, the circular plate is divided into eight curved elements. The upper part of the circular plate is a cone composed of four curved tetrahedral elements. The lower part is a conical groove composed of four curved pyramidal elements. The common boundary of the two parts is a conical surface. It can be seen that there are two curved surfaces for a single pyramidal element. One is a conical surface and the other one is a cylindrical surface.

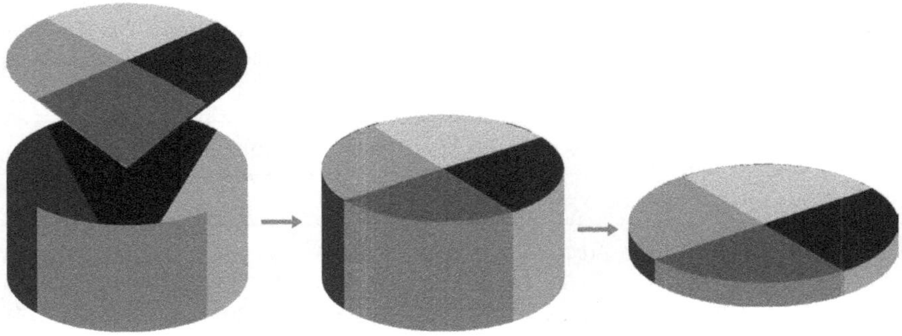

Fig. 5.31. The mesh of a circular plate.

In addition, nodal distribution for geometric mapping should be emphasized again. A certain number of nodes should be assigned for curved surfaces to ensure the geometric accuracy. While only the endpoints are needed for linear boundaries. Usually, dense mapping nodes are set in the area with large curvature, while sparse nodes are set in the area with small curvature. The distribution of the mapping nodes in this example is shown in Fig. 5.32.

The circular plate has thickness h, radius R, and relative thickness $h/R = 0.2$. The boundary conditions are completely free. Dimensionless frequency parameter $\lambda = \omega R\sqrt{\rho/G}$ is used. Due to the particularity of the circular plate, only the first six symmetric modes are given in Table 5.10. The results obtained by the DQHFEM wedge elements are also included in the table. By comparing with DQFEM, 3D FEM and 3D Ritz method, one can see that the results obtained by the mixed curved DQHFEM elements still will agree with those obtained by other methods.

Now, the HQEM on pyramidal domain and several other 3D HQEM elements will be used to solve 3D structural vibration problems. The number of nodes on the edge, face and body of an element is defined by $[n_e, n_f, n_h]$. A model looks like a rocket as shown in Fig. 5.33 is considered. The geometry of the model consists of two wedge elements, one pyramidal element and three hexahedral elements. The material constants are density $\rho = 7722.7\,\text{kg/m}^3$, elastic modulus $E = 190\,\text{GPa}$, and Poisson's ratio $v = 0.3$. One can see from the results in Table 5.11 that the HQEM results still have a

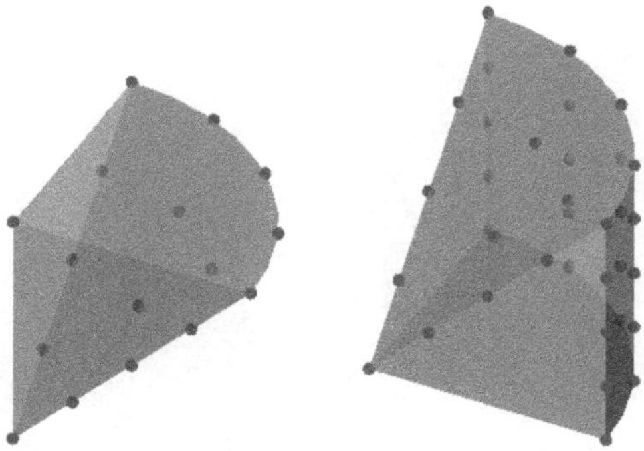

Fig. 5.32. The distribution of mapping nodes for the elements of the circular plate.

Table 5.10. The dimensionless frequency parameter $\lambda = \omega R\sqrt{\rho/G}$ of a free circular plate, $h/R = 0.2$.

		Frequency sequence					
Methods	Element order	1	2	3	4	5	6
DQHFEM	(2, 2, 1)	1.3024	2.029	2.5323	2.8803	3.0551	3.9757
	(3, 3, 2)	0.5919	0.8943	1.5580	2.3858	2.4256	2.8927
	(4, 4, 3)	0.5271	0.8372	1.2163	1.9484	2.0985	2.7379
	(5, 5, 4)	0.5045	0.8329	1.1250	1.7931	1.8856	2.7341
	(6, 6, 5)	0.5006	0.8320	1.1099	1.7704	1.8545	2.7164
	(7, 7, 6)	0.4998	0.8317	1.1072	1.7655	1.8450	2.6819
	(8, 8, 7)	0.4996	0.8316	1.1067	1.7634	1.8432	2.6748
	(9, 9, 8)	0.4995	0.8315	1.1065	1.7625	1.8428	2.6735
	(10, 10, 9)	0.4995	0.8314	1.1065	1.7622	1.8427	2.6732
	(11, 11, 10)	0.4995	0.8314	1.1064	1.7621	1.8427	2.6731
	(12, 12, 11)	0.4995	0.8314	1.1064	1.7621	1.8427	2.6731
	(13, 13, 12)	0.4995	0.8314	1.1064	1.7621	1.8426	2.6731
DQHFEM wedge element		0.4995	0.8314	1.106	1.762	1.843	2.673
DQFEM solutions[32]		0.4995	0.8314	1.106	1.762	1.843	2.673
3D FEM solutions[349]		0.4997	0.8316	1.107	1.763	1.844	2.677
3D Ritz solutions[350]		0.4995	0.8314	1.106	1.762	1.843	2.673

fast convergence rate and good agreement with the results obtained by ANSYS. Figure 5.34 presents the mode shapes obtained by HQEM and ANSYS. One can see that the modes obtained by HQEM agree well with those obtained by ANSYS.

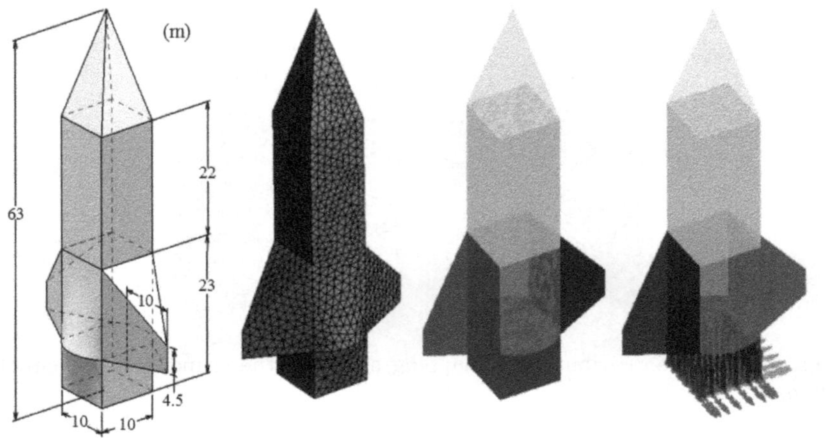

(a) (b) (c) (d)

Fig. 5.33. A rocket model and its mesh: (a) dimensions; (b) ANSYS meshes; (c) HQEM meshes; (d) boundary conditions.

Table 5.11. The frequencies of the rocket model under clamped boundary conditions at the bottom.

		Frequency sequence					
Methods	Element order	1	2	3	4	5	6
HQEM	(2, 2, 1)	3.5784	3.7206	15.6509	15.6810	16.4131	25.9542
	(3, 3, 2)	3.4761	3.6069	15.0136	15.1583	15.6175	25.7387
	(4, 4, 3)	3.4463	3.5848	14.9414	15.0896	15.5484	25.6247
	(5, 5, 4)	3.4338	3.5760	14.9059	15.0647	15.4911	25.5906
	(6, 6, 5)	3.4279	3.5724	14.8935	15.0531	15.4787	25.5643
	(9, 9, 8)	3.4244	3.5700	14.8843	15.0465	15.4664	25.5545
	(10, 10, 9)	3.4222	3.5688	14.8802	15.0425	15.4626	25.5446
	(13, 13, 12)	3.4179	3.5664	14.8712	15.0352	15.4530	25.5301
	(14, 14, 13)	3.4176	3.5662	14.8707	15.0347	15.4525	25.5286
	Number of elements						
FEM	17683	3.4195	3.5061	14.8782	15.0839	15.4604	25.4155
	24060	3.419	3.505	14.8737	15.0832	15.4594	25.413

5.5.3 *The interaction of grains and grain boundaries of metals*

This section uses the DQHFEM to analyze the interaction of grains and grain boundaries of metals in 3D scale which is closer to the fact. It should be noted that the length scale of the grain will not

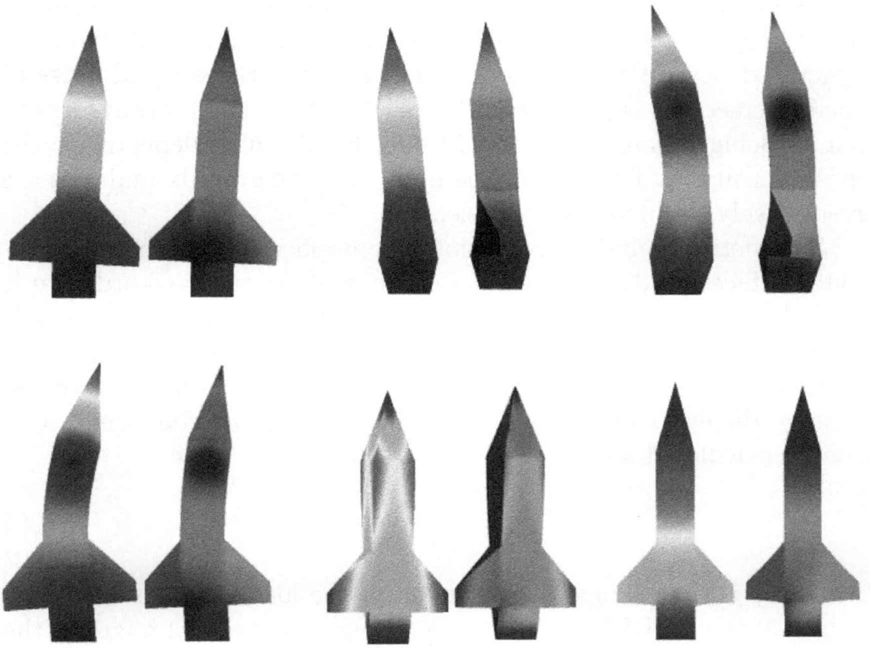

Fig. 5.34. Mode shapes obtained by HQEM (left) and ANSYS (right).

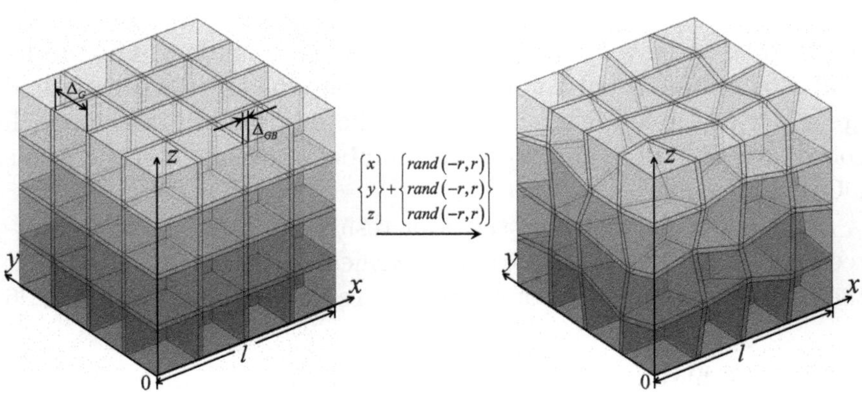

Fig. 5.35. The method of generating a metallographic structure model.

change the result of the modulus computation of the metallographic structure. For simplicity, a unit cubic representative volume element (RVE) will be considered. The method of generating a metallographic structure model is given in Fig. 5.35. Firstly, a cubic RVE is divided into regular grids. Δ_G and Δ_{GB} are the initial lengths of the grain and the grain boundary. Then a metallographic structure model is generated by adding a random number to the three coordinates of each intersection of grain boundaries. But Δ_{GB}/Δ_G is always a constant. Being different from traditional FEM which depends on the mesh density, by DQHFEM, the grain and the grain boundaries are respectively simulated by one element.

Symmetric boundary conditions are applied on the three coordinate surfaces, and the uniaxial tension σ along the x co-ordinate is loaded on the free y–z surface. According to Ref. [351], the Young's modulus of the grain boundary E_{GB} is taken as 75% of that of the grain E_G. The Young's modulus of the structure $\bar{E}$ is equal to the average displacement of the loaded face divided by the stress, so it can be calculated as

$$\bar{E} = \frac{\sigma l}{\bar{u}_x} \tag{5.99}$$

where $\bar{u}_x$ is the average displacement of the loaded face and l is the length of the DQHFEM model. According to the rule of mixture, the formula of approximated global Young's modulus can be obtained as

$$\tilde{E} = R_{GB}E_{GB} + (1 - R_{GB})E_G \tag{5.100}$$

where R_{GB} is the volume fraction of the grain boundary. The results of approximated global Young's modulus obtained by the two method are given in Table 5.12, where N_G is the number of grains in each coordinate direction, and N_E is the number of elements in the whole DQHFEM model.

By comparing with two series of results in Table 5.12, the correctness of the DQHFEM result is verified. It should be noted that the no-dimensional forms of the main stress components in Fig. 5.36

Table 5.12. The results of approximated global Young's modulus obtained by two methods.

N_G	N_E	DOFs	$\tilde{E}$	$\bar{E}$	R_{GB}
3	125	3630	0.9826 E_G	0.9856 E_G	5.768%
4	343	9450	0.9810 E_G	0.9839 E_G	6.457%
5	729	19494	0.9802 E_G	0.9828 E_G	6.868%

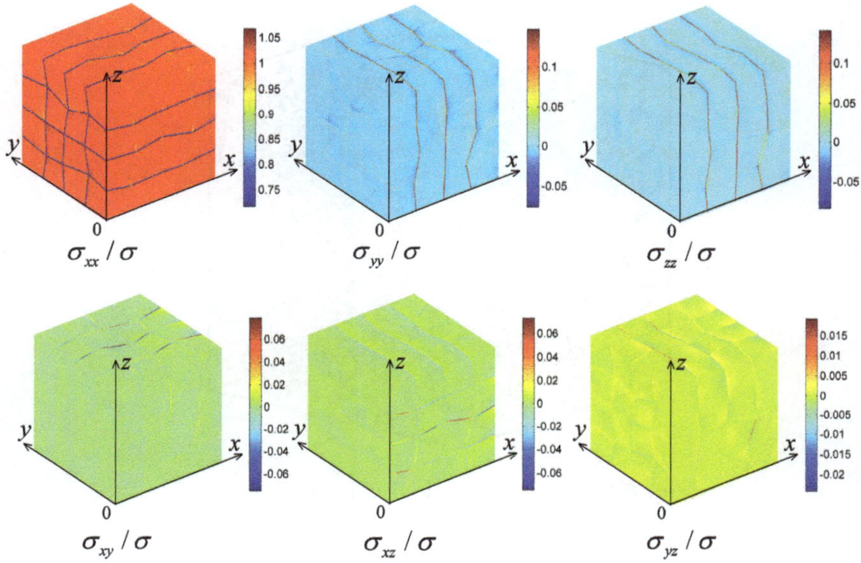

Fig. 5.36. The stress components of DQHFEM model.

are close to 1 and those of the other stress components are very small, which is consistent with the tensile regulation. Due to the approximate symmetry of the model and the symmetric boundary conditions, σ_{yy} and σ_{zz} are nearly equal. The same result can be found between σ_{xy} and σ_{xz}.

5.5.4 *Three-dimensional analysis of nanoparticulate polymer nanocomposites*

Figures 5.37 and 5.38 show a periodic polymer–nanoparticle composite (PNPC) molecular model and the continuum-based analytical model for the representative volume element (RVE) which is widely used by the previous study. Here, the-three dimensional DQHFEM is used to analyze the continuum-based analytical model without any assumed displacement continuous on the interface.

The DQHFEM model consisted of 12 DQHFEM elements as shown in Fig. 5.39. The particle core and the interface are respectively simulated by only one tetrahedron element and one triangular prism element. The matrix is represented by tetrahedron elements.

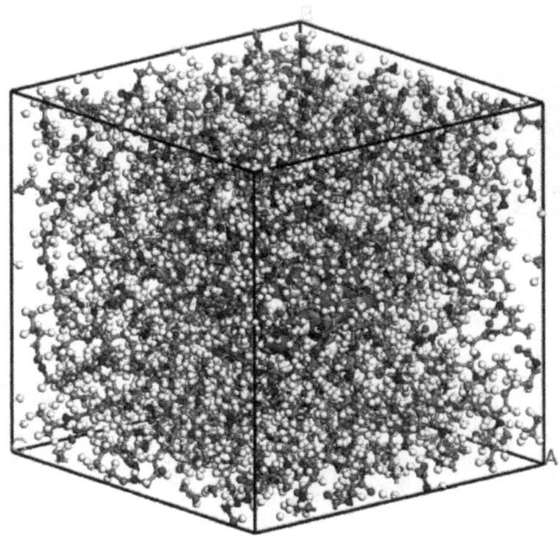

Fig. 5.37. A periodic PNPC molecular model.

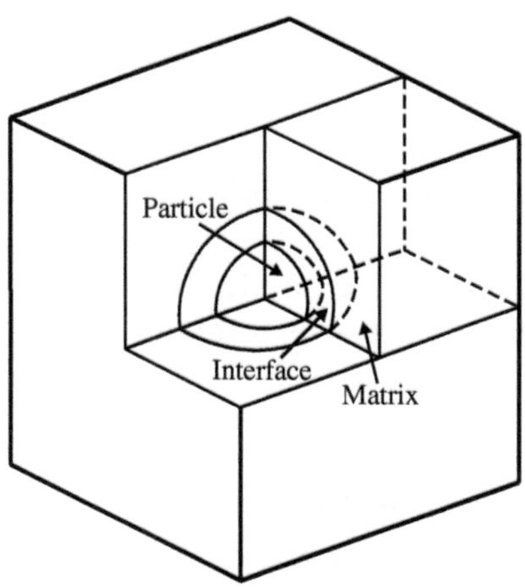

Fig. 5.38. The continuum-based PNPC model.

The radius of the particle $r = 0.5$, and the thickness of the interface $d = r/10$. It should be noted that this is a challenging two-scale problem because d is far less than r and l.

In order to verify the DQHFEM, a homogeneous cube is analyzed first. The Young's modulus and the Poisson's ratio of all elements are all respectively taken as 195 GPa and 0.3 to analyze a free vibration problem. The results of the DQHFEM and the FEM are

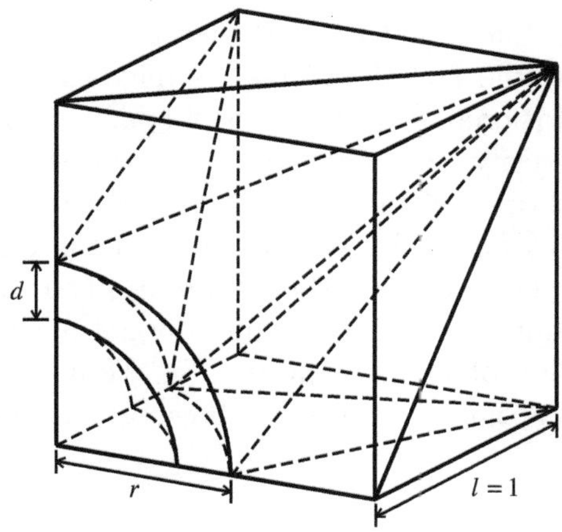

Fig. 5.39. The DQHFEM model of PNPC.

Table 5.13. Comparison of frequency parameters $\lambda = \omega/2\pi$ for the cube with free boundary condition.

Solution method	Element degree/ Mesh size	DOFs	Mode sequence number 1	2	3	4	5	6
DQHFEM	4	426	1475.1	1475.7	1935.1	1967.3	1971.2	1971.5
	5	1023	1422.5	1422.6	1911.6	1914.3	1914.5	1947.3
	6	2166	1417.8	1417.9	1908.1	1908.6	1908.6	1940.6
	7	4071	1416.7	1416.7	1907.6	1907.6	1907.6	1940.1
	8	6954	1416.6	1416.6	1907.5	1907.6	1907.6	1939.9
	9	**11031**	**1416.5**	**1416.5**	**1907.5**	**1907.5**	**1907.5**	**1939.9**
	10	16518	1416.5	1416.5	1907.5	1907.5	1907.5	1939.9
	11	23631	1416.5	1416.5	1907.5	1907.5	1907.5	1939.9
FEM	0.05–0.01	103488	1410.8	1410.8	1662.3	1665.0	1684.0	1709.1
	0.03–0.01	165498	1414.1	1414.1	1902.6	1902.6	1903.3	1939.4
	0.02–0.01	392592	1415.3	1415.3	1905.1	1905.1	1905.4	1939.6
	0.016–0.01	836181	1415.9	1415.9	1906.2	1906.2	1906.2	1939.8
	0.014–0.007	1283244	1416.0	1416.0	1906.5	1906.5	1906.6	1939.8
	0.012–0.007	1911180	1416.2	1416.2	1906.8	1906.8	1906.8	1939.9

shown in Table 5.13. Using the DQHFEM, the results are converged when the elements' degrees are 9 and the degrees of freedom are only 11031. On the contrary, the FEM results by using commercial software ABAQUS-6.14 are still not converged when the degrees of freedom are 1911180, which reveals that the two-scale analysis may not be an easy task for the FEM even with high computational cost. It is obvious that the DQHFEM results have excellent convergence with low computational cost although in sparse grids.

In order to verify the static property of the DQHFEM, the Young's modulus of all elements in the DQHFEM model (Fig. 5.39) are taken as 2.44 GPa, and the shear modulus is taken as 0.88 GPa [338]. Symmetric boundary conditions are applied on the three cube surfaces which contain the grain. The uniaxial normal displacement u is loaded on one of the free faces. The Young's modulus of the structure can be obtained from the following equation:

$$\bar{E} = \frac{\bar{\sigma} l}{u} \tag{5.101}$$

where $\bar{\sigma}$ is the average stress of the loaded face. The relative error of the Young's modulus calculated by this method is 2.48×10^{-6}, which shows the great performance of the DQHFEM in the two-scale problem.

Then the Young's modulus of the particle is taken as 104 GPa, and the shear modulus is taken as 37 GPa [338]. The Young's modulus obtained by the DQHFEM is 2.60 GPa. Due to the symmetry of the model and the symmetric boundary conditions, σ_{xx} and σ_{yy} are symmetric in Fig. 5.40. The same result can be found between σ_{yz} and σ_{xz}. When the Young's modulus of the interface is taken as 0.244 GPa, the Young's modulus given by the DQHFEM is 2.42 GPa. It is obvious that the strength of the interface is not negligible.

5.5.5 *Thermo-mechanical analysis of cross-ply laminated plates*

In this section, 3D vibration analyses, static analyses and thermal-mechanical responses are respectively carried out for cross-ply and angle-ply laminated plates and shells. For the wedge elements, the number of nodes on the edges of the two triangular faces is taken as the same and denoted as N_e, the number of nodes on the interiors of the two triangular faces is also taken as the same and denoted as N_f, the number of nodes on the thickness direction of the quadrilateral faces on edges and interiors are also taken as the same and

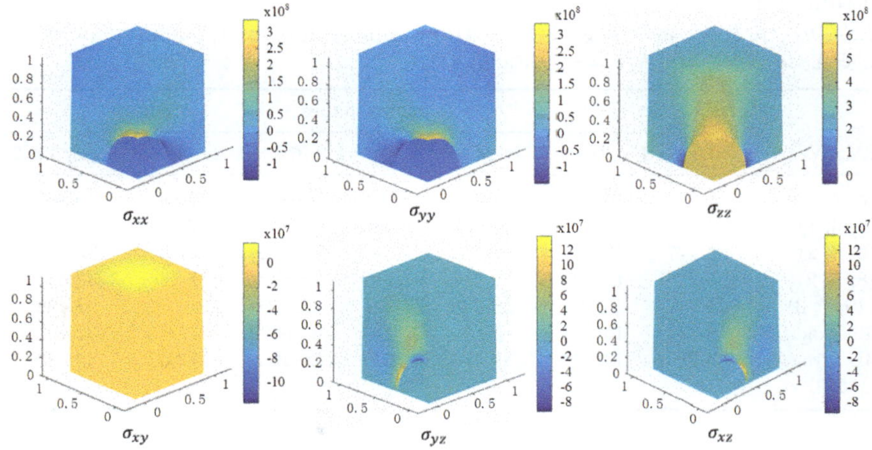

Fig. 5.40. The stress components of DQHFEM model of PNPC.

denoted as N_t. The number of nodes inside the wedge elements correlates with the face and edge nodes by keeping the maximum polynomial degree unchanged. The geometries were first constructed from NURBS and then expressed by the DQH bases. To avoid comparing with round-off errors that are unrealistic, the natural frequencies are expressed as the same frequency parameters as used in literatures.

First, an isotropic simply supported square plate is the analyzed in Table 5.14. Here, simply supported condition means that the tangent displacements on edges equal zeros and the normal displacement on edges is free. The square plate is discretized as eight wedge elements on each layer are two layers are used for analyses (Fig. 5.41). One can see from Table 5.14 that the present method using only a few nodes can provide results with almost all digits agreeing with the exact results. Especially, the wedge element can only use a few nodes on the thickness direction. One can see from Table 5.14 that the present method with one node on thickness direction (without the two ends) can provide results with two digits agreeing with exact ones. Therefore, 3D analyses using wedge DQHFEM elements have great similarities with high-order shear deformation theory of plates, that is, the present DQHFEM using the computational cost

Table 5.14. Comparison of non-dimensional frequency parameters $\Omega = (\omega b^2/h)\sqrt{\rho/E}$ for a homogeneous isotropic simply supported square plate with $\upsilon = 0.3, a/b = 1$ and $h/b = 0.1$ (Fig. 5.41).

	Mode shapes						
(N_e, N_f, N_t)	1×1	$1 \times 2/2 \times 1$	In-plane	2×2	$1 \times 3/3 \times 1$	In-plane	$2 \times 3/3 \times 2$
(6, 3, 0)	5.948	14.251	19.483	22.001	26.824	27.554	33.904
(6, 3, 1)	5.778	13.815	19.483	21.293	25.938	27.554	32.747
(6, 3, 2)	5.777	13.808	19.483	21.278	25.915	27.554	32.710
(8, 5, 2)	5.777	13.805	19.483	21.215	25.870	27.554	32.494
(10, 7, 3)	5.777	13.805	19.483	21.214	25.870	27.554	32.492
3D exact [352]	5.777	13.805	N/A	21.214	25.869	27.553	32.491
3D layerwise [353]	5.785	13.871	19.483	21.300	26.420	27.662	32.930

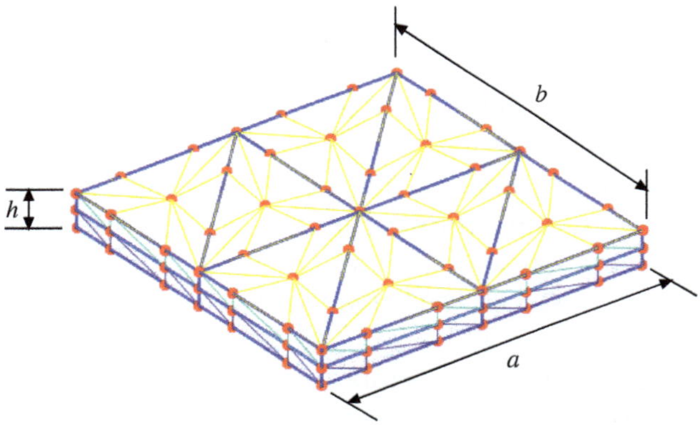

Fig. 5.41. A square plate discretized as eight wedge elements: the bold blue lines denote element boundaries; the red dots denote nodes of elements; the thin lines are used to draw the geometry.

of shear deformation theories can reach the accuracy of 3D elasticity theory.

In Tables 5.15 and 5.16, vibration of a simply supported cross-ply square plate with several orthotropic layers is analyzed. The material properties for Table 5.15 are $E_1/E_2 = 40$, $E_3/E_2 = 1$, $G_{12}/E_2 = G_{13}/E_2 = 0.6$, $G_{23}/E_2 = 0.5$, $\upsilon_{12} = \upsilon_{23} = \upsilon_{13} = 0.25$, and $\rho = 1$. The material properties for Table 5.3 are different from those for Table 5.15 and are $G_{12}/E_2 = G_{13}/E_2 = 0.5$, and $G_{23}/E_2 = 0.6$. The stacking sequence includes $[0°/90° \cdots 0°/90°]$ and $[0°/90°/90°/0°]$.

Table 5.15. Comparison of non-dimensional fundamental frequency parameters $\Omega = \omega\sqrt{\rho h^2/E_2}$ for a simply supported cross-ply square plate with orthotropic layers $[0\circ/90\circ\cdots 0\circ/90\circ]$ and $h/b = 0.5$ (Fig. 5.41).

No. of layers	(N_e, N_f, N_t)	E_1/E_2 3	10	20	30	40
2	(5, 2, 1)	0.2496	0.2794	0.3086	0.3304	0.3476
	(5, 2, 2)	0.2493	0.2783	0.3057	0.3257	0.3411
	(7, 4, 2)	0.2493	0.2782	0.3057	0.3257	0.3411
	3D elasticity [354]	0.25031	0.27938	0.30698	0.32705	0.34250
	HOST [355]	0.24782	0.27764	0.30737	0.33003	0.34810
4	(5, 2, 0)	0.2616	0.3269	0.3794	0.4116	0.4335
	(5, 2, 1)	0.2602	0.3238	0.3740	0.4043	0.4248
	(7, 4, 1)	0.2602	0.3238	0.3740	0.4042	0.4248
	3D elasticity [354]	0.26182	0.32578	0.37622	0.40660	0.42719
	HOST [355]	0.25997	0.32486	0.37801	0.41041	0.43240

Table 5.16. Comparison of normalized fundamental frequency parameters $\Omega = (\omega b^2/h)\sqrt{\rho/E_2}$ for a simply supported cross-ply square plate with orthotropic layers $[0°/90°/90°/0°]$ and $h/b = 0.2$ (Fig. 5.41).

(N_e, N_f, N_t)	E_1/E_2 10	20	30	40
(5, 2, 1)	8.2490	9.5391	20.2725	10.7572
(5, 2, 2)	8.2483	9.5352	10.2623	10.7485
(7, 4, 2)	8.2474	9.5344	10.2655	10.7478
(9, 6, 2)	8.2474	9.5344	10.2655	10.7478
FOST [356]	8.3094	9.5698	10.3224	10.8471

The ratio of E_1/E_2 for Tables 5.15 and 5.16 varies between 3 and 40, and the numbers of layers are 2 and 4. The plate is still discretized as eight wedge elements, as shown in Fig. 5.41. From Tables 5.15 and 5.16, one can again see the fast convergence and high accuracy of the present wedge DQHFEM for cross-plate laminates.

In Tables 5.17 and 5.18, a clamped anti-symmetric skew $[0°/90°/0°/90°]$ and $[45°/-45°/45°/-45°]$ composite laminates with relative thickness $h/b0.1$ are analyzed by the present DQHFEM using

Table 5.17. Non-dimensional frequency parameters $\Omega = (\omega b^2/\pi^2 h)\sqrt{\rho/E_2}$ for clamped anti-symmetric cross-ply skew composite laminates [0°/90°/0°/90°] with $h/b = 0.1$ (Fig. 5.42).

Skew angle (θ)	(N_e, N_f, N_t)	Mode sequence number							
		1	2	3	4	5	6	7	8
0°	(8, 5, 0)	2.3075	3.8097	3.8128	4.9607	5.8069	5.8626	6.6259	6.9234
	(12, 9, 0)	2.3056	3.8050	3.8050	4.8871	5.7424	5.7503	6.5318	6.5327
	(15, 12, 0)	2.3056	3.8048	3.8048	4.8867	5.7420	5.7499	6.5312	6.5312
	(15, 12, 3)	2.2555	3.7171	3.7171	4.7749	5.6052	5.6129	6.3793	6.3793
	(15, 12, 5)	2.2555	3.7170	3.7170	4.7746	5.6049	5.6126	6.3790	6.3790
	HOST [357]	2.2990	3.7880	3.7880	4.8610	5.7129	5.7203	6.4927	6.4927
	FOST [357]	2.3947	3.9533	3.9533	5.0666	5.9685	5.9762	6.7684	6.7684
	FOST [359]	2.3947	3.9532	3.9532	5.0665	5.9680	5.9757	6.7679	6.7679
30°	(8, 5, 0)	2.6783	4.0084	4.7539	5.2123	6.5152	6.5178	7.0897	7.8618
	(12, 9, 0)	2.6764	4.0045	4.7488	5.2035	6.4815	6.4890	7.0186	7.7661
	(15, 12, 0)	2.6763	4.0044	4.7486	5.2032	6.4811	6.4887	7.0182	7.7656
	(15, 12, 3)	2.6163	3.9118	4.6383	5.0815	6.3283	6.3360	6.8521	7.5828
	HOST [358]	2.6521	3.9657	4.7014	5.1513	6.4224	6.4149	6.9457	7.6849
	HOST [357]	2.6666	3.9851	4.7227	5.1752	6.4445	6.4510	6.9755	7.7218
	FOST [357]	2.7798	4.1566	4.9240	5.3989	6.7216	6.7248	7.2738	8.0459
	FOST [359]	2.7796	4.1564	4.9237	5.3983	6.7204	6.7240	7.2729	8.0467
45°	(8, 5, 0)	3.3197	4.6597	5.8857	6.0497	7.1615	7.8054	8.5058	9.1165
	(12, 9, 0)	3.3178	4.6560	5.8778	6.0438	7.1211	7.7782	8.3168	9.0446
	(15, 12, 0)	3.3177	4.6559	5.8777	6.0435	7.1209	7.7780	8.3163	9.0442
	(15, 12, 3)	3.2406	4.5475	5.7395	5.9043	6.9525	7.5990	8.1186	8.8349
	HOST [358]	3.2853	4.6090	5.8173	5.9813	7.0465	7.6971	8.2283	8.9487
	HOST [357]	3.3015	4.6290	5.8423	6.0039	7.0792	7.7269	8.2726	8.9874
	FOST [357]	3.4434	4.8223	6.0858	6.2421	7.3717	8.0226	8.6118	9.3262
	FOST [359]	3.4430	4.8219	6.0850	6.2414	7.3720	8.0239	8.6233	9.3320

wedge elements. The skew angle (θ) of the rhombic plate varies from 0° to 45° (Fig. 5.42). The material properties for Tables 5.17 and 5.18 are $E_1/E_2 = 40$, $E_3/E_2 = 1$, $G_{12}/E_2 = G_{13}/E_2 = 0.6$, $G_{23}/E_2 = 0.5$, $\upsilon_{12} = \upsilon_{23} = \upsilon_{13} = 0.25$, and $\rho = 1$. One can again find the fast convergence of the present DQHFEM. One can see that the DQHFEM with no node (without the two ends) on the thickness direction agree with the high-order shear deformation theory (HOST) [357, 358] for about two digits. If the number of nodes of the DQHFEM on the thickness direction increases, the final converged results agree with the HOST [357, 358] results for 1–2 digits.

In Tables 5.19–5.21, a classical example given by Srinivas [360] is analyzed. Only six elements are used in analyses. The square plate

Table 5.18. Non-dimensional frequency parameters $\Omega = (\omega b^2/\pi^2 h)\sqrt{\rho/E_2}$ for clamped anti-symmetric angle-ply skew composite laminates [45°/−45°/45°/−45°] with h/b = 0.1 (Fig. 5.42).

Skew angle (θ)	(N_e, N_f, N_t)	Mode sequence number							
		1	2	3	4	5	6	7	8
0°	(8, 5, 0)	2.2207	3.7550	3.7559	5.1529	5.6104	5.7129	6.7973	7.0165
	(12, 9, 0)	2.2183	3.7492	3.7494	5.0855	5.5417	5.6103	6.6838	6.6840
	(15, 12, 0)	2.2181	3.7490	3.7491	5.0850	5.5412	5.6099	6.6827	6.6828
	(15, 12, 3)	2.1729	3.6644	3.6645	4.9611	5.4138	5.4794	6.5188	6.5188
	(15, 12, 5)	2.1728	3.6643	3.6644	4.9610	5.4136	5.4792	6.5186	6.5187
	HOST [357]	2.2119	3.7339	3.7339	5.0618	5.5129	5.5810	6.6473	6.6473
	FOST [357]	2.2965	3.8921	3.8921	5.3045	5.7455	5.8202	6.9654	6.9654
	FOST [359]	2.2964	3.8919	3.8919	5.3042	5.7449	5.8196	6.9647	6.9647
30°	(8, 5, 0)	2.6443	3.9779	4.7436	5.2469	6.4246	6.5663	7.0857	7.8557
	(12, 9, 0)	2.6423	3.9738	4.7380	5.2381	6.3957	6.5318	7.0181	7.7625
	(15, 12, 0)	2.6422	3.9737	4.7377	5.2379	6.3954	6.5314	7.0177	7.7620
	(15, 12, 3)	2.5839	3.8828	4.6280	5.1140	6.2475	6.3745	6.8513	7.5792
	HOST [358]	2.6184	3.9353	4.6905	5.1852	6.3301	6.4638	6.9447	7.6810
	HOST [357]	2.6325	3.9549	4.7125	5.2107	6.3577	6.4954	6.9760	7.7176
	FOST [357]	2.7418	4.1221	4.9130	5.4402	6.6192	6.7854	7.2763	8.0421
	FOST [359]	2.7416	4.1219	4.9126	5.4395	6.6183	6.7842	7.2753	8.0427
45°	(8, 5, 0)	3.3197	4.6597	5.8857	6.0497	7.1615	7.8054	8.5058	9.1165
	(12, 9, 0)	3.3178	4.6560	5.8778	6.0438	7.1211	7.7782	8.3168	9.0446
	(15, 12, 0)	3.3177	4.6559	5.8777	6.0435	7.1209	7.7780	8.3163	9.0442
	(15, 12, 3)	3.2406	4.5475	5.7395	5.9043	6.9525	7.5990	8.1186	8.8349
	HOST [358]	3.2853	4.6090	5.8173	5.9813	7.0465	7.6971	8.2283	8.9487
	HOST [357]	3.3015	4.6290	5.8423	6.0039	7.0792	7.7269	8.2726	8.9874
	FOST [357]	3.4434	4.8223	6.0858	6.2421	7.3717	8.0226	8.6118	9.3262
	FOST [359]	3.4430	4.8219	6.0850	6.2414	7.3720	8.0239	8.6233	9.3320

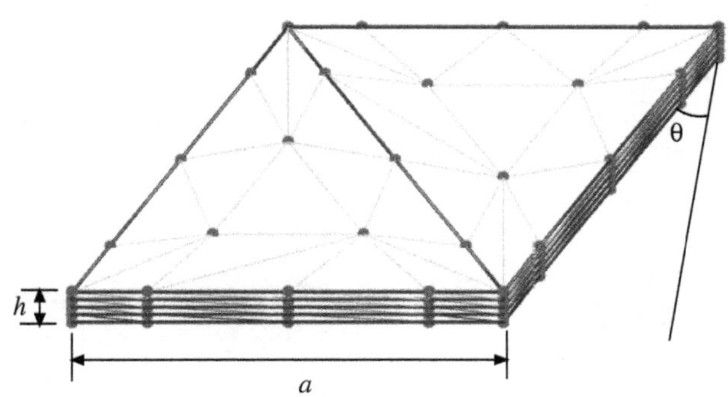

Fig. 5.42. A rhombic plate with four layers; each layer is discretized as two wedge elements.

Table 5.19. Square laminated plate under uniform load with $R = 5$ (Fig. 5.43).

(N_e, N_f, N_t)	$\bar{w}$	$\bar{\sigma}_x^1$	$\bar{\sigma}_x^2$	$\bar{\sigma}_x^3$	$\bar{\sigma}_y^1$	$\bar{\sigma}_y^2$	$\bar{\sigma}_y^3$	$\bar{\tau}_{xz}^1$	$\bar{\tau}_{yz}^1$
HSDT [364]	256.13	62.38	46.91	9.382	38.93	30.33	6.065	3.089	2.566
FSDT [364]	236.10	61.87	49.50	9.899	36.65	29.32	5.864	3.313	2.444
CLT [364]	216.94	61.141	48.623	9.783	36.622	29.297	5.860	4.5899	3.386
Ferreira [362]	258.74	59.21	45.61	9.122	37.88	29.59	5.918	3.593	3.593
Ferreira [363]-15	257.38	58.725	46.980	9.396	37.643	27.714	4.906	3.848	2.839
HSDT [365]-11	253.671	59.6447	46.4292	9.2858	38.0694	29.9313	5.9863	3.8449	1.9650
HSDT [365]-15	256.239	60.1834	46.8581	9.3716	38.3592	30.1642	6.0328	4.2768	2.2227
HSDT [365]-21	257.110	60.3660	47.0028	9.4006	38.4563	30.2420	6.0484	4.5481	2.3910
Ferreira [361]-11	252.084	58.8628	45.4232	9.8846	37.6901	29.4765	5.8953	3.8311	2.5319
Ferreira [361]-15	255.920	59.6503	46.0366	9.2073	38.1408	29.8296	5.9659	3.9773	2.5375
Ferreira [361]-21	257.523	59.9675	46.2906	9.2581	38.3209	29.9740	5.9948	4.0463	2.3901
DQFEM [366]-11	258.828	60.2396	46.5201	9.3040	38.4895	30.1130	6.0226	4.1153	3.4118
DQFEM [366]-15	258.833	60.2466	46.5151	9.3030	38.4914	30.1118	6.0224	4.1080	3.4028
DQFEM [366]-25	258.835	60.2540	46.5098	9.3020	38.4951	30.1091	6.0218	4.1076	3.3994
(10, 7, 2)	259.160	60.3654	46.5612	9.3122	38.5663	30.1466	6.0293	4.4001	3.2403
(14, 11, 4)	259.167	60.3423	46.5682	9.3136	38.5488	30.1465	6.0293	4.3714	3.2633
(18, 15, 6)	259.169	60.3388	46.5766	9.3153	38.5477	30.1513	6.0303	4.3623	3.2307
(20, 17, 6)	259.170	60.3417	46.5806	9.3161	38.5530	30.1553	6.0311	4.3631	3.2100
(22, 19, 6)	259.170	60.3403	46.5794	9.3159	38.5496	30.1537	6.0307	4.3640	3.1966
(24, 21, 6)	259.170	60.3403	46.5793	9.3159	38.5513	30.1540	6.0308	4.3649	3.1855
Exact [360]	258.97	60.353	46.623	9.340	38.491	30.097	6.161	4.3641	3.2675

Table 5.20. Square laminated plate under uniform load with $R = 10$ (Fig. 5.43).

(N_e, N_f, N_t)	$\bar{w}$	$\bar{\sigma}_x^1$	$\bar{\sigma}_x^2$	$\bar{\sigma}_x^3$	$\bar{\sigma}_y^1$	$\bar{\sigma}_y^2$	$\bar{\sigma}_y^3$	$\bar{\tau}_{xz}^1$	$\bar{\tau}_{yz}^1$
HSDT [364]	152.33	64.65	51.31	5.131	42.83	33.97	3.397	3.147	2.587
FSDT [364]	131.095	67.80	54.24	4.424	40.10	32.08	3.208	3.152	2.676
CLT [364]	118.87	65.332	48.857	5.356	40.099	32.079	3.208	4.3666	3.7075
Ferreira [362]	159.402	64.16	47.72	4.772	42.970	42.900	3.290	3.518	3.518
Ferreira [363]-15	158.55	62.723	50.16	5.01	42.565	34.052	3.400	3.596	3.053
HSDT [365]-11	153.008	64.7415	49.4716	4.9472	42.8860	33.3524	3.3352	2.7780	1.8207
HSDT [365]-15	154.249	65.2223	49.8488	4.9849	43.1521	33.5663	3.3566	3.1925	2.1360
HSDT [365]-21	154.658	65.3809	49.9729	4.9973	43.2401	33.6366	3.3637	3.5280	2.3984
Ferreira [361]-11	155.037	63.5984	47.4765	4.7476	42.6696	32.7369	3.2737	3.7016	3.3051
Ferreira [361]-15	157.374	64.4828	48.1544	4.8154	43.1887	33.1392	3.3139	3.8447	3.2183
Ferreira [361]-21	158.380	64.8462	48.4434	4.8443	43.3989	33.3062	3.3306	3.9237	2.8809
DQFEM [366]-11	159.396	65.2015	48.7577	4.8758	43.6378	33.5098	3.3510	4.0040	3.3982
DQFEM [366]-15	159.403	65.2178	48.744	4.8744	43.6425	33.5061	3.3506	3.9957	3.3862
DQFEM [366]-25	159.406	65.2306	48.733	4.8733	43.6494	33.5002	3.3500	3.9957	3.3829
(10, 7, 2)	159.474	65.3100	48.7567	4.8757	43.6994	33.5181	3.3518	4.1195	3.4893
(14, 11, 4)	159.484	65.2827	48.7810	4.8781	43.6774	33.5290	3.3529	4.0984	3.4709
(18, 15, 6)	159.487	65.2900	48.7879	4.8788	43.6834	33.5347	3.3535	4.0943	3.4323
(20, 17, 6)	159.487	65.2998	48.7849	4.8785	43.6950	33.5330	3.3533	4.0955	3.4186
(22, 19, 6)	159.487	65.2959	48.7874	4.8787	43.6892	33.5349	3.3535	4.0965	3.4105
(24, 21, 6)	159.487	65.2966	48.7854	4.8785	43.6914	33.5334	3.3533	4.0970	3.4048
Exact [360]	159.38	65.332	48.857	4.903	43.566	33.413	3.500	4.0959	3.5154

Table 5.21. Square laminated plate under uniform load with $R = 10$ (Fig. 5.43).

(N_e, N_f, N_t)	$\bar{w}$	$\bar{\sigma}_x^1$	$\bar{\sigma}_x^2$	$\bar{\sigma}_x^3$	$\bar{\sigma}_y^1$	$\bar{\sigma}_y^2$	$\bar{\sigma}_y^3$	$\bar{\tau}_{xz}^1$	$\bar{\tau}_{yz}^1$
HSDT [364]	110.43	66.62	51.97	3.465	44.92	35.41	2.361	3.035	2.691
FSDT [364]	90.85	70.04	56.03	3.753	41.39	33.11	2.208	3.091	2.764
CLT [364]	81.768	69.135	55.308	3.687	41.410	33.128	2.209	4.2825	3.8287
Ferreira [362]	121.821	65.650	47.09	3.140	45.850	34.420	2.294	3.466	3.466
Ferreira [363]-15	121.184	63.214	50.571	3.371	45.055	36.044	2.400	3.466	3.099
HSDT [365]-11	113.594	66.3646	49.8957	3.3264	45.2979	34.9096	2.3273	2.1686	1.5578
HSDT [365]-15	114.387	66.7830	50.2175	3.3478	45.5427	35.1057	2.3404	2.6115	1.9271
HSDT [365] -21	114.644	66.9196	50.3230	3.3549	45.6229	35.1696	2.3446	3.0213	2.2750
Ferreira [361]-11	118.298	64.9159	46.8241	3.1216	45.4432	34.2237	2.2816	3.6123	3.8412
Ferreira [361]-15	120.077	65.8418	47.5260	3.1684	46.0049	34.6566	2.3104	3.7556	3.6695
Ferreira [361]-21	120.988	66.2911	47.8992	3.1933	46.2924	34.8898	2.3260	3.8311	3.2562
DQFEM [366]-11	121.764	66.6355	48.2074	3.2138	46.5281	35.0953	2.3397	3.9264	3.4062
DQFEM [366]-15	121.774	66.6633	48.1818	3.2121	46.5367	35.0878	2.3392	3.9190	3.3932
DQFEM [366]-25	121.777	66.6800	48.1664	3.2111	46.5463	35.0790	2.3386	3.9192	3.3580
(10, 7, 2)	121.777	66.7313	48.1855	3.2124	46.5789	35.0898	2.3393	3.9803	3.5793
(14, 11, 4)	121.789	66.7088	48.2186	3.2146	46.5572	35.1094	2.3406	3.9642	3.5371
(18, 15, 6)	121.792	66.7287	48.2174	3.2145	46.5718	35.1113	2.3408	3.9628	3.5025
(20, 17, 6)	121.792	66.7402	48.2107	3.2140	46.5858	35.1056	2.3404	3.9640	3.4933
(22, 19, 6)	121.792	66.7366	48.2143	3.2143	46.5800	35.1091	2.3406	3.9648	3.4882
(24, 21, 6)	121.792	66.7369	48.2120	3.2141	46.5816	35.1071	2.3405	3.9652	3.4847
Exact [360]	121.72	66.787	48.299	3.238	46.424	34.955	2.494	3.9638	3.5768

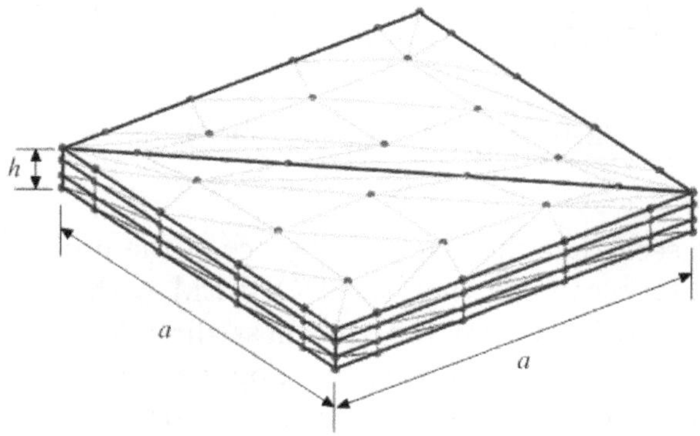

Fig. 5.43. A sandwich plate is discretized as six wedge elements.

was simply supported and subjected to a uniform transverse load. The material property matrix of the core layer is as follows:

$$Q_{\text{core}} = \begin{bmatrix} 0.999781 & 0.231192 & 0 & 0 & 0 & 0 \\ 0.231192 & 0.524886 & 0 & 0 & 0 & 0 \\ 0 & 0 & 0.524886 & 0 & 0 & 0 \\ 0 & 0 & 0 & 0.266810 & 0 & 0 \\ 0 & 0 & 0 & 0 & 0.159914 & 0 \\ 0 & 0 & 0 & 0 & 0 & 0.262931 \end{bmatrix} \tag{5.102}$$

Skin material properties are related to core properties by a factor R as follows:

$$Q_{\text{skin}} = RQ_{\text{core}} \tag{5.103}$$

Transverse displacement and stresses are normalized through the following factors:

$$\bar{w} = w(a/2, a/2, 0)\frac{0.999781}{hq}$$

$$\bar{\sigma}_x^1 = \frac{\sigma_x^{(1)}(a/2, a/2, -h/2)}{-q}, \quad \bar{\sigma}_x^2 = \frac{\sigma_x^{(1)}(a/2, a/2, -2h/5)}{-q}$$

$$\bar{\sigma}_x^3 = \frac{\sigma_x^{(2)}(a/2, a/2, -2h/5)}{-q}, \quad \bar{\sigma}_y^1 = \frac{\sigma_y^{(1)}(a/2, a/2, -h/2)}{-q}$$

$$\bar{\sigma}_y^2 = \frac{\sigma_y^{(1)}(a/2, a/2, -2h/5)}{-q}, \quad \bar{\sigma}_y^3 = \frac{\sigma_y^{(2)}(a/2, a/2, -2h/5)}{-q}$$

$$\bar{\tau}_{xz}^1 = \frac{\tau_{xz}^{(2)}(0, a/2, 0)}{q}, \quad \bar{\tau}_{yz}^1 = \frac{\tau_{yz}^{(2)}(a/2, 0, 0)}{q} \tag{5.104}$$

It can be seen that the fast convergence of the present DQHFEM is achieved. One can see that the DQHFEM with only two nodes (without the two ends) on the thickness direction agree with the exact 3D theory of Srinivas [354] for about three digits. So, it has the potential to obtain accurate result with a low computational cost. The convergence does not have obvious fluctuations when skin properties are quite different from the core properties, which is a typical industrial case. Compared with high-order formulation presented by Ferreira *et al.* [361–363], the present results are closer to the exact three-dimensional results.

In Tables 5.22–5.24, the examples presented by Bhimaraddi [367] are analyzed. Isotropic and orthotropic plates and shells are subjected to sinusoidal distributed lateral load on the surface of the plates and shells.

$$p_z = P \sin\left(\frac{\pi x}{a}\right) \sin\left(\frac{\pi y}{a}\right) \tag{5.105}$$

The Poisson's ratio has been assumed to be 0.3 in the case of an isotropic material. The orthotropic material properties are as follows:

$$\begin{aligned} &\frac{E_x}{E_y} = 25, \quad \frac{E_z}{E_y} = 1, \quad \frac{G_{xz}}{E_y} = \frac{G_{xy}}{E_y} = \frac{1}{2}, \quad \frac{G_{yz}}{E_y} = \frac{1}{2} \\ &\upsilon_{xy} = 0.25, \quad \upsilon_{zx} = 0.03, \quad \upsilon_{yz} = 0.4, \quad \text{hence } \upsilon_{xz} = 0.75 \end{aligned} \tag{5.106}$$

The central deflection is normalized through the following factor:

$$\bar{w} = w(a/2, a/2, 0)\frac{E_y}{q} \tag{5.107}$$

In Table 5.22, only six wedge elements have been used. In the isotropic plate condition, when $h/a(= 0.01)$ is low, the present DQHFEM results in Table 5.22 are similar to those of 2D theories and 3D elastic theories. With the number h/a increasing, the results of 2D theories become increasingly different from those of the present

Table 5.22. Comparison of central deflection for homogeneous square plate with different h/a (Fig. 5.43).

	Isotropic plate			Orthotropic plate		
(N_e, N_f, N_t)	$h/a = 0.01$	$h/a = 0.1$	$h/a = 0.15$	$h/a = 0.01$	$h/a = 0.1$	$h/a = 0.15$
3D elastic [367]	29504	29.440	9.2352	4343.0	6.3343	2.5879
PSD [367]	28041	29.606	9.3562	4333.5	6.3709	2.6196
CST [367]	28026	28.026	8.3040	4312.5	4.3125	1.2778
FOST [339]	28042	29.607	9.3578	4333.5	6.3830	2.6370
HOST9 [339]	28042	29.606	9.3560	4333.5	6.3713	2.6210
HOST11 [339]	28043	29.725	9.4365	4333.7	6.4014	2.6433
HOST12 [339]	28043	29.725	9.4365	4333.7	6.4014	2.6433
(10, 7, 2)	28040	29.425	9.2334	4333.1	6.3250	2.5856
(14, 11, 4)	28040	29.425	9.2330	4333.1	6.3246	2.5851
(16, 13, 4)	28040	29.425	9.2330	4333.1	6.3246	2.5851

DQHFEM, especially the errors achieved 2.2% when h/a is 0.15. The present results of DQHFEM are hardly affected, i.e., the errors only vary from 0.024% to 0.051%. In the orthotropic plate condition, when the $h/a(= 0.01)$ is low, compared with 3D elastic solutions, both the errors of 2D theory (HOST12 [339]) and present DQHFEM are quite small, i.e., less than 0.23%. With the increase in h/a, the errors of HOST12 (1–2.1%) will increase while the errors of the present DQHFEM just vary from 0.11% to 0.15%. In Tables 5.23 and 5.24, only three hexahedron elements have been used. When $h/a(= 0.15)$ is high, the results of the present DQHFEM can maintain good agreement with the results of 3D elastic theories (the errors only vary from 0.1398% to 1.056%) while the errors of the results of 2D theories vary from 0.4113% to 1.523%. From the above discussions, one can conclude that the results of the present DQHFEM are the best either for the isotropic or the orthotropic cases.

In Tables 5.25–5.28, the plate is discretized as six wedge elements, and the shells are discretized as three hexahedron elements. The geometric and material properties are the same as those used in Table 5.22. The linear coefficients of thermal expansion are as follows:

$$\text{Face sheets:} \quad \alpha_1 = 0.1 \times 10^{-5}/^\circ\text{C}, \quad \alpha_2 = 2 \times 10^{-5}/^\circ\text{C}, \quad \alpha_3 = \alpha_1$$

$$\text{Core:} \quad \alpha_1 = 0.1 \times 10^{-6}/^\circ\text{C}, \quad \alpha_2 = 0.2 \times 10^{-5}/^\circ\text{C}, \quad \alpha_3 = \alpha_1$$

Table 5.23. Comparison of central deflection for isotropic spherical shell with different h/a and R/a ratios (Fig. 5.44(b)).

	$R_1 = R_2 = R, R/a = 5$			$R/a = 10$			$R/a = 20$		
(N_e, N_f, N_t)	$h/a = 0.01$	0.1	0.15	0.01	0.1	0.15	0.01	0.1	0.15
3D elastic [367]	2301.4	27.061	9.0755	7383.1	28.910	9.2505	16499	29.356	9.2666
PSD [367]	2295.4	26.471	8.8589	7371.3	28.754	9.2267	16485	29.388	9.3235
CST [367]	2295.3	25.201	7.9099	7370.2	27.262	8.2019	16479	27.831	8.2783
FOST [339]	2296.3	26.598	8.9052	7373.9	28.790	9.2404	16488	29.399	9.3282
HOST9 [339]	2296.3	26.597	8.9039	7373.9	28.790	9.2389	16488	29.398	9.3266
HOST11 [339]	2296.8	26.749	8.9964	7375.1	28.921	9.3225	16490	29.520	9.4077
HOST12 [339]	2296.8	26.749	8.9966	7375.1	28.921	9.3225	16490	29.520	9.4077
(10, 7, 2)	2308.8	26.916	8.9797	7387.7	28.816	9.2159	16500	29.342	9.2800
(14, 11, 4)	2309.2	26.917	8.9796	7388.0	28.816	9.2155	16500	29.342	9.2795
(16, 13, 4)	2309.3	26.917	8.9797	7388.0	28.816	9.2155	16500	29.342	9.2796

Table 5.24. Comparison of central deflection for orthotropic spherical shell with different h/a and R/a ratios (Fig. 5.44(b)).

	$R_1 = R_2 = R, R/a = 5$			$R/a = 10$			$R/a = 20$		
(N_e, N_f, N_t)	$h/a = 0.01$	0.1	0.15	0.01	0.1	0.15	0.01	0.1	0.15
3D elastic [367]	1325.5	6.3332	2.6494	2767.7	6.3593	2.6256	3802.5	6.3532	2.6022
PSD [367]	1314.8	6.1629	2.5662	2753.2	6.1376	2.6061	3789.7	6.3575	2.6162
CST [367]	1312.9	4.2161	1.2649	2744.7	4.2880	1.2745	3773.6	4.3063	1.2770
FOST [339]	1317.8	6.2187	2.6016	2756.5	6.3411	2.6281	3791.2	6.3725	2.6348
HOST9 [339]	1317.8	6.2080	2.5860	2756.5	6.3297	2.6120	3791.3	6.3610	2.6185
HOST11 [339]	1318.8	6.2534	2.6156	2757.7	6.3638	2.6363	3791.9	6.3919	2.6416
HOST12 [339]	1318.8	6.2540	2.6159	2757.7	6.3639	2.6364	3791.9	6.3920	2.6416
(10, 7, 2)	1347.2	6.4165	2.6386	2786.7	6.3658	2.6132	3805.1	6.3546	2.6074
(14, 11, 4)	1347.4	6.4177	2.6387	2786.8	6.3658	2.6128	3805.1	6.3543	2.6070
(16, 13, 4)	1347.5	6.4176	2.6387	2786.8	6.3658	2.6128	3805.1	6.3543	2.6070

The temperature rise is assumed to be sinusoidally distributed as

$$\Delta T = \frac{z}{h} \sin\left(\frac{\pi x}{a}\right) \sin\left(\frac{\pi y}{a}\right) \tag{5.108}$$

The central deflection is normalized through the following factors:

$$\bar{w} = \frac{10hw(a/2, a/2, 0)}{a^2 \alpha_1} \tag{5.109}$$

From the results of sinusoidal transverse load in Table 5.25, one can see that the effects of HOST12 correspond to those of 3D elastic

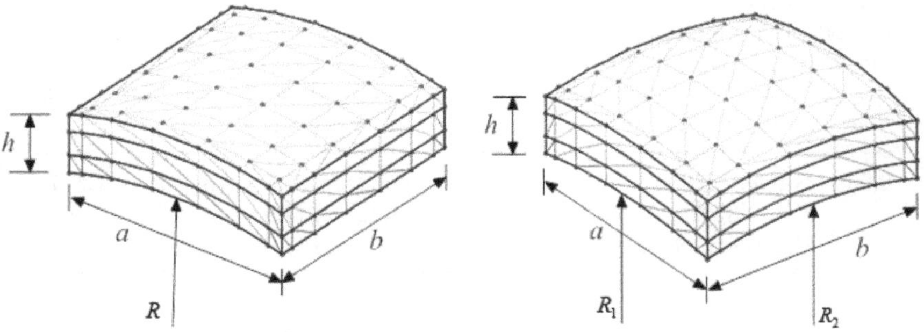

Fig. 5.44. A cylindrical shell (a) and a spherical shell (b) are respectively discretized as three hexahedron elements.

Table 5.25. Comparison of central deflection for orthotropic sandwich square plate with different h/a (Fig. 5.43).

	Sinusoidal transverse load			Sinusoidal temperature load		
(N_e, N_f, N_t)	$h/a = 0.01$	$h/a = 0.1$	$h/a = 0.25$	$h/a = 0.01$	$h/a = 0.1$	$h/a = 0.25$
FOST [339]	0.88522	1.5604	4.7666	1.78397	2.15428	3.27843
HOST9 [339]	0.89103	2.0848	7.1538	1.78798	2.47955	4.23433
HOST11 [339]	0.88818	2.0823	7.1794	1.82208	2.51464	4.24478
HOST12 [339]	0.88818	2.0823	7.1794	1.82208	2.51464	4.24478
Pagano [368]	0.892	2.20	7.596	—	—	—
Reddy [369]	0.8924	2.20046	7.5965	—	—	—
(10, 7, 2)	0.89239	2.20048	7.59675	1.82652	2.61240	4.48137
(14, 11, 4)	0.89239	2.20047	7.59668	1.82652	2.61240	4.48143
(16, 13, 4)	0.89239	2.20047	7.59668	1.82652	2.61240	4.48143

results only when h/a is low. With the increase in h/a, the errors will increase from 0.43% ($h/a = 0.01$) to 5.5% ($h/a = 0.25$). While the results of the present DQHFEM highly correspond to those of 3D elastic results for all cases, i.e., the relative errors are less than 0.044%. Under sinusoidal temperature load, the results of the present DQHFEM correspond to those of HOST12 better when h/a is low. With the increase in h/a, the gaps between the present DQHFEM results and those of HOST12 increase. It achieves 5.3% when h/a is 0.25.

Cross-ply square sandwich plates subjected to sinusoidal temperature load are analyzed in Table 5.26. One can see that the results of

Table 5.26. Comparison of displacements and stresses for cross-ply (0°/90°/0°) square plate subjected to sinusoidal temperature load with different relative thicknesses h/a (Fig. 5.43).

h/a	(N_e, N_f, N_t)	$\bar{u}$	$\bar{v}$	$\bar{w}$	$\bar{\tau}_{xy}$	$\bar{\tau}_{xz}$	$\bar{\tau}_{yz}$
0.25	Navier [370]	17.87	81.24	42.34	155.7	83.27	−129.7
	$\Gamma = 0$[370]	14.29	77.43	26.22	144.1	87.97	−12.9
	Elasticity [368]	18.11	81.83	42.69	157.0	84.81	−128.7
	(6, 4, 4)	18.08	81.72	42.61	155.4	84.25	−128.2
	(8, 6, 6)	18.11	81.83	42.69	157.0	84.82	−128.7
	(10, 8, 6)	18.11	81.83	42.69	157.0	84.81	−128.7
	(12, 10, 6)	18.11	81.83	42.69	157.0	84.81	−128.7
0.1	Navier [370]	16.59	31.92	17.37	76.20	60.45	−66.05
	$\Gamma = 0$ [370]	15.90	31.23	14.65	74.03	60.88	−65.99
	Elasticity [368]	16.61	31.95	17.39	76.29	60.54	−66.01
	(6, 4, 4)	16.58	31.90	17.36	75.57	59.62	−66.94
	(8, 6, 6)	16.61	31.95	17.39	76.30	60.55	−65.99
	(10, 8, 6)	16.61	31.95	17.39	76.29	60.54	−66.01
	(12, 10, 6)	16.61	31.95	17.39	76.29	60.54	−66.01
0.05	Navier [370]	16.16	20.34	12.12	57.34	33.98	−34.77
	$\Gamma = 0$[370]	15.99	20.16	11.43	56.78	34.03	−34.76
	Elasticity [368]	16.17	20.34	12.12	57.35	33.98	−34.76
	(6, 4, 4)	16.13	20.31	12.09	56.93	32.49	−35.47
	(8, 6, 6)	16.17	20.34	12.12	57.36	34.01	−34.75
	(10, 8, 6)	16.17	20.34	12.12	57.35	33.98	−34.76
	(12, 10, 6)	16.17	20.34	12.12	57.35	33.98	−34.76
0.02	Navier [370]	16.02	16.71	10.50	51.40	14.07	−14.13
	$\Gamma = 0$[370]	15.99	16.68	10.39	51.31	14.09	−14.13
	Elasticity [368]	16.02	16.71	10.50	51.41	14.07	−14.13
	(6, 4, 4)	15.98	16.67	10.46	51.21	11.51	−14.47
	(8, 6, 6)	16.02	16.71	10.50	51.41	14.12	−14.12
	(10, 8, 6)	16.02	16.71	10.50	51.41	14.07	−14.13
	(12, 10, 6)	16.02	16.71	10.50	51.41	14.07	−14.13

the present DQHFEM are highly consistent with the exact elasticity results [368] in all cases of h/a. The displacements can converge to four digits when N_e is only 8, while the stresses can also converge to the exact results when N_e, N_f and N_t are respectively 10, 8 and 6.

Symmetric and anti-symmetric cross-ply sandwich shells are analyzed in Tables 5.27 and 5.28. One can also see that the present DQHFEM has good convergence. It can converge up to four digits when N_e, N_f and N_t are only 16, 14 and 4, respectively.

Table 5.27. Comparison of central deflection for orthotropic sandwich shell subjected to sinusoidal temperature load with different h/a and R/a ratios (Fig. 5.44).

	$R/a = 5$			$R/a = 10$			$R/a = 20$		
(N_e, N_f, N_t)	$h/a = 0.01$	0.1	0.25	0.01	0.1	0.25	0.01	0.1	0.25
Cylindrical shell with $a/b = 1$, $R_1 = \infty$, $R_2 = R$									
FOST [339]	1.40652	2.14244	3.26804	1.67189	2.15131	3.27583	1.75457	2.15354	3.27778
HOST9 [339]	1.40776	2.46217	4.21900	1.67497	2.47518	4.23050	1.75833	2.47845	4.23337
HOST11 [339]	1.43231	2.49650	4.22787	1.60612	2.51009	4.24055	1.79164	2.51350	4.24373
HOST12 [339]	1.43467	2.49668	4.22802	1.70695	2.51013	4.24059	1.79187	2.51351	4.24374
(10, 7, 2)	1.41416	2.53223	4.35234	1.67708	2.54302	4.36447	1.75952	2.54574	4.36751
(14, 11, 4)	1.41421	2.53385	4.36716	1.67709	2.54406	4.37288	1.75953	2.54650	4.37272
(16, 13, 4)	1.41421	2.53389	4.36736	1.67709	2.54408	4.37297	1.75953	2.54650	4.37276
Spherical shell with $a/b = 1$, $R_1 = R_2 = R$									
FOST [339]	0.86243	2.12390	3.26282	1.40854	2.14663	3.27453	1.67258	2.15236	3.27746
HOST9 [339]	0.86148	2.42979	4.19525	1.40977	2.46695	4.22452	1.67565	2.42639	4.23188
HOST11 [339]	0.87444	2.46345	4.20297	1.43447	2.50169	4.23429	1.70685	2.51139	4.24216
HOST12 [339]	0.87796	2.46373	4.20316	1.43679	2.50176	4.23434	1.70767	2.51141	4.24217
(10, 8, 2)	0.86445	2.52785	4.34561	1.41609	2.54194	4.36268	1.67893	2.54547	4.36706
(14, 12, 4)	0.86461	2.52880	4.35786	1.41613	2.54261	4.36994	1.67894	2.54603	4.37171
(16, 14, 4)	0.86464	2.52882	4.35808	1.41614	2.54261	4.37002	1.67895	2.54603	4.37175

Table 5.28. Comparison of central deflection for anti-symmetric cross-ply sandwich (0°/core/90°) shell subjected to sinusoidal temperature load with different h/a and R/a ratios (Fig. 5.44).

	$R/a = 5$			$R/a = 10$			$R/a = 20$		
(N_e, N_f, N_t)	$h/a = 0.01$	0.1	0.25	0.01	0.1	0.25	0.01	0.1	0.25
Cylindrical shell with $a/b = 1$, $R_1 = \infty$, $R_2 = R$									
FOST [339]	2.47964	6.47765	6.62009	4.64214	6.53324	6.58232	5.93417	6.54297	6.56187
HOST9 [339]	2.47910	6.47152	6.57582	4.64179	6.52688	6.53658	5.93405	6.53587	6.51479
HOST11 [339]	2.56368	6.57865	6.64635	4.75799	6.63061	6.60227	6.05340	6.63771	6.57810
HOST12 [339]	2.56432	6.57874	6.64641	4.75854	6.63063	6.60229	6.05362	6.63772	6.57810
(10, 8, 2)	2.47873	6.41396	6.43198	4.63882	6.49774	6.45268	5.93243	6.51946	6.45946
(14, 12, 4)	2.47914	6.41501	6.43269	4.63895	6.49860	6.45488	5.93247	6.52031	6.46299
(16, 14, 4)	2.47923	6.41510	6.43273	4.63897	6.49862	6.45488	5.93248	6.52032	6.46298
Spherical shell with $a/b = 1$, $R_1 = R_2 = R$									
FOST [339]	0.86296	6.12662	6.45714	2.47738	6.43204	6.51931	4.63950	6.51294	6.53504
HOST9 [339]	0.86271	6.10588	6.39433	2.47684	6.42031	6.46685	4.63899	6.50379	6.48527
HOST11 [339]	0.88472	6.20118	6.45132	2.53425	6.51905	6.52684	4.72977	6.60336	6.54600
HOST12 [339]	0.88504	6.20150	6.45154	2.53488	6.51914	6.52690	4.73032	6.60338	6.54602
(10, 8, 2)	0.87920	6.11472	6.38462	2.48236	6.41892	6.44343	4.64009	6.50000	6.45854
(14, 12, 4)	0.87954	6.11596	6.38637	2.48260	6.41969	6.44384	4.64019	6.50078	6.46071
(16, 14, 4)	0.87962	6.11619	6.38650	2.48266	6.41975	6.44386	4.64021	6.50080	6.46072

The difference between the DQHFEM results and the HOST12 results respectively reach 3.55% and 3.43% for the orthotropic and the anti-symmetric cross-ply shell, which means that the present 3D results are necessary especially when h/a is high.

To summarize, in various conditions, the present DQHFEM can provide quite precise results. Compared with the results of 2D theories and high-order theories, the present DQHFEM can maintain quite small errors with the increase in h/a. Due to the flexibility of the present wedge and hexahedron DQHFEM elements, we can use less points on the thickness direction which decreases the computational cost greatly to obtain results with satisfactory accuracy at the same time.

5.6 Comments on Gauss Integration

When the Gauss integration scheme is used, one has to decide how many Gauss points should be used. Theoretically, for a one-dimensional integral, using m points can give the exact solution for the integral of a polynomial integrand of up to an order of $(2m-1)$. As a general rule of thumb, more points should be used for a higher order of elements. It is also noted that using a smaller number of Gauss points tends to counteract the *over-stiff behavior* associated with the displacement-based finite element method.

This over-stiff behavior of the displacement-based finite element method comes about primarily because of the use of the shape function. As discussed, the displacement in an element is assumed using shape functions interpolated from the nodal displacements. This implies that the deformation of the element is actually prescribed in the fashion of the shape function. This gives a constraint to the element, and thus the element behaves more stiffly than it should. It is often observed that higher-order elements are usually softer than lower-order ones. This is because the use of more nodes decreases the constraint on the element.

Coming back to the Gauss integration issue, two Gauss points for linear elements, and about two or three Gauss points in each direction for quadratic elements, should be sufficient for most cases. For high-order elements, the number of Gauss points depends on the order of the bases used. Many of the explicit FEM codes based on explicit formulation tend to use one-point integration to achieve the best performance in saving CPU time.

5.7 Conclusions

This chapter presents 3D solid elements. All the elements are C^0, so they have some degree of similarities. Three-dimensional elements of various shapes are presented. Theoretically, quadrilateral and hexahedral elements are respectively the most efficient as well as the most accurate and the simplest 2D and 3D elements. Therefore, whenever possible, we should use the two types of elements for 2D or 3D analyses. However, the techniques of generating such meshes are not fully developed so far. Commercial software presently only supports automatic generation of triangular and tetrahedral meshes. Thus, to improve computational efficiency, we may have to use hexa-dominated meshes. Some software today supports such meshes. Therefore, besides tetrahedral and hexahedral elements, transition elements like wedge and pyramid elements are necessary for a hexa-dominated mesh. Unlike other types of elements that have many symmetries, the pyramid element is very special and needs rational vertex bases. Once the shape functions are constructed, the procedures left in FEM are more or less the same. To demonstrate the performance of the elements, some examples are presented for all elements.

Chapter 6

The Hierarchical Quadrature Element Method for Kirchhoff Plates

This chapter focuses on the construction of p-version finite elements *that have curve boundaries* for C^1 problems. Both triangular and quadrilateral elements are constructed based on the C^1-version blending function interpolation methods. Orthogonal hierarchical bases are constructed and subsequently transformed into interpolative nodal bases to facilitate the imposition of boundary conditions and the implementation of C^1 conformity on curvilinear domains. Nodal collocation strategies are also studied for improving the numerical performance and novel non-uniformly distributed nodes; namely, Gauss–Jacobi points, are proposed. For parallelograms and straight-sided triangular elements, C^1 continuity is exactly satisfied between neighboring elements. The difficulty of C^1 conformity for elements that have curved boundaries is circumvented by interpolating the normal derivatives at Gauss–Lobatto nodes. Moreover, with the help of the blending function interpolation method, the bases on edges and the internal modes can differ in terms of approximation order. Therefore, local p-refinements (Fig. 6.1) can be easily performed by these elements. Numerical results demonstrated that these elements are computationally inexpensive and converge fast for problems with regular and irregular domains.

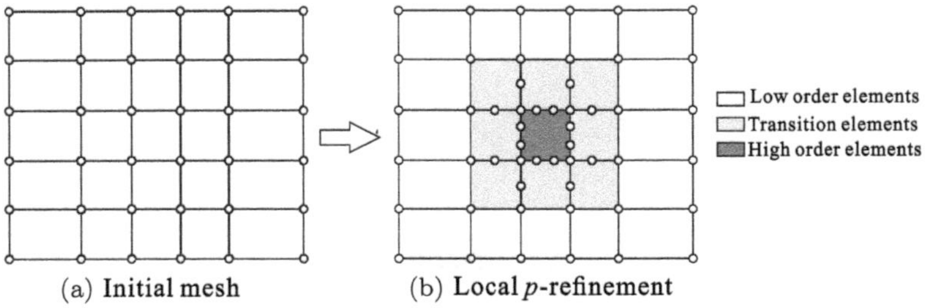

Fig. 6.1. Local p-refinement.

6.1 Hermite Blending Function Interpolation

The finite element method (FEM) shape functions can be constructed intuitively if one is familiar with the interpolative properties of the shape functions. It is easy to construct these functions when the topology of the elements and the conformity requirements are simple. However, the construction of C^1 elements is not easy since high-order derivatives are often involved in the nodal variables. Therefore, the Hermite blending function interpolation method, which is introduced in this section, will be used to construct the shape functions (the vertex and edge bases only; see the following section). The formulae for a unit square domain can be found in [371], which were used by Watkins [372] to construct the conforming rectangular elements. They are included here for completeness. Inspired by Gordon [371, 373], formulae for a triangular domain are derived by the authors of this work.

6.1.1 *Blending function interpolation on a unit square domain*

As shown in Fig. 6.2(a), consider a smooth function $F(\xi, \eta)$ that is defined on a unit square domain with the following boundary functions of values and derivatives:

$$\begin{aligned} f_1(\xi) &= F(\xi, -1), & f_2(\eta) &= F(1, \eta) \\ f_3(\xi) &= F(\xi, 1), & f_4(\eta) &= F(-1, \eta) \\ g_1(\xi) &= F_\eta(\xi, -1), & g_2(\eta) &= F_\xi(1, \eta) \\ g_3(\xi) &= F_\eta(\xi, 1), & g_4(\eta) &= F_\xi(-1, \eta) \end{aligned} \tag{6.1}$$

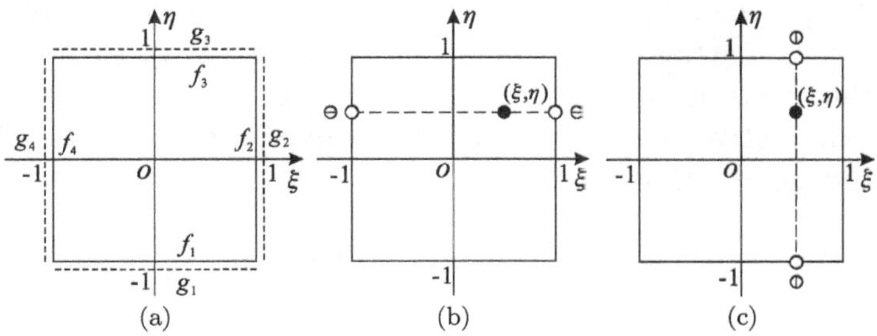

Fig. 6.2. Blending function interpolation on a unit square domain: (a) boundary functions of values and derivatives; (b) interpolation along the ξ-direction; (c) interpolation along the η-direction.

where F_ξ and F_η denote the first-order partial derivatives with respect to ξ and η. The Hermite blending function interpolation can be used to find an approximation function, denoted as $\tilde{F}(\xi, \eta)$, that has the same boundary functions of values and derivatives as $F(\xi, \eta)$. To this end, the following third-order Hermite interpolation bases are often used:

$$h_1(\xi) = \frac{1}{4}(\xi + 1)(\xi - 1)^2, \quad h_2(\xi) = \frac{1}{4}(\xi + 2)(\xi - 1)^2$$
$$h_3(\xi) = \frac{1}{4}(2 - \xi)(\xi + 1)^2, \quad h_4(\xi) = \frac{1}{4}(\xi - 1)(\xi + 1)^2 \quad (6.2)$$

where h_1 and h_4 are attached to the derivatives at the two ends of $[-1, 1]$, and h_2 and h_3 are attached to the function values at the ends. By interpolating F along the ξ-direction (see Fig. 6.2(b)), one can obtain

$$P_\xi[F] = g_4(\eta)h_1(\xi) + f_4(\eta)h_2(\xi) + f_2(\eta)h_3(\xi) + g_2(\eta)h_4(\xi) \quad (6.3)$$

where $P_\xi[\cdot]$ is the projection operator [371] that is related to the ξ-direction. $P_\xi[F]$ has the same boundary values and derivatives as F on edges $\xi = \pm 1$. Therefore, the interpolation residual, which is denoted as R and defined as

$$R = F - P_\xi[F] \quad (6.4)$$

has zero values and normal derivatives on $\xi = \pm 1$. By interpolating R in the η-direction (see Fig. 6.2(c)), one can obtain

$$P_\eta[R] = R(\xi, -1)h_1(\eta) + R_\eta(\xi, -1)h_2(\eta)$$
$$+ R(\xi, 1)h_3(\eta) + R_\eta(\xi, 1)h_4(\eta) \quad (6.5)$$

where $P_\eta[\cdot]$ is the projection operator that is related to the η-direction and R_η is the partial derivative of R with respect to η. $P_\eta[R]$ also has zero values and normal derivatives on $\xi = \pm 1$, while the values and normal derivatives on $\eta = \pm 1$ are the same as those of R. Finally, one can conclude that

$$\tilde{F} = P_\xi[F] + P_\eta[R] \tag{6.6}$$

is the desired approximation that has the same boundary values and derivatives as F. According to Eq. (6.3), the projection operators are linear. It follows that

$$P_\eta[R] = P_\eta[F - P_\xi[F]] = P_\eta[F] - P_\eta[P_\xi[F]] \tag{6.7}$$

The projection operators are also commutative [371], which implies $P_\eta[P_\xi[F]] = P_\xi[P\eta[F]]$ or, for simplicity, $P_\eta P_\xi[F] = P_\xi P\eta[F]$. Substituting Eq. (6.7) into Eq. (6.6), the approximation function can be expressed as

$$\tilde{F} = P_\xi[F] + P_\eta[F] - P_\xi P_\eta[F] \tag{6.8}$$

where

$$\begin{aligned}
P_\xi[F] &= g_4(\eta)h_1(\xi) + f_4(\eta)h_2(\xi) + f_2(\eta)h_3(\xi) + g_2(\eta)h_4(\xi) \\
P_\eta[F] &= g_1(\xi)h_1(\eta) + f_1(\xi)h_2(\eta) + f_3(\xi)h_3(\eta) + g_3(\xi)h_4(\eta) \\
P_\xi P_\eta[F] &= [g_1'(-1)h_1(\eta) + f_1'(-1)h_2(\eta) + f_3'(-1)h_3(\eta) \\
&\quad + g_3'(-1)h_4(\eta)]h_1(\xi) + [g_1(-1)h_1(\eta) + f_1(-1)h_2(\eta) \\
&\quad + f_3(-1)h_3(\eta) + g_3(-1)h_4(\eta)]h_2(\xi) \\
&\quad + [g_1(1)h_1(\eta) + f_1(1)h_2(\eta) + f_3(1)h_3(\eta) \\
&\quad + g_3(1)h_4(\eta)]h_3(\xi) + [g_1'(1)h_1(\eta) + f_1'(1)h_2(\eta) \\
&\quad + f_3'(1)h_3(\eta) + g_3'(1)h_4(\eta)]h_4(\xi)
\end{aligned} \tag{6.9}$$

For more details about the blending function interpolation, one may refer to [371]. An example of verifying the interpolation properties of the method is shown in Fig. 6.3. The errors at all boundaries vanish; hence, all the boundary conditions in Eq. (6.1) are well satisfied.

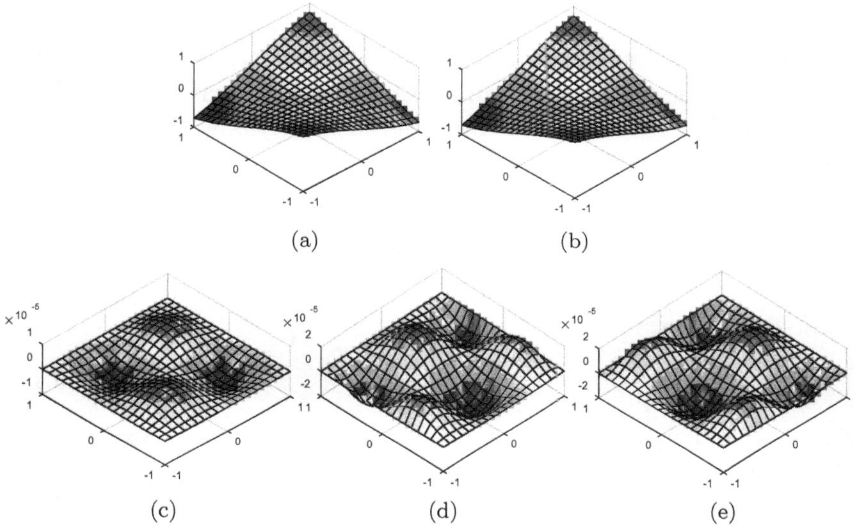

Fig. 6.3. Interpolation on a unit square domain: (a) $F = \sin(\xi)\sin(\eta)$; (b) $\tilde{F}$; (c) $F - \tilde{F}$; (d) $\partial F/\partial\xi - \partial\tilde{F}/\partial\xi$; (e) $\partial F/\partial\eta - \partial\tilde{F}/\partial\eta$.

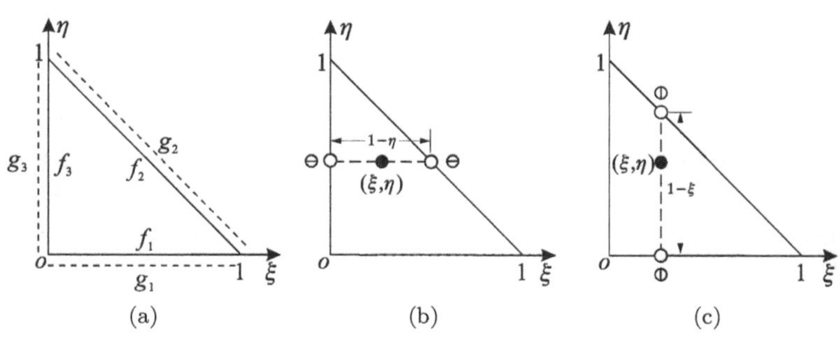

Fig. 6.4. Blending function interpolation on a unit triangular domain: (a) boundary functions of values and derivatives; (b) interpolation along the ξ-direction; (c) interpolation along the η-direction.

6.1.2 *Blending function interpolation on a unit triangular domain*

Consider a two-dimensional function, namely, $F(\xi, \eta)$, that is defined on a unit triangular domain (see Fig. 6.4(a)) and has the following boundary functions of values and derivatives:

$$\begin{aligned} &f_1(\xi) = F(\xi, 0),\ \ f_2(\eta) = F(1-\eta, \eta),\ \ f_3(\eta) = F(0, \eta) \\ &g_1(\xi) = F_\eta(\xi, 0),\ g_2(\eta) = F_\xi(1-\eta, \eta),\ g_3(\eta) = F_\xi(0, \eta) \end{aligned} \tag{6.10}$$

Function g_2 is no longer the normal derivative of F on the bevel edge. In contrast to the square domain, the Hermite interpolation functions on the triangular domain should be rational because of the bevel edge. For example, the interpolation functions, namely, ϕ_i, for the ξ-direction (see Fig. 6.4(b)) are expressed as

$$\begin{aligned}\phi_1(\xi,\eta) &= \frac{\xi(\xi+\eta-1)^2}{(1-\eta)^2}, \quad \phi_2(\xi,\eta) = \frac{(\xi+\eta-1)^2(2\xi+1-\eta)}{(1-\eta)^3}\\ \phi_3(\xi,\eta) &= \frac{\xi^2(2\xi-3+3\eta)}{(\eta-1)^3}, \quad \phi_4(\xi,\eta) = \frac{(\xi+\eta-1)\xi^2}{(1-\eta)^2}\end{aligned} \tag{6.11}$$

where ϕ_1 and ϕ_4 are attached to the derivatives at the two ends of a line segment in the ξ-direction and ϕ_2 and ϕ_3 are attached to the function values at the two ends. Similarly, the interpolation functions, namely, ψ_i, for the η-direction (see Fig. 6.4(c)) can be easily obtained by exchanging the variables of ϕ_i in Eq. (6.11):

$$\begin{aligned}\psi_1(\xi,\eta) &= \phi_1(\eta,\xi), \quad \psi_2(\xi,\eta) = \phi_2(\eta,\xi)\\ \psi_3(\xi,\eta) &= \phi_3(\eta,\xi), \quad \psi_4(\xi,\eta) = \phi_4(\eta,\xi)\end{aligned} \tag{6.12}$$

where ψ_1 and ψ_4 are attached to the derivatives (with respect to η) at the two ends of a line segment in the η-direction, and ψ_2 and ψ_3 are related to the function values at the two ends. Following the same procedure as in Section 6.1.1 and interpolating F in the ξ-direction yields

$$P_\xi[F] = g_3(\eta)\phi_1 + f_3(\eta)\phi_2 + f_2(\eta)\phi_3 + g_2(\eta)\phi_4 \tag{6.13}$$

Then, the residual, namely, R, is expressed as

$$R = F - P_\xi[F] \tag{6.14}$$

and

$$P_\eta[R] = R_\eta(\xi,0)\psi_1 + R(\xi,0)\psi_2 + R(\xi,1-\xi)\psi_3 + R_\eta(\xi,1-\xi)\psi_4 \tag{6.15}$$

Since $P_\xi[F]$ and F share the same function value and partial derivative with respect to ξ on the bevel edge, it follows that $P_\xi[F]$ and F also have the same function value and partial derivative with respect to η on that edge, which implies $R(\xi, 1-\xi) = R_\eta(\xi, 1-\xi) = 0$ in Eq. (6.15). Define another projection operator of R in the η-direction:

$$\tilde{P}_\eta[R] = R_\eta(\xi, 0)\psi_1 + R(\xi, 0)\psi_2 \tag{6.16}$$

Then,

$$P_\eta[R] = \tilde{P}_\eta[R] = \tilde{P}_\eta[F] - \tilde{P}_\eta P_\xi[F] \tag{6.17}$$

and the final approximation function, namely, $\tilde{F}$, can be expressed as

$$\tilde{F} = P_\xi[F] + P_\eta[R] = P_\xi[F] + \tilde{P}_\eta[F] - \tilde{P}_\eta P_\xi[F] \tag{6.18}$$

where

$$\begin{aligned}
P_\xi[F] &= g_3(\eta)\phi_1 + f_3(\eta)\phi_2 + f_2(\eta)\phi_3 + g_2(\eta)\phi_4 \\
P_\eta[F] &= g_1(\xi)\psi_1 + f_1(\xi)\psi_2 \\
\tilde{P}_\eta P_\xi[F] &= \psi_1 \begin{bmatrix} g_3'(0)\phi_1(\xi,0) + f_3'(0)\phi_2(\xi,0) + f_2'(0)\phi_3(\xi,0) \\ +g_2'(0)\phi_4(\xi,0) + g_3(0)\phi_{1,\eta}'(\xi,0) + f_3(0)\phi_{2,\eta}'(\xi,0) \\ +f_2(0)\phi_{3,\eta}'(\xi,0) + g_2(0)\phi_{4,\eta}'(\xi,0) \end{bmatrix} \\
&\quad + \psi_2[g_3(0)\phi_1(\xi,0) + f_3(0)\phi_2(\xi,0) \\
&\quad + f_2(0)\phi_3(\xi,0) + g_2(0)\phi_4(\xi,0)]
\end{aligned} \tag{6.19}$$

As verified in Fig. 6.5, the approximation function $\tilde{F}$ has the same function values and two partial derivatives on the boundaries as the specified function F.

Using the interpolation method in Eqs. (6.8) and (6.18), one can construct a two-dimensional function with prescribed boundary functions, which is the main strategy for the construction of shape functions in this work. The blending function interpolation is valid only if $F$is sufficiently smooth; hence, the boundary functions should be sufficiently consistent. For example, on the corner ($\xi = 0$, $\eta = 0$), the specified functions should satisfy $f_1(0) = f_3(0)$, $f_3'(0) = g_1(0)$, etc. Evaluating the consistency is tedious. However, this difficulty will not be encountered during element construction in this work since the boundary functions are interpolated by the nodal variables and the consistency is satisfied automatically.

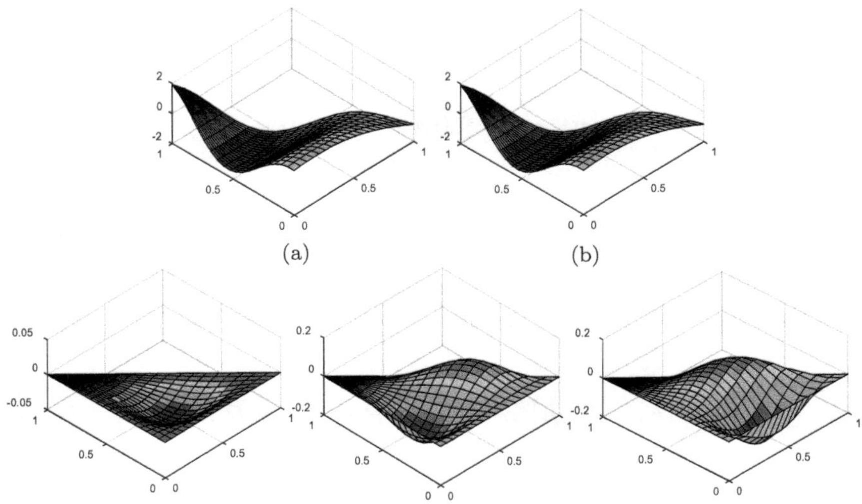

(a) (b) (c) (d) (e)

Fig. 6.5. Interpolation on a unit triangular domain: (a) $F = (1 + \xi + \eta)\cos(2\xi)\cos(6\eta)$; (b) $\tilde{F}$; (c) $F - \tilde{F}$; (d) $\partial F/\partial \xi - \partial \tilde{F}/\partial \xi$; (e) $\partial F/\partial \eta - \partial \tilde{F}/\partial \eta$.

6.2 Hierarchical Bases for Quadrilateral Elements

In this section, hierarchical bases for quadrilateral elements will be developed. The boundary shape functions, including vertex and edge functions, are derived using the blending function interpolation in Section 6.1.1, while the face functions are constructed via tensor production of one-dimensional bases in the ξ- and η-directions. Function F in Section 6.1 is replaced by displacement variable w in the following discussions (including Section 6.2).

6.2.1 *Vertex functions*

To derive the shape functions using the blending function interpolation method, the boundary functions in Eq. (6.1) should be determined first. As shown in Fig. 6.6, six nodal variables are used on each vertex. The boundary normal derivatives, which are denoted as w_n (namely, g_i in Eq. (6.1)), can be interpolated by the third-order Hermite bases in Eq. (6.2), along with two variables on each end. For example, the variables for the interpolation of g_1 on vertices 1 and 2 are $\partial w/\partial \eta$ and $\partial^2 w/\partial \xi \partial \eta$. Similarly, the boundary value of

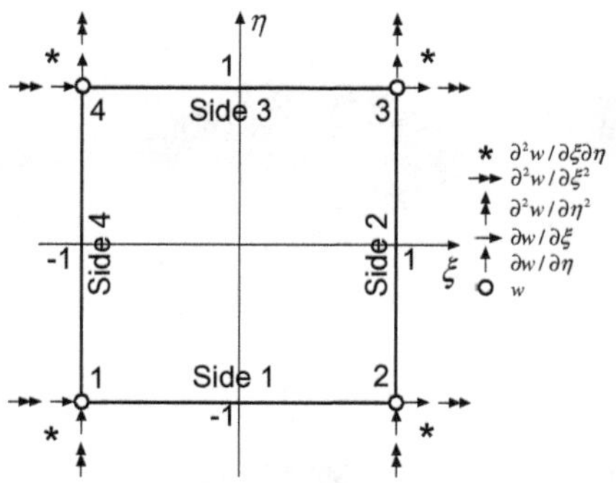

Fig. 6.6. Vertex DOF collocation for a unit quadrilateral element.

w (namely, f_i in Eq. (6.1)) on each side can be interpolated by the fifth-order Hermite bases that are expressed in Eq. (6.20) and three variables are used on each end. For example, the related variables for the interpolation of f_1 are w, $\partial w/\partial\xi$ and $\partial^2 w/\partial\xi^2$ on vertices 1 and 2.

$$\begin{aligned}
H_1^{(2)} &= (1-\xi)^3(1+\xi)^2/16, \quad H_2^{(2)} = (1+\xi)^3(1-\xi)^2/16 \\
H_1^{(1)} &= (1-\xi)^3(1+\xi)(5+3\xi)/16 \\
H_2^{(1)} &= (1+\xi)^3(\xi-1)(5-3\xi)/16 \\
H_1 &= (1-\xi)^3(3\xi^2+9\xi+8)/16 \\
H_2 &= (1+\xi)^3(3\xi^2-9\xi+8)/16
\end{aligned} \tag{6.20}$$

After the boundary functions in Eq. (6.1) have been determined, the shape functions can be obtained directly via Eqs. (6.8) and (6.9). To determine the shape function for each nodal variable, one can set the variable to 1 and all other nodal variables to 0 and the corresponding shape function can be obtained via the blending function interpolation procedure. For example, to determine the shape function that is attached to variable w on vertex 1 (denoted by w^{V1}), we set $w^{V1} = 1$ and set all other nodal variables on the vertex and other vertices to zero. Then, the boundary functions in Eq. (6.1) can be interpolated as

$$f_1 = H_1(\xi); \quad f_4 = H_1(\eta); \quad f_2 = f_3 = g_1 = \cdots = g_4 = 0 \tag{6.21}$$

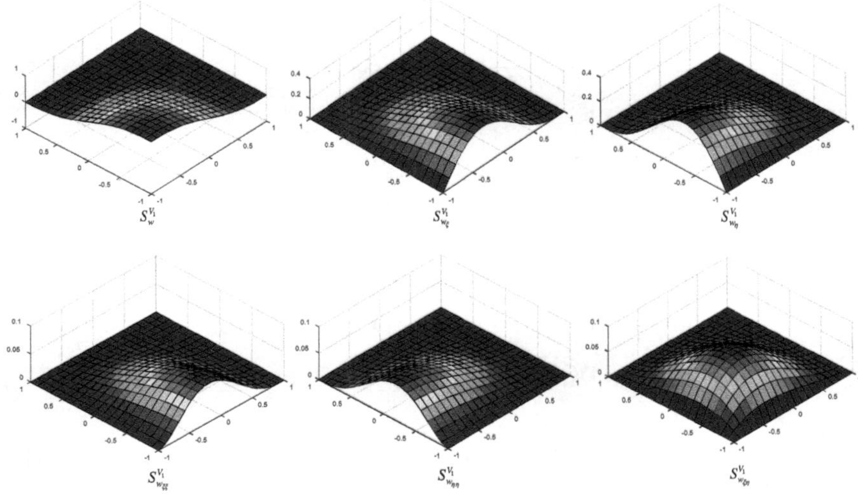

Fig. 6.7. Vertex functions for a unit quadrilateral element on vertex 1.

Using the blending function interpolation in Eq. (6.8), the corresponding shape function can be obtained:

$$S_w^{V_1} = H_1(\eta)h_2(\xi) + H_1(\xi)h_2(\eta) - h_2(\xi)h_2(\eta) \tag{6.22}$$

Using similar procedures, one can obtain the other shape functions on the remaining vertices. The shape functions on vertex 1 are plotted in Fig. 6.7.

6.2.2 *Edge functions*

The edge functions for C^1 elements can be divided into two parts: (1) bases for the enrichment of boundary values, namely, w, that have zero normal derivatives: (2) bases for the enrichment of boundary normal derivatives, namely, w_n,that have zero function values. According to Section 6.2.1, the boundary functions for w and w_n are complete up to the fifth and third orders, respectively. Therefore, the enriching bases for w and w_n should start from the sixth and fourth orders. Taking side 1 as an example, we let

$$f_1 = (1+\xi)^3(1-\xi)^3 J_i^{(\alpha,\beta)}(\xi), \quad i = 0, 1, \ldots \tag{6.23}$$

where $J_i^{(\alpha,\beta)}$, $i = 0, 1, \ldots$ are the Jacobi polynomials and the weights (α, β) are set as (4, 4) in accordance with the internal modes (see Section 6.2.3) so that the bases can be calculated together. By forcing the other boundary functions in Eq. (6.1) to be zero and using Eq. (6.8), one can obtain the edge functions for w on side 1 as

$$S_{w,i}^{S_1} = (1+\xi)^3(1-\xi)^3 J_i^{(4,4)}(\xi) h_2(\eta), \quad i = 0, 1, \ldots \tag{6.24}$$

To improve the computational efficiency, using the recursion formulae of Jacobi polynomials to calculate the bases and their derivatives is highly recommended:

$$a_n J_n^{(\alpha,\beta)}(x) = (b_n x + c_n) J_{n-1}^{(\alpha,\beta)}(x) - d_n J_{n-2}^{(\alpha,\beta)}(x) \tag{6.25}$$

$$\frac{d}{dx} J_n^{(\alpha,\beta)}(x) = \frac{n+\alpha+\beta+1}{2} J_{n-1}^{(\alpha+1,\beta+1)}(x), \quad J_{-1}^{(\alpha,\beta)}(x) = 0 \tag{6.26}$$

where

$$\begin{aligned} a_n &= 2n(\alpha+\beta+n)(\alpha+\beta+2n-2) \\ b_n &= (\alpha+\beta+2n-2)(\alpha+\beta+2n-1)(\alpha+\beta+2n) \\ c_n &= (\alpha+\beta+2n-1)(\alpha^2-\beta^2) \\ d_n &= 2(\alpha+n-1)(\beta+n-1)(\alpha+\beta+2n) \end{aligned} \tag{6.27}$$

These formulae can be found in mathematics handbooks. They are included here for completeness. Similarly, the edge functions that correspond to the normal derivatives are expressed as

$$S_{w_n,i}^{S_1} = (1+\xi)^2(1-\xi)^2 J_i^{(4,4)}(\xi) h_1(\eta), \quad i = 0, 1, 2, \ldots \tag{6.28}$$

The edge functions on other sides can be derived via similar approaches. The numbers of edge functions on the four sides can differ; therefore, joining elements that have different polynomial degrees is possible. Moreover, the numbers of edge functions for w and w_n on the same side can also differ; hence, the approximation orders for w and w_n on the boundaries are independent. As will be shown in Section 6.5.3, this property is beneficial for enhancing the C^1 conformity on curved edges. Figure 6.8 shows the first three edge functions for w and w_n.

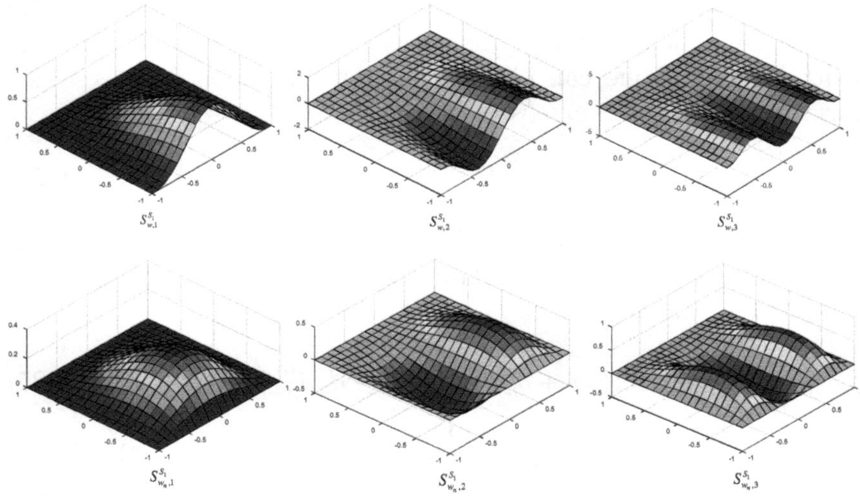

Fig. 6.8. Edge functions for a unit quadrilateral element.

6.2.3 *Face functions*

The face functions of quadrilateral elements can be obtained by simply calculating the tensor product of one-dimensional C^1 hierarchical bases as follows:

$$S^F_{mn} = C_{mn}(1-\xi^2)^2(1-\eta^2)^2 J^{(4,4)}_{m-1}(\xi) J^{(4,4)}_{n-1}(\eta),$$
$$m = 1, 2, \ldots, H_\xi, \quad n = 1, 2, \ldots, H_\eta$$
$$C_{mn} = \frac{1}{2^9}\sqrt{(2m+7)(2n+7)\prod_{i=4}^{7}(m+i)\prod_{j=4}^{7}(n+j)\Big/\prod_{i=0}^{3}(m+i)\prod_{j=0}^{3}(n+i)} \tag{6.29}$$

where m and n are the indices, and H_ξ and H_η are the numbers of bases along each direction of the natural coordinates. The coefficients, which are denoted as C_{mn}, of the Jacobi polynomials are determined by the orthogonality requirement:

$$\int_{-1}^{1}\int_{-1}^{1} S^F_{pq} S^F_{mn} d\xi d\eta = \delta_{pm}\delta_{qn} \tag{6.30}$$

where δ_{pm} and δ_{qn} are Kronecker deltas. The first three face functions are shown in Fig. 6.9. According to Eq. (6.29), both the function

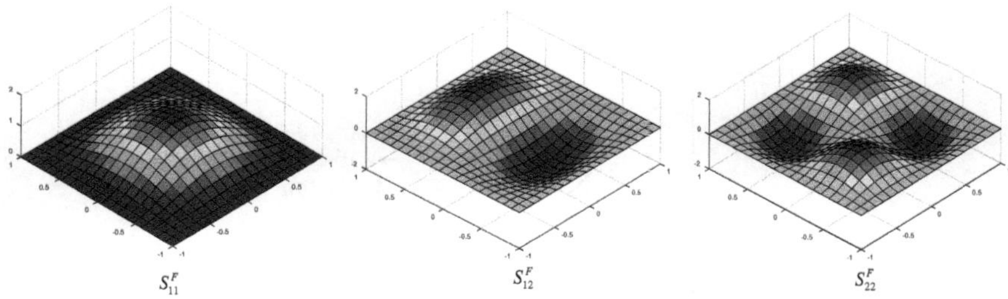

Fig. 6.9. First three face functions for a unit quadrilateral element.

values and the normal derivatives of the bases vanish on the element boundaries. Appendix 6.1 summarizes all the shape functions for the quadrilateral elements.

6.3 Hierarchical Bases for Triangular Elements

6.3.1 *Vertex functions*

A process similar to that in Section 6.2.1 can be used to obtain the shape functions on a unit triangular domain. However, the vertex functions on vertices 2 and 3 should be determined carefully since the angles are no longer right angles as in quadrilateral elements (see Fig. 6.10). To obtain the boundary functions, the fifth-order Hermite polynomials in Eq. (6.20), which are defined in $[-1, 1]$, are shifted to $[0, 1]$:

$$\begin{aligned} \tilde{H}_1^{(2)}(\eta) &= (1-\eta)^3\eta^2/2, & \tilde{H}_2^{(2)}(\eta) &= \eta^3(1-\eta)^2/2 \\ \tilde{H}_1^{(1)}(\eta) &= (1-\eta)^3\eta(3\eta+1), & \tilde{H}_2^{(1)}(\eta) &= \eta^3(\eta-1)(4-3\eta) \\ \tilde{H}_1(\eta) &= (1-\eta)^3(6\eta^2+3\eta+1), & \tilde{H}_2(\eta) &= \eta^3(6\eta^2-15\eta+10) \end{aligned} \tag{6.31}$$

We derive the shape functions for w_η (which abbreviates $\partial w/\partial\eta$) on vertex 3 (denoted by $w_\eta^{V_3}$) as an example. First, let $w_\eta^{V_3} = 1$ and the remaining nodal variables be zero. Then, the boundary functions on sides 1 and 3 can be interpolated directly as

$$f_3(\eta) = \tilde{H}_2^{(1)}(\eta), \quad g_3(\eta) = f_1(\eta) = g_1(\eta) = 0 \tag{6.32}$$

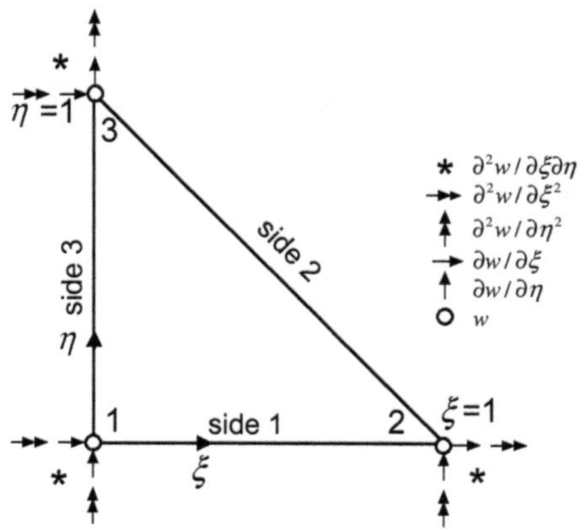

Fig. 6.10. Vertex DOF collocation of a unit triangular element.

To determine the boundary functions on side 2, the following relations (which follow from Eq. (6.10)) should be used:

$$\begin{gathered} f_2(1) = w^{V_3}, f_2'(1) = w_\eta^{V_3} - w_\xi^{V_3}, f_2''(1) = w_{\xi\xi}^{V_3} - 2w_{\xi\eta}^{V_3} + w_{\eta\eta}^{V_3} \\ g_2(1) = w_\xi^{V_3}, g_2'(1) = w_{\xi\eta}^{V_3} - w_{\xi\xi}^{V_3} \end{gathered} \tag{6.33}$$

As $w_\eta^{V_3}$ is assumed to be 1 and all other variables are assumed to be zero, Eq. (6.33) becomes

$$f_2(1) = 0, \quad f_2'(1) = 1, \quad f_2''(1) = 0, \quad g_2(1) = 0, \quad g_2'(1) = 0 \tag{6.34}$$

On vertex 2, we have $f_2(0) = f_2'(0) = f_2''(0) = g_2(0) = g_2'(0) = 0$. Finally, boundary functions f_2 and g_2 can be interpolated by fifth- and third-order Hermite bases as

$$f_2(\eta) = \tilde{H}_2^{(1)}(\eta), \quad g_2(\eta) = 0 \tag{6.35}$$

Substituting Eqs. (6.32) and (6.35) into Eq. (6.18), one can obtain the shape function for w_η^{V3} as

$$S_{w_\eta}^{V_3} = H_2^{(1)}(\eta)[\phi_2(\xi,\eta) + \phi_3(\xi,\eta)] \tag{6.36}$$

which can be simplified to

$$S_{w_\eta}^{V_3} = -\eta^3(3\eta^2 - 7\eta + 4) \tag{6.37}$$

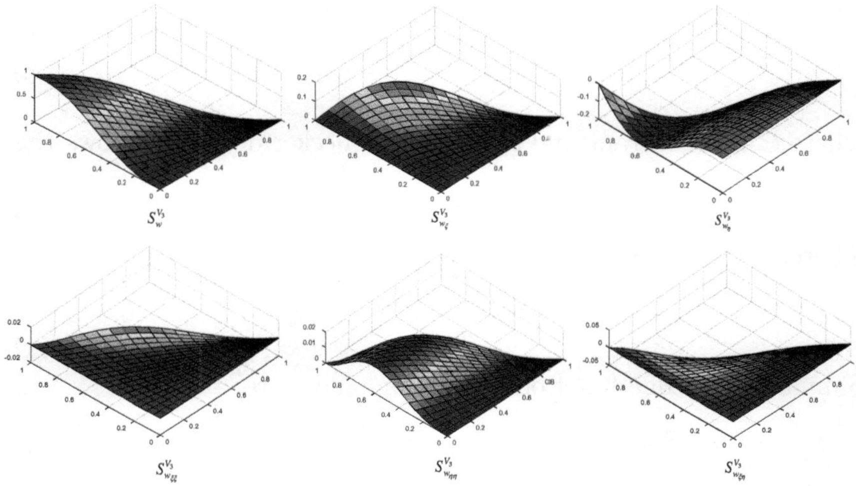

Fig. 6.11. Vertex functions for a unit triangular element on vertex 3.

Similarly, one can derive the vertex functions for other nodal variables. The shape functions on vertex 3 are plotted in Fig. 6.11. Equation (6.33) is essential for constructing shape functions that have consistent boundary functions and it is even more important for the construction of shape functions for nodal variables such as $w_{\xi\xi}$, $w_{\xi\eta}$, and $w_{\eta\eta}$ on vertices 2 and 3 since these variables are coupled with both boundary values and the derivatives (see Eq. (6.10)).

6.3.2 *Edge functions*

The derivation of edge functions is also similar. For example, to derive the edge functions for $w$on side 1, one can consider

$$f_1 = \xi^3(1-\xi)^3 P_i(\xi), \quad i = 0, 1, \ldots \tag{6.38}$$

where $P_i(\xi)$ is an unknown polynomial of order i. The other boundary functions in Eq. (6.10) are set to zero. Then, the corresponding edge functions can be expressed as

$$S_{w,i}^{S_1}(\xi) = \xi^3(1-\xi)^3 P_i(\xi)\psi_2(\xi,\eta), \quad i = 0, 1 \ldots \tag{6.39}$$

Theoretically, P_i can be any group of complete polynomials. To improve the numerical properties of the basis transformation (see Section 6.4.1), P_i is determined by the orthogonality condition

$$\int_A S_{w,i}^{S_1} S_{w,j}^{S_1} dA = C\delta_{ij} \tag{6.40}$$

where C is a constant and δ_{ij} is the Kronecker delta. It follows that

$$P_i(\xi) = J_i^{(7,6)}(2\xi - 1). \tag{6.41}$$

Therefore, the edge functions for w on side 1 can be expressed as

$$S_{w,i}^{S_1}(\xi) = \xi^3(\xi + \eta - 1)^2(2\eta + 1 - \xi)J_i^{(7,6)}(2\xi - 1), \quad i = 0, 1, 2, \ldots \tag{6.42}$$

Similarly, the edge functions for w_η on side 1 are

$$S_{w_\eta,i}^{S_1}(\xi) = \xi^2\eta(\xi + \eta - 1)^2 J_i^{(7,4)}(2\xi - 1), \quad i = 0, 1, 2, \ldots \tag{6.43}$$

The edge functions on other sides can be derived in similar ways. The first three edge functions for w and w_η on side 1 are plotted in Fig. 6.12.

6.3.3 *Face functions*

The orthogonal face functions of these triangular elements are developed from the techniques for deriving the C^0 bases in [54]. Neglecting

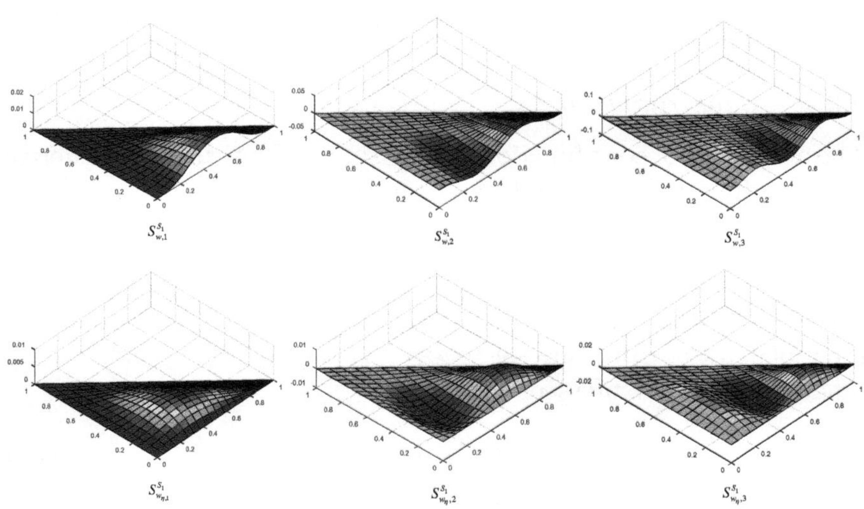

Fig. 6.12. Edge functions for a unit triangular element on side 1.

the derivation steps, they are expressed as

$$S_{pn}^F(\xi,\eta) = C_{pn}(1-\xi-\eta)^2\xi^2\eta^2(1-\eta)^{p-n-6}J_{p-n-6}^{(4,4)} \times \left(\frac{2\xi}{1-\eta}-1\right) J_n^{(2p-2n-3,4)}(2\eta-1) \tag{6.44}$$

where $p(\geq 6)$ is the polynomial order and $n = 0, 1, \ldots, p-6$, are the indices of the pth-order face functions. The constant, namely, C_{pn}, is determined from the normalization condition in Eq. (6.30) as

$$C_{pn} = \sqrt{(2p-2n-3)(2p+2)\prod_{i=1}^{4}(p-n-2+i)(2p-n-3+i) \Big/ \prod_{i=1}^{4}(p-n-6+i)(n+i)} \tag{6.45}$$

The number of linearly independent p^{th}-order face functions is $p-5$. For a specified order $p \geq 6$, the number of complete face functions is $N_f = (p-4)(p-5)/2$. The rational term in Eq. (6.44) is eliminated. The first three face functions are expressed as

$$\begin{aligned} S_{60}^F &= C_{60}\xi^2\eta^2(1-\xi-\eta)^2 \\ S_{70}^F &= C_{70}5\xi^2\eta^2(1-\xi-\eta)^2(2\xi+\eta-1) \\ S_{71}^F &= C_{71}5\xi^2\eta^2(1-\xi-\eta)^2(3\eta-1) \end{aligned} \tag{6.46}$$

They are plotted in Fig. 6.13. Both the function values and the normal derivatives of the bases vanish on the boundaries. Appendix 6.1 summarizes all the hierarchical bases for triangular elements.

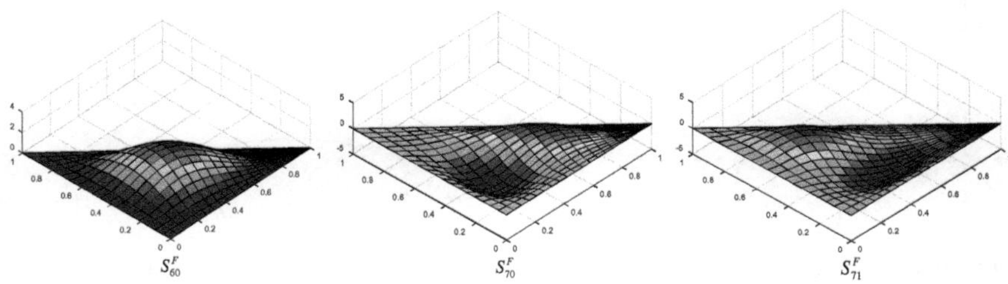

Fig. 6.13. First three face functions for a unit triangular element.

6.4 Numerical Implementation

6.4.1 *Basis transformation*

The hierarchical bases that are developed above using Jacobi polynomials can be calculated efficiently via the recursion formulae in Eqs. (6.25) and (6.26). However, since they are defined in natural coordinates, they cannot be directly applied to elements that have curved boundaries because of the difficulties of satisfying the C^1 conformity. To overcome this problem, a basis transformation is invoked to transform them into nodal bases. For a quadrilateral element, as shown in Fig. 6.14(a), the displacement field, namely, w, on the unit square domain can be approximated by the hierarchical bases as

$$w(\xi,\eta) = \boldsymbol{N}^T\boldsymbol{a} \tag{6.47}$$

where $\boldsymbol{N}$ is a vector of the hierarchical bases and $\boldsymbol{a}$ is the generalized displacement vector, which are expressed as

$$\begin{aligned}
\boldsymbol{N}^T = [&S_w^{V_1},\ldots,S_{w_{\xi\eta}}^{V_1},\ldots,S_w^{V_4},\ldots,S_{w_{\xi\eta}}^{V_4},\\
&S_{w,1}^{S_1},\ldots,S_{w,M}^{S_1},\ldots,S_{w,1}^{S_4},\ldots,S_{w,Q}^{S_4},\\
&S_{w_n,1}^{S_1},\ldots,S_{w_n,M+\Delta}^{S_1},\ldots,S_{w_n,1}^{S_4},\ldots,S_{w_n,Q+\Delta}^{S_4},S_{11}^{F},\ldots,S_{H_\xi H_\eta}^{F}]\\
\boldsymbol{a}^T = [&w^{V_1},\ldots,w_{\xi\eta}^{V_1},\ldots,w^{V_4},\ldots,w_{\xi\eta}^{V_4},a_1,\ldots,a_{N_w},\\
&b_1,\ldots,b_{N_d},c_1,\ldots,c_{N_f}]
\end{aligned} \tag{6.48}$$

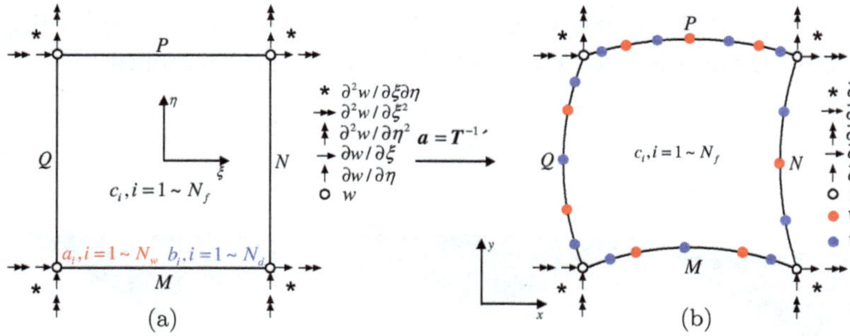

Fig. 6.14. Basis transformation for quadrilateral elements: (a) coefficients of hierarchical bases; (b) variables of nodal bases, where $M = 2$, $N = 1$, $P = 3$, $Q = 2$ and $\Delta = 1$.

where M, N, P and Q are the numbers of edge functions for w; V_i ($i = 1, 2, 3, 4$) denotes vertex i; and S_i ($i = 1, 2, 3, 4$) denotes side i. The numbers of edge functions for normal derivatives w_n on each side are set as $M + \Delta$, $N + \Delta$, $P + \Delta$ and $Q + \Delta$, respectively. The increment Δ is typically set to 1 for straight-sided elements to ensure the C^1 conformity. However, more nodes can be used to improve the interpolation accuracy of normal derivatives (namely, the C^1 conformity) in elements that have curved boundaries. Accordingly, the total number of edge functions is $N_w = M + N + P + Q$ for w and $N_d = M+N+P+Q+4\Delta$ for w_n, while the number of face functions is $N_f = H_\xi H_\eta$.

Using proper mapping functions, the unit domain in Fig. 6.14(a) can be mapped into a curved domain in global coordinates, as shown in Fig. 6.14(b). In this work, the C^0 blending functions [48, 374] are employed as the exact mapping functions for our quadrilateral and triangular elements. The transformed nodal variables are also expressed in Fig. 6.14(b). The vertex unknowns in natural coordinates have been transformed into variables with respect to global coordinates. The nodal variables, namely, w and w_n (deflection and normal derivative), are also allocated on each edge in x–y coordinates and the numbers of these variables are the same as those of the hierarchical edge functions for w and w_n. By evaluating the nodal values of Eq. (6.47), one can obtain the following relation:

$$\boldsymbol{\delta} = \boldsymbol{T}\boldsymbol{a} \tag{6.49}$$

where $\boldsymbol{\delta}$ contains all the transformed nodal variables and the unchanged coefficients c_i that are attached to the face functions and $\boldsymbol{T}$ is the transition matrix. Consequently, the displacement field can be rewritten as

$$w(\xi, \eta) = \boldsymbol{N}^T \boldsymbol{T}^{-1} \boldsymbol{\delta} = \tilde{\boldsymbol{N}}^T \boldsymbol{\delta} \tag{6.50}$$

where $\tilde{\boldsymbol{N}}$ contains the interpolative nodal bases and the face functions. Similarly, the basis transformation for triangular elements is illustrated in Fig. 6.15. Due to the collocation property of $\tilde{\boldsymbol{N}}$, the element assembly and boundary condition imposition can be easily performed via FEM, which is the typically used approach. Moreover, the C^1 conformity of these elements is exactly satisfied in parallelograms and straight-sided triangular elements since the Jacobian

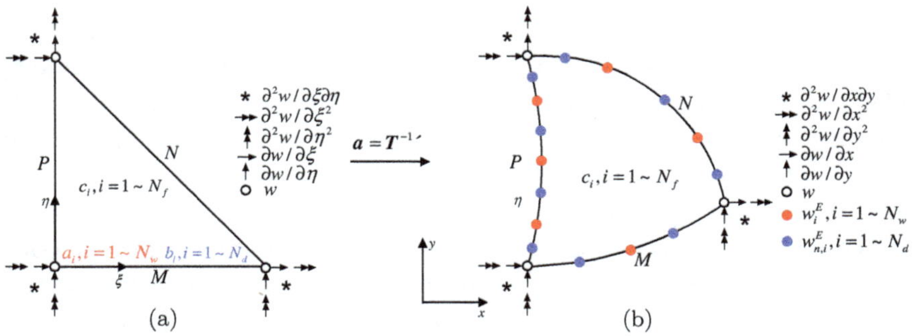

Fig. 6.15. Basis transformation for triangular elements: (a) coefficients of hierarchical bases; (b) variables of nodal bases, where $M = 1$, $N = 2$, $P = 3$ and $\Delta = 1$.

is constant. For elements that have curved boundaries, only the C^0 continuity of field variable w is exactly satisfied; the continuity of the normal derivative is satisfied on the nodes. In this case, these elements are referred to as *quasi-conforming* elements in this work. However, as will be shown, fast convergence can still be realized by using Gauss–Lobatto points as the nodes for w_n.

6.4.2 *Node collocation*

In conventional h-version Lagrange elements, equally spaced nodes are widely used in elements of moderately high order. For example, three and four nodes are used on each edge of quadratic and cubic elements. However, for p-version approximation, more nodes may be required on element boundaries when the element order is high. In such cases, equally spaced nodes often result in poorly conditioned matrices because of the significant numerical oscillations when many nodes are used (see Figs. 6.16(a) and (b)). Therefore, hierarchical bases that are based on orthogonal polynomials are widely used in p-version finite elements [375]. However, this may also render inconvenience in the imposition of non-homogeneous boundary conditions and element assembly. To solve this problem, inspired by the differential quadrature method [376], Liu *et al.* [35, 301] developed a p-version finite element in which Lagrange functions that were based on Gauss–Lobatto quadrature points were used as the edge functions. Via this approach, the numerical oscillations were eliminated, while boundary condition imposition and element assembly could be easily

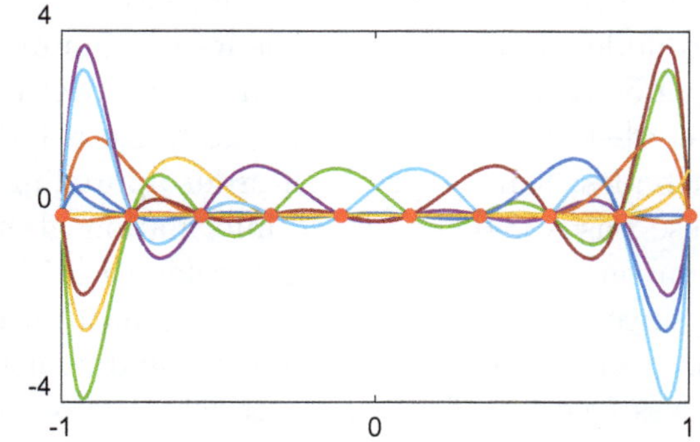

(a) 10 equally spaced nodes

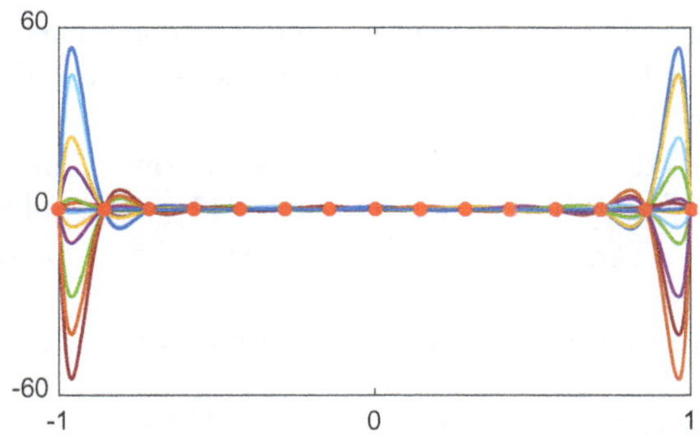

(b) 15 equally spaced nodes

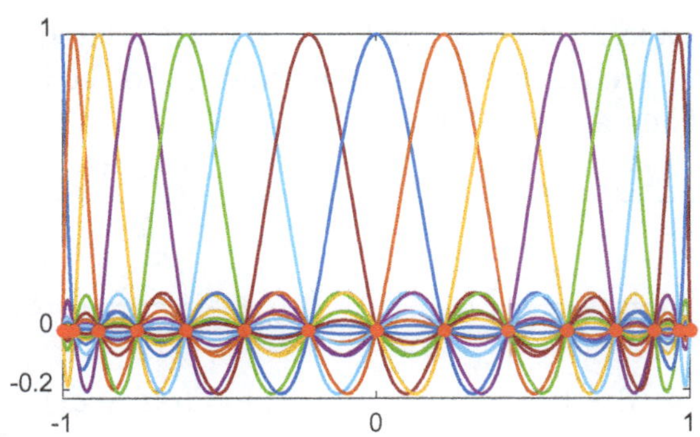

(c) 15 Gauss–Lobatto nodes

Fig. 6.16. Comparison of Lagrange bases that were obtained by various nodes.

performed. Figure 6.16(c) shows the Lagrange bases that are based on Gauss–Lobatto nodes. Each basis attains its maximum value on the corresponding node and the function value varies over a small range of approximately $[-0.2, 1]$. In addition, this range is independent of the number of nodes; hence, very high-order approximation is possible using these bases [34]. For more information about Lagrange node collocation on surfaces and bodies, refer to [377–379]. In the present work, since the transformed bases (in natural coordinates) for w and w_n on element boundaries are the one-dimensional C^2 and C^1 Hermite bases (see Fig. 6.17), respectively, the optimized nodes that have small numerical oscillations are no longer Gauss–Lobatto nodes. However, they can be derived via a similar approach by letting each basis reach its maximum value at the attached node. In the following, we will derive the optimized nodes for C^1 Hermite bases as an example.

As illustrated in Fig. 6.18, consider a one-dimensional function, namely, $f(\xi)$, that is defined in $[-1, 1]$ with values on nodes $[\xi_1 = -1, \xi_2, \ldots, \xi_{N-1}, \xi_N = 1]$ and first-order derivatives at two ends.
The C^1 Hermite interpolation of f can be expressed as

$$\tilde{f}(\xi) = h_1^{(1)}(\xi) f'(-1) + h_N^{(1)}(\xi) f'(1) + \sum_{i=1}^{N} h_i(\xi) f(\xi_i) \tag{6.51}$$

where $h_1^{(1)}(\xi)$ and $h_N^{(1)}(\xi)$ are the bases that are related to the derivatives and $h_i(\xi)$ are the bases that are attached to the function values on node ξ_i. They are defined as

$$\begin{aligned} h_1^{(1)}(\xi) &= \frac{1-\xi^2}{2} L_1(\xi); & h_N^{(1)}(\xi) &= \frac{\xi^2-1}{2} L_N(\xi) \\ h_1(\xi) &= (c_1\xi + c_2)\frac{1-\xi}{2} L_1(\xi); & h_i(\xi) &= \frac{1-\xi^2}{1-\xi_i^2} L_i(\xi) \\ j &= 2 \sim N-1; & h_N(\xi) &= (c_3\xi + c_4)\frac{1+\xi}{2} L_N(\xi) \end{aligned} \tag{6.52}$$

where $L_i(\xi)$, $i = 12, \ldots, N$ are the Lagrange functions, which are defined as

$$L_i(\xi) = \prod_{k=1, k\neq i}^{N} \frac{\xi - \xi_k}{\xi_i - \xi_k} \tag{6.53}$$

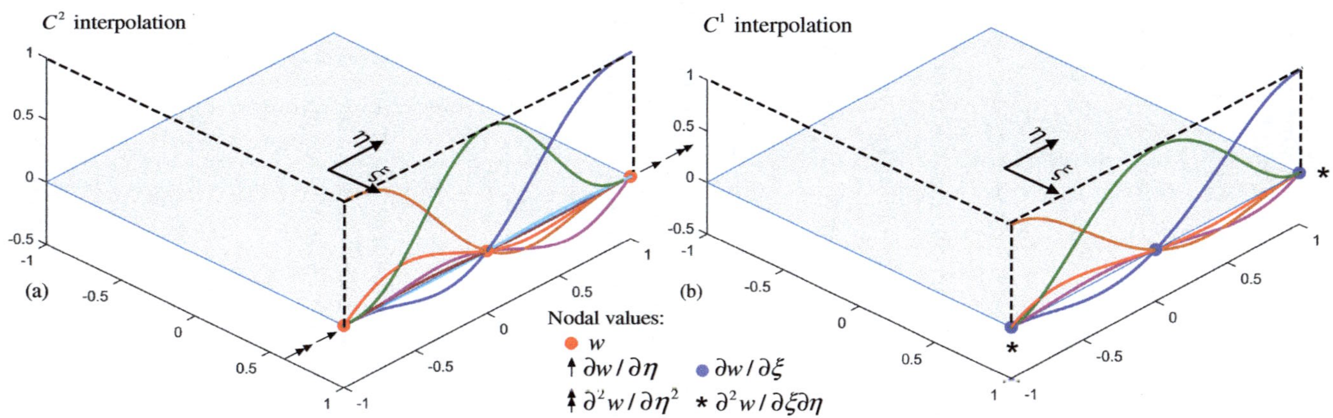

Fig. 6.17. Transformed bases on an element boundary (side 2) with three nodes: (a) C^2 Hermite bases for w; (b) C^1 Hermite bases for $\partial w/\partial \xi$.

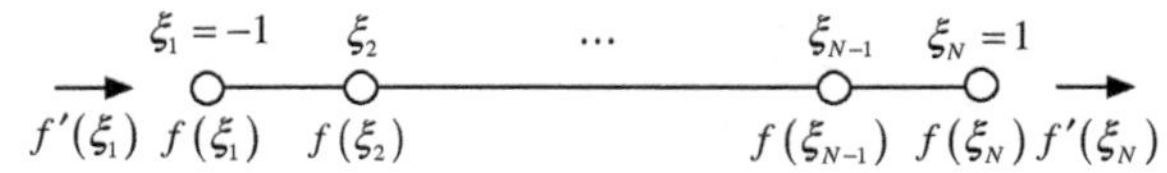

Fig. 6.18. C^1 Hermite interpolation with N nodes.

and the coefficients c_i, $i = 1, 2, 3, 4$, which are expressed as

$$\begin{aligned} c_1 &= \frac{1}{2} - L_1'(\xi_1), \qquad c_2 = \frac{3}{2} - L_1'(\xi_1) \\ c_3 &= -\frac{1}{2} - L_N'(\xi_N), \quad c_4 = \frac{3}{2} + L_N'(\xi_N) \end{aligned} \tag{6.54}$$

can be determined from the interpolation requirements

$$h_1(-1) = 1; \quad h_1'(-1) = 0; \quad h_N(1) = 1; \quad h_N'(1) = 0 \tag{6.55}$$

The Hermite bases that are attached to the middle nodes have the following properties:

$$h_i(\xi_j) = \delta_{ij} \quad \text{and} \quad h_i'(\pm 1) = 0, \quad i = 2, 3, \ldots, N-1 \tag{6.56}$$

To obtain the optimized nodes, we force the middle bases, namely, $h_i(\xi)$, $i = 2, 3, \ldots, N-1$, to attain their maximum values at nodes ξ_i as the Lagrange bases that are based on Gauss–Lobatto nodes, which results in the following equations:

$$\begin{gathered} g_i(\xi) = \frac{dh_i(\xi_i)}{d\xi} = \frac{2\xi_i}{\xi_i^2 - 1} + \frac{dL_i(\xi_i)}{d\xi} = 0 \\ \xi = [\xi_2, \ldots, \xi_{N-1}] \quad i = 2, 3, \ldots, N-1 \end{gathered} \tag{6.57}$$

The $N-2$ middle nodes can be determined from the system of $N-2$ nonlinear equations. The Newton–Raphson method can be used to solve these equations efficiently and the required Jacobian matrix can be computed from

$$\frac{\partial g_i(\xi)}{\partial \xi_j} = \begin{cases} -\dfrac{2(1+\xi_j^2)}{(1-\xi_j^2)^2} - \displaystyle\sum_{k=1, k\neq j}^{N} \dfrac{1}{(\xi_j - \xi_k)^2}, & (i = j) \\ \dfrac{1}{(\xi_j - \xi_i)^2}, & (i \neq j) \end{cases} \tag{6.58}$$

The Gauss–Lobatto nodes can be used as the initial guess. The solutions of Eq. (6.57) are the zeros of Jacobi polynomial $J_{N-2}^{(3,3)}(\xi)$;

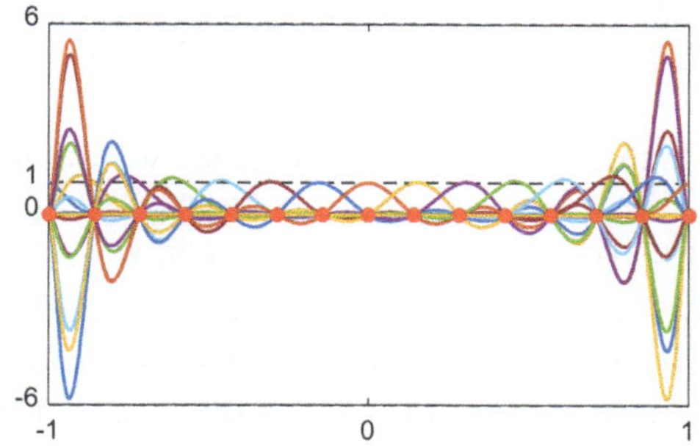

(a) 15 equally spaced nodes

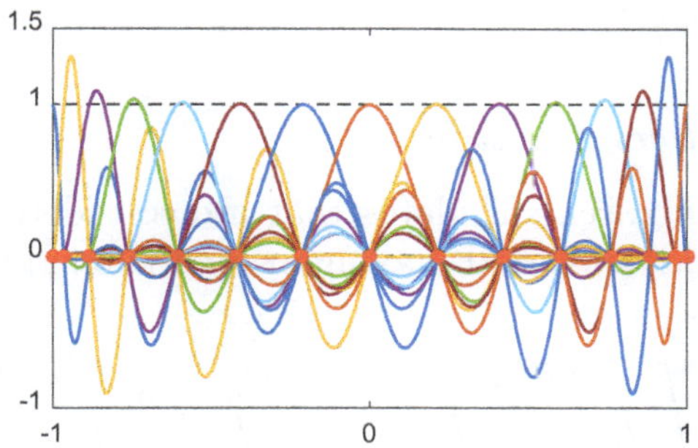

(b) 15 Gauss–Lobatto nodes

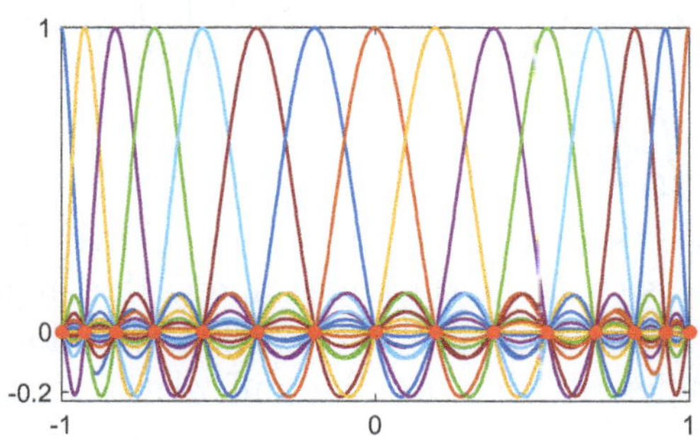

(c) 15 Gauss–Jacobi (3, 3) nodes

Fig. 6.19. Comparison of C^1 Hermite bases that were obtained by various different nodes.

hence, they are referred to as Gauss–Jacobi (GJ) nodes in this work. Figure 6.19 compares the C^1 Hermite bases that were obtained by nodes of various types. The numerical oscillations are more profound in bases that use equally spaced nodes than in the bases that are generated by Gauss–Jacobi nodes.

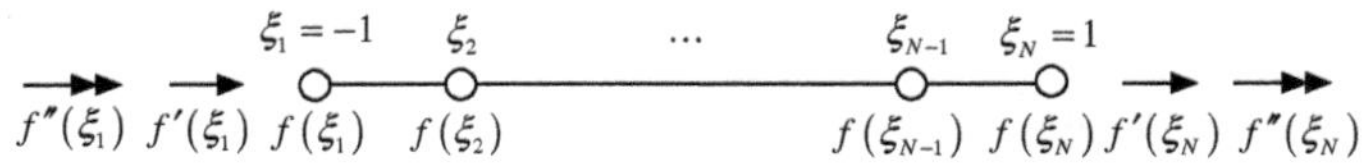

Fig. 6.20. C^2 Hermite interpolation with N nodes.

Figure 6.20 illustrates the C^2 Hermite interpolation, where the derivative up to second order at each end should be interpolated (the fifth order Hermite bases in Eq. (6.20) correspond to the special case of $N = 2$). The derivation of the optimized C^2 Gauss–Jacobi nodes is similar to the derivation above. They are the zeros of Jacobi polynomial $J_{N-2}^{(5,5)}(\xi)$.

6.4.3 *FEM discretization*

For the isotropic Kirchhoff plates that are considered in this work, the potential functional and maximum kinetic energy are expressed as

$$\begin{aligned}\Pi = \frac{D}{2}\int\int_\Omega \left(\frac{\partial^2 w}{\partial x^2}\right)^2 + \left(\frac{\partial^2 w}{\partial y^2}\right)^2 + 2\upsilon\frac{\partial^2 w}{\partial x^2}\frac{\partial^2 w}{\partial y^2}\\ + 2(1-\upsilon)\left(\frac{\partial^2 w}{\partial x\partial y}\right)^2 dxdy - \int\int_\Omega qwdxdy \end{aligned}\tag{6.59}$$

$$T_{\max} = \int\int_\Omega \frac{1}{2}\rho h\omega^2 w^2 dxdy$$

where υ is Poisson's ratio, $D = Eh^3/12(1-\upsilon^2)$ the flexural rigidity, ρ the volume density, h the thickness, and ω the angular frequency. For numerical discretization, we define the following weighting matrices for differentiation as the differential quadrature (DQ) method:

$$\begin{aligned}\mathbf{D}_0 = \begin{bmatrix} N^T(\xi_1,\eta_1) \\ \vdots \\ N^T(\xi_{N_\xi},\eta_{N_\eta}) \end{bmatrix}, \quad \mathbf{D}_{xx} = \begin{bmatrix} N_{,xx}^{\prime\prime T}(\xi_1,\eta_1) \\ \vdots \\ N_{,xx}^{\prime\prime T}(\xi_{N_\xi},\eta_{N_\eta}) \end{bmatrix}\\ \mathbf{D}_{yy} = \begin{bmatrix} N_{,yy}^{\prime\prime T}(\xi_1,\eta_1) \\ \vdots \\ N_{xx}^{\prime\prime T}(\xi_{N_\xi},\eta_{N_\eta}) \end{bmatrix}, \quad \mathbf{D}_{xy} = \begin{bmatrix} N_{,xy}^{\prime\prime T}(\xi_1,\eta_1) \\ \vdots \\ N_{,xy}^{\prime\prime T}(\xi_{N_\xi},\eta_{N_\eta}) \end{bmatrix}\end{aligned}\tag{6.60}$$

where (ξ_i, η_j) are the Gauss–Lobatto quadrature points [35], and N_ξ and N_η are the numbers of quadrature points in the two directions. The following chain rules of differentiation are needed in the computation of Eq. (6.60):

$$\begin{bmatrix} \dfrac{\partial w}{\partial x} \\ \dfrac{\partial w}{\partial y} \end{bmatrix} = \frac{1}{|\mathbf{J}|} \begin{bmatrix} J_{22} & -J_{12} \\ -J_{21} & J_{11} \end{bmatrix} \begin{bmatrix} \dfrac{\partial w}{\partial \xi} \\ \dfrac{\partial w}{\partial \eta} \end{bmatrix} \tag{6.61}$$

$$\begin{bmatrix} \dfrac{\partial^2 w}{\partial x^2} \\ \dfrac{\partial^2 w}{\partial y^2} \\ \dfrac{\partial^2 w}{\partial x \partial y} \end{bmatrix} = \frac{1}{|\mathbf{J}|^2} \begin{bmatrix} J_{22}^2 & J_{12}^2 & -2J_{12}J_{22} \\ J_{21}^2 & J_{11}^2 & -2J_{11}J_{21} \\ -J_{21}J_{22} & -J_{11}J_{12} & J_{11}J_{22} + J_{12}J_{21} \end{bmatrix}$$
$$\times \begin{bmatrix} \dfrac{\partial^2 w}{\partial \xi^2} - \dfrac{\partial w}{\partial x}\dfrac{\partial^2 x}{\partial \xi^2} - \dfrac{\partial w}{\partial y}\dfrac{\partial^2 y}{\partial \xi^2} \\ \dfrac{\partial^2 w}{\partial \eta^2} - \dfrac{\partial w}{\partial x}\dfrac{\partial^2 x}{\partial \eta^2} - \dfrac{\partial w}{\partial y}\dfrac{\partial^2 y}{\partial \eta^2} \\ \dfrac{\partial^2 w}{\partial \eta \partial \xi} - \dfrac{\partial w}{\partial x}\dfrac{\partial^2 x}{\partial \eta \partial \xi} - \dfrac{\partial w}{\partial y}\dfrac{\partial^2 y}{\partial \eta \partial \xi} \end{bmatrix} \tag{6.62}$$

where $x(\xi, \eta)$ and $y(\xi, \eta)$ are the geometry mapping functions and $\mathbf{J}$ is the Jacobian matrix, which is defined as

$$\mathbf{J} = \begin{bmatrix} J_{11} & J_{12} \\ J_{21} & J_{22} \end{bmatrix} = \begin{bmatrix} \dfrac{\partial x}{\partial \xi} & \dfrac{\partial y}{\partial \xi} \\ \dfrac{\partial x}{\partial \eta} & \dfrac{\partial y}{\partial \eta} \end{bmatrix} \tag{6.63}$$

Using Eqs. (6.50) and (6.60), the functional in Eq. (6.59) can be discretized as

$$\begin{aligned} \Pi &= \frac{1}{2}\boldsymbol{\delta}^T \mathbf{T}^{-T}\mathbf{H}^T\mathbf{B}\mathbf{H}\mathbf{T}^{-1}\boldsymbol{\delta} - \mathbf{q}^T\mathbf{C}\mathbf{D}_0\mathbf{T}^{-1}\boldsymbol{\delta} \\ T_{\max} &= \frac{1}{2}\omega^2 \rho h \boldsymbol{\delta}^T \mathbf{T}^{-T}\mathbf{D}_0^T\mathbf{C}\mathbf{D}_0\mathbf{T}^{-1}\boldsymbol{\delta} \end{aligned} \tag{6.64}$$

where

$$\mathbf{B} = D\begin{bmatrix} \mathbf{C} & \upsilon\mathbf{C} & \mathbf{0} \\ \upsilon\mathbf{C} & \mathbf{C} & \mathbf{0} \\ \mathbf{0} & \mathbf{0} & 2(1-\upsilon)\mathbf{C} \end{bmatrix}, \quad \mathbf{H} = \begin{bmatrix} \mathbf{D}_{xx} \\ \mathbf{D}_{yy} \\ \mathbf{D}_{xy} \end{bmatrix}$$

$$\mathbf{q} = [q(\xi_1, \eta_1), \ldots, q(\xi_{N_\xi}, \eta_{N_\eta})]^T$$

$$\mathbf{C} = \mathrm{diag}\{J_{ij} C_i^\xi C_j^\eta\}, \quad J_{ij} = |\mathbf{J}(\xi_i, \eta_j)| \tag{6.65}$$

in which C_i^ξ and C_j^η are the quadrature weightings and J_{ij} are the values of the Jacobian at quadrature points. Finally, the stiffness and mass matrices and load vector can be expressed as

$$\mathbf{K} = D\mathbf{T}^{-T}[\mathbf{D}_{xx}^T\mathbf{C}\mathbf{D}_{xx} + \mathbf{D}_{yy}^T\mathbf{C}\mathbf{D}_{yy} + \upsilon(\mathbf{D}_{xx}^T\mathbf{C}\mathbf{D}_{yy} + \mathbf{D}_{yy}^T\mathbf{C}\mathbf{D}_{yy})$$
$$+ 2(1-\upsilon)\mathbf{D}_{xy}^T\mathbf{C}\mathbf{D}_{xy}]\mathbf{T}^{-1}$$
$$\mathbf{M} = \rho h\mathbf{T}^{-T}\mathbf{D}_0^T\mathbf{C}\mathbf{D}_0\mathbf{T}^{-1}, \quad \mathbf{F} = \mathbf{T}^{-T}\mathbf{D}_0^T\mathbf{C}\mathbf{q}. \tag{6.66}$$

The boundary condition imposition of present elements can be found in the early literature [380] and also more recently [381]; hence, it is omitted in this work due to space limitations.

6.5 Results and Discussion

In this section, numerical examples for vibration and deformation analyses of Kirchhoff plates are presented for evaluating the computational performance of these quadrilateral (abbreviated as *Quad.*) and triangular (*Tri.*) elements. First, the complete order and the computational efficiency of the elements are analyzed. Then, the computational performance of fully C^1 conforming elements (straight-sided triangles and parallelograms) is evaluated via bending and free vibration analyses of thin plates. A singular problem is employed to demonstrate the advantages of these elements in local p-refinement. The performance of elements that have curved boundaries is studied in the third part.

6.5.1 *Complete element order and computational efficiency*

For simplicity, unless otherwise stated, all the numbers of middle-side nodes have the same values of $N \geq 0$ for w and $N + 1$ for

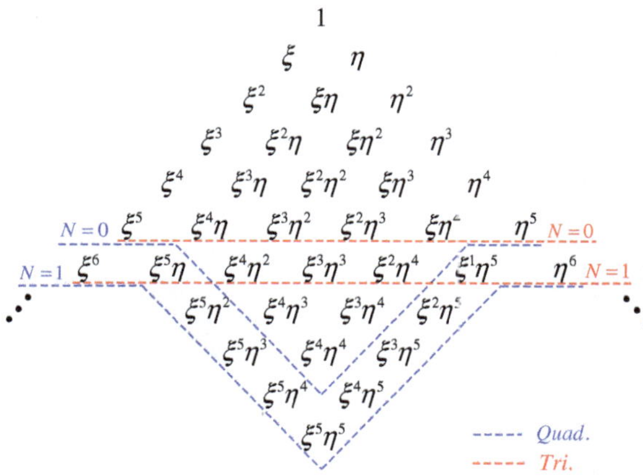

Fig. 6.21. Complete order of the quadrilateral and triangular elements.

w_n in this work, while the numbers of face functions are set as $N_f = H_\xi H_\eta = (N+1)^2$ for quadrilateral elements and $N_f = (N+1)N/2$ for triangular elements. Therefore, for a specified value of N, there are $N^2+10N+29$ bases for a quadrilateral element and $(N+6)(N+7)/2$ bases for a triangular element.

The edge and vertex bases in this work are constructed via the blending function method instead of tensor product 1D bases for local p-refinement. The projection test is used to analyze the complete order of the elements and the results are shown in Fig. 6.21. The complete order of both quadrilateral and triangular elements is $N + 5$. Incomplete high-order monomials, along with two additional polynomials, are included in the quadrilateral elements. However, these incomplete terms do not appear in our triangular elements. Although the number of nodes, namely, N, is set as 0 and 1 in the figure, the regularity of the complete order can be extended to any value of N.

In p-version FEM, computational costs are mainly incurred in the generation of element matrices, especially when the element order is very high. To evaluate the computational efficiency of these elements, Fig. 6.22 presents the run time for the generation of stiffness matrices of present quadrilateral and triangular elements at various values of complete order p. The total number of shape functions for each p is also specified in the figure. The number of quadrature points is set as $p \times p$ for each case. The MATLAB codes were run on a personal computer with Core i5 4200U @ 1.60 GHz CPU. The

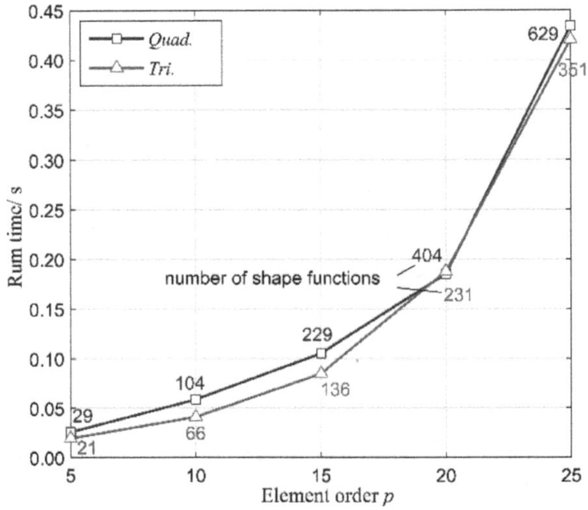

Fig. 6.22. Run time of stiffness matrix generation versus complete order p.

computation process includes the calculation of the high-order bases and their up-to-second-order derivatives, the computation of the up-to-second-order Jacobian, the transformation in Eq. (6.50), and the linear summation of the Gauss–Lobatto quadrature. The run time for both types of elements is less than half a second, even when $p = 25$.

6.5.2 *Analysis using conforming elements*

As discussed previously, if the elements are parallelograms or straight-sided triangles, the C^1 continuity can be exactly satisfied. In this section, the computational performance of these conforming elements is studied. The developed Gauss–Jacobi points are used as the nodes for w and w_n on element boundaries to improve the numerical performance.

6.5.2.1 *Bending of a square plate*

The first example, which is illustrated in Fig. 6.23, is a simply supported square plate that is under a sinusoidally distributed load. Both h- and p-refinements will be performed in the analyses. Figure 6.24

shows the coarsest and finest meshes for h-refinement using non-distorted quadrilateral and triangular elements. Distorted quadrilateral and triangular meshes for h-refinement are shown in Figs. 6.25 and 6.26, respectively, while the meshes for p-refinement are shown in Fig. 6.27. The quadrilateral elements in h-refinement are generated by forcing the numbers of nodes to satisfy $M = N = P = Q = N_f = 0$; these elements are the same as Watkins's 28-DOF conforming rectangular elements. The triangular elements in h-refinement are generated by forcing the numbers of nodes to satisfy $M = N = P = N_f = 0$; they are the same as the TUBA-21 elements. The distorted quadrilateral elements in Fig. 6.25 are indeed non-conforming; they are studied here to demonstrate the influence of C^1 conformity on the convergence rate. Nevertheless, our p-version distorted quadrilateral elements perform very well, as shown in the next part. The energy norm of the displacement error [374] and the relative error of the central point moment M_x, which are defined as follows, will be used to study the convergence rate:

$$\|e\|_E = \left[D \int\int_\Omega \left(\frac{\partial^2 e}{\partial x^2}\right)^2 + \left(\frac{\partial^2 e}{\partial y^2}\right)^2 + 2v\frac{\partial^2 e}{\partial x^2}\frac{\partial^2 e}{\partial y^2} + 2(1-v)\left(\frac{\partial^2 e}{\partial x \partial y}\right)^2 dxdy\right]^{1/2} \tag{6.67}$$

$$\|e_r\|_{M_x} = \frac{|M_{x,app} - M_{x,Exact}|}{M_{x,Exact}}, M_x = -D\left(\frac{\partial^2 w}{\partial x^2} + v\frac{\partial^2 w}{\partial y^2}\right) \tag{6.68}$$

where $e = w^* - w_{app}$ is the difference between the exact solution, namely, w^*, and the approximated numerical solution, namely, w_{app}, $M_{x,app}$ is the approximated numerical solution of moment M_x; and $M_{x,Exact}$ denotes the exact solution.

The errors are presented in logarithmic scale in Fig. 6.28, where N is set as 0, 2, 4, 6, 8 for the p-refinement in Figs. 6.27(a), (c), (d) and 1, 3, 5, 7, 9 in Fig. 6.27(b) so that the numbers of DOFs are similar for comparison. According to Fig. 6.28(a), the convergence rate of the non-distorted quadrilateral elements in h-refinement is higher than that of the non-distorted triangular elements. This is because the number of triangular elements is higher than the number of quadrilateral elements if the numbers of DOFs of the two

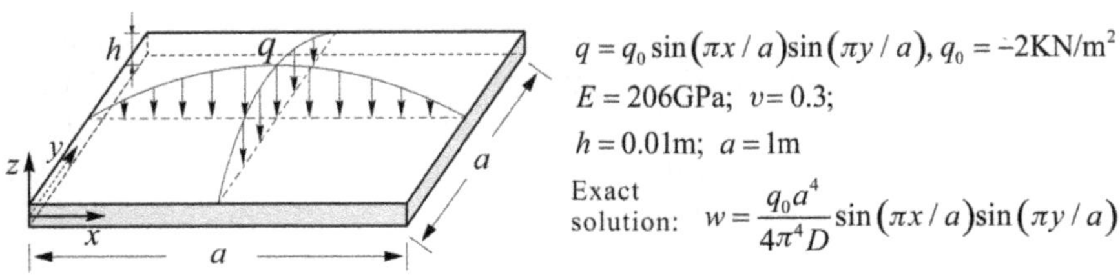

Fig. 6.23. Simply supported square plate under a sinusoidal load.

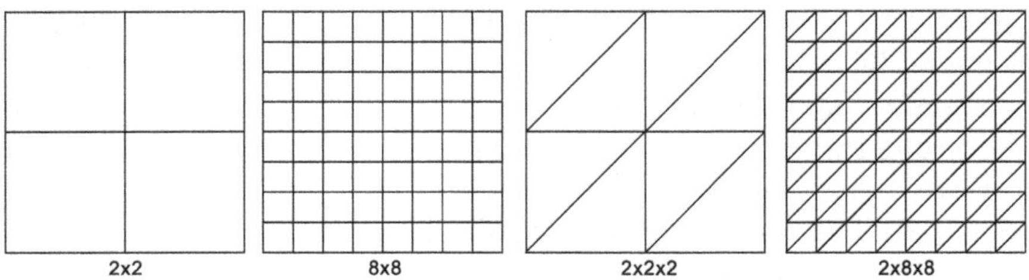

Fig. 6.24. Non-distorted meshes for h-refinement.

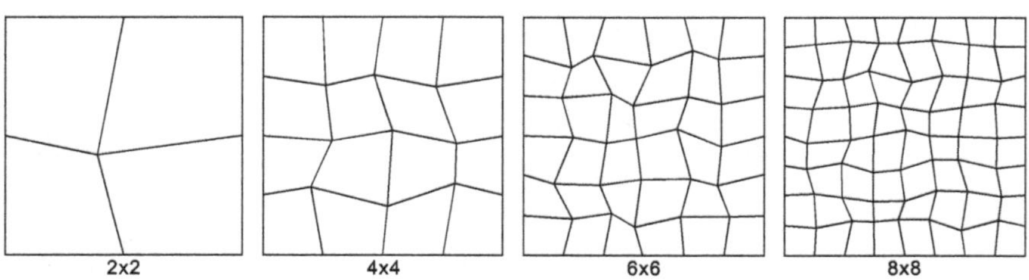

Fig. 6.25. Distorted quadrilateral meshes for h-refinement.

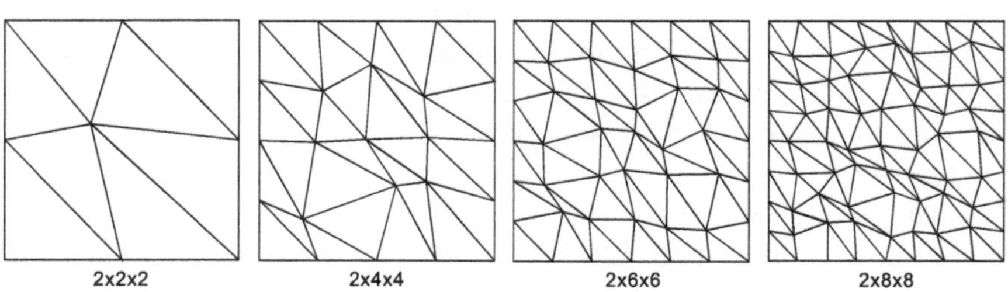

Fig. 6.26. Distorted triangular meshes for h-refinement.

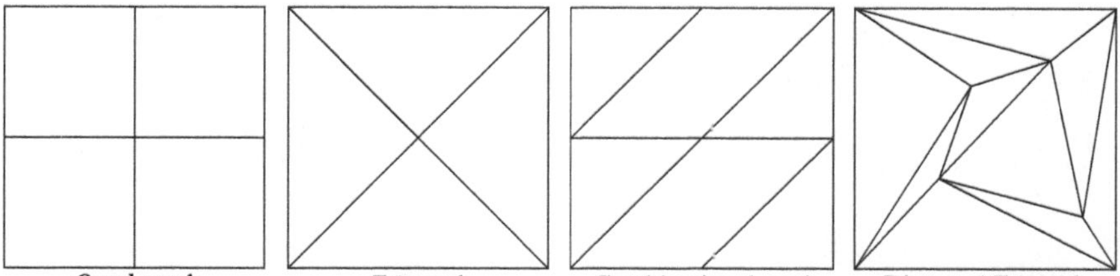

Fig. 6.27. Meshes for p-refinement.

types of elements are the same. As a result, the global continuity of the solution space of triangular elements is poorer than that of quadrilateral elements. The convergence rate of distorted triangular elements in h-refinement is slightly lower than that of the non-distorted mesh. The convergence of distorted quadrilateral elements in h-refinement is poor due to the loss of C^1 conformity, which is a common problem in h-version FEM. Exponential rates of convergence are observed in all cases of p-refinement. The convergence rate of non-distorted triangular elements in p-refinement is higher than that of non-distorted quadrilateral elements because the complete order of triangular elements is higher than that of quadrilateral elements when the same number of DOFs is used (Fig. 6.21). The convergence rate of the combinational case is lower compared to the cases with a single type of elements. This phenomenon is expected because the global continuity in the mesh (Fig. 6.27(c)) is worse than in the other cases due to the use of more elements. However, fast convergence is also realized when more DOFs are used. Satisfactory performance can still be realized when distorted triangular elements are used. The convergence rate is very close to that of the combinational mesh in which the element distortion is less severe. Fast convergence of the central point moment is also observed in Fig. 6.28(b).

6.5.3 *Plate with a singularity*

In the second example, a fully simply supported L-shaped plate under a uniformly distributed flexural load of $q = -1,000\,\text{N/m}^2$ is employed to study the performance of these elements in local p-refinement. The moment is singular at the inner corner. The dimensions of the plate are specified in Fig. 6.29(a). While the material constants are taken

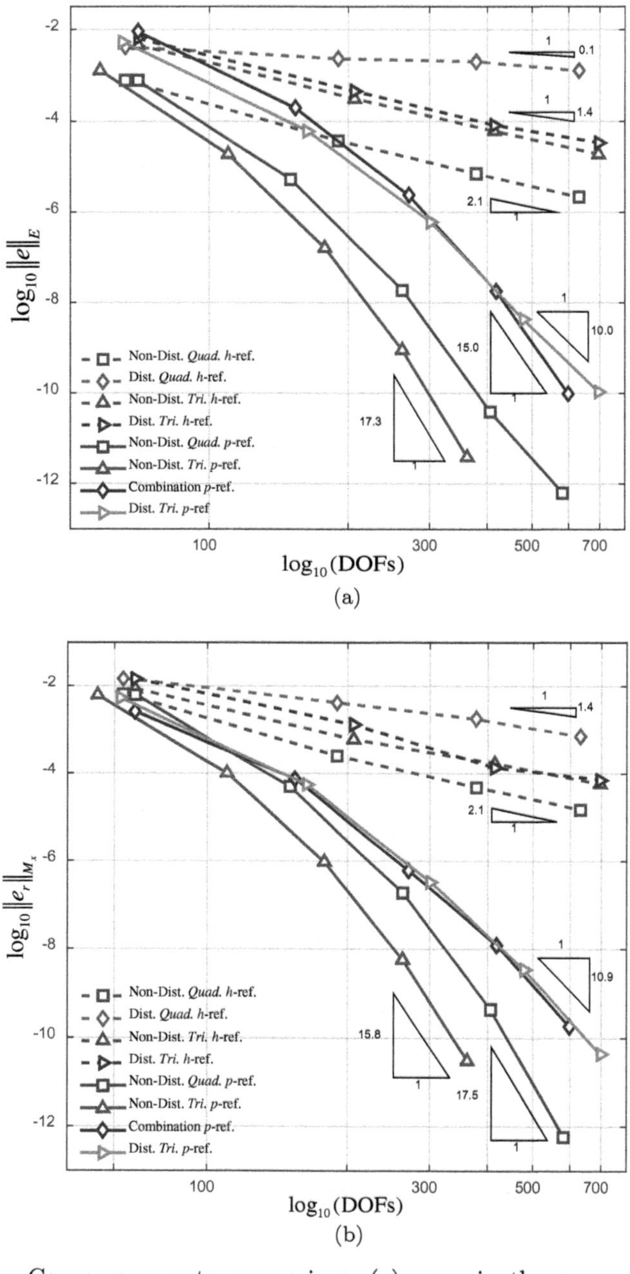

Fig. 6.28. Convergence rate comparison: (a) error in the energy norm and (b) relative error in the logarithm of central-point moment M_x.

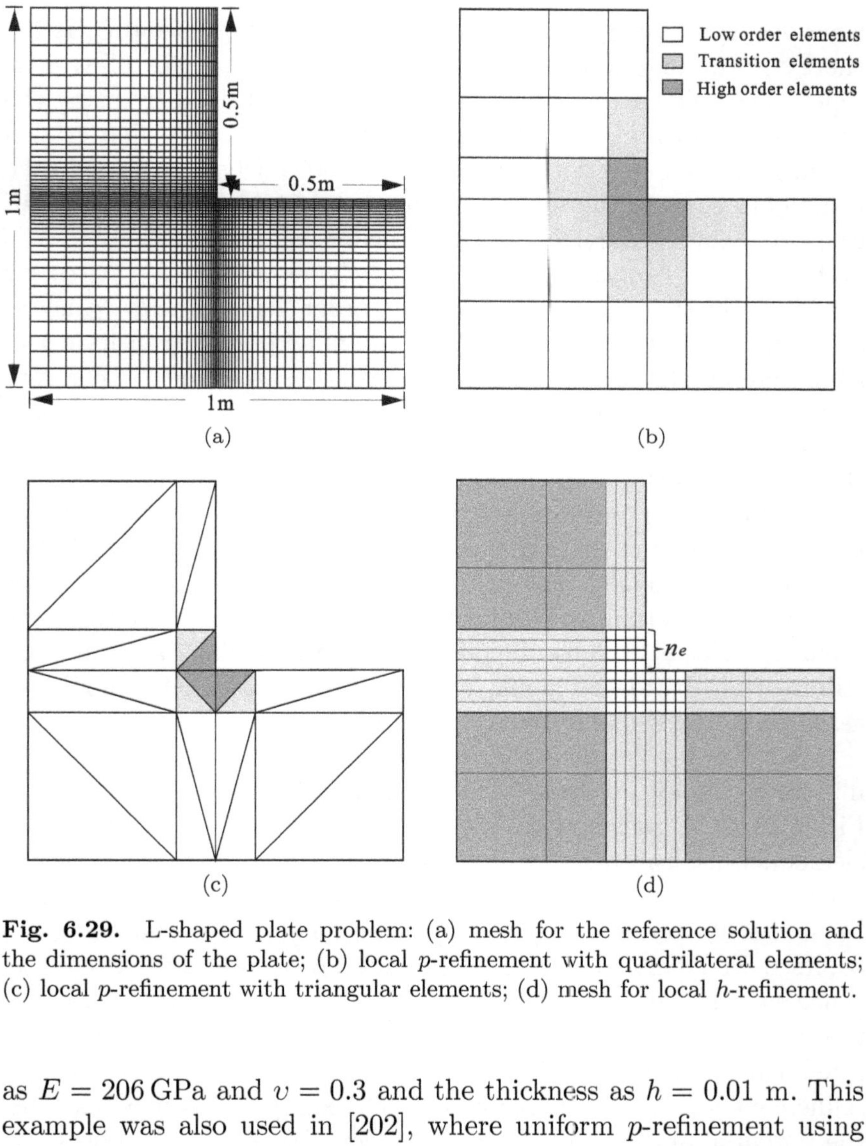

Fig. 6.29. L-shaped plate problem: (a) mesh for the reference solution and the dimensions of the plate; (b) local p-refinement with quadrilateral elements; (c) local p-refinement with triangular elements; (d) mesh for local h-refinement.

as $E = 206\,\text{GPa}$ and $\upsilon = 0.3$ and the thickness as $h = 0.01$ m. This example was also used in [202], where uniform p-refinement using elements with the same order at each step was considered. To assess the convergence rate, the relative error of the energy norm, which is defined as follows, is used for comparison [202]:

$$E_r = \frac{|U_r - U_i|}{U_r} \times 100\% \tag{6.69}$$

where U_r is the energy norm of the reference solution and U_i is the energy norm that is obtained via refinement strategies of various types. The reference solution, namely, U_r, is obtained by using the 28-DOF conforming rectangular elements with a very fine graded mesh, as shown in Fig. 6.29(a), where 15094 DOFs are used (the results are close to the solutions that were obtained by the 28-DOF elements using a uniform mesh with nearly 1,00,000 DOFs).

In Figs. 6.29(b) and (c), high-order elements with N_1 nodes are used near the singular corner, while low-order elements with N_2 nodes are used in the area away from the singular point. Transition elements are employed to connect the two types of elements. For simplicity, the numbers of node are set as $N_2 = 0, 2, 4, 6, 8,$ and 10 and $N_1 = N_2 + 10$ during the process of local p-refinement. In Fig. 6.30, the results of local p-refinement are compared with the solutions that were obtained via uniform p-refinement with $N = 7, 9, 11, 13,$ and 15 in a quadrilateral mesh (Fig. 6.29(b)) and $N = 9, 11, 13, 15,$ and 17 in a triangular mesh (Fig. 6.29(c)). In both cases, the local p-refinements use fewer DOFs than the uniform p-refinements to achieve the same accuracy. This is because the domain around the singular point contributes mostly to the global energy; thus, more DOFs should be assigned to the singular area.

We have also studied the local h-refinement strategy, which is illustrated in Fig. 6.29(d). The mesh profile is similar to that in Fig. 6.29(b). However, only mesh refinement with fixed fifth-order elements is used around the singular corner. In the area away from the singular corner, high-order elements with $p = 7$ are used (which realize sufficient accuracy). By setting the number of elements, namely, n_e, (see Fig. 6.29(d)) as 6, 8, 10 and 12, we obtain the convergence curve that is shown in Fig. 6.30. The convergence rate of the local h-refinement strategy is close to the rate of the uniform p-refinement strategy using quadrilateral elements. More efficient refinement strategies are possible. This example also demonstrates the potential application of these elements for adaptive analysis.

6.5.4 *Free vibration*

In the third example, vibration analysis of a simply supported square plate is employed to study the effects of the node type on the condition numbers of the element matrices and the accuracy of the results.

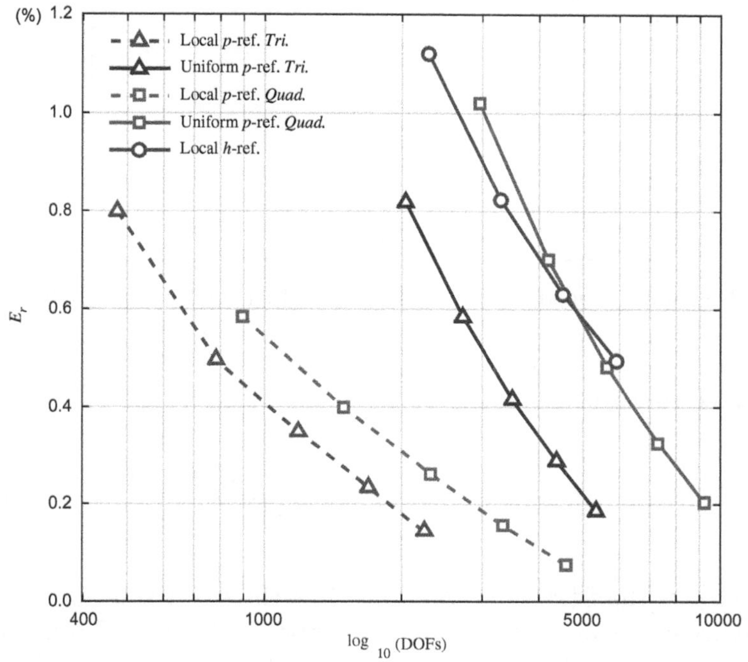

Fig. 6.30. Convergence of the relative error of the energy norm, which is expressed as a percentage.

The expression for the natural frequency parameters, which can be found in textbooks, is as follows:

$$\omega_{ij}^* = (i^2 + j^2)(\pi/a)^2\sqrt{D/\rho h} \tag{6.70}$$

where i and j are the wave numbers along each direction, a is the side length, and D, ρ and h are as defined previously. Six types of nodes are considered here: equally spaced (ES) nodes, the Gauss–Jacobi (GJ) nodes that are developed in this work, Gauss–Lobatto (GL) nodes, the first type of Chebyshev nodes (Cheby1), the second type of Chebyshev nodes (Cheby2) and Chebyshev-Gauss–Lobatto (CGL) nodes. The last four types of non-equally spaced nodes are widely used in DQM; they are defined in [382]. Two types of meshes, which are illustrated in Fig. 6.31, are used in the computation. As a measure of accuracy, we define the following relative error, which is

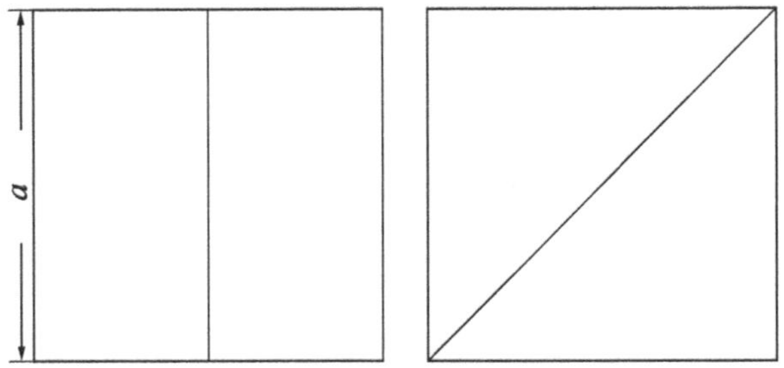

Fig. 6.31. Quadrilateral mesh (left) and triangular mesh (right).

expressed in logarithmic form:

$$\log_{10} \|e_r\| = \log_{10} \left| \frac{\omega_{ij} - \omega_{ij}^*}{\omega_{ij}} \right| \tag{6.71}$$

where ω_{ij} is the FEM solution and ω_{ij}^* is the exact solution.

Figure 6.32 shows the normalized errors of the first 10 frequencies that were obtained by these quadrilateral and triangular elements with $p = 10$, 20 and 30. When $p = 10$, the relative errors of the results that were obtained by the various types of nodes are almost the same. This is because the nodal bases of the nodes of all the types are linearly equivalent and the corresponding algebraic system is well conditioned. When $p = 20$, more accurate results are obtained using non-uniformly distributed nodes. However, the accuracies of the first two frequencies that were obtained by equally spaced nodes decrease from magnitude 10^{-10} to 10^{-8}. When $p = 30$, the accuracy of the equally spaced nodes decreases dramatically and becomes unacceptable, while the results that are obtained using non-uniform nodes are still highly accurate. In Fig. 6.33, the corresponding condition numbers of the element matrices that are generated for the various node types are plotted. The condition numbers of these matrices are high since very high-order bases are used. Nevertheless, the bases can still produce highly accurate results (Fig. 6.32) and the results are stable. Moreover, compared with nodes of other types, the Gauss–Jacobi nodes that are developed in this work have the lowest condition numbers for all cases, while the condition numbers for the equally spaced nodes are the largest.

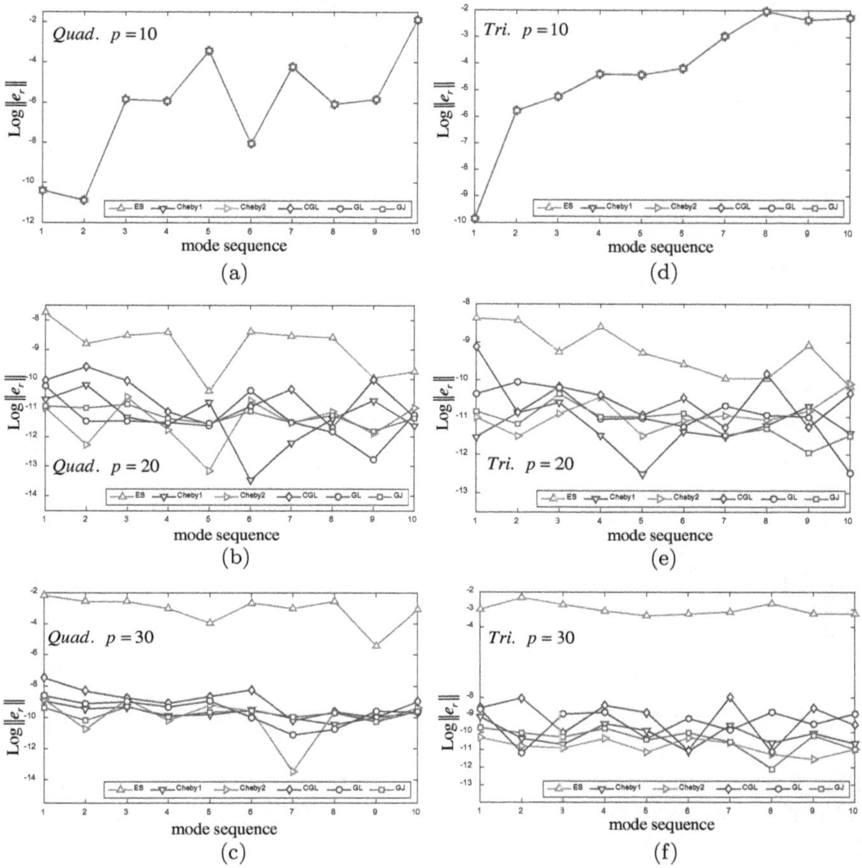

Fig. 6.32. Relative errors of the first 10 frequencies with various orders: (a) *Quad.* $p = 10$; (b) *Quad.* $p = 20$; (c) *Quad.* $p = 30$; (d) *Tri.* $p = 10$; (e) *Tri.* $p = 20$; (f) *Tri.* $p = 30$.

6.5.5 *Analysis using quasi-conforming elements*

For elements that have curved boundaries (including distorted straight-sided quadrilateral elements), the continuity of the displacement, namely, w, is exactly satisfied on the boundaries, while the normal derivative, namely, w_n, is continuous on nodes only. Although C^1 conformity is not fully satisfied, it will be demonstrated in the following that the convergence rate of these elements is still very high. In Section 6.5.2, the number of nodes of w_n is set as one more than that of w on each edge. However, as discussed in Section 6.2.2, the

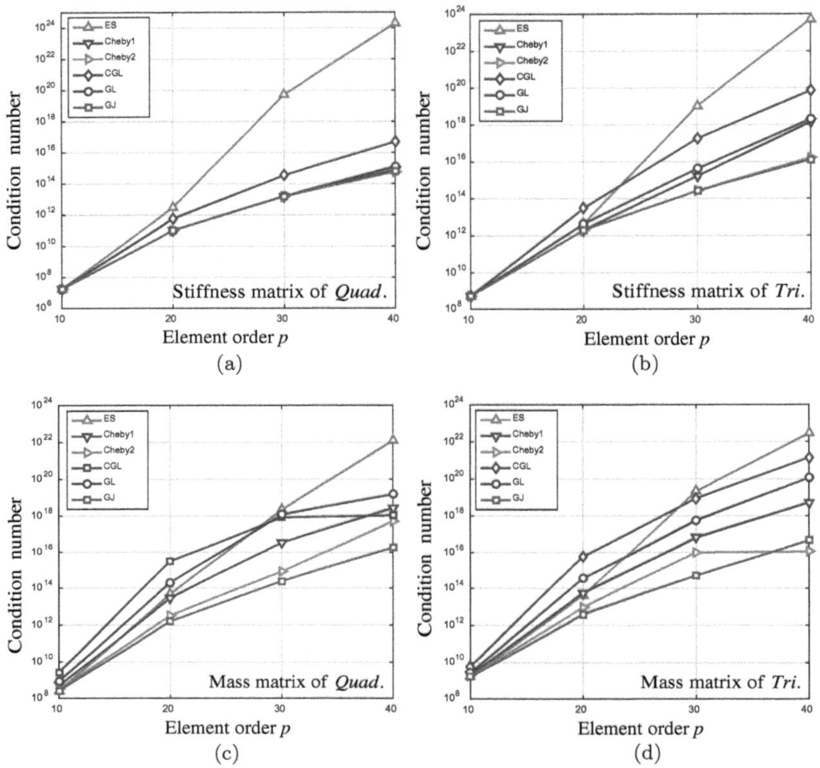

Fig. 6.33. Condition numbers of element matrices versus the element order: (a) stiffness matrix of *Quad*; (b) stiffness matrix of *Tri*; (c) mass matrix of *Quad*; (d) mass matrix of *Tri*.

number of nodes for w_n can be selected freely because of the inherent characteristics of blending function interpolation. This property is very useful for improving the convergence rate, especially for elements that have curved boundaries. In the following, the number of nodes for w on each edge is denoted by N, while the number of nodes for w_n is set as $N + \Delta$. The face function numbers are as previously defined.

6.5.6 *Bending of a square plate*

The square plate that was analyzed in Section 6.4.2 will be considered again in this part with meshes that are generated by elements that have curved boundaries (see Fig. 6.34). The convergence curves

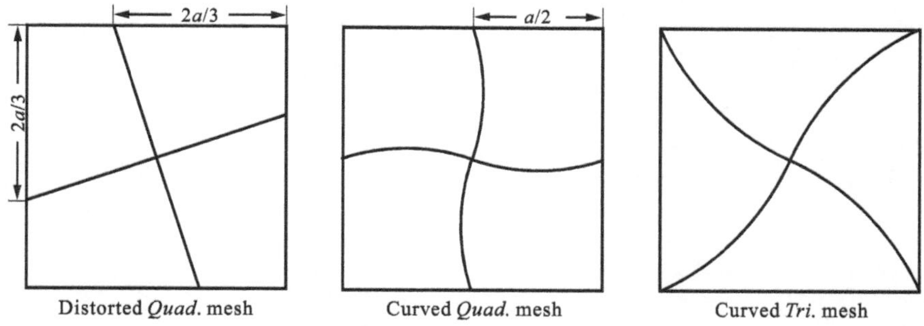

Fig. 6.34. Three meshes for the square plate that use quasi-conforming elements.

of the error energy norm for distorted quadrilateral meshes with $N = 0, 2, 4, 6$, and 8 (p from 5 to 13) are shown in Fig. 6.35(a). Gauss–Lobatto points are used as the nodes for w_n (the nodes for w are still Gauss–Jacobi nodes with weights (5, 5)) and Δ ranges from 1 to 3. The results that are obtained by rectangular conforming elements (Fig. 6.27(a)) are also included in the figure for comparison. An exponential rate of convergence is observed in all cases with various values of the node-number increment, which is denoted as Δ. Improved convergence rates are obtained when Δ is increased because C^1 continuity is refined as the number of nodes that are used for w_n increases. It is surprising that when $\Delta = 3$, the convergence curve that is obtained by our distorted elements is very close to the results of conforming elements using non-distorted meshes. To examine the influence of the node type on the convergence rate, Fig. 6.35(b) presents the convergence curves that were obtained by various node types for the distorted quadrilateral mesh with $\Delta = 3$. An exponential convergence rate is only obtained when Gauss–Lobatto nodes are used. A linear convergence rate is observed when other non-uniform nodes are used, while the results of equally spaced nodes seem to diverge. The convergence curves of the error energy norm and the central point moment that are obtained by various meshes with $\Delta = 3$ are compared in Figs. 6.35(c) and (d). Fast convergence is observed in all cases.

6.5.7 *Bending of a circular plate*

The second example is a clamped circular plate that is subjected to a uniformly distributed load (see Fig. 6.36). The exact solution can

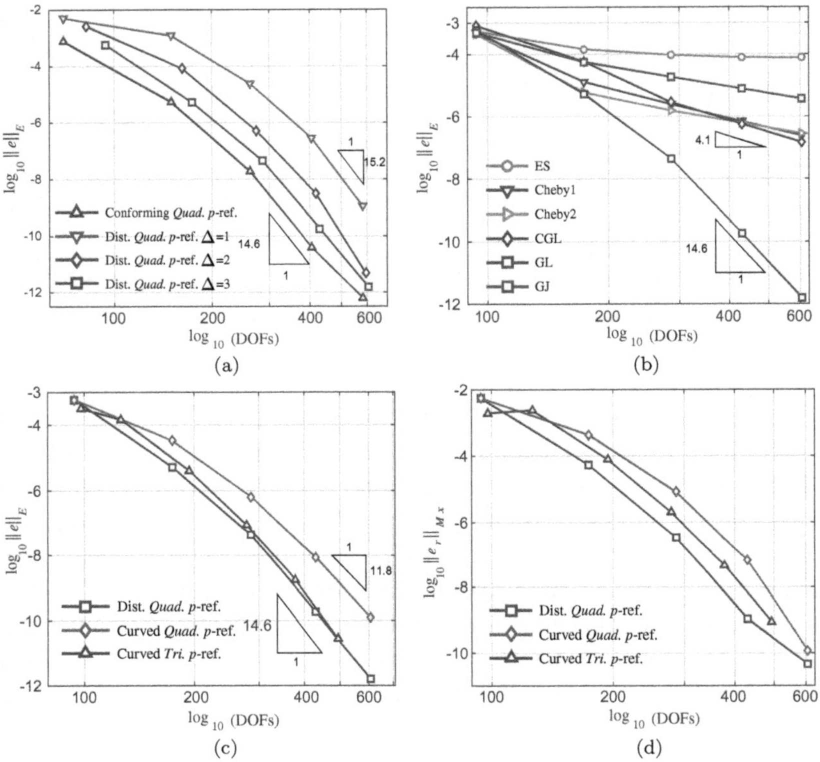

Fig. 6.35. Convergence comparison: (a) convergence of the error energy norm for distorted quadrilateral meshes with $\Delta = 1, 2$, and 3; (b) convergence of the error energy norm using distorted quadrilateral meshes with various node types and $\Delta = 3$; (c) convergence of the error energy norm using various meshes with $\Delta = 3$; (d) convergence of the relative error of central-point moment M_x.

be found in [183]. Three types of meshes, namely, curved quadrilateral mesh, curved triangular mesh and blended mesh, are used to discretize the domain. Gauss–Lobatto nodes with $\Delta = 5$ are used for w_n. The convergence curves of the error energy norm and the central point moment error are shown in Fig. 6.37. Satisfactory convergence is observed. Similar to the results in Fig. 6.28, the results for the triangular mesh converge faster than those for the quadrilateral mesh and the blended mesh because the complete order of the triangular elements is higher compared to the others when the same DOFs are used.

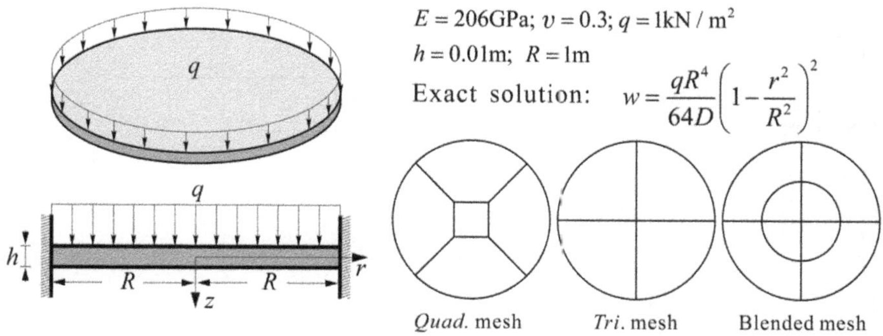

Fig. 6.36. Clamped circular plate under a uniformly distributed load.

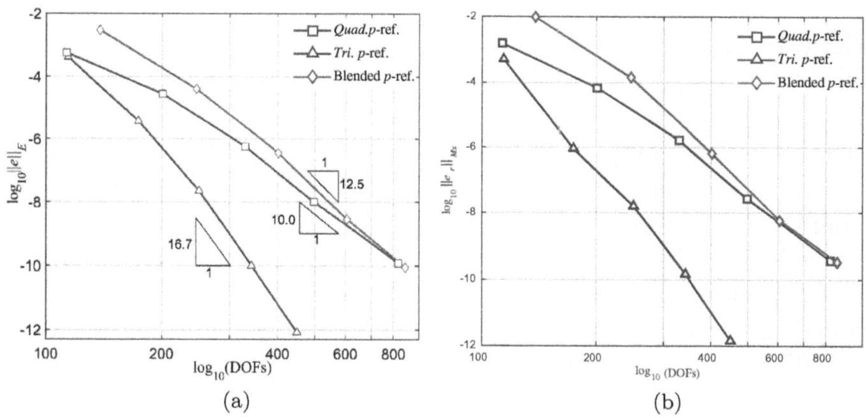

Fig. 6.37. Convergence comparison: (a) convergence of the error energy norm using various meshes; (b) convergence of the relative error of the central-point moment.

6.5.8 *Plate with an irregular cutout*

To evaluate the computational performance of the considered elements in problems with irregular domains, free vibration analysis of a square plate with an irregular cutout (Fig. 6.38(a)) is employed as our last example. The same mesh as IGA [209] is used for domain decomposition (Fig. 6.38(b)). The first 10 non-dimensional frequency parameters, namely, $\Omega = (\omega a^4 \rho h/D)^{1/4}$, for three types of boundary conditions are presented in Table 6.1. Satisfactory convergence is observed and the results agree well with the solutions that are obtained via the edge-based smoothing radial point interpolation

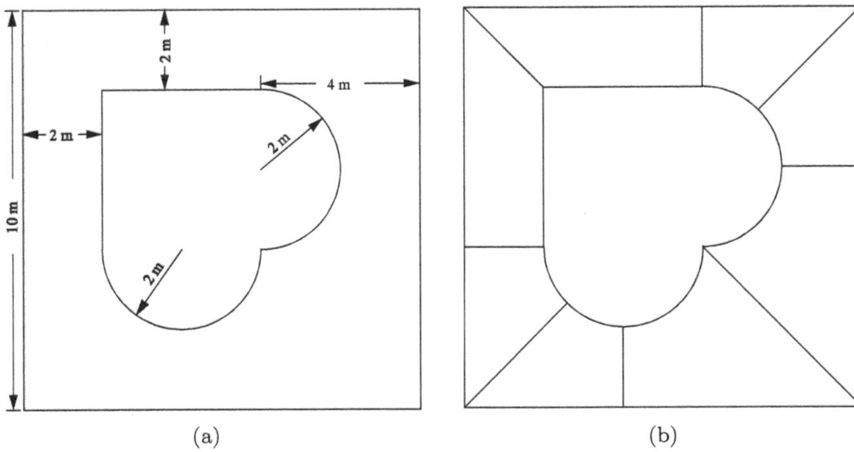

Fig. 6.38. Square plate with a complicated cutout: (a) geometric dimensions; (b) the mesh.

method (ES-RPIM) and the nodal-based smoothing radial point interpolation method (NS-RPIM) [383]. These results differ slightly from the IGA results [209], where the bending strip method was used to deal with the C^1 conformity. The first six mode shapes of the simply supported plate are presented in Fig. 6.39, where only three nodes for w are used on each edge and the results exhibit satisfactory smoothness on the element boundaries.

6.6 Conclusions

Both quadrilateral and triangular hierarchical p-version C^1 elements that have curved boundaries are developed via the blending function interpolation method. The features of the elements can be summarized as follows:

(1) The hierarchical bases are constructed using the Hermite blending function interpolation methods, along with the Jacobi polynomials. Because of the advantages of blending function interpolation, the approximation orders of the edge bases and internal modes in these elements can differ from each other; hence, local h- and p-refinements can be performed conveniently.
(2) For convenience in imposing essential boundary conditions and enforcing the C^1 conformity on curved edges, the orthogonal

Table 6.1. First 10 non-dimensional frequency parameters for a square plate with an irregular cutout.

	Mode sequence									
N	1	2	3	4	5	6	7	8	9	10
					Simply supported					
3	4.913	6.389	6.758	8.557	8.973	10.655	10.884	11.600	12.805	13.159
5	4.912	6.388	6.754	8.557	8.966	10.648	10.884	11.599	12.804	13.149
7	4.912	6.388	6.752	8.556	8.963	10.646	10.883	11.599	12.804	13.145
12	4.912	6.388	6.750	8.556	8.960	10.643	10.883	11.599	12.804	13.142
15	4.912	6.388	6.750	8.556	8.960	10.642	10.883	11.599	12.804	13.141
ES-RPIM [383]	4.905	6.389	6.753	8.574	8.986	10.685	10.897	11.594	12.856	13.223
NS-RPIM [383]	4.919	6.398	6.775	8.613	9.016	10.738	10.930	11.601	12.903	13.283
IGA [209]	5.193	6.579	6.597	7.819	8.812	9.42	10.742	10.776	11.919	13.200
					Clamped					
3	7.437	9.819	9.838	10.943	11.157	12.340	12.832	13.446	14.433	14.712
5	7.437	9.817	9.838	10.942	11.153	12.332	12.828	13.438	14.433	14.700
7	7.437	9.817	9.838	10.942	11.151	12.329	12.828	13.438	14.433	14.695
12	7.437	9.816	9.838	10.942	11.149	12.326	12.828	13.438	14.432	14.691
15	7.437	9.816	9.838	10.942	11.148	12.325	12.828	13.438	14.432	14.690
ES-RPIM [383]	7.423	9.770	9.797	10.927	11.137	12.363	12.822	13.428	14.508	14.789
NS-RPIM [383]	7.410	9.726	9.764	10.896	11.114	12.353	12.781	13.368	14.485	14.766
IGA [209]	7.621	9.810	9.948	11.135	11.216	12.482	12.872	13.650	14.676	14.738
					Free					
3	3.223	3.825	4.658	5.473	5.526	6.922	7.331	8.077	8.353	8.664
5	3.222	3.824	4.658	5.473	5.526	6.922	7.328	8.077	8.353	8.664
7	3.222	3.824	4.658	5.473	5.525	6.922	7.327	8.077	8.353	8.664
12	3.222	3.824	4.658	5.473	5.525	6.922	7.326	8.076	8.353	8.664
15	3.222	3.824	4.658	5.473	5.525	6.922	7.325	8.076	8.353	8.664
ES-RPIM [383]	3.225	3.823	4.646	5.473	5.524	6.902	7.303	8.060	8.349	8.623
NS-RPIM [383]	3.234	3.846	4.684	5.497	5.547	6.929	7.350	8.103	8.392	8.667
IGA [209]	3.482	3.968	5.216	5.805	6.015	7.100	7.485	8.114	8.853	9.001

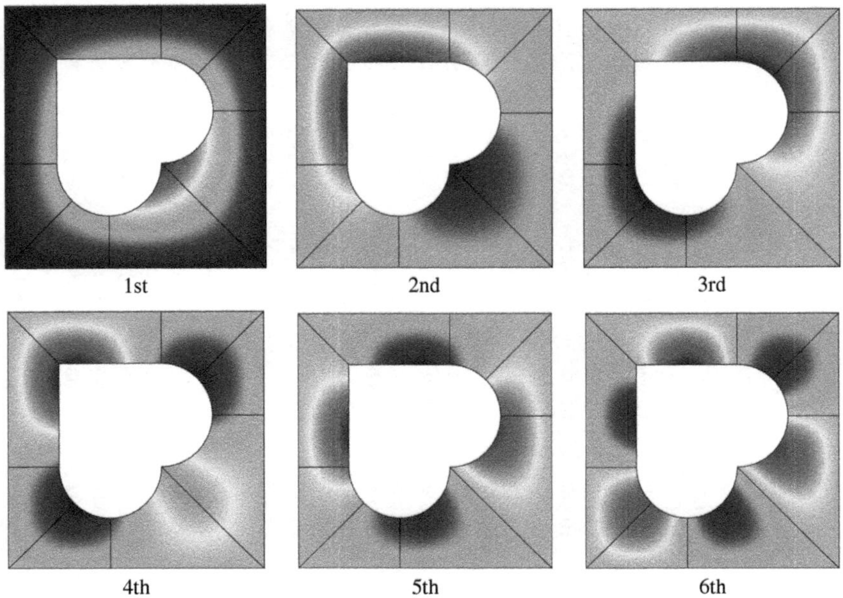

Fig. 6.39. First six mode shapes of a simply supported square plate with an irregular cutout ($N = 3$ and $\Delta = 1$).

hierarchical bases on the edges of elements are transformed into interpolative nodal bases. Nodal collocation strategies are studied for improving the matrix conditioning. Stable results are obtained, even when very high-order bases are used.

(3) For straight-sided triangular elements and parallelograms, the C^1 conformity is fully satisfied. For elements that have curved boundaries, the C^1 conformity is satisfied on nodes and enhanced by Gauss–Lobatto nodes. Numerical results demonstrate that fast convergence can be realized in both fully conforming elements and elements that have curved boundaries.

(4) The developed p-version quadrilateral and triangular elements can be used cooperatively. They are expected to be used in problems with more complicated domains.

In our future work, these elements will be extended to thin-shell analysis with an exact geometry that is created using NURBS or other parameterization methods. The developed elements that have curved boundaries will be very useful not only for the accurate modeling of trimmed surfaces but also for local refinements around the trimmed areas.

Appendix 6.1: Shape Functions

	Quadrilateral elements	Triangular elements
Vertex 1	$S_w^{V_1} = h_2(\xi)H_1(\eta) + H_1(\xi)h_2(\eta) - h_2(\xi)h_2(\eta)$ $S_{w_\xi}^{V_1} = H_1^{(1)}(\xi)h_2(\eta); S_{w,\eta}^{V_1} = h_2(\xi)H_1^{(1)}(\eta)$ $S_{w_{\xi\xi}}^{V_1} = H_1^{(2)}(\xi)h_2(\eta); S_{w,\eta\eta}^{V_1} = h_2(\xi)H_1^{(2)}(\eta);$ $S_{w,\xi\eta}^{V_1} = h_1(\xi)h_1(\eta)$	$S_w^{V_1} = 6\zeta^5 - 5\zeta^2(3\zeta^2 - 2\zeta - 6\xi\eta + 6\xi\eta\zeta), \zeta = 1 - \xi - \eta$ $S_{w_\xi}^{V_1} = \xi(3\xi + 1)\zeta^2(3\eta + \zeta); S_{w_\eta}^{V_1} = \eta(3\eta + 1)\zeta^2(3\xi + \zeta)$ $S_{w_{\xi\xi}}^{V_1} = \xi^2\zeta^2(3\eta + \zeta)/2; S_{w_{\eta\eta}}^{V_1} = \eta^2\zeta^2(3\xi + \zeta)/2; S_{w_{\xi\eta}}^{V_1} = \xi\eta\zeta^2$
Vertex 2	$S_w^{V_2} = h_3(\xi)H_1(\eta) + H_2(\xi)h_2(\eta) - h_3(\xi)h_2(\eta)$ $S_{w_\xi}^{V_2} = H_2^{(1)}(\xi)h_2(\eta); S_{w_\eta}^{V_2} = h_3(\xi)H_1^{(1)}(\eta)$ $S_{w_{\xi\xi}}^{V_2} = H_2^{(2)}(\xi)h_2(\eta); S_{w_{\eta\eta}}^{V_2} = h_3(\xi)H_1^{(2)}(\eta);$ $S_{w_{\xi\eta}}^{V_2} = h_4(\xi)h_1(\eta)$	$S_w^{V_2} = \xi^3(6\xi^2 - 15\xi + 10); S_{w_\xi}^{V_2} = -\xi^3(3\xi^2 - 7\xi + 4)$ $S_{w_\eta}^{V_2} = \xi^2\eta(9\xi\eta - 9\eta - 2\xi + 6\eta^2 + 3); S_{w_{\xi\xi}}^{V_2} = \xi^3(\xi - 1)^2/2$ $S_{w_{\eta\eta}}^{V_2} = \xi^2\eta^2(3\xi + 2\eta - 2)/2;$ $S_{w_{\xi\eta}}^{V_2} = -\xi^2\eta(3\xi\eta - 3\eta - \xi + 2\eta^2 + 1)$
Vertex 3	$S_w^{V_3} = h_3(\xi)H_2(\eta) + H_2(\xi)h_3(\eta) - h_3(\xi)h_3(\eta)$ $S_{w_\xi}^{V_3} = H_2^{(1)}(\xi)h_3(\eta); S_{w_\eta}^{V_3} = h_3(\xi)H_2^{(1)}(\eta)$ $S_{w_{\xi\xi}}^{V_3} = H_2^{(2)}(\xi)h_3(\eta); S_{w,\eta\eta}^{V_3} = h_3(\xi)H_2^{(2)}(\eta); S_{w,\xi\eta}^{V_3} = h_4(\xi)h_4(\eta)$	$S_w^{V_3} = \eta^3(6\eta^2 - 15\eta + 10); S_{w_\xi}^{V_3} = \eta^2\xi(9\xi\eta - 9\xi - 2\eta + 6\xi^2 + 3)$ $S_{w_\eta}^{V_3} = -\eta^3(3\eta^2 - 7\eta + 4); S_{w_{\xi\xi}}^{V_3} = \eta^2\xi^2(3\eta + 2\xi - 2)/2$ $S_{w_{\eta\eta}}^{V_3} = \eta^3(1 - \eta)^2/2; S_{w_{\xi\eta}}^{V_3} = -\eta^2\xi(3\xi\eta - 3\xi - \eta + 2\xi^2 + 1)$
Vertex 4	$S_w^{V_4} = h_2(\xi)H_2(\eta) + H_1(\xi)h_3(\eta) - h_2(\xi)h_3(\eta)$ $S_{w_\xi}^{V_4} = H_1^{(1)}(\xi)h_3(\eta); S_{w_\eta}^{V_4} = h_2(\xi)H_2^{(1)}(\eta);$ $S_{w_{\xi\xi}}^{V_4} = H_1^{(2)}(\xi)h_3(\eta); S_{w_{\eta\eta}}^{V_4} = h_2(\xi)H_2^{(2)}(\eta); S_{w_{\xi\eta}}^{V_4} = h_1(\xi)h_4(\eta)$	—
Edge 1	$S_{w,i}^{S_1} = (1 + \xi)^3(1 - \xi)^3 J_i^{(4,4)}(\xi)h_2(\eta);\ i = 0 \sim M - 1$ $S_{w_n,i}^{S_1} = (1 + \xi)^2(1 - \xi)^2 J_i^{(4,4)}(\xi)h_1(\eta);\ i = 0 \sim M$	$S_{w,i}^{S_1} = \xi^3\zeta^2(3\eta + \zeta)J_i^{(7,6)}(2\xi - 1),\ i = 0 \sim M - 1$ $S_{w_\eta,i}^{S_1} = \xi^2\eta\zeta^2 J_i^{(7,4)}(2\xi - 1),\ i = 0 \sim M$
Edge 2	$S_{w,i}^{S_2} = (1 - \eta)^3(1 + \eta)^3 J_i^{(4,4)}(\eta)h_3(\xi);\ i = 0 \sim N - 1$ $S_{w_n,i}^{S_2} = (1 - \eta)^2(1 + \eta)^2 J_i^{(4,4)}(\eta)h_4(\xi);\ i = 0 \sim N$	$S_{w,i}^{S_2} = -\xi^2\eta^3(\xi + 3\zeta)J_i^{(7,6)}(2\eta - 1),\ i = 0 \sim N - 1$ $S_{w_\xi,i}^{S_1} = -\xi^2\eta^2\zeta J_i^{(7,4)}(2\eta - 1),\ i = 0 \sim N$
Edge 3	$S_{w,i}^{S_3} = (1 - \xi)^3(1 + \xi)^3 J_i^{(4,4)}(\xi)h_3(\eta);\ i = 0 \sim P - 1$ $S_{w_n,i}^{S_3} = (1 - \xi)^2(1 + \xi)^2 J_i^{(4,4)}(\xi)h_4(\eta);\ i = 0 \sim P$	$S_{w,i}^{S_3} = \eta^3\zeta^2(3\xi + \zeta)J_i^{(7,6)}(2\eta - 1),\ i = 0 \sim P - 1$ $S_{w_\xi,i}^{S_3} = \xi\eta^2\zeta^2 J_i^{(7,4)}(2\eta - 1),\ i = 0 \sim P$
Edge 4	$S_{w,i}^{S_4} = (1 - \eta)^3(1 + \eta)^3 J_i^{(4,4)}(\eta)h_2(\xi);\ i = 0 \sim Q - 1$ $S_{w_n,i}^{S_4} = (1 - \eta)^2(1 + \eta)^2 J_i^{(4,4)}(\eta)h_1(\xi);\ i = 0 \sim Q$	—
Face	$S_{mn}^F = C_{mn}(1 - \xi^2)^2(1 - \eta^2)^2 J_{m-1}^{(4,4)}(\xi)J_{n-1}^{(4,4)}(\eta);$ $m = 1 \sim H_\xi, n = 1 \sim H_\eta$	$S_{pn}^F = C_{pn}\zeta^2\xi^2\eta^2(1 - \eta)^{p-n-6} J_{p-n-6}^{(4,4)}\left(\frac{2\xi}{1-\eta} - 1\right)$ $J_n^{(2p-2n-3,4)}(2\eta - 1)$ $p \geq 6, n = 0 \sim p - 6$

Chapter 7

The Hierarchical Quadrature Element Method for Shells in Orthogonal Curvilinear Coordinate System

This chapter applies the differential quadrature hierarchical finite element method (DQHFEM) to deep shells in orthogonal curvilinear coordinate system. The dynamic analyses of functionally graded material (FGM) sandwich shallow shells in thermal and non-thermal environments are taken as an example. A layerwise theory based on the first-order shear deformation theory (FSDT) for each layer is adopted. Effective material properties of the FGM are estimated according to Voigt's rule of mixture (ROM) and/or Mori–Tanaka (MT) scheme. For the shells in thermal environment, a nonlinear temperature distribution in thickness direction is considered and the elastic properties are assumed to be temperature-dependent. The results obtained from the proposed formulation are validated with those available in the literatures. Natural frequencies obtained from Sander's, Love's and Donnell's shell theories for different geometric and boundary conditions are compared with each other first to assess the performance of different shell theories for FGM sandwich shells under non-thermal environment. Then the effects of volume fraction index, core thickness and temperature gradient on natural frequencies of FGM sandwich shells are investigated. The presented DQHFEM is much like the fixed interface mode synthesis method but does not need modal analysis and is of high accuracy. This work is the first application of the method to functionally graded sandwich shells in thermal environments.

7.1 Element Energy Functions for Deep Shell Element

For ease of understanding, the orthogonal curvilinear coordinate system is used for a rough idea of deep shell element in this section. Assuming that a normal to the middle surface remains straight during deformation but not normal to the middle surface, the displacements can be written as

$$\begin{aligned} u_1(\alpha,\beta,\gamma) &= u(\alpha,\beta)+\gamma\psi_\alpha(\alpha,\beta) \\ u_2(\alpha,\beta,\gamma) &= v(\alpha,\beta)+\gamma\psi_\beta(\alpha,\beta) \\ u_3(\alpha,\beta,\gamma) &= w(\alpha,\beta) \end{aligned} \tag{7.1}$$

where u, v and w are middle surface displacements of the shell, and ψ_α and ψ_β are middle surface rotations. The third of the above equations yields the strain component $e_\gamma = 0$. The Lame coefficients of an orthogonal curvilinear coordinate system can be expressed by middle surface variables as

$$H_1 = A(1+k_1\gamma), H_2 = B(1+k_2\gamma) \tag{7.2}$$

where k_1 and k_2 are respectively the principal curvatures of middle surface with respect to α and β; and A and B are respectively the Lame coefficients of middle surface with respect to α and β. They are defined by

$$\begin{aligned} A &= \left[\left(\frac{\partial x}{\partial \alpha}\right)^2+\left(\frac{\partial y}{\partial \alpha}\right)^2+\left(\frac{\partial z}{\partial \alpha}\right)^2\right]^{\frac{1}{2}} \\ B &= \left[\left(\frac{\partial x}{\partial \beta}\right)^2+\left(\frac{\partial y}{\partial \beta}\right)^2+\left(\frac{\partial z}{\partial \beta}\right)^2\right]^{\frac{1}{2}} \end{aligned} \tag{7.3}$$

on middle surface ($\gamma = 0$). The strains at any point in the shells can then be written in terms of middle surface strains and curvature changes as [384, 385]

$$e_1 = e_1^0 + \gamma\chi_1, \quad e_2 = e_2^0 + \gamma\chi_2, \quad e_{12} = e_{12}^0 + \gamma\chi_{12} \tag{7.4}$$

$$\begin{aligned} e_{23} &= \frac{1}{H_2}\left(B\psi_\beta + \frac{\partial w}{\partial \beta} - B\frac{v}{R_2}\right) \\ e_{31} &= \frac{1}{H_1}\left(A\psi_\alpha + \frac{\partial w}{\partial \alpha} - A\frac{u}{R_1}\right) \end{aligned} \tag{7.5}$$

where

$$\left.\begin{aligned} e_1^0 &= \frac{1}{H_1}\left(\frac{\partial u}{\partial \alpha} + \frac{v}{H_2}\frac{\partial H_1}{\partial \beta} + A\frac{w}{R_1}\right) \\ e_2^0 &= \frac{1}{H_2}\left(\frac{\partial v}{\partial \beta} + \frac{u}{H_1}\frac{\partial H_2}{\partial \alpha} + B\frac{w}{R_2}\right) \\ e_{12}^0 &= \frac{H_2}{H_1}\frac{\partial}{\partial \alpha}\left(\frac{v}{H_2}\right) + \frac{H_1}{H_2}\frac{\partial}{\partial \beta}\left(\frac{u}{H_1}\right) \end{aligned}\right\} \tag{7.6}$$

$$\left.\begin{aligned} \chi_1 &= \frac{1}{H_1}\left(\frac{\partial \psi_\alpha}{\partial \alpha} + \frac{1}{H_2}\frac{\partial H_1}{\partial \beta}\psi_\beta\right) \\ \chi_2 &= \frac{1}{H_2}\left(\frac{\partial \psi_\beta}{\partial \beta} + \frac{1}{H_1}\frac{\partial H_2}{\partial \alpha}\psi_\alpha\right) \\ \chi_{12} &= \frac{H_2}{H_1}\frac{\partial}{\partial \alpha}\left(\frac{\psi_\beta}{H_2}\right) + \frac{H_1}{H_2}\frac{\partial}{\partial \beta}\left(\frac{\psi_\alpha}{H_1}\right) \end{aligned}\right\} \tag{7.7}$$

For Kirchhoff–Love thin shell theory, we have $e_{23} = e_{31} = 0$. This leads to the following results from Eq. (7.5):

$$\begin{aligned} \psi_\alpha &= \frac{u}{R_1} - \frac{1}{A}\frac{\partial w}{\partial \alpha} \\ \psi_\beta &= \frac{v}{R_2} - \frac{1}{B}\frac{\partial w}{\partial \beta} \end{aligned} \tag{7.8}$$

where $R_1 = 1/k_1$ and $R_2 = 1/k_2$ are radii of curvatures. Considering the following facts:

$$\frac{\gamma}{R_1} \ll 1, \quad \frac{\gamma}{R_2} \ll 1 \tag{7.9}$$

Replacing H_1 and H_2 respectively by A and B, we have

$$\left.\begin{aligned} e_1^0 &= \frac{1}{A}\frac{\partial u}{\partial \alpha} + \frac{v}{AB}\frac{\partial A}{\partial \beta} + \frac{w}{R_1} \\ e_2^0 &= \frac{1}{B}\frac{\partial v}{\partial \beta} + \frac{u}{AB}\frac{\partial B}{\partial \alpha} + \frac{w}{R_2} \\ e_{12}^0 &= \frac{B}{A}\frac{\partial}{\partial \alpha}\left(\frac{v}{B}\right) + \frac{A}{B}\frac{\partial}{\partial \beta}\left(\frac{u}{A}\right) \end{aligned}\right\} \tag{7.10}$$

$$\left.\begin{aligned} \chi_1 &= \frac{1}{A}\left(\frac{\partial \psi_\alpha}{\partial \alpha} + \frac{1}{B}\frac{\partial A}{\partial \beta}\psi_\beta\right) \\ \chi_2 &= \frac{1}{B}\left(\frac{\partial \psi_\beta}{\partial \beta} + \frac{1}{A}\frac{\partial B}{\partial \alpha}\psi_\alpha\right) \\ \chi_{12} &= \frac{B}{A}\frac{\partial}{\partial \alpha}\left(\frac{\psi_\beta}{B}\right) + \frac{A}{B}\frac{\partial}{\partial \beta}\left(\frac{\psi_\alpha}{A}\right) \end{aligned}\right\} \tag{7.11}$$

If we further take

$$\psi_\alpha = -\frac{1}{A}\frac{\partial w}{\partial \alpha}, \quad \psi_\beta = -\frac{1}{B}\frac{\partial w}{\partial \beta} \tag{7.12}$$

we get the Donnell thin shell theory or the engineering shell theory.

The stress–strain relations of isotropic materials are

$$\begin{gathered} \sigma_1 = \frac{E}{1-\upsilon^2}(e_1 + \upsilon e_2), \quad \sigma_2 = \frac{E}{1-\upsilon^2}(e_2 + \upsilon e_1) \\ \tau_{12} = Ge_{12} \end{gathered} \tag{7.13}$$

$$\tau_{13} = Ge_{13}, \quad \tau_{23} = Ge_{23} \tag{7.14}$$

Substitution of Eq. (7.4) into Eq. (7.13) leads to

$$\begin{aligned} \sigma_1 &= \frac{E}{1-\upsilon^2}[(e_1^0 + \upsilon e_2^0) + \gamma(\chi_1 + \upsilon\chi_2)] \\ \sigma_2 &= \frac{E}{1-\upsilon^2}[(e_2^0 + \upsilon e_1^0) + \gamma(\chi_2 + \upsilon\chi_1)] \\ \tau_{12} &= \frac{E}{2(1+\upsilon)}(e_{12}^0 + \gamma\chi_{12}) \end{aligned} \tag{7.15}$$

The moments M_1, M_2, M_{21} and M_{12}, shear forces Q_1 and Q_2, and membrane forces N_1, N_2, N_{21} and N_{12} can be expressed by stresses as

$$
\begin{aligned}
\begin{bmatrix} N_1 \\ N_{12} \\ Q_1 \end{bmatrix} &= \int_{-h/2}^{h/2} \begin{bmatrix} \sigma_1 \\ \tau_{12} \\ \tau_{13} \end{bmatrix} (1+\gamma/R_2)\mathrm{d}\gamma \\
\begin{bmatrix} N_2 \\ N_{21} \\ Q_2 \end{bmatrix} &= \int_{-h/2}^{h/2} \begin{bmatrix} \sigma_2 \\ \tau_{12} \\ \tau_{23} \end{bmatrix} (1+\gamma/R_1)\mathrm{d}\gamma
\end{aligned} \tag{7.16}
$$

$$
\begin{aligned}
\begin{bmatrix} M_1 \\ M_{12} \end{bmatrix} &= \int_{-h/2}^{h/2} \begin{bmatrix} \sigma_1 \\ \tau_{12} \end{bmatrix} (1+\gamma/R_2)\gamma\mathrm{d}\gamma \\
\begin{bmatrix} M_2 \\ M_{21} \end{bmatrix} &= \int_{-h/2}^{h/2} \begin{bmatrix} \sigma_2 \\ \tau_{12} \end{bmatrix} (1+\gamma/R_1)\gamma\mathrm{d}\gamma
\end{aligned} \tag{7.17}
$$

These forces and moments are shown in Figs. 7.1 and 7.2.

Substituting Eqs. (7.15) and (7.14) into Eqs. (7.16) and (7.17) leads to

$$
\left.
\begin{aligned}
N_1 &= \frac{Eh}{1-\upsilon^2}\left[(e_1^0+\upsilon e_2^0)+\frac{h^2}{12R_2}(\chi_1+\upsilon\chi_2)\right] \\
N_{12} &= \frac{Eh}{2(1+\upsilon)}\left(e_{12}^0+\frac{h^2}{12R_2}\chi_{12}\right) \\
N_2 &= \frac{Eh}{1-\upsilon^2}\left[(e_2^0+\upsilon e_1^0)+\frac{h^2}{12R_1}(\chi_2+\upsilon\chi_1)\right] \\
N_{21} &= \frac{Eh}{2(1+\upsilon)}\left(e_{12}^0+\frac{h^2}{12R_1}\chi_{12}\right)
\end{aligned}
\right\} \tag{7.18}
$$

$$
\left.
\begin{aligned}
M_1 &= \frac{Eh^3}{12(1-\upsilon^2)}\left[(\chi_1+\upsilon\chi_2)+\frac{1}{R_2}(e_1^0+\upsilon e_2^0)\right] \\
M_{12} &= \frac{Eh^3}{24(1+\upsilon)}\left(\chi_{12}+\frac{1}{R_2}e_{12}^0\right) \\
M_2 &= \frac{Eh^3}{12(1-\upsilon^2)}\left[(\chi_2+\upsilon\chi_1)+\frac{1}{R_1}(e_2^0+\upsilon e_1^0)\right] \\
M_{21} &= \frac{Eh^3}{24(1+\upsilon)}\left(\chi_{12}+\frac{1}{R_1 e_{12}^0}\right)
\end{aligned}
\right\} \tag{7.19}
$$

and

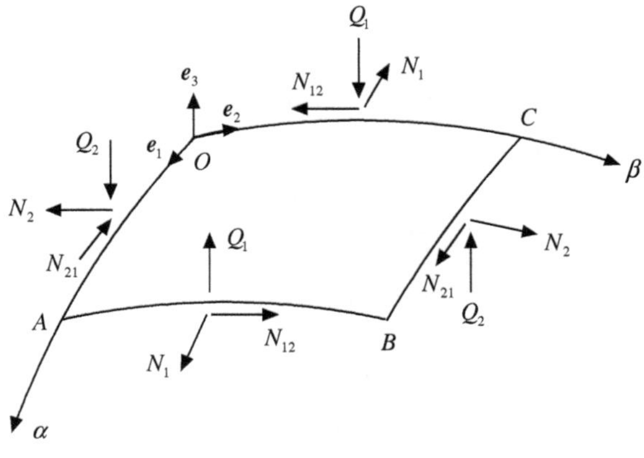

Fig. 7.1. Inner forces of a shell.

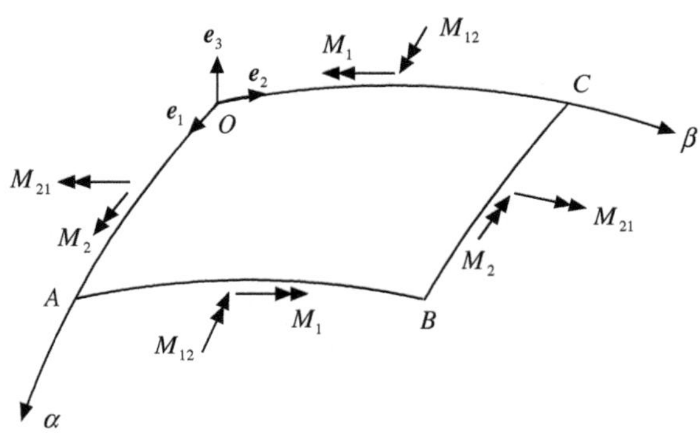

Fig. 7.2. Inner moments of a shell.

$$Q_1 = \frac{Eh}{2(1+\upsilon)} e_{13}, \quad Q_2 = \frac{Eh}{2(1+\upsilon)} e_{23} \tag{7.20}$$

For Kirchhoff–Love and Donnell thin shell theories, shear forces $Q1$ and $Q2$ are calculated from equilibrium equations, and the terms $1+\gamma/R_1$ and $1+\gamma/R_2$ should be omitted, so we have

$$\begin{aligned} N_1 &= C(e_1^0 + \upsilon e_2^0), & N_{12} &= \upsilon_1 C e_{12}^0 \\ N_2 &= C(e_2^0 + \upsilon e_1^0), & N_{21} &= \upsilon_1 C e_{12}^0 \\ M_1 &= D(\chi_1 + \upsilon \chi_2), & M_{12} &= \upsilon_1 D \chi_{12} \\ M_2 &= D(\chi_2 + \upsilon \chi_1), & M_{21} &= \upsilon_1 D \chi_{12} \end{aligned} \tag{7.21}$$

where

$$C = \frac{Eh}{1-\upsilon^2}, \quad D = \frac{Eh^3}{12(1-\upsilon^2)}, \quad \upsilon_1 = \frac{1-\upsilon}{2} \tag{7.22}$$

The strain energy stored in the element is given by

$$U = \frac{1}{2}\int_\alpha\int_\beta\int_{-h/2}^{+h/2} \{\sigma_1 e_1 + \sigma_2 e_2 + \tau_{12}e_{12} + \tau_{13}e_{13} + \tau_{23}e_{23}\} \times AB\mathrm{d}\alpha\mathrm{d}\beta\mathrm{d}\gamma \tag{7.23}$$

which can be expressed by middle surface variables as

$$U = \frac{1}{2}\int_\alpha\int_\beta \{N_1 e_1^0 + N_2 e_2^0 + N_{12}e_{12}^0 + M_1\chi_1 + M_2\chi_2 + M_{12}\chi_{12} + M_{21}\chi_{21} + Q_1 e_{13}^0 + Q_2 e_{23}^0\} AB\mathrm{d}\alpha\mathrm{d}\beta \tag{7.24}$$

The kinetic energy of the shell is given by

$$\begin{aligned} T &= \frac{1}{2}\int_V \{\dot{u}_1^2 + \dot{u}_2^2 + \dot{u}_3^2\} dV \\ &= \frac{1}{2}\int_\alpha\int_\beta\int_{-h/2}^{+h/2} (\dot{u}^2 + \dot{v}^2 + \dot{w}^2 + \gamma^2\dot{\psi}_\alpha^2 + \gamma^2\dot{\psi}_\beta^2 \\ &\quad + 2\gamma\dot{u}\dot{\psi}_\alpha + 2\gamma\dot{v}\dot{\psi}_\beta) \\ &\quad \times \left(1 + \frac{\gamma}{R_1}\right)\left(1 + \frac{\gamma}{R_2}\right) AB\mathrm{d}\alpha\mathrm{d}\beta\mathrm{d}\gamma \end{aligned} \tag{7.25}$$

The virtual work is given by

$$\delta W = \int_\alpha\int_\beta \{q_1\delta u + q_2\delta v + q_3\delta w + m_1\delta\psi_\alpha + m_2\delta\psi_\beta\} AB\mathrm{d}\alpha\mathrm{d}\beta \tag{7.26}$$

where m_1 and m_2 are distributed momentums.

The equilibrium equations for a shell is given by

$$\left.\begin{aligned}
&\frac{\partial(BN_1)}{\partial\alpha}+\frac{\partial(AN_{21})}{\partial\beta}+N_{12}\frac{\partial A}{\partial\beta}-N_2\frac{\partial B}{\partial\alpha}\\
&\quad+AB\frac{Q_1}{R_1}+ABq_1=AB\rho h\frac{\partial^2 u}{\partial t^2}\\
&\frac{\partial(BN_{12})}{\partial\alpha}+\frac{\partial(AN_2)}{\partial\beta}+N_{21}\frac{\partial B}{\partial\alpha}-N_1\frac{\partial A}{\partial\beta}\\
&\quad+AB\frac{Q_2}{R_2}+ABq_2=AB\rho h\frac{\partial^2 v}{\partial t^2}\\
&\frac{\partial(BQ_1)}{\partial\alpha}+\frac{\partial(AQ_2)}{\partial\beta}-AB\left(\frac{N_1}{R_1}+\frac{N_2}{R_2}\right)\\
&\quad+ABq_3=AB\rho h\frac{\partial^2 w}{\partial t^2}
\end{aligned}\right\}\tag{7.27}$$

$$\left.\begin{aligned}
&\frac{\partial(BM_1)}{\partial\alpha}+\frac{\partial(AM_{21})}{\partial\beta}+M_{12}\frac{\partial A}{\partial\beta}-M_2\frac{\partial B}{\partial\alpha}\\
&\quad-ABQ_1+ABm_1=AB\rho\frac{h^3}{12}\frac{\partial^2\psi_\alpha}{\partial t^2}\\
&\frac{\partial(BM_{12})}{\partial\alpha}+\frac{\partial(AM_2)}{\partial\beta}+M_{21}\frac{B}{\partial\alpha}-M_1\frac{\partial A}{\partial\beta}\\
&\quad-ABQ_2+ABm_2=AB\rho\frac{h^3}{12}\frac{\partial^2\psi_\beta}{\partial t^2}
\end{aligned}\right\}\tag{7.28}$$

7.2 The Configuration of Double-Curved Sandwich Shell

The central plane configuration of double-curved shell studied in this paper is shown in Fig. 7.3. The projection of the central plane in the Cartesian coordinate system is a rectangle, its length and width are respectively a and b, and the thickness is h. Three coordinate systems, namely, plane coordinate system $x-y-z$ and curved coordinate system $\alpha-\beta-z$ and polar coordinate system $\phi_x-\phi_y-z$ are shown in Fig. 7.3. The relations between these coordinate systems are as follows:

$$\begin{aligned}
&x=R_x\sin\phi_x,\quad y=R_y\sin\phi_y\\
&z=R_x(\cos\phi_x-1)+R_y(\cos\phi_y-1)
\end{aligned}\tag{7.29}$$

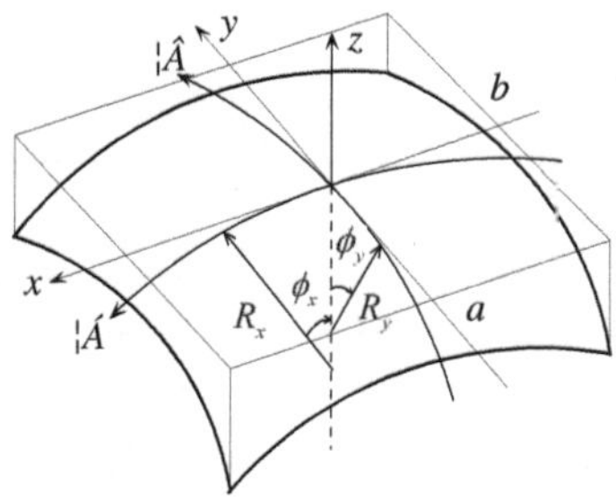

Fig. 7.3. A double-curved shell with rectangular planform.

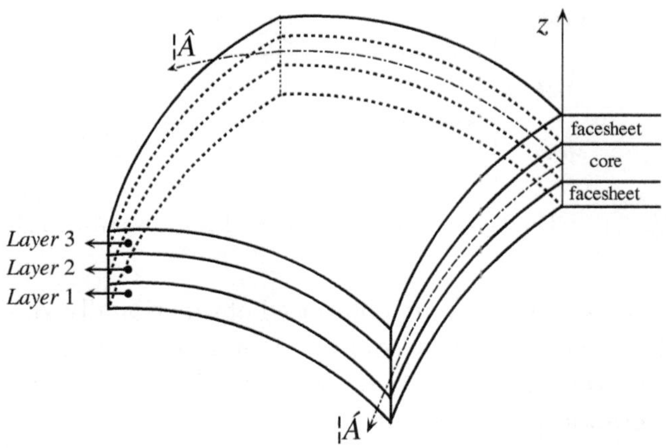

Fig. 7.4. The layered configuration of the sandwich shell.

$$\begin{aligned} \alpha &= R_x \phi_x \\ \beta &= R_y \phi_y \end{aligned} \tag{7.30}$$

We can find that the middle surface of the shell is simplified to a sphere $x^2 + y^2 + (z + R)^2 = R^2$ when $R_x = R_y = R$.

The model studied in this work is a sandwich shell, as shown in Fig. 7.4. It is composed of three layers with different material properties. The third and first layers are respectively called upper and lower facesheet layers, and the second layer is called the core layer. The above geometric parameter relations are defined in Fig. 7.3.

Two types of functionally graded sandwich shells are studied in this work: Type A and Type B configurations. The upper and lower

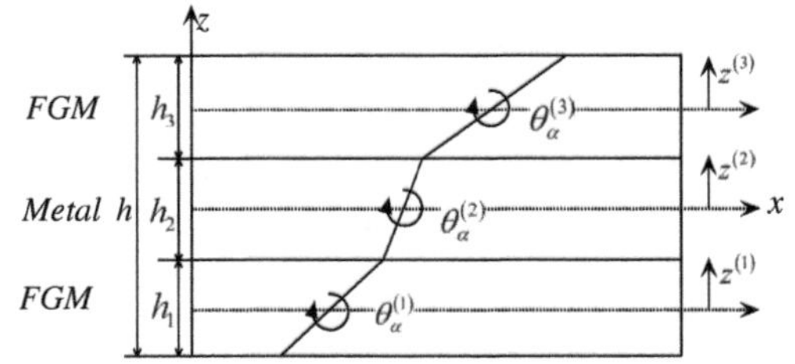

Fig. 7.5. The layered configuration of Type A sandwich shell.

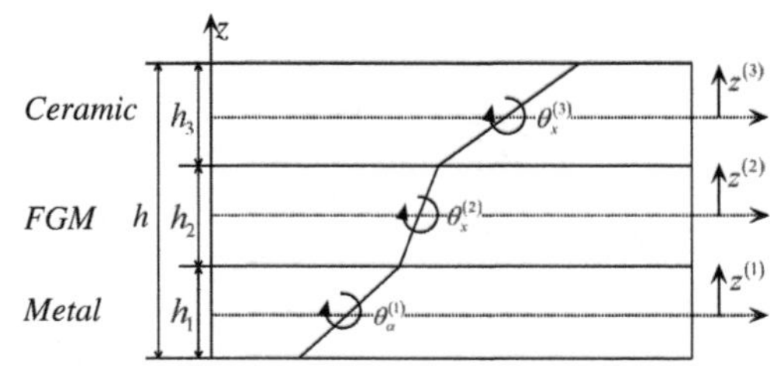

Fig. 7.6. The layered configuration of Type B sandwich shell.

facesheet layers of Type A are composed of functionally graded materials, and the core layer is composed of pure metal, as shown in Fig. 7.5. For the functionally graded material of the lower layer, it has pure ceramic properties at $z^{(1)} = -h_1/2$ and pure metal properties at $z^{(1)} = h_1/2$. For the functionally graded material of the upper layer, it has pure metal properties at $z^{(3)} = -h_3/2$ and pure ceramic properties at $z^{(3)} = h_3/2$. The upper facesheet layer of Type B is composed of pure ceramic material, the lower facesheet layer is composed of pure metal, and the core layer is functionally graded material, as shown in Fig. 7.6. The material properties of the whole shell are continuous. That is to say, for the core layer, it has the properties of pure metal at $z^{(2)} = -h_2/2$ and the properties of pure ceramic at $z^{(2)} = h_2/2$.

7.3 Estimation of Material Properties

We know that functionally graded materials are composed of two kinds of material compositions which are generally metal and ceramic. Functionally graded shells form a shell structure in which the physical properties change continuously in the direction of thickness. Estimation of the effective physical attributes is extremely important for knowing the mechanical properties of functionally graded shells. There are generally two kinds of models in the estimation of material properties. One of the most simple estimation models is the ROM model. The other one is the Mori–Tanaka model. The Mori–Tanaka model is more effective for the estimation of functionally graded materials with discontinuous particulate phases. Both models are used in the two functionally graded sandwich shell configurations of this work. The volume fraction $\lambda^{(i)}$ of materials for each shell is defined as follows:

$$\lambda^{(i)} = \left(\frac{z_i + h_i/2}{h_i}\right)^n z_i \in \left\{\frac{-h_i}{2}, \frac{h_i}{2}\right\} \tag{7.31}$$

in which the index i is the number of layers.

7.3.1 *ROM model*

According to the ROM model, physical properties such as elastic modulus E, Poisson's ratio v, shear modulus G, thermal expansion coefficient α, thermal conductivity K and density ρ can be estimated according to the following formula:

$$P_{\mathit{eff}} = P_1 + (P_2 - P_1)\lambda \tag{7.32}$$

where P_1 and P_2 are respectively the physical properties of component 1 and component 2. P can represent any one of the above physical properties. It deserves to be noted that λ is the volume fraction of component 2 according to the above expression. If one wants to represent the volume fraction of component 1, one just needs to exchange the subscripts 1 and 2. P_{eff} is the effective physical property.

Similarly, the effective material properties of each layer of Type A shell can be estimated as follows from the above definition:

$$\begin{aligned} P_{\mathit{eff}}^{(1)} &= P_c + (P_m - P_c)\lambda^{(1)} \\ P_{\mathit{eff}}^{(2)} &= P_m + (P_m - P_m)\lambda^{(2)} \\ P_{\mathit{eff}}^{(3)} &= P_m + (P_c - P_m)\lambda^{(3)} \end{aligned} \tag{7.33}$$

It can be observed that $\lambda^{(i)}$ is the volume fraction of metal for layers 1 and 2, and the volume fraction of ceramic for layer 3. There is a similar expression for type B.

$$\begin{aligned} P_{eff}^{(1)} &= P_m + (P_m - P_m)\lambda^{(1)} \\ P_{eff}^{(2)} &= P_m + (P_c - P_m)\lambda^{(2)} \\ P_{eff}^{(3)} &= P_c + (P_c - P_c)\lambda^{(3)} \end{aligned} \tag{7.34}$$

where $\lambda^{(1)}$ is the volume fraction of the metal, and $\lambda^{(2)}$ and $\lambda^{(3)}$ are the volume fractions of ceramic. In fact, for the second layers of Type A shell, as well as the first and third layers of Type B shell, they are not functionally graded materials and can be written directly as $P_{eff-A}^{(2)} = P_m$, $P_{eff-B}^{(1)} = P_m$, $P_{eff-B}^{(3)} = P_c$. But we still express them as Eqs. (7.33) and (7.34) because Type A and Type B shells can be obtained by changing the subscript of Eqs. (7.33) and (7.34), which is convenient for programming.

7.3.2 *Mori–Tanaka model*

It is assumed that the two constituents of functionally graded materials are respectively material 1 and material 2. According to the Mori–Tanaka model [386], the effective volume modulus B and effective shear modulus G can be expressed as

$$\frac{B - B_2}{B_1 - B_2} = \frac{\lambda}{1 + [1 - \lambda]\frac{B_1 - B_2}{B_2 + 4/3G_2}} \tag{7.35}$$

$$\frac{G - G_2}{G_1 - G_2} = \frac{\lambda}{1 + [1 - \lambda]\frac{G_1 - G_2}{G_2 + f}} \tag{7.36}$$

where

$$f = \frac{G_2(9B_2 + 8G_2)}{6(B_2 + 2G_2)} \tag{7.37}$$

The effective elastic modulus E and Poisson's ratio υ of the functionally graded shell are obtained according to the following formula:

$$E = \frac{9BG}{3B + G} \tag{7.38}$$

$$\upsilon = \frac{0.5(3B - 2G)}{3B + G} \tag{7.39}$$

Because the Mori–Tanaka model does not provide specific estimation formula for thermal expansion coefficient α, thermal conductivity coefficient K and density ρ, their estimation is still based on the ROM model. When we consider the effective material properties of each layer of Type A and Type B shells, the component subscripts 1 and 2 of Eqs. (7.35) to (7.39) are replaced by specific material properties that are the same as Eqs. (7.33)–(7.34). The specific expressions are as follows:

The first layer of Type A: replace components 1 and 2 respectively with ceramic and metal.

The second layer of Type A: replace components 1 and 2 with metal.

The third layer of Type A: replace components 1 and 2 respectively with metal and ceramic.

The first layer of Type B: replace components 1 and 2 with metal.

The second layer of Type B: replace components 1 and 2 respectively with metal and ceramic.

The third layer of Type B: replace components 1 and 2 with ceramic.

7.4 Solution of Temperature Field

Generally speaking, temperature has influence on the physical properties of any material. In the case of a small change in temperature, we generally assume that the physical properties do not change with temperature. However, functionally graded shells often need to work in high-temperature environment. Therefore, we need to consider their material property variation with temperature. Different materials and different properties have different sensitivities to temperature. In this paper, we assume that only the elastic modulus E, Poisson's ratio v and the thermal expansion coefficient α change with respect to temperature, while thermal conductivity K and density ρ do not change with temperature. For these three special physical properties, the literature [387] gives an approximate function model of nonlinear change with temperature:

$$P_{eff} = P_0(P_{-1}/T + 1 + P_1 T + P_2 T^2 + P_3 T^3) \qquad (7.40)$$

where P_i ($i = -1, 0, 1, 2$ and 3) is the temperature coefficient of a certain physical property of material. It is unique for each physical property. The above temperature coefficients are obtained for pure materials through experiments. The material properties of the pure components need to be obtained first, then the component properties are used in ROM or MT models to estimate the effective properties of FGM. According to Eq. (7.40), it is also necessary to know the temperature field distribution in the structure. We know that when the thermal conductivity of a material is a constant, a linear temperature field is obtained from the one-dimensional steady-state equation of heat conduction. However, the thermal conductivity is not a constant but a function of the thickness coordinate z of the functionally graded shell. The temperature field satisfies the following one-dimensional heat conduction equation:

$$-\frac{\mathrm{d}}{\mathrm{d}z_i}\left[K^{(i)}(z_i)\frac{\mathrm{d}T^{(i)}}{\mathrm{d}z_i}\right] = 0 \tag{7.41}$$

For Type A sandwich shell, it is not only necessary to satisfy the temperature boundary conditions of the upper and lower boundaries as follows:

$$\begin{aligned} T^{(1)}\left(-\frac{h_1}{2}\right) &= T_b \\ T^{(3)}\left(\frac{h_3}{2}\right) &= T_t \end{aligned} \tag{7.42}$$

but also required to satisfy the temperature continuity and heat flow continuity conditions as follows for the two internal interfaces:

$$\begin{aligned} T^{(2)}\left(-\frac{h_2}{2}\right) &= T^{(1)}\left(\frac{h_1}{2}\right) \\ T^{(2)}\left(\frac{h_2}{2}\right) &= T^{(3)}\left(-\frac{h_3}{2}\right) \end{aligned} \tag{7.43}$$

According to the ROM model, the thermal conductivity $K^{(i)}(z)$ of each layer should satisfy

$$\begin{aligned} K^{(1)}(z_1) &= K_c + (K_m - K_c)\lambda^{(1)}(z_1) \\ K^{(2)}(z_2) &= K_m \\ K^{(3)}(z_3) &= K_c + (K_m - K_c)\lambda^{(3)}(z_3) \end{aligned} \tag{7.44}$$

with $\lambda^{(i)}(z^{(i)})$ $(i = 1, 3)$ defined by Eq. (7.31)

$$\begin{aligned} K^{(2)}(z_2)\frac{dT^{(2)}}{dz_2}\bigg|_{z_2=-\frac{h_2}{2}} &= K^{(1)}\frac{dT^{(1)}}{dz_1}\bigg|_{z_1=\frac{h_1}{2}} \\ K^{(2)}(z_2)\frac{dT^{(2)}}{dz_2}\bigg|_{z_2=\frac{h_2}{2}} &= K^{(3)}\frac{dT^{(3)}}{dz_3}\bigg|_{z_3=-\frac{h_3}{2}} \end{aligned} \tag{7.45}$$

In this work, the numerical method is not used to solve the above nonlinear heat conduction differential equation. The approximate theoretical solution is given by [388]

$$\begin{aligned} T^{(1)}(z_1) &= T_b + (T_{m1} - T_b)\eta^{(1)}(z_1) \\ T^{(2)}(z_2) &= T_{m1} + (T_{m2} - T_{m1})\frac{z_2}{h_2} \\ T^{(3)}(z_1) &= T_{m2} + (T_t - T_{m2})\eta^{(3)}(z_3) \end{aligned} \tag{7.46}$$

where T_{m1} is the interfacial temperature of the first and second layers, T_{m2} is the interfacial temperature of the second and third layers, $\eta^{(1)}$ and $\eta^{(3)}$ are calculated according to Eqs. (7.47)–(7.49).

$$\begin{aligned} \eta^{(1)}(z_1) = \frac{1}{C}\Bigg[&\left(\frac{z_1 + h_1/2}{h_1}\right) - \frac{Kk_{mc}}{(n-1)K_c}\left(\frac{z_1 + h_1/2}{h_1}\right)^{n+1} \\ &+ \frac{K_{mc}^2}{(2N+1)K_c^2}\left(\frac{z_1 + h_1/2}{h_1}\right)^{2n+1} \\ &- \frac{K_{mc}^3}{(3N+1)K_c^3}\left(\frac{z_1 + h_1/2}{h_1}\right)^{3n+1} \\ &+ \frac{K_{mc}^4}{(4N+1)K_c^4}\left(\frac{z_1 + h_1/2}{h_1}\right)^{4n+1} \\ &- \frac{K_{mc}^5}{(5N+1)K_c^5}\left(\frac{z_1 + h_1/2}{h_1}\right)^{5n+1}\Bigg] \end{aligned} \tag{7.47}$$

$$\eta^{(3)}(z_3) = \frac{1}{C}\left[\left(\frac{h_3/2 - z_3}{h_3}\right) - \frac{Kk_{mc}}{(n-1)K_c}\left(\frac{h_3/2 - z_3}{h_3}\right)^{n+1}\right.$$
$$+\frac{K_{mc}^2}{(2N+1)K_c^2}\left(\frac{h_3/2 - z_3}{h_3}\right)^{2n+1}$$
$$-\frac{K_{mc}^3}{(3N+1)K_c^3}\left(\frac{h_3/2 - z_3}{h_3}\right)^{3n+1}$$
$$+\frac{K_{mc}^4}{(4N+1)K_c^4}\left(\frac{h_3/2 - z_3}{h_3}\right)^{4n+1}$$
$$\left.-\frac{K_{mc}^5}{(5N+1)K_c^5}\left(\frac{h_3/2 - z_3}{h_3}\right)^{5n+1}\right] \tag{7.48}$$

where

$$C = 1 - \frac{K_{mc}}{(N+1)K_c} + \frac{K_{mc}^2}{(2N+1)K_c^2} - \frac{K_{mc}^3}{(3N+1)K_c^3}$$
$$+ \frac{K_{mc}^4}{(4N+1)K_c^4} - \frac{K_{mc}^5}{(5N+1)K_c^5} \tag{7.49}$$
$$G = 1 - \frac{K_{mc}}{K_c} + \frac{K_{mc}^2}{K_c^2} - \frac{K_{mc}^3}{K_c^3} + \frac{K_{mc}^4}{K_c^4} - \frac{K_{mc}^5}{K_c^5}$$
$$K_{mc} = K_m - K_c$$

The two internal interface temperatures are calculated according to Eq. (7.50)

$$T_{m1} = \frac{\left(\frac{K_m}{h_2} + \frac{K_m G}{h_1 C}\right)T_t + \frac{K_m}{h_2}T_b}{\frac{K_m G}{h_1 C} + \frac{2K_m}{h_2}}$$
$$T_{m2} = \frac{\left(\frac{K_m}{h_2} + \frac{K_m G}{h_3 C}\right)T_b + \frac{K_m}{h_2}T_t}{\frac{K_m G}{h_3 C} + \frac{2K_m}{h_2}} \tag{7.50}$$

For Type B sandwich shell, it is assumed that the temperature fields in the first and third layers are constant. The calculation of the temperature field in the second layer is similar to that in Eq. (7.46).

The specific form is as follows:

$$\begin{aligned} T^{(1)}(z_1) &= T_b \\ T^{(2)}(z_2) &= T_b + (T_t - T_b)\eta^{(2)}(z_2) \\ T^{(3)}(z_3) &= T_t \end{aligned} \tag{7.51}$$

where the calculation method of $\eta^{(2)}(z_2)$is similar to Eq. (7.47). As long as h_2 is replaced by h_1, the positions of subscripts m and c (metal and ceramic) are exchanged.

7.5 Layerwise Theory of Functionally Graded Shells

In this chapter, the layerwise shell theory based on the first-order shear deformation theory [389] will be employed. The theory requires that interfaces between shells satisfy displacement continuity. According to the first-order shear deformation theory, the displacement of the ith layer can be expressed as

$$\begin{aligned} u^{(i)}(\alpha,\beta,z) &= u_0^{(i)}(\alpha,\beta) + z^{(i)}\theta_\alpha^{(i)} \\ v^{(i)}(\alpha,\beta,z) &= v_0^{(i)}(\alpha,\beta) + z^{(i)}\theta_\beta^{(i)} \\ w^{(i)}(\alpha,\beta,z) &= w(\alpha,\beta) \end{aligned} \tag{7.52}$$

where u_0 and v_0 are the displacement in the middle of the shell, and θ_α and θ_β are respectively the rotation angles along α and β directions. According to the displacement continuity requirements between interfaces, the mid-plane displacements of the ith layer and the $(i+1)$th layer satisfy the following relation:

$$\begin{aligned} u_0^{(i+1)}(\alpha,\beta) &= u_0^{(i)}(\alpha,\beta) + \frac{h_i}{2}\theta_\alpha^{(i)} + \frac{h_{i+1}}{2}\theta_\alpha^{(i+1)} \\ v_0^{(i+1)}(\alpha,\beta) &= v_0^{(i)}(\alpha,\beta) + \frac{h_i}{2}\theta_\beta^{(i)} + \frac{h_{i+1}}{2}\theta_\beta^{(i+1)} \end{aligned} \tag{7.53}$$

In the hierarchical finite element method [390], each layer of sandwich shell can be taken as an element. Then the assembly [390] of elements of adjacent layers can be carried out according to the displacement continuity condition given by the iterative formula of Eq. (7.53). By using the p-version method, we only need to calculate the stiffness matrix of three elements corresponding to the layers. This greatly reduces the computational complexity.

7.5.1 *Linear strain energy*

The linear strain–displacement relation is the same for each shell, so we ignore the superscript i. According to the literature [391], the linear strain–displacement relation of the shell can be written in the following form:

$$\begin{Bmatrix} \varepsilon_{\alpha\alpha} \\ \varepsilon_{\beta\beta} \\ \gamma_{\alpha\beta} \\ \gamma_{\beta z} \\ \gamma_{z\alpha} \end{Bmatrix} = \begin{bmatrix} D_\alpha & 0 & zD_\alpha & 0 & k_x \\ 0 & D_\beta & 0 & zD_\beta & k_y \\ (1+zC_0)D_\beta & (1-zC_0)D_\alpha & zD_\beta & zD_\alpha & 0 \\ 0 & -C_1k_y & 0 & 1 & D_\beta \\ -C_1k_x & 0 & 1 & 0 & D_\alpha \end{bmatrix} \begin{Bmatrix} u_0 \\ v_0 \\ \theta_\alpha \\ \theta_\beta \\ w \end{Bmatrix} \tag{7.54}$$

which can also be written as $\varepsilon_L = \boldsymbol{Bu}$. In Eq. (7.54), $D_\alpha = \partial/\partial\alpha$, $D_\beta = \partial/\partial\beta$, $k_x = 1/R_x$, $k_y = 1/R_y$ and $C_0 = 0.5C_2(k_x - k_y)$. When $k_x = k_y = 0$, the double-curved shell is reduced to a plate. C_1 and C_2 are the coefficients used to distinguish several shell theories. Different values of them correspond to different shell theories. When $C_1 = C_2 = 1$, it is Sander's shell theory. When $C_1 = 1$ and $C_2 = 0$, it is Love's shell theory. When $C_1 = C_2 = 0$, it is Donnell's shell theory.

The stress–strain relationship is also linear. The thermal strain needs to be introduced due to the temperature change as follows:

$$\begin{Bmatrix} \sigma_{11} \\ \sigma_{22} \\ \tau_{12} \\ \tau_{23} \\ \tau_{31} \end{Bmatrix} = \begin{bmatrix} D_{11} & D_{12} & 0 & 0 & 0 \\ D_{12} & D_{22} & 0 & 0 & 0 \\ 0 & 0 & D_{66} & 0 & 0 \\ 0 & 0 & 0 & D_{44} & 0 \\ 0 & 0 & 0 & 0 & D_{55} \end{bmatrix} \times \left\{ \begin{pmatrix} \varepsilon_{11} \\ \varepsilon_{22} \\ \gamma_{12} \\ \gamma_{23} \\ \gamma_{31} \end{pmatrix} - \begin{pmatrix} \alpha\Delta T \\ \alpha\Delta T \\ 0 \\ 0 \\ 0 \end{pmatrix} \right\} \tag{7.55}$$

which can also be written as $\sigma = \boldsymbol{D}(\varepsilon_L - \varepsilon_0)$, where $\varepsilon_0 = (\alpha\Delta T, \alpha\Delta T, 0, 0, 0)^T$. The subscripts 1 and 2 in Eq. (7.55) respectively correspond to α and β coordinates in the plane. The subscript 3 corresponds to the normal direction of the α $-\beta$ tangent plane. For isotropic materials, D_{ij}, are defined as

$$\begin{aligned} D_{11} &= D_{22} = \frac{E(z)}{1 - \upsilon(z)^2}, \quad D_{12} = \upsilon(z) D_{11} \\ D_{66} &= D_{44} = D_{55} = G = \frac{E(z)}{2(1 + \upsilon(z))} \end{aligned} \tag{7.56}$$

Similar to the higher-order shear deformation theory, the lamination theory does not need shear correction factor. The linear strain energy is computed by

$$U_L = \int_V \varepsilon_L^T \bar{\sigma} dV = \int_V \boldsymbol{u}^T \boldsymbol{B}^T \boldsymbol{D} \boldsymbol{B} \boldsymbol{u} dV \tag{7.57}$$

It needs to be discretized to obtain the linear stiffness matrix. The nonlinear strain energy will be considered in the Section 7.5.2.

7.5.2 *Nonlinear strain energy*

In this paper, we use the nonlinear strain–displacement relation of Sander [392]. The nonlinear strain can be obtained as follows:

$$\begin{aligned} \varepsilon_{\alpha\alpha nl} &= \frac{1}{2}\left(\frac{\partial u_0}{\partial \alpha}\right)^2 + \frac{1}{2}\left(\frac{\partial v_0}{\partial \alpha}\right)^2 + \frac{1}{2}\left(\frac{\partial w}{\partial \alpha} - k_x u_0\right)^2 \\ &\quad + \frac{1}{2} z^2 \left[\left(\frac{\partial \theta_\alpha}{\partial \alpha}\right)^2 + \left(\frac{\partial \theta_\beta}{\partial \alpha}\right)^2\right] \\ \varepsilon_{\beta\beta nl} &= \frac{1}{2}\left(\frac{\partial u_0}{\partial \beta}\right)^2 + \frac{1}{2}\left(\frac{\partial v_0}{\partial \beta}\right)^2 + \frac{1}{2}\left(\frac{\partial w}{\partial \beta} - k_y v_0\right)^2 \\ &\quad + \frac{1}{2} z^2 \left[\left(\frac{\partial \theta_\alpha}{\partial \beta}\right)^2 + \left(\frac{\partial \theta_\beta}{\partial \beta}\right)^2\right] \end{aligned}$$

$$\begin{aligned}\varepsilon_{\alpha\beta nl} = & \left(\frac{\partial u_0}{\partial \alpha}\right)\left(\frac{\partial u_0}{\partial \beta}\right) + \left(\frac{\partial v_0}{\partial \alpha}\right)\left(\frac{\partial v_0}{\partial \beta}\right) \\ & + \left(\frac{\partial w}{\partial \alpha} - k_x u_0\right)\left(\frac{\partial w}{\partial \beta} - k_y v_0\right) \\ & + z^2\left[\left(\frac{\partial \theta_\alpha}{\partial \alpha}\right)\left(\frac{\partial \theta_\alpha}{\partial \beta}\right) + \left(\frac{\partial \theta_\beta}{\partial \alpha}\right)\left(\frac{\partial \theta_\beta}{\partial \beta}\right)\right]\end{aligned} \tag{7.58}$$

The nonlinear strain energy is expressed as follows according to the nonlinear shell theory:

$$U_{NL} = \frac{1}{2}\int_V \boldsymbol{\varepsilon}_{nl}^T \boldsymbol{\sigma}_0 \mathrm{d}V \tag{7.59}$$

where $\boldsymbol{\varepsilon}_{nl} = [\varepsilon_{\alpha\alpha nl}, \varepsilon_{\beta\beta nl}, \varepsilon_{\alpha\beta nl}, 0, 0]^T$ is the nonlinear strain vector; and $\boldsymbol{\sigma} = [\sigma^0_{\alpha\alpha}, \sigma^0_{\beta\beta}, \sigma^0_{\alpha\beta}, \sigma^0_{\beta z}, \sigma^0_{\alpha z}]^T$ is the initial stress vector. In this paper, the initial thermal stresses caused by temperature will rise. They can be easily obtained according to the thermal elasticity theory. After solving the following governing equation to obtain the thermal deformation displacement field $\boldsymbol{u}_0$, the initial thermal stress σ_0 can be obtained according to Eq. (7.55):

$$\int_V \boldsymbol{B}^T \boldsymbol{D}\boldsymbol{B}\boldsymbol{u}_0 dV = \int_V \boldsymbol{B}^T \boldsymbol{D}\boldsymbol{\varepsilon}_0 dV \tag{7.60}$$

Equation (7.59) is expressed in matrix form to facilitate programming. The geometric stiffness matrix is derived by constructing the following vector:

$$\begin{aligned}\boldsymbol{g} = & \left[\frac{\mathrm{d}u_0}{\mathrm{d}\alpha}, \frac{\mathrm{d}u_0}{\mathrm{d}\beta}, \frac{\mathrm{d}v_0}{\mathrm{d}\alpha}, \frac{\mathrm{d}v_0}{\mathrm{d}\beta}, \frac{\mathrm{d}w}{\mathrm{d}\alpha} - \frac{u_0}{R_x}, \frac{\mathrm{d}w}{\mathrm{d}\beta}\right. \\ & \left. -\frac{v_0}{R_y}, \frac{\mathrm{d}\theta_\alpha}{\mathrm{d}\alpha}, \frac{\mathrm{d}\theta_\alpha}{\mathrm{d}\beta}, \frac{\mathrm{d}\theta_\beta}{\mathrm{d}\alpha}, \frac{\mathrm{d}\theta_\beta}{\mathrm{d}\beta}\right]^T\end{aligned} \tag{7.61}$$

Then the nonlinear strain $\boldsymbol{\varepsilon}_{nl}$ can be written in matrix form as

$$\boldsymbol{\varepsilon}_{nl} = \boldsymbol{E}\boldsymbol{g} \tag{7.62}$$

where

$$\boldsymbol{E}=\begin{bmatrix} \frac{1}{2}\frac{\partial u_0}{\partial \alpha} & 0 & \frac{1}{2}\frac{\partial v_0}{\partial \alpha} & 0 & \frac{1}{2}\left(\frac{\partial w}{\partial \alpha}-\frac{u_0}{R_x}\right) & 0 & \frac{1}{2}z^2\frac{\partial \theta_\alpha}{\partial \alpha} & 0 & \frac{1}{2}z^2\frac{\partial \theta_\beta}{\partial \alpha} & 0 \\ 0 & \frac{1}{2}\frac{\partial u_0}{\partial \beta} & 0 & \frac{1}{2}\frac{\partial v_0}{\partial \beta} & 0 & \frac{1}{2}\left(\frac{\partial w}{\partial \beta}-\frac{v_0}{R_y}\right) & 0 & \frac{1}{2}z^2\frac{\partial \theta_\alpha}{\partial \beta} & 0 & \frac{1}{2}z^2\frac{\partial \theta_\beta}{\partial \beta} \\ \frac{1}{2}\frac{\partial u_0}{\partial \beta} & \frac{1}{2}\frac{\partial u_0}{\partial \alpha} & \frac{1}{2}\frac{\partial v_0}{\partial \beta} & \frac{1}{2}\frac{\partial v_0}{\partial \alpha} & \frac{1}{2}\left(\frac{\partial w}{\partial \beta}-\frac{v_0}{R_y}\right) & \frac{1}{2}\left(\frac{\partial w}{\partial \alpha}-\frac{u_0}{R_x}\right) & \frac{1}{2}z^2\frac{\partial \theta_\alpha}{\partial \beta} & \frac{1}{2}z^2\frac{\partial \theta_\alpha}{\partial \beta} & \frac{1}{2}z^2\frac{\partial \theta_\beta}{\partial \beta} & \frac{1}{2}z^2\frac{\partial \theta_\beta}{\partial \alpha} \end{bmatrix} \tag{7.63}$$

The column vector $\boldsymbol{g}$ can be expressed by the differential operator as

$$\boldsymbol{g}=\boldsymbol{B}_g\boldsymbol{u} \tag{7.64}$$

where

$$\boldsymbol{B}_g=\begin{bmatrix} D_\alpha & 0 & 0 & 0 & 0 \\ D_\beta & 0 & 0 & 0 & 0 \\ 0 & D_\alpha & 0 & 0 & 0 \\ 0 & D_\beta & 0 & 0 & 0 \\ -k_x & 0 & 0 & 0 & 0 \\ 0 & -k_y & 0 & 0 & 0 \\ 0 & 0 & D_\alpha & 0 & 0 \\ 0 & 0 & D_\beta & 0 & 0 \\ 0 & 0 & 0 & D_\alpha & 0 \\ 0 & 0 & 0 & D_\beta & 0 \end{bmatrix} \tag{7.65}$$

Substituting Eq. (7.62) into Eq. (7.59), we get

$$U_{nl}=\frac{1}{2}\int_V \boldsymbol{g}^T(\boldsymbol{E}^T\boldsymbol{\sigma_0})\mathrm{d}V \tag{7.66}$$

Transform the above equation through

$$\boldsymbol{E}^T\boldsymbol{\sigma}_0=\boldsymbol{S}\boldsymbol{g} \tag{7.67}$$

where

$$\boldsymbol{S}=\begin{bmatrix} s_1 & \boldsymbol{o} & \boldsymbol{o} & \boldsymbol{o} & \boldsymbol{o}\\ \boldsymbol{o} & s_1 & \boldsymbol{o} & \boldsymbol{o} & \boldsymbol{o}\\ \boldsymbol{o} & \boldsymbol{o} & s_1 & \boldsymbol{o} & \boldsymbol{o}\\ \boldsymbol{o} & \boldsymbol{o} & \boldsymbol{o} & s_2 & \boldsymbol{o}\\ \boldsymbol{o} & \boldsymbol{o} & \boldsymbol{o} & \boldsymbol{o} & s_2 \end{bmatrix} \tag{7.68}$$

where $\boldsymbol{s}_1$ and $\boldsymbol{s}_2$ are defined as

$$\boldsymbol{s}_1=\begin{bmatrix} \frac{1}{2}\sigma^0_{\alpha\alpha} & \frac{1}{2}\sigma^0_{\alpha\beta}\\ \frac{1}{2}\sigma^0_{\alpha\beta} & \frac{1}{2}\sigma^0_{\beta\beta} \end{bmatrix} \quad \boldsymbol{s}_2=\begin{bmatrix} \frac{1}{2}z^2\sigma^0_{\alpha\alpha} & \frac{1}{2}z^2\sigma^0_{\alpha\beta}\\ \frac{1}{2}z^2\sigma^0_{\alpha\beta} & \frac{1}{2}z^2\sigma^0_{\beta\beta} \end{bmatrix} \tag{7.69}$$

Substituting Eq. (7.67) into Eq. (7.66) and combining with Eq. (7.64), we obtain

$$U_{NL}=\frac{1}{2}\int_V \boldsymbol{g}^T\boldsymbol{S}\boldsymbol{g}\mathrm{d}V=\frac{1}{2}\int_V \boldsymbol{u}^T\boldsymbol{B}_g^T\boldsymbol{S}\boldsymbol{B}_g\boldsymbol{u}\mathrm{d}V \tag{7.70}$$

This is the nonlinear strain energy. It will be discretized to obtain the nonlinear stiffness matrix.

7.5.3 *The kinetic energy*

In dynamic analysis, we also need to know the kinetic energy of structure:

$$T=\frac{1}{2}\int_V \rho[\dot{u}^2+\dot{v}^2+\dot{w}^2]\mathrm{d}V \tag{7.71}$$

According to Eq. (7.52), we have

$$\begin{bmatrix} \dot{u}\\ \dot{v}\\ \dot{w} \end{bmatrix}=\begin{bmatrix} 1 & 0 & z & 0 & 0\\ 0 & 1 & 0 & z & 0\\ 0 & 0 & 0 & 0 & 1 \end{bmatrix}\begin{bmatrix} \dot{u}_0\\ \dot{v}_0\\ \dot{\theta}_\alpha\\ \dot{\theta}_\beta\\ \dot{w} \end{bmatrix} \tag{7.72}$$

Equation (7.71) is then expressed in matrix form as

$$T=\int_V \rho\dot{\boldsymbol{u}}^T\boldsymbol{m}^T\boldsymbol{m}\dot{\boldsymbol{u}}dV \tag{7.73}$$

where $\dot{\boldsymbol{u}} = [\dot{u}_0, \dot{v}_0, \dot{\theta}_\alpha, \dot{\theta}_\beta, \dot{w}_0]^{\mathrm{T}}$ and

$$\boldsymbol{m} = \begin{bmatrix} 1 & 0 & z & 0 & 0 \\ 0 & 1 & 0 & z & 0 \\ 0 & 0 & 0 & 0 & 1 \end{bmatrix} \tag{7.74}$$

With the strain energy and kinetic energy, we can then derive the governing equations through the variational principle.

7.5.4 *The governing equation*

Finally, the free vibration equation of functionally graded shell is derived by the Hamiltonian variational principle [393] as follows:

$$\int_{t_1}^{t_2} \delta[T - U_L - U_{NL}]dt = 0 \tag{7.75}$$

where T, U_L and U_{NL} are respectively calculated according to the expressions in Eqs. (7.57), (7.70) and (7.73).

7.6 The Differential Quadrature Hierarchical Finite Element Method

7.6.1 *Approximation of the displacement field*

Now, we introduce the differential quadrature hierarchical finite element method (DQHFEM) that is an improved version of the hierarchical finite element method (HFEM). The sandwich shell model in this paper is a two-dimensional model. The displacement field is approximated by

$$w\left[\alpha(\xi,\eta),\beta(\xi,\eta)\right] \approx \sum_{k=1}^{K} S_k(\xi,\eta)w_k + \sum_{m=1}^{H_\xi}\sum_{n=1}^{H_\eta} H_{mn}(\xi,\eta)a_{mn} \tag{7.76}$$

First, we need to map the geometry domain $(\alpha, \beta) = [0, a] \times [0, b]$ to the parameter domain $(\xi, \eta) = [-1, 1] \times [-1, 1]$, as shown in Fig. 7.7. The mapping is simple but is necessary in order to use the standard integration points and weights. In Eq. (7.76), w_k is the

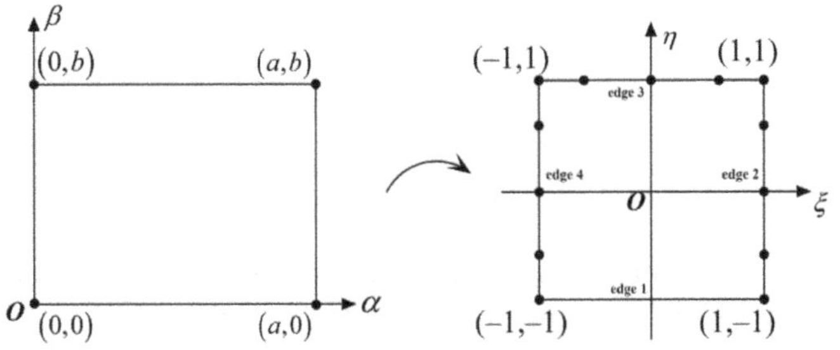

Fig. 7.7. The geometric mapping from a unit square domain to an arbitrary rectangular domain.

nodal deflection on the boundary, S_k is the corresponding boundary nodal shape functions (serendipity functions), and H_{mn} is the hierarchical shape functions. A brief introduction to the two types of bases are presented in the following sections.

7.6.1.1 *The boundary shape functions*

The boundary shape functions are the serendipity functions. The number of boundary nodes on the edge can be set freely. In order to avoid the Runge phenomenon, the non-uniform nodes are generally used. The Gauss–Lobatto nodes are used in the work. We assume the number of nodes on edges 1, 2, 3 and 4 are respectively M, N, M and N. The specific form of the shape function is as follows:

(a) Corner point shape function ($\xi = -1$, $\eta = -1$)

$$S_1 = \frac{1-\eta}{2} L_1^M(\xi) + \frac{1-\xi}{2} L_1^N(\eta) - \frac{(1-\xi)(1-\eta)}{4} \tag{7.77}$$

The shape functions for the rest of the corners are similar.

(b) Edge shape function (edge 1: $\eta = -1$)

$$S_i = \frac{1-\eta}{2} L_i^M(\xi), \quad i = 2, \ldots, M-1 \tag{7.78}$$

The shape functions for the remaining edges can be obtained in a similar way. The order of the edge shape functions can also be increased by increasing the number of nodes. They do not satisfy the hierarchical property but bring convenience in boundary condition imposition and element assembly.

7.6.1.2 The hierarchical shape functions

The construction of the hierarchical shape functions is important for the completeness of the bases and for achieving self-adaptive analysis. The hierarchical shape functions are expressed by tensor product of orthogonal polynomials whose boundary values are 0:

$$H_{mn}(\xi,\eta)=\varphi_m(\xi)\varphi_n(\eta) \tag{7.79}$$

where $\varphi_n(x)$ is the integral form of Legendre's orthogonal polynomials as follows:

$$\varphi_m(\xi)=\int_{-1}^{\xi}L_m(\xi)\mathrm{d}\xi=\frac{(\xi^2-1)}{m(m+1)}\frac{dL_m(\xi)}{\mathrm{d}\xi},\xi\in[-1,1],\quad m=1,2,\ldots \tag{7.80}$$

Figure 7.8 shows a corner shape function, an edge shape function and two hierarchical shape functions. It can be observed that the shape functions on the boundary satisfies the interpolation characteristics. The hierarchical shape functions vanish on the boundaries. Therefore, they do not affect the application of boundary conditions.

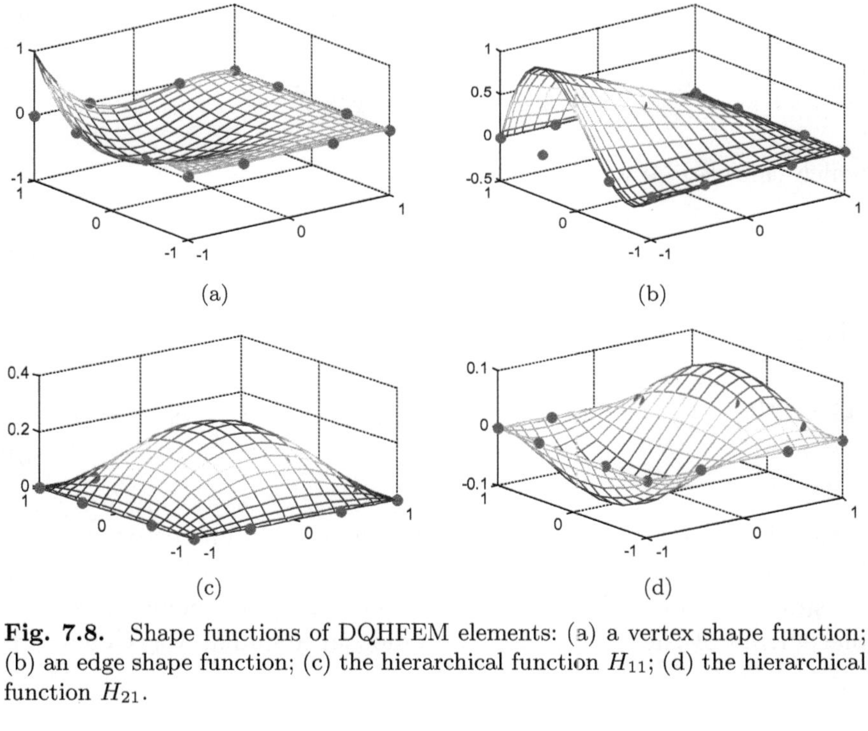

Fig. 7.8. Shape functions of DQHFEM elements: (a) a vertex shape function; (b) an edge shape function; (c) the hierarchical function H_{11}; (d) the hierarchical function H_{21}.

For convenience of forming element matrices for the computations in this work, we rewrite Eq. (48) in matrix form:

$$\mathbf{w} = \mathbf{N}^T \bar{\mathbf{w}} \tag{7.81}$$

where

$$\begin{aligned}\mathbf{N}^T &= [S_1, \ldots, S_K, H_{11}, \ldots, H_{1m}, H_{21}, \ldots, H_{mn}] \\ \bar{\mathbf{w}}^T &= [w_1, \ldots, w_K, a_{11}, \ldots, a_{1m}, a_{21}, \ldots, a_{mn}]\end{aligned} \tag{7.82}$$

In this chapter, vectors of displacements, strains or stresses are in italic bold form, as shown in Section 7.5. Their discretized forms are in bold form as in this section. Equation (7.81) is discretized by $N = pq$ integration points as

$$\begin{aligned}&\mathbf{w}(\xi_1, \eta_1) = \mathbf{N}^T(\xi_1, \eta_1)\bar{\mathbf{w}} \\ &\cdots \\ &\mathbf{w}(\xi_i, \eta_j) = \mathbf{N}^T(\xi_i, \eta_j)\bar{\mathbf{w}} \\ &\cdots \\ &\mathbf{w}(\xi_p, \eta_q) = \mathbf{N}^T(\xi_p, \eta_q)\bar{\mathbf{w}}\end{aligned} \tag{7.83}$$

which can also be written as

$$\mathbf{w} = \mathbf{G}\bar{\mathbf{w}} \tag{7.84}$$

where

$$\mathbf{G} = [\mathbf{N}(\xi_1, \eta_1) \ \cdots \ \mathbf{N}(\xi_i, \eta_j) \ \cdots \ \mathbf{N}(\xi_p, \eta_q)]^T, \ i \in [1, p], \ j \in [1, q] \tag{7.85}$$

is the discretized displacement matrix, $\bar{\mathbf{w}}$ is a column vector of shape function coefficients, and $\mathbf{w}$ is a column vector of displacement at integration points. Formulations for the other four generalized displacements of the shell can be obtained in a similar way. They are expressed as

$$\begin{aligned}
\mathbf{u}_0 &= \mathbf{G}\bar{\mathbf{u}}_0 \\
\mathbf{v}_0 &= \mathbf{G}\bar{\mathbf{v}}_0 \\
\boldsymbol{\theta}_\alpha &= \mathbf{G}\bar{\boldsymbol{\theta}}_\alpha \\
\boldsymbol{\theta}_\beta &= \mathbf{G}\bar{\boldsymbol{\theta}}_\beta
\end{aligned} \tag{7.86}$$

The discretized values of derivatives of each generalized displacement can be obtained in a similar way as

$$\begin{aligned}
\frac{\partial \mathbf{u}_0}{\partial \alpha} &= \mathbf{G}_\alpha \bar{\mathbf{u}}_0 & \frac{\partial \mathbf{u}_0}{\partial \beta} &= \mathbf{G}_\beta \bar{\mathbf{u}}_0 \\
\frac{\partial \mathbf{v}_0}{\partial \alpha} &= \mathbf{G}_\alpha \bar{\mathbf{v}}_0 & \frac{\partial \mathbf{v}_0}{\partial \beta} &= \mathbf{G}_\beta \bar{\mathbf{v}}_0 \\
\frac{\partial \boldsymbol{\theta}_\alpha}{\partial \alpha} &= \mathbf{G}_\alpha \bar{\boldsymbol{\theta}}_\alpha & \frac{\partial \boldsymbol{\theta}_\alpha}{\partial \beta} &= \mathbf{G}_\beta \bar{\boldsymbol{\theta}}_\alpha \\
\frac{\partial \boldsymbol{\theta}_\beta}{\partial \alpha} &= \mathbf{G}_\alpha \bar{\boldsymbol{\theta}}_\beta & \frac{\partial \boldsymbol{\theta}_\beta}{\partial \beta} &= \mathbf{G}_\beta \bar{\boldsymbol{\theta}}_\beta
\end{aligned} \tag{7.87}$$

where

$$\begin{aligned}
\mathbf{G}_\alpha &= [\mathbf{N}_\alpha(\xi_1, \eta_1) \cdots \mathbf{N}_\alpha(\xi_i, \eta_j) \cdots \mathbf{N}_\alpha(\xi_p, \eta_q)]^T, \\
&\quad i \in [1, p],\ j \in [1, q] \\
\mathbf{G}_\beta &= \left[\mathbf{N}_\beta(\xi_1, \eta_1) \cdots \mathbf{N}_\beta(\xi_i, \eta_j) \cdots \mathbf{N}_\beta(\xi_p, \eta_q)\right]^T, \\
&\quad i \in [1, p],\ j \in [1, q]
\end{aligned} \tag{7.88}$$

are weighting coefficient matrices for discretized vectors of derivatives of the generalized displacements.

7.6.2 *Integral scheme*

In this work, the Gauss–Lobatto integration that contains the end-points will be used. The one-dimensional formula is as follows:

$$f(x) = \sum_{i=1}^{N} C_i f(x_i) \tag{7.89}$$

where the weights C_j of Gauss–Lobatto integration are given by

$$C_1 = C_n = \frac{2}{n(n-1)}, \quad C_j = \frac{2}{n(n-1)[P_{n-1}(\xi_j)]^2} \ (j \neq 1, n) \tag{7.90}$$

where ξ_j is the $(j-1)$th zero of $P'_{n-1}(\xi_j)$. Liu *et al.* [34] presented the detail of computing roots for Legendre polynomials using the recursion formula of Legendre polynomials if more than 40 roots are required.

7.6.3 *Linear stiffness matrix*

The above matrices and vectors will be used to discretize the strain energy and kinetic energy. The strain column vector in Eq. (7.54) can be discretized as

$$\begin{Bmatrix} \boldsymbol{\varepsilon}_{\alpha\alpha} \\ \boldsymbol{\varepsilon}_{\beta\beta} \\ \boldsymbol{\gamma}_{\alpha\beta} \\ \boldsymbol{\gamma}_{\beta z} \\ \boldsymbol{\gamma}_{z\alpha} \end{Bmatrix} = \begin{bmatrix} \mathbf{G}_\alpha & \mathbf{0} & z\mathbf{G}_\alpha & \mathbf{0} & k_x\mathbf{G} \\ \mathbf{0} & \mathbf{G}_\beta & \mathbf{0} & z\mathbf{G}_\beta & k_y\mathbf{G} \\ (1+zC_0)\mathbf{G}_\beta & (1-zC_0)\mathbf{G}_\alpha & z\mathbf{G}_\beta & z\mathbf{G}_\alpha & \mathbf{0} \\ \mathbf{0} & -C_1k_y\mathbf{G} & \mathbf{0} & \mathbf{G} & \mathbf{G}_\beta \\ -C_1k_x\mathbf{G} & \mathbf{0} & \mathbf{G} & \mathbf{0} & \mathbf{G}_\alpha \end{bmatrix} \times \begin{Bmatrix} \bar{\mathbf{u}}_0 \\ \bar{\mathbf{v}}_0 \\ \bar{\boldsymbol{\theta}}_\alpha \\ \bar{\boldsymbol{\theta}}_\beta \\ \bar{\mathbf{w}} \end{Bmatrix} \tag{7.91}$$

which can be expressed in a compact matrix form as

$$\boldsymbol{\varepsilon} = \mathbf{B}\bar{\mathbf{u}} \tag{7.92}$$

Note that $\mathbf{B}$ is a function of z.The elastic matrix with consideration of the integration weighting coefficients is

$$\mathbf{D}(z) = \begin{bmatrix} D_{11}\mathbf{C} & D_{12}\mathbf{C} & \mathbf{0} & \mathbf{0} & \mathbf{0} \\ D_{12}\mathbf{C} & D_{22}\mathbf{C} & \mathbf{0} & \mathbf{0} & \mathbf{0} \\ \mathbf{0} & \mathbf{0} & D_{66}\mathbf{C} & \mathbf{0} & \mathbf{0} \\ \mathbf{0} & \mathbf{0} & \mathbf{0} & D_{44}\mathbf{C} & \mathbf{0} \\ \mathbf{0} & \mathbf{0} & \mathbf{0} & \mathbf{0} & D_{55}\mathbf{C} \end{bmatrix} \tag{7.93}$$

where

$$\mathbf{C} = \mathrm{diag}[C_{\xi_1} C_{\eta_1} |J|_{\xi_1\eta_1}, \ldots, C_{\xi_p} C_{\eta_q} |J|_{\xi_p\eta_q}] \tag{7.94}$$

where C_{ξ_i}, C_{η_i} and $|J|_{\xi_i\eta_i}$ are respectively the integration weights and the Jacobian determinant at the integral points. By using numerical integration similar to isoparametric elements, linear strain energy Eq. (7.57) can be expressed as

$$U_L = \frac{1}{2}\bar{\mathbf{u}}^T \mathbf{K}_L \bar{\mathbf{u}} \tag{7.95}$$

where $\mathbf{K}_L$ is the linear stiffness matrix and can be expressed as

$$\mathbf{K}_L = \sum_{i=1}^{N_z} C_{z_i} \mathbf{B}^T(z_i)\mathbf{D}(z_i)\mathbf{B}(z_i) \tag{7.96}$$

where C_{zi} is the integration weighting coefficients in the z-direction.

7.6.4 *Geometric stiffness matrix*

In the same way as above, Eq. (7.65) can be discretized by the weighting coefficient matrices of values and derivatives as

$$\mathbf{B}_g = \begin{bmatrix} \mathbf{G}_\alpha & \mathbf{0} & \mathbf{0} & \mathbf{0} & \mathbf{0} \\ \mathbf{G}_\beta & \mathbf{0} & \mathbf{0} & \mathbf{0} & \mathbf{0} \\ \mathbf{0} & \mathbf{G}_\alpha & \mathbf{0} & \mathbf{0} & \mathbf{0} \\ \mathbf{0} & \mathbf{G}_\beta & \mathbf{0} & \mathbf{0} & \mathbf{0} \\ -k_x\mathbf{G} & \mathbf{0} & \mathbf{0} & \mathbf{0} & \mathbf{0} \\ \mathbf{0} & -k_y\mathbf{G} & \mathbf{0} & \mathbf{0} & \mathbf{0} \\ \mathbf{0} & \mathbf{0} & \mathbf{G}_\alpha & \mathbf{0} & \mathbf{0} \\ \mathbf{0} & \mathbf{0} & \mathbf{G}_\beta & \mathbf{0} & \mathbf{0} \\ \mathbf{0} & \mathbf{0} & \mathbf{0} & \mathbf{G}_\alpha & \mathbf{0} \\ \mathbf{0} & \mathbf{0} & \mathbf{0} & \mathbf{G}_\beta & \mathbf{0} \end{bmatrix} \tag{7.97}$$

Similarly, the nonlinear strain energy is discretized as

$$U_L = \frac{1}{2}\bar{\mathbf{u}}^T \mathbf{K}_g \bar{\mathbf{u}} \tag{7.98}$$

where $\mathbf{K}_g$ is the geometric stiffness matrix that can be expressed as

$$\mathbf{K}_g = \sum_{i=1}^{N_z} C_{z_i} \mathbf{B}_g^T(z_i)\mathbf{S}(z_i)\mathbf{B}_g(z_i) \tag{7.99}$$

The discretized expression of $\boldsymbol{S}$ is similar to the discrete expression of the elastic matrix $\boldsymbol{D}$, and therefore it is not presented here.

7.6.5 *Mass matrix*

Using the same way as above, the mass matrix given by Eq. (7.74) can be discretized as

$$\mathbf{m} = \begin{bmatrix} \mathbf{G} & \mathbf{0} & z\mathbf{G} & \mathbf{0} & \mathbf{0} \\ \mathbf{0} & \mathbf{G} & \mathbf{0} & z\mathbf{G} & \mathbf{0} \\ \mathbf{0} & \mathbf{0} & \mathbf{0} & \mathbf{0} & \mathbf{G} \end{bmatrix} \tag{7.100}$$

Thus, the kinetic energy can be discretized as

$$T = \frac{1}{2}\dot{\mathbf{u}}^{\mathbf{T}}\mathbf{M}\dot{\mathbf{u}} \tag{7.101}$$

where

$$\mathbf{M} = \sum_{i=1}^{N_z} C_{z_i}\rho(z_i)\mathbf{m}^T(z_i)\mathbf{m}(z_i) \tag{7.102}$$

The element matrices above are derived for each layer. We will discuss their assembly in Section 7.6.6.

7.6.6 *Element assembly*

According to the laminated shell theory, it can be found that describing the displacement field of each layer of the shell needs 5 DoFs. If the continuity between layers is not considered, then a three-layer shell needs 15 DoFs.

$$\bar{\boldsymbol{u}} = \Big[u_0^{(1)}, v_0^{(1)}, \theta_\alpha^{(1)}, \theta_\beta^{(1)}, w^{(1)}, u_0^{(2)}, v_0^{(2)}, \theta_\alpha^{(2)}, \theta_\beta^{(2)}, w^{(2)}, u_0^{(3)}, v_0^{(3)}, \theta_\alpha^{(3)}, \theta_\beta^{(3)}, w^{(3)}\Big] \tag{7.103}$$

Considering displacement continuity given by Eq. (7.53), one needs only 9 DOFs for the whole sandwich shell:

$$\tilde{\bar{\boldsymbol{u}}} = \left[\theta_\alpha^{(1)}, \theta_\beta^{(1)}, \theta_\alpha^{(2)}, \theta_\beta^{(2)}, \theta_\alpha^{(3)}, \theta_\beta^{(3)}, u_0^{(2)}, v_0^{(2)}, w\right] \tag{7.104}$$

The relationship between Eqs. (7.103) and (7.104) according to (7.53) is as follows:

$$\begin{Bmatrix} u_0^{(1)} \\ v_0^{(1)} \\ \psi_\alpha^{(1)} \\ \vdots \\ u_0^{(3)} \\ v_0^{(3)} \\ \psi_\alpha^{(3)} \\ \psi_\beta^{(3)} \\ w \end{Bmatrix} = \begin{bmatrix} -\frac{1}{2}h_1 & 0 & -\frac{1}{2}h_2 & 0 & 0 & 0 & 1 & 0 & 0 \\ 0 & -\frac{1}{2}h_1 & 0 & -\frac{1}{2}h_2 & 0 & 0 & 0 & 1 & 0 \\ 1 & 0 & 0 & 0 & 0 & 0 & 0 & 0 & 0 \\ \vdots & \vdots & \vdots & \vdots & \vdots & \vdots & \vdots & \vdots & \vdots \\ 0 & 0 & \frac{1}{2}h_2 & 0 & \frac{1}{2}h_3 & 0 & 1 & 0 & 0 \\ 0 & 0 & 0 & \frac{1}{2}h_2 & 0 & \frac{1}{2}h_3 & 0 & 1 & 0 \\ 0 & 0 & 0 & 0 & 1 & 0 & 0 & 0 & 0 \\ 0 & 0 & 0 & 0 & 0 & 1 & 0 & 0 & 0 \\ 0 & 0 & 0 & 0 & 0 & 0 & 0 & 0 & 1 \end{bmatrix}$$

$$\times \begin{Bmatrix} \psi_\alpha^{(1)} \\ \psi_\beta^{(1)} \\ \psi_\alpha^{(2)} \\ \psi_\beta^{(2)} \\ \psi_\alpha^{(3)} \\ \psi_\beta^{(3)} \\ u_0^{(2)} \\ v_0^{(2)} \\ w \end{Bmatrix} \tag{7.105}$$

which can be written as

$$\bar{\boldsymbol{u}} = \boldsymbol{H}\tilde{\bar{\boldsymbol{u}}} \tag{7.106}$$

The discretization of Eq. (7.106) leads to

$$\bar{\mathbf{u}} = \mathbf{H}\tilde{\bar{\mathbf{u}}} \tag{7.107}$$

where $\mathbf{H} = \text{kron}(\boldsymbol{H}, \boldsymbol{I})$, in which $\boldsymbol{I}$ is an $N_b \times N_b$ unit matrix and N_b is the number of shape function. According to the above definition, we can do elements assembly as follows:

$$\mathbf{K} = \mathbf{H}^T \begin{bmatrix} \mathbf{K}^{(1)} & \mathbf{0} & \mathbf{0} \\ \mathbf{0} & \mathbf{K}^{(2)} & \mathbf{0} \\ \mathbf{0} & \mathbf{0} & \mathbf{K}^{(3)} \end{bmatrix} \mathbf{H} \tag{7.108}$$

$$\mathbf{K}_g = \mathbf{H}^T \begin{bmatrix} \mathbf{K}_g^{(1)} & \mathbf{0} & \mathbf{0} \\ \mathbf{0} & \mathbf{K}_g^{(2)} & \mathbf{0} \\ \mathbf{0} & \mathbf{0} & \mathbf{K}_g^{(3)} \end{bmatrix} \mathbf{H} \tag{7.109}$$

$$\mathbf{M} = \mathbf{H}^T \begin{bmatrix} \mathbf{M}^{(1)} & \mathbf{0} & \mathbf{0} \\ \mathbf{0} & \mathbf{M}^{(2)} & \mathbf{0} \\ \mathbf{0} & \mathbf{0} & \mathbf{M}^{(3)} \end{bmatrix} \mathbf{H} \tag{7.110}$$

With the global element matrices above, we can solve the dynamic problem. The derivation of the governing equations for the dynamic problem will be presented in Section 7.6.7.

7.6.7 *The dynamic equation*

After discretizing the linear and nonlinear strain energies and the kinetic energy, the governing equation given by Eq. (7.75) is transformed into the following eigenvalue problem:

$$(\mathbf{K} + \mathbf{K}_g)\omega^2 + \mathbf{M} = \mathbf{0} \tag{7.111}$$

We can now solve the equations to analyze free vibration of functionally graded sandwich shells in thermal environments.

7.7 Results and Discussion

In this section, extensive comparison and parametric studies are carried out for free vibration of FGM shells with or without thermal environments. The natural frequencies obtained from the present layerwise formulations are compared with the results available in the literatures in order to validate the correctness of the present formulation. Shells with two different types of boundary conditions, namely, simply supported (SSSS) and clamped (CCCC) boundaries, will be considered. An SSSS shell implies that the degrees of freedom

Table 7.1. Convergence study of non-dimensional frequency parameters $\Omega = \omega(a^2/h)\sqrt{\rho_m/E_m}$ of a square Al/SiC SPH single-layered FGM shells with CCCC boundaries ($R_y/a = 5$, $R_x = R_y$, $a/h = 5$, $n = 0.5$).

Kapuria [394]	Pandey [393]		DQFEM [395]		Present (DQHFEM)	
15.9159	Mesh	Result	Node numbers	Result	Bases numbers	Result
	4×4	15.6950	5×5	15.6121	$4+2\times2$	18.8932
	6×6	15.6576	8×8	15.6382	$4+3\times3$	15.6718
	8×8	15.6516	10×10	15.6375	$4+4\times4$	15.6617
	10×10	15.6514	15×15	15.6374	$4+5\times5$	15.6485
					$4+6\times6$	15.6484
					$4+7\times7$	15.6477
					$4+8\times8$	15.6477

(DOFs) v_0, w_0, $\theta_y^{(1)}$, $\theta_y^{(2)}$ and $\theta_y^{(3)}$ are zeros along edges $x = 0$, a or the DOFs u_0, w_0, $\theta_x^{(1)}$, $\theta_x^{(2)}$ and $.\theta_x^{(3)}$ are zeros along edges $y = 0$, b. A CCCC shell means all 9 DoFs are zero along all of the four edges.

7.7.1 *Free vibration of functionally graded single-layer shell in non-thermal environment*

In this section, the free vibration of a single-layer functionally graded square cylindrical shell in non-thermal environment is studied. First, in Table 7.1, the metal composition of the single-layer functionally graded shell is aluminum (Al), whose elastic modulus $E_m = 70\,\text{GPa}$, density $\rho_m = 2702\,\text{kg/m}^3$ and Poisson's ratio $v_m = 0.3$; the ceramic composition is silicon carbide (SiC), whose elastic modulus $E_c = 427\,\text{GPa}$, density $\rho_c = 3100\,\text{kg/m}^3$ and Poisson's ratio $v_c = 0.17$. The microscopic mechanical equivalence model ROM with metal volume fraction λ is employed. Thus, when $n = 0$, the functionally graded shell is reduced to a pure metal shell, and when $n = \infty$, it is reduced to a pure ceramic shell. When n is equal to other values, the lower surface of the shell ($z = -h_1/2$) is pure ceramic and the upper surface ($z = h_1/2$) is pure metal. For single-layer shells, the lamination theory used in this paper is reduced to the first-order shear deformation theory (FSDT). The shear correction factor is taken as $k = 5/6$. The dimensionless frequency parameters in Table 7.1 are defined by $\Omega = \omega(a^2/h)\sqrt{\rho_m/E_m}$. The geometric parameters are set

Table 7.2. Convergence study of non-dimensional frequency parameters $\Omega = \omega(a^2/h)\sqrt{\rho_m/E_m}$ of a square Al/SiC SPH single-layered FGM shells with SSSS boundaries ($R_y/a = 5$, $R_x = R_y$, $a/h = 5$, $n = 0.5$).

Bases numbers	$4+2\times2$	$8+2\times2$	$12+2\times2$	$12+4\times4$	$12+5\times5$	$12+6\times6$	$16+6\times6$	$20+7\times7$
DQHFEM	18.8932	10.2779	10.2779	10.0061	10.0057	10.0057	9.9919	9.9919

to $R_y/a = 5$, $a/b = 1$, and $a/h = 5$. Donnell's shell theory is used in this example.

The differential quadrature hierarchical finite element method (DQHFEM) is a typical p-version finite element method, whose result gradually converges to the exact result with the elevation of the order of the shape functions. The shape functions of the two-dimensional DQHFEM can be divided into boundary shape functions and face shape functions. In order to compare with other methods, convergence studies of other methods are also included in Table 7.1. We can see that the results of the DQHFEM highly agree with those of Kapuria [394] and the DQFEM [395]. When the number of boundary nodes is 4 and the number of hierarchical functions is 7 in each direction, the DQHFEM converges. The total number of degrees of freedom of the DQHFEM is only $4 + 7 \times 7 = 53$. When Pandey's method using the 8-node quadrilateral finite elements converges, it needs $10\times10 + 11\times11 = 221$ (the number of 8-node element is 10×10 [393]) DoFs. When the DQFEM converges, it needs $15 \times 15 = 225$ DoFs. Therefore, the DQHFEM has the highest convergence rate.

According to the properties of the DQHFEM basis functions, the number of boundary nodes has no influence on the results in Table 7.1 because of the clamped boundary conditions. In order to simplify the computation, only the shape functions of four corner points are used. In order to study the influence of the boundary shape functions, the boundary conditions are changed to simply supported (SSSS) in Table 7.2. Other parameters do not change. We can find that the DQHFEM elements not only can be assembled as the FEM but also has the hierarchical characteristics. Moreover, the number of boundary nodes can be assigned freely. For clamped boundary conditions, the number of boundary DOFs can be greatly reduced. In the following examples, in order to make sure the results have converged, the number of basis functions is taken as $28 + 8 \times 8 = 92$.

Table 7.3. Comparison study of non-dimensional frequencies $\Omega = \omega(a^2/h)\sqrt{\rho_m/E_m}$ of square Al/SiC SPH single-layered FGM shells with SSSS boundary conditions ($R_y/a = 2$, $R_y = R_x$).

	a/h	n	Kapuria *et al.* [394]	Pandey *et al.* [393]	DQFEM [395]	Sander (current)	Donnell (current)
ROM	5	0.5	10.8079	10.3851	10.5452	10.5614	10.7478
		1	9.5173	9.2030	9.2754	9.2929	9.4807
		2	8.3062	8.0760	8.1172	8.1344	8.3185
		5	7.3351	7.1920	7.2739	7.2851	7.4595
	10	1	12.6834	12.5339	12.4847	12.4955	12.6304
		5	9.7248	9.6171	9.6718	9.6787	9.8065
	20	1	20.0728	20.0347	19.9198	19.9277	20.0028
		5	15.0152	14.9544	15.016	15.0208	15.0928
MT	5	0.5	9.1723	9.0079	8.9942	9.0082	9.1909
		1	8.1694	8.0353	8.0192	8.0351	8.2124
		2	7.4340	7.3156	7.3260	7.3410	7.5095
		5	6.8286	6.6959	6.7794	6.7886	6.9370
	10	1	10.8526	10.8743	10.745	10.7548	10.8821
		5	8.9674	8.9016	8.8990	8.9048	9.0199
	20	1	16.9993	17.1575	16.9713	16.9728	17.0489
		5	13.5566	13.5672	13.5363	13.5404	13.6094

Next, comparison study of natural frequencies for a square SSSS single-layered spherical FGM shell with $R_y/a = 2$ and 5 are presented in Table 7.3. The material properties are computed by both ROM and MT micromechanical models. The non-dimensional frequency parameters and the material properties used in this example are the same as those used above. Since the shell has equal radio of curvature in x- and y-directions and has square planform, the tracer C_0 becomes zero. Hence, the results obtained from Sander's and Love's shell theories are the same and so the results of Love's shell theories is not presented. One can see from Table 7.3 that the results obtained by the present finite element formulation are in good agreement with those of Kapuria *et al.* [394], Pandey *et al.* [393] and the DQFEM [395]. The results of Pandey were also based on Sanders's first-order approximate shell theory. In addition, we can find that the results obtained by the MT model are relatively smaller than those of the ROM model. Such phenomenon was also observed in the results of Kapuria *et al.* [394] and Pandey *et al.* [393].

7.7.2 *Vibration of functionally graded sandwich shells in non-thermal environment*

In this section, free vibration of functionally graded sandwich shell in non-thermal environment is studied. The metal composition of the single-layer functionally graded shell is aluminum (Al), whose physical properties are elastic modulus $E_m = 70\,\text{GPa}$, density $\rho_m = 2707\,\text{kg/m}^3$ and Poisson's ratio $v_m = 0.3$. The ceramic composition is alumina (Al_2O_3), whose physical properties are elastic modulus $E_c = 380\,\text{GPa}$, density $\rho_c = 3800\,\text{kg/m}^3$ and Poisson's ratio $v_c = 0.3$. Its configuration is similar to Type B. The upper and lower facesheets are made of pure material. The middle layer is functionally graded material. The total thicknesses is h. The thicknesses of the upper and lower panels are $h_1 = h_3 = 0.1h$. The thickness of the middle layer is $h_2 = 0.8h$. Its difference compared with Type B is that the upper layer is pure metal material and the lower layer is pure ceramic material. The λ of the functionally graded layer is the volume fraction of the ceramic component. When n is 0 or ∞, the functionally graded layer is respectively reduced to pure metal and pure ceramic material. When n is other value, the lower boundary ($z_2 = -h_2/2$) of the functional gradient layer is pure ceramic, while the upper boundary ($z_2 = h_2/2$) is pure metal. The value of R_y/a is 0.5 or 5, while $R_x = \infty$. Length–width ratio is $a/h = 10$. The non-dimensional frequency parameter is $\Omega = \omega(a^2/h)\sqrt{\rho_0/E_0}$. The density and the elasticity modulus are respectively $\rho_0 = 1\,\text{kg/m}^3$ and $E_0 = 1\,\text{GPa}$. Sander's shell theory is employed. The results of Pandey [393] and the DQFEM [395] indicate that the result using Sander's theory is much closer to the three-dimensional solution of Fazzolari and Carrera [396]. It can be observed from Table 7.4 that the DQHFEM is highly consistent with the results calculated by Fazzolari and Carrera [396], Pandey [393], and the DQFEM [395].

7.7.3 *Free vibration of functionally graded single-layer shell in thermal environment*

In this section, the influence of thermal environment is studied. In order to make the thermal effect more obvious, it is necessary to select materials that are sensitive to temperature for the ceramic and metal components of FGM. According to the introduction of the

Table 7.4. Comparison of non-dimensional frequencies of square SSSS 1-8-1 Al2O3/Al sandwich CYL shells with FGM core.

R_y/a	Model	Theory	$n = 0.5$	$n = 1$	$n = 5$	$n = 10$
0.5	ROM	Fazzolari and Carrera [396]	2.2828	2.4567	2.7545	2.8199
		Pandey *et al.* [393]	2.2983	2.4681	2.7618	2.8271
		DQFEM [395]	2.2973	2.4673	2.7608	2.8259
		Sander's (Current)	2.2944	2.4643	2.7582	2.8236
	MT	Pandey *et al.* [393]	2.1886	2.3608	2.7152	2.8023
		DQFEM [395]	2.1782	2.3508	2.7060	2.7934
		Sander's (Current)	2.1751	2.3479	2.7304	2.7912
5	ROM	Fazzolari and Carrera [396]	1.3186	1.3721	1.5186	1.5748
		Pandey *et al.* [393]	1.3197	1.3721	1.5180	1.5749
		DQFEM [395]	1.3189	1.3714	1.5174	1.5744
		Sander's (Current)	1.3136	1.3666	1.5123	1.5688
	MT	Pandey *et al.* [393]	1.2806	1.3255	1.4812	1.5519
		DQFEM [395]	1.2753	1.3196	1.4729	1.5434
		Sander's (Current)	1.2697	1.3149	1.4680	1.5380

Table 7.5. Temperature coefficients of Si_3N_4 and SUS304 [387].

	Properties	P_0	P_{-1}	P_1	P_2	P_3
Si_3N_4	E_c (Pa)	348.43×10^9	0	-3.070×10^{-4}	2.160×10^{-7}	-8.946×10^{-11}
	α_c (1/K)	5.8723×10^{-6}	0	9.095×10^{-4}	0	0
	v_c	0.24	0	0	0	0
SUS304	E_m (Pa)	201.04×10^9	0	3.079×10^{-4}	-6.534×10^{-7}	0
	α_m (1/K)	12.33×10^{-6}	0	8.086×10^{-4}	0	0
	v_m	0.3262	0	-2.002×10^{-4}	3.797×10^{-7}	0

literature [387], two temperature-dependent materials are adopted in this work. The ceramic and metal components are respectively silicon nitride (Si_3N_4) and stainless steel (SUS304). The temperature-dependent model of their physical properties is defined by Eq. (7.40). Their temperature coefficients are listed in Table 7.5.

The free vibration of single-layer functionally graded shell in thermal environment is studied in this example. Three material properties in this example are assumed to be temperature-dependent: the elastic modulus, the coefficients of thermal expansion and the Poisson's ratio. The remaining material properties are assumed to be temperature-independent and they are $\rho_c = 2370\,\text{kg/m}^3$,

$K_c = 9.19\,\text{W/mK}$, $\rho_m = 8166\,\text{kg/m}^3$ and $K_m = 12.04\,\text{W/mK}$. The formulation of non-dimensional frequency parameter is $\Omega = \omega a^2\sqrt{\rho_m h/D_m^*}$, where $D_\text{m}^* = E_\text{m}h^3/12(1v_\text{m}^2)$, in which ρ_m, E_m and v_m are physical properties of pure metallic materials at a reference temperature 300 K. The geometric parameters are $a/h = 10$, $R_x = 0$ and $R_y/a = 10$. For the functionally graded shell, λ is the volume fraction of metal component. It means that when $n = 0$ or ∞, the functionally graded shell is respectively reduced to pure ceramic or pure metal shell. When n is other value, the upper surface of the shell is pure ceramic, and the lower surface is pure metal. The boundary conditions of temperature are that the upper surface and lower surface respectively have constant temperature $T_t = 600\,\text{K}$ and $T_b = 300\,\text{K}$. The boundary conditions are clamped (CCCC).

In order to ensure convergence and improve the convergence speed, we choose the number of basis functions to be $4 + 8 \times 8 = 68$. The results are compared with those of Yang *et al.* [397]. For convenience of comparisons, Sander's shell theory is used, as shown in Table 7.6. One can see that the results of the DQHFEM are highly consistent with those in the literatures. The difference between the MT model and the ROM model in non-thermal environment is also reflected in thermal environment. One can see that the results of Pandey [393] with $n = 0$ or ∞ obtained by MT and ROM model are inconsistent. This is not reasonable since the MT and ROM models should have no difference in the case of $n = 0$ or ∞ according to the theory in Section 3. The result of the DQHFEM satisfies this condition. Therefore, it is reasonable to think that Pandey's results should have some error. Figure 7.9 shows some mode shapes of a case in Table 7.6.

7.7.4 *Free vibration of functionally graded sandwich shells in thermal environment*

First, free vibration of functionally graded sandwich plate in thermal environment is studied. The curvature of the plate is equal to 0. Other parameters of the plate are the same as those of Type A sandwich shell. The metal and ceramic components are still respectively SUS304 and Si_3N_4, but the heat conduction coefficients are different from those in Section 7.7.3. They are $K_m = 15.379\,\text{W/mK}$ and $K_c = 13.923\,\text{W/mK}$. The formulation of non-dimensional frequency

Table 7.6. Comparison of non-dimensional frequency parameters of a square Si3N4/SUS304 CYL single-layered FGM shell in thermal environment with CCCC boundary conditions ($T_t = 600\,\text{K}$, $T_b = 300\,\text{K}$, $R_y/a = 10$, $a/h = 10$, $R_x = \infty$).

	n							
	SiC		2		10		SUS304	
Model	ROM	MT	ROM	MT	ROM	MT	ROM	MT
Yang and Shen [397]	72.4460		39.3690		34.5710		31.5110	
Pandey [393]	72.1210	72.1176	39.5420	39.3146	34.2512	33.9864	33.9093	31.7104
Sander's (current)	72.3970	72.3970	39.3213	39.1711	33.9154	33.8052	31.3436	31.3436

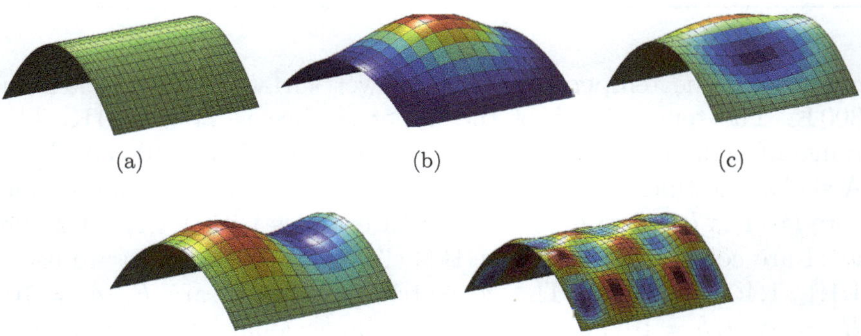

Fig. 7.9. The first three-order and high-order modal shapes of a single-layer functionally graded shell: (a) initial shell configuration; (b) $\Omega_1 = 39.3213$; (c) $\Omega_2 = 75.1068$; (d) $\Omega_3 = 75.2095$; (e) $\Omega_{50} = 395.8740$.

parameters is $\Omega = \omega(a^2/h)\sqrt{\rho_m/E_m}$. Other physical properties are consistent with those in Section 7.3, and the relative thickness is taken as $a/h = 10$. Two kinds of temperature boundary conditions are considered: $T_b = 600\,\text{K}$ or $300\,\text{K}$, and $T_t = 300\,\text{K}$. The results are compared with those of Pandey [393] and Wang *et al.* [398]. One can see from Table 7.7 that the results obtained by the DQHFEM are in good agreement with the results in the literatures.

Now, the influence of configuration, temperature and volume fraction on the frequency parameters is studied. Sander's theory will be

Table 7.7. Comparison of non-dimensional frequency parameters of square Si3N4/SUS304 sandwich plates with FGM facesheet of Type A in thermal environment and SSSS boundary conditions ($a/h = 10$).

			Wang and Shen [398]	Pandey [393]	Sander (current)	Wang and Shen [398]	Pandey [393]	Sander (current)
Thickness	Model	n	$T_t = 300\,\mathrm{K}$, $T_b = 300\,\mathrm{K}$			$T_t = 300\,\mathrm{K}$, $T_b = 600\,\mathrm{K}$		
$h = 10\,\mathrm{mm}$	ROM	0.2	5.9984	9.9673	5.9736	5.7821	5.7401	5.7909
$h_c/h_f = 8$		2	6.5705	6.5517	6.5396	6.3503	6.3376	6.3516
	MT	0.2	5.9848	5.9539	5.9603	5.7686	5.7251	5.7777
		2	6.5571	6.5380	6.5265	6.3371	6.3224	6.3388
$h = 6\,\mathrm{mm}$	ROM	0.2	6.1144	6.0820	6.0861	5.8976	5.8284	5.8977
$h_c/h_f = 4$		2	7.0417	7.0278	7.0048	6.8161	6.7894	6.8017
	MT	0.2	6.0938	6.0614	6.0659	5.8771	5.8056	5.8777
		2	7.0233	7.0091	6.9868	6.7979	6.7691	6.7691

employed. The temperature of the lower surface is $T_b = 600\,\mathrm{K}$ or $300\,\mathrm{K}$. The temperature of the upper surface is $T_t = 300\,\mathrm{K}$. The range of volume fraction index is $n = 0.2$, 0.5, 2 and 10. For Type A shells, the thickness ratio of upper and lower facesheet layers and core layer is $h_f : h_c : h_f$, for which three cases with 1:1:1, 2:1:2 and 4:1:4 are considered. For Type B shells, other three ratios are used: 1:1:1, 1:4:1 and 1:8:1. The geometric parameters are $R_y/a = 10$, $R_x = \infty$, $a/h = 10$ and $a/b = 1$. Non-dimensional frequency parameters are taken similar to those in Table 7.6. The results are listed in Tables 7.8 and 7.9. We can see that the results obtained by the DQHFEM are in good agreement with those obtained by other methods. For Type A shells, the increase of core thickness will reduce the non-dimensional frequency. Theoretically, the frequency is mainly determined by the stiffness and mass of the structure. The increase of core thickness results in the increase of metal composition of laminated shell. Because the elastic modulus of metal (SUS304) is smaller than that of ceramic (Si_3N_4), the average modulus of elasticity will decrease, and thus the overall stiffness will decrease. Moreover, the density of metal (SUS304) is greater than that of ceramic (Si_3N_4), and thus the overall mass will increase. These two factors result in a reduction in non-dimensional frequency parameters. In Type A shells, we also find that the non-dimensional frequency parameters

Table 7.8. Non-dimensional frequency parameters of square Si3N4/SUS304 sandwich CYL Type A shells with SSSS boundary conditions in thermal environment ($R_y/a = 10$, $a/h = 10$, $R_x = \infty$).

Temperature			$T_b = 600\,\text{K}$, $T_t = 300\,\text{K}$				$T_b = 400\,\text{K}$, $T_t = 300\,\text{K}$			
Model		n	0.2	0.5	2	10	0.2	0.5	2	10
ROM	1:1:1	Pandey [393]	20.1415	22.1142	26.3347	29.7082	20.7953	22.7517	26.9591	30.3472
		Sander (current)	20.2350	22.1074	26.1178	29.3363	20.7345	22.6198	26.6803	29.9610
	2:1:2	Pandey [393]	20.4867	22.7969	28.0059	32.5648	21.1225	23.4412	28.6542	33.2513
		Sander (current)	20.5559	22.7782	27.7627	32.1483	21.0638	23.3058	28.3630	32.8416
	4:1:4	Pandey [393]	20.6770	23.2491	29.2059	34.8407	21.3338	23.8979	29.8721	35.5662
		Sander (current)	20.7653	23.2256	28.9512	34.3987	21.2779	23.7628	29.5784	35.1466
MT	1:1:1	Pandey [393]	20.0235	21.9666	26.2576	29.6966	20.6831	22.6101	26.8826	30.3344
		Sander (current)	20.1286	21.9752	26.0480	29.3240	20.6295	22.4885	26.6096	29.9489
	2:1:2	Pandey [393]	20.33360	22.6374	27.9318	32.5576	20.9976	23.2869	28.5786	33.2418
		Sander (current)	20.4377	22.6344	27.6935	32.1393	20.9469	23.1625	28.2926	32.8322
	4:1:4	Pandey [393]	20.5387	23.0831	29.1350	34.8365	21.2015	23.7366	29.7981	35.5590
		Sander (current)	20.6401	23.0753	28.8831	34.3914	21.1541	23.6129	29.5091	35.7390

Table 7.9. Non-dimensional frequency parameters of square Si3N4/SUS304 sandwich CYL Type A shells with SSSS boundary conditions in thermal environment ($R_y/a = 10$, $a/h = 10$, $R_x = \infty$).

Temperature			$T_b = 600\,\mathrm{K}$, $T_t = 300\,\mathrm{K}$				$T_b = 400\,\mathrm{K}$, $T_t = 300\,\mathrm{K}$			
Model		n	0.2	0.5	2	10	0.2	0.5	2	10
ROM	1:1:1	Pandey [393]	26.3572	25.4236	23.8102	22.8077	27.5772	26.6004	24.9169	23.8764
		Sander (current)	27.3374	26.3406	24.6186	23.5220	27.8774	26.9017	25.1938	24.0830
	1:4:1	Pandey [393]	29.4383	27.0050	23.4738	21.5653	30.4658	27.9338	24.2720	22.3061
		Sander(current)	30.1035	27.6250	24.0028	21.9458	30.6612	28.1801	24.5315	22.4331
	1:8:1	Pandey [393]	31.0678	27.7802	23.3370	21.0359	32.0297	28.6142	24.0132	21.6442
		Sander (current)	31.5535	28.1993	23.6644	21.2382	32.1303	28.7668	24.1923	21.7126
MT	1:1:1	Pandey [393]	26.3441	25.4086	23.8002	22.7978	27.5602	26.5801	24.9025	23.8654
		Sander (current)	27.3325	26.3341	24.6091	23.5150	27.8705	26.8921	25.1823	24.0759
	1:4:1	Pandey [393]	29.3511	26.9101	23.3971	21.4775	30.3660	27.8248	24.1883	22.2200
		Sander (current)	30.0577	27.5777	23.9699	21.9199	30.6157	28.1315	24.4959	22.4063
	1:8:1	Pandey [393]	30.9347	27.6297	23.2087	20.8907	31.8732	28.4390	23.8718	21.4970
		Sander (current)	31.4700	28.1168	23.6175	21.2000	32.0492	28.6846	24.1423	21.6732

will increase with the increase of metal volume fraction index. This effect can also be attributed to the reduction in metal composition. So, the stiffness goes up, and the mass goes down. Similarly, in Type B shells, the change in frequency can also be attributed to the above reasons.

7.8 Conclusions

In this chapter, DQHFEM is used to analyze the free vibration of single-layer and sandwich functionally graded shells in thermal and non-thermal environments. In all examples, the results are in good agreement with the results in the literatures. In this paper, the DQHFEM combined with a lamination theory is used to analyze the FGM sandwich shell in thermal environments for the first time. The effective material properties of FGM are estimated according to two micromechanical models: Voigt's rule of mixing (ROM) and the Mori–Tanaka (MT) model. The present results are based on the LW theory and the DQHFEM, and are compared with the results in the literatures. Comparisons show that all of the results of the present method are of high accuracy, which indicates that the DQHFEM is a simple, effective and accurate numerical method for the analysis of thick/thin FGM sandwich composite structures.

Chapter 8

The Hierarchical Quadrature Element Method for Isotropic and Composite Laminated General Shells

In this chapter, a p-version curved composite laminated shell element has been developed using the modified high-order bases of differential quadrature hierarchical finite element method (DQHFEM). The theoretical model of the shell is based on the layerwise theory with linear expansion in each layer. Exact geometry is established using the CAD technique of non-uniform rational B-splines (NURBS). As a result, the FEM discretization errors of geometry are eliminated. To solve the coupling difficulty of elements with different parameterizations commonly used in the complicated model, a novel method based on interpolation on arc length coordinates is proposed in this work. Even though the computation efficiency based on layerwise theory is not as fast as those based on equivalent single-layer theory, they produce as accurate results as the 3D theory and use less DOFs as well as less input data compared to the 3D model. Additionally, because of the fast convergence of p-version FEM and the exact representation of the geometric model, the present elements are expected to produce more accurate results than the conventional h-version flat shell elements with the same number of DOFs. Numerical examples are provided to illustrate the accuracy as well as versatility of the present elements.

8.1 Geometry Representation

Consider a shell body (Fig. 8.1) with uniform thickness h (for non-uniform case, h is a function of ξ^1 and ξ^2), geometry description of which can be expressed as

$$\boldsymbol{X}(\xi^1,\xi^2,\xi^3) = \boldsymbol{S}(\xi^1,\xi^2) + \xi^3\boldsymbol{n}(\xi^1,\xi^2) \tag{8.1}$$

where $\boldsymbol{X}$ denotes the coordinates of material points in global x–y coordinates, $(\xi^1,\xi^2,\xi^3) \in [0,1]\times[0,1]\times[-h,h]$ is the parametric coordinates, $\boldsymbol{S}$ denotes the middle surface point, and $\boldsymbol{n}$ is the unit normal director, see Fig. 8.1. The middle surface $\boldsymbol{S}$ is exactly expressed by NURBS

$$\boldsymbol{S}(\xi^1,\xi^2) = \sum_{i=1}^{m}\sum_{j=1}^{n} R_{ij}(\xi^1,\xi^2)\boldsymbol{P}_{ij} \tag{8.2}$$

where $\boldsymbol{P}_{ij}$, $i = 1 \sim m$, $j = 1 \sim n$, is the control point and R_{ij} is the corresponding basis defined as

$$R_{ij} = \frac{N_{i,p}(\xi^1)N_{j,q}(\xi^2)w_{ij}}{\sum\limits_{i=1}^{m}\sum\limits_{i=1}^{n} N_{i,p}(\xi^1)N_{j,q}(\xi^2)w_{ij}} \tag{8.3}$$

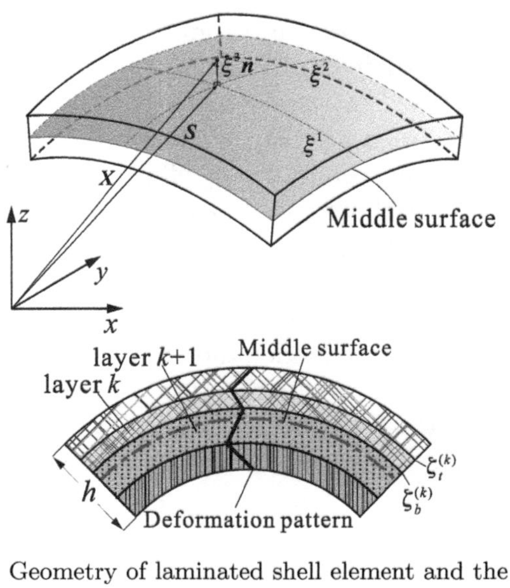

Fig. 8.1. Geometry of laminated shell element and the notations.

where w_{ij} is the weighting coefficient and $N_{i,p}$ and $N_{j,q}$ are the piecewise B-splines of orders p and q in directions of ξ^1 and ξ^2. The recursive formulae for B-splines are given by

$$
\begin{aligned}
p = 0 : N_{i,0}(\xi) &= \begin{cases} 1 & \xi_i \leq \xi < \xi_{i+1} \\ 0 & \text{otherwise} \end{cases} \\
p > 0 : N_{i,p}(\xi) &= \frac{\xi - \xi_i}{\xi_{i+p} - \xi_i} N_{i,p-1}(\xi) \\
&\quad + \frac{\xi_{i+p+1} - \xi}{\xi_{i+p+1} - \xi_{i+1}} N_{i+1,p-1}(\xi)
\end{aligned} \tag{8.4}
$$

where $\xi_i \in \Xi = [\xi_1, \xi_1, \ldots, \xi_{n+p+1}]$ is the knot and Ξ is the knot vector. An example of a NURBS surface along with its control points is shown in Fig. 8.2. The NURBS technique has become a standard tool in describing and modeling free form curves and surfaces in computer-aided design (CAD) and computer graphics; one may refer to [399, 400] for more details.

Using Eqs. (8.1) and (8.2), we defined the following covariant vectors:

$$
\boldsymbol{G}_i = \frac{\partial \mathbf{X}}{\partial \xi^i}, \ i = 1, 2, 3; \quad \boldsymbol{g}_\alpha = \frac{\partial \boldsymbol{S}}{\partial \xi^\alpha}, \quad \alpha = 1, 2 \tag{8.5}
$$

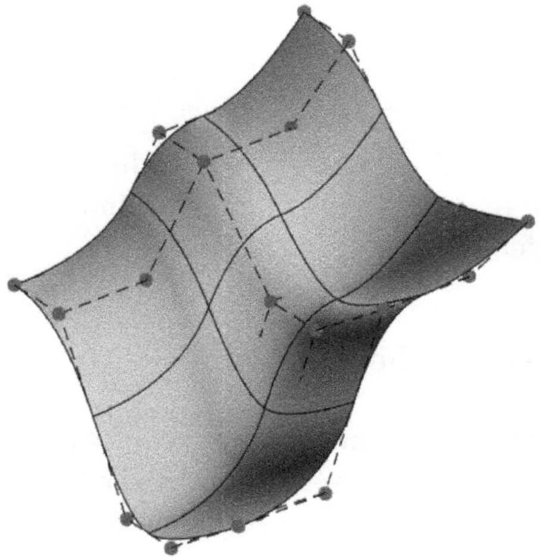

Fig. 8.2. A NURBS surface: (•) the control points.

It follows that

$$\boldsymbol{G}_\alpha = \boldsymbol{g}_\alpha + \xi^3 \frac{\partial \boldsymbol{n}}{\partial \xi^\alpha}; \quad \alpha = 1, 2; \quad \boldsymbol{G}_3 = \boldsymbol{n} \tag{8.6}$$

where the normal director can be expressed as

$$\boldsymbol{n} = \frac{\boldsymbol{g}_1 \times \boldsymbol{g}_2}{|\boldsymbol{g}_1 \times \boldsymbol{g}_2|} \tag{8.7}$$

To apply the assumptions of present elements, we also defined the local Cartesian coordinates

$$\boldsymbol{i}_1 = \frac{\boldsymbol{g}_1}{|\boldsymbol{g}_1|}; \quad \boldsymbol{i}_3 = \boldsymbol{n}; \quad \boldsymbol{i}_2 = \boldsymbol{i}_3 \times \boldsymbol{i}_1 \tag{8.8}$$

In this manner, $\boldsymbol{i}_1$ and $\boldsymbol{i}_2$ are always located in the tangent plane of the shell. For planar shell, $\boldsymbol{i}_1$ and $\boldsymbol{i}_2$ can be simply taken as form unit vectors of global coordinates. The required differential relations of Eq. (8.8) are given by

$$\begin{aligned}
\frac{\partial \boldsymbol{i}_1}{\partial \xi^\alpha} &= \frac{1}{\sqrt{g_{11}}}\left(\Gamma^2_{1\alpha}\frac{\boldsymbol{g}^2}{g^{22}} + b_{1\alpha}\boldsymbol{i}_3\right)\\
\frac{\partial \boldsymbol{i}_2}{\partial \xi^\alpha} &= \frac{1}{\sqrt{g_{22}}}\left(\Gamma^1_{2\alpha}\frac{\boldsymbol{g}^1}{g^{11}} + b_{2\alpha}\boldsymbol{i}_3\right)\\
\frac{\partial \boldsymbol{i}_3}{\partial \xi^\alpha} &= -b_{\alpha\beta}\boldsymbol{g}^\beta = -b_{\alpha\beta}g^{\beta\gamma}\boldsymbol{g}_\gamma = -b_{\alpha}^{\cdot\beta}\boldsymbol{g}_\beta = -b_{\beta\alpha}\boldsymbol{g}^\beta = -b^{\beta}_{\cdot\alpha}\boldsymbol{g}_\beta
\end{aligned} \tag{8.9}$$

where $g_{\alpha\beta} = \boldsymbol{g}_\alpha \cdot \boldsymbol{g}_\beta$ and $\boldsymbol{g}^\alpha$ is the contravariant bases of $\boldsymbol{g}_\alpha$ determined by $\boldsymbol{g}_\alpha \cdot g^\beta = \delta^\beta_\alpha$; $g^{\alpha\beta} = \boldsymbol{g}^\alpha \cdot \boldsymbol{g}^\beta$ and Γ^k_{ij} is the Christoffel. In the following discussions, the unified symbols ξ, η, ζ commonly used in FEM will be employed to denote ξ^1, ξ^2, ξ^3, respectively.

8.2 The Layerwise Shell Model

In present shell elements, the kinematics of each layer is based on the assumptions of the Reissner–Mindlin shell theory: (1) the normal strain is assumed to be zero, i.e., the transverse normal is inextensible; (2) the influence of normal stress on deformations is neglected. Besides, all layers are bonded perfectly at the interfaces during the deformation. For a single layer k, according to the first assumption, the displacement field can be expressed as

$$\boldsymbol{u}^{(k)} = \boldsymbol{u}_R^{(k)} + \boldsymbol{\Omega}^{(k)} \times (\zeta - \zeta_R^{(k)})\boldsymbol{n}; \quad \zeta \in [\zeta_b^{(k)}, \zeta_t^{(k)}] \tag{8.10}$$

where $\boldsymbol{u}_R^{(k)}$ is the displacement vector of the reference surface (need not be the middle surface of the layer), $\boldsymbol{n}$ is the unit normal director defined in Eq. (8.7); vector $\boldsymbol{\Omega}^{(k)}$ defines the rotation of the normal fiber in layer k, and it will be expanded in terms of $\boldsymbol{g}_\alpha$, $\alpha = 1, 2$, $\zeta_R^{(k)}$ is the thickness coordinate of the reference surface while $\zeta_b^{(k)}$ and $\zeta_t^{(k)}$ are respectively the thickness coordinates of the bottom and top surfaces of layer k. The thickness of layer k is denoted by $h_k = \zeta_t^{(k)} - \zeta_b^{(k)}$.

For example, the displacement field of an n-layered shell element can be determined by the following steps. First, the original reference surface (ORS) displacement vector should be designated. The term ORS means that, as it will show, the displacement fields for other layers will be constructed with reference to the ORS. In our model, this surface can be specified arbitrarily, which means the ORS can be located in any layer, for example, layer k, with an arbitrarily given ζ coordinate, i.e., $\zeta_R^{(k)}$. Then the displacement in layer k is determined by Eq. (8.10). For layer k+1, see Fig. 8.1, the reference surface is specified as the bottom surface of layer k, then we have

$$\begin{cases} \zeta_R^{(k+1)} = \zeta_b^{(k)} \\ \boldsymbol{u}_R^{(k+1)} = \boldsymbol{u}^{(k)}(\xi, \eta, \zeta_b^{(k)}) \end{cases} \tag{8.11}$$

In this manner, the inter-lamina continuity of displacement is ensured. Using Eqs. (8.10) and (8.11), the displacement field of layer $k+1$ is given by

$$\boldsymbol{u}^{(k+1)} = \boldsymbol{u}_R^{(k+1)} + \boldsymbol{\Omega}^{(k+1)} \times (\zeta - \zeta_b^{(k)})\boldsymbol{n}; \quad \zeta \in [\zeta_b^{(k+1)}, \zeta_t^{(k+1)}] \tag{8.12}$$

where the only new added variable is the rotation vector $\Omega^{(k+1)}$. Using the same process, the displacement field for all layers below the original layer k is determined. Similarly, the displacement field of the layers above layer k is also determined by letting the top surface of layer k be the reference surface of layer $k-1$ and so forth. Consequently, a zigzag deformation model is introduced in present elements, see Fig. 8.1.

The constitutive law for present element is very similar to [401]. Using Eq. (8.10), the global Cartesian components of infinitesimal strain for each layer, say k, can be expressed as

$$\begin{gathered}\varepsilon_{11}=\frac{\partial \boldsymbol{u}^{(k)}}{\partial x}\cdot\boldsymbol{e}_1,\quad \varepsilon_{22}=\frac{\partial \boldsymbol{u}^{(k)}}{\partial y}\cdot\boldsymbol{e}_2,\quad \varepsilon_{33}=\frac{\partial \boldsymbol{u}^{(k)}}{\partial z}\cdot\boldsymbol{e}_3\\ \gamma_{12}=2\varepsilon_{12}=\frac{\partial \boldsymbol{u}^{(k)}}{\partial x}\cdot\boldsymbol{e}_2+\frac{\partial \boldsymbol{u}^{(k)}}{\partial y}\cdot\boldsymbol{e}_1\\ \gamma_{23}=2\varepsilon_{23}=\frac{\partial \boldsymbol{u}^{(k)}}{\partial y}\cdot\boldsymbol{e}_3+\frac{\partial \boldsymbol{u}^{(k)}}{\partial z}\cdot\boldsymbol{e}_2\\ \gamma_{31}=2\varepsilon_{31}=\frac{\partial \boldsymbol{u}^{(k)}}{\partial x}\cdot\boldsymbol{e}_3+\frac{\partial \boldsymbol{u}^{(k)}}{\partial z}\cdot\boldsymbol{e}_1\end{gathered}\tag{8.13}$$

where $\boldsymbol{e}_i$ represent the unit orthogonal bases of global axes. Using Eq. (8.8), one can obtain the local strain components $\tilde{\varepsilon}_{ij}$ by

$$\boldsymbol{\varepsilon}=\tilde{\varepsilon}_{ij}\boldsymbol{i}_i\otimes\boldsymbol{i}_j=\varepsilon_{rs}(\boldsymbol{e}_r\cdot\boldsymbol{i}_i)(\boldsymbol{e}_s\cdot\boldsymbol{i}_j)\boldsymbol{i}_i\otimes\boldsymbol{i}_j\tag{8.14}$$

One can verify that the normal strain component $\tilde{\varepsilon}_{33}$ in each layer is automatically vanished. Then the constitutive equation of isotropic elasticity in local coordinates is given in a matrix form as

$$\begin{aligned}\tilde{\boldsymbol{\sigma}}=\begin{bmatrix}\tilde{\sigma}_{11}\\ \tilde{\sigma}_{22}\\ \tilde{\tau}_{12}\\ \tilde{\tau}_{23}\\ \tilde{\tau}_{31}\end{bmatrix}&=\frac{E}{1-\nu^2}\begin{bmatrix}1&\nu&&&\\ \nu&1&&&\\ &&(1-\nu)/2&&\\ &&&\kappa(1-\nu)/2&\\ &&&&\kappa(1-\nu)/2\end{bmatrix}\\ &\times\begin{bmatrix}\tilde{\varepsilon}_{11}\\ \tilde{\varepsilon}_{22}\\ \tilde{\gamma}_{12}\\ \tilde{\gamma}_{23}\\ \tilde{\gamma}_{31}\end{bmatrix}=\tilde{\mathbf{D}}\tilde{\boldsymbol{\varepsilon}}\end{aligned}\tag{8.15}$$

where $\tilde{\gamma}_{ij}=2\tilde{\varepsilon}_{ij}$ is the engineering strain, E is the Young's modulus, ν is the Poison's ration, κ is the shear correction factor, and $\tilde{\sigma}_{ij}$ is the stress component in local coordinates. Note that the normal stress $\tilde{\sigma}_{33}$ should not be determined by the constitutive equation.

Additionally, for orthotropic material, the stress–strain relation in material coordinates is expressed as

$$\hat{\boldsymbol{\sigma}} = \begin{bmatrix} \hat{\sigma}_{11} \\ \hat{\sigma}_{22} \\ \hat{\tau}_{12} \\ \hat{\tau}_{23} \\ \hat{\tau}_{31} \end{bmatrix} = \begin{bmatrix} D_{11} & D_{12} & & & \\ D_{21} & D_{22} & & & \\ & & G_{12} & & \\ & & & G_{23} & \\ & & & & G_{31} \end{bmatrix} \begin{bmatrix} \hat{\varepsilon}_{11} \\ \hat{\varepsilon}_{22} \\ \hat{\gamma}_{12} \\ \hat{\gamma}_{23} \\ \hat{\gamma}_{31} \end{bmatrix} = \hat{\mathbf{D}}\hat{\boldsymbol{\varepsilon}} \tag{8.16}$$

where the hat symbol '^' implies that components are defined in material coordinates, subscripts 1 and 2 are respectively the direction of the fiber and its normal in the tangent plane, see Fig. 8.3. The constants G_{12}, G_{23}, G_{31} are the shear moduli and

$$\begin{aligned} D_{11} &= \frac{E_1}{1-\nu_{12}\nu_{21}}, \quad D_{22} = \frac{E_2}{1-\nu_{12}\nu_{21}} \\ D_{12} &= \frac{E_1\nu_{21}}{1-\nu_{12}\nu_{21}} = \frac{E_2\nu_{12}}{1-\nu_{12}\nu_{21}} = D_{21} \end{aligned} \tag{8.17}$$

where E_1 and E_2 are the Young's moduli in the fiber and transverse directions, and ν_{12} and ν_{21} are the Poisson's ratios in the tangent plane of orthotropic layer. Using the same relation as Eq. (8.14), the transition matrix from material coordinates to the local Cartesian coordinates is given by

$$\mathbf{T} = \begin{bmatrix} \cos^2\theta & \sin^2\theta & -2\sin\theta\cos\theta & & \\ \sin^2\theta & \cos^2\theta & 2\sin\theta\cos\theta & & \\ \sin\theta\cos\theta & -\sin\theta\cos\theta & \cos^2\theta-\sin^2\theta & & \\ & & & \cos\theta & \sin\theta \\ & & & -\sin\theta & \cos\theta \end{bmatrix} \tag{8.18}$$

where θ is the angle between the fiber direction and the local axis $\boldsymbol{i}_1$, see Fig. 8.3. Then the constitutive relation in local coordinates is obtained as

$$\tilde{\boldsymbol{\sigma}} = \mathbf{T}\hat{\boldsymbol{\sigma}}; \quad \hat{\boldsymbol{\varepsilon}} = \mathbf{T}^{\mathrm{T}}\tilde{\boldsymbol{\varepsilon}}; \quad \tilde{\boldsymbol{\sigma}} = \mathbf{T}\hat{\mathbf{D}}\mathbf{T}^{\mathrm{T}}\tilde{\boldsymbol{\varepsilon}} = \tilde{\mathbf{D}}\tilde{\boldsymbol{\varepsilon}} \tag{8.19}$$

With the geometry and layerwise model defined above, they will be used to discretize the potential functional together with the finite element method given in the following section.

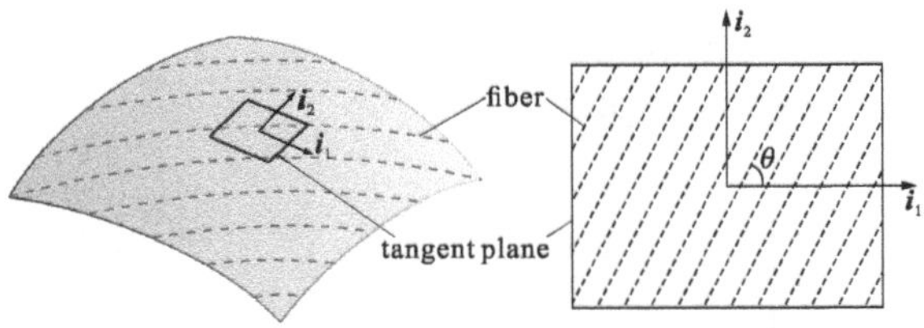

Fig. 8.3. A sketch of fibers in local coordinates.

8.3 The Hierarchical Quadrature Elements

This section presents the details of the present p-version composite laminated shell elements. The first part introduces the modified high-order bases of present elements. Patch coupling of NURBS patches with different parameterizations is also considered. The discretization process is then presented in the second part.

8.3.1 *Modified high-order bases*

The high-order bases of present elements are derived from the C^0 bases in DQHFEM [402] with consideration of patch coupling. The bases in [402] are summarized in Table 8.1 (shifted in span $[0,1] \times [0,1]$), where M, P, Q, N denote the node numbers at four edges; $L_i^M(\xi)$ is the ith Lagrange function of order M and so forth, H_ξ and H_η are the numbers of internal hierarchical bases referring to ξ- and η-directions, and $P_n(\xi)$ denotes the Legendre polynomial of order n. Non-uniformly distributed Gauss–Lobatto nodes are used as the element nodes on each side to improve the matrix conditioning. The C^0 conformity is ensured if the parameterization of the boundary curves of two adjacent elements is the same. This is easy to be satisfied in conventional isoparametric mapping or blending function mapping; however, to make full use of the advantages of NURBS technique in geometry representation, we need to consider the general case that the element geometry is represented by NURBS patches with different parameterizations, see Fig. 8.4(b). In this case, the inter-element conformity will not be satisfied directly by using the bases in Table 8.1.

To solve the problem, modifications of the vertex and edge functions have been made in Table 8.2, where the Lagrange functions on each side have been modified as the function of the corresponding arc length s_i, $i = 1$–4. In this way, the conformity of the middle surface

Table 8.1. High-order bases of the DQHFEM [402].

	Vertex bases	Edge bases	Face bases
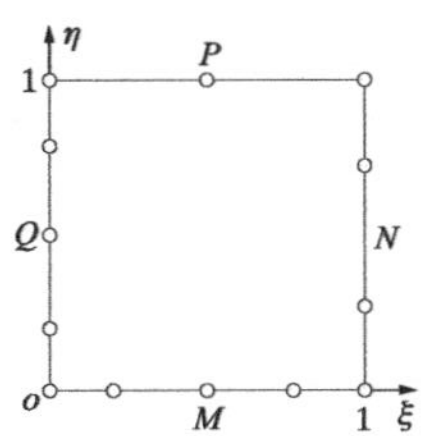	$S^{\mathrm{V1}} = (1-\eta)L_1^M(\xi) + (1-\xi)L_1^Q(\eta) - (1-\xi)(1-\eta)$ $S^{\mathrm{V2}} = (1-\eta)L_M^M(\xi) + \xi L_1^P(\eta) - \xi(1-\eta)$ $S^{\mathrm{V3}} = \xi L_N^N(\eta) + \eta L_P^P(\xi) - \xi\eta$ $S^{\mathrm{V4}} = (1-\xi)L_Q^Q(\eta) + \eta L_1^P(\xi) - (1-\xi)\eta$	$S_i^{E1} = (1-\eta)L_i^M(\xi),\ i = 2 \sim M-1$ $S_i^{E2} = \xi L_i^N(\eta),\ i = 2 \sim N-1$ $S_i^{E3} = \eta L_i^P(\xi),\ i = 2 \sim P-1$ $S_i^{E4} = (1-\xi)L_i^Q(\eta),\ i = 2 \sim Q-1$	$S_{mn}^F = \phi_m(\xi)\phi_n(\eta),$ $m = 1 \sim H_\xi; n = 1 \sim H_\eta$ $\phi_n(\xi) = \dfrac{[(2\xi-1)^2-1]}{n(n+1)} \left.\dfrac{\mathrm{d}P_n(x)}{\mathrm{d}x}\right\|_{x=2\xi-1}$

Table 8.2. Modified high-order bases.

Vertex	Side	Face
$S^{\mathrm{V1}} = (1-\eta)\tilde{L}_1^M(s_1) + (1-\xi)\tilde{L}_1^Q(s_4) - (1-\xi)(1-\eta)$	$S_i^{\mathrm{E1}} = (1-\eta)\tilde{L}_i^M(s_1), i = 2 \sim M-1$	$S_{mn}^F = \phi_m(\xi)\phi_n(\eta),$
$S^{\mathrm{V2}} = (1-\eta)\tilde{L}_M^M(s_1) + \xi\tilde{L}_1^P(s_2) - \xi(1-\eta)$	$S_i^{\mathrm{E2}} = \xi\tilde{L}_i^N(s_2), i = 2 \sim N-1$	$m = 1 \sim H_\xi; n = 1 \sim H_\eta$
$S^{\mathrm{V3}} = \xi\tilde{L}_N^N(s_2) + \eta\tilde{L}_P^P(s_3) - \xi\eta$	$S_i^{\mathrm{E3}} = \eta\tilde{L}_i^P(s_3), i = 2 \sim P-1$	$\phi_n(\xi) = \frac{[(2\xi-1)^2-1]}{n(n+1)} \left.\frac{\mathrm{d}P_n(x)}{\mathrm{d}x}\right\vert_{x=2\xi-1}$
$S^{\mathrm{V4}} = (1-\xi)\tilde{L}_Q^Q(s_4) + \eta\tilde{L}_1^P(s_3) - (1-\xi)\eta$	$S_i^{\mathrm{E4}} = (1-\xi)\tilde{L}_i^Q(s_4), i = 2 \sim Q-1$	

$$\tilde{L}_i^N(s) = \prod_{j=1;j\neq i}^{N} \frac{s-s_j}{s_i-s_j}; \quad \begin{aligned} s_1(\xi) &= \int_0^\xi |g_1(\xi,0)|\,\mathrm{d}\xi; s_2(\eta) = \int_0^\eta |g_2(1,\eta)|\mathrm{d}\eta \\ s_3(\xi) &= \int_0^\xi |g_1(\xi,1)|\,\mathrm{d}\xi; s_4(\eta) = \int_0^\eta |g_2(0,\eta)|\mathrm{d}\eta \end{aligned}$$

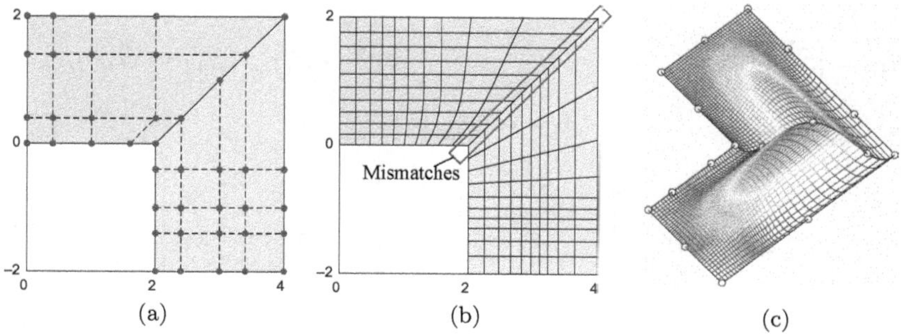

Fig. 8.4. Conformity of modified bases on the common edge of two NURBS patches with different parametrizations: (a) NURBS patches and the control points '•'; (b) mismatched isoparametric curves of two NURBS patches; (c) C^0 conformity on the common edge: '○' denotes the element nodes.

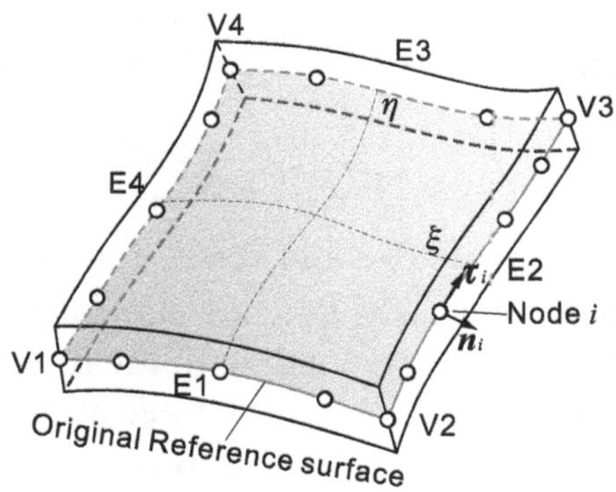

Fig. 8.5. Node collocation on original reference surface.

displacement should be achieved even when different parameterizations exist. The internal bases in Table 8.1 are retained since they do not affect the conformity. As an example, Fig. 8.4(c) shows the conformity on the common edge of two NURBS patches with different parameterizations.

8.3.2 *FEM discretization*

According to Section 8.2, the basic variables are the displacements of the original reference surface (ORS) and the rotations of each layer. Therefore, the element nodes are collocated on the edge of ORS, see Fig. 8.5. Then the displacement of ORS can be approximated as

$$\mathbf{u}_{\mathrm{R}} \approx \mathbf{N}_{\mathrm{R}}\tilde{\mathbf{u}} \tag{8.20}$$

where

$$\mathbf{N}_{\mathrm{R}} = [S^{\mathrm{V1}} \sim S^{\mathrm{V4}}, S_2^{E1} \sim S_{M-1}^{E1}, \ldots, S_2^{E4} \sim S_{Q-1}^{E4}, S_{11}^{F} \sim S_{H_\xi H_\eta}^{F}] \otimes \mathbf{I}$$

$$\mathbf{I} = \mathrm{diag}\{[1, 1, 1]\}$$

$$\tilde{\mathbf{u}} = \begin{bmatrix} \mathbf{u}^E \\ \mathbf{u}^F \end{bmatrix}, \quad \mathbf{u}^{\mathrm{E}} = \begin{bmatrix} \mathbf{u}_1^{\mathrm{E}} \\ \vdots \\ \mathbf{u}_{N_d}^{\mathrm{E}} \end{bmatrix}, \quad \mathbf{u}_i^{\mathrm{E}} = \begin{bmatrix} u_i \\ v_i \\ w_i \end{bmatrix} \tag{8.21}$$

$$\mathbf{u}^{\mathrm{F}} = \begin{bmatrix} \mathbf{u}_{11}^{\mathrm{F}} \\ \vdots \\ \mathbf{u}_{H_\xi H_\eta}^{\mathrm{F}} \end{bmatrix}, \quad \mathbf{u}_{ij}^{\mathrm{F}} = \begin{bmatrix} u_{ij} \\ v_{ij} \\ w_{ij} \end{bmatrix}$$

where $N_d = M + N + P + Q - 4$ is the total number of edge and vertex bases. Similarly, the rotation field for each layer, k, can be approximated as

$$\boldsymbol{\Omega}^{(k)} \approx \mathbf{N}^{\mathrm{E}} \tilde{\boldsymbol{\omega}}_E^{(k)} + \mathbf{N}^{\mathrm{F}} \tilde{\boldsymbol{\omega}}_F^{(k)} \tag{8.22}$$

where $\mathbf{N}^E$ is the matrix of edge and vertex functions and $\mathbf{N}^F$ is the face function matrix. Obviously, they are submatrices of $\mathbf{N}_R$; $\tilde{\boldsymbol{\omega}}_E^{(k)}$ is the nodal rotation vector corresponding to the edge functions, and it can be expanded by the unit boundary tangent and normal vectors (see Fig. 8.5); $\tilde{\boldsymbol{\omega}}_F^{(k)}$ is the generalized rotation vector related to the face function, and it can be expanded by the covariant bases. They are expressed as

$$\begin{aligned} \tilde{\boldsymbol{\omega}}_E^{(k)} &= \begin{bmatrix} \vdots \\ \alpha_i^{(k)} \boldsymbol{n}_i + \beta_i^{(k)} \boldsymbol{\tau}_i \\ \vdots \end{bmatrix}, \quad i = 1 \sim N_{\mathrm{d}}; \\ \tilde{\boldsymbol{\omega}}_F^{(k)} &= \begin{bmatrix} \vdots \\ \alpha_{ij}^{(k)} \boldsymbol{g}_1 + \beta_{ij}^{(k)} \boldsymbol{g}_2 \\ \vdots \end{bmatrix}, \quad i = 1 \sim H_\xi, \ j = 1 \sim H_\eta \end{aligned} \tag{8.23}$$

Note that $\boldsymbol{n}_{\mathrm{i}}$ and $\boldsymbol{\tau}_i$ are the same for all layers, while the unknowns, $\alpha_i^{(k)}$, $\beta_i^{(k)}$, $\alpha_{ij}^{(k)}$, $\beta_{ij}^{(k)}$, vary from layer to layer. For convenience of the

FEM implementation, Eq. (8.23) can be rewritten in terms of the unknowns

$$\mathbf{\Omega}^{(k)} \approx \mathbf{N}_\Omega \tilde{\boldsymbol{\omega}}^{(k)} \tag{8.24}$$

where

$$\begin{aligned}
&\mathbf{N}_\Omega = [\mathbf{N}_\Omega^1\ \mathbf{N}_\Omega^2\ \mathbf{N}_\Omega^3\ \mathbf{N}_\Omega^4];\\
&\mathbf{N}_\Omega^1 = \mathbf{N}^{\mathrm{E}} \begin{bmatrix} \mathbf{n}_1 & & \\ & \ddots & \\ & & \mathbf{n}_{N_\mathrm{d}} \end{bmatrix}, \quad \mathbf{N}_\Omega^2 = \mathbf{N}^{\mathrm{E}} \begin{bmatrix} \boldsymbol{\tau}_1 & & \\ & \ddots & \\ & & \boldsymbol{\tau}_{Nd} \end{bmatrix}\\
&\mathbf{N}_\Omega^3 = \mathbf{N}^{\mathrm{F}} \begin{bmatrix} \boldsymbol{g}_1 & & \\ & \ddots & \\ & & \boldsymbol{g}_1 \end{bmatrix}, \quad \mathbf{N}_\Omega^4 = \mathbf{N}^{\mathrm{F}} \begin{bmatrix} \boldsymbol{g}_2 & & \\ & \ddots & \\ & & \boldsymbol{g}_2 \end{bmatrix}\\
&\tilde{\omega}^{(k)} = \begin{bmatrix} \boldsymbol{\omega}_1^{(k)} \\ \boldsymbol{\omega}_2^{(k)} \\ \boldsymbol{\omega}_3^{(k)} \\ \boldsymbol{\omega}_4^{(k)} \end{bmatrix}, \quad \boldsymbol{\omega}_1^{(k)} = \begin{bmatrix} \vdots \\ \alpha_i^{(k)} \\ \vdots \end{bmatrix}, \quad \boldsymbol{\omega}_2^{(k)} = \begin{bmatrix} \vdots \\ \beta_i^{(k)} \\ \vdots \end{bmatrix}\\
&\boldsymbol{\omega}_3^{(k)} = \begin{bmatrix} \vdots \\ \alpha_{ij}^{(k)} \\ \vdots \end{bmatrix}, \quad \boldsymbol{\omega}_4^{(k)} = \begin{bmatrix} \vdots \\ \beta_{ij}^{(k)} \\ \vdots \end{bmatrix}
\end{aligned} \tag{8.25}$$

As an example, we will derive the FEM matrices for a three-layered laminated shell element in the following. Firstly, the principle of virtual work is expressed as

$$\int_\Omega \boldsymbol{\sigma} : \delta\boldsymbol{\varepsilon} \mathrm{d}\Omega + \int_\Omega \rho\ddot{\boldsymbol{u}} \cdot \delta\boldsymbol{u} \mathrm{d}\Omega = \int_\Omega \boldsymbol{p} \cdot \delta\boldsymbol{u} \mathrm{d}\Omega + \int_\Gamma \boldsymbol{t} \cdot \delta\boldsymbol{u} \mathrm{d}\Gamma \tag{8.26}$$

where $\boldsymbol{\sigma}$ is the stress tensor; $\delta\boldsymbol{\varepsilon}$ is the virtual strain, $\delta\boldsymbol{u}$ is the virtual displacement, $\boldsymbol{p}$ is the body force and $\boldsymbol{t}$ is the traction load. Suppose that the ORS is located in layer 2 with thickness coordinate $\zeta_R^{(2)}$, then using Eqs. (26) and (30), the displacement field of layer 2 can be expressed as

$$\mathbf{u}^{(2)} = \mathbf{N}_{\mathrm{R}}^{(2)}\tilde{\mathbf{u}} + \boldsymbol{\Omega}^{(2)} \times (\zeta - \zeta_R^{(2)})\boldsymbol{n}; \quad \zeta \in [z_b^{(2)},\ z_t^{(2)}] \tag{8.27}$$

where $\mathbf{N}_{\mathrm{R}}^{(2)} = \mathbf{N}_{\mathrm{R}}$ because the ORS is located in layer 2. Define a skew-symmetric matrix

$$\boldsymbol{\Lambda} = \begin{bmatrix} 0 & n_z & -n_y \\ -n_z & 0 & n_x \\ n_y & -n_x & 0 \end{bmatrix} \tag{8.28}$$

where n_x, n_y and n_z are the Cartesian components of the normal director $\boldsymbol{n}$. Then Eq. (27) can be rewritten as

$$\begin{aligned} \mathbf{u}^{(2)} &= \tilde{\mathbf{N}}^{(2)}\tilde{\mathbf{d}}^{(2)} \\ \tilde{\mathbf{N}}^{(2)} &= [\mathbf{N}_{\mathrm{R}}^{(2)}, (\zeta - \zeta_R^{(2)})\boldsymbol{\Lambda}\mathbf{N}_{\Omega}]; \quad \zeta \in [\zeta_b^{(2)}, \zeta_t^{(2)}]; \\ \tilde{\mathbf{d}}^{(2)} &= \begin{bmatrix} \tilde{\mathbf{u}} \\ \tilde{\boldsymbol{\omega}}^{(2)} \end{bmatrix} \end{aligned} \tag{8.29}$$

Then the virtual displacement is given by

$$\delta\mathbf{u}^{(2)} = \tilde{\mathbf{N}}^{(2)}\delta\tilde{\mathbf{d}}^{(2)} \tag{8.30}$$

Define the following strain matrices:

$$\boldsymbol{\varepsilon} = [\varepsilon_{11}, \varepsilon_{22}, \varepsilon_{33}, \gamma_{12}, \gamma_{23}, \gamma_{31}]^{\mathrm{T}}; \quad \tilde{\boldsymbol{\varepsilon}} = [\tilde{\varepsilon}_{11}, \tilde{\varepsilon}_{22}, \tilde{\gamma}_{12}, \tilde{\gamma}_{23}, \tilde{\gamma}_{31}]^{\mathrm{T}}$$

where ε denotes the global strain components and $\tilde{\varepsilon}$ denotes the local strain components (note that $\tilde{\varepsilon}_{33} = 0$). According to Eq. (8.14), they are related by a transition matrix $\mathbf{Q}$ through

$$\tilde{\boldsymbol{\varepsilon}} = \mathbf{Q}\boldsymbol{\varepsilon} \tag{8.31}$$

Using Eqs. (8.13) and (8.29), ε and its variation $\delta\varepsilon$ can be expressed as

$$\boldsymbol{\varepsilon} = \mathbf{B}^{(2)}\tilde{\mathbf{d}}^{(2)}; \ \delta\boldsymbol{\varepsilon} = \mathbf{B}^{(2)}\delta\tilde{\mathbf{d}}^{(2)} \tag{8.32}$$

where

$$\mathbf{B}^{(2)} = [\mathbf{B}_{11}^{\mathrm{T}}\ \mathbf{B}_{22}^{\mathrm{T}}\ \mathbf{B}_{33}^{\mathrm{T}}\ \mathbf{B}_{12}^{\mathrm{T}}\ \mathbf{B}_{23}^{\mathrm{T}}\ \mathbf{B}_{31}^{\mathrm{T}}]$$

$$\begin{gathered}
\mathbf{B}_{11} = \mathbf{e}_1^{\mathrm{T}} \cdot \frac{\partial \tilde{\mathbf{N}}^{(2)}}{\partial x}; \quad \mathbf{B}_{22} = \mathbf{e}_2^{\mathrm{T}} \cdot \frac{\partial \tilde{\mathbf{N}}^{(2)}}{\partial y}; \quad \mathbf{B}_{33} = \mathbf{e}_3^{\mathrm{T}} \cdot \frac{\partial \tilde{\mathbf{N}}^{(2)}}{\partial z} \\
\mathbf{B}_{12} = \mathbf{e}_2^{\mathrm{T}} \cdot \frac{\partial \tilde{\mathbf{N}}^{(2)}}{\partial x} + \mathbf{e}_1^{\mathrm{T}} \cdot \frac{\partial \tilde{\mathbf{N}}^{(2)}}{\partial y} \\
\mathbf{B}_{23} = \mathbf{e}_3^{\mathrm{T}} \cdot \frac{\partial \tilde{\mathbf{N}}^{(2)}}{\partial y} + \mathbf{e}_2^{\mathrm{T}} \cdot \frac{\partial \tilde{\mathbf{N}}^{(2)}}{\partial z} \\
\mathbf{B}_{31} = \mathbf{e}_1^{\mathrm{T}} \cdot \frac{\partial \tilde{\mathbf{N}}^{(2)}}{\partial z} + \mathbf{e}_3^{\mathrm{T}} \cdot \frac{\partial \tilde{\mathbf{N}}^{(2)}}{\partial x}
\end{gathered} \tag{8.33}$$

Using Eq. (8.19), the functional in Eq. (8.26) can be discretized as

$$\delta \tilde{\mathbf{d}}^{\mathrm{T}} (\mathbf{K}^{(2)} \tilde{\mathbf{d}}^{(2)} + \mathbf{M}^{(2)} \ddot{\tilde{\mathbf{d}}}^{(2)}) = \delta \tilde{\mathbf{d}}^{\mathrm{T}} \mathbf{F}^{(2)} \tag{8.34}$$

where the stiffness, mass matrix and load vector are given by

$$\begin{aligned}
\mathbf{K}^{(2)} &= \int_{\Omega^{(2)}} \mathbf{B}^{(2)} \mathbf{T} \mathbf{Q}^{\mathrm{T}} \tilde{\mathbf{D}} \mathbf{Q} \mathbf{B}^{(2)} \mathrm{d}\Omega \\
\mathbf{M}^{(2)} &= \int_{\Omega^{(2)}} \rho \tilde{\mathbf{N}}^{(2)\mathrm{T}} \tilde{\mathbf{N}}^{(2)} \mathrm{d}\Omega \\
\mathbf{F}^{(2)} &= \int_{\Omega^{(2)}} \tilde{\mathbf{N}}^{(2)\mathrm{T}} \mathbf{p} \mathrm{d}\Omega + \int_{\Gamma^{(2)}} \tilde{\mathbf{N}}^{(2)\mathrm{T}} \bar{\mathbf{t}} \mathrm{d}\Gamma
\end{aligned} \tag{8.35}$$

For layers 1 and 3, the top and bottom faces of layer 2 are used as their reference surface. According to Eq. (8.29), the displacement fields of reference surface in layers 1 and 3 can be approximated as

$$\begin{aligned}
\mathbf{u}_R^{(1)} &= \mathbf{N}_R^{(1)} \tilde{\mathbf{d}}^{(2)}; \quad \mathbf{N}_R^{(1)} = [\mathbf{N}_R^{(2)}, (\zeta_t^{(2)} - \zeta_R^{(2)}) \mathbf{\Lambda} \mathbf{N}_\Omega] \\
\mathbf{u}_R^{(3)} &= \mathbf{N}_R^{(3)} \tilde{\mathbf{d}}^{(2)}; \quad \mathbf{N}_R^{(3)} = [\mathbf{N}_R^{(2)}, (\zeta_b^{(2)} - \zeta_R^{(2)}) \mathbf{\Lambda} \mathbf{N}_\Omega]
\end{aligned} \tag{8.36}$$

Then the displacement for layers 1 and 3 can be expressed by

$$\begin{aligned}
\mathbf{u}^{(1)} &= \mathbf{N}_{\mathrm{R}}^{(1)} \mathbf{d}^{(2)} + (\zeta - \zeta_R^{(1)}) \mathbf{\Lambda} \mathbf{N}_\Omega \tilde{\omega}^{(1)} = \tilde{\mathbf{N}}^{(1)} \tilde{\mathbf{d}}^{(1)} \\
\mathbf{u}^{(3)} &= \mathbf{N}_{\mathrm{R}}^{(3)} \mathbf{d}^{(2)} + (\zeta - \zeta_R^{(3)}) \mathbf{\Lambda} \mathbf{N}_\Omega \tilde{\omega}^{(3)} = \tilde{\mathbf{N}}^{(3)} \tilde{\mathbf{d}}^{(3)}
\end{aligned} \tag{8.37}$$

where

$$\tilde{\mathbf{N}}^{(1)} = [\mathbf{N}_{\mathrm{R}}^{(1)}, (\varsigma - \varsigma_R^{(1)})\mathbf{\Lambda}\mathbf{N}_\Omega]; \tilde{\mathbf{N}}^{(3)} = [\mathbf{N}_{\mathrm{R}}^{(3)}, (\varsigma - \varsigma_R^{(3)})\mathbf{\Lambda}\mathbf{N}_\Omega] \quad (8.38)$$

$$\tilde{\mathbf{d}}^{(1)} = \begin{bmatrix} \tilde{\mathbf{d}}^{(2)} \\ \tilde{\omega}^{(1)} \end{bmatrix}; \quad \tilde{\mathbf{d}}^{(3)} = \begin{bmatrix} \tilde{\mathbf{d}}^{(2)} \\ \tilde{\omega}^{(3)} \end{bmatrix} \quad (8.39)$$

Using a similar process, one can easily obtain the corresponding element matrices. Apparently, for the three-layered shell element, the gross unknowns are

$$\tilde{\mathbf{d}} = \begin{bmatrix} \tilde{\mathbf{u}} \\ \tilde{\omega}^{(1)} \\ \tilde{\omega}^{(2)} \\ \tilde{\omega}^{(3)} \end{bmatrix} \quad (8.40)$$

Then, according to Eqs. (8.29) and (8.39), it follows that

$$\tilde{\mathbf{d}}^{(1)} = \mathbf{T}_1\tilde{\mathbf{d}}; \quad \tilde{\mathbf{d}}^{(2)} = \mathbf{T}_2\tilde{\mathbf{d}}; \quad \tilde{\mathbf{d}}^{(3)} = \mathbf{T}_3\tilde{\mathbf{d}} \quad (8.41)$$

where $\mathbf{T}_i$, $i = 1, 2, 3$, is the transition matrix. Finally, the matrices for the three-layered elements can be obtained as

$$\mathbf{K} = \sum_{i=1}^{3} \mathbf{T}_i^{\mathrm{T}}\mathbf{K}^{(i)}\mathbf{T}_i; \quad \mathbf{M} = \sum_{i=1}^{3} \mathbf{T}_i^{\mathrm{T}}\mathbf{M}^{(i)}\mathbf{T}_i; \quad \mathbf{F} = \sum_{i=1}^{3} \mathbf{T}_i^{\mathrm{T}}\mathbf{F}^{(i)} \quad (8.42)$$

Using the above processes, one can obtain the FEM matrices of laminated shell element with arbitrary number of layers.

8.4 Numerical Examples

In this section, numerical examples including static and free vibration analyses of isotropic and laminated composite plates and shells are presented to check the computation performance of present elements. Convergence studies are made and the results are compared with those in the literatures and obtained by the authors using ABAQUS. For simplicity if not mentioned otherwise, the element node at each edge is taken as the same N as the face function number in each direction.

8.4.1 *Analyses of plates*

The first example for the analyses of composite structures is the bending problem of a simply supported sandwich plate under uniform distributed load, see Fig. 8.6. Orthotropic material with the same axes of orthotropy is used in the two skin layers and the core. The material parameters E and G of the skin are R times stronger than those of the core. This example is used to show that the present elements can produce very accurate stresses even when the material properties vary severely. We note that the present layerwise model can easily degenerate into FSDT model by forcing the rotations in each layer, i.e., $\tilde{\omega}^{(k)}$ in Eq. (8.24), to take the same value; then shear correction factors κ_{11} and κ_{22} are required to improve the accuracy. In Table 8.3, we present the normalized central point deflection and stresses obtained by both layerwise model and FSDT model, and compare them with the solutions obtained by 3D theory [403] and other finite element solutions [404, 405] based on assumed strain method and FSDT. As it shows, the present FSDT results converge slightly faster than the LW results. When R is small, both FSDT theory and LW theory can produce sufficiently accurate results, and the results agree well with the other numerical solutions. However, when $R = 50$, significant deviations of stresses from the 3D reference solutions are observed in FSDT and other results, while the LW results are still very accurate. Figure 8.7 shows the distribution of normalized stress $\bar{\sigma}_x$ and the displacement u obtained by FSDT theory and LW model with different ratios R. As it shows, a piecewise linear warping is allowed in the LW model.

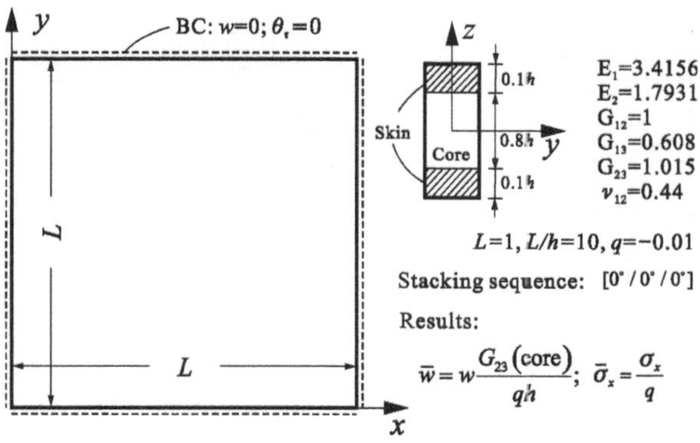

Fig. 8.6. A simply supported sandwich plate.

Table 8.3. Convergence of displacement and stresses at the center of simply supported sandwich plate.

R	Theory	N	$\bar{w}$ $(z = 0)$	$\bar{\sigma}_x$ $(z = -0.4h$, core)	$\bar{\sigma}_x$ $(z = -0.4h$, skin)	$\bar{\sigma}_x$ $(z = -0.5h)$
1	FSDT	5	181.32	28.50	28.50	35.63
	$\kappa_{11} = \kappa_{22} = 0.8333$	7	181.33	28.49	28.49	35.62
		11	181.33	28.50	28.50	35.62
	Layerwise	5	180.57	28.44	28.44	35.92
		7	180.58	28.43	28.43	35.92
		11	180.58	28.44	28.44	35.92
		15	180.58	28.44	28.44	35.92
	3D elasticity [403]	—	181.05	28.45	28.45	35.94
	Heterosis [405]	—	183.99	27.48	27.48	34.34
	QLLL [404]	—	180.05	29.10	29.10	36.37
10	FSDT	5	41.99	5.05	50.49	63.11
	$\kappa_{11} = \kappa_{22} = 0.3521$	7	41.99	5.05	50.50	63.12
		11	41.99	5.05	50.50	63.13
	Layerwise	5	41.92	4.87	48.74	65.17
		7	41.93	4.87	48.70	65.25
		11	41.93	4.87	48.71	65.25
		15	41.93	4.87	48.72	65.23

	3D elasticity [403]	—	41.91	4.86	48.61	65.08
	Heterosis [405]	—	41.92	4.87	48.73	65.23
	QLLL [404]	—	39.50	5.40	54.37	67.90
50	FSDT	5	16.84	0.93	46.35	57.93
	$\kappa_{11} = \kappa_{22} = 0.0938$	7	16.84	0.92	46.39	57.99
		11	16.84	0.93	46.40	58.00
	Layerwise	5	16.75	0.75	37.59	66.65
		7	16.79	0.74	37.03	67.30
		11	16.78	0.74	37.22	67.13
		15	16.78	0.74	37.27	67.08
	3D elasticity [403]	—	16.75	0.74	37.15	66.90
	Heterosis [405]	—	16.85	0.93	46.65	58.31
	QLLL [404]	—	16.25	1.18	59.04	73.08

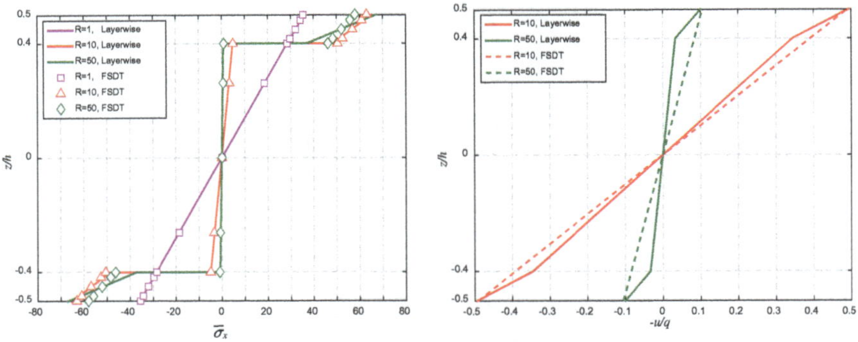

Fig. 8.7. Distribution of $\bar{\sigma}_x(L/2, L/2)$ and $u(0, L/2)/q$ through thickness.

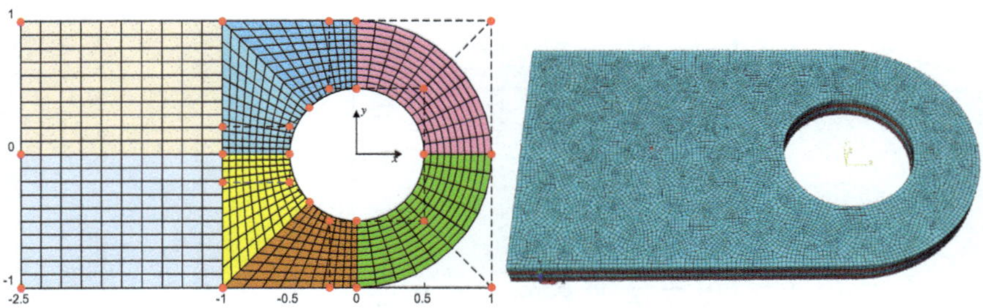

Fig. 8.8. The meshes of straight lug: (a) meshed by 8 p-elements (• the control points); (b) meshes in ABAQUS.

In the second example, we consider the free vibration analysis of a laminated straight lug. This problem is employed as a case of plates with complicated shape. The geometric dimensions and meshes are shown in Fig. 8.8, where eight elements (identified by colors) are used to part the irregular domain. The patch coupling is considered in this example as one may notice the mismatch of isoparametric curves on $x = -1$. The material parameters are given as follows: (I) graphite–epoxy: $E_1 = 137.9\,\text{GPa}$, $E_2 = 8.96\,\text{GPa}$, $G_{12} = G_{13} = 7.1\,\text{GPa}$, $G_{23} = 6.21\,\text{GPa}$, $\nu_{12} = 0.3$, $\rho = 1450\,\text{kg/m}^3$; (II) glass–epoxy: $E_1 = 53.78\,\text{GPa}$, $E_2 = 13.93\,\text{GPa}$, $G_{12} = G_{13} = 8.96\,\text{GPa}$, $G_{23} = 3.45\,\text{GPa}$, $\nu_{12} = 0.25$, $\rho = 1900\,\text{kg/m}^3$. Table 8.4 presents the convergence of first five natural frequencies of the lug with different stacking sequences and ply thickness. We compare our

Table 8.4. First five natural frequencies of laminated straight lug with different stacking sequences.

Thickness	Stacking sequence	N	Frequencies (Hz) 1	2	3	4	5
$h = 0.1$	$[45°/0°/-45°]$	7	6.4333	26.705	31.597	75.290	82.284
$[h/4, h/2, h/4]$	[I/II/I]	9	6.4324	26.701	31.591	75.278	82.272
		11	6.4322	26.700	31.589	75.274	82.270
		ABAQUS	6.4588	26.684	31.672	75.215	82.340
	$[90°/0°/90°]$	7	4.8906	19.066	24.665	56.765	67.907
	[I/II/I]	9	4.8905	19.061	24.663	56.755	67.904
		11	4.8905	19.061	24.663	56.754	67.903
		ABAQUS	4.9292	19.107	24.833	56.910	68.397
$h = 0.2$	$[0°/90°/0°/90°]$	7	20.362	43.743	91.282	124.70	146.30
$[h/4, h/4,$	[I/II/I/II]	9	20.362	43.741	91.278	124.69	146.29
$h/4, h/4]$		11	20.362	43.741	91.278	124.69	146.29
		ABAQUS	20.338	43.323	89.902	124.60	143.72
	$[-45°/45°/45°/-45°]$	7	12.529	51.201	61.187	105.69	135.60
	[I/II/II/I]	9	12.523	51.195	61.164	105.62	135.59
		11	12.520	51.193	61.156	105.59	135.59
		ABAQUS	12.528	50.900	60.861	105.98	134.56

results with the solutions obtained by ABAQUS continuum shell elements (SC8R) with very fine meshes (Fig. 8.8(b)). One can see that the present results converge very well, and they are in good agreement with the ABAQUS results. Fig. 8.9 plots the first five mode shapes of the laminated straight lug with material [I/II/II/I] and stacking sequence $[-45°/45°/45°/-45°]$.

In the third example, we consider the free vibration of a stiffened isotropic square plate shown in Fig. 8.10(a). This example is used to show the flexibility of present elements in modeling complex structures. The plate is modeled by five elements shown in Fig. 8.10(a). Each stiffener is modeled by one narrow element with 2–4 layers shown in Fig. 8.10(b), while elements with single layer are used in other area. The node number in width direction of the stiffener is taken as 3 as the width is small, while the node number in length direction is denoted by the same N as the single-layer elements. Because of the flexibility of present elements in choosing the ORS, the elements in different area can be assembled compatibly, see Fig. 8.10(b). The results of first four natural frequencies

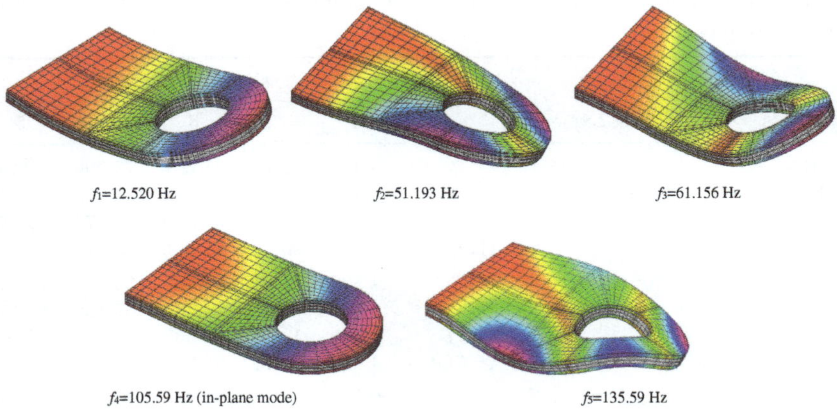

Fig. 8.9. First five mode shapes of [−45°/45°/45°/−45°] laminated straight lug.

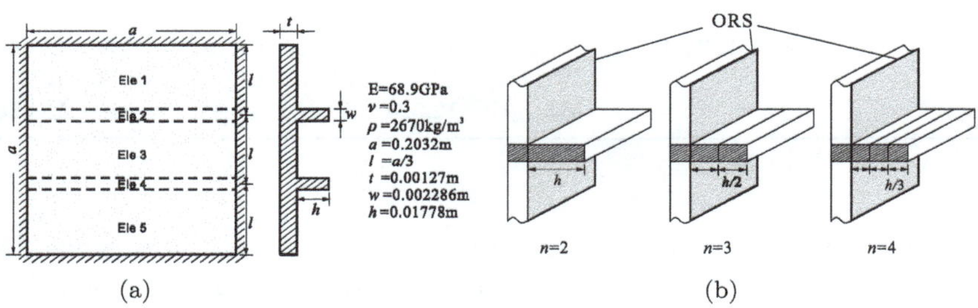

Fig. 8.10. Double ribbed plate with clamped boundaries: (a) the geometric dimensions and material constants; (b) modeling of stiffener using an n-layered element.

are listed in Table 8.5 and compared with other solutions in the literatures. The data in brackets are the relative error of our best results ($n = 4$, $N = 11$). As it can be observed, with the increase of layers in thickness direction of the stiffeners, the frequencies are decreased accordingly. This is because the adding of layers is equivalent to the mesh refinement in the thickness direction. Since this example is not a standard problem, small difference of the results from those obtained by different authors is observed. However, the relative errors refer to different results still under the level of 5%. As it shows, the present solutions are very close to the differential quadrature solutions except the basic frequency. Figure 8.11 shows the mode shapes of the stiffened plate.

Table 8.5. Convergence comparison of first four frequencies of stiffened plate.

Layers of stiffener (n)	N	Frequencies (Hz) 1	2	3	4
2	5	951.3	1252.3	1426.4	1754.6
	7	944.8	1247.4	1353.8	1422.4
	9	943.7	1246.5	1352.4	1421.4
	11	943.5	1246.2	1351.5	1420.9
3	5	950.2	1250.6	1421.9	1753.9
	7	943.8	1245.8	1352.2	1417.8
	9	942.8	1245.0	1351.0	1416.7
	11	942.6	1244.7	1350.1	1416.2
4	5	949.7	1249.9	1420.0	1753.4
	7	943.3	1245.0	1351.5	1415.7
	9	942.3	1244.2	1350.2	1414.6
	11	942.1	1243.9	1349.4	1414.1
Semi-analytical [406]		931.5(1.14%)	1220.9(1.88%)	1331.8(1.34%)	1403.3(0.77%)
FEM [407]		965.3(−2.40%)	1272.3(−2.23%)	1364.3(−1.09%)	1418.1(−0.28%)
DQ [408]		915.9(2.86%)	1242.2(0.14%)	1344.4(0.37%)	1414.1(0.00%)

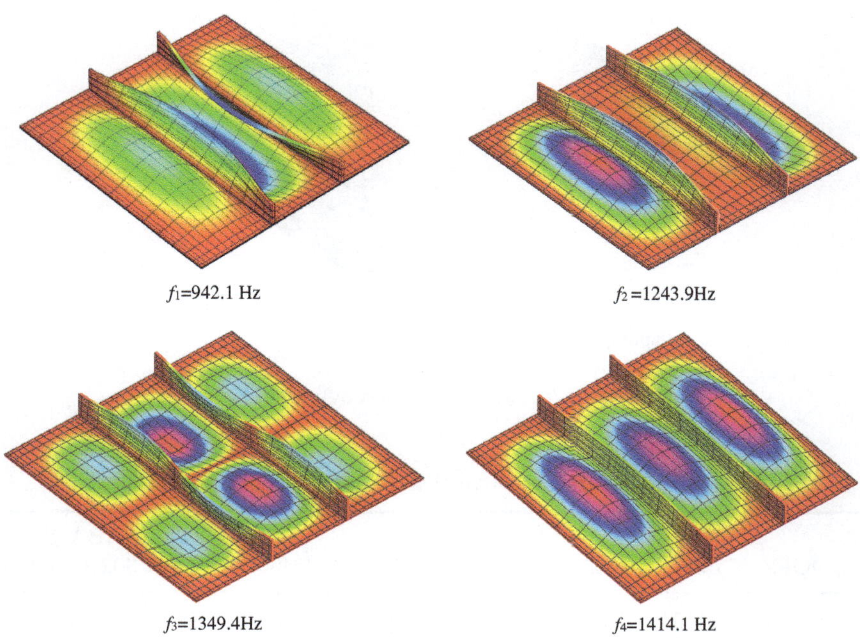

Fig. 8.11. First four mode shapes of stiffened square plates with clamped boundary ($n = 4$).

8.4.2 Analyses of shells

The first example of the shell analysis is the pinched cylinder problem with a unit concentrated load applied on the center (Fig. 8.12). Both ends of the cylinder are covered with rigid diaphragms which allow displacement only in the axial direction of the cylinder. Due to the symmetry, 1/8 of the shell is modeled by one DQHFEM element. This example is a classical problem to check the shear locking as the shell is very thin with ratio $h/R = 100$. Table 8.6 presents the numerical results obtained by present elements and compared with the solutions given by Roh [409] using assumed strain method and the ABAQUS results given by the author using 70×70 S4R elements (30246DOFs). As it can be seen, the present results agree well with the reference solutions. Moreover, as shown in Fig. 8.13, compared with the h-version S4R elements, the present DQHFEM

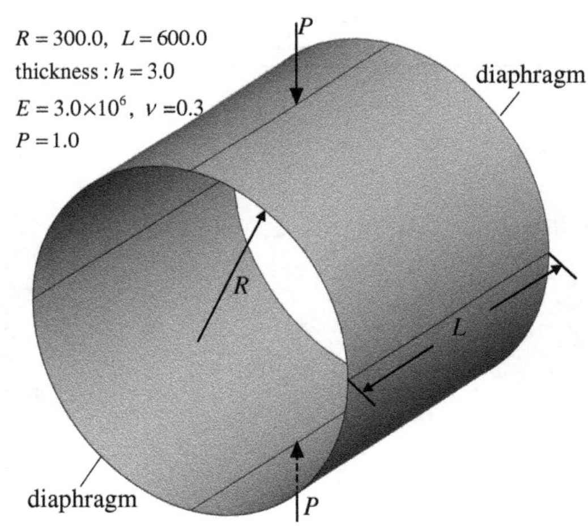

Fig. 8.12. A pinched cylinder.

Table 8.6. Convergence of vertical displacement at loading point.

N/DOFs	7/365	9/565	11/805	13/1085	15/1405	Roh [409]	ABAQUS 30246 DOFs
w (e$^{\cdot -05}$)	1.3116	1.7677	1.8274	1.8409	1.8464	1.8541	1.8447

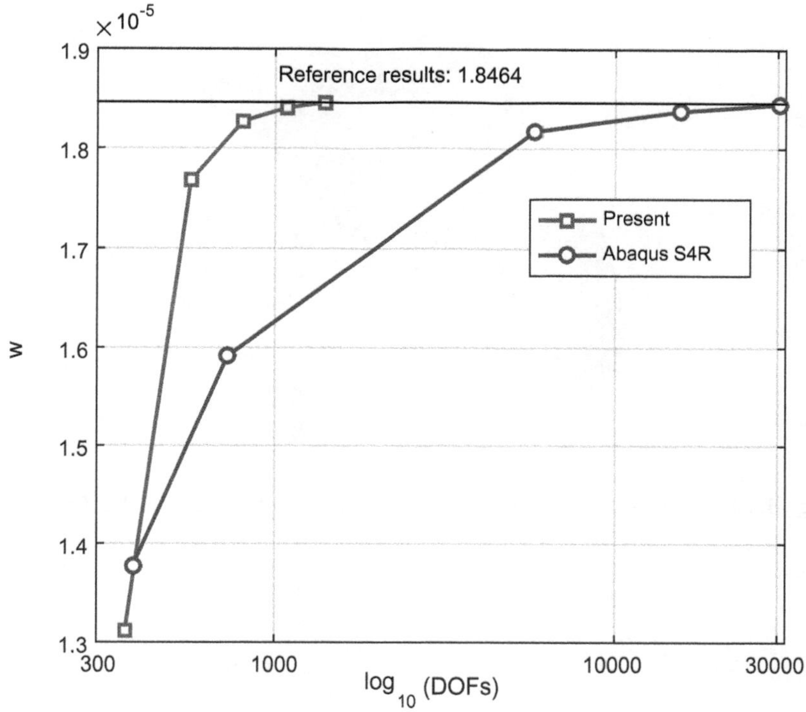

Fig. 8.13. Convergence comparison of vertical displacement at loading point.

uses far less DOFs to achieve the same accuracy. Importantly, this example also indicates that the present elements are insensitive to the shear locking. Figure 8.14 shows the vertical displacement field on the 1/8 shell.

The second example is the classical Scordelis–Lo roof. The geometry and material constants as well as loadings are shown in Fig. 8.15. This example is used to show that the present elements are free from membrane locking. The coupling problem of elements with different parameterizations is also studied. As shown in Fig. 8.16, two elements with equal (Fig. 8.16(a)) and different (Fig. 8.16(b)) parameterizations are used to model the roof. The results of vertical displacement at point "A" are presented in Table 8.7 and compared with the numerical solutions given by Scordelis [410] and Macneal [411]. Good agreement is observed again in both kinds of meshes, and the results

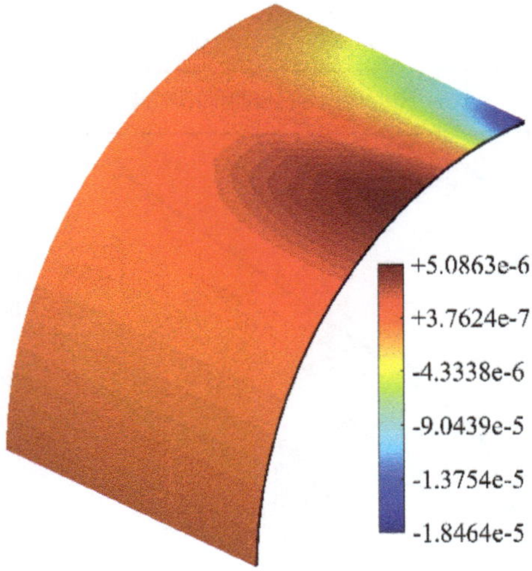

Fig. 8.14. The vertical displacement field obtained by DQHFEM with $N = 15$.

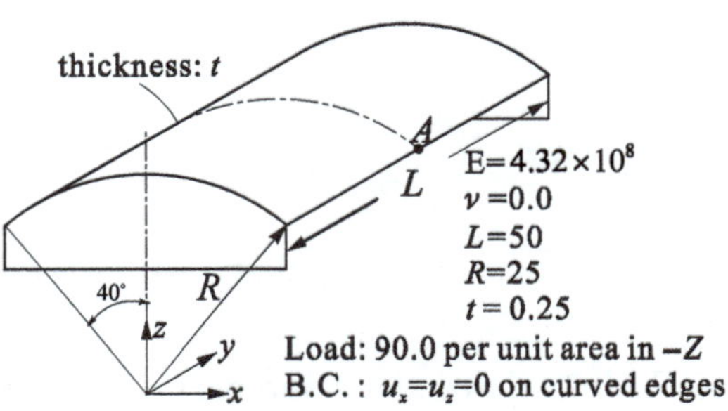

Fig. 8.15. Scordelis–Lo roof.

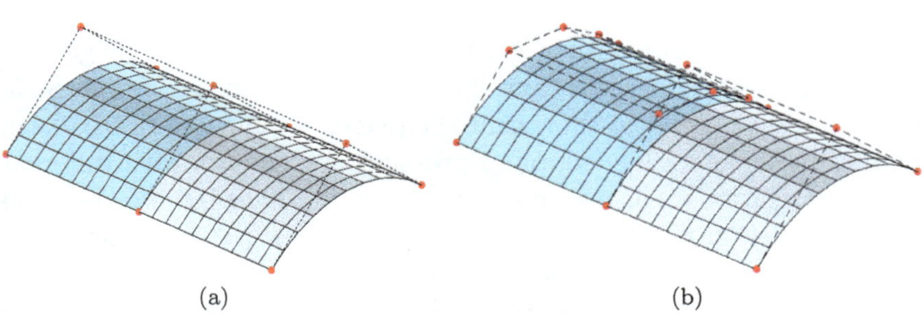

Fig. 8.16. The Scordelis–Lo roof modeled by two elements (identified by color): (a) two elements with the same parameterization; (b) two elements with different parameterization.

Table 8.7. Convergence comparison of the vertical displacement at point A.

N	5	7	9	11	13	15	17	Scordelis [410]	Macneal [411]
Mesh (a)	0.2965	0.3093	0.3076	0.3078	0.3080	0.3081	0.3083	0.3086	0.3024
Mesh (b)	0.2964	0.3095	0.3077	0.3079	0.3080	0.3082	0.3083		

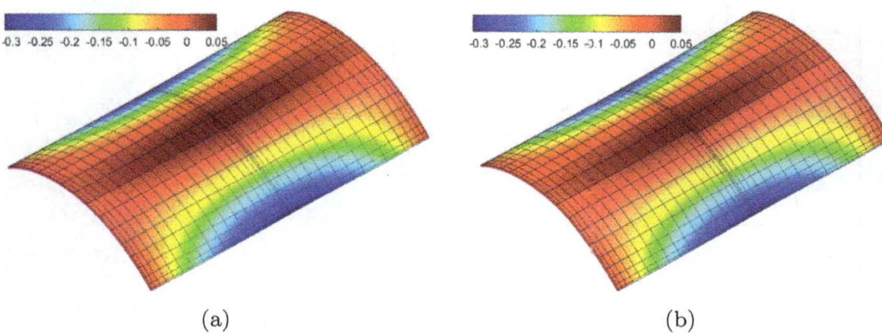

Fig. 8.17. Comparison of the vertical displacement field: (a) two elements with the same parameterization; (b) two elements with different parameterization.

obtained by different meshes are very close when the same number of nodes are used, which shows the coupling problem of elements with different parameterizations is well solved. Additionally, it also shows that sufficient accurate results can be obtained when a few nodes are used and no membrane locking is observed. Figure 8.17 verifies the good agreement of the overall vertical displacement field obtained by the two meshes.

In the third example, the static analysis of an isotropic hook is performed. This example is used to show the advantages of NURBS in modeling complicated structures. The geometric dimensions and the load are shown in Fig. 8.18(a), while the DQHFEM meshes and the ABAQUS meshes are shown in Figs. 8.18(b) and (c). The DQHFEM results of top surface displacement magnitude as well as two in-plane stresses S_{11} and S_{22} are shown in Fig. 8.19 and compared with the ABAQUS solutions obtained by S4R elements with very fine meshes. S_{11} denotes the stress in the projected direction of z-axis on the surface (1-direction), while S_{22} denotes the stress in the

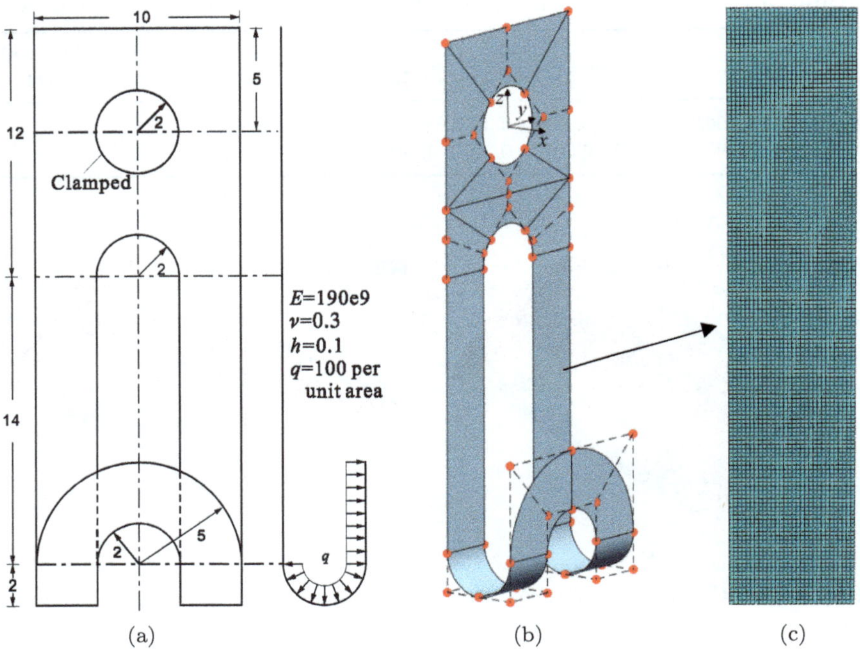

Fig. 8.18. A hook: (a) dimensions of the hook; (b) NURBS representation and the DQHFEM meshes; (c) ABAQUS meshes on a portion of the hook.

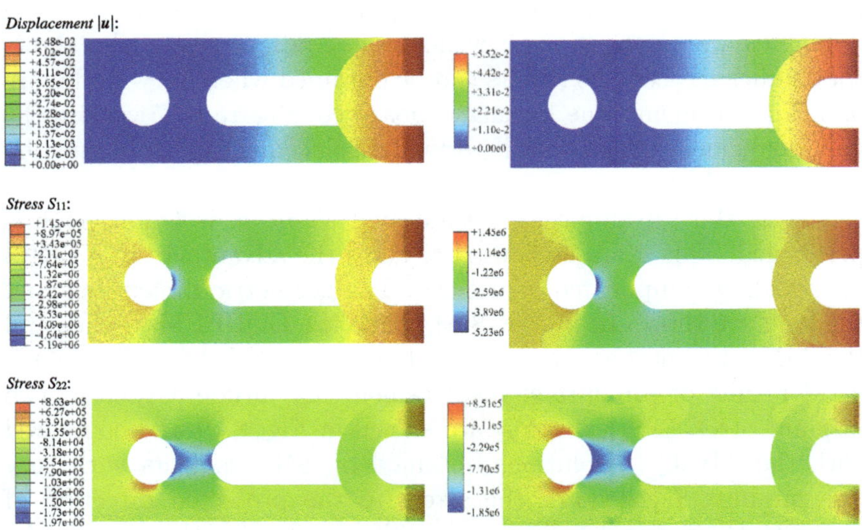

Fig. 8.19. Results comparison: (left) ABAQUS results using 135594 DOFs; (right) DQHFEM results using 6320 DOFs ($N = 7$ for each element).

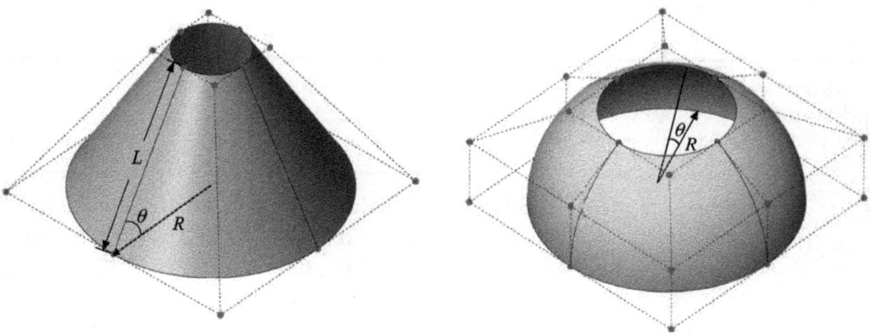

Fig. 8.20. Geometry and control points of conic and hemispherical shells.

direction which is perpendicular to the 1-direction and the normal direction. As it shows, good agreement of the DQHFEM results and the ABAQUS results is observed. The DQHFEM results of stress also exhibit good continuity at the element boundary as the smoothed ABAQUS results. Moreover, because of the advantages of high-order approximation, the DQHFEM uses far less DOFs, nearly one half, than the *h*-version S4R elements to obtain the given results.

In the last two examples, we consider the free vibration analysis of two revolutionary laminated shells, namely, the conic and the hemispherical shells as shown in Fig. 8.20. Each of them will be modeled by four elements. The geometry parameters of the conic shell are $R = 1\,\text{m}$, $L = 3\,\text{m}$, $h = 0.3\,\text{m}$, $\theta = \pi/6$, while those for the hemispherical shell, are $R = 1\,\text{m}$, $h = 0.1\,\text{m}$ and $\theta = \pi/3$. The material parameters for each layer of the conic shell are $E_1 = 137.9\,\text{GPa}$, $E_2 = 8.96\,\text{GPa}$, $\nu_{12} = 0.25$, $G_{12} = G_{13} = 7.1\,\text{GPa}$, $G_{23} = 6.21\,\text{GPa}$, $\rho = 1450\,\text{kg/m}^3$, while those for each layer of the hemispherical shell are $E_2 = 10\,\text{GPa}$, $E_1/E_2 = 15$, $\nu_{12} = 0.25$, $G_{12} = G_{13} = 0.6E_2$, $G_{23} = 0.5E_2$, $\rho = 1500\,\text{kg/m}^3$.

Table 8.8 presents the convergence of the first six natural frequencies of the $[45°/-45°]$ conic shell with C–C (clamped at two ends) and C-F (clamped at the bottom and free at the top) boundary conditions. The ply thickness for each layer is $h/2$. The results are compared with the solutions obtained by variational formulation based on FSDT [412] and GDQ solutions [413], based on the second-order shear deformation theory (SSDT). Slight discrepancies between the

Table 8.8. Convergence of first six natural frequencies for [45°/−45°] conic shell.

Boundary Condition (B. C.)	Method	N	Frequencies (Hz) 1	2	3	4	5	6
C-F	Present	7	179.820	253.2285	269.395	333.460	350.3324	393.693
		9	179.708	252.718	269.360	333.163	347.887	393.257
		11	179.691	252.680	269.356	333.142	347.772	393.230
		13	179.689	252.674	269.356	333.139	347.768	393.227
	FSDT [412]	—	181.574	252.191	267.253	326.712	350.024	396.493
	SSDT [413]	—	177.579	250.953	267.719	329.643	346.569	390.134
C-C	Present	7	302.275	316.312	361.577	375.792	404.221	442.346
		9	302.571	316.142	358.889	375.591	403.690	441.082
		11	302.539	316.130	358.802	375.586	403.677	441.056
		13	302.535	316.129	358.798	375.586	403.677	441.056
	FSDT [413]	—	305.960	316.531	363.613	370.638	395.490	432.104

present results and the reference solutions are observed. This is probably due to the strain–displacement relation. In [412] and [413], the strain–displacement relations are expressed in orthogonal curvilinear coordinates and simplified by the geometry assumptions $h/R_\alpha \ll 1$ and $h/R_\beta \ll 1$, where h is the thickness, and R_α, R_β are the radii of curvatures. However, these assumptions are not employed in our elements as the isoparametric curves of the NURBS patch are generally non-orthogonal. Nevertheless, one can verify that the relative errors of the results are still very small. Figure 8.21 shows the first six mode shapes of the C–F conic shell.

Table 8.9 presents the five picked frequency parameters $\Omega = \omega R\sqrt{\rho/E_2}$ of the hemispherical shell with four equally thick plies stacked in a sequence of [0°/90°/0°/90°]. The results are compared with the Fourier solutions based on FSDT [414]. As it can be seen, the present DQHFEM results converge very fast even with four layers analyzed in this example. It also shows that the present results based on LW model are slightly lower than those based on FSDT. This is because the deformation of normal fibers in present elements is allowed to be piecewise linear, which is more flexible than the FSDT model. The corresponding mode shapes for the C–F shell are presented in Fig. 8.22.

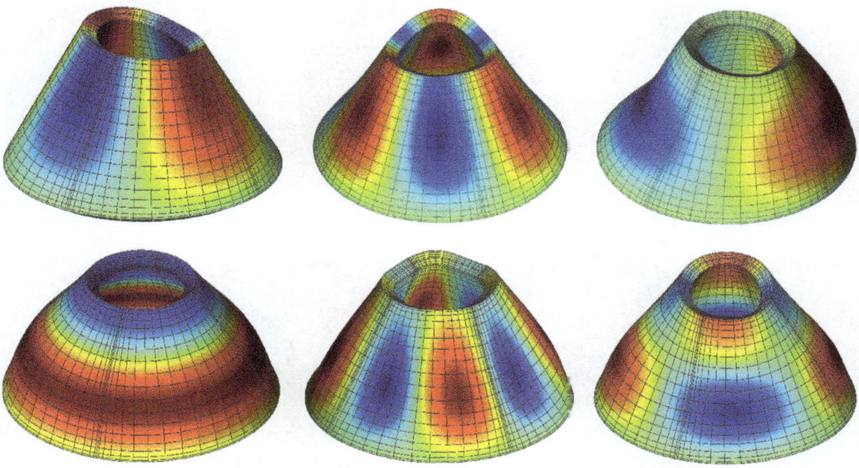

Fig. 8.21. The first six mode shapes of the [45°/−45°] conic shell with C–F boundary.

Table 8.9. Convergence of the five picked frequency parameters of [0°/90°/0°/90°] hemispherical shell.

B. C.	Method	N	Frequencies ($\Omega = \omega R\sqrt{\rho/E_2}$) 1	2	3	4	5
C–F	Present	7	0.79103	1.02562	1.40688	2.17008	2.18520
		9	0.79034	1.02540	1.40340	2.15866	2.18324
		11	0.79023	1.02538	1.40309	2.15785	2.18298
		13	0.79021	1.02537	1.40306	2.15776	2.18294
	FSDT [414]	—	0.80084	1.03778	1.42610	2.19843	2.21868
C–C	Present	7	2.39715	2.52436	2.64800	2.87486	3.07681
		9	2.39656	2.52207	2.64197	2.86392	3.07191
		11	2.39655	2.52203	2.64185	2.86362	3.07093
		13	2.39655	2.52203	2.64184	2.86361	3.07077
	FSDT [414]	—	2.43409	2.56885	2.69277	2.92038	3.15755

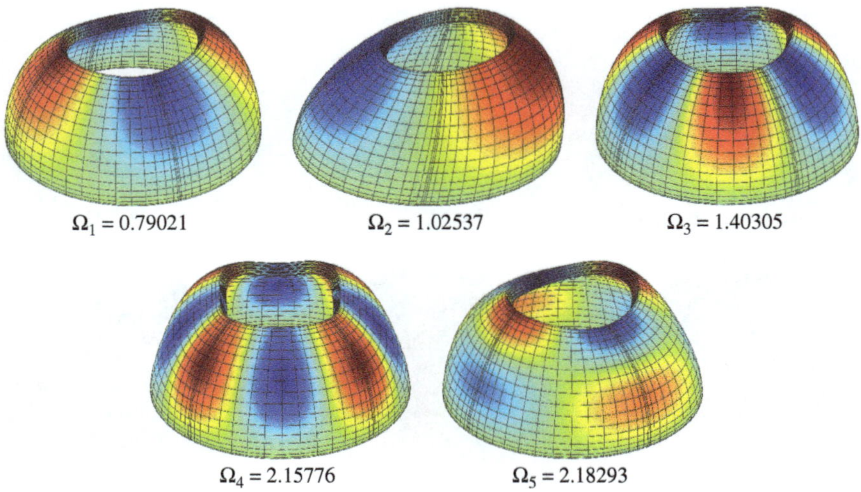

Fig. 8.22. The five picked mode shapes of [0°/90°/0°/90°] hemispherical shell with C–F boundary.

8.5 Conclusions

In this chapter, both isotropic and composite laminated plates and shells have been analyzed by a *p*-version differential quadrature hierarchical finite element method (DQHFEM). The theoretical model of the shell is based on the layerwise theory with linear expansion across the thickness in each lamina. Therefore, sandwich structures with significant material property variation can be analyzed accurately, and no shear correction factors are required. Moreover, to avoid geometry discretization errors that usually arose in conventional flat shell elements, the CAD technique NURBS is used to exactly represent the middle surface of the shell. In this way, complicated models can be easily constructed without losing precisions. As the complicated models are often composed by patches with different parameterization, the coupling difficulties on element (or patch) boundaries are also solved in this work by a novel method of interpolation on arc length coordinates. Therefore, the present elements can be used to model complicated structures. A lot of numerical examples verified the high accuracy, fast convergence, as well as the versatilities of the present element in analyzing the isotropic and composite structures with simple and complicated shapes.

Chapter 9

Hierarchical Quadrature Element Method for Geometrically Nonlinear Problems

The high-order finite element method has the advantages of fast convergence, insensitivity to mesh distortion, and convenience in realizing adaptive analysis. As a high-order scheme, the hierarchical quadrature element method (HQEM) has already been applied to linear finite element analysis and is found with high accuracy and efficiency. This paper carries out a set of geometrically nonlinear analyses using the HQEM. Under the total Lagrange framework, three types of nonlinear elements are constructed: two-dimensional, three-dimensional and Mindlin hierarchical quadrature elements. Based on the Green–Lagrange strain tensor and the second Piola–Kirchhoff stress tensor under the reference configuration, the virtual work equation is constructed and discretized in the space domain. Two iteration solution techniques, the Newton–Raphson and the arclength methods, are used for geometrically nonlinear quasi-static analyses based on the constructed elements. The results are first compared with the solutions of similar units in the commercial software ABAQUS to verify the correctness and accuracy of the presented method. Then the convergence characteristics are investigated. Some numerical examples used single element as well as multiple element assembly for analysis to verify the ease of assembly of the presented formulation.

Compared with low-order schemes, the presented method requires far fewer degrees of freedom when analyzing the same problem. It shows a promising application prospect in geometrically nonlinear finite element analysis which requires repeated iteration and update of the stiffness matrix.

9.1 The Hierarchical Quadrature Element Method

The HQEM is essentially a high-order finite element method. It uses weak form equations to get rid of the difficulty of analyzing problems in complex domains and uses non-uniformly distributed Gauss–Lobatto integration points to avoid the Runge phenomenon on the model boundary, while ensuring the accuracy of analyses. The HQEM inherits the advantages of HFEM, that is, the number of internal basis functions is independent of the number of bases on the boundaries, so local order elevation will not propagate from patch to patch. The interpolation nodes on the boundary of HQEM elements avoid the difficulty of HFEM in element assembly and boundary condition imposition.

9.1.1 *Shape functions for quadrilateral elements*

In this chapter, two-dimensional and the Mindlin shallow shell hierarchical quadrature elements are constructed as a quadrilateral element, as shown in Fig. 9.1. The serendipity interpolation shape functions based on non-uniform Gauss–Lobatto nodes are used on the boundary of the quadrilateral element, and the hierarchical shape functions in tensor product form is used inside the quadrilateral element.

First, the displacement field composed of interpolation bases on the boundary of an element and hierarchical bases inside the domain can be approximated as follows:

$$u[x(\xi,\eta),y(\xi,\eta)]=\sum_{k=1}^{K}S_k(\xi,\eta)u_k+\sum_{m=1}^{H_\xi}\sum_{n=1}^{H_\eta}\varphi_m(\xi)\varphi_n(\eta)a_{mn} \quad (9.1)$$

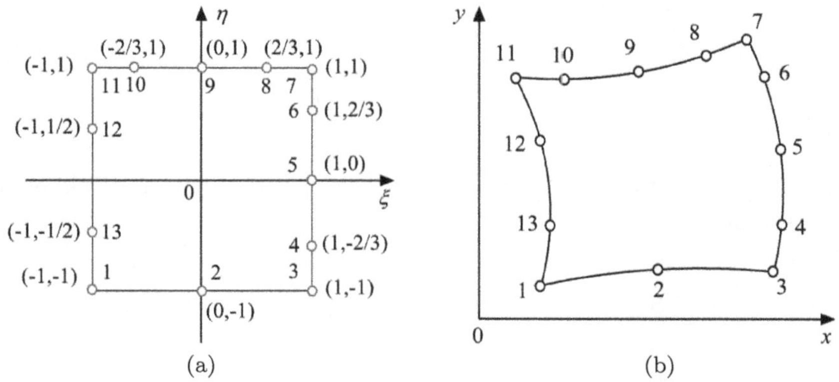

Fig. 9.1. A quadrilateral element: (a) parametric domain; (b) geometric domain.

where K is the number of nodes on the boundary of the quadrilateral element, S_K is the serendipity interpolation shape function on the boundary, $H_\xi H_\eta$ are the number of hierarchical shape functions along the two orthogonal directions inside the quadrilateral element, and φ_m, φ_n are the corresponding one-dimensional hierarchical quadrature shape functions. The one-dimensional hierarchical quadrature bases are usually constructed by integration of Legendre polynomials:

$$\varphi_m(\xi) = \int_{-1}^{\xi} L_m(\xi)\mathrm{d}\xi = \frac{(\xi^2-1)}{m(m+1)}\frac{\mathrm{d}L_m(\xi)}{\mathrm{d}\xi},$$
$$\xi \in [-1,1],\ m = 1, 2, \ldots \tag{9.2}$$

Therefore, the four shape functions at the corners of the quadrilateral element are as follows:

$$S^{\mathrm{V1}} = \frac{(1-\eta)}{2}L_1^M(\xi) + \frac{(1-\xi)}{2}L_1^N(\eta) - \frac{(1-\xi)(1-\eta)}{4}$$

$$S^{\mathrm{V2}} = \frac{(1-\eta)}{2}L_M^M(\xi) + \frac{\xi}{2}L_1^N(\eta) - \frac{\xi(1-\eta)}{4}$$

$$
\begin{aligned}
S^{V3} &= \frac{\xi}{2}L_N^N(\eta) + \frac{\eta}{2}L_M^M(\xi) - \frac{\xi\eta}{4} \\
S^{V4} &= \frac{(1-\xi)}{2}L_N^N(\eta) + \frac{\eta}{2}L_1^M(\xi) - \frac{(1-\xi)\eta}{4}
\end{aligned}
\tag{9.3}
$$

The shape functions on the four edges of the quadrilateral element are as follows:

$$
\begin{aligned}
S_i^{E1} &= \frac{(1-\eta)}{2}L_i^M(\xi), \quad && 2 \le i \le M-1 \\
S_i^{E2} &= \frac{\xi}{2}L_i^N(\eta), && 2 \le i \le N-1 \\
S_i^{E3} &= \frac{\eta}{2}L_i^M(\xi), && 2 \le i \le M-1 \\
S_i^{E4} &= \frac{(1-\xi)}{2}L_i^N(\eta), && 2 \le i \le N-1
\end{aligned}
\tag{9.4}
$$

where M is the number of nodes along the two edges in the ξ-direction and N is the number of nodes along the two edges in the η-direction. L_i^M is the Lagrange shape function corresponding to the ith node on an edge along the ξ-direction:

$$
L_i^M(\xi) = \prod_{j=1, j\neq i}^{M} \frac{\xi - \xi_j}{\xi_i - \xi_j} \tag{9.5}
$$

The shape function inside the quadrilateral element is

$$
S_{m,n} = \varphi_m(\xi)\varphi_n(\eta) \tag{9.6}
$$

With the displacement field at hand, the potential functional of the nonlinear elastic problem will be discretized.

9.1.2 *Shape functions for hexahedral elements*

The three-dimensional element in this paper is a curved hexahedral element. The numbering pattern of its eight vertices and 12 edges is shown in Fig. 9.2.

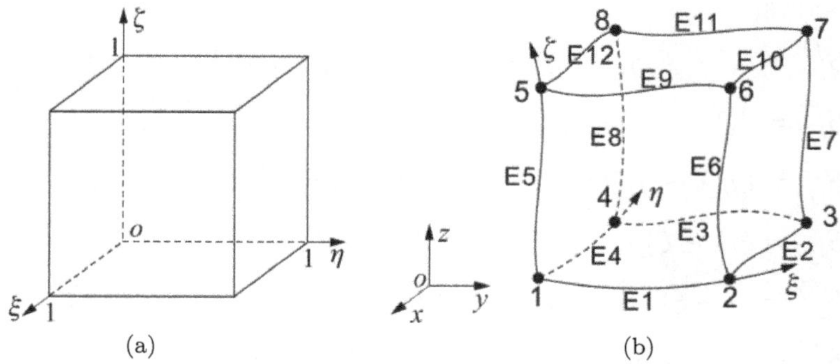

Fig. 9.2. A hexahedral element: (a) parameter domain; (b) geometric domain.

The displacement field composed of interpolation vertex bases and orthogonal edge, face and body bases can be described as

$$\begin{aligned} & u[x(\xi,\eta,\zeta), y(\xi,\eta,\zeta), z(\xi,\eta,\zeta)] \\ & \quad = \sum_{i=1}^{6} S^{v,i}(\xi,\eta,\zeta)u_i + \sum_{i=1}^{L_k}\sum_{k=1}^{12} S_i^{e,k}(\xi,\eta,\zeta)a_i^{e,k} \\ & \quad + \sum_{i=1}^{M_k}\sum_{j=1}^{N_k}\sum_{k=1}^{6} S_{ij}^{f,k}(\xi,\eta,\zeta)a_{ij}^{f,k} \\ & \quad + \sum_{i=1}^{L}\sum_{j=1}^{M}\sum_{k=1}^{N} S_{ijk}^{f}(\xi,\eta,\zeta)a_{ijk}^{b} \end{aligned} \tag{9.7}$$

where the interpolative vertex bases are

$$\begin{aligned} & S_1^v = (1-\xi)(1-\eta)(1-\zeta), && S_2^v = \xi(1-\eta)(1-\zeta) \\ & S_3^v = \xi\eta(1-\zeta), && S_4^v = (1-\xi)\eta(1-\zeta) \\ & S_5^v = (1-\xi)(1-\eta)\zeta, && S_6^v = \xi(1-\eta)\zeta \\ & S_7^v = \xi\eta\zeta, && S_8^v = (1-\xi)\eta\zeta \end{aligned} \tag{9.8}$$

We set the faces surrounded by edge 1-2-3-4, 1-2-5-6, 2-3-6-7, 3-4-7-8, 1-4-5-8, and 5-6-7-8 as F1, F2, F3, F4, F5 and F6, respectively. The shape functions of the hexahedral element can be constructed as follows:

$$\begin{aligned}
S_{ij}^{f1} &= \xi(1-\xi)\eta(1-\eta)(1-\zeta)J_i^{(2,2)}(2\xi-1)J_j^{(2,2)}(2\eta-1)\\
S_{ij}^{f2} &= \xi(1-\xi)(1-\eta)\zeta(1-\zeta)J_i^{(2,2)}(2\xi-1)J_j^{(2,2)}(2\zeta-1)\\
S_{ij}^{f3} &= \xi\eta(1-\eta)\zeta(1-\zeta)J_i^{(2,2)}(2\eta-1)J_j^{(2,2)}(2\zeta-1)\\
S_{ij}^{f4} &= \xi(1-\xi)\eta\zeta(1-\zeta)J_i^{(2,2)}(2\xi-1)J_j^{(2,2)}(2\zeta-1)\\
S_{ij}^{f5} &= (1-\xi)\eta(1-\eta)\zeta(1-\zeta)J_i^{(2,2)}(2\eta-1)J_j^{(2,2)}(2\zeta-1)\\
S_{ij}^{f6} &= \xi(1-\xi)\eta(1-\eta)\zeta J_i^{(2,2)}(2\xi-1)J_j^{(2,2)}(2\eta-1)\\
i,j &= 0,1,2,\ldots
\end{aligned} \tag{9.9}$$

where $J_i^{(2,2)}$ is the ith Jacobi orthogonal polynomial with weighting coefficients (2, 2). Finally, the body shape functions of the hexahedral element are

$$S_{ijk}^{b} = \xi(1-\xi)\eta(1-\eta)\zeta(1-\zeta)B_{ijk}(\xi,\eta,\zeta), \quad i,j,k = 0,1,2,\ldots \tag{9.10}$$

Equation (9.10) is a general form that vanishes at all boundary nodes of the hexahedral element, where B_{ijk} are polynomials to be determined. Normally, we construct them through Jacobi orthogonal polynomials.

The unit hexahedral domain in Fig. 9.2(a) and the parametric domain of Jacobi orthogonal polynomials have the following mapping relationships:

$$\xi = \frac{(1+r)}{2}, \quad \eta = \frac{(1+s)}{2}, \quad \zeta = \frac{(1+t)}{2} \tag{9.11}$$

The orthogonal conditions of the body bases S_{ijk}^{b} of the hexahedral element can be written as

$$\int_V S_{lmn}^{b} S_{ijk}^{b} \mathrm{d}V = \int_0^1\int_0^1\int_0^1 S_{lmn}^{b} S_{ijk}^{b} \mathrm{d}\xi\mathrm{d}\eta\mathrm{d}\zeta = \delta_{li}\delta_{mj}\delta_{nk}C_{ijk} \tag{9.12}$$

Using Eq. (9.11), it can be further rewritten as

$$\int_V S_{lmn}^{b} S_{ijk}^{b} \mathrm{d}V = \int_{-1}^1\int_{-1}^1\int_{-1}^1 \bar{J} B_{lmn}(r,s,t)B_{ijk}(r,s,t)\mathrm{d}r\mathrm{d}s\mathrm{d}t \tag{9.13}$$

where

$$\bar{J} = [\xi(1-\xi)\eta(1-\eta)\zeta(1-\zeta)]^2 \left| \frac{\partial(\xi,\eta,\zeta)}{\partial(r,s,t)} \right|$$
$$= \frac{(1-r)^2(1+r)^2(1-s)^2(1+s)^2(1-t)^2(1+t)^2}{2^{15}} \qquad (9.14)$$

Therefore, using the orthogonality of the Jacobi polynomial, we can take B_{ijk} in the following form:

$$B_{ijk}(r,s,t) = J_i^{(2,2)}(r)J_j^{(2,2)}(s)J_k^{(2,2)}(t) \qquad (9.15)$$

Replacing Eq. (9.15) into Eq. (9.10), the body bases of the hexahedral element can be obtained as

$$S_{ijk}^b = \xi(1-\xi)\eta(1-\eta)\zeta(1-\zeta)J_i^{(2,2)}(2\xi-1)J_j^{(2,2)}(2\eta-1)$$
$$J_k^{(2,2)}(2\zeta-1) \quad i,j,k = 0,1,2,\ldots \qquad (9.16)$$

The procedure above can also be used for triangular and tetrahedral HQEM elements [76].

9.2 Measures of Stress and Strain

In continuous media, measurements of stress and strain must be defined in a clear and physically meaningful manner. The Lagrange description is a common method of describing statics and kinematics of a structure when geometrical nonlinearity occurred. In this method, the static and kinematic quantities must be referred to some previous configuration or reference configuration. We assume that the position of the studied structure in space in the reference configuration (usually the undeformed configuration) is $\boldsymbol{\xi}$, the so-called material coordinates (ξ_1, ξ_2, ξ_3), and its position in space in the current configuration (usually the deformed configuration) is $\mathbf{x}$, the so-called space coordinates (x_1, x_2, x_3). Usually when describing large deformation problems, the former coordinate system is called Lagrange coordinates, and the latter is called Euler coordinates. Obviously, the two are not independent: $\mathbf{x} = \mathbf{x}(\boldsymbol{\xi})$.

Firstly, the deformation gradient tensor can be obtained by differentiating the spatial coordinates with respect to the material coordinates:

$$\mathbf{F} = \frac{\partial \mathbf{x}}{\partial \boldsymbol{\xi}} \tag{9.17}$$

A deformation measurement that is relatively easy to calculate is right Cauchy–Green deformation tensor $\mathbf{C}$, which is obtained using the square difference of $\mathrm{d}\boldsymbol{\xi}$ and $\mathrm{d}\mathbf{x}$:

$$\mathrm{d}\mathbf{x} \cdot \mathrm{d}\mathbf{x} - \mathrm{d}\boldsymbol{\xi} \cdot \mathrm{d}\boldsymbol{\xi} = (\mathbf{F} \cdot \mathrm{d}\boldsymbol{\xi}) \cdot (\mathbf{F} \cdot \mathrm{d}\boldsymbol{\xi}) - \mathrm{d}\boldsymbol{\xi} \cdot \mathrm{d}\boldsymbol{\xi} = \mathrm{d}\boldsymbol{\xi} \cdot (\mathbf{F}^{\mathrm{T}}\mathbf{F} - \mathbf{I}) \cdot \mathrm{d}\boldsymbol{\xi} \tag{9.18}$$

where

$$\mathbf{C} = \mathbf{F}^{\mathrm{T}} \cdot \mathbf{F} \tag{9.19}$$

Therefore, the Green–Lagrange strain tensor is defined as

$$\boldsymbol{\gamma} = \frac{1}{2}(\mathbf{C} - \mathbf{I}) \tag{9.20}$$

It is defined in the reference configuration and degenerated into Euler strain tensor when the deformation gradient is small.

Like the measure of strain, the use of an auxiliary stress measure in reference configuration is needed. An auxiliary stress often used in nonlinear analysis is the second Piola–Kirchhoff stress $\boldsymbol{\tau}$, which possesses the following relationship with Cauchy stress tensor:

$$\boldsymbol{\sigma} = \frac{\rho}{\rho_0}\mathbf{F} \cdot \boldsymbol{\tau} \cdot \mathbf{F}^{\mathrm{T}} \tag{9.21}$$

Other stress and strain measures will not be used in this work and therefore they are not defined here.

9.3 Geometrically Nonlinear Formulation of Hierarchical Quadrature Elements

9.3.1 *Formulation for two- and three-dimensional elements*

In this section, we derive the geometric nonlinear finite element formulation for two- and three-dimensional elements based on the total Lagrange description. In this description, the undeformed configuration is set as the reference configuration. According to the strain and stress measure we introduced in Section 9.2, we can obtain the virtual work principle under the reference configuration:

$$\int_{V_0} \delta\boldsymbol{\gamma}^{\mathrm{T}}\boldsymbol{\tau}^{t+\Delta t}\mathrm{d}V = \int_{S_0} \delta\mathbf{u}^{\mathrm{T}}\mathbf{t}_0\mathrm{d}S + \int_{V_0} \rho_0\delta\mathbf{u}^{\mathrm{T}}\mathbf{g}\mathrm{d}V \tag{9.22}$$

where $\mathbf{t}_0$ is the force divided by a surface in the reference configuration or the nominal traction. The stress $\boldsymbol{\tau}^{t+\Delta t}$ is yet unknown, and it is decomposed into a known stress $\boldsymbol{\tau}^t$ which is evaluated at the start of the current load step and a stress increment $\Delta\boldsymbol{\tau}$. Replacing the decomposing into Eq. (9.22), we have

$$\int_{V_0} \delta\boldsymbol{\gamma}^{\mathrm{T}}\Delta\boldsymbol{\tau}\mathrm{d}V + \int_{V_0} \delta\boldsymbol{\gamma}^{\mathrm{T}}\boldsymbol{\tau}^t\mathrm{d}V = \int_{S_0} \delta\mathbf{u}^{\mathrm{T}}\mathbf{t}_0\mathrm{d}S + \int_{V_0} \rho_0\delta\mathbf{u}^{\mathrm{T}}\mathbf{g}\mathrm{d}V \tag{9.23}$$

The material used in this paper is considered as linear elastic. Using Hooke's law,

$$\Delta\boldsymbol{\tau} = \mathbf{D}\Delta\gamma \tag{9.24}$$

Eq. (9.23) can be re-written as

$$\int_{V_0} \delta\gamma^{\mathrm{T}}\mathbf{D}\Delta\gamma\mathrm{d}V + \int_{V_0} \delta\gamma^{\mathrm{T}}\boldsymbol{\tau}^t\mathrm{d}V = \int_{S_0} \delta u^{\mathrm{T}}\mathbf{t}_0\mathrm{d}S + \int_{V_0} \rho_0\delta\mathbf{u}^{\mathrm{T}}\mathbf{g}\mathrm{d}V \tag{9.25}$$

The strain increment $\Delta\boldsymbol{\gamma}$ consists of a linear contribution $\Delta\mathbf{e}$ and a non-linear contribution $\Delta\boldsymbol{\eta}$ to displacement increment $\Delta\mathbf{u}$ or in the following decomposition:

$$\Delta\gamma = \Delta\mathbf{e} + \Delta\boldsymbol{\eta} \tag{9.26}$$

Substitution of the strain decomposition and the variation of the Green–Lagrange strain tensor $\delta\gamma^{t+\Delta t} = \delta\Delta\gamma$ into Eq. (9.25) gives

$$\begin{aligned}
&\int_{V_0} \delta\Delta\mathbf{e}^{\mathrm{T}}\mathbf{D}\Delta\mathbf{e}\mathrm{d}V + \int_{V_0} \delta\Delta\mathbf{e}^{\mathrm{T}}\mathbf{D}\Delta\boldsymbol{\eta}\mathrm{d}V + \int_{V_0} \delta\Delta\boldsymbol{\eta}^{\mathrm{T}}\mathbf{D}\Delta\mathbf{e}\mathrm{d}V \\
&\quad + \int_{V_0} \delta\Delta\boldsymbol{\eta}^{\mathrm{T}}\mathbf{D}\Delta\boldsymbol{\eta}\mathrm{d}V + \int_{V_0} \delta\boldsymbol{\eta}^{\mathrm{T}}\boldsymbol{\tau}^t\mathrm{d}V \\
&= \int_{S_0} \delta\mathbf{u}^{\mathrm{T}}\mathbf{t}_0\mathrm{d}S + \int_{V_0} \rho_0\delta\mathbf{u}^{\mathrm{T}}\mathbf{g}\mathrm{d}V - \int_{V_0} \delta\mathbf{e}^{\mathrm{T}}\boldsymbol{\tau}^t\mathrm{d}V
\end{aligned} \tag{9.27}$$

The expressions in Eq. (9.27) can be classified and handled by its degree with respect to the displacement increment $\Delta\mathbf{u}$, respectively. First, $\int(\delta\mathbf{e})^{\mathrm{T}}\boldsymbol{\tau}^t\mathrm{d}V$ is of degree zero, thus it is brought to the right-hand side of Eq. (9.27) and constructed as internal force vector. The expressions $\int \delta\Delta\mathbf{e}^{\mathrm{T}}\mathbf{D}\Delta\mathbf{e}\mathrm{d}V$ and $\int \delta\boldsymbol{\eta}^{\mathrm{T}}\boldsymbol{\tau}^{t}\mathrm{d}V$ is of degree one, or it is linear in displacement increment. Thus, these are the contributions constructed as the tangential stiffness matrix. Finally, the expressions that are of a degree higher than one are erased as linearization of the virtual work principle. The above classification and disposition give

$$\int_{V_0} \delta\Delta\mathbf{e}^{\mathrm{T}}\mathbf{D}\Delta\mathbf{e}\mathrm{d}V + \int_{V_0} \delta\boldsymbol{\eta}^{\mathrm{T}}\boldsymbol{\tau}^{\mathrm{t}}\mathrm{dV}$$
$$= \int_{S_0} \delta\mathbf{u}^{\mathrm{T}}t_0\mathrm{d}S + \int_{V_0} \rho_0\delta\mathbf{u}^{\mathrm{T}}\mathbf{g}\mathrm{d}V - \int_{V_0} \delta\mathbf{e}^{\mathrm{T}}\tau^{t}\mathrm{d}V \quad (9.28)$$

The linear part of strain $\Delta\mathbf{e}$ is related to the displacement increment $\Delta\mathbf{u}$ in a straight manner:

$$\Delta\mathbf{e} = \mathbf{L}\Delta\mathbf{u} \quad (9.29)$$

where $\mathbf{L}$ is a differential matrix constructed by structural configurations. In a two-dimensional configuration, $\mathbf{L}$ is constructed as

$$L = \begin{bmatrix} F_{11}\dfrac{\partial}{\partial\xi_1} & F_{21}\dfrac{\partial}{\partial\xi_1} \\ F_{12}\dfrac{\partial}{\partial\xi_2} & F_{22}\dfrac{\partial}{\partial\xi_2} \\ F_{11}\dfrac{\partial}{\partial\xi_2} + F_{12}\dfrac{\partial}{\partial\xi_1} & F_{21}\dfrac{\partial}{\partial\xi_2} + F_{22}\dfrac{\partial}{\partial\xi_1} \end{bmatrix} \quad (9.30)$$

In a three-dimensional configuration, $\mathbf{L}$ reads

$$\mathbf{L} = \begin{bmatrix} F_{11}\dfrac{\partial}{\partial\xi_1} & F_{21}\dfrac{\partial}{\partial\xi_1} & F_{31}\dfrac{\partial}{\partial\xi_1} \\ F_{12}\dfrac{\partial}{\partial\xi_2} & F_{22}\dfrac{\partial}{\partial\xi_2} & F_{32}\dfrac{\partial}{\partial\xi_2} \\ F_{13}\dfrac{\partial}{\partial\xi_3} & F_{23}\dfrac{\partial}{\partial\xi_3} & F_{33}\dfrac{\partial}{\partial\xi_3} \\ F_{11}\dfrac{\partial}{\partial\xi_2} + F_{12}\dfrac{\partial}{\partial\xi_1} & F_{21}\dfrac{\partial}{\partial\xi_2} + F_{22}\dfrac{\partial}{\partial\xi_1} & F_{31}\dfrac{\partial}{\partial\xi_2} + F_{32}\dfrac{\partial}{\partial\xi_1} \\ F_{12}\dfrac{\partial}{\partial\xi_3} + F_{13}\dfrac{\partial}{\partial\xi_2} & F_{22}\dfrac{\partial}{\partial\xi_3} + F_{23}\dfrac{\partial}{\partial\xi_2} & F_{32}\dfrac{\partial}{\partial\xi_3} + F_{33}\dfrac{\partial}{\partial\xi_2} \\ F_{13}\dfrac{\partial}{\partial\xi_1} + F_{11}\dfrac{\partial}{\partial\xi_3} & F_{23}\dfrac{\partial}{\partial\xi_1} + F_{21}\dfrac{\partial}{\partial\xi_3} & F_{33}\dfrac{\partial}{\partial\xi_1} + F_{31}\dfrac{\partial}{\partial\xi_3} \end{bmatrix} \quad (9.31)$$

The continuum finite element requires interpolation of the continuous displacement field. The matrix $\mathbf{H}$ containing the interpolation functions h_1, h_2, h_3, $h_4, \ldots$ is written as

$$\mathbf{H} = \begin{bmatrix} h_1 & 0 & 0 & h_2 & 0 & 0 & \cdots & h_n & 0 & 0 \\ 0 & h_1 & 0 & 0 & h_2 & 0 & \cdots & 0 & h_n & 0 \\ 0 & 0 & h_1 & 0 & 0 & h_2 & \cdots & 0 & 0 & h_n \end{bmatrix} \tag{9.32}$$

The relationship between continuous displacement field and discrete nodal displacements can be written as $\mathbf{u} = \mathbf{Ha}$ or in incremental form as $\Delta\mathbf{u} = \mathbf{H}\Delta\mathbf{a}$. Define $\mathbf{B}_L = \mathbf{LH}$, then the relation between linear strain increment and continuous displacement field is

$$\Delta\mathbf{e} = \mathbf{B}_L\Delta\mathbf{a} \tag{9.33}$$

For a two-dimensional configuration,

$$\mathbf{B}_L = \begin{bmatrix} F_{11}\dfrac{\partial h_1}{\partial \xi_1} & F_{21}\dfrac{\partial h_1}{\partial \xi_1} & \cdots & \cdots \\ F_{12}\dfrac{\partial h_1}{\partial \xi_2} & F_{22}\dfrac{\partial h_1}{\partial \xi_2} & \cdots & \cdots \\ F_{11}\dfrac{\partial h_1}{\partial \xi_2} + F_{12}\dfrac{\partial h_1}{\partial \xi_1} & F_{21}\dfrac{\partial h_1}{\partial \xi_2} + F_{22}\dfrac{\partial h_1}{\partial \xi_1} & \cdots & \cdots \end{bmatrix} \tag{9.34}$$

and for a three-dimensional configuration,

$$\mathbf{B}_L = \begin{bmatrix} F_{11}\dfrac{\partial h_1}{\partial \xi_1} & F_{21}\dfrac{\partial h_1}{\partial \xi_1} & F_{31}\dfrac{\partial h_1}{\partial \xi_1} & \cdots & \cdots \\ F_{12}\dfrac{\partial h_1}{\partial \xi_2} & F_{22}\dfrac{\partial h_1}{\partial \xi_2} & F_{32}\dfrac{\partial h_1}{\partial \xi_2} & \cdots & \cdots \\ F_{13}\dfrac{\partial h_1}{\partial \xi_3} & F_{23}\dfrac{\partial h_1}{\partial \xi_3} & F_{33}\dfrac{\partial h_1}{\partial \xi_3} & \cdots & \cdots \\ F_{11}\dfrac{\partial h_1}{\partial \xi_2} + F_{12}\dfrac{\partial h_1}{\partial \xi_1} & F_{21}\dfrac{\partial h_1}{\partial \xi_2} + F_{22}\dfrac{\partial h_1}{\partial \xi_1} & F_{31}\dfrac{\partial h_1}{\partial \xi_2} + F_{32}\dfrac{\partial h_1}{\partial \xi_1} & \cdots & \cdots \\ F_{12}\dfrac{\partial h_1}{\partial \xi_3} + F_{13}\dfrac{\partial h_1}{\partial \xi_2} & F_{22}\dfrac{\partial h_1}{\partial \xi_3} + F_{23}\dfrac{\partial h_1}{\partial \xi_2} & F_{32}\dfrac{\partial h_1}{\partial \xi_3} + F_{33}\dfrac{\partial h_1}{\partial \xi_2} & \cdots & \cdots \\ F_{13}\dfrac{\partial h_1}{\partial \xi_1} + F_{11}\dfrac{\partial h_1}{\partial \xi_3} & F_{23}\dfrac{\partial h_1}{\partial \xi_1} + F_{21}\dfrac{\partial h_1}{\partial \xi_3} & F_{33}\dfrac{\partial h_1}{\partial \xi_1} + F_{31}\dfrac{\partial h_1}{\partial \xi_3} & \cdots & \cdots \end{bmatrix} \tag{9.35}$$

Using the linear strain matrix $\mathbf{B}_L$, the linear contribution of the tangential stiffness matrix can be constructed using the following conversion of the first term in Eq. (9.28):

$$\int_{V_0} (\delta\Delta\mathbf{e})^{\mathrm{T}}\mathbf{D}\Delta\mathbf{e}dV = (\delta\Delta\mathbf{a})^{\mathrm{T}}\mathbf{K}_L\Delta\mathbf{a} \tag{9.36}$$

where $\mathbf{K}_L$ is the linear contribution of the tangential stiffness matrix:

$$\mathbf{K}_L = \int_{V_0} \mathbf{B}_L^{\mathrm{T}}\mathbf{D}\mathbf{B}_L\mathrm{d}V \tag{9.37}$$

The geometrical contribution of the tangential stiffness matrix can be constructed likewise, and the second term of Eq. (9.28) can be converted as:

$$\int_{V_0} (\delta\Delta\boldsymbol{\eta})^{\mathrm{T}}\boldsymbol{\tau}^t dV = (\delta\Delta\mathbf{a})^{\mathrm{T}}\mathbf{K}_{NL}\Delta\mathbf{a} \tag{9.38}$$

where $\mathbf{K}_{NL}$ is the geometric contribution of the tangential stiffness matrix:

$$\mathbf{K}_{NL} = \int_{V_0} \mathbf{B}_{NL}^{\mathrm{T}}\mathbf{T}^t\mathbf{B}_{NL}\mathrm{d}V \tag{9.39}$$

The second Piola–Kirchhoff stress and the nonlinear strain–stress matrix are defined according to structural configurations. For two-dimensional configuration,

$$\mathbf{T} = \begin{bmatrix} \tau_{xx} & \tau_{xy} & 0 & 0 \\ \tau_{xy} & \tau_{yy} & 0 & 0 \\ 0 & 0 & \tau_{xx} & \tau_{xy} \\ 0 & 0 & \tau_{xy} & \tau_{yy} \end{bmatrix} \tag{9.40}$$

and

$$\mathbf{B}_{NL} = \begin{bmatrix} \dfrac{\partial h_1}{\partial \xi_1} & 0 & \dfrac{\partial h_2}{\partial \xi_1} & 0 & \cdots & \cdots \\ \dfrac{\partial h_1}{\partial \xi_2} & 0 & \dfrac{\partial h_2}{\partial \xi_2} & 0 & \cdots & \cdots \\ 0 & \dfrac{\partial h_1}{\partial \xi_1} & 0 & \dfrac{\partial h_2}{\partial \xi_1} & \cdots & \cdots \\ 0 & \dfrac{\partial h_1}{\partial \xi_2} & 0 & \dfrac{\partial h_2}{\partial \xi_2} & \cdots & \cdots \end{bmatrix} \tag{9.41}$$

For three-dimensional configuration:

$$\mathbf{T} = \begin{bmatrix} \tau_{xx} & \tau_{xy} & \tau_{zx} & 0 & 0 & 0 & 0 & 0 & 0 \\ \tau_{xy} & \tau_{yy} & \tau_{yz} & 0 & 0 & 0 & 0 & 0 & 0 \\ \tau_{zx} & \tau_{yz} & \tau_{zz} & 0 & 0 & 0 & 0 & 0 & 0 \\ 0 & 0 & 0 & \tau_{xx} & \tau_{xy} & \tau_{zx} & 0 & 0 & 0 \\ 0 & 0 & 0 & \tau_{xy} & \tau_{yy} & \tau_{yz} & 0 & 0 & 0 \\ 0 & 0 & 0 & \tau_{zx} & \tau_{yz} & \tau_{zz} & 0 & 0 & 0 \\ 0 & 0 & 0 & 0 & 0 & 0 & \tau_{xx} & \tau_{xy} & \tau_{zx} \\ 0 & 0 & 0 & 0 & 0 & 0 & \tau_{xy} & \tau_{yy} & \tau_{yz} \\ 0 & 0 & 0 & 0 & 0 & 0 & \tau_{zx} & \tau_{yz} & \tau_{zz} \end{bmatrix} \tag{9.42}$$

and

$$\mathbf{B}_{NL} = \begin{bmatrix} \frac{\partial h_1}{\partial \xi_1} & 0 & 0 & \frac{\partial h_2}{\partial \xi_1} & 0 & 0 & \cdots & \cdots \\ \frac{\partial h_1}{\partial \xi_2} & 0 & 0 & \frac{\partial h_2}{\partial \xi_2} & 0 & 0 & \cdots & \cdots \\ \frac{\partial h_1}{\partial \xi_3} & 0 & 0 & \frac{\partial h_2}{\partial \xi_3} & 0 & 0 & \cdots & \cdots \\ 0 & \frac{\partial h_1}{\partial \xi_1} & 0 & 0 & \frac{\partial h_2}{\partial \xi_1} & 0 & \cdots & \cdots \\ 0 & \frac{\partial h_1}{\partial \xi_2} & 0 & 0 & \frac{\partial h_2}{\partial \xi_2} & 0 & \cdots & \cdots \\ 0 & \frac{\partial h_1}{\partial \xi_3} & 0 & 0 & \frac{\partial h_2}{\partial \xi_3} & 0 & \cdots & \cdots \\ 0 & 0 & \frac{\partial h_1}{\partial \xi_1} & 0 & 0 & \frac{\delta h_2}{\partial \xi_1} & \cdots & \cdots \\ 0 & 0 & \frac{\partial h_1}{\partial \xi_2} & 0 & 0 & \frac{\delta h_2}{\partial \xi_2} & \cdots & \cdots \\ 0 & 0 & \frac{\partial h_1}{\partial \xi_3} & 0 & 0 & \frac{\partial h_2}{\partial \xi_3} & \cdots & \cdots \end{bmatrix} \tag{9.43}$$

Substituting the tangential stiffness matrix, Eq. (9.37) and Eq. (9.39), into Eq. (9.28) gives

$$(\delta \Delta \mathbf{a})^{\mathrm{T}} (\mathbf{K}_L + \mathbf{K}_{NL}) \Delta \mathbf{a} = (\delta \Delta \mathbf{a})^{\mathrm{T}} (\mathbf{f}_{\mathrm{ext}}^{t+\Delta t} + \mathbf{f}_{\mathrm{int}}^{t}) \tag{9.44}$$

where the external force vector and the internal force vector are defined as

$$\begin{cases} \mathbf{f}_{\text{ext}}^{t+\Delta t} = \int_{S_0} \mathbf{H}^{\mathrm{T}} \mathbf{t}_0 \mathrm{d}S + \int_{V_0} \rho_0 \mathbf{H}^{\mathrm{T}} \mathbf{g} \mathrm{d}V \\ \mathbf{f}_{\text{int}}^{t} = \int_{V_0} \mathbf{B}_L^{\mathrm{T}} \boldsymbol{\tau}^t \mathrm{d}V \end{cases} \tag{9.45}$$

Equation (9.44) holds for any virtual displacement increment $\delta\Delta\mathbf{a}$, and immediately we have the linearized equation set:

$$(\mathbf{K}_L + \mathbf{K}_{NL})\Delta\mathbf{a} = \mathbf{f}_{\text{ext}}^{t+\Delta t} + \mathbf{f}_{\text{int}}^{t} \tag{9.46}$$

The above derivation applies to two- and three-dimensional continuum elements. The original, undeformed configuration has been chosen as the reference configuration, so that formally we have a total Lagrange formulation.

9.3.2 *Formulation for shallow shell elements*

The hierarchical quadrature shallow shell element is based on the Mindlin formulation, which was originally proposed for medium-thickness plates considering transverse shear deformation. Thus, the rotation angle of the mid-plane is not equal to the rotation angle of the normal.

Assume the section remains planar, the strain $\boldsymbol{\varepsilon}^{\mathrm{T}} = (\varepsilon_{xx}, \varepsilon_{yy}, \varepsilon_{xy})$ of a microbody with distance of l from the middle surface of the shallow shell can be described as

$$\boldsymbol{\varepsilon} = \boldsymbol{\varepsilon}_l + z_l \boldsymbol{\chi} \tag{9.47}$$

where the subscript l refers to the reference plane. It was taken as the middle surface here but not necessary.

Assume that the magnitude of in-plane strain remains small compared with out-of-plane strain $\left(\frac{\partial u}{\partial x}\right)^2 \leq \left(\frac{\partial w}{\partial x}\right)^2$ and out-of-plane

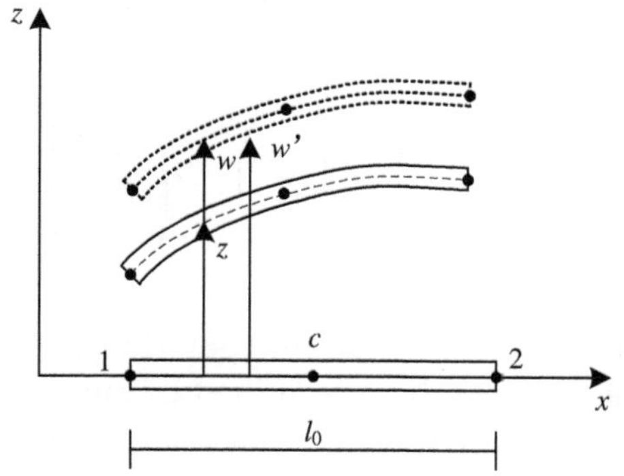

Fig. 9.3. Definition of shallow shell geometry.

normal strain ε_{zz} vanishes. Thus, the in-plane strain of mid-surface gives

$$\boldsymbol{\varepsilon}_l = \begin{pmatrix} \dfrac{\partial u_l}{\partial x} \\ \dfrac{\partial u_l}{\partial y} \\ \dfrac{\partial u_l}{\partial y} + \dfrac{\partial u_l}{\partial x} \end{pmatrix} + \begin{pmatrix} \dfrac{1}{2}\left(\dfrac{\partial w'}{\partial x}\right)^2 - \dfrac{1}{2}\left(\dfrac{\partial z}{\partial x}\right)^2 \\ \dfrac{1}{2}\left(\dfrac{\partial w'}{\partial y}\right)^2 - \dfrac{1}{2}\left(\dfrac{\partial z}{\partial x}\right)^2 \\ \dfrac{\partial w'}{\partial x}\dfrac{\partial w'}{\partial y} - \dfrac{\partial z}{\partial x}\dfrac{\partial z}{\partial y} \end{pmatrix} \tag{9.48}$$

where z is the original vertical coordinates of shell mid-surface, or initial deflection, w is the vertical displacement of the shell according to the undeformed configuration, and we have $w' = w + z$. The relations are shown in Fig. 9.3.

In the Mindlin formulation, the curvatures are defined as

$$\boldsymbol{\chi} = \begin{pmatrix} \chi_x \\ \chi_y \\ \chi_z \end{pmatrix} = \begin{pmatrix} \dfrac{\partial \theta_x}{\partial x} \\ \dfrac{\partial \theta_y}{\partial y} \\ \dfrac{\partial \theta_x}{\partial y} + \dfrac{\partial \theta_y}{\partial x} \end{pmatrix} \tag{9.49}$$

where θ_x and θ_y are the rotation angles of the mid-surface of the shell.

The out-of-plane shear strain is defined in the following form:

$$\boldsymbol{\gamma} = \begin{pmatrix} \gamma_{xz} \\ \gamma_{yz} \end{pmatrix} = \begin{pmatrix} \theta_x \\ \theta_y \end{pmatrix} + \begin{pmatrix} \dfrac{\partial w}{\partial x} \\ \dfrac{\partial w}{\partial y} \end{pmatrix} \tag{9.50}$$

Equations (9.47) and (9.50) indicate that the strain inside the shell microbody can be completely determined by the strain of the middle surface of the shell, which greatly facilitates the analysis of shell structural problems.

In the Mindlin formulation, we have five non-vanishing stress components: the in-plane normal stress σ_{xx} and σ_{yy}, the in-plane shear stress σ_{xy}, and the out-of-plane shear stress σ_{xz} and σ_{yz}. Integrating these stresses along the thickness direction gives the generalized stresses, so we have normal forces as

$$\mathbf{N} = \begin{pmatrix} N_x \\ N_y \\ N_{xy} \end{pmatrix} = \int_{-h/2}^{h/2} \begin{pmatrix} \sigma_{xx}(z_l) \\ \sigma_{yy}(z_l) \\ \sigma_{xy}(z_l) \end{pmatrix} \mathrm{d}z_l \tag{9.51}$$

the bending moment as

$$\mathbf{M} = \begin{pmatrix} M_x \\ M_y \\ M_{xy} \end{pmatrix} = \int_{-h/2}^{h/2} \begin{pmatrix} \sigma_{xx}(z_l) \\ \sigma_{yy}(z_l) \\ \sigma_{xy}(z_l) \end{pmatrix} z_l \mathrm{d}z_l \tag{9.52}$$

and out-of-plane shear forces as

$$\mathbf{Q} = \begin{pmatrix} Q_x \\ Q_y \end{pmatrix} = \int_{-h/2}^{h/2} \begin{pmatrix} \sigma_{xz}(z_l) \\ \sigma_{yz}(z_l) \end{pmatrix} \mathrm{d}z_l \tag{9.53}$$

Based on the assumption that the strain remains small, we can use Hooke's law to relate the in-plane strain increment (which can be considered as a degenerate form of Green–Lagrange strain) with the in-plane stress increment:

$$\begin{pmatrix} \dot{\sigma}_{xx} \\ \dot{\sigma}_{yy} \\ \dot{\sigma}_{xy} \end{pmatrix} = \mathbf{D} \begin{pmatrix} \dot{\varepsilon}_{xx} \\ \dot{\varepsilon}_{yy} \\ \dot{\varepsilon}_{xy} \end{pmatrix} \tag{9.54}$$

where

$$\mathbf{D} = \frac{E}{1 - v^2} \begin{bmatrix} 1 & v & 0 \\ v & 1 & 0 \\ 0 & 0 & (1-v)/2 \end{bmatrix} \tag{9.55}$$

and E is the Young's modulus, and v is the Poisson's ratio.

Therefore, the increase (or rate) of the in-plane generalized forces $\mathbf{M}$ and $\mathbf{N}$ and the out-of-plane shear stress $\mathbf{Q}$ reads:

$$\dot{\mathbf{N}} = \int_{-h/2}^{h/2} \mathbf{D}\dot{\varepsilon}_l \mathrm{d}z_l = \mathbf{D}h\dot{\varepsilon}_l = \mathbf{D}_{\mathrm{m}}\dot{\varepsilon}_l \tag{9.56}$$

$$\dot{\mathbf{M}} = \int_{-h/2}^{h/2} \mathbf{D}\dot{\chi} z_l^2 \mathrm{d}z_l = \mathbf{D}\frac{h^3}{12}\dot{\chi} = \mathbf{D}_{\mathrm{b}}\dot{\chi} \tag{9.57}$$

$$\dot{\mathbf{Q}} = \int_{-h/2}^{h.2} \mathbf{G}\dot{\gamma} \mathrm{d}z_l = \mathbf{G}h\dot{\gamma} = \mathbf{D}_{\mathrm{s}}\dot{\gamma} \tag{9.58}$$

where

$$\mathbf{G} = \frac{5}{6}\begin{bmatrix} \mu & 0 \\ 0 & \mu \end{bmatrix} \tag{9.59}$$

$\mathbf{D}_{\mathrm{m}}$ and $\mathbf{D}_{\mathrm{b}}$ are the tangential constitutive matrices for membrane action and for bending, respectively. $\mathbf{D}_{\mathrm{s}}$ is the tangential constitutive matrix of material response to shear deformation, and μ is the shear modulus. Combining the above generalized stresses, we get the stress–strain relationship of all components:

$$\begin{pmatrix} \dot{\mathbf{N}} \\ \dot{\mathbf{M}} \\ \dot{\mathbf{Q}} \end{pmatrix} = \begin{bmatrix} \mathbf{D}_{\mathrm{m}} & \mathbf{0} & \mathbf{0} \\ \mathbf{0} & \mathbf{D}_{\mathrm{b}} & \mathbf{0} \\ \mathbf{0} & \mathbf{0} & \mathbf{D}_{\mathrm{s}} \end{bmatrix} \begin{pmatrix} \dot{\boldsymbol{\varepsilon}}_l \\ \dot{\boldsymbol{\chi}} \\ \dot{\boldsymbol{\gamma}} \end{pmatrix} \tag{9.60}$$

Now, the geometrical nonlinear finite element formulation can be derived. In the Mindlin formulation, there are five non-vanishing basic unknowns u_l, v_l, w, θ_x and θ_y. They are collected in nodal displacement vectors as

$$\mathbf{a} = \begin{pmatrix} (u_l)_1 \\ (v_l)_1 \\ \cdots \\ \cdots \\ (u_l)_n \\ (v_l)_n \end{pmatrix}, \quad \mathbf{w} = \begin{pmatrix} w_1 \\ w_2 \\ \cdots \\ \cdots \\ w_n \end{pmatrix}, \quad \theta = \begin{pmatrix} (\theta_x)_1 \\ (\theta_y)_1 \\ \cdots \\ \cdots \\ (\theta_x)_n \\ (\theta_y)_n \end{pmatrix} \tag{9.61}$$

Defining

$$\mathbf{H} = \begin{bmatrix} h_1 & 0 & h_2 & 0 & \cdots & \cdots & h_n & 0 \\ 0 & h_1 & 0 & h_2 & \cdots & \cdots & 0 & h_n \end{bmatrix} \tag{9.62}$$

and

$$\mathbf{h} = [h_1 \; h_2 \; \cdots \; h_n]^{\mathrm{T}} \tag{9.63}$$

we then have the following discretization of the displacement field:

$$\begin{pmatrix} u_l \\ v_l \end{pmatrix} = \mathbf{Ha}, \quad w = \mathbf{h}^{\mathrm{T}}\mathbf{w}, \quad \begin{pmatrix} \theta_x \\ \theta_y \end{pmatrix} = \mathbf{H}\theta \tag{9.64}$$

Thus the in-plane strain, Eq. (9.47), is processed in a variational and discretized form:

$$\delta\boldsymbol{\varepsilon}_l = \mathbf{B}\delta\mathbf{a} + (\mathbf{B}\mathbf{w}')\mathbf{B}_w\delta\mathbf{w} \tag{9.65}$$

where

$$\mathbf{B} = \begin{bmatrix} \dfrac{\partial h_1}{\partial x} & 0 & \cdots & \cdots & \dfrac{\partial h_n}{\partial x} & 0 \\ 0 & \dfrac{\partial h_1}{\partial y} & \cdots & \cdots & 0 & \dfrac{\partial h_n}{\partial y} \\ \dfrac{\partial h_1}{\partial y} & \dfrac{\partial h_1}{\partial x} & \cdots & \cdots & \dfrac{\partial h_n}{\partial y} & \dfrac{\partial h_n}{\partial x} \end{bmatrix} \tag{9.66}$$

$$\mathbf{w}' = \begin{bmatrix} w_1' & 0 \\ 0 & w_1' \\ \cdots & \cdots \\ \cdots & \cdots \\ w_n' & 0 \\ 0 & w_n' \end{bmatrix} \tag{9.67}$$

$$\mathbf{B}_w = \begin{bmatrix} \dfrac{\partial h_1}{\partial x} & \cdots & \dfrac{\partial h_n}{\partial x} \\ \dfrac{\partial h_1}{\partial y} & \cdots & \dfrac{\partial h_n}{\partial y} \end{bmatrix} \tag{9.68}$$

Similar processing can be done to curvature $\boldsymbol{\chi}$ and out-of-plane shear strain $\boldsymbol{\gamma}$:

$$\delta\boldsymbol{\chi} = \mathbf{B}\delta\boldsymbol{\theta} \tag{9.69}$$

$$\delta\boldsymbol{\gamma} = \mathbf{H}\delta\boldsymbol{\theta} + \mathbf{B}_w\delta\mathbf{w} \tag{9.70}$$

For Mindlin formulation, the virtual principle at time $t + \Delta t$ reads

$$\int_{V_0} (\delta\boldsymbol{\varepsilon}^{t+\Delta t})^{\mathrm{T}}\boldsymbol{\sigma}^{t+\Delta t}\mathrm{d}V = (\delta\mathbf{u}^{t+\Delta t})^{\mathrm{T}}\mathbf{f}_{\mathrm{ext}}^{t+\Delta t} \tag{9.71}$$

The above variation of strains (Eqs. (9.65), (9.69) and (9.70)) and generalized stresses (Eqs. (9.56)–(9.58)) can be brought into Eq. (9.71):

$$\begin{aligned} \int_A [\delta\mathbf{a}^{\mathrm{T}}\mathbf{B}^{\mathrm{T}}\mathrm{d}\mathbf{N} + \delta\boldsymbol{\theta}^{\mathrm{T}}\mathbf{B}^{\mathrm{T}}\mathrm{d}\mathbf{M} + (\delta\boldsymbol{\theta}^{\mathrm{T}}\mathbf{H}^{\mathrm{T}} + \delta\mathbf{w}^{\mathrm{T}}\mathbf{B}_w^{\mathrm{T}})d\mathbf{Q} \\ + \delta\mathbf{w}^{\mathrm{T}}\mathbf{B}_w^{\mathrm{T}}(\mathbf{B}d\mathbf{w}')^{\mathrm{T}}\mathbf{N}]\mathrm{d}A = \delta\mathbf{a}^{\mathrm{T}}(\mathbf{f}_{\mathrm{ext}}^{\mathrm{a}} - \mathbf{f}_{\mathrm{int}}^{\mathrm{a}}) \\ + \delta\mathbf{w}^{\mathrm{T}}(\mathbf{f}_{\mathrm{ext}}^{\mathrm{w}} - \mathbf{f}_{\mathrm{int}}^{\mathrm{w}}) + \delta\boldsymbol{\theta}^{\mathrm{T}}(\mathbf{f}_{\mathrm{ext}}^{\theta} - \mathbf{f}_{\mathrm{int}}^{\theta}) \end{aligned} \tag{9.72}$$

Replace the constitutive relation Eq. (9.60) into Eq. (9.72), and note that Eq. (9.72) is satisfied for arbitrary virtual displacement $(\delta\mathbf{a}, \delta\mathbf{w}, \delta\boldsymbol{\theta})$, then we have

$$\begin{bmatrix} \mathbf{K}_{\mathrm{aa}} & \mathbf{K}_{\mathrm{aw}} & \mathbf{K}_{\mathrm{a}\theta} \\ \mathbf{K}_{\mathrm{aw}}^{\mathrm{T}} & \mathbf{K}_{\mathrm{ww}} & \mathbf{K}_{\mathrm{w}\theta} \\ \mathbf{K}_{\mathrm{a}\theta}^{\mathrm{T}} & \mathbf{K}_{\mathrm{w}\theta}^{\mathrm{T}} & \mathbf{K}_{\theta\theta} \end{bmatrix} \begin{pmatrix} \mathrm{d}\mathbf{a} \\ \mathrm{d}\mathbf{w} \\ \mathrm{d}\boldsymbol{\theta} \end{pmatrix} = \begin{pmatrix} \mathbf{f}_{\mathrm{ext}}^{\mathrm{a}} - \mathbf{f}_{\mathrm{int}}^{\mathrm{a}} \\ \mathbf{f}_{\mathrm{ext}}^{\mathrm{w}} - \mathbf{f}_{\mathrm{int}}^{\mathrm{w}} \\ \mathbf{f}_{\mathrm{ext}}^{\theta} - \mathbf{f}_{\mathrm{int}}^{\theta} \end{pmatrix} \tag{9.73}$$

where the internal force vector is defined as

$$\mathbf{f}_{\text{int}}^{\text{a}} = \int_A \mathbf{B}^{\text{T}}\mathbf{N}\text{d}A \tag{9.74}$$

$$\mathbf{f}_{\text{int}}^{\text{w}} = \int_A \mathbf{B}_w^{\text{T}}((\mathbf{B}\mathbf{w}')^{\text{T}}\mathbf{N} + \mathbf{Q})\text{d}A \tag{9.75}$$

$$\mathbf{f}_{\text{int}}^{\theta} = \int_A (\mathbf{B}^{\text{T}}\mathbf{M} + \mathbf{H}^{\text{T}}\mathbf{Q})\text{d}A \tag{9.76}$$

and the format of each sub-item of the tangent stiffness matrix is as follows:

$$\begin{aligned}
\mathbf{K}_{\text{aa}} &= \int_A \mathbf{B}^{\text{T}}\mathbf{D}_{\text{m}}\mathbf{B}\text{d}A \\
\mathbf{K}_{\text{aw}} &= \int_A \mathbf{B}^{\text{T}}\mathbf{D}_{\text{m}}(\mathbf{B}\mathbf{w}')\mathbf{B}_w\text{d}A \\
\mathbf{K}_{\text{a}\theta} &= \mathbf{0} \\
\mathbf{K}_{\text{ww}} &= \int_A \mathbf{B}_w^{\text{T}}\mathbf{D}_{\text{s}}\mathbf{B}_w\text{d}A + \int_A \mathbf{B}_w^{\text{T}}(\mathbf{B}\mathbf{w}')^{\text{T}}\mathbf{D}_{\text{m}}(\mathbf{B}\mathbf{w}')\mathbf{B}_w\text{d}A \\
&\quad + \int_A \mathbf{B}_w^{\text{T}}\boldsymbol{\mathcal{N}}\mathbf{B}_w\text{d}A \\
\mathbf{K}_{\text{w}\theta} &= \int_A \mathbf{B}_w^{\text{T}}\mathbf{D}_{\text{s}}\mathbf{H}\text{d}A + \int_A \mathbf{B}_w^{\text{T}}(\mathbf{B}\mathbf{w}')^{\text{T}}\mathbf{D}_{\text{c}}\mathbf{B}\text{d}A \\
\mathbf{K}_{\theta\theta} &= \int_A \mathbf{B}^{\text{T}}\mathbf{D}_{\text{b}}\mathbf{B}\text{d}A + \int_A \mathbf{H}^{\text{T}}\mathbf{D}_{\text{s}}\mathbf{H}\text{d}A
\end{aligned} \tag{9.77}$$

where the third term of $\mathbf{K}_{\text{ww}}$ is considered as a geometric nonlinear contribution and $\boldsymbol{\mathcal{N}}$ is a matrix composed of normal forces:

$$\boldsymbol{\mathcal{N}} = \begin{pmatrix} N_x & N_{xy} \\ N_{xy} & N_y \end{pmatrix} \tag{9.78}$$

9.3.3 *Solution of the system of nonlinear equations*

The solution of nonlinear equations requires the use of iterative solving techniques. Such techniques prevent the departure of solution from the repeated linearization of the governing equations, like Eq. (9.46) for two- and three-dimensional configurations and Eq. (9.73) for shallow shell configuration.

In a load step of full Newton–Raphson method, the first evaluation of the displacement increment is given by

$$\Delta \mathbf{a}_1 = \mathbf{K}_0^{-1}\mathbf{r}_0 \tag{9.79}$$

where

$$\mathbf{r}_0 = \mathbf{f}_{\text{ext}}^{t+\Delta t} - \mathbf{f}_{\text{int},0} \tag{9.80}$$

In each iteration, the correction of displacement increment is calculated by the updated tangential stiffness matrix according to the last evolution of displacement increment:

$$\mathrm{d}\mathbf{a}_2 = \mathbf{K}_1^{-1}\mathbf{r}_1 \tag{9.81}$$

and added to the last displacement increment evolution:

$$\Delta \mathbf{a}_2 = \Delta \mathbf{a}_1 + \mathrm{d}\mathbf{a}_2 \tag{9.82}$$

until the residual vector $\mathbf{r}_n$ in nth iteration is a small value.

The arc length method follows a similar manner but regards the load increment factor $\Delta\lambda$ as an additional unknown quantity, and the n-dimensional unknown node displacement solution space stored in vector $\mathbf{a}$ is expanded to $n+1$ dimension. Thus, a supplementary constraint is

$$g(\mathbf{a}_0, \lambda_0, \Delta\mathbf{a}, \Delta\lambda, \Delta l) = 0 \tag{9.83}$$

and the expanded equation set is

$$\begin{bmatrix} \mathbf{K} & -\hat{\mathbf{f}}_{\text{ext}} \\ \mathbf{h}^{\mathrm{T}} & s \end{bmatrix} \begin{bmatrix} \mathrm{d}\mathbf{a}_{j+1} \\ \mathrm{d}\lambda_{j+1} \end{bmatrix} = \begin{bmatrix} \mathbf{r}_j \\ -g_j \end{bmatrix} \tag{9.84}$$

The latter can easily handle the problem of "limited points" or "snap-back behavior" while the former failed at such situations. For details of the Newton–Raphson method and the arc length method, refer to the book by Borst *et al.* [415].

9.4 Numerical Tests

In this section, some representative numerical examples are carried out for the introduced two-dimensional, three-dimensional, and shallow shell hierarchical quadrature elements, respectively. A code in C++ is developed and the visualization in post-processing is handled by ParaView (version 5.5.2). The results obtained by the C++ code are compared with those obtained by commercial finite element software (ABAQUS/CAE 2018).

9.4.1 *A cantilever beam in planar configuration*

In this section, a cantilever beam with tip load (see Fig. 9.4) is discussed to validate the procedure of the geometrically nonlinear analysis and the C++ code. The beam is assigned a length $l = 8.0\,\text{mm}$, a height $h = 0.5\,\text{mm}$, a Young's modulus $E = 100\,\text{N/mm}^2$, and a Poisson's ratio $\upsilon = 0.3$. The beam is loaded by a concentrated load $F = 0.2\,\text{N}$, which is applied in 20 steps of $\Delta F = 0.01\,\text{N}$.

First, the convergence characteristics of the HQEM elements in our C++ code are compared with the CPS4 element used in ABAQUS. Then the load–displacement curves of the beam tip by the Newton–Raphson method and the arc length method obtained by both C++ code and ABAQUS are plotted. A displacement contour is shown in the end to give an intuitionistic vision of the deformed cantilever beam.

A fast convergence character of two-dimensional HQEM can be found in Fig. 9.5, where the convergent results can be obtained with tens of DOFs and ABAQUS CPS4 element is not able to converge until the number of DOFs rise to 1,000. The Newton–Raphson method and the arc length method are both implemented with two-dimensional HQEM and ABAQUS CPS4 element, and coincident load–displacement curves are obtained as show in Fig. 9.6.

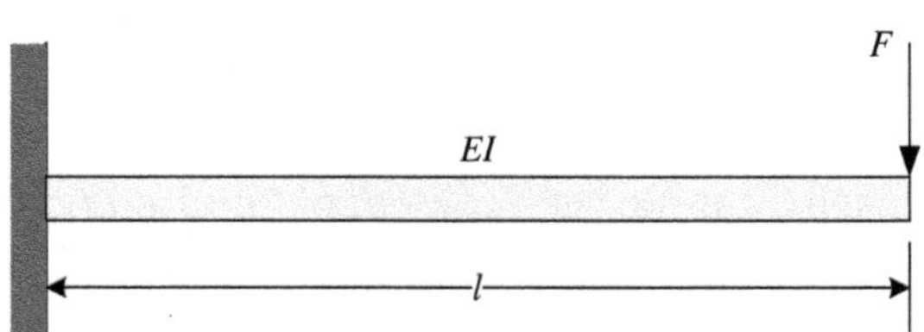

Fig. 9.4. A cantilever beam.

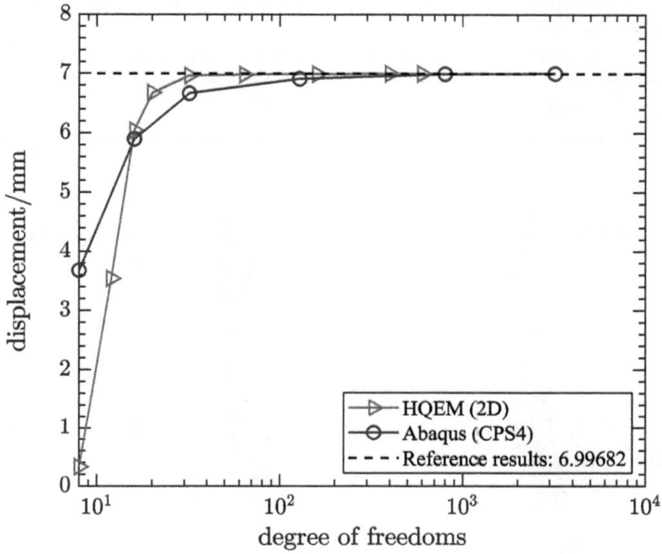

Fig. 9.5. Tip deflection convergence characteristics of HQEM (2D) and ABAQUS (CPS4).

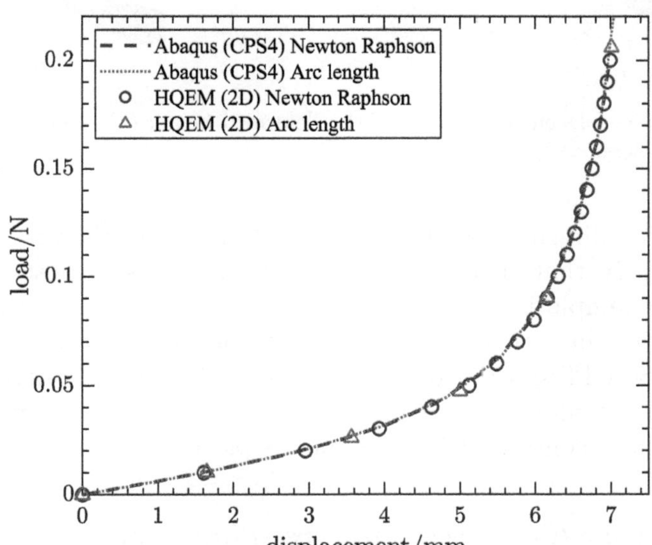

Fig. 9.6. Load–displacement curves obtained by HQEM (2D) and ABAQUS (CPS4).

Table 9.1. Tip deflection of deformed cantilever beam.

Method	HQEM (2D)	HQEM (2D, double elements)	ABAQUS (CPS4)
Results/mm	6.99682	6.9967	6.993

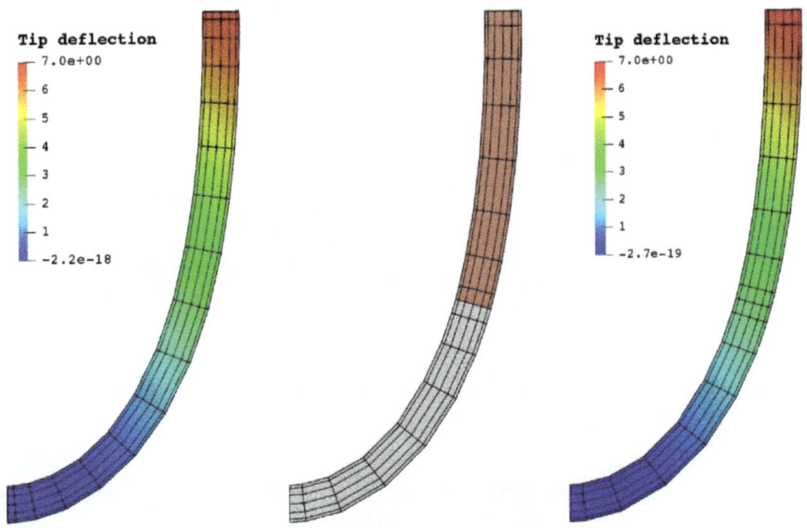

(a) (b)

Fig. 9.7. Displacement contour of the deformed beam: (a) single element; (b) double elements.

The two-dimensional HQEM elements used in this work have C^0 continuity so that element assembly can be easily disposed. A two-element example is carried out below and the results of tip deflection of two-dimensional HQEM (single and double element(s)) and ABAQUS (CPS4) are listed in Table 9.1. The beam is divided into two elements along length direction, as Fig. 9.7 shows, where the displacement contour of HQEM code is plotted.

9.4.2 *A cantilever beam in three-dimensional configuration*

The cantilever beam problem (Fig. 9.4) is reconsidered in three-dimensions in this section to verify the correctness of the nonlinear

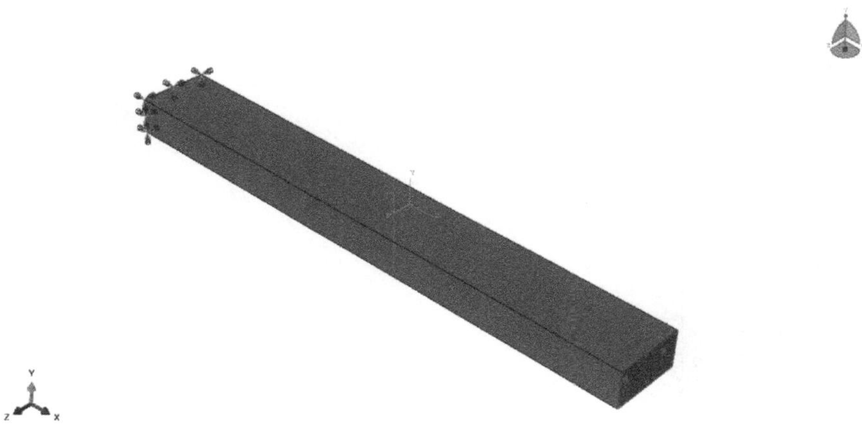

Fig. 9.8. Tip force is applied as surface traction in three-dimensional numerical tests.

three-dimensional HQEM C++ code. The beam's geometry and material properties are the same and it is omitted here for the sake of brevity. However, the tip force is applied on the right tip rectangular section of the beam (since it is more reasonable for a three-dimensional structure), as shown in Fig. 9.8.

The results of tip deflection obtained by three-dimensional HQEM using C++ code are compared with those obtained by ABAQUS CAE 2018 with C3D8 element, an 8-noded linear hexahedron element. First, the convergence characters are investigated in Fig. 9.9, which illustrates that the HQEM (p-FEM) still converges much faster than C38D (h-FEM) in a three-dimensional scenario. In this case, the three-dimensional HQEM converges at 540 DOFs of 540 while C3D8 converges with 3,528 DOFs (with criterion that the second decimal place remains unchanged). Second, the load–displacement curves obtained by both three-dimensional HQEM and ABAQUS C3D8 with two different nonlinear solution approaches (Newton–Raphson method and arc length method) are investigated in Fig. 9.10. It can be seen clearly that the four sets of results coincide with each other. Table 9.2 lists the exact tip deflection results obtained by three-dimensional HQEM and ABAQUS C3D8, and the results coincide as in the two-dimensional case. The result of triple element example is also listed here and will be discussed later in this section.

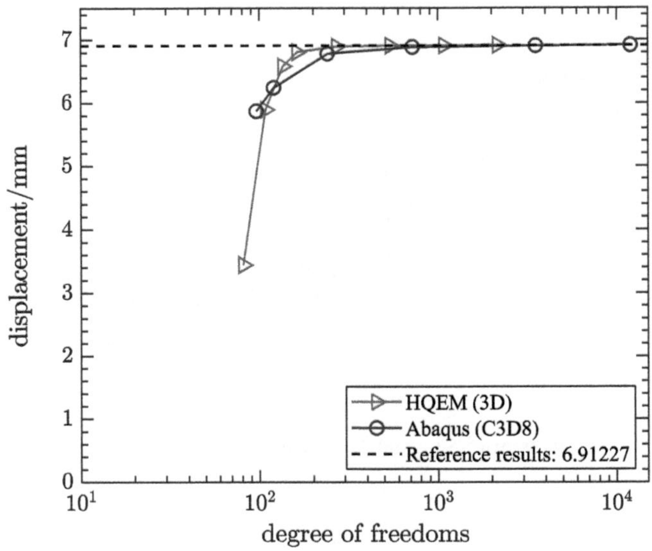

Fig. 9.9. Tip deflection convergence characteristics of HQEM (3D) and ABAQUS (C3D8).

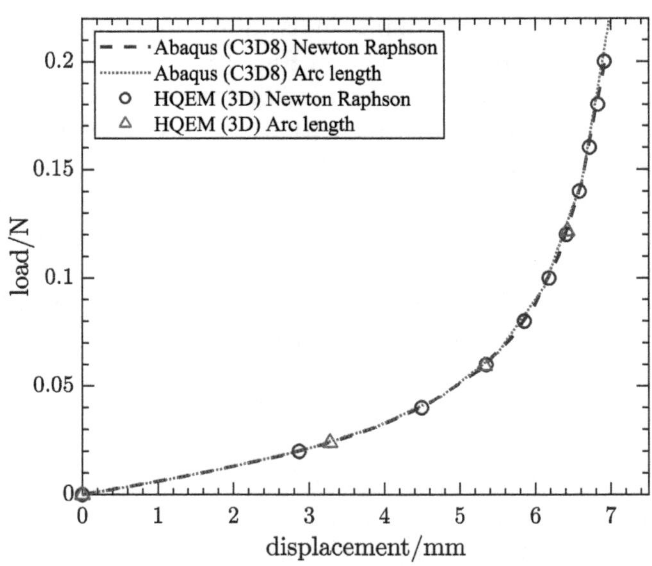

Fig. 9.10. Load–displacement curves gained by HQEM (3D) and ABAQUS (C3D8).

Table 9.2. Tip deflection of deformed cantilever beam.

Method	HQEM (3D)	HQEM (3D, triple elements)	ABAQUS (C3D8)
Tip deflection (mm)	6.91227	6.91406	6.913

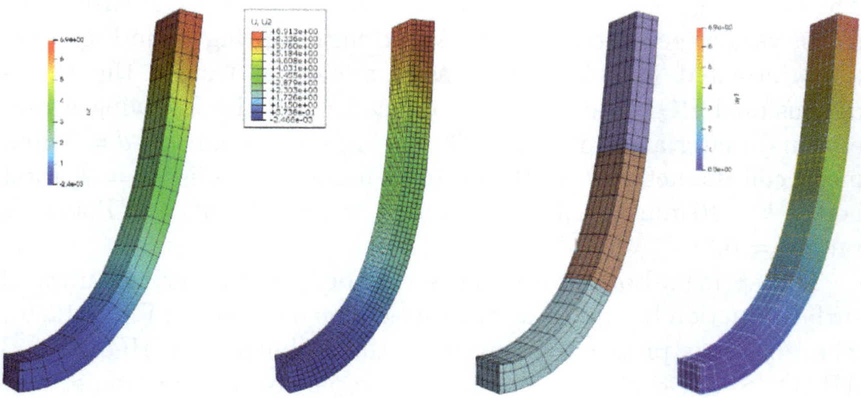

(a) (b) (c)

Fig. 9.11. Displacement contour of the deformed beam: (a) HQEM 3D; (b) ABAQUS C3D8; (c) HQEM 3D triple elements.

Finally, the displacement contour obtained by three-dimensional HQEM and ABAQUS C3D8 are plotted in Fig. 9.11. The contours' spectrums are both set in the style of 'Rainbow' and the data range of '0–6.913'. Note that the result difference between two- and three-dimensional scenarios is caused by the difference in load conditions (concentrated force in the two-dimensional case and surface traction in the three-dimensional case). Even though the 2D element assembly property has been tested in Section 3.1 with the plane problem, HQEM 3D elements contain face bases with difficulty to assemble directly due to the lack of interpolation characters. Thus, an interpolation transportation needs to be applied to implement the element assembly. The element assembly properties of the HQEM 3D are investigated as follows. This time the cantilever beam is divided into three elements along the length direction, thus the surface traction is applied on the rightmost element and the clamped boundary conditions are applied on the leftmost element. Figure 9.11(c) shows the displacement contour of the assembly problem. The relationship between the three elements can be seen on the left and the displacement continuity can be seen on the right.

9.4.3 *An extension spring in three-dimensional configuration*

The advantage of three-dimensional HQEM is that it can easily represent various geometric shapes. An extension spring (as in Fig. 9.12) is investigated in this section as a numerical test of the three-dimensional HQEM code and is defined using the following geometry and material parameters. The spring's wire diameter $d = 1\,\text{mm}$, mean coil diameter $D = 10\,\text{mm}$, total number of coils $N_t = 2$, total height $h = 10\,\text{mm}$, Young's modulus $E = 100\,\text{N/mm}^2$, and Poisson's ratio $\upsilon = 0.3$.

The spring's bottom section is clamped, and a vertical upward surface traction is applied on the top section as shown in Fig. 9.12(b). The nonlinear problem is solved by three-dimensional HQEM and ABAQUS C3D8 element so that a comparison can be made. The results obtained by HQEM 3D and C3D8 are listed in Table 9.3 and displacement contours (both are set in style of 'Rainbow' and range of '0–10.65') in front view, side view, vertical view, and perspective view of the two sets of solutions are shown in Fig. 9.13.

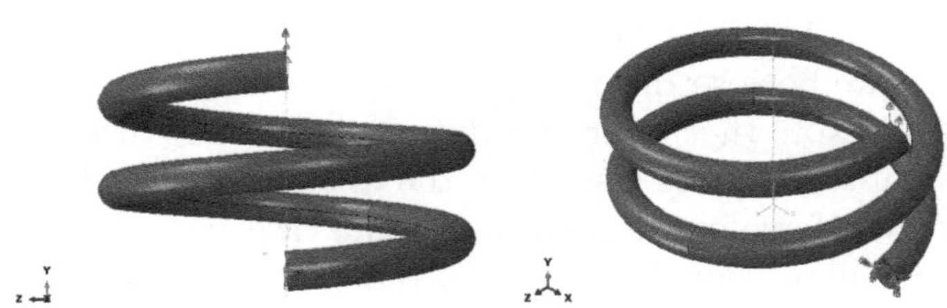

(a) (b)

Fig. 9.12. Geometry of the extension spring: (a) front view; (b) perspective view.

Table 9.3. Vertical extension of the spring.

Method	HQEM (3D)	ABAQUS (C3D8)
Tip deflection/mm	10.6139	10.65

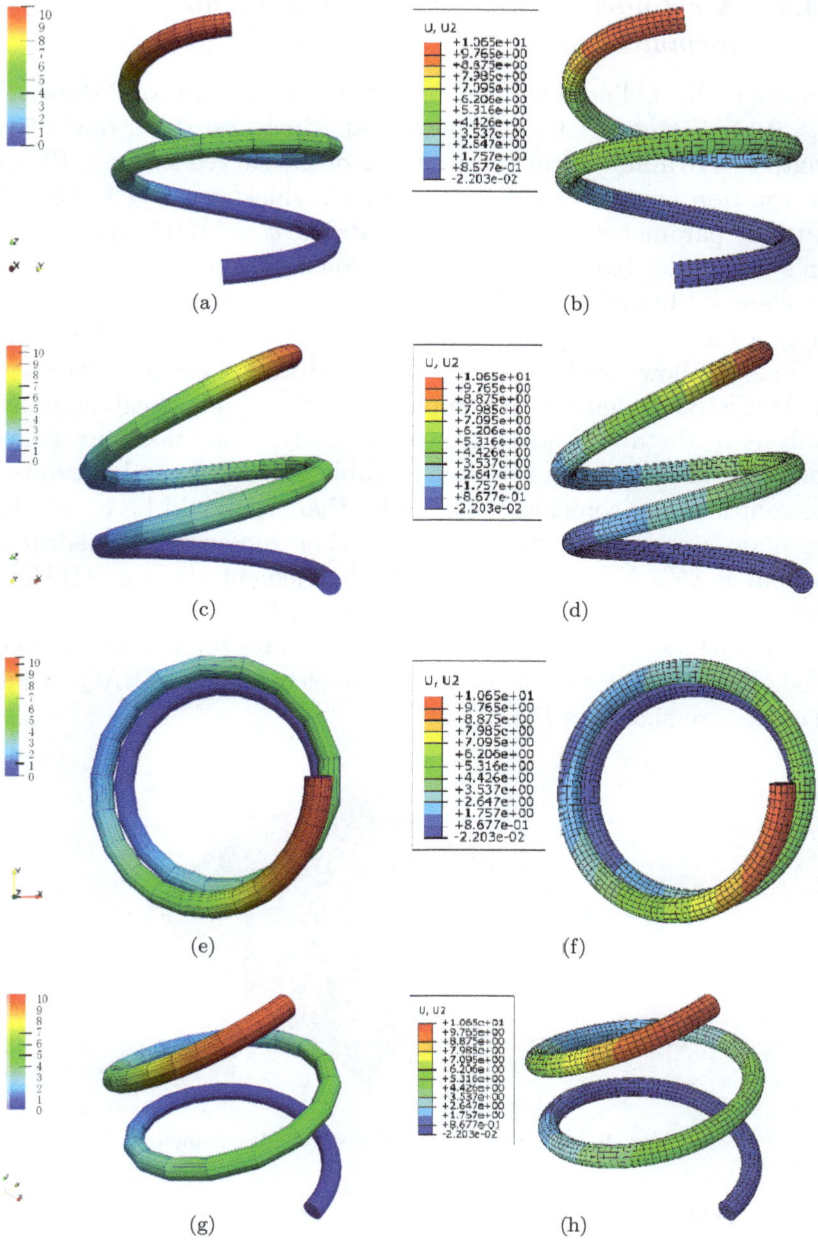

Fig. 9.13. Displacement contours of the extension spring: (a) HQEM 3D; (b) Abaqus C3D8.

9.4.4 A cylindrical shallow shell in Mindlin formulation

Consider the cylindrical shell mid-plane configuration, shown in Fig. 9.14, the geometric parameters of which are the straight side length $l = 10\,\text{mm}$, the curvature radius of the curved side $r = 10\,\text{mm}$, the rotation angle $\theta = \pi/3$, and the shell thickness $h = 1\,\text{mm}$. The material parameters are Young's modulus $E = 100\,\text{N/mm}^2$, Poisson's ratio $\upsilon = 0.3$ and shear correction factor $\kappa = 5/6$. Consider the clamped boundary conditions on the four sides of the shell, and apply a uniform load $p = 1\,\text{N/mm}^2$ on the top surface of the shell.

The shallow shell hierarchical quadrature element and the ABAQUS S8R element are applied for the geometrical nonlinear analysis of the above structure, respectively. The maximum deflection of the shell structure is listed in Table 9.4, and it can be seen that the solutions are coincided with each other. It should be noted that the node arrangement of the shallow shell hierarchical quadrature element is $[N_x, N_y] = [10, 10]$, while the element size for ABAQUS S8R is $h = 0.5\,\text{mm}$ and the node number is 462.

The displacement contours plotted based on the results obtained by shallow shell hierarchical quadrature element and ABAQUS S8R elements are shown in Fig. 9.15.

Fig. 9.14. The geometry of a cylindrical shell.

Table 9.4. The maximum deflection of the cylindrical shell.

Method	HQEM (Mindlin)	ABAQUS S8R
Maximum deflection/mm	1.32689	1.327

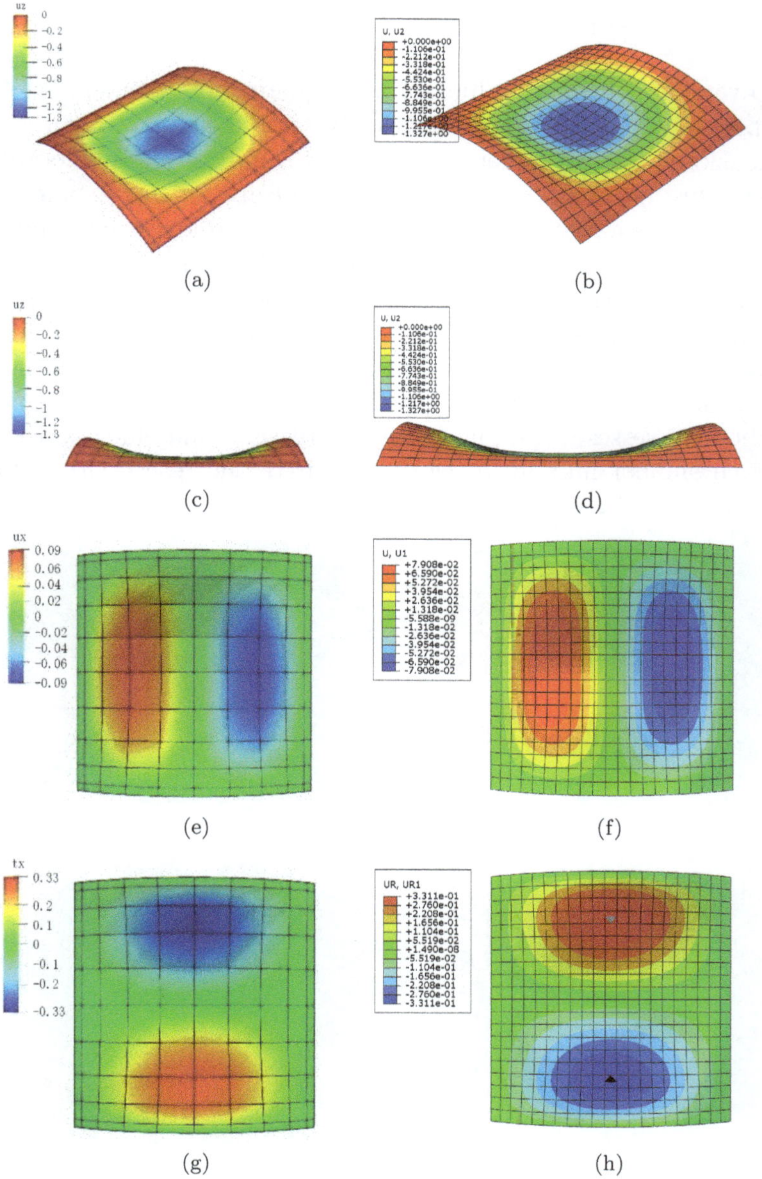

Fig. 9.15. The displacement contours of the cylindrical shell: (a) HQEM Mindlin u_z (perspective view); (b) ABAQUS S8R u_z (perspective view); (c) HQEM Mindlin u_z (front view); (d) ABAQUS S8R u_z (front view); (e) HQEM Mindlin u_x; (f) ABAQUS S8R u_x; (g) HQEM Mindlin θ_x; (h) ABAQUS S8R θ_x.

9.5 Conclusions

A p-version FEM, the hierarchical quadrature element method, is presented for the geometrically nonlinear problems. Two-dimensional, three-dimensional, and Mindlin shell hierarchical quadrature elements are constructed and applied for geometrically nonlinear quasi-static analysis of planar, complex space, and shallow shell structure, respectively. Both the Newton–Raphson method and the arc length methods are used. The displacement responses of some common structures are calculated and compared with those obtained by the commercial software. It is shown that the results of the presented elements coincide with those of ABAQUS and the method possesses advantages of high precision and efficiency. Due to a much smaller number of DOFs required, the results are obtained with a reduced computational cost.

Chapter 10

The Hierarchical Quadrature Element Method for Incremental Elasto-Plastic Analysis

The hierarchical quadrature element method (HQEM) combines the hierarchical finite element method and differential quadrature method. In this chapter, the implementation of the HQEM to the J_2 flow theory with isotropic hardening for small displacements and small strains is presented. Based on higher-order quadrilateral and hexahedral element formulations, a Newton–Raphson iteration scheme combined with a return-mapping algorithm is applied to find approximate solutions for the underlying physically nonlinear model in the two- as well as three-dimensional problems. The accuracy and convergence of the HQEM are validated through comparison with h-version approximate results calculated by commercial codes.

10.1 Classical J_2 Flow Theory with Nonlinear Isotropic Hardening

10.1.1 *Classical three-dimensional elasto-plastic theory*

The elasto-plastic problem to be considered in this chapter is based on the J_2 flow theory for small strains with isotropic hardening [144–146, 168]. The classic three-dimensional elasto-plastic model is used to solve the three-dimensional and plane strain problems:

(1) Additive decomposition

Under the assumption of small deformation, strain and strain rate can be decomposed into elastic and plastic parts, where the elastic component determines the stress and the plastic component affects the yield stress:

$$\boldsymbol{\varepsilon} = \boldsymbol{\varepsilon}_e + \boldsymbol{\varepsilon}_p \; \dot{\boldsymbol{\varepsilon}} = \dot{\boldsymbol{\varepsilon}}_e + \dot{\boldsymbol{\varepsilon}}_p \tag{10.1}$$

where superscripts e and p denote elastic and plastic parts, respectively. The superposed 'dot' denotes the rate of a quantity. It is reminded that in static problems, the rate is equivalent to the strain increment.

(2) Strain energy density

The following form of strain energy density can be considered:

$$W(\boldsymbol{\varepsilon}_e) = \frac{1}{2}\boldsymbol{\varepsilon}_e : \mathbf{D} : \boldsymbol{\varepsilon}_e = \frac{1}{2}(\boldsymbol{\varepsilon} - \boldsymbol{\varepsilon}_p) : \mathbf{D} : (\boldsymbol{\varepsilon} - \boldsymbol{\varepsilon}_p) \tag{10.2}$$

where $\mathbf{D}$ is a fourth-order constitutive tensor:

$$\mathbf{D} = (\lambda + (2/3)\mu)\mathbf{1} \otimes \mathbf{1} + 2\mu\mathbf{I}_{\text{dev}} \tag{10.3}$$

where $\mathbf{1}$ is the second-order unit tensor, $\mathbf{I}_{\text{dev}} = \mathbf{I} - (1/3)\mathbf{1} \otimes \mathbf{1}$, and $\mathbf{I}$ is the fourth-order unit tensor, λ and μ are the Lame constants. By differentiating the above definition, stress can be related to the elastic strain as

$$\boldsymbol{\sigma} = \frac{\partial W(\boldsymbol{\varepsilon}_e)}{\partial \boldsymbol{\varepsilon}_e} = \mathbf{D} : \boldsymbol{\varepsilon}_e = \mathbf{D} : (\boldsymbol{\varepsilon} - \boldsymbol{\varepsilon}_p) \tag{10.4}$$

(3) Yield function

For metal plasticity, the *von Mises* yield criterion with the associative flow rule is commonly used to describe material behavior after elastic deformation. Accordingly, the yield criterion or yield function is formulated as

$$f(\boldsymbol{\sigma}, \alpha) \equiv \|\text{dev}[\boldsymbol{\sigma}]\| - \sqrt{\frac{2}{3}}\kappa(\alpha) = \|\mathbf{s}\| - \sqrt{\frac{2}{3}}\kappa(\alpha) \leq 0 \tag{10.5}$$

where $\| \cdot \| = \sqrt{(\cdot : \cdot)}$ is the Euclidean norm of a tensor, $\text{dev}[\cdot] = (\cdot) - 1/3\text{tr}[\cdot]$ is the deviatoric part of a tensor and $\text{tr}[\cdot]$ denotes the trace operator, $\mathbf{s}$ is the deviatoric part of stress and is the center of the yield surface (the elastic domain), and $\kappa(\alpha)$ is the radius of the elastic domain determined by the isotropic hardening model. An internal variable α being often referred to as equivalent plastic strain describes the nonlinear isotropic hardening:

$$\kappa(\alpha) = \sigma_0 + h\alpha + (\sigma_\infty - \sigma_0)(1 - \exp(-\omega\alpha)) \tag{10.6}$$

which is composed by a linear and an exponential function, where σ_0 is the initial yield stress, h the linear hardening parameter, σ_∞ the saturation stress and ω the hardening exponent.

(4) Associative flow rule

The flow rule determines the evolution of the plastic strain $\boldsymbol{\varepsilon}_p$. For multidimensional plasticity, since $\boldsymbol{\varepsilon}_p$ is a tensor, it is necessary to determine its magnitude as well as its direction. Thus, a general form of the flow rule can be written as

$$\dot{\boldsymbol{\varepsilon}}_p = \gamma \mathbf{r}(\boldsymbol{\sigma}, \alpha) \tag{10.7}$$

where γ is called a plastic consistency parameter. The expression of $\mathbf{r}(\boldsymbol{\sigma}, \alpha)$ depends on the plasticity model. It is often assumed that there is a flow potential (or plastic potential), g, such that the plastic strain evolves in the direction normal to the flow potential:

$$\dot{\boldsymbol{\varepsilon}}_p = \gamma \frac{\partial g(\boldsymbol{\sigma}, \boldsymbol{\xi})}{\partial \boldsymbol{\sigma}} \tag{10.8}$$

where $g(\boldsymbol{\sigma}, \alpha)$ is the flow potential. When the flow potential is the same as the yield function, the plastic model is called associative. Thus,

$$\dot{\boldsymbol{\varepsilon}}_p = \gamma \frac{\partial f(\mathbf{s}, \alpha)}{\partial \mathbf{s}} = \gamma \frac{\mathbf{s}}{\|\mathbf{s}\|} = \gamma \mathbf{N} \tag{10.9}$$

where $\mathbf{N}$ is a unit deviatoric tensor normal to the yield surface and γ is a plastic consistency parameter, which is non-negative. If the material status is elastic, γ must be 0, but if it is plastic, then γ must be positive. Thus, the plastic strain increases in the direction normal to the yield surface and has the magnitude of plastic consistency parameter γ.

(5) Plastic consistency parameter

As mentioned before, γ is 0 when the material is elastic ($f < 0$) and positive when it is plastic ($f < 0$). In optimization, this is called the Kuhn–Tucker condition and can be written as

$$\gamma \geq 0, \quad f \leq 0, \quad \gamma f = 0 \tag{10.10}$$

When the stress is on the yield surface, i.e., in the plastic state, Eq. (10.10) is satisfied because $f = 0$. However, when the state varies, it is possible to have three different cases:

(a) elastic unloading: $\dot{f} < 0$, $\gamma = 0 \Rightarrow \gamma\dot{f} = 0$;
(b) neutral loading: $\dot{f} = 0$, $\gamma = 0 \Rightarrow \gamma\dot{f} = 0$;
(c) plastic loading: $\dot{f} = 0$, $\gamma > 0 \Rightarrow \gamma\dot{f} = 0$.

Thus, $\gamma f = 0$ in Eq. (10.10) is equivalent to $\gamma\dot{f} = 0$ when the stress is on the yield surface. Therefore, the rate form of the Kuhn–Tucker condition can be used in calculating the plastic consistency parameter.

10.1.2 *Numerical algorithm for three dimensional elasto-plastic problems*

Since constitutive relations and evolution of plastic variables are in the form of rates in the elasto-plastic model, they need to be integrated over time (or load) increments. In static problems, the time increment should be understood as a load increment. The full magnitude of load is first divided by N increments, and the structural equilibrium at each increment is sought with the incremental force method. It is assumed that the solutions and the status of material at time t_n are known, which includes stress and plastic variables. Then, at time t_{n+1}, the Newton–Raphson method solves for the incremental displacements during the convergence iteration. Thus, the objective is to update stress and plastic variables from time t_n to t_{n+1} using the given displacement increments or equivalently using the given strain increments.

Note that the structural equilibrium is only satisfied at the discrete set of time increments. Thus, it is possible that there might be discretization error in time, especially when the status of material changes within a time increment. If a smaller size of time increment is used, the error will be reduced.

In this chapter, the elasto-plastic model with rate form is transformed into incremental form by backward implicit Euler time integration method, and the plastic problem is solved by return-mapping algorithm. In the return-mapping algorithm, a two-step method is often used. First, the elastic trial status is computed, in which all strain increments are purely elastic. If the trial stress resides outside the elastic domain, then the trial stress is projected onto the yield surface, which is a convex set. This step is called the return mapping to the yield surface. The process of the algorithm is as follows.

The first step is called the elastic predictor and uses this incremental strain. The stress and hardening parameters are predicted elastically as

$$\mathbf{s}^{tr} = \mathbf{s}^n + 2\mu\Delta\mathbf{e}, \quad \alpha^{tr} = \alpha^n \tag{10.11}$$

where the right superscript n denotes the time t_n and "tr" denotes the trial status. If the trial stress is within the elastic domain, i.e.,$f(\boldsymbol{\sigma}^{tr}, \alpha^{tr}) \leq 0$, then the status of the material is elastic, and the stress and plastic variables are updated using the trial predictors as

$$\mathbf{s}^{n+1} = \mathbf{s}^{tr}, \quad \alpha^{n+1} = \alpha^{tr} \tag{10.12}$$

If the trial stress is outside the elastic domain, i.e.,$f(\boldsymbol{\sigma}^{tr}, \alpha^{tr}) > 0$, then the status of the material becomes plastic, and the plastic correction step needs to be carried out to find the plastic status of the material:

$$\begin{aligned} \mathbf{s}^{n+1} &= \mathbf{s}^{tr} - 2\mu\Delta\varepsilon^p = \mathbf{s}^{tr} - 2\mu\hat{\gamma}\mathbf{N} \\ \alpha^{n+1} &= \alpha^n + \sqrt{\frac{2}{3}}\hat{\gamma} \end{aligned} \tag{10.13}$$

where $\hat{\gamma} = \gamma\Delta t$ is the plastic consistency parameter and $\mathbf{N} = \mathbf{s}^{tr}/\|\mathbf{s}^{tr}\|$ is a unit deviatoric tensor, normal to the yield surface at time t_{n+1}. Note that stress and back stress are corrected in the parallel direction to $\mathbf{N}$. This fact makes it convenient for finding the updated stress on the updated yield surface.

Thus, the plastic correction step condenses to determine the plastic consistency parameter $\hat{\gamma}$, from which the plastic strain increment can be obtained. The basic idea is to make the yield function satisfy the yield condition at the updated state. Thus, at the return-mapped point, the following yield condition must be satisfied:

$$\begin{aligned} f(\mathbf{s}^{n+1}, \alpha^{n+1}) &= \|\mathbf{s}^{n+1}\| - \sqrt{\frac{2}{3}}\kappa(\alpha^{n+1}) \\ &= \|\mathbf{s}^{tr}\| - (2\mu + H(\alpha^{n+1}))\hat{\gamma} \\ &\quad -\sqrt{\frac{2}{3}}\kappa(\alpha^{n+1}) = 0 \end{aligned} \tag{10.14}$$

which is a nonlinear equation in terms of $\hat{\gamma}$. Equation (10.14) can be solved for $\hat{\gamma}$ using the local Newton–Raphson method. Note that this is different from the Newton–Raphson method for the convergence iteration in which equilibrium between internal and external forces is sought. This is a local iteration to find the stress point on the yield surface. Thus, the solution procedure for elasto-plastic systems has double iteration loops. The inside, local iteration loop usually converges quickly within five or six iterations, but it needs to be performed at every integration point that has plastic deformation.

The next step is to update stress and plastic variables. After $\hat{\gamma}$ is found, the deviatoric stress can be updated at time t_{n+1} by

$$\mathbf{s}^{n+1} = \mathbf{s}^n + 2\mu\Delta\mathbf{e} - 2\mu\hat{\gamma}\mathbf{N} \tag{10.15}$$

The stress and plastic variables are updated as follows:

$$\boldsymbol{\sigma}^{n+1} = \boldsymbol{\sigma}^n + \Delta\boldsymbol{\sigma} = \boldsymbol{\sigma}^n + (\mathbf{D} : \Delta\boldsymbol{\varepsilon} - 2\mu\hat{\gamma}\mathbf{N}) \tag{10.16}$$

$$\alpha^{n+1} = \alpha^n + \sqrt{\frac{2}{3}}\hat{\gamma} \tag{10.17}$$

Note that the stress and back stress increments corresponding to the plastic correction component in Eqs. (10.15) and (10.17) are in the same direction as $\mathbf{N}$, which is a radial direction of the yield surface.

When the continuous tangential stiffness is used, the Newton–Raphson iteration may not achieve quadratic convergence because the continuous tangential stiffness cannot be consistent with the time integration algorithm, so it is necessary to derive the consistent tangential stiffness to match the time integration algorithm.

The expression of consistent tangential stiffness is

$$\mathbf{D}^{\text{alg}} = \frac{\partial \Delta \boldsymbol{\sigma}}{\partial \Delta \boldsymbol{\varepsilon}} = \mathbf{D} - \frac{4\mu^2 \mathbf{N} \otimes \mathbf{N}}{2\mu + \frac{2}{3}\kappa_\alpha} - \frac{4\mu^2}{\|\mathbf{s}\|}[\mathbf{I}_{\text{dev}} - \mathbf{N} \otimes \mathbf{N}] \tag{10.18}$$

under the condition of nonlinear isotropic hardening, the expression becomes

$$\begin{aligned}\mathbf{D}^{\text{alg}} = \mathbf{D} &- \frac{4\mu^2 \mathbf{N} \otimes \mathbf{N}}{2\mu + \frac{2}{3}[h + \omega(\sigma_\infty - \sigma_0)(1 - \exp(-\omega\alpha))]} \\ &- \frac{4\mu^2 \hat{\gamma}}{\|\mathbf{s}\|}[\mathbf{I}_{\text{dev}} - \mathbf{N} \otimes \mathbf{N}]\end{aligned} \tag{10.19}$$

At each integration point of the element, the stress and plastic variables are determined using the return-mapping algorithm, which is shown in the following flowchart:

(a) Unit tensor $\mathbf{1} = [1, 1, 1, 0, 0, 0]^T$.
(b) Trial stress $\boldsymbol{\sigma}^{tr} = \boldsymbol{\sigma}^n + \mathbf{D} \cdot \Delta\boldsymbol{\varepsilon}$.
(c) Trace of stress $\mathbf{s}^{tr} = \boldsymbol{\sigma}^{tr} - (1/3)\text{tr}(\boldsymbol{\sigma}^{tr})\mathbf{1}$.
(d) Norm
$\|\mathbf{s}^{tr}\| = \sqrt{(s_{11}^{tr})^2 + (s_{22}^{tr})^2 + (s_{33}^{tr})^2 + 2[(s_{12}^{tr})^2 + (s_{23}^{tr})^2 + (s_{13}^{tr})^2]}$.
(e) Yield function $f = \|\mathbf{s}^{tr}\| - \sqrt{2/3}[\sigma_0 + h\alpha + (\sigma_\infty - \sigma_0)(1 - \exp(-\omega\alpha))]$.
(f) Check for the yield status.
If $f < 0$, then the material is elastic:

$$\boldsymbol{\sigma}^{n+1} = \boldsymbol{\sigma}^{tr}$$

$$\mathbf{D}^{\text{alg}} = \mathbf{D} = \begin{bmatrix} \lambda + 2\mu & \lambda & \lambda & 0 & 0 & 0 \\ \lambda & \lambda + 2\mu & \lambda & 0 & 0 & 0 \\ \lambda & \lambda & \lambda + 2\mu & 0 & 0 & 0 \\ 0 & 0 & 0 & \mu & 0 & 0 \\ 0 & 0 & 0 & 0 & \mu & 0 \\ 0 & 0 & 0 & 0 & 0 & \mu \end{bmatrix}$$

End of process.

(g) Solve nonlinear equation to get consistency parameter:

$$f(\hat{\gamma}) = 0 = \|\sigma^{tr}\| - 2\mu\hat{\gamma} - \sqrt{\frac{2}{3}}\left(\sigma_0 + h\left(\alpha^n + \sqrt{\frac{2}{3}}\hat{\gamma}\right)\right.$$

$$\left. + (\sigma_\infty - \sigma_0)\left(1 - \exp\left(-\omega\left(\alpha^n + \sqrt{\frac{2}{3}}\hat{\gamma}\right)\right)\right)\right)$$

(h) Unit deviatoric vector $\mathbf{N} = \mathbf{s}^{tr}/\|\mathbf{s}^{tr}\|$.

(i) Update stress and plastic variables:

$$\boldsymbol{\sigma}^{n+1} = \boldsymbol{\sigma}^{tr} - 2\mu\hat{\gamma}\mathbf{N}, \alpha^{n+1} = \alpha^n + \sqrt{\frac{2}{3}}\hat{\gamma}$$

(j) Calculate consistent tangent stiffness

$$c_1 = \frac{4\mu^2}{2\mu + \frac{2}{3}[h + \omega(\sigma_\infty - \sigma_0)(1 - \exp(-\omega\alpha))]}, c_2 = \frac{4\mu^2\hat{\gamma}}{\|\boldsymbol{\sigma}^{tr}\|}$$

$$\mathbf{I}^{\text{dev}} = \begin{bmatrix} \frac{2}{3} & -\frac{1}{3} & -\frac{1}{3} & 0 & 0 & 0 \\ -\frac{1}{3} & \frac{2}{3} & -\frac{1}{3} & 0 & 0 & 0 \\ -\frac{1}{3} & -\frac{1}{3} & \frac{2}{3} & 0 & 0 & 0 \\ 0 & 0 & 0 & \frac{1}{2} & 0 & 0 \\ 0 & 0 & 0 & 0 & \frac{1}{2} & 0 \\ 0 & 0 & 0 & 0 & 0 & \frac{1}{2} \end{bmatrix}$$

$$D_{ij}^{\text{alg}} = D_{ij} - (c_1 - c_2)N_iN_j - c_2 \times I_{ij}^{\text{dev}}$$

Note that the above calculation must be performed for each integration point of an element. Accordingly, the stress and plastic variables at each integration point must be stored and updated. There is no need to store the stress and plastic variables during unconverged iteration. Once the Newton–Raphson method is converged at load increment t_{n+1}, they are stored, and next load increment starts.

10.1.3 *Return-mapping algorithm for plane stress elasto-plastic problems*

The idea of solving plane stress elasto-plastic problems is to execute return mapping directly in constrained plane stress subspace through the basic equations in the previous section. Under the condition of isotropic hardening, the theoretical solution is reduced to the iterative solution of nonlinear equation for plane stress problems. The return-mapping algorithm of plane stress elasto-plastic problem is shown in the following flowchart:

(a) Calculate trial stress

$$\boldsymbol{\sigma}^{tr} = \boldsymbol{\sigma}^n + \mathbf{D}\Delta\boldsymbol{\varepsilon}$$

(b) Check for the yield status $f(\mathbf{s}^{tr}, \alpha^{tr})$;
If $f < 0$, then the material is elastic:

$$\boldsymbol{\sigma}^{n+1} = \boldsymbol{\sigma}^{tr}$$

$$\mathbf{D}^{\text{alg}} = \mathbf{D} = \begin{bmatrix} \lambda + 2\mu & \lambda & 0 \\ \lambda & \lambda + 2\mu & 0 \\ 0 & 0 & \mu \end{bmatrix}$$

End of process.

(c) Solve nonlinear equation to get consistency parameter:

$$f^2(\hat{\gamma}) := \frac{1}{2}\bar{f}^2(\hat{\gamma}) - R^2(\hat{\gamma}) \equiv 0$$

$$\bar{f}^2(\hat{\gamma}) := \frac{1}{2}\frac{\frac{1}{3}(\sigma_{11}^{tr} + \sigma_{22}^{tr})^2}{\left\{1 + \frac{E}{3(1-\nu)}\hat{\gamma}\right\}^2} + \frac{\frac{1}{2}(\sigma_{11}^{tr} - \sigma_{22}^{tr})^2 + 2(\sigma_{12}^{tr})^2}{\{1 + 2\mu\hat{\gamma}\}^2}$$

$$R^2(\Delta\gamma) := \frac{1}{3}\kappa^2\left[\alpha^n + \sqrt{\frac{2}{3}}\Delta\gamma\bar{f}(\hat{\gamma})\right]$$

where $\bar{f} = \sqrt{\boldsymbol{\sigma}\mathbf{P}\boldsymbol{\sigma}}$ and

$$\mathbf{P} := \frac{1}{3}\begin{bmatrix} 2 & -1 & 0 \\ -1 & 2 & 0 \\ 0 & 0 & 6 \end{bmatrix}$$

(d) Calculate the correction tangential stiffness $\boldsymbol{\Xi} := [\mathbf{D}^{-1} + \hat{\gamma}\mathbf{P}]^{-1}$

(e) Update stress and plastic variables:

$$\boldsymbol{\sigma}^{n+1} = \boldsymbol{\Xi}(\hat{\gamma})(\mathbf{D}^{-1})\boldsymbol{\sigma}^{tr}$$

$$\alpha^{n+1} = \alpha^{n} + \sqrt{\frac{2}{3}}\hat{\gamma}\bar{f}(\hat{\gamma})$$

(f) Calculate consistent tangent stiffness

$$\left.\frac{\mathrm{d}\boldsymbol{\sigma}}{\mathrm{d}\boldsymbol{\varepsilon}}\right|_{n+1} = \boldsymbol{\Xi} - \frac{[\boldsymbol{\Xi}\mathbf{P}\boldsymbol{\sigma}^{n+1}][\boldsymbol{\Xi}\mathbf{P}\boldsymbol{\sigma}^{n+1}]^{\mathrm{T}}}{(\boldsymbol{\sigma}^{n+1})^{\mathrm{T}}\mathbf{P}\boldsymbol{\Xi}\mathbf{P}\boldsymbol{\sigma}^{n+1} + \overline{\beta}^{n+1}}$$

$$\theta := 1 - \frac{2}{3}\kappa'_{n+1}\hat{\gamma}$$

$$\overline{\beta}^{n+1} := \frac{2}{3}\frac{1}{\theta}(\kappa'_{n+1})(\boldsymbol{\sigma}^{n+1})^{\mathrm{T}}\mathbf{P}\boldsymbol{\sigma}^{n+1}$$

10.1.4 *Numerical calculation process of elasto-plastic problems*

The numerical solution process of elasto-plastic problems can be summarized as follows: The global incremental iterative form combines with the HQEM and the return-mapping algorithm. In each iteration of the incremental step, it is necessary to apply the return-mapping algorithm to each integral point to determine the elasto-plastic state. Figure 10.1 shows the overall flowchart of this chapter for calculating the elasto-plastic problem. Figure 10.2 shows the application process of the return-mapping algorithm at each integral point under non-linear isotropic hardening conditions.

10.2 The Hierarchical Quadrature Element Method

The HQEM formulations for 2D and 3D solids have been presented in Chapters 4 and 5. The hexahedral element is taken as an example here. The displacement field of a hexahedron element can be expressed as

$$\mathbf{u} = \tilde{\mathbf{S}}^{T}\bar{\mathbf{u}} \tag{10.20}$$

The value of displacement field $\mathbf{u}$ at the integral point is

$$\hat{\mathbf{u}} = \mathbf{G}\bar{\mathbf{u}} \tag{10.21}$$

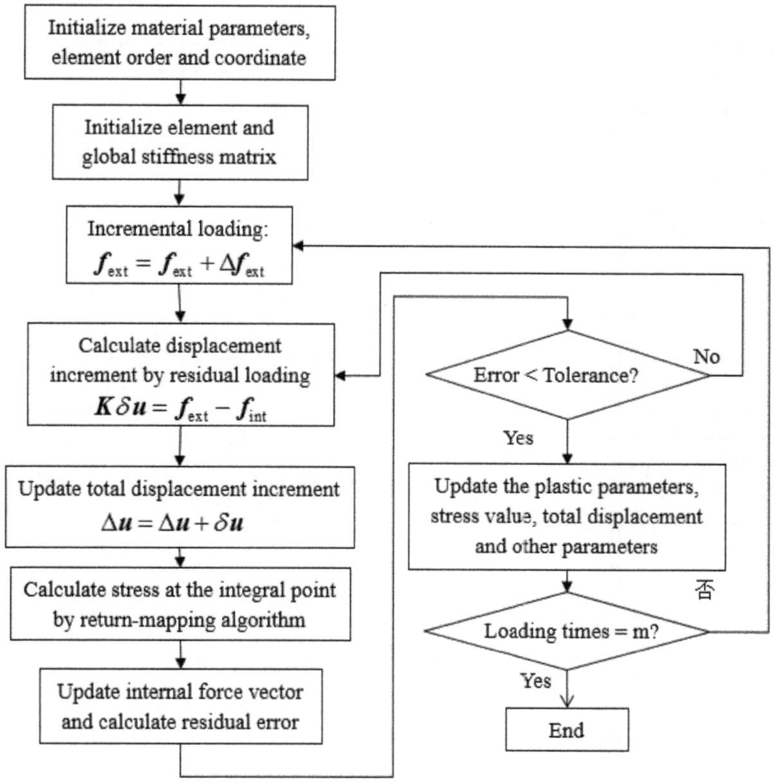

Fig. 10.1. Flowchart of elasto-plastic numerical solution.

where

$$\begin{aligned}\mathbf{G}^T &= [\mathbf{S}(\xi_1,\eta_1,\zeta_1)\ \mathbf{S}(\xi_2,\eta_2,\zeta_2)\cdots\mathbf{S}(\xi_K,\eta_K,\zeta_K)]\\ \bar{\mathbf{u}} &= [u(\xi_1,\eta_1,\zeta_1)\ u(\xi_2,\eta_2,\zeta_2)\cdots u(\xi_K,\eta_K,\zeta_K)]^T\end{aligned} \tag{10.22}$$

Using the following chain rule, we can get the derivative of the displacement field u with respect to global coordinates x, y and z:

$$\begin{bmatrix}\dfrac{\partial u}{\partial x}\\ \dfrac{\partial u}{\partial y}\\ \dfrac{\partial u}{\partial z}\end{bmatrix} = \mathbf{J}^{-1}\begin{bmatrix}\dfrac{\partial u}{\partial \xi}\\ \dfrac{\partial u}{\partial \eta}\\ \dfrac{\partial u}{\partial \zeta}\end{bmatrix} \tag{10.23}$$

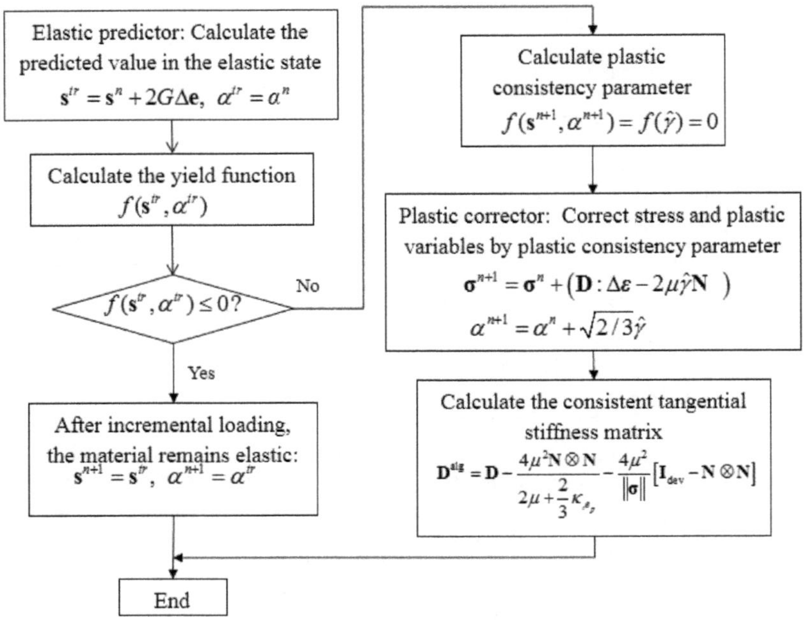

Fig. 10.2. Flowchart of return-mapping algorithm.

where $\mathbf{J}$ is Jacob determinant:

$$\mathbf{J} = \begin{vmatrix} \dfrac{\partial x}{\partial \xi} & \dfrac{\partial y}{\partial \xi} & \dfrac{\partial z}{\partial \xi} \\ \dfrac{\partial x}{\partial \eta} & \dfrac{\partial y}{\partial \eta} & \dfrac{\partial z}{\partial \eta} \\ \dfrac{\partial x}{\partial \zeta} & \dfrac{\partial y}{\partial \zeta} & \dfrac{\partial z}{\partial \zeta} \end{vmatrix} \tag{10.24}$$

Then, the derivatives at integral points $x_{ijk} = x(\xi_i, \eta_j, \zeta_k)$, $y_{ijk} = y(\xi_i, \eta_j, \zeta_k)$, $z_{ijk} = z(\xi_i, \eta_j, \zeta_k)$ are expressed as:

$$\left(\frac{\partial u}{\partial x}\right)_{ijk} = \frac{1}{|\mathbf{J}|_{ijk}} \begin{bmatrix} \left(\dfrac{\partial y}{\partial \eta}\dfrac{\partial z}{\partial \zeta} - \dfrac{\partial y}{\partial \zeta}\dfrac{\partial z}{\partial \eta}\right)_{ijk} \\ -\left(\dfrac{\partial y}{\partial \xi}\dfrac{\partial z}{\partial \zeta} - \dfrac{\partial y}{\partial \zeta}\dfrac{\partial z}{\partial \xi}\right)_{ijk} \\ \left(\dfrac{\partial y}{\partial \xi}\dfrac{\partial z}{\partial \eta} - \dfrac{\partial y}{\partial \eta}\dfrac{\partial z}{\partial \xi}\right)_{ijk} \end{bmatrix} \begin{bmatrix} \mathbf{S}_\xi^{\mathrm{T}}(\xi_i, \eta_j, \zeta_k) \\ \mathbf{S}_\eta^{\mathrm{T}}(\xi_i, \eta_j, \zeta_k) \\ \mathbf{S}_\zeta^{\mathrm{T}}(\xi_i, \eta_j, \zeta_k) \end{bmatrix} \mathbf{u}$$

$$\left(\frac{\partial u}{\partial y}\right)_{ijk} = \frac{1}{|\mathbf{J}|_{ijk}} \begin{bmatrix} -\left(\dfrac{\partial x}{\partial \eta}\dfrac{\partial z}{\partial \zeta} - \dfrac{\partial x}{\partial \zeta}\dfrac{\partial z}{\partial \eta}\right)_{ijk} \\ \left(\dfrac{\partial x}{\partial \xi}\dfrac{\partial z}{\partial \zeta} - \dfrac{\partial x}{\partial \zeta}\dfrac{\partial z}{\partial \xi}\right)_{ijk} \\ -\left(\dfrac{\partial x}{\partial \xi}\dfrac{\partial z}{\partial \eta} - \dfrac{\partial x}{\partial \eta}\dfrac{\partial z}{\partial \xi}\right)_{ijk} \end{bmatrix} \begin{bmatrix} \mathbf{S}_\xi^{\mathrm{T}}(\xi_i, \eta_j, \zeta_k) \\ \mathbf{S}_\eta^{\mathrm{T}}(\xi_i, \eta_j, \zeta_k) \\ \mathbf{S}_\zeta^{\mathrm{T}}(\xi_i, \eta_j, \zeta_k) \end{bmatrix} \mathbf{u}$$

$$\left(\frac{\partial u}{\partial z}\right)_{ijk} = \frac{1}{|\mathbf{J}|_{ijk}} \begin{bmatrix} -\left(\dfrac{\partial x}{\partial \eta}\dfrac{\partial y}{\partial \zeta} - \dfrac{\partial x}{\partial \zeta}\dfrac{\partial y}{\partial \eta}\right)_{ijk} \\ \left(\dfrac{\partial x}{\partial \xi}\dfrac{\partial y}{\partial \zeta} - \dfrac{\partial x}{\partial \zeta}\dfrac{\partial y}{\partial \xi}\right)_{ijk} \\ -\left(\dfrac{\partial x}{\partial \xi}\dfrac{\partial y}{\partial \eta} - \dfrac{\partial x}{\partial \eta}\dfrac{\partial y}{\partial \xi}\right)_{ijk} \end{bmatrix} \begin{bmatrix} \mathbf{S}_\xi^{\mathrm{T}}(\xi_i, \eta_j, \zeta_k) \\ \mathbf{S}_\eta^{\mathrm{T}}(\xi_i, \eta_j, \zeta_k) \\ \mathbf{S}_\zeta^{\mathrm{T}}(\xi_i, \eta_j, \zeta_k) \end{bmatrix} \mathbf{u} \tag{10.25}$$

For clearance, we write the partial derivatives of the base function **S** and displacement field **u** as

$$\begin{aligned} \mathbf{S}_\xi &= \frac{\partial \mathbf{S}}{\partial \xi}, \quad \mathbf{S}_\eta = \frac{\partial \mathbf{S}}{\partial \eta}, \quad \mathbf{S}_\zeta = \frac{\partial \mathbf{S}}{\partial \zeta} \\ \mathbf{u}_x &= \frac{\partial \mathbf{u}}{\partial x}, \quad \mathbf{u}_y = \frac{\partial \mathbf{u}}{\partial y}, \quad \mathbf{u}_z = \frac{\partial \mathbf{u}}{\partial z} \end{aligned} \tag{10.26}$$

and the derivatives, Eq. (10.25), are shortened as

$$\bar{\mathbf{u}}_x = \mathbf{G}_x\mathbf{u}, \quad \bar{\mathbf{u}}_y = \mathbf{G}_y\mathbf{u}, \quad \bar{\mathbf{u}}_z = \mathbf{G}_z\mathbf{u} \tag{10.27}$$

Similarly, the derivatives with respect to v, w can be given as

$$\begin{aligned} \bar{\mathbf{v}}_x &= \mathbf{G}_x\mathbf{v}, \quad \bar{\mathbf{v}}_y = \mathbf{G}_y\mathbf{v}, \quad \bar{\mathbf{v}}_z = \mathbf{G}_z\mathbf{v} \\ \bar{\mathbf{w}}_x &= \mathbf{G}_x\mathbf{w}, \quad \bar{\mathbf{w}}_y = \mathbf{G}_y\mathbf{w}, \quad \bar{\mathbf{w}}_z = \mathbf{G}_z\mathbf{w} \end{aligned} \tag{10.28}$$

Consider discretized geometric equation $\boldsymbol{\epsilon} = \mathbf{B}\mathbf{u}$ or

$$\begin{Bmatrix} \boldsymbol{\varepsilon}_{xx} \\ \boldsymbol{\varepsilon}_{yy} \\ \boldsymbol{\varepsilon}_{zz} \\ \boldsymbol{\gamma}_{yz} \\ \boldsymbol{\gamma}_{xz} \\ \boldsymbol{\gamma}_{xy} \end{Bmatrix} = \begin{bmatrix} \mathbf{G}_x & \mathbf{0} & \mathbf{0} \\ \mathbf{0} & \mathbf{G}_y & \mathbf{0} \\ \mathbf{0} & \mathbf{0} & \mathbf{G}_z \\ \mathbf{0} & \mathbf{G}_z & \mathbf{G}_y \\ \mathbf{G}_z & \mathbf{0} & \mathbf{G}_x \\ \mathbf{G}_y & \mathbf{G}_x & \mathbf{0} \end{bmatrix} \begin{bmatrix} \mathbf{u} \\ \mathbf{v} \\ \mathbf{w} \end{bmatrix} \tag{10.29}$$

Combine with the following weight coefficient matrix defined in the domain of the hexahedron element:

$$\mathbf{C} = \mathrm{diag}\left\{|J|_{111}\, C_1^{\xi} C_1^{\eta} C_1^{\varsigma}, \ldots, |J|_{N_\xi 11}\, C_{N_\xi}^{\xi} C_1^{\eta} C_1^{\varsigma}, \ldots, |J|_{N_\xi N_\eta 1} \right.$$
$$\left. C_{N_\xi}^{\xi} C_{N_\eta}^{\eta} C_1^{\varsigma}, \ldots, |J|_{N_\xi N_\eta N_\varsigma}\, C_{N_\xi}^{\xi} C_{N_\eta}^{\eta} C_{N_\varsigma}^{\varsigma} \right\} \tag{10.30}$$

then the format of energy functional is obtained as follows:

$$\delta\Pi = \int_{\Omega} (\sigma_{xx}\delta\varepsilon_{xx} + \sigma_{yy}\delta\varepsilon_{yy} + \sigma_{zz}\delta\varepsilon_{zz} + \tau_{yz}\delta\gamma_{yz}$$
$$+ \tau_{zx}\delta\gamma_{zx} + \tau_{xy}\delta\gamma_{xy})\mathrm{d}\Omega \tag{10.31}$$

in which its value is taken as 0 to let the energy functional take the stagnation value and we can get

$$\mathbf{K}_L\mathbf{U} + \mathbf{M}\ddot{\mathbf{U}} = \mathbf{R} \tag{10.32}$$

Thus, the linear stiffness matrix $\mathbf{K}_L$ and the mass matrix $\mathbf{M}$ can be obtained using the variational principle and are listed as follows:

$$\mathbf{K}_L = \mathbf{B}^T\mathbf{D}\mathbf{B}$$
$$\mathbf{M} = \rho \begin{bmatrix} \mathbf{G}^T\mathbf{C}\mathbf{G} & & \\ & \mathbf{G}^T\mathbf{C}\mathbf{G} & \\ & & \mathbf{G}^T\mathbf{C}\mathbf{G} \end{bmatrix} \tag{10.33}$$

When the element is in elastic state, $\mathbf{D}$ matrix is elastic matrix $\mathbf{D}_e$, which has the following form:

$$\mathbf{D}_e = \frac{E(1-\upsilon)}{(1+\upsilon)(1-2\upsilon)}$$
$$\times \begin{bmatrix} \mathbf{C} & \frac{\upsilon}{1-\upsilon}\mathbf{C} & \frac{\upsilon}{1-\upsilon}\mathbf{C} & \mathbf{0} & \mathbf{0} & \mathbf{0} \\ \frac{\upsilon}{1-\upsilon}\mathbf{C} & \mathbf{C} & \frac{\upsilon}{1-\upsilon}\mathbf{C} & \mathbf{0} & \mathbf{0} & \mathbf{0} \\ \frac{\upsilon}{1-\upsilon}\mathbf{C} & \frac{\upsilon}{1-\upsilon}\mathbf{C} & \mathbf{C} & \mathbf{0} & \mathbf{0} & \mathbf{0} \\ \mathbf{0} & \mathbf{0} & \mathbf{0} & \frac{1-2\upsilon}{2(1-\upsilon)}\mathbf{C} & \mathbf{0} & \mathbf{0} \\ \mathbf{0} & \mathbf{0} & \mathbf{0} & \mathbf{0} & \frac{1-2\upsilon}{2(1-\upsilon)}\mathbf{C} & \mathbf{0} \\ \mathbf{0} & \mathbf{0} & \mathbf{0} & \mathbf{0} & \mathbf{0} & \frac{1-2\upsilon}{2(1-\upsilon)}\mathbf{C} \end{bmatrix} \tag{10.34}$$

When the element is in the elasto-plastic state, the **D** matrix is the elasto-plastic matrix $\mathbf{D}_{ep}$, which is assembled by the consistent tangent stiffness matrix and integral weight at each integral point of the element. The elasto-plastic matrix of the element can be expressed as follows:

$$\mathbf{D}_{ep} = \begin{bmatrix} \mathbf{D}_{11} & \mathbf{D}_{12} & \mathbf{D}_{13} & \mathbf{D}_{14} & \mathbf{D}_{15} & \mathbf{D}_{16} \\ \mathbf{D}_{21} & \mathbf{D}_{22} & \mathbf{D}_{23} & \mathbf{D}_{24} & \mathbf{D}_{25} & \mathbf{D}_{23} \\ \mathbf{D}_{31} & \mathbf{D}_{32} & \mathbf{D}_{33} & \mathbf{D}_{34} & \mathbf{D}_{35} & \mathbf{D}_{36} \\ \mathbf{D}_{41} & \mathbf{D}_{42} & \mathbf{D}_{43} & \mathbf{D}_{44} & \mathbf{D}_{45} & \mathbf{D}_{46} \\ \mathbf{D}_{51} & \mathbf{D}_{52} & \mathbf{D}_{53} & \mathbf{D}_{54} & \mathbf{D}_{55} & \mathbf{D}_{56} \\ \mathbf{D}_{61} & \mathbf{D}_{62} & \mathbf{D}_{63} & \mathbf{D}_{64} & \mathbf{D}_{65} & \mathbf{D}_{66} \end{bmatrix} \tag{10.35}$$

where

$$\mathbf{D}_{ij} = \begin{bmatrix} C_{111}d_{ij}^{(111)} & 0 & 0 & \cdots & 0 \\ 0 & C_{211}d_{ij}^{(211)} & 0 & \cdots & 0 \\ \vdots & \vdots & \vdots & \vdots & \vdots \\ 0 & 0 & 0 & \cdots & C_{N_\xi N_\eta N_\zeta}d_{ij}^{(N_\xi N_\eta N_\zeta)} \end{bmatrix}$$

$$\mathbf{D}_{ep}^{(klm)} = \begin{bmatrix} d_{11}^{(klm)} & d_{12}^{(klm)} & d_{13}^{(klm)} & d_{14}^{(klm)} & d_{15}^{(klm)} & d_{16}^{(klm)} \\ d_{21}^{(klm)} & d_{22}^{(klm)} & d_{23}^{(klm)} & d_{24}^{(klm)} & d_{25}^{(klm)} & d_{26}^{(klm)} \\ d_{31}^{(klm)} & d_{32}^{(klm)} & d_{33}^{(klm)} & d_{34}^{(klm)} & d_{35}^{(klm)} & d_{36}^{(klm)} \\ d_{41}^{(klm)} & d_{42}^{(klm)} & d_{43}^{(klm)} & d_{44}^{(klm)} & d_{45}^{(klm)} & d_{46}^{(klm)} \\ d_{51}^{(klm)} & d_{52}^{(klm)} & d_{53}^{(klm)} & d_{54}^{(klm)} & d_{55}^{(klm)} & d_{56}^{(klm)} \\ d_{61}^{(klm)} & d_{62}^{(klm)} & d_{63}^{(klm)} & d_{64}^{(klm)} & d_{65}^{(klm)} & d_{66}^{(klm)} \end{bmatrix}$$

$$klm = 111, 211, \ldots N_\xi 11, \ldots N_\xi N_\eta N_\zeta \tag{10.36}$$

The consistent tangent stiffness matrix at the integral point is derived from the return-mapping algorithm given in the previous section.

10.3 Numerical Examples and Discussions

This section is concerned with the illustration of the advantages of the present method in 2D and 3D elasto-plastic analysis and reveals its accuracy and performance. To that purpose, we thus consider three numerical examples and then investigate the numerical results computed, addressing the effectiveness of the HQEM in modeling inelastic material behaviors. The results obtained by HQEM are compared with analytical solutions and other results of FEM. In this chapter, the commercial software ABAQUS is picked up for the h-version FEM analysis.

Two different sets of material parameters are to be considered (see Table 10.1). The first set corresponds to an elastic perfectly plastic behavior whereas the second one also includes nonlinear isotropic hardening. Example in Section 10.3.1 considers a thick-walled tube under plane strain condition with internal pressure and the perfect elasto-plastic model is used. Example in Section 10.3.2 considers a thin-walled plate with circular hole under plane stress condition. As a 3D elasto-plastic problem, Example in Section 10.3.3 considers a thick-walled plate with circular hole. The nonlinear isotropic hardening model is included in these two examples.

10.3.1 *A thick-walled tube under uniform internal pressure*

First, we study an example for which an analytical solution is available: the thick-walled tube under internal pressure. The inner and outer radii of the thick-walled tube are set to be $a = 1.5\,\mathrm{m}$ and $b = 4\,\mathrm{m}$. The inner surface is loaded with uniform pressure

Table 10.1. Material parameters.

Number	Material parameter	Set 1	Set 2
1	Bulk modulus	166.7 (MPa)	164206.0 (MPa)
2	Shear modulus	3.36 (MPa)	80193.8 (MPa)
3	Initial yield stress	58.0 (KPa)	450.0 (MPa)
4	Saturation stress	0.0 (MPa)	715.0 (MPa)
5	Linear hardening	0.0 (MPa)	129.24 (MPa)
6	Hardening exponent	0.0 (dimensionless)	16.93 (dimensionless)

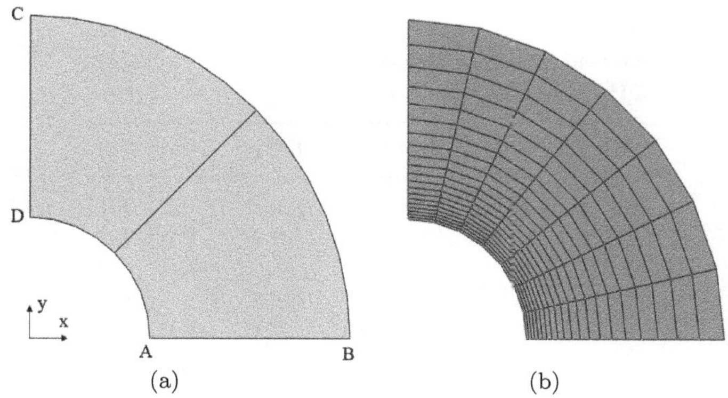

Fig. 10.3. Mesh discretization for a quarter of a thick-walled tube: (a) mesh used in the HQEM analysis; (b) mesh used in the h-version FEM analysis.

$P = 6.2 \times 10^4$ Pa. Due to the symmetry of the structure, a quarter of the structure is used in the computation.

In Fig. 10.3(a), the mesh used for the thick-walled tube in the p-version scheme, which consists of two elements, is shown. The boundary conditions are $u_y = 0$ on edge A–B, $u_x = 0$ on edge C–D, and normal pressure $t_n = P$ on A–D. The comparison results derived from the h-version FEM use a fine mesh of 140 CPE4R 4-noded elements with a total number of 336 DOFs (see Fig. 10.3(b)). In this example, the implementation is tackled by applying a uniform load with 25 steps. The analytical solution is derived under plane strain and incompressibility conditions, which means the Poisson's ratio is assumed to be 0.5. Although the Poisson ratio used in this example is 0.49, the unrealistic assumption of incompressibility in the elastic region does not significantly affect the validity of the analytical solution for the practical case [132].

Table 10.2 shows the comparison of the results of HQEM and ABAQUS using different degrees of freedom. It can be seen that by using HQEM, more accurate results can be obtained, and the convergence speed of HQEM is significantly faster than that of the h-method (the convergence of two digits can be achieved at 104 DOFs).

The displacement u_r of the inner surface calculated by HQEM ($p = 8 \times 8$) and ABAQUS (140 elements) is depicted in Fig. 10.4 in comparison with the analytical solutions, which will be given in the following. When the applied load exceeds 105 MPa, the radial

Table 10.2. Convergence speed comparison between HQEM and ABAQUS in the computation of thick-walled tube problem.

Method	$p(m \times n)$	DOFs	Node A u_x
Analytical result			0.032201
HQEM	4×4	104	0.032357
	6×6	212	0.032433
	8×8	352	0.032158
	10×10	524	0.032232
ABAQUS		60	0.030955
		176	0.031718
		336	0.032211
		682	0.032324

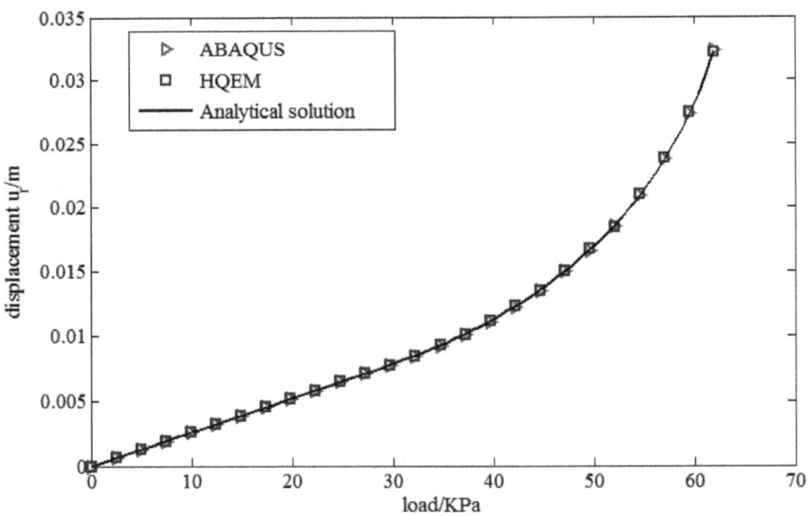

Fig. 10.4. Comparison of the displacement u_r between the HQEM, ABAQUS and the analytical solutions.

displacement u_r behavior of the inner surface is no longer linear, so it can be concluded that the elastic limit load is approximately 27 KPa. Interestingly, the elastic limit load, on the other hand, estimated through the analytical solution, i.e., $P_e = \sigma_s/\sqrt{3}(1 - a^2/b^2)$ is about 28 KPa, which clearly reveals that the elastic limit load valuated by HQEM is in good agreement with the analytical solution.

The plastic limit load estimated for the hollow cylinder is $P_p = 2\sigma_s/\sqrt{3}\ln(b/a) = 65.7\,\text{KPa}$. When the thick-walled tube is in an elasto-plastic state, which means it consists of an elastic part and a plastic one, the radius of the elasto-plastic interface r_p can be obtained by the following formula:

$$P = \frac{\sigma_s}{\sqrt{3}}\left(2\ln\frac{r_p}{a} + 1 - \frac{r_p^2}{b^2}\right) \tag{10.37}$$

By applying Eq. (10.37), the radius of the elasto-plastic interface is $r_p = 3.101\,\text{m}$, while that estimated by HQEM is $r_p = 3.049\,\text{m}$, which is represented by the shaded area in Fig. 10.5 as the yielded area.

The analytical solutions of the stresses for the elastic part of the thick-walled tube, i.e., $r \in [r_p, b]$, can be derived from the literature under plane strain and incompressibility condition, which has the following forms:

$$\begin{aligned}
\sigma_r &= \frac{\sigma_s}{\sqrt{3}}\frac{r_p^2}{b^2}\left(1 - \frac{b^2}{r^2}\right)\\
\sigma_\theta &= \frac{\sigma_s}{\sqrt{3}}\frac{r_p^2}{b^2}\left(1 + \frac{b^2}{r^2}\right)\\
\sigma_z &= \frac{1}{\sqrt{3}}\frac{\sigma_s r_p^2}{b^2}
\end{aligned} \tag{10.38}$$

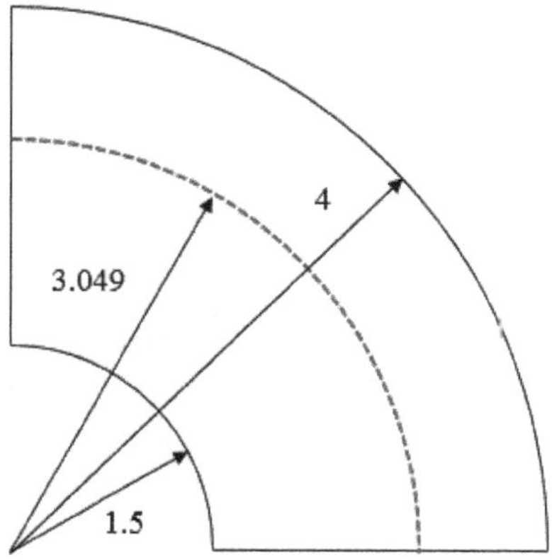

Fig. 10.5. Schematic diagram of the yielded area of the quarter of a thick-walled tube.

Those for the plastic part of the thick-walled tube, i.e., $r \in [a, r_p]$ are as follows:

$$\begin{aligned}\sigma_r &= -\frac{\sigma_s}{\sqrt{3}}\left[1 + 2\ln\frac{r_p}{r} - \frac{r_p^2}{b^2}\right]\\ \sigma_\theta &= \frac{\sigma_s}{\sqrt{3}}\left[1 - 2\ln\frac{r_p}{r} + \frac{r_p^2}{b^2}\right]\\ \sigma_z &= \frac{\sigma_s}{\sqrt{3}}\left[\frac{r_p^2}{b^2} - 2\ln\frac{r_p}{r}\right]\end{aligned} \tag{10.39}$$

The radial displacement along the thickness can be obtained by the same formula in the elastic and plastic regions:

$$u = \frac{\sqrt{3}\sigma_s r_p^2}{2Er} \tag{10.40}$$

Figure 10.6 shows a comparison of the radial displacement results along the thickness among the HQEM, FEM (ABAQUS) and the analytical methods. Results indicate an excellent agreement among considered methods as the obtained curves are almost identical. The radial displacement is found to be decreased along the thickness

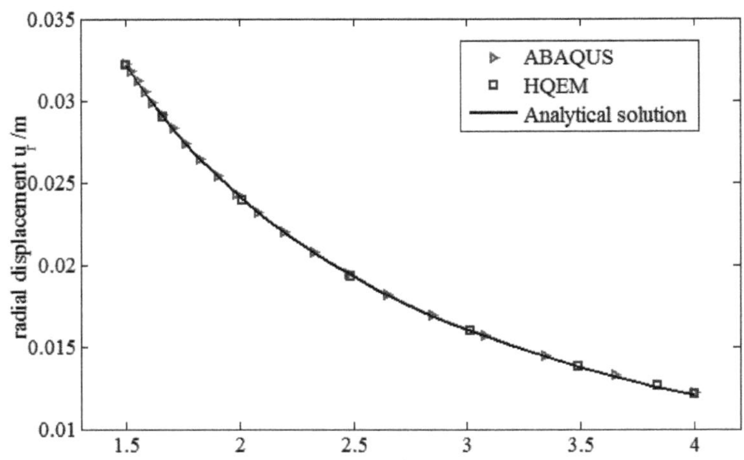

Fig. 10.6. Comparison of the radial displacement along the thickness of the thick-walled tube among HQEM, FEM (ABAQUS) and the analytical solutions.

direction of the hollow cylinder. It is clear that, due to the applied pressure, the deformation of the inner surface of the thick-walled tube is larger than that of the outer one.

Figure 10.7 shows the radial stress (a), hoop stress (b) and axial stress (c) in which different solutions derived from HQEM, FEM (ABAQUS) and the analytical solutions are compared with each other. As expected, the stresses calculated by HQEM are in good agreement with the analytical solutions, and it offers a more accurate result than the FEM solutions. Additionally, it is interesting to see the *von Mises* stress distributions on the thick-walled tube. To this end, the *von Mises* stress calculated by HQEM and FEM (ABAQUS) are hence depicted in Fig. 10.8, showing a good agreement between each other.

10.3.2 *Perforated square plate under plane stress condition*

The second numerical example to be considered is a perforated square plate under plane stress condition and a nonlinear isotropic hardening model is used. The material parameters corresponding to set 2 are listed in Table 10.1. Due to the symmetry of the structure, a quarter of the structure is used in the computation (see Fig. 10.9). The edge 2–3 is loaded with uniform pressure $P = 4.5 \times 10^8$ Pa.

For HQEM, the mesh consists of two uniform elements used in the computation and a series of computations varying the polynomial degree uniformly from $p = 4$ up to $p = 12$ was performed (see Fig. 10.10(a)). The comparison results derived from the h-version FEM used a series of meshes of CPS4R 4-noded elements. Figure 10.10(b) shows one of the meshes used in the ABAQUS analysis. In this example, the implementation is tackled by applying a uniform load with 50 steps.

Table 10.3 shows the comparison of the results of HQEM and ABAQUS under different degrees of freedom. Taking the result calculated by ABAQUS with 72782 DOFs as reference, the results are plotted in the form of relative errors in double logarithmic scale in Fig. 10.11. It can be seen that by using HQEM, more accurate results can be obtained, and the convergence speed of HQEM is significantly faster than that of the h-method (the convergence of two digits can be achieved at 622 DOFs).

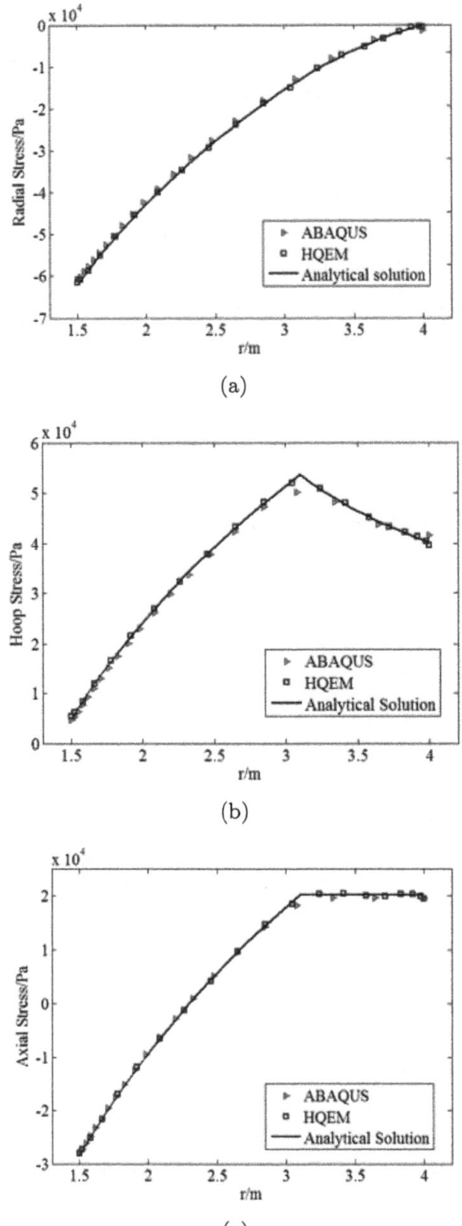

Fig. 10.7. Comparison of stresses for the quarter of a thick-walled tube: (a) radial stress; (b) hoop stress; (c) axial stress among HQEM, FEM (ABAQUS) and the analytical solutions.

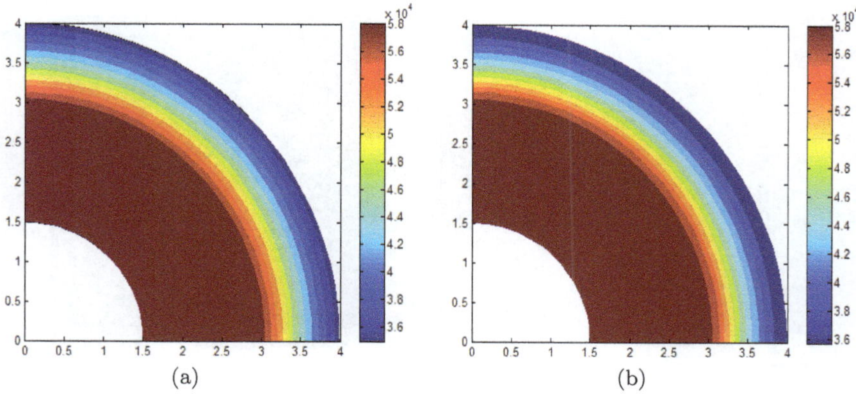

Fig. 10.8. Comparison results of the *von Mises* stresses of the quarter of a thick-walled tube using HQEM (a) and FEM (ABAQUS) (b). The red color represents the plastic region.

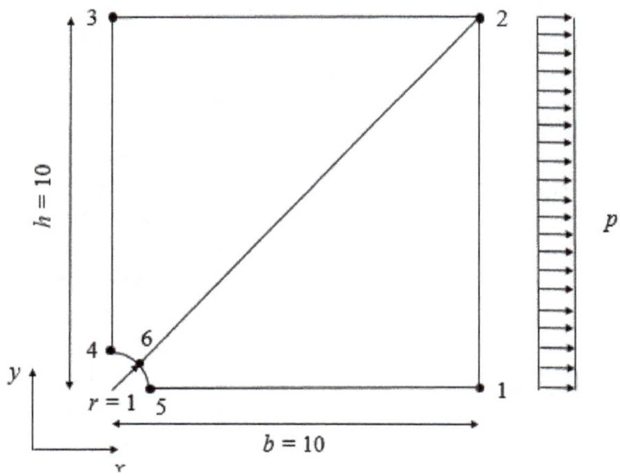

Fig. 10.9. A quarter of the perforated square plate.

The displacement u_y in node 4 calculated by HQEM ($p = 12\times12$) and ABAQUS (72782 elements) is depicted in Fig. 10.12. It can be seen that although the number of DOFs is not large, HQEM can still achieve a good agreement with the reference solution, which shows the high efficiency of HQEM in calculating the elasto-plastic problems.

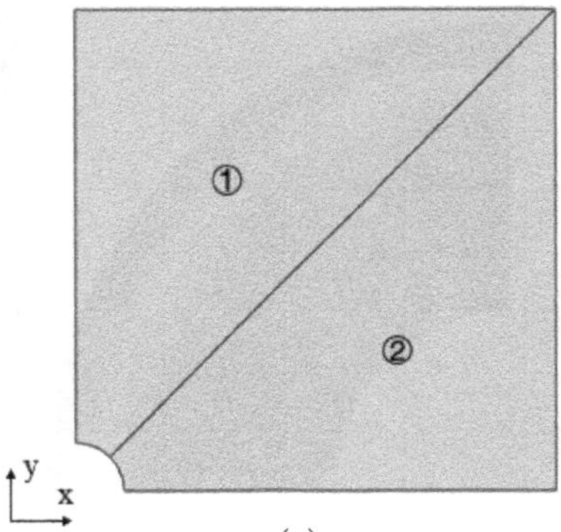

(a)

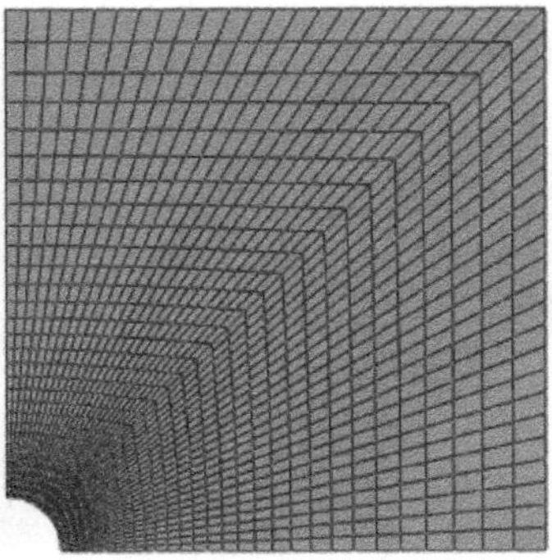

(b)

Fig. 10.10. Mesh discretization for a quarter of a perforated square plate: (a) mesh used in the HQEM analysis; (b) mesh used in the h-version FEM analysis.

Table 10.3. Convergence speed comparison between HQEM and ABAQUS in the computation of the perforated square plate problem.

Method	$p(m \times n)$	DOFs	Node 4 u_y
HQEM	4×4	104	−0.009155
	6×6	212	−0.008546
	8×8	352	−0.007432
	10×10	524	−0.006769
	11×11	622	−0.006692
	12×12	726	−0.006613
ABAQUS		108	−0.007801
		374	−0.007253
		800	−0.006982
		1386	−0.006835
		3038	−0.006703
		5330	−0.006660
		72782	−0.006621

Figure 10.13 shows the stress in x-direction in which different solutions derived from HQEM, the reference solutions ((ABAQUS 72782 DOFs), are compared among each other. As expected, the stresses calculated by HQEM are in good agreement with the reference solutions. Figure 10.14 shows the *von Mises* stress distributions

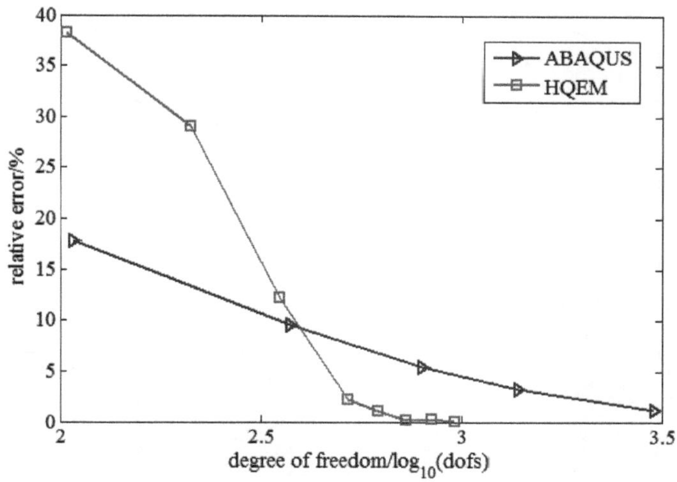

Fig. 10.11. Convergence speed comparison between HQEM and ABAQUS in the computation of the perforated square plate problem.

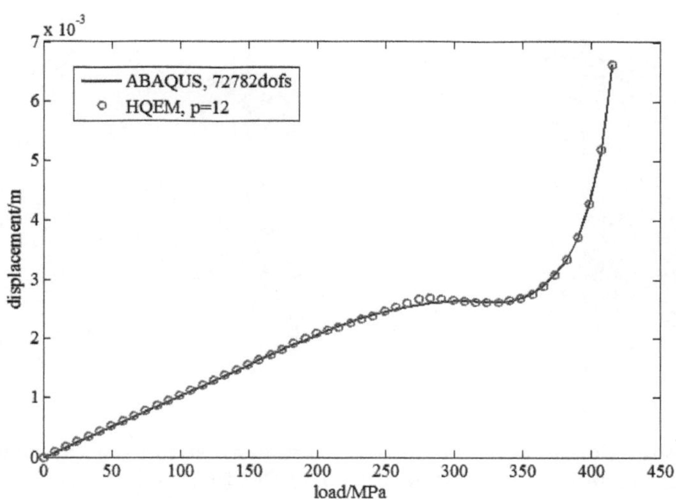

Fig. 10.12. Comparison of the displacement u_y in node 4 among HQEM, the h-version FEM (ABAQUS).

on the perforated square plate which are calculated by HQEM and the h-version FEM (ABAQUS 72782 DOFs), showing a good agreement between each other.

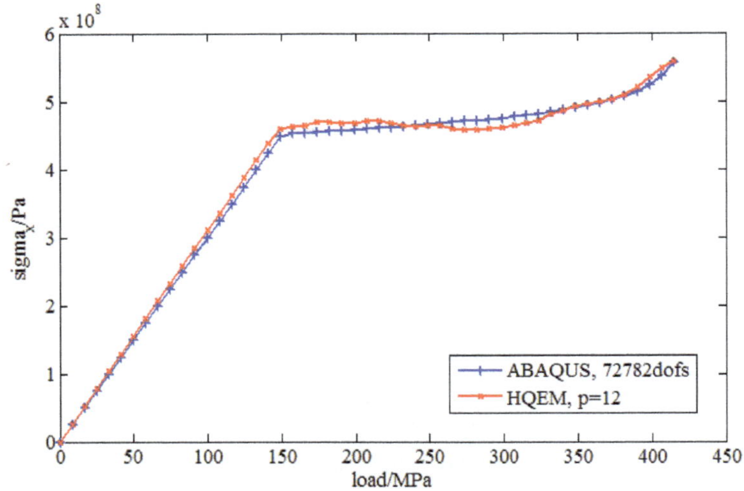

Fig. 10.13. Comparison of the stress in x-direction of node 4 among HQEM and the h-version FEM (ABAQUS).

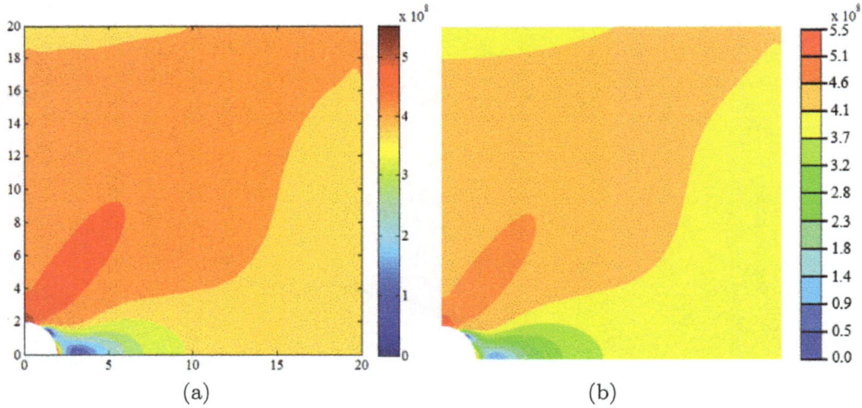

Fig. 10.14. Comparison results of the *von Mises* stresses of the quarter of a perforated square plate using HQEM (a) and FEM (ABAQUS) (b).

10.3.3 *Thick perforated square plate*

The last example deals with a thick perforated square plate under the nonlinear isotropic hardening model (as shown in Table 10.1, set 2), which is a three-dimensional elasto-plastic problem. Similarly, an eighth of the system is used in the computation due to

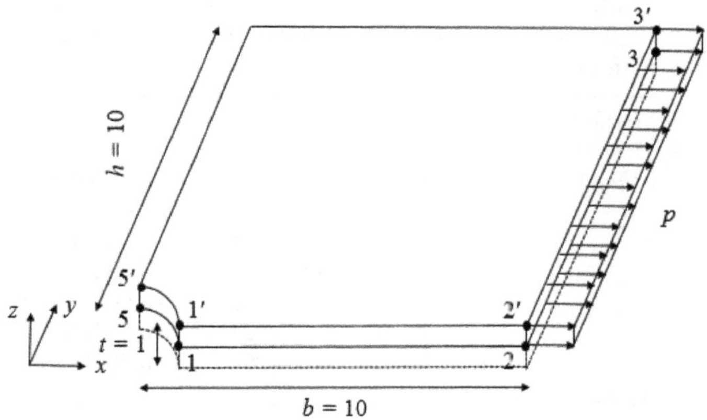

Fig. 10.15. An eighth of the thick perforated square plate.

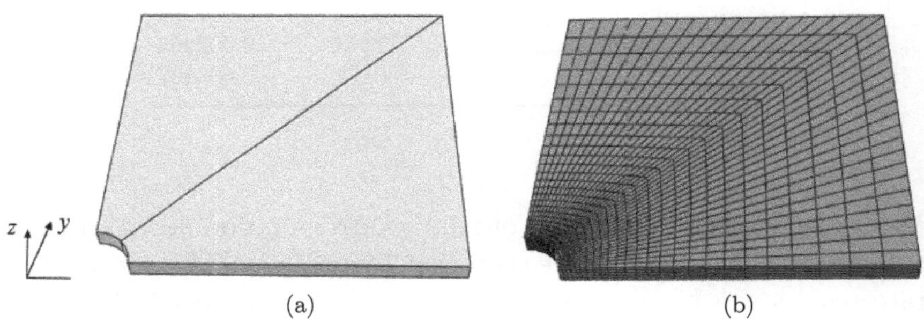

Fig. 10.16. Mesh discretization for an eighth of the thick perforated square plate: (a) mesh used in the HQEM analysis; (b) mesh used in the h-version FEM analysis.

the symmetry of the structure (see Fig. 10.15). The face 2-2′–3-3′ is loaded with uniform pressure $P = 4.5 \times 10^8$ Pa.

Similarly, a mesh consisting of two uniform elements is used for the computation of HQEM and the polynomial degree ranges from $p = 4$ up to $p = 10$ are performed (Fig. 10.16(a)). For the h-version FEM, a series of meshes of 8-noded elements is taken. Figure 10.16(b) shows one of the meshes used in the ABAQUS analysis. In this example, the implementation is tackled by applying a uniform load with 50 steps.

Table 10.4 provides the comparison of results calculated by HQEM and ABAQUS using different degrees of freedom. Take the result calculated by ABAQUS with 130174 DOFs as reference; the

Table 10.4. Convergence speed comparison between HQEM and ABAQUS in the computation of the thick perforated square plate problem.

Method	$p(l \times m \times n)$	DOFs	Node 5 u_y
HQEM	$4 \times 4 \times 4$	1188	−0.003048
	$5 \times 5 \times 5$	1911	−0.002729
	$6 \times 6 \times 6$	2474	−0.002474
	$7 \times 7 \times 7$	4131	−0.002289
	$8 \times 8 \times 8$	5700	−0.002195
	$9 \times 9 \times 9$	7623	−0.002173
	$10 \times 10 \times 8$	7008	−0.002180
ABAQUS		1584	−0.002530
		2970	−0.002471
		4788	−0.002369
		6048	−0.002323
		8100	−0.002297
		11718	−0.002263
		130174	−0.002167

results are plotted in the form of relative errors in double logarithmic scale in Fig. 10.17. It can be seen that by using HQEM, more accurate results can be obtained, and the convergence speed of HQEM is significantly faster than that of the h-method (the convergence of two digits can be achieved at 5700 DOFs).

The displacement u_y in node 5 calculated by HQEM ($p = 8 \times 8 \times 3$) and ABAQUS (130174 elements) is depicted in Fig. 10.18. It can be seen that although the number of DOFs is not large, HQEM can still achieve a good agreement with the reference solution, which shows the high efficiency of HQEM in calculating the elasto-plastic problems.

10.4 Conclusions

In this chapter, a hierarchical quadrature element method (HQEM) based on the J_2 flow theory with isotropic hardening is used to analyze two- as well as three-dimensional elasto-plastic problems. The backward Euler time integration method is applied to the classical

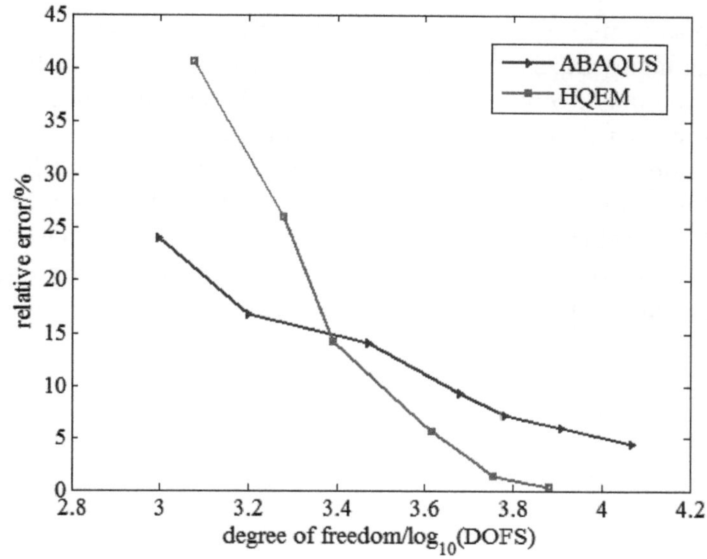

Fig. 10.17. Convergence speed comparison between HQEM and ABAQUS of the thick perforated square plate problem.

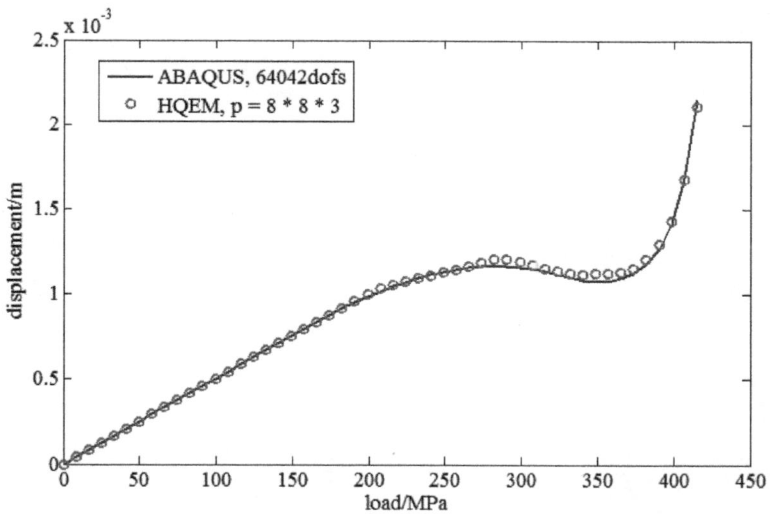

Fig. 10.18. Comparison of the displacement u_y in node 5 among HQEM, the h-version FEM (ABAQUS) of the thick perforated square plate problem.

elasto-plastic model which is in the rate form and then the return-mapping algorithm is introduced to treat the J_2 plasticity by determining the plastic consistency parameter. The local return-mapping algorithm for the pointwise integration of the constitutive equations is embedded in a global Newton–Raphson iteration. For the investigated elasto-plastic problems, it is numerically demonstrated that HQEM can achieve fast convergence and only needs a few degrees of freedom to obtain highly accurate results, which is conducive to reducing the amount of calculation in the elasto-plastic analysis.

Chapter 11

Curved p-version C^1 Finite Elements for the Finite Deformation Analysis of Isotropic and Composite Laminated Thin Shells

In this chapter, we generalize the p-version curved C^1 plate elements in chapter 6 to the analysis of general isotropic and laminated thin shells with arbitrary curvature and shape. The kinematic formulations are based on the geometrically exact nonlinear thin shell model where the Kirchhoff–Love hypothesis is strictly implemented during the deformation. Since high-accurate curved coarse meshes are required in the implementation of p-version finite element method, the quasi-regional mapping technique is employed and modified to obtain the high-order meshes with fitted boundaries. Through this method, shells with irregular shapes can be easily modeled. To avoid dealing with the difficult problem of constructing fully C^1 conforming displacement fields on shell surface, the C^1 conformity of present elements is implemented on the selected Gauss–Lobatto nodes, which had been found to be able to produce very fast convergence. New interpolation variables on surfaces are derived and used for element assembly. In this manner, direct construction of displacement fields on shell mid-surfaces defined by the quasi-regional mapping method is allowed. Several benchmark problems of large deformation analysis of isotropic and laminated thin shells are presented to illustrate the accuracy as well as robustness of the present method.

11.1 Thin Shell Model

In this section, we give a brief review of the thin shell model by following the formulations introduced in [172]. This model is essentially a geometrically exact shell model [416] where the shell body is defined by a reference surface and an attached director field. Since there is no restriction on the rotation angle of director field, the formulations are thus suitable for the fully geometrically nonlinear problems.

11.1.1 *Kinematics*

As shown in Fig. 11.1, the three-dimensional (3D) shell body is represented as a two-dimensional (2D) surface enhanced by the normal directors. Therefore, the reference configuration of the thin shell is defined by

$$\boldsymbol{\Phi}(\xi_1, \xi_2, \xi_3) = \boldsymbol{X}(\xi_1, \xi_2) + \xi_3 \boldsymbol{A}_3(\xi_1, \xi_2), \quad \xi_3 \in [-h/2, h/2] \tag{11.1}$$

where ξ_i are the convected coordinates, h is the thickness, and $\boldsymbol{X}(\xi_\alpha)$ denotes the parameterization of the reference mid-surface, which comes along with the covariant bases $\boldsymbol{A}_\alpha$ and the unit normal director $\boldsymbol{A}_3$ defined by

$$\boldsymbol{A}_\alpha := \boldsymbol{X}_{,\alpha} \quad \text{and} \quad \boldsymbol{A}_3 := \frac{\boldsymbol{A}_1 \times \boldsymbol{A}_2}{\|\boldsymbol{A}_1 \times \boldsymbol{A}_2\|} \tag{11.2}$$

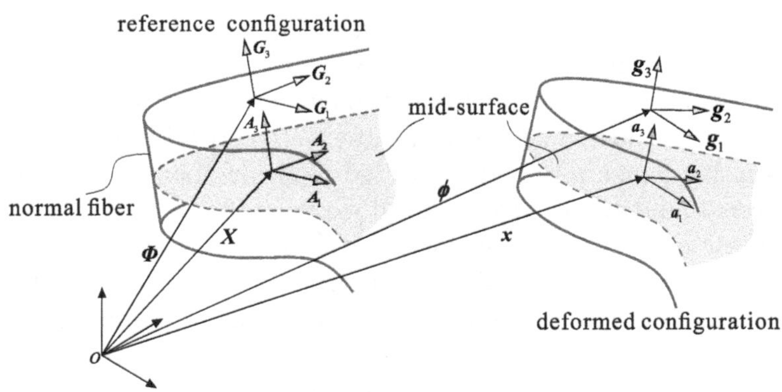

Fig. 11.1. Shell geometry in the reference and deformed configurations.

where the subscript comma implies partial derivative. The metrics and the contravariant bases of mid-surface can be obtained as

$$
\begin{aligned}
A_{\alpha\beta} &= \boldsymbol{A}_\alpha \cdot \boldsymbol{A}_\beta \\
\boldsymbol{A}^\alpha &= A^{\alpha\beta}\boldsymbol{A}_\beta, \quad [A^{\alpha\beta}] = [A_{\alpha\beta}]^{-1}
\end{aligned}
\tag{11.3}
$$

where repeated indices imply summarization, and the square brackets indicate the component matrix. Similarly, the covariant bases and metrics of the 3D shell body are given by

$$
\begin{aligned}
\boldsymbol{G}_\alpha &= \boldsymbol{A}_\alpha + \xi_3\boldsymbol{A}_{3,\alpha} = \boldsymbol{A}_\alpha - \xi_3\boldsymbol{B}\cdot\boldsymbol{A}_\alpha \\
\boldsymbol{G}_3 &= \boldsymbol{A}_3 \\
G_{ij} &= \boldsymbol{G}_i \cdot \boldsymbol{G}_j
\end{aligned}
\tag{11.4}
$$

where $\boldsymbol{B}$ is the mid-surface curvature tensor defined by

$$
\boldsymbol{B} = B_{\alpha\beta}\boldsymbol{A}^\alpha \otimes \boldsymbol{A}^\beta, \quad B_{\alpha\beta} = \boldsymbol{A}_{\alpha,\beta}\cdot\boldsymbol{A}_3 \tag{11.5}
$$

According to the Kirchhoff–Love assumptions, it is assumed that the shell normal fibers remain inextensible and normal to its mid-surface throughout the deformation. Consequently, the motion of the shell body can be fully described by the motion of its mid-surface. Assume the mid-surface displacement is $\boldsymbol{u}$, then the deformed configuration of shell body can be expressed as

$$
\begin{aligned}
\boldsymbol{\phi}(\xi_1,\xi_2,\xi_3) &= \boldsymbol{x}(\xi_1,\xi_2) + \xi_3\boldsymbol{a}_3(\xi_1,\xi_2) \\
\boldsymbol{x} &= \boldsymbol{X} + \boldsymbol{u}
\end{aligned}
\tag{11.6}
$$

where $\boldsymbol{\phi}$ is the position vector of material point, and $\boldsymbol{x}$ and a_3 are the deformed shell mid-surface and its unit normal vector defined by

$$
\begin{aligned}
\boldsymbol{a}_3 &= \frac{\boldsymbol{a}_1 \times \boldsymbol{a}_2}{\|\boldsymbol{a}_1 \times \boldsymbol{a}_2\|} \\
\boldsymbol{a}_\alpha &= \boldsymbol{x}_{,\alpha} = \boldsymbol{A}_a + \boldsymbol{u}_{,\alpha}
\end{aligned}
\tag{11.7}
$$

where $\boldsymbol{a}_\alpha$ are the covariant bases of deformed mid-surface. Likewise, the metrics and contravariant bases of deformed mid-surface can be given by

$$
\begin{aligned}
a_{\alpha\beta} &= \boldsymbol{a}_\alpha \cdot \boldsymbol{a}_\beta \\
\boldsymbol{a}^\alpha &= a^{\alpha\beta}\boldsymbol{a}_\beta, \quad [a^{\alpha\beta}] = [a_{\alpha\beta}]^{-1}
\end{aligned}
\tag{11.8}
$$

Then the covariant bases and metrics of deformed shell body can be expressed as

$$
\begin{aligned}
\boldsymbol{g}_\alpha &= \boldsymbol{a}_\alpha + \xi_3 \boldsymbol{a}_{3,\alpha} = \boldsymbol{a}_\alpha - \xi_3 \boldsymbol{b} \cdot \boldsymbol{a}_\alpha \\
\boldsymbol{g}_3 &= \boldsymbol{a}_3 \\
g_{ij} &= \boldsymbol{g}_i \cdot \boldsymbol{g}_j
\end{aligned}
\tag{11.9}
$$

where $\boldsymbol{b}$ is the curvature tensor of deformed mid-surface defined by

$$
\boldsymbol{b} = b_{\alpha\beta}\boldsymbol{a}^\alpha \otimes \boldsymbol{a}^\beta, \quad b_{\alpha\beta} = \boldsymbol{a}_{\alpha,\beta} \cdot \boldsymbol{a}_3 \tag{11.10}
$$

Since the present elements are formulated in the total Lagrange (TL) form, the required Green–Lagrange strain of the shell body is given by

$$
\begin{aligned}
\boldsymbol{E} &= \frac{1}{2}(\boldsymbol{F}^T\boldsymbol{F} - \boldsymbol{I}) = E_{ij}\boldsymbol{G}^i \otimes \boldsymbol{G}^j \\
\boldsymbol{F} &= \frac{\partial \boldsymbol{\phi}}{\partial \boldsymbol{\Phi}} = \boldsymbol{g}_i \otimes \boldsymbol{G}^i \\
E_{ij} &= \frac{1}{2}(g_{ij} - G_{ij})
\end{aligned}
\tag{11.11}
$$

where $\boldsymbol{F}$ denotes the deformation gradient. With the Kirchhoff–Love assumptions, it can be verified that the transverse components $E_{\alpha 3}$ are reduced to zero. Since the transverse normal stress is neglected in thin shells, only the in-plane components of $\boldsymbol{E}$ are required to be considered in the strain energy. Substituting Eqs. (11.4) and (11.9) into (11.11), the in-plane components of $\boldsymbol{E}$ can be expressed as

$$
\begin{aligned}
E_{\alpha\beta} &= (g_{\alpha\beta} - G_{\alpha\beta})/2 \\
g_{\alpha\beta} &= a_{\alpha\beta} - 2\xi_3 b_{\alpha\beta} + \xi_3^2 \boldsymbol{a}_{3,\alpha} \cdot \boldsymbol{a}_{3,\beta} \\
G_{\alpha\beta} &= A_{\alpha\beta} - 2\xi_3 B_{\alpha\beta} + \xi_3^2 \boldsymbol{A}_{3,\alpha} \cdot \boldsymbol{A}_{3,\beta}
\end{aligned}
\tag{11.12}
$$

As the thickness is small in shells, the quadratic terms of ξ_3 are neglected. Finally, the in-plane Green strain components can be further reduced to

$$\begin{aligned} E_{\alpha\beta} &= \varepsilon_{\alpha\beta} + \xi_3 \kappa_{\alpha\beta} \\ \varepsilon_{\alpha\beta} &= (a_{\alpha\beta} - A_{\alpha\beta})/2 \\ \kappa_{\alpha\beta} &= B_{\alpha\beta} - b_{\alpha\beta} \end{aligned} \tag{11.13}$$

where $\varepsilon_{\alpha\beta}$ are the membrane strains, and $\kappa_{\alpha\beta}$ are the bending and twisting strains.

11.1.2 *Weak form*

The principle of virtual work is generally given by

$$\delta W^{\text{int}} - \delta W^{\text{ext}} = 0 \tag{11.14}$$

where δW^{int} and δW^{ext} are respectively the internal and external virtual work. Using the Green–Lagrange strain and the second Piola–Kirchhoff (PK2) stress as energy conjugate pair, the internal virtual work can be expressed as

$$\delta W^{\text{int}} = \int_\Omega \boldsymbol{S} : \delta \boldsymbol{E} \, \mathrm{d}\Omega \tag{11.15}$$

where Ω denotes the reference configuration of shell body, $\boldsymbol{S}$ is the PK2 stress, and $\delta\boldsymbol{E}$ is the variation of Green strain. Define the following shifter tensor:

$$\boldsymbol{Z} = \boldsymbol{G}^i \otimes \boldsymbol{A}_i \tag{11.16}$$

The 3D integration in Eq. (11.15) can be expressed by

$$\begin{aligned} \delta W^{\text{int}} &= \int_\Omega \boldsymbol{S} : \delta \boldsymbol{E} \, \mathrm{d}\Omega = \int_{\Omega_0} \int_{-h/2}^{h/2} \delta E_{ij} S^{ij} \det(\boldsymbol{Z}^{-1}) \, \mathrm{d}\Omega_0 \\ &\approx \int_{\Omega_0} \int_{-h/2}^{h/2} \delta E_{\alpha\beta} S^{\alpha\beta} \, \mathrm{d}\Omega_0 \end{aligned} \tag{11.17}$$

where $\det(\boldsymbol{Z}^{-1})$ is the determinant of $\boldsymbol{Z}$, and Ω_0 denotes the mid-surface of reference configuration. The last equation is obtained by neglecting the transverse components and using the truth that

$\det(\boldsymbol{Z}^{-1}) \approx 1$ when the thickness is small. Substituting Eq. (11.13) into Eq. (11.17), the internal virtual work can be expressed as

$$\begin{aligned}\delta W^{\text{int}} &= \int_{\Omega_0} \boldsymbol{m} : \delta\boldsymbol{\kappa} + \boldsymbol{n} : \delta\boldsymbol{\varepsilon}\, \mathrm{d}\Omega_0 \\ &= \int_{\Omega_0} m^{\alpha\beta}\delta\kappa_{\alpha\beta} + n^{\alpha\beta}\delta\varepsilon_{\alpha\beta}\, \mathrm{d}\Omega_0 \end{aligned} \tag{11.18}$$

where $\boldsymbol{\varepsilon}$ and $\boldsymbol{\kappa}$ are the membrane and bending strain tensors, and $\boldsymbol{m}$ and $\boldsymbol{n}$ are the stress resultant tensors of bending moments and membrane forces. They are expressed as

$$\begin{aligned} &\boldsymbol{\varepsilon} = \varepsilon_{\alpha\beta}\boldsymbol{A}^{\alpha} \otimes \boldsymbol{A}^{\beta}, \qquad \boldsymbol{\kappa} = \kappa_{\alpha\beta}\boldsymbol{A}^{\alpha} \otimes \boldsymbol{A}^{\beta} \\ &\boldsymbol{n} = n^{\alpha\beta}\boldsymbol{A}_{\alpha} \otimes \boldsymbol{A}_{\beta}, \qquad \boldsymbol{m} = m^{\alpha\beta}\boldsymbol{A}_{\alpha} \otimes \boldsymbol{A}_{\beta} \\ &n^{\alpha\beta} = \int_{h/2}^{h/2} S^{\alpha\beta}\,\mathrm{d}\xi_3, \quad m^{\alpha\beta} = \int_{-h/2}^{h/2} \xi_3 S^{\alpha\beta}\,\mathrm{d}\xi_3 \end{aligned} \tag{11.19}$$

The external virtual work in Eq. (11.14) is given by

$$\delta W^{\text{ext}} = \int_{\Gamma_t} \boldsymbol{t} \cdot \delta\boldsymbol{u}\, \mathrm{d}\Gamma + \int_{\Omega_0} \boldsymbol{p} \cdot \delta\boldsymbol{u}\, \mathrm{d}\Omega_0 \tag{11.20}$$

where Γ_t is the traction edge, $\boldsymbol{t}$ is the line traction, and $\boldsymbol{p}$ is the surface pressure or surface traction.

11.1.3 *Constitutive equation*

In the material primary Cartesian coordinates, denoted by $\hat{\mathbf{e}}_i$, the simplified plane stress constitutive equation for orthotropic elastic material is given by

$$\hat{\mathbf{S}} = \begin{bmatrix} \hat{S}_{11} \\ \hat{S}_{22} \\ \hat{S}_{12} \end{bmatrix} = \begin{bmatrix} D_{11} & D_{12} & \\ D_{21} & D_{22} & \\ & & G_{12} \end{bmatrix} \begin{bmatrix} \hat{E}_{11} \\ \hat{E}_{22} \\ 2\hat{E}_{12} \end{bmatrix} = \hat{\mathbf{C}}\hat{\mathbf{E}} \tag{11.21}$$

where $\hat{\mathbf{C}}$ is the elastic matrix, G_{12} is the in-plane shear modulus, and $D_{\alpha\beta}$ are defined as

$$\begin{aligned} &D_{11} = \frac{E_1}{1 - \upsilon_{12}\upsilon_{21}}, \qquad D_{22} = \frac{E_2}{1 - \upsilon_{12}\upsilon_{21}}, \\ &D_{12} = \frac{E_1\upsilon_{21}}{1 - \upsilon_{12}\upsilon_{21}} = \frac{E_2\upsilon_{12}}{1 - \upsilon_{12}\upsilon_{21}} = D_{21} \end{aligned} \tag{11.22}$$

where E_1 and E_2 are the Young's moduli in the fiber and transverse directions and υ_{12} and υ_{21} are Poisson's ratios. These material parameters may vary from layer to layer when multilayer laminates are considered.

In order to define the material coordinates $\hat{\mathbf{e}}_i$, an auxiliary field of reference frame $\boldsymbol{e}_i$ is generally specified. A common practice is to use the following projection method:

$$\boldsymbol{e}_3 = \boldsymbol{A}_3, \quad \boldsymbol{e}_1 = \frac{\boldsymbol{v} - (\boldsymbol{v}\cdot\boldsymbol{e}_3)\boldsymbol{e}_3}{\|\boldsymbol{v} - (\boldsymbol{v}\cdot\boldsymbol{e}_3)\boldsymbol{e}_3\|}, \quad \boldsymbol{e}_2 = \boldsymbol{e}_3 \times \boldsymbol{e}_1 \tag{11.23}$$

where $\boldsymbol{v}$ is a user-specified reference vector. Then the material coordinates can be defined by a proper rotation of $\boldsymbol{e}_i$ about axis $\boldsymbol{e}_3$:

$$\hat{\mathbf{e}}_i = \boldsymbol{R}(\theta)\boldsymbol{e}_i \tag{11.24}$$

where $\boldsymbol{R}$ is the rotation tensor, and θ is the angle. Note that in thickness integration of multilayered structures, see Eq. (11.29), the reference frame is shared by all layers, while the rotation angle is specified independently for each layer, depending on the stacking sequence.

Since the Green strain components are expressed in curvilinear coordinates, the following transformation is required to transform the curvilinear coordinate strain components into the components expressed in Cartesian material coordinates:

$$\begin{aligned}\boldsymbol{E} &= E_{\alpha\beta}\boldsymbol{G}^\alpha\boldsymbol{G}^\beta \approx E_{\alpha\beta}\boldsymbol{A}^\alpha\boldsymbol{A}^\beta = E_{\alpha\beta}(\boldsymbol{A}^\alpha\cdot\hat{\mathbf{e}}_\gamma)(\boldsymbol{A}^\beta\cdot\hat{\mathbf{e}}_\lambda)\hat{\mathbf{e}}_\gamma\hat{\mathbf{e}}_\lambda \\ &= \hat{E}_{\gamma\lambda}\hat{\mathbf{e}}_\gamma\hat{\mathbf{e}}_\lambda\end{aligned} \tag{11.25}$$

In matrix form, it is given by

$$\hat{\mathbf{E}} = \begin{bmatrix}\bar{E}_{11}\\ \bar{E}_{22}\\ 2\bar{E}_{12}\end{bmatrix} = \begin{bmatrix} T_1^1T_1^1 & T_1^2T_1^2 & T_1^1T_1^2 \\ T_2^1T_2^1 & T_2^2T_2^2 & T_2^1T_2^2 \\ 2T_1^1T_2^1 & 2T_1^2T_2^2 & T_1^1T_2^2 + T_1^2T_2^1\end{bmatrix}\begin{bmatrix}E_{11}\\ E_{22}\\ 2E_{12}\end{bmatrix} = \mathbf{TE} \tag{11.26}$$

where $T_\alpha^\beta := \bar{\boldsymbol{G}}^\beta \cdot \hat{\mathbf{e}}_\alpha$. Accordingly, the transformation of stress components is expressed by

$$\boldsymbol{S} \approx S^{\alpha\beta}\boldsymbol{A}_\alpha\boldsymbol{A}_\beta = \hat{S}_{ij}(\boldsymbol{A}^\alpha\cdot\hat{\mathbf{e}}_i)(\boldsymbol{A}^\beta\cdot\hat{\mathbf{e}}_j)\bar{\boldsymbol{G}}_\alpha\bar{\boldsymbol{G}}_\beta \tag{11.27}$$

In matrix form, it is

$$\mathbf{S} = \begin{bmatrix} S^{11} \\ S^{22} \\ S^{12} \end{bmatrix} = \begin{bmatrix} T_1^1 T_1^1 & T_2^1 T_2^1 & 2T_1^1 T_2^1 \\ T_1^2 T_1^2 & T_2^2 T_2^2 & 2T_1^2 T_2^2 \\ T_1^1 T_1^2 & T_2^1 T_2^2 & T_1^1 T_2^2 + T_2^1 T_1^2 \end{bmatrix} \begin{bmatrix} \hat{S}_{11} \\ \hat{S}_{22} \\ \hat{S}_{12} \end{bmatrix} = \mathbf{T}^{\mathrm{T}} \hat{\mathbf{S}} \tag{11.28}$$

Substituting Eq. (11.26) and (11.28) into Eq. (11.19), we can obtain the following stress-resultant-expressed constitutive equations:

$$\begin{aligned} \mathbf{n} &= \int_{h/2}^{h/2} \mathbf{T}^{\mathrm{T}} \hat{\mathbf{C}} \mathbf{T} \mathbf{E} \, \mathrm{d}\xi_3 = \int_{h/2}^{h/2} \mathbf{C} \mathbf{E} \, \mathrm{d}\xi_3 \\ \mathbf{m} &= \int_{h/2}^{h/2} \xi_3 T^{\mathrm{T}} \hat{\mathbf{C}} \mathbf{T} \mathbf{E} \, \mathrm{d}\xi_3 = \int_{h/2}^{h/2} \xi_3 \mathbf{C} \mathbf{E} \, \mathrm{d}\xi_3 \end{aligned} \tag{11.29}$$

where $\mathbf{C}$ is the elastic matrix in terms of curvilinear coordinates, and the vectors $\mathbf{m}$, $\mathbf{n}$ and $\mathbf{E}$ are expressed by

$$\begin{aligned} \mathbf{m} &= \begin{bmatrix} m^{11} \\ m^{22} \\ m^{12} \end{bmatrix}, \quad \mathbf{n} = \begin{bmatrix} n^{11} \\ n^{22} \\ n^{12} \end{bmatrix} \\ \mathbf{E} &= \begin{bmatrix} \varepsilon_{11} \\ \varepsilon_{22} \\ 2\varepsilon_{12} \end{bmatrix} + \xi_3 \begin{bmatrix} \kappa_{11} \\ \kappa_{22} \\ 2\kappa_{12} \end{bmatrix} = \boldsymbol{\varepsilon} + \xi_3 \boldsymbol{\kappa} \end{aligned} \tag{11.30}$$

For single-layered shells, the elastic matrix $\mathbf{C}$ is constant along the thickness direction. In this case, Eq. (11.29) can be further reduced to

$$\mathbf{n} = h\mathbf{C}\boldsymbol{\varepsilon}, \quad \mathbf{m} = \frac{h^3}{12}\mathbf{C}\boldsymbol{\kappa} \tag{11.31}$$

which indicates that the bending part and the membrane part are decoupled. For multilayered structures, numerical integration is generally implemented layer by layer. In this case, the bending and membrane forces are affected by both membrane and bending strains, which is known as the coupling effect.

11.2 Mesh Generation

As high-order bases are used in the p-version finite elements, using coarse mesh is sufficient to get highly accurate results during the analysis. Unlike the h-version method, the solution error of p-version method is controlled by increasing the polynomial degree (p) of elements on a fixed mesh pattern. Therefore, it is very important to ensure that the representation of the domain (i.e., the coarse mesh) is sufficiently accurate for the solution error to be controlled by p rather than the errors of geometry discretization [188–190].

For 2D flat domains, the blending function method is widely employed for p-version mesh generation [190, 213, 417]. In this method, parametric descriptions of domain boundary curves are incorporated into the construction of element mapping, thus the geometry errors on domain boundaries are fully removed. As an example, Fig. 11.2(b) shows a curved triangular mesh generated by the blending function mapping on triangular domain:

$$\boldsymbol{P}(\xi,\eta) = \frac{\zeta}{1-\xi}\boldsymbol{s}_1 + \frac{\xi}{1-\eta}\boldsymbol{s}_2 + \frac{\zeta}{1-\eta}\boldsymbol{s}_3 - \zeta\boldsymbol{P}_1 - \frac{\xi\zeta}{1-\xi}\boldsymbol{P}_2,$$
$$\zeta = 1-\xi-\eta \tag{11.32}$$

where $\boldsymbol{s}_i$, $i = 1,2,3$ are the boundary curves, and $\boldsymbol{P}_1$ and $\boldsymbol{P}_2$ are element vertices, see Fig. 11.3. In practical implementation, the conventional mesh generation algorithm is firstly applied to get the straight-side triangular mesh shown in Fig. 11.2(a).

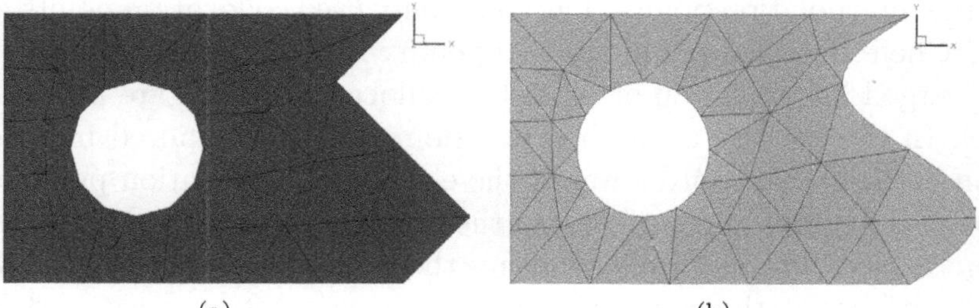

Fig. 11.2. High-order triangular mesh generated by blending function mapping on 2D flat domain: (a) linear mesh with geometry discretization error on curved domain boundaries; (b) exact blending function mesh.

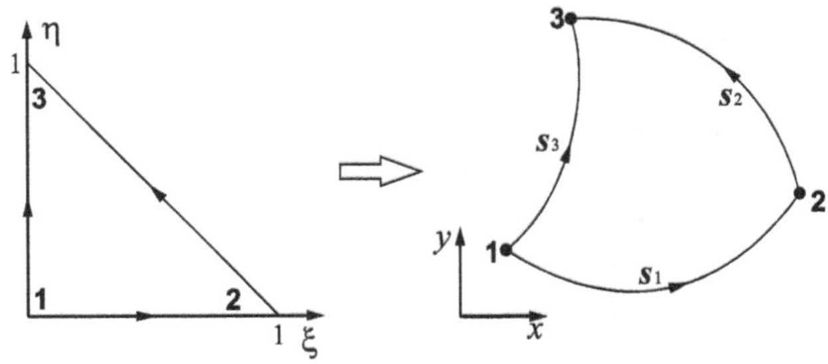

Fig. 11.3. Blending function mapping on triangular domain.

Then a reparameterization process is implemented for the elements on curved boundaries by using the blending function mapping in conjunction with the boundary geometry information. Through this two-step procedure, curved mesh can be generated.

For shell analysis using p-version FEM, curved surface mesh in 3D space is required. However, the blending function method cannot be directly applied to this kind of mesh generation since only the edge curves are considered into the mapping functions. A possible approach is to generate the 2D mesh first on parametric domain and then map it onto the surface. However, using this method, the mesh quality may be badly affected by mapping function when the surface is defined by a complicated parametric function, and the existing surface mesh generation algorithms may not be fully exploited. With this in mind, the quasi-regional mapping (QRM) method [188] is chosen for the surface mesh generation of the present study.

The QRM method is very similar to the isoparametric mapping method but distinct from it in that special collocation points are used as the interpolation points. The commonly used collocation points are the Chen–Babuška points, see Appendix A.2, which can produce a converged interpolation error as the collocation points are increased [89]. In the process of mesh generation, the only required information is the global coordinates of the elaborated collocation points on boundary curves and surfaces. In the original paper [188], these coordinates are obtained by evaluating the mapping function of curves and surfaces at the collocation points on parametric space, and the parametric collocation points are generated by a local linear mapping obtained by the interpolation of vertex parametric coordinates. However, this method is not suitable for surfaces with irregular cutout

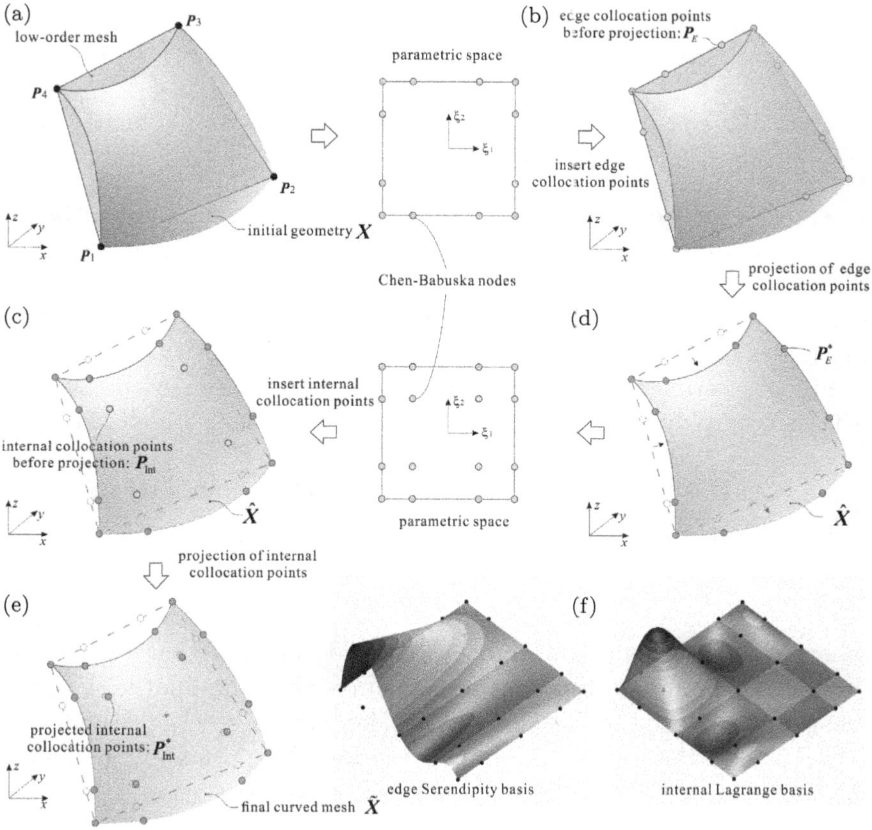

Fig. 11.4. The quasi-regional mapping method: (a) initial low-order mesh; (b) assignment of edge collocation points; (c) assignment of internal collocation points; (d) projection of edge points; (e) projection of internal points and the final surface; (f) the geometry interpolation bases.

where the cut boundary curves on parametric space are not straight lines. Considering this limitation, a slight modification based on a two-stage point-to-surface projection is proposed in the following to obtain the global coordinates of collocation points.

Take the quadrilateral mesh generation as an example. As illustrated in Fig. 11.4, in order to obtain the curved high-order mesh for the surface defined by $\boldsymbol{X}$ (the mid-surface of shell), coarse facet low-order mesh is firstly generated by the classical algorithm. For each facet element, pre-projected edge points are generated by the following linear interpolation:

$$\boldsymbol{P}_{\mathrm{E},i} = \sum_{J=1}^{4} N_J(\xi_{1i}, \xi_{2i})\boldsymbol{P}_J, \quad i = 1, 2, \ldots, n_e \tag{11.33}$$

where $\boldsymbol{P}_J$ are the four global vertex points, N_J are the bilinear bases on the standard parametric domain, (ξ_{1i}, ξ_{2i}) are the parametric coordinates of Chen–Babuška nodes on element boundaries, n_e is the edge node number; and $\boldsymbol{P}_{E,i}$ are the obtained pre-projected boundary nodes in global coordinates. Then projection algorithms based on minimum distance to surface or minimum distance to surface boundaries are employed to get the projected edge nodes P_E^* defined by

$$\boldsymbol{P}_{E,i}^* := \begin{cases} \boldsymbol{P} \in X, \|\boldsymbol{P} - \boldsymbol{P}_{E,i}\| \to \min, & \text{if } \boldsymbol{P}_{E,i} \text{ is an internal point.} \\ \boldsymbol{P} \in \partial\boldsymbol{X}, \|\boldsymbol{P} - \boldsymbol{P}_{E,i}\| \to \min, & \text{if } \boldsymbol{P}_{E,i} \text{ is an edge point.} \end{cases} \tag{11.34}$$

where $\partial\boldsymbol{X}$ denotes the boundary of $\boldsymbol{X}$. Using different projection criteria for points on different area ensures that the final high-order mesh can well fit the domain boundaries, see Figs. 11.5(c) and (d). After we get the projected boundary nodes $\boldsymbol{P}_E^*$, a mapping surface $\hat{\boldsymbol{X}}$ can be constructed using the serendipity interpolation bases:

$$\hat{\boldsymbol{X}} = \sum_{J=1}^{n_e} S_J(\xi, \eta)\boldsymbol{P}_{E,J}^* \tag{11.35}$$

where S_J are the high-order serendipity shape functions [35, 418] based on boundary Chen–Babuska nodes, see Fig. 11.4(f); $\boldsymbol{P}_{E,J}^*$ are the projected boundary nodes. Clearly, according to the property of serendipity interpolation, surface $\hat{\boldsymbol{X}}$ shall fit well with $\boldsymbol{X}$ on all of the element boundaries but may not well on the internal points. To further improve the accuracy, internal collocation points are then added in, see Fig. 11.4(d), by

$$\boldsymbol{P}_{\mathrm{Int},i} = \hat{\boldsymbol{X}}(\xi_i, \eta_i) \tag{11.36}$$

where (ξ_i, η_i) denote the internal Chen–Babuška nodes, and $\boldsymbol{P}_{\mathrm{Int},i}$ are the pre-projected internal points on $\hat{\boldsymbol{X}}$. Similarly, the projection

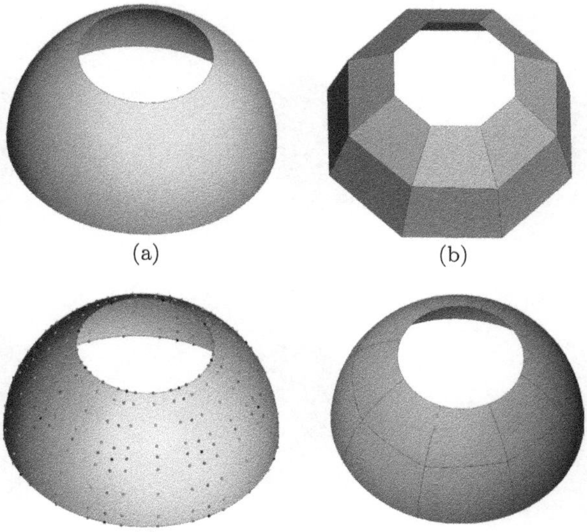

(a) (b) (c) (d)

Fig. 11.5. Curved mesh generation on a hemispherical surface: (a) original geometry; (b) coarse low-order mesh; (c) projected collocation points; (d) final curved mesh.

algorithm based on minimum distance to surface is used to obtain the projected internal points $\boldsymbol{P}^*_{\text{Int}}$:

$$\boldsymbol{P}^*_{\text{Int},i} := \{\boldsymbol{P} \in \boldsymbol{X} \text{ and } \|\boldsymbol{P} - \boldsymbol{P}_{\text{Int},i}\| \to \min\} \tag{11.37}$$

Finally, the revised mapping function can be expressed by

$$\tilde{\boldsymbol{X}} = \hat{\boldsymbol{X}} + \Delta, \quad \Delta = \sum_{i=1}^{n_{\text{Int}}} \hat{L}_i(\xi, \eta)(\boldsymbol{P}^*_{\text{Int},i} - \boldsymbol{P}_{\text{Int},i}) \tag{11.38}$$

where Δ is the correction term, n_{Int} is the internal node number, and $\hat{L}_i(\xi, \eta)$ are the Lagrange bases based on internal Chen–Babuška points, see Fig. 11.4(f). As for illustration, Fig. 11.5 shows the generated curved mesh for a hemispherical surface.

11.3 Finite Element Implementation

11.3.1 *Hierarchical bases*

In Chapter 6, the p-version C^1 hierarchical bases for both quadrilateral and triangular plate elements were constructed based on a

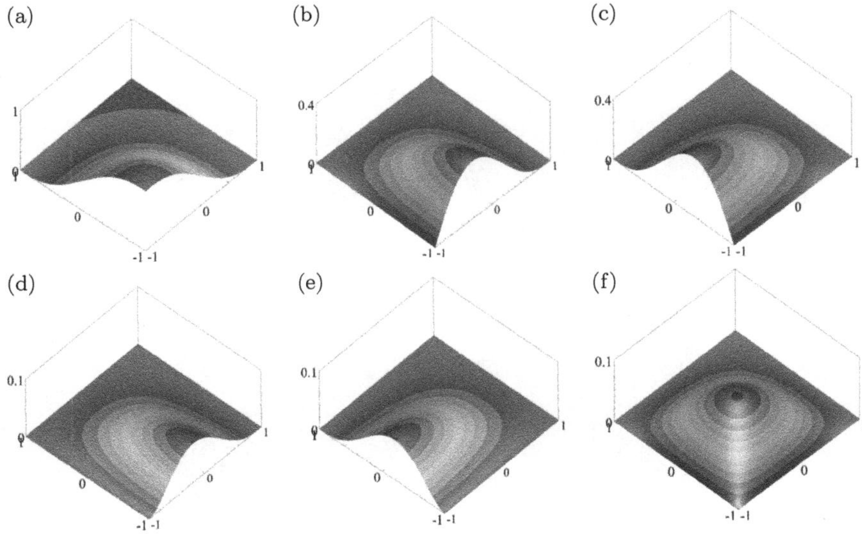

Fig. 11.6. Base functions on vertex 1 ($\xi = -1, \eta = -1$): (a) H^{V1}; (b) H_{ξ}^{V1}; (c) H_{η}^{V1}; (d) $H_{\xi\xi}^{\mathrm{V1}}$; (e) $H_{\eta\eta}^{\mathrm{V1}}$; (f) $H_{\xi\eta}^{\mathrm{V1}}$.

framework of Hermite blending function interpolation. In this work, only the bases of quadrilateral elements are used in the shell element construction as the formulations are equally suitable for the bases of triangular elements. Generally, the hierarchical bases are classified into vertex, edge and face functions according to their attached topology entities. Figure 11.6 shows the six vertex functions of our quadrilateral elements. These bases are the same as the fifth-order h-version rectangular elements [372] where 6 DOFs including the second-order derivatives are used as nodal variables. The edge functions at one boundary are shown in Fig. 11.7). These bases can be divided into two categories: one has non-zero boundary values and zero normal derivatives, and the other has zero boundary values and non-zero boundary normal derivatives. The internal face functions are shown in Fig. 11.8). It can be observed that these bases have both zero boundary values and first-order derivatives. For convenience of reference, detailed expressions of these bases are given in Appendix A.1. One can observe that these bases are of a separated variable form and are formulated in terms of Jacobi orthogonal polynomials. These properties ensures that the evaluation of these bases can be calculated efficiently by using the recursion formula [213]. Since the number of nodal bases is fixed during the p-refinement, the lowest

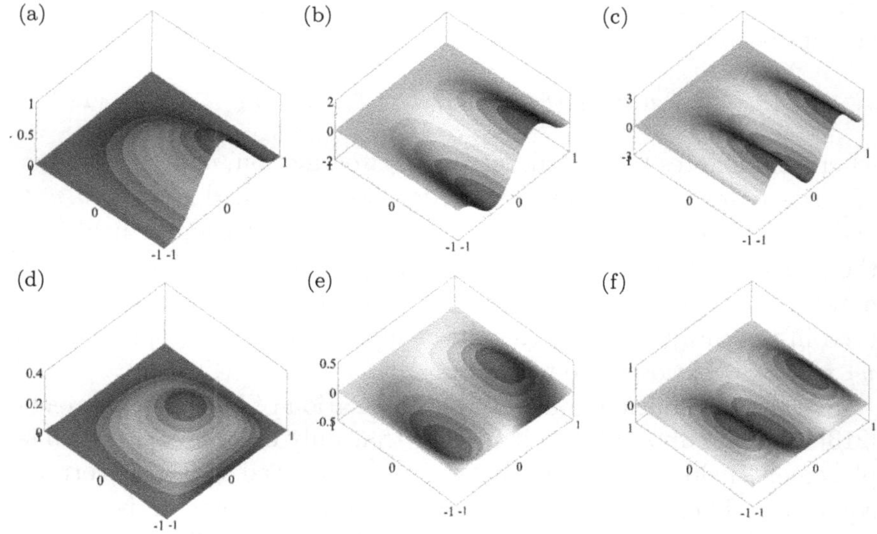

Fig. 11.7. Base functions on edge 1 ($\eta = -1$): (a) H_1^{E1}; (b) H_2^{E1}; (c) H_3^{E1}; (d) $H_{\text{d},1}^{\text{E1}}$; (e) $H_{\text{d},2}^{\text{E1}}$; (f) $H_{\text{d},3}^{\text{E1}}$.

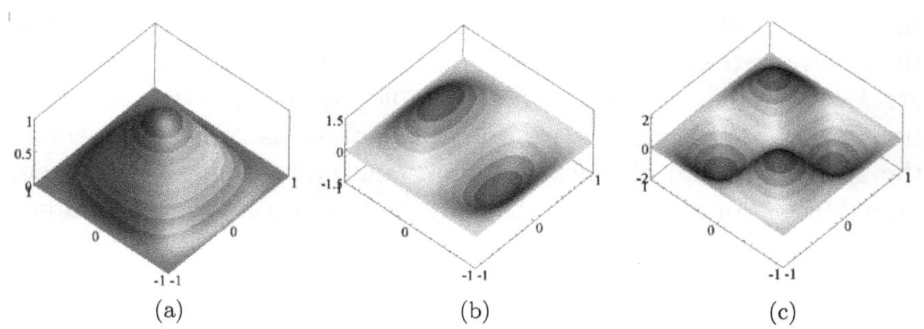

Fig. 11.8. Internal base functions: (a) H_{11}^{F}; (b) H_{12}^{F}; (c) H_{22}^{F}.

order of present elements is $p = 5$. Moreover, use of different number of edge functions is allowed in the implementation of present elements, which means that local p-refinement can be easily performed by our elements.

11.3.2 *Nodal variable collocation*

As mentioned above, the independent field variables of the thin shell model are the mid-surface displacements in the three global directions. Using the hierarchical bases, the displacement field can be expressed as

$$\boldsymbol{u} = H_I(\xi_1, \xi_2)\tilde{\boldsymbol{u}}_I,\ I = 1, 2, \ldots, m \tag{11.39}$$

where H_I denotes the Ith hierarchical base function, m is the number of bases determined by element order, see [213], and $\boldsymbol{u} = [u, v, w]^{\mathrm{T}}$ and $\tilde{\boldsymbol{u}}_I = [\tilde{u}_I, \tilde{v}_I, \tilde{w}_I]^{\mathrm{T}}$ are respectively the displacement field and the generalized nodal variables. Since the hierarchical bases H_I are non-interpolative, the related nodal variables $\tilde{u}_I$ cannot be directly used in assembly. To solve this problem, shifted nodal displacements $\boldsymbol{u}_I$ will be derived in the following.

We start from the nodal variable collocation of our plate elements [213]. For Kirchhoff plates, it is clear that only one independent displacement field, i.e., the deflection w, is involved in the governing equations. Therefore, the key point of the plate elements is to construct a C^1 displacement field w on the flat domain. To this end, we used six nodal variables including the complete second-order derivatives with respect to the global x- and y-axis on the element vertices, see Fig. 11.9(a), while the normal derivatives and the node deflections were used on the middle edge nodes. In practical computation, the generalized nodal variables related to the hierarchical edge bases are transformed into the nodal variables by multiplying a transition matrix at element level. For most solutions of C^1 conformity, the specified nodal displacements are continuous along the element boundaries. For this reason, they can be used directly in assembly.

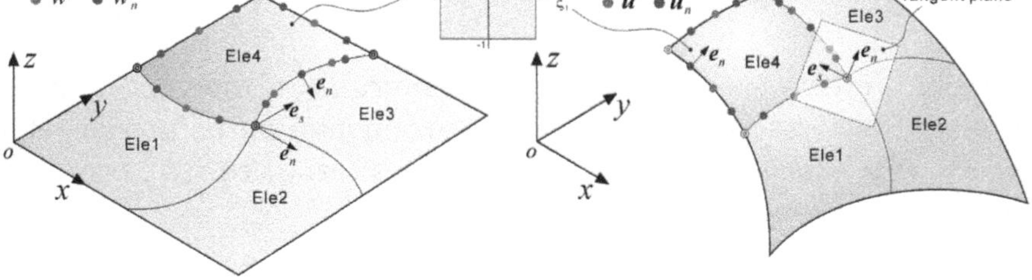

Fig. 11.9. Nodal variable collocation on plate (a) and shell (b) elements.

However, the generalization of plate nodal variable to the curved shell elements is not straightforward. A commonly used method is to construct a conforming solution first on the flat parametric domain of the shell using the plate element and then map them onto the shell surface [227, 419]. However, as in these works, the nodal variables are defined on the parametric space, it is difficult for this method to be applied to shells defined by multipatches with difference parameterizations. In the present study, this limitation will be eliminated as the deriving nodal variables are defined directly on shell surfaces.

In order to generalize the node variable definition of flat plates to curved thin shells, the crucial issue concerns the definition of vertex variables where high-order derivatives are involved. To this end, we firstly specify two unit vectors $\boldsymbol{e}_n$ and $\boldsymbol{e}_s$ on each vertex. As illustrated in Fig. 11.9(b), these two vectors are perpendicular to each other and lie on the tangent plane of shell mid-surface in reference configuration. Of course, the definition of $\boldsymbol{e}_n$ and $\boldsymbol{e}_s$ is not unique. A simple way is by the cross product of normal director $\boldsymbol{A}_3$ and proper selected global axis. However, when the vertex lies on the interface of different material regions, $\boldsymbol{e}_n$ is specified as the normal of that interface, which is required in the relaxation of the over continuity of second normal derivatives, see the following. With these two vectors, we can obtain the following chain rules:

$$\begin{aligned}\begin{bmatrix}\mathrm{d}\xi_1\\ \mathrm{d}\xi_2\end{bmatrix} &= \begin{bmatrix}\boldsymbol{A}^1\cdot\boldsymbol{e}_n & \boldsymbol{A}^1\cdot\boldsymbol{e}_s\\ \boldsymbol{A}^2\cdot\boldsymbol{e}_n & \boldsymbol{A}^2\cdot\boldsymbol{e}_s\end{bmatrix}\begin{bmatrix}\mathrm{d}n\\ \mathrm{d}s\end{bmatrix}\\ &= \begin{bmatrix}\partial\xi_1/\partial n & \partial\xi_1/\partial s\\ \partial\xi_2/\partial n & \partial\xi_2/\partial s\end{bmatrix}\begin{bmatrix}\mathrm{d}n\\ \mathrm{d}s\end{bmatrix}\end{aligned} \tag{11.40}$$

which can be easily derived from the differential relation $\mathrm{d}\boldsymbol{r} = \boldsymbol{A}_\alpha\,\mathrm{d}\xi_\alpha = \boldsymbol{e}_n\,\mathrm{d}n + \boldsymbol{e}_s\,\mathrm{d}s$, where $\mathrm{d}\boldsymbol{r}$ is an infinitesimal line segment on the mid-surface. Finally, the partial derivatives with respect to $\boldsymbol{e}_n$ and $\boldsymbol{e}_s$ can be defined as

$$\boldsymbol{u}_n = \frac{\partial\boldsymbol{u}}{\partial n} = \frac{\partial\boldsymbol{u}}{\partial\xi_\alpha}\frac{\partial\xi_\alpha}{\partial n}$$

$$\boldsymbol{u}_s = \frac{\partial\boldsymbol{u}}{\partial s} = \frac{\partial\boldsymbol{u}}{\partial\xi_\alpha}\frac{\partial\xi_\alpha}{\partial s}$$

$$\boldsymbol{u}_{nn} = \frac{\partial^2 \boldsymbol{u}}{\partial n^2} = \frac{\partial}{\partial \xi_\beta}\left[\frac{\partial \boldsymbol{u}}{\partial \xi_\alpha}\frac{\partial \xi_\alpha}{\partial n}\right]\frac{\partial \xi_\beta}{\partial n}$$

$$\boldsymbol{u}_{ss} = \frac{\partial^2 \boldsymbol{u}}{\partial s^2} = \frac{\partial}{\partial \xi_\beta}\left[\frac{\partial \boldsymbol{u}}{\partial \xi_\alpha}\frac{\partial \xi_\alpha}{\partial s}\right]\frac{\partial \xi_\beta}{\partial s}$$

$$\boldsymbol{u}_{ns} = \frac{\partial^2 \boldsymbol{u}}{\partial n \partial s} = \frac{\partial}{\partial \xi_\beta}\left[\frac{\partial \boldsymbol{u}}{\partial \xi_\alpha}\frac{\partial \xi_\alpha}{\partial n}\right]\frac{\partial \xi_\beta}{\partial s} \tag{11.41}$$

where repeated indices imply summation. The above equations are indeed a generalization of conventional concept of directional derivative on flat 2D domain to the 3D surface. If the surface is a plane, and $\boldsymbol{e}_n$ and $\boldsymbol{e}_s$ are aligned along the global x- and y-axis, Eq. (11.41) can be degenerated to the nodal variable definition of the flat thin plates. In our thin shell elements, the derivatives in Eq. (11.41) along with the node displace $\boldsymbol{u}$ are used as the nodal variables on element vertices. Since three components are contained in the displacement vector, there are hence 18 DOFs on each vertex. Similar to the plate elements, nodal variables on middle edges are still the node displacement $\boldsymbol{u}$ and normal derivative $\boldsymbol{u}_n$. For the definition of $\boldsymbol{u}_n$, director $\boldsymbol{e}_n$ is specified as the vector which is normal to the attached edge and tangent to the mid-surface, see Fig. 11.9(b). In practical implementation, one should ensure that $\boldsymbol{e}_n$ and $\boldsymbol{e}_s$ of adjacent elements are in the same direction.

With the usage of nodal variables, the finite element displacement field can be reexpressed as

$$\boldsymbol{u} = H_I \tilde{\boldsymbol{u}}_I = H_I Q_{IJ}^{-1} \boldsymbol{u}_J = N_J \boldsymbol{u}_J, \quad I, J = 1, 2, \ldots, m \tag{11.42}$$

where $\boldsymbol{u}_J$ denotes the Jth nodal variables (note that the generalized nodal variables related to the internal bases remain unchanged), and N_J is the Jth shifted basis, and Q_{IJ}^{-1} are items of the inverse of transition matrix $[Q_{IJ}]$ which can be calculated by evaluating the Ith nodal variable of H_J. To make it clear, assume that the first six nodal variables are defined as

$$\begin{aligned} &\boldsymbol{u}_1 = \boldsymbol{u}^{V1}, \quad \boldsymbol{u}_2 = \boldsymbol{u}_n^{V1}, \quad \boldsymbol{u}_3 = \boldsymbol{u}_s^{V1} \\ &\boldsymbol{u}_4 = \boldsymbol{u}_{nn}^{V1}, \quad \boldsymbol{u}_5 = \boldsymbol{u}_{ss}^{V1}, \quad \boldsymbol{u}_6 = \boldsymbol{u}_{ns}^{V1} \end{aligned} \tag{11.43}$$

where $\boldsymbol{u}^{V1} = [u, v, w]_{V1}$ is the node displacement on Vertex 1; $\boldsymbol{u}_n^{V1} = [u_n, v_n, w_n]_{V1}$ is the first-order derivative with respect to $\boldsymbol{e}_n$

on vertex 1, and the like. Then the corresponding components in $[Q_{IJ}]$ can be expressed as

$$\begin{cases} Q_{1J} = H_J|_{V1}, \quad Q_{2J} = \left.\dfrac{\partial H_J}{\partial n}\right|_{V1}, \quad Q_{3J} = \left.\dfrac{\partial H_J}{\partial s}\right|_{V1} \\ Q_{4J} = \left.\dfrac{\partial^2 H_J}{\partial n^2}\right|_{V1}, \quad Q_{5J} = \left.\dfrac{\partial^2 H_J}{\partial s^2}\right|_{V1}, \quad Q_{6J} = \left.\dfrac{\partial^2 H_J}{\partial n \partial s}\right|_{V1} \end{cases}$$

$$J = 1 \sim m \tag{11.44}$$

where subscript V1 means evaluation on vertex 1. Similarly, items of $[Q_{IJ}]$ corresponding to the other nodal variables can be easily derived.

Using the shifted nodal variables $\boldsymbol{u}_J$, the C^1 conformity of field variable can be enforced by the usual assembly process of FEM. Strictly speaking, the C^1 conformity is not fully satisfied in advance on the curved shell element boundaries. Indeed, it is satisfied on element nodes. The relaxation of C^1 conformity may occasionally result in lower convergence rate or even divergent results. However, as indicated in [213], by choosing the Gauss–Lobatto points, see Appendix A.2, as the boundary nodes, the exponential rate of convergence can still be achieved even when curved elements are used. This strategy is also employed in the present work to enhance the convergence performance of our shell elements.

Additionally, as in our plate elements, the continuity of present shell elements is indeed C^2 on the vertices. This over conformity (compared with the strict C^1 conformity) will not cause problems in the analysis of structures with continuous material property and curvatures because the bending and twisting strains are continuous due to the continuous of stresses. However, in the case where material or curvature discontinuity is involved, the continuity of second order normal derivatives $\boldsymbol{u}_{nn}$ ($\boldsymbol{e}_n$ is normal to the discontinuity interface) should be relaxed. This can be easily performed by sharing the five vertex displacement in the assembly process [420]. Numerical example in Section 11.4.5 verifies the effectiveness of this method. Moreover, in the present study, we limit our goals to the shells with at least C^1 continuity in geometry. Shells with C^0 lines on the area of sharp folds, creases and stiffener junctions will be considered in the ongoing work as special treatments are required.

11.3.3 Discretization and linearization

Substituting Eq. (11.42) into Eq. (11.18), the internal work can be expressed as

$$\begin{aligned}
&\int_{\Omega_0} \boldsymbol{m} : \delta\boldsymbol{\kappa} + \boldsymbol{n} : \delta\boldsymbol{\varepsilon}\, \mathrm{d}\Omega_0 \\
&\quad = \delta u_{iJ} \cdot \int_{\Omega_0} \boldsymbol{m} : \frac{\partial \boldsymbol{\kappa}}{\partial u_{iJ}} + \boldsymbol{n} : \frac{\partial \boldsymbol{\varepsilon}}{\partial u_{iJ}}\, \mathrm{d}\Omega_0 \\
&\quad = \delta u_{iJ} F_{iJ},\ i = 1 \sim 3,\ J = 1 \sim m
\end{aligned} \tag{11.45}$$

where u_{iJ} is the ith component of nodal displacement $\boldsymbol{u}_J$ and F_{iJ} is the corresponding internal force. Then the general nonlinear algebraic system can be expressed as

$$\mathbf{f}^{\mathrm{int}}(\mathbf{u}) = \mathbf{f}^{\mathrm{ext}} \tag{11.46}$$

where $\mathbf{f}^{\mathrm{int}}$ is the internal force vector with items F_{iJ}, $\mathbf{f}^{\mathrm{ext}}$ is the external force vector, and $\mathbf{u}$ is the unknown displacement vector. To solve this equation, iteration methods such as the load-incremental Newton–Raphson method or the arc length method are frequently used [421], where the required tangent stiffness can be obtained by differentiating F_{iJ} as

$$\begin{aligned}
\mathrm{d}F_{iJ} &= \mathrm{d}u_{kL} \cdot \underbrace{\int_{\Omega_0} \frac{\partial \boldsymbol{m}}{u_{kL}} : \frac{\partial \boldsymbol{\kappa}}{\partial u_{iJ}} + \boldsymbol{m} : \frac{\partial^2 \boldsymbol{\kappa}}{\partial u_{iJ} \partial u_{kL}} + \frac{\partial \boldsymbol{n}}{\partial u_{kL}} : \frac{\partial \boldsymbol{\varepsilon}}{\partial u_{iJ}} + \boldsymbol{n} : \frac{\partial^2 \boldsymbol{\varepsilon}}{\partial u_{iJ} \partial u_{kL}}\, \mathrm{d}\Omega_0}_{K_{iJkL}} \\
&= K_{iJkL}\, \mathrm{d}u_{kL}
\end{aligned} \tag{11.47}$$

where K_{iJkL} are the components of tangent stiffness matrix and the relevant derivatives with respect to the unknowns are summarized in Table 11.1 [231].

11.3.4 Boundary condition imposition

For a simply supported boundary Γ, the displacement and the tangent derivatives are zeros, which can be expressed by

$$\boldsymbol{u} = \boldsymbol{0}, \quad \frac{\partial \boldsymbol{u}}{\partial s} = \boldsymbol{0}, \quad \frac{\partial^2 \boldsymbol{u}}{\partial s^2} = \boldsymbol{0} \quad \text{on } \Gamma \tag{11.48}$$

Table 11.1. Derivatives required for the linearization.

(1)	Derivatives of displacement $\boldsymbol{u}$:
	$\boldsymbol{u}_{,iJ} = N^J \boldsymbol{e}_i, \quad \boldsymbol{u}_{,iJkL} = \boldsymbol{0}$
(2)	Derivatives of covariant bases:
	$\boldsymbol{a}_{\alpha,iJ} = N^J_{,\alpha} \boldsymbol{e}_i, \quad \boldsymbol{a}_{\alpha,iJkL} = \boldsymbol{0}$
	$\boldsymbol{a}_{\alpha,\beta,iJ} = N^J_{,\alpha\beta} \boldsymbol{e}_i, \ \boldsymbol{a}_{\alpha,\beta,iJkL} = \boldsymbol{0}$
(3)	Derivatives of metric tensor components:
	$\boldsymbol{a}_{\alpha\beta,iJ} = \boldsymbol{a}_{\alpha,iJ} \cdot \boldsymbol{a}_\beta + \boldsymbol{a}_\alpha \cdot \boldsymbol{a}_{\beta,iJ}$
	$\boldsymbol{a}_{\alpha\beta,iJkL} = \boldsymbol{a}_{\alpha,iJ} \cdot \boldsymbol{a}_{\beta,kL} + \boldsymbol{a}_{\alpha,kL} \cdot \boldsymbol{a}_{\beta,iJ}$
(4)	Derivatives of membrane strains:
	$\varepsilon_{\alpha\beta,iJ} = \frac{1}{2} a_{\alpha\beta,iJ}, \ \varepsilon_{\alpha\beta,iJkL} = \frac{1}{2} a_{\alpha\beta,iJkL}$
(5)	Derivatives of $\tilde{\boldsymbol{a}}_3 = \boldsymbol{a}_1 \times \boldsymbol{a}_2$ and $\bar{a} = \Vert \tilde{a}_3 \Vert$:
	$\tilde{\boldsymbol{a}}_{3,iJ} = \boldsymbol{a}_{1,iJ} \times \boldsymbol{a}_2 + \boldsymbol{a}_1 \times \boldsymbol{a}_{2,iJ}$
	$\tilde{\boldsymbol{a}}_{3,iJkL} = \boldsymbol{a}_{1,iJ} \times \boldsymbol{a}_{2,kL} + \boldsymbol{a}_{1,kL} \times \boldsymbol{a}_{2,iJ}$
	$\bar{a}_{,iJ} = \boldsymbol{a}_3 \cdot \tilde{\boldsymbol{a}}_{3,iJ}$
	$\bar{a}_{,iJkL} = \frac{1}{\bar{a}} [\tilde{\boldsymbol{a}}_{3,iJ} \cdot \tilde{\boldsymbol{a}}_{3,kL} + \tilde{\boldsymbol{a}}_3 \cdot \tilde{\boldsymbol{a}}_{3,iJkL} - (\boldsymbol{a}_3 \cdot \tilde{\boldsymbol{a}}_{3,iJ})(\boldsymbol{a}_3 \cdot \tilde{\boldsymbol{a}}_{3,kL})]$
(6)	Derivatives of $\boldsymbol{a}_3$:
	$\boldsymbol{a}_3 = \tilde{\boldsymbol{a}}_3 / \bar{a}$
	$\boldsymbol{a}_{3,iJ} = \bar{a}^{-1} (\tilde{\boldsymbol{a}}_{3,iJ} - \bar{a}_{,iJ} \boldsymbol{a}_3)$
	$\boldsymbol{a}_{3,iJkL} = \bar{a}^{-1} (\tilde{\boldsymbol{a}}_{3,iJkL} - \bar{a}_{,iJkL} \boldsymbol{a}_3) + \bar{a}^{-2} (2 \bar{a}_{,iJ} \bar{a}_{,kL} \boldsymbol{a}_3 - \bar{a}_{,kL} \tilde{\boldsymbol{a}}_{3,iJ} - \bar{a}_{,iJ} \tilde{\boldsymbol{a}}_{3,kL})$
(7)	Derivatives of curvature tensor:
	$b_{\alpha\beta,iJ} = \boldsymbol{a}_{\alpha,\beta,iJ} \cdot \boldsymbol{a}_3 + \boldsymbol{a}_{\alpha,\beta} \cdot \boldsymbol{a}_{3,iJ}$
	$b_{\alpha\beta,iJkL} = \boldsymbol{a}_{\alpha,\beta,iJ} \cdot \boldsymbol{a}_{3,kL} + \boldsymbol{a}_{\alpha,\beta,kL} \cdot \boldsymbol{a}_{3,iJ} + \boldsymbol{a}_{\alpha,\beta} \cdot \boldsymbol{a}_{3,iJkL}$
(8)	Derivatives of bending strains:
	$\kappa_{\alpha\beta,iJ} = -b_{\alpha\beta,iJ}, \ \kappa_{\alpha\beta,iJkL} = -b_{\alpha\beta,iJkL}$

where the second and third equations are the direct results of the first equation. For a clamped boundary Γ, the constraint equations are given by

$$\boldsymbol{u} = \boldsymbol{0}, \quad \frac{\partial \boldsymbol{u}}{\partial s} = \boldsymbol{0}, \frac{\partial^2 \boldsymbol{u}}{\partial s^2} = \boldsymbol{0}, \frac{\partial \boldsymbol{u}}{\partial n} \cdot \boldsymbol{A}_3 = 0, \frac{\partial}{\partial s} \left(\frac{\partial \boldsymbol{u}}{\partial n} \cdot \boldsymbol{A}_3 \right) = 0 \text{ on } \Gamma \tag{11.49}$$

where the first three are the same as the simply supported case, the fourth equation is derived from the condition that normal fibers remain unchanged on the clamped boundary, and the last equation is derived from the fourth equation.

As high-order derivatives are used as element DOFs on vertices, the imposition of essential boundary conditions then becomes difficult. In Chapter 6, exact imposition techniques for these two types of boundary conditions and their mixed cases were thoroughly discussed for the flat plate elements with arbitrary curved boundaries. Theoretically, these methods can be extended to the boundary condition imposition of our shell elements. However, the formulations would be very complicated when the shell boundaries are 3D spatial curves with varied normal direction A_3. With this in mind, we use the more convenient penalty method to deal with the boundary condition imposition on general curved boundaries. For example, the penalty terms for a clamped boundary are given by

$$\delta W^p = \int_\Gamma \alpha_1 \boldsymbol{u} \cdot \delta \boldsymbol{u} \, \mathrm{d}\Gamma + \int_\Gamma \alpha_2 \left(\frac{\partial \boldsymbol{u}}{\partial n} \cdot \boldsymbol{A}_3 \right) \delta \left(\frac{\partial \boldsymbol{u}}{\partial n} \cdot \boldsymbol{A}_3 \right) \mathrm{d}\Gamma \quad (11.50)$$

where α_1, α_2 are the penalty factors, and α_2 is taken as zero when the boundary is simply supported. The associated penalty stiffness is expressed as

$$K^p_{iJkL} = \alpha_1 \int_\Gamma \frac{\partial \boldsymbol{u}}{\partial u_{iJ}} \cdot \frac{\partial \boldsymbol{u}}{\partial u_{kL}} \mathrm{d}\Gamma + \int_\Gamma \alpha_2 \left(\frac{\partial \boldsymbol{u}_n}{\partial u_{iJ}} \cdot \boldsymbol{A}_3 \right) \left(\frac{\partial \boldsymbol{u}_n}{\partial u_{kL}} \cdot \boldsymbol{A}_3 \right) \mathrm{d}\Gamma \quad (11.51)$$

Clearly, the penalty stiffness is independent of the nodal displacements, and it is calculated only once.

11.4 Numerical Examples

In this section, we use several benchmark problems of finite deformation analysis of single-layered isotropic and multilayered laminated thin shells to test the performance of our p-version thin shell elements. Mesh distortion tests are implemented to emphasize the capability of present C^1 elements in modeling problems with curved boundaries. Unless stated otherwise, uniform p-refinement is used by default in all examples, and the number of quadrature points is specified as $p + 5$ on each direction of parametric domain. The high-order meshes are all generated by using the QRM method introduced above. The convergence criteria of nonlinear iteration are simultaneous 0.01 displacement tolerance and 0.005 force tolerance.

11.4.1 *Cantilever beam*

In the first problem, we consider the bending analysis of an isotropic beam subjected to end moment. The geometry dimensions, material parameters and the load value are shown in Fig. 11.10. Three mesh patterns with different extents of distortions are used to check the computation performance. The load–deflection curves of tip point A are shown in Fig. 11.11 and compared with the reference results given by Sze using ABAQUS S4R elements [422]. Where the load factor is defined as ratio $M/M_{\max}$. As it can be observed, for rectangular mesh (mesh 1), using the lowest order ($p = 5$) is sufficient to capture the deformation features of entire loading process. For quadrilateral mesh (mesh 2), using elements with $p = 5$ is then become insufficient to capture the final deformation feature due to the mesh distortion. However, with slight increase in element order, converged results are obtained again. In mesh 3, elements with curved boundary are used for discretization. The distortion is severer than the quadrilateral case, see Fig. 11.12. In this case, using elements with $p = 5$ and 6 is unable to give acceptable results, while the results of $p = 8$ are still

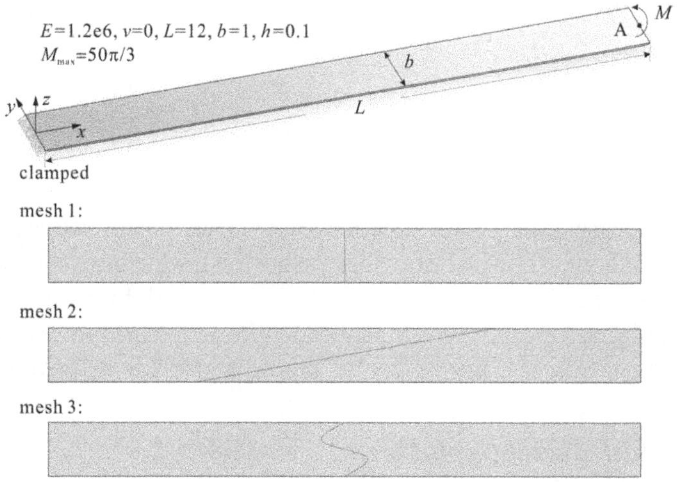

Fig. 11.10. Cantilever beam subjected to end moment and the mesh patterns.

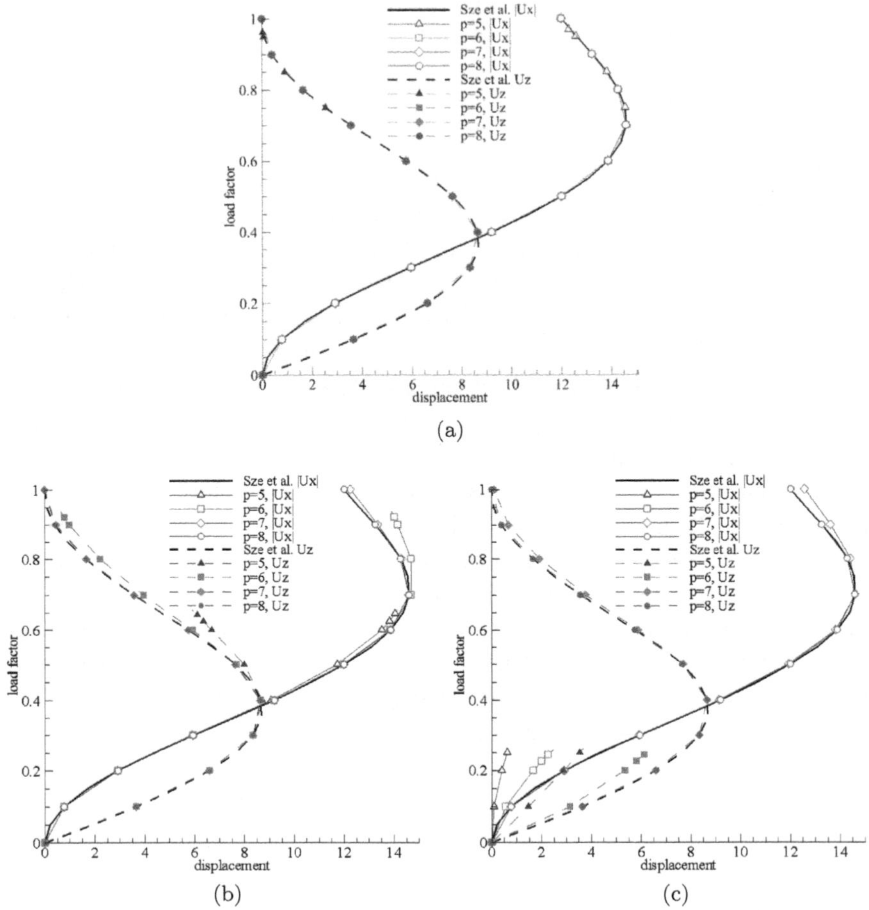

Fig. 11.11. Convergence comparison of load–displacement curves obtained by different meshes: (a) mesh 1; (b) mesh 2; (c) mesh 3.

highly accurate and agree excellently well with the reference solutions. This indicates that the high-order elements are less sensitive to the mesh distortion. Figure 11.13 shows the deformed configurations calculated by the three meshes at different load factors, denoted by λ. In all cases, the final circular configuration is achieved.

11.4.2 *Slit annular plate*

In the second example, the isotropic slit annular plate problem is considered. The geometry dimensions and material constants are shown

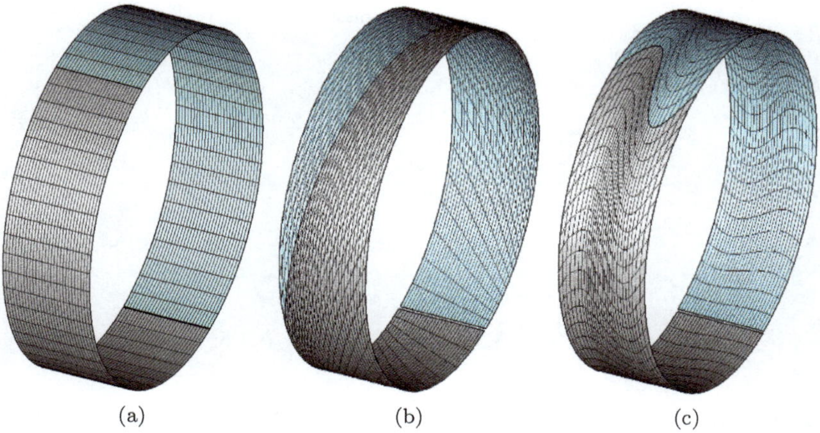

Fig. 11.12. Comparison of element distortion shown by parametric curves on the final configuration: (a) mesh 1; (b) mesh 2; (c) mesh 3.

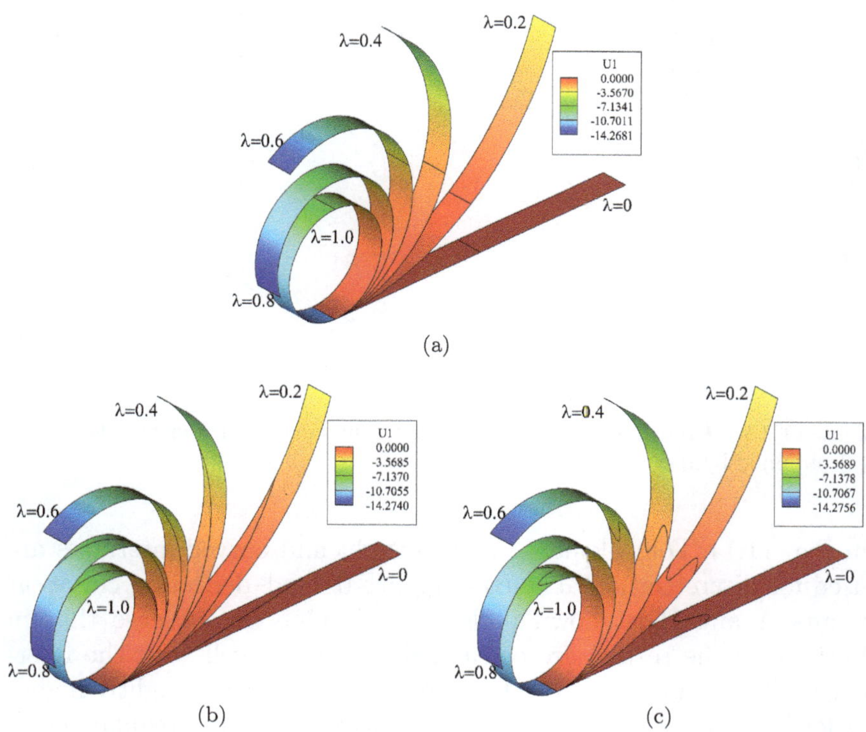

Fig. 11.13. Configurations at different load steps obtained by the three meshes: (a) mesh 1; (b) mesh 2; (c) mesh 3.

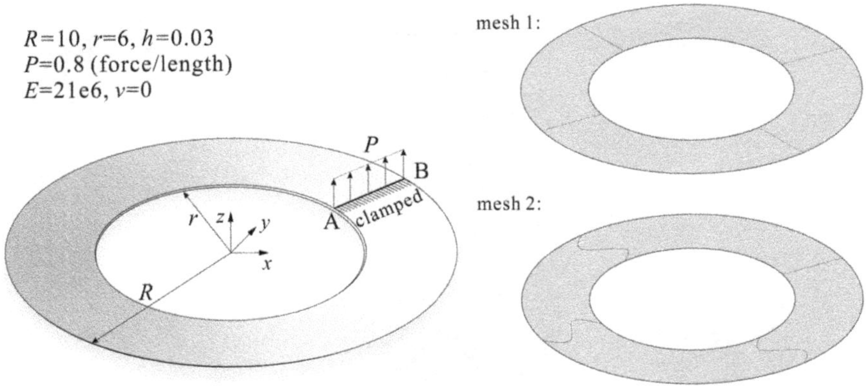

Fig. 11.14. An annular plate subjected to shear force and the mesh patterns.

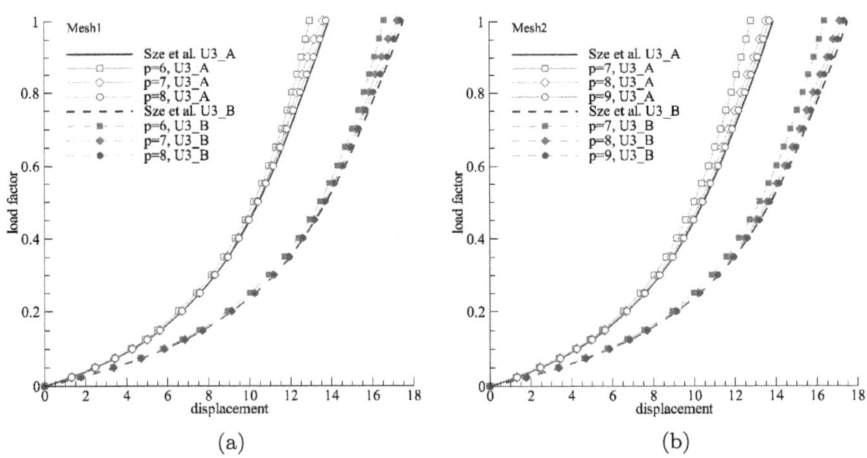

Fig. 11.15. Convergence of load–displacement curves for the annular plate: (a) obtained by mesh 1; (b) obtained by mesh 2.

in Fig. 11.14. Four elements with straight and curved interfaces are used for discretization. The convergence of load–deflection curves at points A and B is shown in Fig. 11.15. One can observe that in both cases the results are converged and agree well with the reference solutions given by Sze. For the distorted mesh, a slightly higher order is required to achieve the same accuracy as the regular mesh. Figure 11.16 shows the deformed configurations obtained by the two meshes at different load levels, and good agreement is verified again.

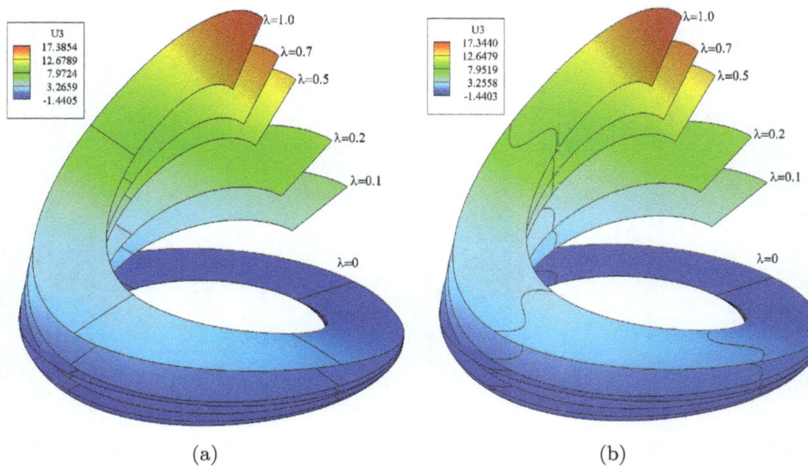

Fig. 11.16. Deformed configurations at different load levels: (a) obtained by mesh 1; (b) obtained by mesh 2.

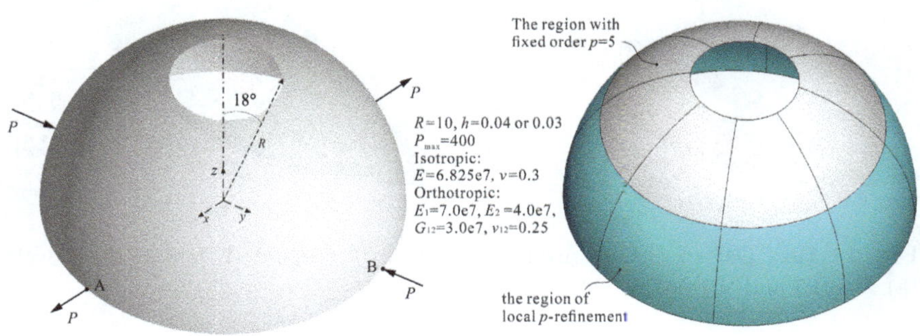

Fig. 11.17. Pinched hemispherical shell and the mesh pattern.

11.4.3 *Pinched hemispherical shell*

In the third example, we consider the hemispherical shell problem. The geometric dimensions and material parameters are shown in Fig. 11.17. In this example, both isotropic and orthotropic laminated materials are considered. The stacking sequence for the laminated shell is [0/90/0], and the reference direction is the projection of global z-axis on the surface. The thickness of isotropic shell is 0.04, while for laminated shell, the thickness is 0.01 for each ply. As mentioned previously, the approximation orders on the boundaries of our elements can be different. This property allows the local p-refinement on areas

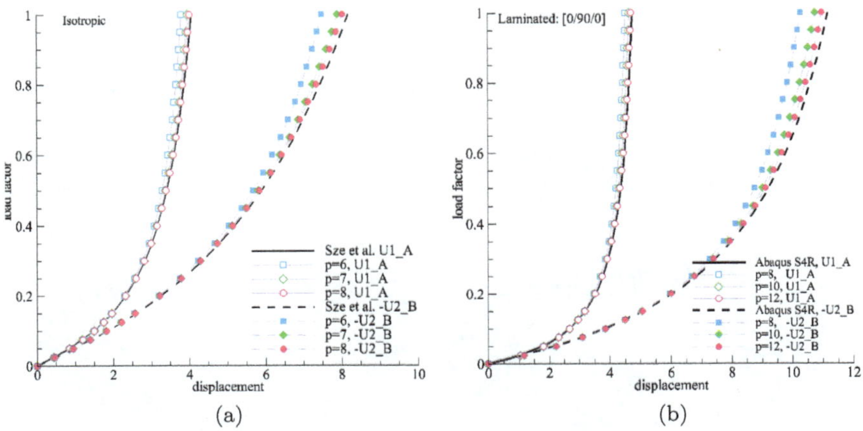

Fig. 11.18. Convergence of load–displacement curves for pinched hemispherical shell: (a) isotropic shell; (b) laminated shell.

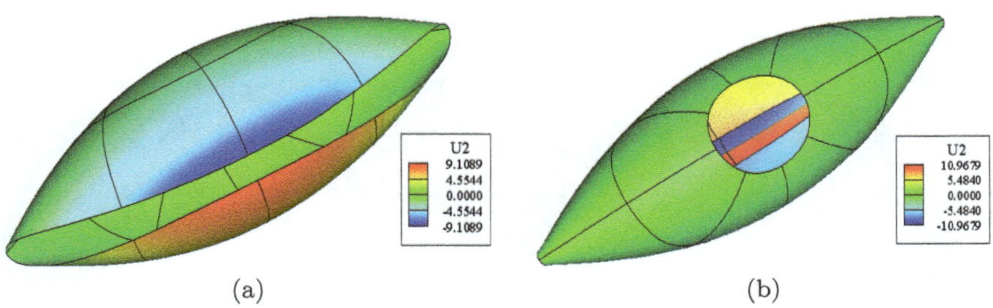

Fig. 11.19. Deformed configurations of [0/90/0] laminated shell at load levels (a) $\lambda = 0.5$ and (b) $\lambda = 1$.

of large deformation gradient, which is helpful for saving the computation costs. For this example, we used low-order elements with fixed $p = 5$ on the top half of the shell because the deformation is smaller. Elements with higher orders are used in the bottom half during the p-convergence study. The results of load–deflection curves are shown in Fig. 11.18 and compared with the solutions by Sze and the ABAQUS S4R results obtained by the present authors. Good agreement is again observed between the two cases, which shows that local p-refinement is sufficient to achieve acceptable results. Figure 11.19 shows the deformation patterns for the laminated shell at load levels $\lambda = 0.5$ and $\lambda = 1$. The smoothness and symmetry of the solutions are observed. One may also note the intersection at the bottom of the shell when load factor $\lambda = 1$. This phenomenon is normal because self-contact is not considered in our model. On the contrary, the intersection verifies the robustness of present elements in analyzing problems with large deformation.

11.4.4 *Post buckling of shallow cylindrical shell*

In the fourth problem, the buckling analysis of shallow cylindrical shells is considered. The geometric dimensions and mesh patterns are shown in Fig. 11.20. Both isotropic and laminated thin shells with different thicknesses and stacking sequences are analyzed. The longitudinal direction is chosen as the reference direction for orthotropic material definition. Regular and curved meshes are used for the domain discretization. Figure 11.21 shows the load–deflection curves for isotropic and laminated shells with thickness $h = 12.7$. The material parameters are $E = 21e6$, $v = 0$ for isotropic material and $E_1 = 3300$, $E_2 = 1100$, $G_{12} = 660$, $v_{12} = 0.25$ for orthotropic material. The maximum pinching load is $P = 3000$. Elements with orders $p = 5, 6, 7$ are used for the computation. As it can be observed, the results obtained by different orders coincide with each other and agree very well with the reference solutions given by Sze. This indicates that the lowest order $p = 5$ is sufficient to capture the entire equilibrium paths of all cases. Moreover, the results obtained by

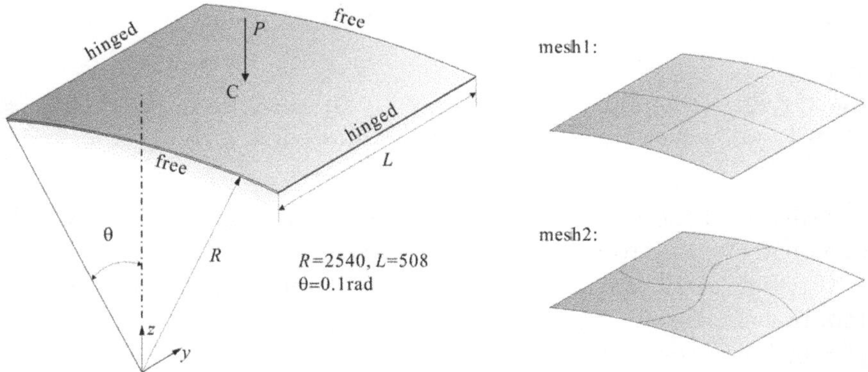

Fig. 11.20. A hinged shallow cylindrical roof subjected to a central pinching load and the mesh patterns.

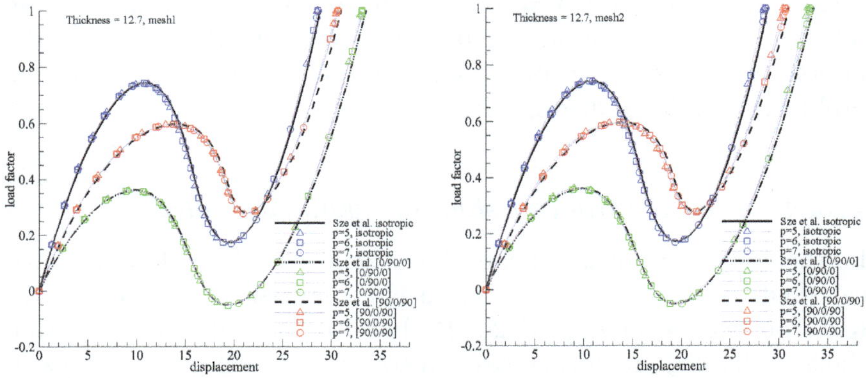

(a) (b)

Fig. 11.21. Convergence of load–deflection (U3) curves for shells with thickness $h = 12.7$: (a) obtained by mesh 1; (b) obtained by mesh 2.

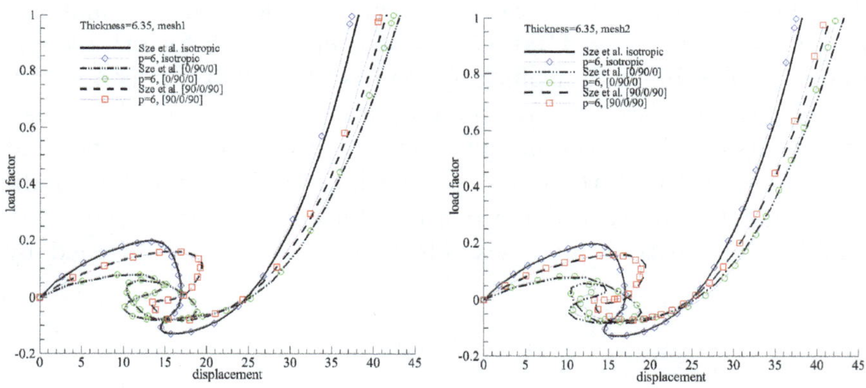

(a) (b)

Fig. 11.22. Comparison of load–deflection (U3) curves for shells of thickness $h = 6.35$: (a) obtained by mesh1; (b) obtained by mesh 2.

the distorted mesh are nearly the same as the regular mesh, which demonstrates the robustness of our curved C^1 elements.

In Fig. 11.22, shells with the same material and load parameters but smaller thickness ($h = 6.35$) are analyzed by elements with $p = 6$. Since complicated snap-through and snap-back behaviors of equilibrium path are involved in this problem, the arc length method [421] is employed to track the solutions. As it shows, decreasing the shell thickness greatly increases the complexity of the load–deflection path. Significantly different behavior of equilibrium path between the

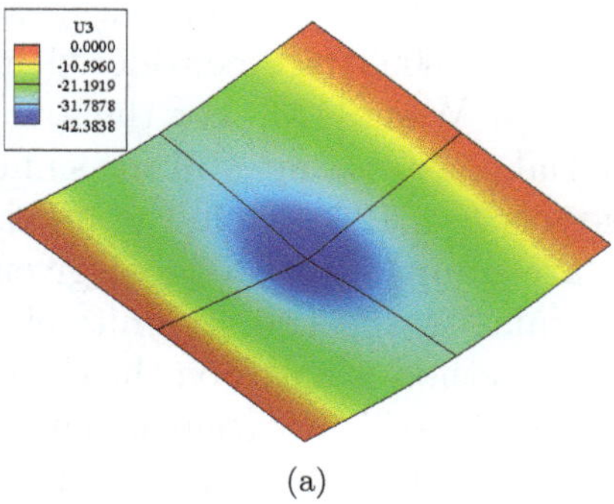

Fig. 11.23. Displacement field comparison of U3 for the shell with stacking sequence [90/0/90] and $h = 6.35$ at load level $P = 3000$.

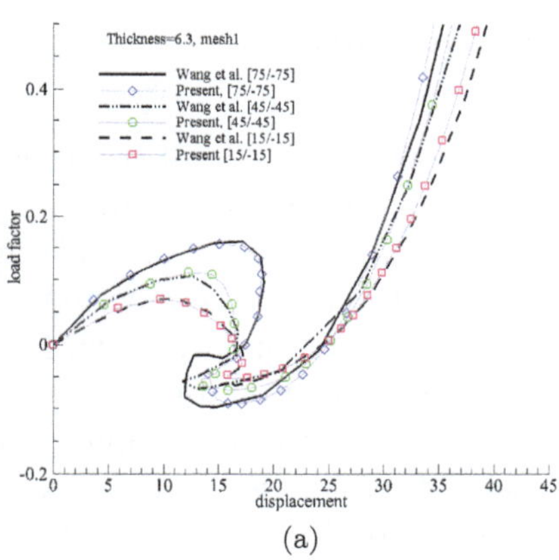

Fig. 11.24. Load–deflection (U3) curves of laminated thin shells with various stacking sequences: (a) two-layer thin shell with thickness $h = 6.3$ and maximum load $P = 0.5 \times 3000$; (b) four-layer thin shell with thickness 12.7 and maximum load $P = 4000$.

[0/90/0] and [90/0/90] laminated shells is observed. Nevertheless, the results obtained by our p-version elements using regular and distorted meshes still show excellent agreement with the reference solutions. Figure 11.23 shows the final deformed configuration of [0/90/0] laminated shell obtained by the regular and distorted elements. Good coincidence is observed again.

In order to test the applicability to arbitrary oblique angle ply laminated composite shells, two- and four-layered laminated shells with equal thickness and various stacking sequences are further analyzed in Fig. 11.24. For the two-layered shells, Fig. 11.24 (a), the material constants are the same as before, while different thickness ($h = 6.3$) and maximum load ($P = 0.5 \times 3000$) are considered in accordance with the reference solutions by Wang [423]. For the four-layered shells, Fig. 11.24(b), the material constants are still the same as mentioned before and the thickness is $h = 12.7$, while the maximum load is $P = 4000$. The results are compared with those given by Payette and Reddy using seven-parameter spectral/*hp* finite element method [219]. Although Payette's method is based on the shear deformable theory, the obtained results are still in excellent agreement with the present results obtained by the thin shell model, which demonstrates the feasibility of KL (or CPT) model in the analysis of multilayered composite thin shells.

11.4.5 *Shell with irregular shape and material discontinuity*

In the last example, we consider the analysis of shells of irregular shape to illustrate the potential of present elements in modeling problems with more complicated domains. Uniform and non-uniform material distribution are also considered in the analysis. The material parameters and problem description are shown in Fig. 11.25. For the uniform case, single-layer isotropic material is used for the entire model. For the non-uniform case, single-layer orthotropic material with ply angle 0° or 90° (the reference direction is along x-axis) is used in region A, while single-layer isotropic material is used

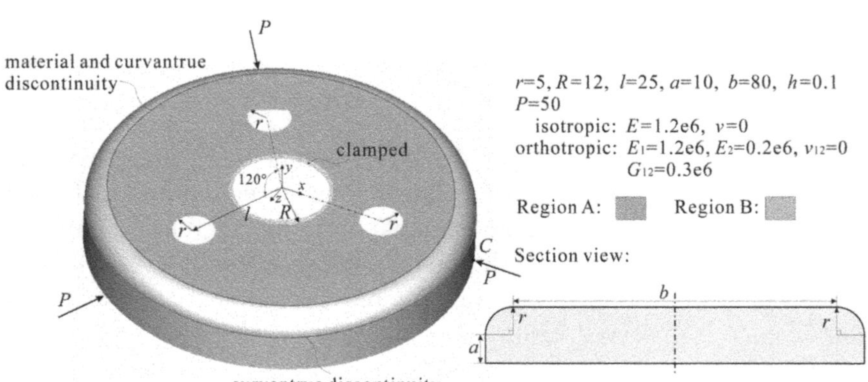

Fig. 11.25. Center clamped shell subjected to symmetric pinching load.

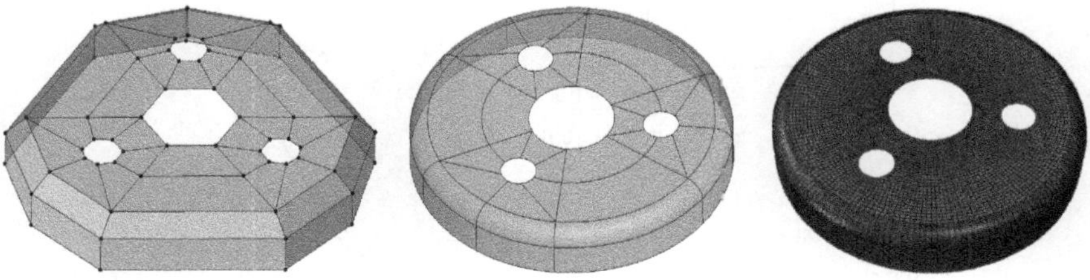

(a) (b) (c)

Fig. 11.26. Mesh patterns: (a) coarse facet mesh before projection; (b) projected high-order curved mesh; (c) ABAQUS S4R mesh.

in region B. The curved mesh with 45 shell elements is shown in Fig. 11.26(b), which is obtained by the QRM projection of the coarse mesh shown in Fig. 11.26(a). Figure 11.26(c) shows the low-order mesh used by ABAQUS S4R elements to obtain the reference solution. In the analysis, the relaxation of second-order normal derivatives is implemented on the interfaces where material property or curvature is discontinuous. It is found that using $p = 9$ with 7600 DOFs in our elements is sufficient to produce converged results for all the material distribution cases, while the reference solution is obtained by 10596 S4R elements with 64998 DOFs.

The load–displacement magnitude curves are shown in Fig. 11.27, where the label of [0] or [90] denotes the non-uniform material case in which the ply angle of orthotropic material in region A is $0°$ or $90°$. One can see good agreement between the present results and the ABAQUS results (note that the S4R elements are C^0 type elements). However, our p-version elements use much less DOFs to give similar results. In Fig. 11.28, contour plots of displacement magnitude obtained by the present elements and S4R elements are displayed. One can see that in all cases the present results are comparable with those obtained by S4R elements. For uniform isotropic material model, the obtained displacement field exhibits good symmetry, and the displacements at three loading points are of the same magnitude. For the non-uniform material model, the displacement magnitude at three loading points are no longer equivalent due to the influence of orthotropic material property. For the case of angle $0°$, the displacement magnitude at point C is slightly larger than the other two loading points, while the opposite situation is observed in the case of angle $90°$. Figure 11.29 shows the contour plots of out-plane

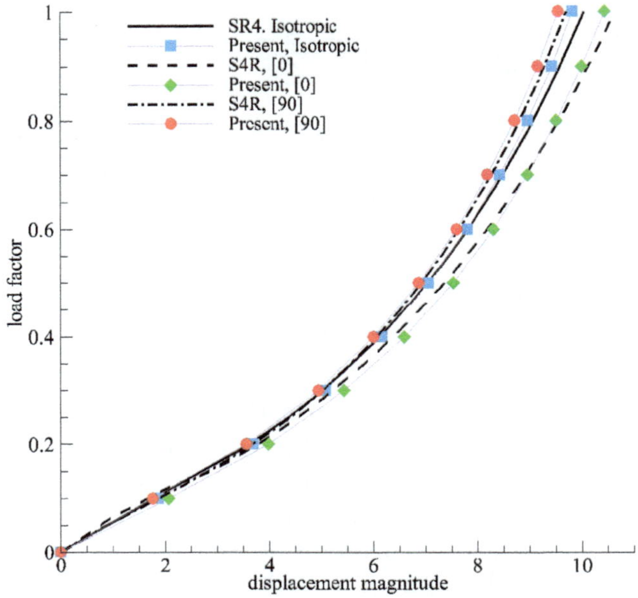

Fig. 11.27. Load–deflection curves at point C for shells with different material distributions.

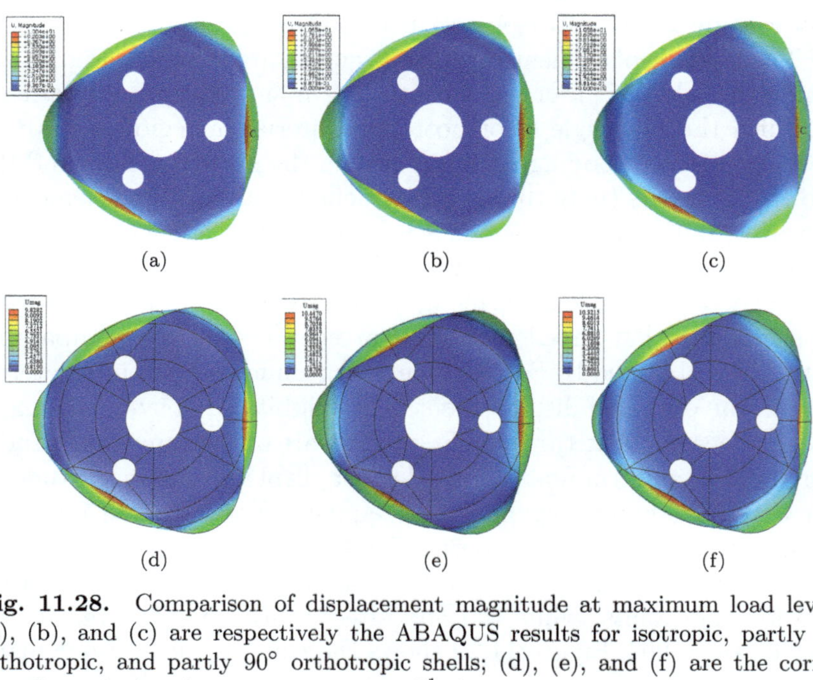

Fig. 11.28. Comparison of displacement magnitude at maximum load level: (a), (b), and (c) are respectively the ABAQUS results for isotropic, partly 0° orthotropic, and partly 90° orthotropic shells; (d), (e), and (f) are the corresponding solutions by present p-version C^1 element with $p = 9$.

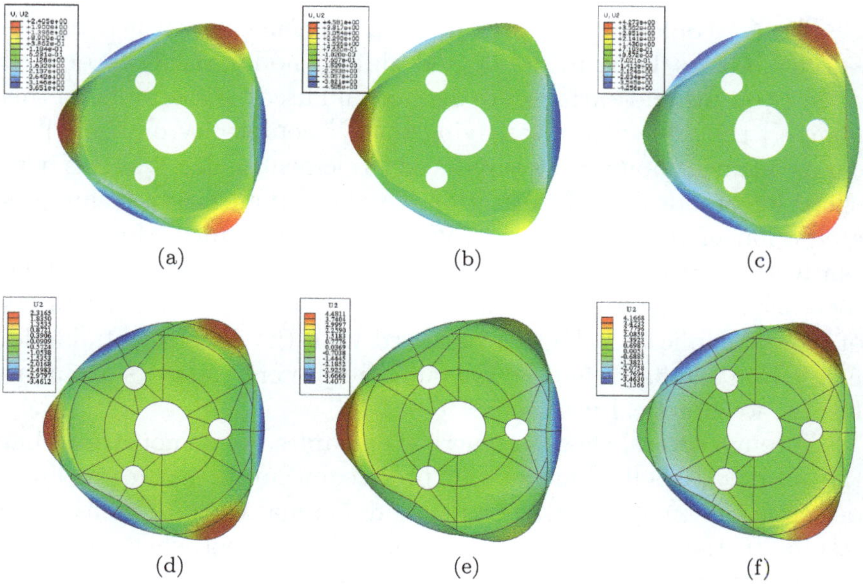

Fig. 11.29. Comparison of displacement U2 at maximum load level: (a), (b), and (c) are respectively the ABAQUS results for isotropic, partly 0° orthotropic, and partly 90° orthotropic shells; (d), (e), and (f) are the corresponding solutions by present p-version C^1 element with $p = 9$.

displacement U2, which is shown to be more sensitive to the material orthotropy. In all cases, the results by our p-version thin shell elements are still in good agreement with the ABAQUS results.

11.5 Conclusions

In this chapter, we propose a curved p-version C^1 finite element formulation for the finite deformation analysis of thin shells with arbitrary curvature and shape. The kinematic model is based on the geometrically exact nonlinear Kirchhoff–Love thin shell model, where a strict normal director field is assumed during the deformation process. The quasi-regional mapping method is employed as a convenient tool to generate the high-order mesh for shells with irregular shape. Modifications based on a two-step projection is implemented to ensure the obtained high-order mesh fits the domain boundaries well.

The C^1 conformity of present shell elements is satisfied on the selected Gauss–Lobatto nodes and shifted nodal variables defined directly on the mid-surfaces are derived and used for assembly. In this manner, the difficulty of satisfying the C^1 conformity on the whole curved element boundaries is reduced to element nodes, which is beneficial for practical implementation. Further enhancement techniques are no longer required to deal with the C^1 conformity. More importantly, the interpolation strategy allows the present elements working together with the quasi-regional mapping method to model problems of more complicated shape. Compared with the C^1 elements implemented on the flat reference domain, the present elements are more flexible for practical use.

Through several chosen numerical examples, we demonstrate that the proposed shell elements can produce accurate results in finite deformation analysis of isotropic and laminated thin shells. The results further provide evidence that the current shell elements are robust even when curved distorted meshes are used. This feature is highly desirable for the p-version C^1 elements because large curved elements are frequently used in domain discretization of p-version finite element method. As expected, the present p-version thin shell elements are able to produce comparable results with those of lower-order h-version elements but uses remarkably less DOFs. Although complete second-order DOFs are used on element vertices, the present elements are still applicable to structures with material discontinuity by relaxing the continuity of second-order normal derivative.

Appendix A.1 Hierarchical Bases of Quadrilateral Elements

Table A.1. Hierarchical bases of quadrilateral elements on $[-1,1]^2$.

Vertex 1 (ξ and η denote ξ_1 and ξ_2, respectively):	Vertex 2:
$H^{\mathrm{V1}} = h_2(\xi)\tilde{h}_1(\eta) + \tilde{h}_1(\xi)h_2(\eta) - h_2(\xi)h_2(\eta)$	$H^{\mathrm{V2}} = h_3(\xi)\tilde{h}_1(\eta) + \tilde{h}_2(\xi)h_2(\eta) - h_3(\xi)h_2(\eta)$
$H^{\mathrm{V1}}_{\xi} = \tilde{h}_1^{(1)}(\xi)h_2(\eta)$ $H^{\mathrm{V1}}_{\eta} = h_2(\xi)\tilde{h}_1^{(1)}(\eta)$	$H^{\mathrm{V2}}_{\xi} = \tilde{h}_2^{(1)}(\xi)h_2(\eta)$ $H^{\mathrm{V2}}_{\eta} = h_3(\xi)\tilde{h}_1^{(1)}(\eta)$
$H^{\mathrm{V1}}_{\xi\xi} = \tilde{h}_1^{(2)}(\xi)h_2(\eta)$ $H^{\mathrm{V1}}_{\eta\eta} = h_2(\xi)\tilde{h}_1^{(2)}(\eta)$	$H^{\mathrm{V2}}_{\xi\xi} = \tilde{h}_2^{(2)}(\xi)h_2(\eta)$ $H^{\mathrm{V2}}_{\eta\eta} = h_3(\xi)\tilde{h}_1^{(2)}(\eta)$
$H^{\mathrm{V1}}_{\xi\eta} = h_1(\xi)h_1(\eta)$	$H^{\mathrm{V2}}_{\xi\eta} = h_4(\xi)h_1(\eta)$
Vertex 3:	Vertex 4:
$H^{\mathrm{V3}} = h_3(\xi)\tilde{h}_2(\eta) + \tilde{h}_2(\xi)h_3(\eta) - h_3(\xi)h_3(\eta)$	$H^{\mathrm{V4}} = h_2(\xi)\tilde{h}_2(\eta) + \tilde{h}_1(\xi)h_3(\eta) - h_2(\xi)h_3(\eta)$
$H^{\mathrm{V3}}_{\xi} = \tilde{h}_2^{(1)}(\xi)h_3(\eta)$ $H^{\mathrm{V3}}_{\eta} = h_3(\xi)\tilde{h}_2^{(1)}(\eta)$	$H^{\mathrm{V4}}_{\xi} = \tilde{h}_1^{(1)}(\xi)h_3(\eta)$ $H^{\mathrm{V4}}_{\eta} = h_2(\xi)\tilde{h}_2^{(1)}(\eta)$
$H^{\mathrm{V3}}_{\xi\xi} = \tilde{h}_2^{(2)}(\xi)h_3(\eta)$ $H^{\mathrm{V3}}_{\eta\eta} = h_3(\xi)\tilde{h}_2^{(2)}(\eta)$	$H^{\mathrm{V4}}_{\xi\xi} = \tilde{h}_1^{(2)}(\xi)h_3(\eta)$ $H^{\mathrm{V4}}_{\eta\eta} = h_2(\xi)\tilde{h}_2^{(2)}(\eta)$
$H^{\mathrm{V3}}_{\xi\eta} = h_4(\xi)h_4(\eta)$	$H^{\mathrm{V4}}_{\xi\eta} = h_1(\xi)h_4(\eta)$

(Continued)

Table A.1. (*Continued*)

Edge 1 ($J_i^{(\alpha,\beta)}(\cdot)$ are Jacobi orthogonal polynomials):	Edge 2:
$H_i^{\mathrm{E1}} = (1+\xi)^3(1-\xi)^3 J_i^{(4,4)}(\xi)h_2(\eta)$	$H_i^{\mathrm{E2}} = (1-\eta)^3(1+\eta)^3 J_i^{(4,4)}(\eta)h_3(\xi)$
$H_{\mathrm{d},i}^{\mathrm{E1}} = (1+\xi)^2(1-\xi)^2 J_i^{(4,4)}(\xi)h_1(\eta);\quad i = 0,1,2,\ldots$	$H_{\mathrm{d},i}^{\mathrm{E2}} = (1-\eta)^2(1+\eta)^2 J_i^{(4,4)}(\eta)h_4(\xi);\quad i = 0,1,2,\ldots$
Edge 3:	Edge 4:
$H_i^{\mathrm{E3}} = (1-\xi)^3(1+\xi)^3 J_i^{(4,4)}(\xi)h_3(\eta)$	$H_i^{\mathrm{E4}} = (1-\eta)^3(1+\eta)^3 J_i^{(4,4)}(\eta)h_2(\xi)$
$H_{\mathrm{d},i}^{\mathrm{E3}} = (1-\xi)^2(1+\xi)^2 J_i^{(4,4)}(\xi)h_4(\eta);\quad i = 0,1,2,\ldots$	$H_{\mathrm{d},i}^{\mathrm{E4}} = (1-\eta)^2(1+\eta)^2 J_i^{(4,4)}(\eta)h_1(\xi);\quad i = 0,1,2,\ldots$

Face functions:

$$H_{mn}^{\mathrm{F}} = C_{mn}(1-\xi^2)^2(1-\eta^2)^2 J_{m-1}^{(4,4)}(\xi) J_{n-1}^{(4,4)}(\eta);\ m,n = 1,2,\ldots$$

Definition of 1D Hermite bases:

$h_1(\xi) = (\xi+1)(\xi-1)^2/4$	$\tilde{h}_1(\xi) = (1-\xi)^3(3\xi^2+9\xi+8)/16$	$\tilde{h}_2(\xi) = (1+\xi)^3(3\xi^2-9\xi+8)/16$
$h_2(\xi) = (\xi+2)(\xi-1)^2/4$	$\tilde{h}_1^{(1)}(\xi) = (1-\xi)^3(1+\xi)(5+3\xi)/16$	$\tilde{h}_2^{(1)}(\xi) = (1+\xi)^3(\xi-1)(5-3\xi)/16$
$h_3(\xi) = (2-\xi)(\xi+1)^2/4$	$\tilde{h}_1^{(2)}(\xi) = (1-\xi)^3(1+\xi)^2/16$	$\tilde{h}_2^{(2)}(\xi) = (1+\xi)^3(1-\xi)^2/16$
$h_4(\xi) = (\xi-1)(\xi+1)^2/4$		

Appendix A.2 The Interpolation Points

Table A.2.* Interpolation nodes on $[-1, 1]$

Node number n	Gauss–Lobatto nodes	Chen–Babuška nodes
2	1.000000000000000	1.000000000000000
3	0.000000000000000, 1.000000000000000	0.000000000000000, 1.000000000000000
4	0.447213595499958, 1.000000000000000	0.417791301355990, 1.000000000000000
5	0.000000000000000, 0.654653670707977	0.000000000000000, 0.620911304689912
	1.000000000000000	1.000000000000000
6	0.285231516480645, 0.765055323929465	0.268907044771973, 0.734126667189175
	1.000000000000000	1.000000000000000
7	0.000000000000000, 0.468848793470714	0.000000000000000, 0.567430602747253
	0.830223896278567, 1.000000000000000	0.803440238269107, 1.000000000000000
8	0.209299217902479, 0.591700181433142	0.199287729905666, 0.567430602747253
	0.871740148509607, 1.000000000000000	0.848871961036656, 1.000000000000000
9	0.000000000000000, 0.363117463826178	0.000000000000000, 0.347787971611667
	0.677186279510738, 0.899757995411460	0.653533479079903, 0.880230852718454
	1.000000000000000	1.000000000000000
10	0.165278957666387, 0.477924949810444	0.158565288657640, 0.460149825922899
	0.738773865105505, 0.919533908166459	0.716613860625308, 0.902770975291773
	1.000000000000000	1.000000000000000

(*Continued*)

Table A.2.*. (*Continued*)

Node number n	Gauss–Lobatto nodes	Chen–Babuska nodes
11	0.000000000000000, 0.295758135586939	0.000000000000000, 0.284888001066926
	0.565235326996205, 0.784483473663144	0.546667696174604, 0.764098454567145
	0.934001430408059, 1.000000000000000	0.919508751794299, 1.000000000000000
12	0.136552932854928, 0.399530940965349	0.131751840053756, 0.386268452294038
	0.632876153031861, 0.819279321644007	0.614435542614339, 0.800682266235608
	0.944899272222882, 1.000000000000000	0.932274783022918, 1.000000000000000
13	0.000000000000000, 0.249286930106240	0.000000000000000, 0.241223569292276
	0.482909821091336, 0.686188469081757	0.468417505900827, 0.668366619463316
	0.846347564651872, 0.953309846642164	0.829435479966906, 0.942231627955178
	1.000000000000000	1.000000000000000
14	0.116331868883704, 0.342724013342713	0.112732706528405, 0.332541822894725
	0.550639402928647, 0.728868599091326	0.535665483103728, 0.711910314047619
	0.867801053830347, 0.959935045267261	0.852427589917411, 0.950146060815101
	1.000000000000000	1.000000000000000
15	0.000000000000000, 0.215353955363794	0.000000000000000, 0.209151011805735
	0.420638054713672, 0.606253205469846	0.409156537764197, 0.591270545747718
	0.763519689951815, 0.885082044222976	0.747528116752139, 0.871091606365657
	0.965245926503839, 1.000000000000000	0.956540263333238, 1.000000000000000

Notes: *Only the positive half is presented, while the negative half can be obtained by symmetry.

Chapter 12

Incorporation of the Hierarchical Quadrature Element Method with Isogeometric Analysis

The concept of isogeometric analysis (IGA) was proposed by Hughes [51,52], who attempted to directly use the CAD geometry and entirely eliminate the finite element polynomial description. A coarse mesh of "non-uniform rational B-splines (NURBS) elements" is used to match the exact CAD geometry. This bridges the gap between geometry and analysis. The solution space for dependent variables is represented in terms of the same functions that represent the geometry. For purposes of analysis, the basis is refined and/or its degree elevated without changing the geometry or its parameterization. This process does not require any further communication with the CAD system, and therefore, IGA is expected to facilitate more widespread adoption in the industry. However, the NURBS are tensor product bases. For triangular and tetrahedral elements, there are singularities on the degenerated edges or faces. IGA for trimmed NURBS geometry needs to use T-spline, etc. The tensor product NURBS bases are not convenient for local adaptive analyses. Due to the boundary representation (B-Rep) nature of CAD geometry, the application of IGA to 3D solids is not so straightforward. We know that the isogeometric concept comes from the isoparametric concept. Application of the isoparametric concept to HQEM makes it possible to apply some techniques of IGA to HQEM. In the generalized isogeometric analysis by HQEM, we can transform exact CAD geometry into high-accuracy DQH elements. The DQH elements can be sub-parametric or super-parametric. Both geometry and analyses can be carried out by using the DQH bases. A whole NURBS element can be modeled by one DQH element. The accuracy of the DQH method is improved through order elevation. These properties are similar to IGA, but the generalized isogeometric analysis by HQEM can also be applied to polar singular meshes, 3D solids, local adaptive analyses, etc. These are areas where the IGA have difficulties. Therefore, the incorporation of HQEM with IGA is necessary and meaningful.

12.1 B-Splines and NURBS

NURBS surfaces are constructed through tensor product of one-dimensional NURBS bases (Fig. 12.1). For convenience of surface intersection, NURBS surfaces are usually first transformed into Bézier surfaces (Fig. 12.2). NURBS surfaces may also need to be transformed into triangular and quadrilateral elements according to parametric domains directly (Figs. 12.3 and 12.4). For completeness, a brief introduction of Bézier and NURBS curves and surfaces [220, 424] will be presented in the following.

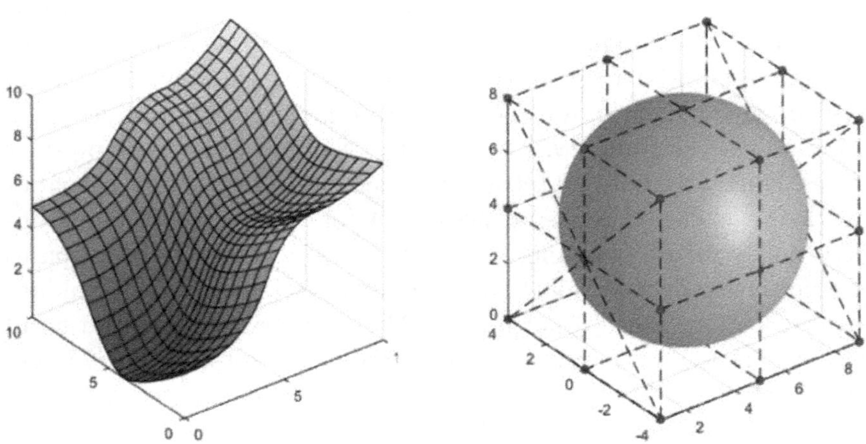

Fig. 12.1. A quadric free surface (left) and a sphere surface and its control points (right).

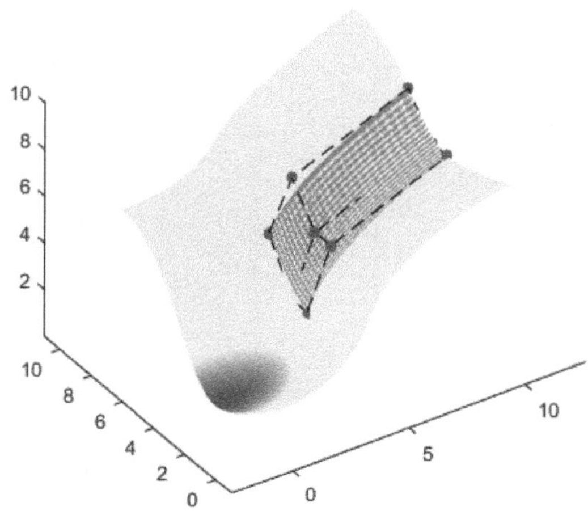

Fig. 12.2. A NURBS surface transformed into Bézier surfaces and one of the Bézier patches and its control points.

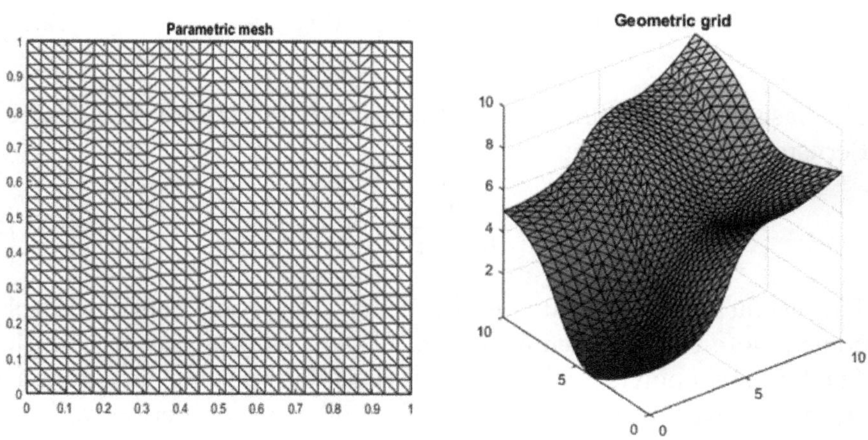

Fig. 12.3. Triangular mesh of a NURBS surface (left: parametric mesh; right: geometric grid).

The Bézier method has the following advantages: (1) It is natural for interactive shape design; the coefficients convey geometric insight about the shape of the curve. A designer can specify end conditions at both ends of the curve. (2) Algorithms for processing Bézier basis have a geometric flavor. (3) Numerically, it is usually well conditioned. Therefore, it is widely used in engineering. An nth-degree

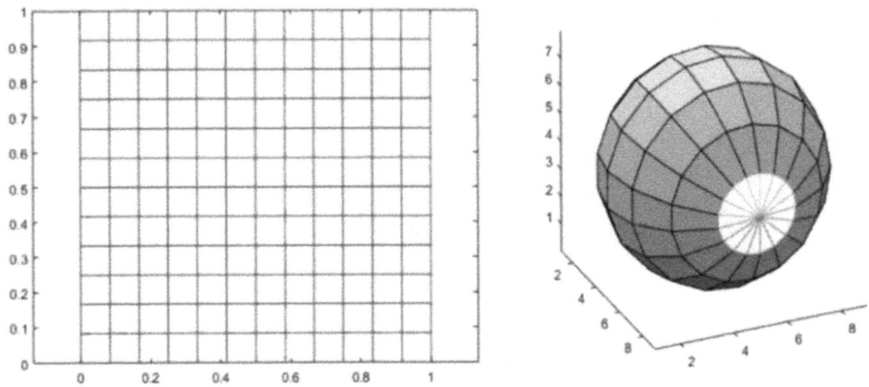

Fig. 12.4. Quadrilateral mesh of a NURBS surface (left: parametric mesh; right: geometric grid).

Bézier curve is defined by

$$\boldsymbol{C}(u) = \sum_{i=0}^{n} B_{i,n}(u)\boldsymbol{P}_i, \quad 0 \le u \le 1 \tag{12.1}$$

The basis (blending) functions, $\{B_{i,n}(u)\}$, are the classical nth-degree Bernstein polynomials given by

$$B_{i,n}(u) = \frac{n!}{i!(n-i)!}u^i(1-u)^{n-i} \tag{12.2}$$

The geometric coefficients of this form, $\{\boldsymbol{P}_i\}$, are called control points. Note that the definition, Eq. (12.1), requires that $u \in [0, 1]$. The Bernstein polynomials have many useful properties, such as non-negativity, partition of unity, etc. The following properties will be used in this work:

(1) *Linear precision*:

$$u = \sum_{i=0}^{n} \frac{i}{n} B_{i,n}(u) \tag{12.3}$$

which implies that the monomial u can be expressed as the weighted sum of Bernstein polynomials of degree n with coefficients evenly spaced in the interval [0, 1].

(2) *Derivatives*:

$$B'_{i,n}(u) = \frac{\mathrm{d}B_{i,n}(u)}{\mathrm{d}u} = n[B_{i-1,n-1}(u) - B_{i,n-1}(u)] \tag{12.4}$$

with $B_{-1,n-1}(u) = B_{n,n-1}(u) \equiv 0$.

(3) *de Casteljau algorithm*: Fixing $u = u_0$ and denoting $\boldsymbol{P}_i$ by $\boldsymbol{P}_{0,i}$, a recursive algorithm for computing the point $\boldsymbol{C}(u_0) = \boldsymbol{P}_{n,0}(u_0)$ on an nth-degree Bezier curve can be given by

$$\boldsymbol{P}_{k,i}(u_0) = (1 - u_0)\boldsymbol{P}_{k-1,i}(u_0) + u_0\boldsymbol{P}_{k-1,i+1}(u_0) \tag{12.5}$$

where $k = 1, \ldots, n$; $i = 0, \ldots, n - k$.

These properties will be used for surface trimming in the following part of this section. The Bézier curves are a special case of the B-spline curves that only have knots on two ends.

The B-spline is defined on knot vectors. A knot vector in one dimension is a non-decreasing set of coordinates in the parameter space, written as $\boldsymbol{E} = \{\xi_0, \xi_1, \ldots, \xi_{m+p+1}\}$, where $\xi_i \in \mathbb{R}$ is the ith knot, $i = 1, 2, \ldots, m + p + 1$ is the knot index, p is the polynomial order, and m is the number of basis functions used to construct the B-spline curve. With a knot vector in hand, the B-spline basis functions are defined recursively starting with piecewise constants ($p = 0$),

$$N_{i,0}(\xi) = \begin{cases} 1 & \text{if } \xi_i \leq \xi \leq \xi_{i+1} \\ 0 & \text{otherwise} \end{cases} \tag{12.6}$$

For $p = 1, 2, 3, \ldots$, they are defined by

$$N_{i,p}(\xi) = \frac{\xi - \xi_i}{\xi_{i+p} - \xi_i} N_{i,p-1}(\xi) + \frac{\xi_{i+p+1} - \xi}{\xi_{i+p+1} - \xi_{i+1}} N_{i+1,p-1}(\xi) \tag{12.7}$$

Note that Eq. (12.7) can yield the quotient 0/0, and we define this quotient to be zero. Figure 12.5 illustrates the B-spline basis for $p = 1, 2, 3$ with seven equally spaced sampling points in [0, 1]. The kth derivative of $N_{i,p}(\xi)$ in terms of the functions

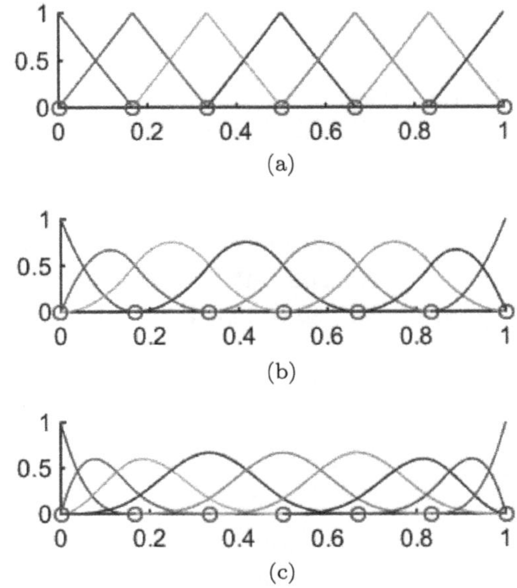

Fig. 12.5. B-spline basis with equally spaced sampling points: (a) $p = 1$; (b) $p = 2$; (c) $p = 3$.

$N_{i,p-k}(\xi), \ldots, N_{i+k,p-k}(\xi)$ is

$$N_{i,p}^{(k)}(\xi) = \frac{\mathrm{d}^k N_{i,p}(\xi)}{\mathrm{d}\xi^k} = \frac{p!}{(p-k)!} \sum_{j=0}^{k} \alpha_{k,j} N_{i+j,p-k}(\xi) \tag{12.8}$$

with

$$\begin{aligned} \alpha_{0,0} &= 1 \\ \alpha_{k,0} &= \frac{\alpha_{k-1,0}}{\xi_{i+p-k+1} - \xi_i} \\ \alpha_{k,j} &= \frac{\alpha_{k-1,j} - \alpha_{k-1,j-1}}{\xi_{i+p+j-k+1} - \xi_{i+j}} \qquad j = 1, \ldots, k-1 \\ \alpha_{k,k} &= \frac{-\alpha_{k-1,k-1}}{\xi_{i+p+1} - \xi_{i+k}} \end{aligned} \tag{12.9}$$

Note that (1) k should not exceed p (all higher derivatives are zero) and (2) the denominators involving knot differences can become zero; the quotient is defined to be zero in this case. A knot vector is said to be open if its first and last knot values appear $p+1$ times. Open knot vectors are the standard in CAD literatures. B-splines formed from open knot vectors are interpolator at the ends of the parameter space interval and have C^0 continuity therein and $p-1$ continuous derivatives elsewhere. The NURBS book [425] has provided efficient algorithms for B-splines.

The B-splines can be regarded as a special case of the NURBS. A pth-degree B-spline curve is defined by

$$\boldsymbol{C}(\xi) = \sum_{i=0}^{m} N_{i,p}(\xi)\boldsymbol{P}_i \tag{12.10}$$

where the $\{\boldsymbol{P}_i\}$ are the control points and the $\{N_{i,p}(\xi)\}$ are the pth-degree B-spline basis functions (Eq. (12.7)) defined on the non-periodic (and non-uniform) knot vector

$$\boldsymbol{U} = \{\underbrace{a,\ldots,a}_{p+1}, \xi_{p+1},\ldots,\xi_{r-p-1}, \underbrace{b,\ldots,b}_{p+1}\} \tag{12.11}$$

($n+1$ knots). The integer $r = m+p+1$. Unless stated otherwise, it is usually assumed that $a = 0$ and $b = 1$. A NURBS curve defined on the knot vector as in Eq. (12.11) can be given as

$$\boldsymbol{C}(\xi) = \frac{\sum_{i=0}^{n} N_{i,p}(\xi) w_i \boldsymbol{P}_i}{\sum_{i=0}^{n} N_{i,p}(\xi) w_i} \qquad a \le \xi \le b \tag{12.12}$$

where $\{w_i\}$ are the weights. It is clear that Eq. (12.12) becomes a B-spline curve due to the partition of unity property of B-splines if $w_i = c$ for any $c \neq 0$. Therefore, when NURBS are referred to, B-splines are included.

A NURBS surface of degree p in the ξ-direction and degree q in the η-direction is a bivariate vector-valued piecewise rational function of the form

$$\boldsymbol{S}(\xi,\eta) = \sum_{i=0}^{m}\sum_{j=0}^{n} R_{i,j}(\xi,\eta)\boldsymbol{P}_{i,j} \quad 0 \le \xi,\eta \le 1 \tag{12.13}$$

where the piecewise rational basis functions

$$R_{i,j}(\xi,\eta) = \frac{N_{i,p}(\xi)N_{j,q}(\eta)w_{i,j}}{\sum_{k=0}^{m}\sum_{l=0}^{n} N_{k,p}(\xi)N_{l,q}(\eta)w_{k,l}} \tag{12.14}$$

$\{\boldsymbol{P}_{i,j}\}$ form a bidirectional control net, $\{w_{i,j}\}$ are the weights, and $\{N_{i,p}(\xi)\}$ and $\{N_{j,q}(\eta)\}$ are the non-rational B-spline basis functions defined on the knot vectors

$$\begin{aligned}\boldsymbol{U} &= \{\underbrace{0,\ldots,0}_{p+1},\xi_{p+1},\ldots,\xi_{r-p-1},\underbrace{1,\ldots,1}_{p+1}\}\\ \boldsymbol{V} &= \{\underbrace{0,\ldots,0}_{q+1},\xi_{q+1},\ldots,\xi_{s-q-1},\underbrace{1,\ldots,1}_{q+1}\}\end{aligned} \tag{12.15}$$

where $r = m + p + 1$ and $s = n + q + 1$. It is clear the Eq. (12.13) becomes a B-spline surface if $w_{i,j} = c$ for any $c \neq 0$. The NURBS can exactly describe conic sections and circles that play a fundamental role in CAD and CAM applications [52, 426].

12.2 Non-uniform Rational Lagrange Functions

NURBS is a standard technology of computer-aided design (CAD). It was also recently used for analysis which led to the famous concept of isogeometric analysis (IGA) to accomplish seamless compatibility between computer-aided design and analysis [51, 52]. However, the NURBS are not interpolation functions, so the imposition of non-homogeneous boundary conditions are not as straightforward as the standard finite element method that uses Lagrange function as basis [427, 428]. This section aims to equivalently transform NURBS geometries into NURL geometries to overcome such difficulties. The B-splines are defined on non-decreasing knot vectors, such as $\boldsymbol{E} = \{\xi_0, \xi_1, \ldots, \xi_m\}$, where $\xi_i \leq \xi_{i+1}$, $i = 0, 1, \ldots, m-1$ [220]. One can see from Eq. (12.7) and Fig. 12.6 that the NURBS curves are piecewise curves. The knots satisfying $\xi_i < \xi_{i+1}$ are known as breakpoints [220]. The points on NURBS curves that correspond to breakpoints are known as Bézier points (Fig. 12.6), e.g., the breakpoint vector for a knot vector $\{0, 0, 0, 1/4, 1/2, 3/4, 3/4, 1, 1, 1\}$ is $\{0, 1/4, 1/2, 3/4, 1\}$. The computation of NURBS needs to determine the knot interval where ξ lies, i.e., to find the interval $[\xi_i, \xi_{i+1})$ that ξ belongs to.

Since the NURBS curves are defined by piecewise polynomials, they can be equivalently transformed to other types of polynomial curves, e.g., the rational Lagrange curves. For a pth-degree NURBS

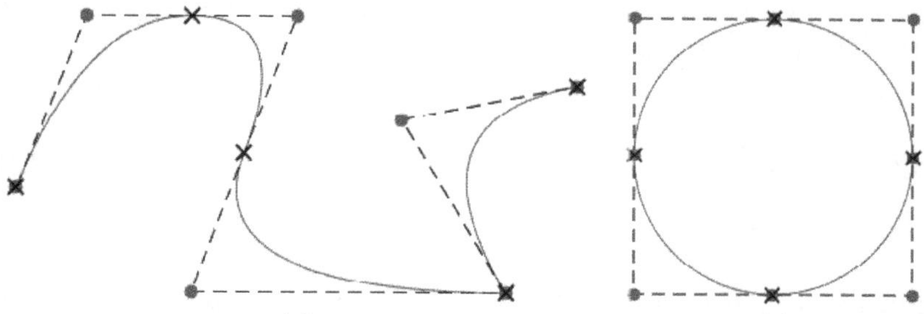

Fig. 12.6. NURBS curves: (a) $\boldsymbol{U} = \{0, 0, 0, 1/4, 1/2, 3/4, 3/4, 1, 1, 1\}$; (b) $\boldsymbol{U} = \{0, 0, 0, 1/4, 1/4, 1/2, 1/2, 3/4, 3/4, 1, 1, 1\}$. The sign $\times$ denotes the Bézier points and $\bullet$ denotes the control points.

curve, each non-empty knot interval of NURBS should have $p+1$ nodes in the equivalent rational Lagrange form, i.e., $p-1$ knots should be inserted into each non-empty knot interval of NURBS in the equivalent rational Lagrange form. This leads to a C^0 basis that is able to represent NURBS as [429]. In order to overcome the shortcomings of global interpolation, one can insert more than $p-1$ knots into each non-empty knot interval of NURBS to obtain the equivalent rational Lagrange form. Due to the uniqueness of polynomials, the pth-degree NURBS curve section at each non-empty knot interval can be exactly interpolated by arbitrary $p+1$ nodes within it. The inserted node intervals are called as 'span' herein in order to be distinguished from the intervals of NURBS. Thus, similar to the finite difference method or the definition of the T-spline [323, 430], local Lagrange interpolation can be defined on interpolation vector $\boldsymbol{E}_k = \{\zeta_0, \zeta_1, \ldots, \zeta_m\}$ for each non-empty knot interval of NURBS, namely, $[\zeta_0, \zeta_m) = [\xi_k, \xi_{k+1}) \in \boldsymbol{E}$, as

$$\boldsymbol{C}(\zeta) = \sum_{i=r}^{r+p} L_{i,p}^r(\zeta)\boldsymbol{P}_i, \quad \zeta_{c_r} \le \zeta \le \zeta_{c_{r+1}}, \quad 0 \le r \le m-p \quad (12.16)$$

where p is the degree of the Lagrange function and r is the starting index of the Lagrange function. The indexes c_r $(r = 0, 1, \ldots, m-p)$ are integers that take values within $[0, m]$ to reach the best interpolation accuracy. In this work, they are taken as

$$c_0 = 0, \quad c_r = r + [p/2] \ (0 < r < m-p), \quad c_{m-p} = m \quad (12.17)$$

where $[\cdot]$ means rounding down decimal places. The Lagrange basis in Eq. (12.16) is defined as

$$L_{i,p}^{r}(\zeta)=\prod_{\substack{j=r\\ j\neq i}}^{r+p}\frac{\zeta-\zeta_j}{\zeta_i-\zeta_j},\quad \zeta_{c_r}\leq\zeta\leq\zeta_{c_{r+1}},\quad 0\leq r\leq m-p \tag{12.18}$$

For a fixed r, Eq. (12.16) becomes a Lagrange interpolation using $p+1$ interpolation points. Figure 12.7 shows how the NURL basis are defined. For the 10 interpolation nodes shown in Fig. 12.7, $r=0$ for points within $[\zeta_0,\ \zeta_2)$; $r=1$ for points within $[\zeta_2,\ \zeta_3)$, and $r=6$ for points within $[\zeta_7,\ \zeta_9)$. Equation (12.16) can be written in the following form for the example in Fig. 12.7:

$$\boldsymbol{C}(\zeta)=\begin{cases}\sum_{i=0}^{3}L_{i,3}^{0}(\zeta)\boldsymbol{P}_i, & \zeta_0\leq\zeta\leq\zeta_2\\ \sum_{i=r}^{r+3}L_{i,3}^{r}(\zeta)\boldsymbol{P}_i, & \zeta_{r+1}\leq\zeta\leq\zeta_{r+2},\ 1\leq r\leq 5\\ \sum_{i=6}^{9}L_{i,3}^{6}(\zeta)\boldsymbol{P}_i, & \zeta_7\leq\zeta\leq\zeta_9\end{cases} \tag{12.19}$$

One can see that the NURL presented in this work differs from the Lagrange extraction and projection for NURBS basis functions [429] only by using Eq. (12.17). This simple change does not increase the computational cost and complexity while it makes the NURL spaces nested, and therefore makes k-refinement possible for the NURL, which has been shown to be more advantageous than the common h- and p-refinement [51, 52]. One can see that the definition of the NURL has similarity to the T-spline [323, 430]. The appended subroutine "spanweak" outputs each interpolation node vector and the corresponding node indexes of discrete values of ξ used for evaluation. One can see that it is very simple since it is a direct usage of Eq. (12.17).

In order to clearly show how the NURL basis functions are nested as defined, Eq. (12.16) is expressed by a single index through a C^0 connection of the local Lagrange basis by sharing the same coefficient

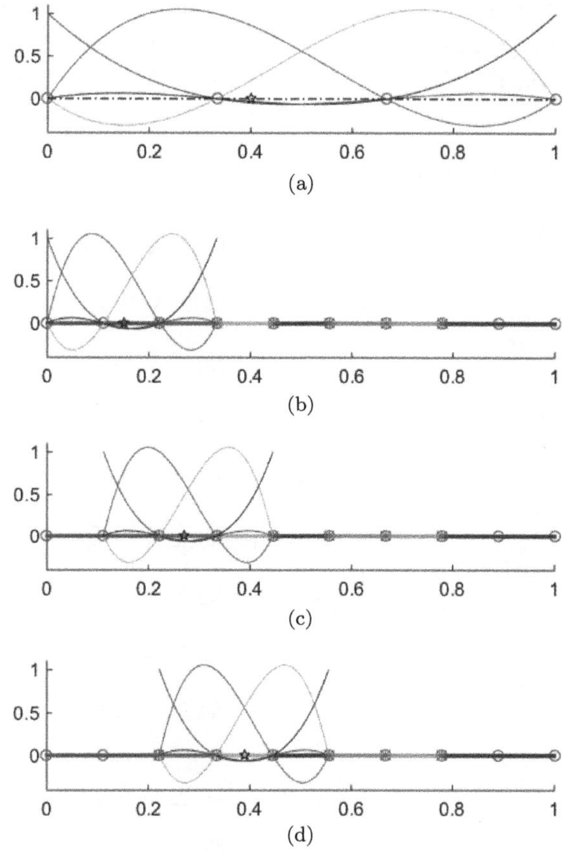

Fig. 12.7. The local basis of NURL with $p = 3$: (a) the number of interpolation nodes equals to $p + 1$; (b–d) the number of interpolation nodes is greater than $p + 1$. The sign "o" denotes the interpolation nodes, the sign * breaks the index c_r, and the sign $\star$ denotes an arbitrary evaluation point.

$\boldsymbol{P}_i$ on different spans as

$$\boldsymbol{C}(\zeta) = \sum_{i=0}^{m} L_{i,p}(\zeta)\boldsymbol{P}_i \tag{12.20}$$

where

$$L_{i,p}(\zeta) = \begin{cases} L_{i,p}^{i-p}(\zeta), & \zeta_{c_{i-p}} \le \zeta \le \zeta_{c_{i-p+1}} \\ \cdots, & \cdots \\ L_{i,p}^{i-1}(\zeta), & \zeta_{c_{i-1}} \le \zeta \le \zeta_{c_i} \\ L_{i,p}^{i}(\zeta), & \zeta_{c_i} \le \zeta \le \zeta_{c_{i+1}} \end{cases} \tag{12.21}$$

where $L_{i,p}^{i-l}(\zeta) = 0$ if $i - l < 0$ or $i - l > m - p$. The local Lagrange interpolation becomes the finite difference method if ζ is an interpolation node, i.e., if $\zeta \in \boldsymbol{E}_k$. The single index basis for the example in Fig. 12.7 is

$$
\begin{aligned}
&L_{0,3}(\zeta) = L_{0,3}^{0}(\zeta), \zeta_0 \le \zeta \le \zeta_2 \\
&L_{1,3}(\zeta) = \begin{cases} L_{1,3}^{0}(\zeta), & \zeta_0 \le \zeta \le \zeta_2 \\ L_{1,3}^{1}(\zeta), & \zeta_2 \le \zeta \le \zeta_3 \end{cases} \\
&L_{2,3}(\zeta) = \begin{cases} L_{2,3}^{0}(\zeta), & \zeta_0 \le \zeta \le \zeta_2 \\ L_{2,3}^{1}(\zeta), & \zeta_2 \le \zeta \le \zeta_3 \\ L_{2,3}^{2}(\zeta), & \zeta_3 \le \zeta \le \zeta_4 \end{cases} \\
&L_{3,3}(\zeta) = \begin{cases} L_{3,3}^{0}(\zeta), & \zeta_0 \le \zeta \le \zeta_2 \\ L_{3,3}^{1}(\zeta), & \zeta_2 \le \zeta \le \zeta_3 \\ L_{3,3}^{2}(\zeta), & \zeta_3 \le \zeta \le \zeta_4 \\ L_{3,3}^{3}(\zeta), & \zeta_4 \le \zeta \le \zeta_5 \end{cases} \\
&\cdots \\
&L_{8,3}(\zeta) = \begin{cases} L_{8,3}^{5}(\zeta), & \zeta_6 \le \zeta \le \zeta_7 \\ L_{8,3}^{6}(\zeta), & \zeta_7 \le \zeta \le \zeta_9 \end{cases} \\
&L_{9,3}(\zeta) = L_{9,3}^{6}(\zeta), \quad \zeta_7 \le \zeta \le \zeta_9
\end{aligned} \tag{12.22}
$$

It deserves to be noted that Eqs. (12.20)–(12.22) are only useful for understanding the NURL basis. The usage of the subroutine "spanweak" (http://nurl.sourceforge.net) coded according to Eq. (12.17) is much simpler than the definition given by Eqs. (12.20)–(12.22).

Equation (12.19) shows that the basis of Eq. (12.22) are composed of Lagrange interpolation function in each span, so the basis $L_{i,p}(\zeta)$ of Eq. (12.22) possesses all the properties of the Lagrange basis except the C^0 continuous and local interpolation features. Figure 12.8 plots the NURL basis with degree $p = 1, 2, 3$ and the NURBS basis with $p = 2$. One can see that the NURBS basis with degree $p \ge 2$ are not interpolators. One can see from Figs. 12.8(a)–(c) that the order

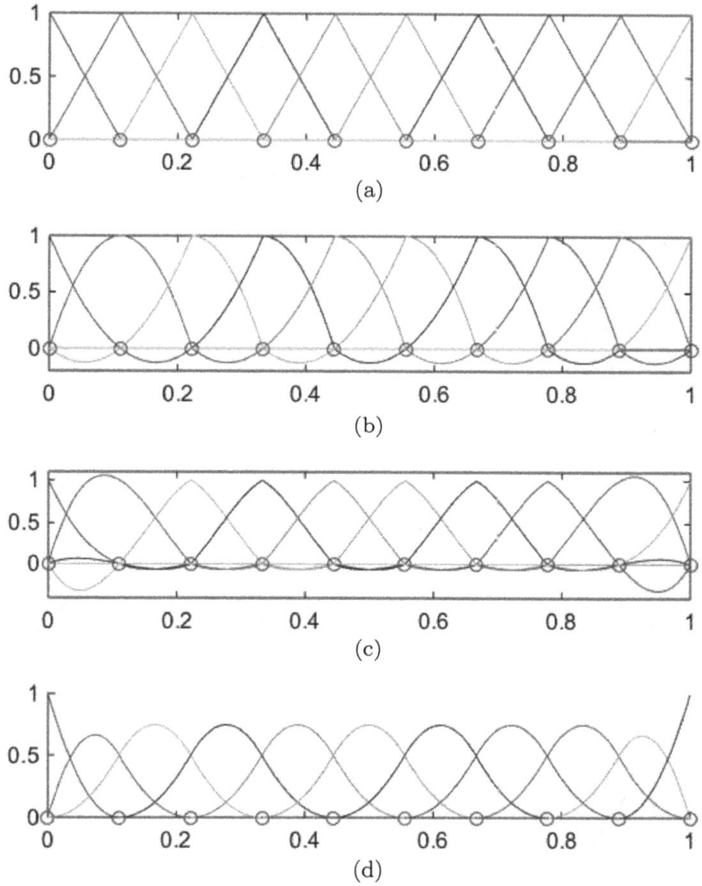

Fig. 12.8. NURL basis with constant weights: (a) $p = 1$; (b) $p = 2$; (c) $p = 3$. NURBS basis: (d) $p = 2$. "○" denotes the interpolation nodes (or knots for NURBS). The order elevation from (a) to (c) is nested like the k-refinement process of IGA using NURBS.

elevation of the NURL is nested like the k-refinement process of IGA using NURBS [51, 52]. Figures 12.9(a) and 12.9(b) respectively plot the first, fifth and ninth NURL basis for $p = 2$ and $p = 3$. One can see that a global NURL basis is a combination of the local Lagrange basis, as shown in Figs. 12.7(b)–12.7(d). One can see from Eq. (12.16) that if $m = p$, one has $r = 0$ and Eq. (12.16) becomes the Lagrange interpolation of a curve, i.e., the Lagrange extraction and projection for NURBS basis functions [429] are obtained. Similar to the NURBS,

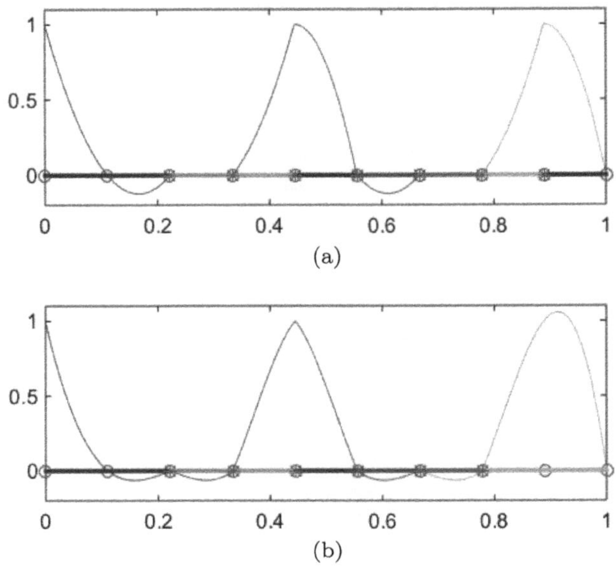

Fig. 12.9. The first, fifth and ninth NURL basis with constant weights for $p = 2$ (a) and $p = 3$ (b). "○" denotes the interpolation nodes and * breaks the index c_r.

a NURL curve can be given as

$$\boldsymbol{C}(\zeta) = \frac{\sum_{i=r}^{r+p} L_{i,p}^{r}(\zeta) w_i \boldsymbol{P}_i}{\sum_{i=r}^{r+p} L_{i,p}^{r}(\zeta) w_i}, \quad \zeta_{c_r} \leq \zeta \leq \zeta_{c_{r+1}}, \quad 0 \leq r \leq m - p \tag{12.23}$$

or

$$\boldsymbol{C}(\zeta) = \sum_{i=r}^{r+p} R_{i,p}^{r}(\zeta) \boldsymbol{P}_i, \quad \zeta_{c_r} \leq \zeta \leq \zeta_{c_{r+1}}, \quad 0 \leq r \leq m - p \tag{12.24}$$

where

$$R_{i,p}^{r}(\zeta) = \frac{L_{i,p}^{r}(\zeta) w_i}{\sum_{i=r}^{r+p} L_{i,p}^{r}(\zeta) w_i} \tag{12.25}$$

The weights w_i in Eq. (12.23) are the values of the denominator of Eq. (12.12) at each node of the NURL if the NURL curve is transformed from a NURBS curve. Note that we used ζ as variable for the NURL curve above to distinguish it from the variable ξ used for NURBS. In the following, ξ will also be used for the NURL since this should not cause confusion in the following. Similar to the NURL curve, a NURL surface can be written as

$$\boldsymbol{S}(\xi,\eta)=\sum_{i=r}^{r+p}\sum_{j=s}^{s+q}R_{i,j;p,q}^{r,s}(\xi,\eta)\boldsymbol{P}_{ij} \tag{12.26}$$

where

$$\begin{aligned}\xi_{c_r}\le\xi\le\xi_{c_{r+1}},\quad \eta_{d_s}\le\eta\le\eta_{d_{s+1}}\\ 0\le r\le m-p,\quad 0\le s\le n-q\end{aligned} \tag{12.27}$$

and

$$R_{i,j;p,q}^{r,s}(\xi,\eta)=\frac{L_{i,p}^{r}(\xi)L_{j,q}^{s}(\eta)w_{ij}}{\sum\limits_{i=r}^{r+p}\sum\limits_{j=s}^{s+q}L_{i,p}^{r}(\xi)L_{j\,q}^{s}(\eta)w_{ij}} \tag{12.28}$$

where p and q are respectively the degrees of the Lagrange function in ξ- and η-directions; and $m+1$ and $n+1$ are respectively the total numbers of nodes in the two directions. Figure 12.10 shows the NURL curves that are equivalent to the NURBS curves given in

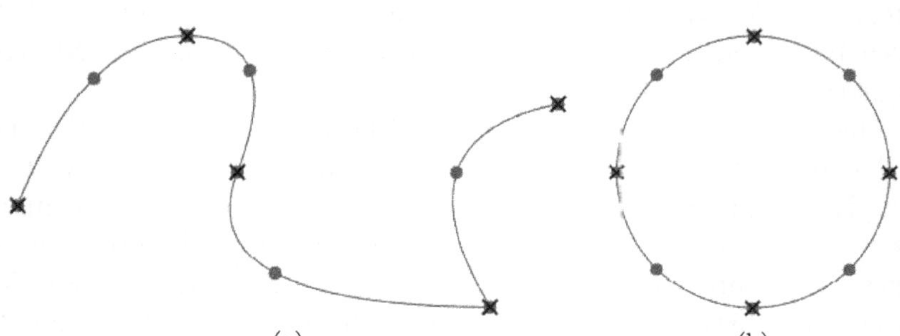

Fig. 12.10. NURL curves: the interpolation vectors for both (a) and (b) is $U=\{0,1/8,1/4,3/8,1/2,5/8,3/4,7/8,1\}$; the breakpoint vectors for both (a) and (b) is $V=\{0,1/4,1/2,3/4,1\}$.

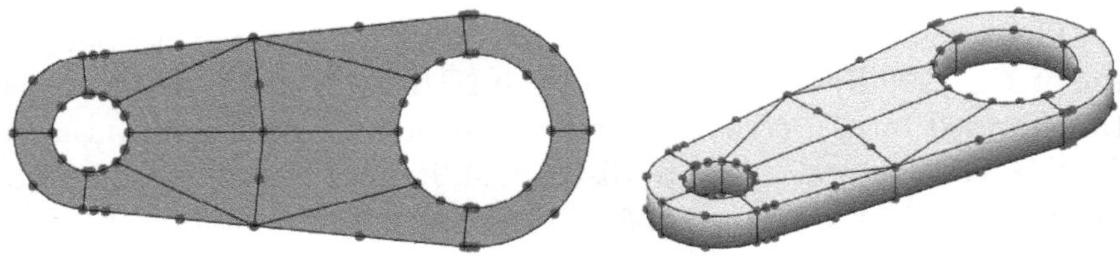

(a) (b)

Fig. 12.11. A NURL surface (a) and a volume (b). • denotes a node.

Fig. 12.6. One can see that the "control" points of a NURL geometry are physical points. NURL surfaces and volumes are tensor products of one-dimensional (1D) NURL basis, as shown in Eq. (12.26). Figure 12.11 shows a NURL surface and a NURL volume.

To transform a NURBS curve to a NURL curve, one only needs to compute the coordinates of NURBS curves on the interpolation nodes of the NURL (used as coefficients of the NURL) and the values of the denominator of the NURBS curve on the interpolation nodes of the NURL (used as weights of the NURL). The same procedure can be used for refinement (node insertion) or degree elevation. Degree elevation indicated that the minimum numbers of nodes are included in each interval of a NURL geometry. The evaluation of a NURL curve needs a breakpoint vector that includes the intervals of the (NURBS) curve and an interpolation vector that includes both the breakpoints and the minimum number points of a pth-degree polynomial in each interval of the NURBS curve. Equation (12.16) and (12.23) should be used within each interval. Using the data structure of the NURBS Toolbox [298], simple MATLAB codes such as the nrb2nrl routine can be utilized to directly transform a NURBS geometry into a NURL geometry.

The NURL geometries in Fig. 12.11 are constructed through NURL patches. One can also construct NURL geometries by trimming NURL patches as shown in Fig. 12.12. The parametric trimmed curves of the two surfaces are shown in Fig. 12.13. The parametric curves are expressed by the NURL with non-uniform interpolation nodes and constant weights. One can see from Fig. 12.13(b) that the parametric domain of the spherical NURL surface has already been trimmed into two quadrilateral domains. The parametric domain of

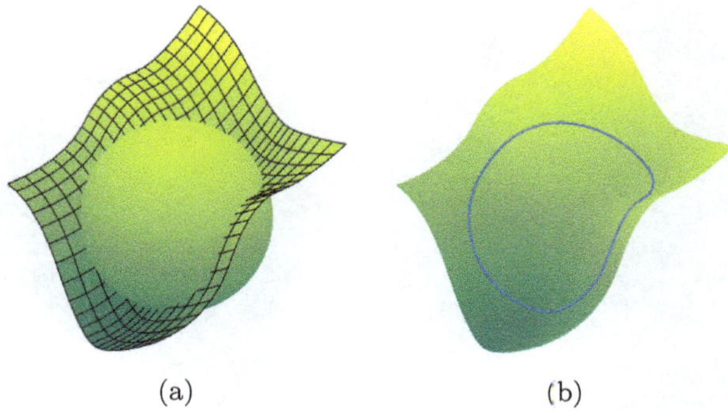

Fig. 12.12. (a) A quadric NURL surface and a spherical NURL surface; (b) intersection curve (blue) of the two surfaces of (a) obtained by the Newton–Raphson method.

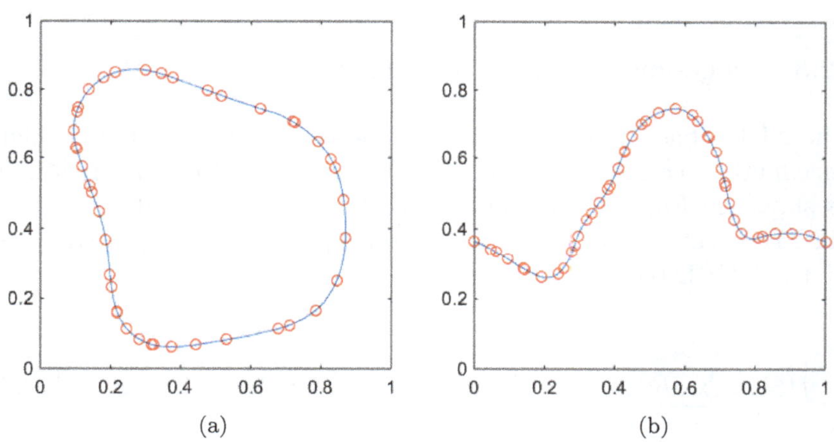

Fig. 12.13. Parametric trimmed curves of the quadric NURL surface (a) and the spherical NURL surface (b). "o" denotes the intersection points obtained by the Newton–Raphson method.

the quadric NURL surface can also be divided into five quadrilateral domains as shown in Fig. 12.14(a). Expressing the quadrilateral elements of parametric domain by NURL, one can obtain a hat-shaped surface as in Fig. 12.14(b) from the trimmed surfaces. Note that the surface in Fig. 12.9(b) is water-tight due to the double parametric approach used herein.

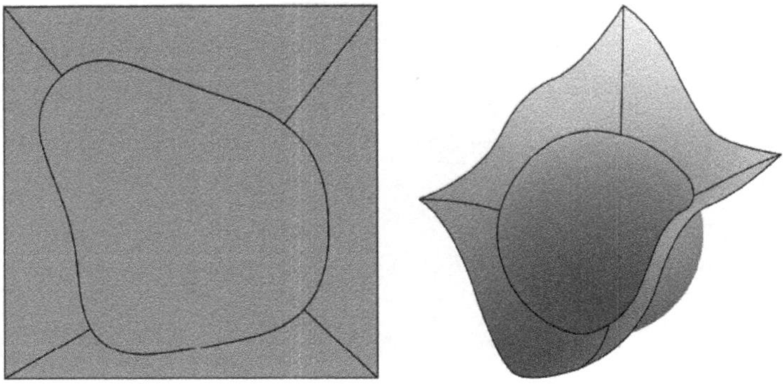

(a) (b)

Fig. 12.14. (a) The parametric domain of the quadric NURL surface (Fig. 8(a)) that was divided into five quadrilateral domains; (b) a hat-shaped surface constructed by trimming the two surfaces in Fig. 7(a).

12.3 Isogeometry Analysis of Rods

The NURL basis functions are constructed through the finite difference method (FDM) in combination with the Lagrange interpolation, as shown in Fig. 12.7(b). Similar to the NURL curve shown in Eq. (12.24), a 1D function $f(\xi)$ defined on $[\xi_0, \xi_m]$ can be approximated by the NURL basis as

$$f(\xi) \approx \sum_{i=r}^{r+p} R_{i,p}^{r}(\xi) f_i, \quad \xi_{c_r} \leq \xi \leq \xi_{c_{r+1}}, \quad 0 \leq r \leq m-p \quad (12.29)$$

where $f_i = f(\xi_i)$ and ξ_i, $i = 0, 1, 2, \ldots, m$ are the interpolation nodes of the function. The kth-order derivatives of the function can be approximated as

$$f^{(k)}(\xi) \approx \sum_{i=r}^{r+p} R_{i,p}^{r,(k)}(\xi) f_i, \quad \xi_{c_r} \leq \xi \leq \xi_{c_{r+1}}, \quad 0 \leq r \leq m-p \quad (12.30)$$

Let $p = 2$ and $m = 4$, the following formulation can be obtained from Eq. (12.30)

$$\begin{Bmatrix} f^{(k)}(\xi_0^*) \\ f^{(k)}(\xi_1^*) \\ f^{(k)}(\xi_2^*) \\ f^{(k)}(\xi_3^*) \\ f^{(k)}(\xi_4^*) \end{Bmatrix} = \begin{bmatrix} R_{0,2}^{0,(k)}(\xi_0^*) & R_{1,2}^{0,(k)}(\xi_0^*) & R_{2,2}^{0,(k)}(\xi_0^*) & 0 & 0 \\ R_{0,2}^{0,(k)}(\xi_1^*) & R_{1,2}^{0,(k)}(\xi_1^*) & R_{2,2}^{0,(k)}(\xi_1^*) & 0 & 0 \\ 0 & R_{1,2}^{1,(k)}(\xi_2^*) & R_{2,2}^{1,(k)}(\xi_2^*) & R_{3,2}^{1,(k)}(\xi_2^*) & 0 \\ 0 & 0 & R_{2,2}^{2,(k)}(\xi_3^*) & R_{3,2}^{2,(k)}(\xi_3^*) & R_{4,2}^{2,(k)}(\xi_3^*) \\ 0 & 0 & R_{2,2}^{2,(k)}(\xi_4^*) & R_{3,2}^{2,(k)}(\xi_4^*) & R_{4,2}^{2,(k)}(\xi_4^*) \end{bmatrix} \times \begin{Bmatrix} f(\xi_0) \\ f(\xi_1) \\ f(\xi_2) \\ f(\xi_3) \\ f(\xi_4) \end{Bmatrix} \tag{12.31}$$

where it is supposed that

$$\xi_i^* < \xi_{i+1}^* \quad \text{and} \quad \xi_i \le \xi_i^* \le \xi_{i+1} \tag{12.32}$$

Obviously, Eq. (12.31) is the well-known formulations of the mid-point finite difference method if $\xi_i^* = \xi_i$, where $i = 1, 2, 3, 4$. Define the following vectors and matrices:

$$\boldsymbol{f} = \{f_0 \ f_1 \ \cdots \ f_m\}^{\mathrm{T}} \tag{12.33}$$

$$\boldsymbol{f}^{(k)} = \{f_1^{(k)} \ f_2^{(k)} \ \cdots \ f_n^{(k)}\}^{\mathrm{T}} \tag{12.34}$$

where $f_j^{(k)} = \mathrm{d}^k f(\xi_j^*)/\mathrm{d}\xi^k$ and .ξ_j^*, $j = 1, 2, \ldots, n$ are the integration nodes. Therefore, the discretized formulation of Eq. (12.30) can be given as

$$\boldsymbol{f}^{(k)} = \bar{\boldsymbol{G}}^{(k)}\boldsymbol{f} \tag{12.35}$$

where the weighting coefficient matrix $\bar{G}^{(k)}$ is determined through derivatives of the basis $R^r_{i,p}(\xi)$ in Eq. (12.29) and the integration nodes. The ith ($i = 0, 1, 2, \ldots, m$) column of the matrix $\bar{G}^{(0)}$ corresponds to the ith NURL basis. One obtains the rational Lagrange interpolation from Eqs. (12.29) and (12.30) if one takes $m = p$ in the equations.

Numerical integration is necessary in the finite element method. This work uses the well-known Gauss–Lobatto quadrature that is a type of Gauss quadrature with two endpoints fixed, so that additional interpolation of the results is not necessary during post-processing. The Gauss–Lobatto quadrature for a function $f(\xi)$ defined on $[a, b]$ can be given as

$$\int_a^b f(\xi)\mathrm{d}\xi = \sum_{j=1}^{n} C_j f(\xi_j^*) \tag{12.36}$$

where ξ_j^* and C_j ($j = 1, 2, \ldots, n$) are respectively the integration nodes and weights. The method of computing Gauss–Lobatto nodes and weights can be found in [34]. Since the NURL basis used in this work involves piecewise polynomials like the B-spline, the integration should be done in each span. For a polynomial of 2pth degree, $p + 2$ integration nodes are required in each span. Let $[a_i, b_i]$ ($i = 1, 2, \ldots, k$) be the ith span. Then Eq. (12.36) can be rewritten as

$$\begin{aligned}\int_a^b f(\xi)\mathrm{d}t &= \int_{a_1}^{b_1} f(\xi)\mathrm{d}\xi + \int_{a_2}^{b_2} f(\xi)\mathrm{d}\xi + \cdots + \int_{a_k}^{b_k} f(\xi)\mathrm{d}\xi \\ &= \sum_{j=1}^{p+2} C_j^1 f(\xi_j^1) + \sum_{j=1}^{p+2} C_j^2 f(\xi_j^2) + \cdots + \sum_{j=1}^{p+2} C_j^k f(\xi_j^k) \\ &= \sum_{j=1}^{(p+1)k+1} C_j f(\xi_j^*)\end{aligned} \tag{12.37}$$

where $C_1 = C_1^1$, $C_{(p+1)k+1} = C_{p+2}^k$. Let $q = (j-1)/(p+1)$. If q is an integer and $j \neq 1$ and $(p+1)k+1$, one has

$$C_j = C_{j-q(p+1)}^q + C_1^{q+1} \tag{12.38}$$

otherwise (q is not an integer and $j \neq 1$, $(p+1)k+1$)

$$C_j = C_{j-[q](p+1)}^{[q]+1} \tag{12.39}$$

where $[\cdot]$ means rounding down decimal places. See the appended GaussLobattoQ function for MATLAB codes.

In order to study the property of the NURL as basis for analysis, vibration of a rod with length L will be considered herein. The strain energy and work potential for vibration of the rod can be discretized as

$$U = \frac{1}{2}\int_0^L ES\left(\frac{\partial u}{\partial x}\right)^2 dx = \frac{ES}{2}\boldsymbol{u}^T\bar{\boldsymbol{G}}^{(1)T}\boldsymbol{C}\bar{\boldsymbol{G}}^{(1)}\boldsymbol{u} \tag{12.40}$$

$$W = -\int_0^L u\left(-\rho S\frac{\partial^2 u}{\partial t^2}\right) dx = \boldsymbol{u}^T(\rho S\bar{\boldsymbol{G}}^{(0)\mathrm{T}}\boldsymbol{C}\bar{\boldsymbol{G}}^{(0)})\ddot{\boldsymbol{u}} \tag{12.41}$$

where $\boldsymbol{C} = \mathrm{diag}\{C_1, C_2, \ldots, C_n\}$, E is the Young's modulus, S is the area of the cross-section, ρ is the density, and $\boldsymbol{u}$ is the discrete value of the displacement at interpolation nodes similar to Eq. (12.33). Therefore, the stiffness matrix $\boldsymbol{K}$ and mass matrix $\boldsymbol{M}$ are

$$\boldsymbol{K} = ES\bar{\boldsymbol{G}}^{(1)T}\boldsymbol{C}\bar{\boldsymbol{G}}^{(1)}, \boldsymbol{M} = \rho S\bar{\boldsymbol{G}}^{(0)\mathrm{T}}\boldsymbol{C}\bar{\boldsymbol{G}}^{(0)} \tag{12.42}$$

Note that for this simple example of rod, the weights in Eq. (12.25) are taken as a constant. Figure 12.15 shows the discrete spectra using FEM, NURBS and NURL. One can see that the cubic NURL solution has the same distribution as the quadratic NURBS solution. Reference [328] reported the same FEM and NURBS results as in Fig. 12.15. Therefore, the NURL results with pth-degree NURL basis provide C^p accuracy although the basis looks like having C^0 continuity as in Fig. 12.8. One can see from Fig. 12.15(a) that the NURL results with even degrees have part of or all discrete spectra smaller than the exact results. It is obvious that the accuracy of the NURL results increases with the increase of the degree of the basis, which validated the correctness of the present results.

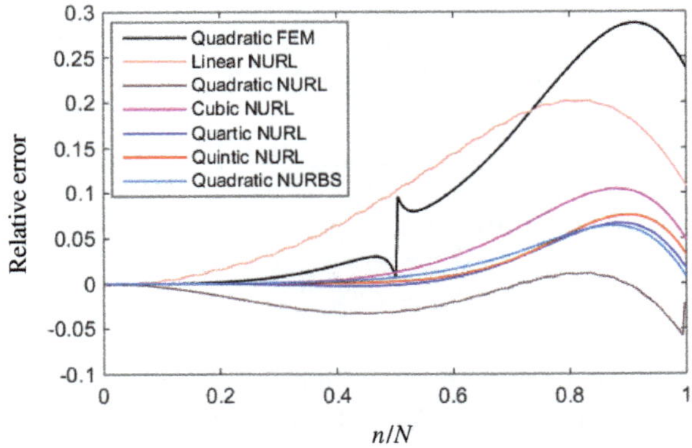

(a)

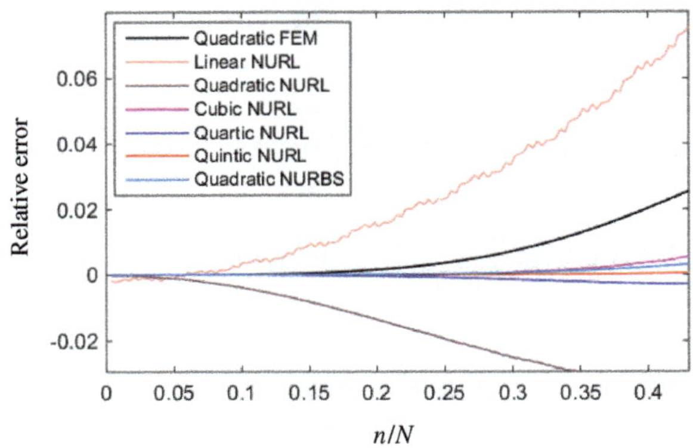

(b)

Fig. 12.15. Fixed rod: (a) relative error of discrete spectra using FEM, NURBS and NURL; (b) partial enlargement of (a).

Results with better accuracy can be obtained if the basis in Eq. (12.29) is changed to be Hermite basis, i.e., a function $f(\xi)$ defined on $[\xi_0, \xi_m]$ is approximated by the Hermite basis as

$$f(\xi) \approx \sum_{i=r}^{r+p} [H_{i,p}^r(\xi) f_i + G_{i,p}^r(\xi) f_i'], \quad \xi_{c_r} \le \xi \le \xi_{c_{r+1}},$$
$$0 \le r \le m-p \tag{12.43}$$

where $f_i = f(\xi_i)$, $f_i' = df(\xi_i)/d\xi$ and ξ_i, $i = 0, 1, 2, \ldots, m$ is the interpolation nodes of the curve. The Hermite polynomials in

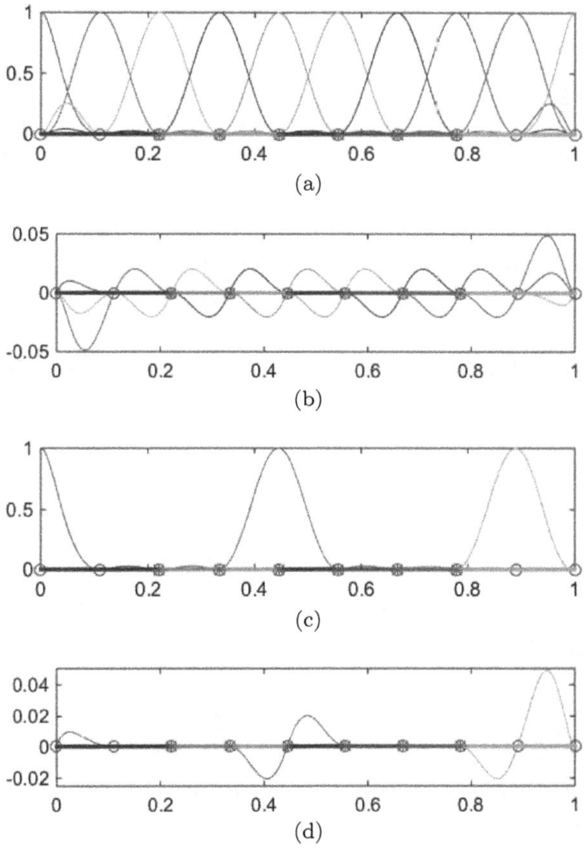

Fig. 12.16. NURH basis with constant weights: (a) $H_{i,3}$; (b) $G_{i,3}$; (c) $H_{1,3}$, $H_{5,3}$,$H_{9,3}$; (d) $G_{1,3}$, $G_{5,3}$,$G_{9,3}$. "o" denotes the interpolation nodes and * breaks the index of c_r.

Eq. (12.43) are

$$\begin{aligned} H^r_{i,p}(\xi) &= [2(\xi_i - \xi)L'^r_{i,p}(\xi_i) + 1][L^r_{i,p}(\xi)]^2 \\ G^r_{i,p}(\xi) &= (\xi - \xi_i)[L^r_{i,p}(\xi)]^2 \end{aligned} \tag{12.44}$$

Figure 12.16 plots the NURL basis formulated into the Hermite basis (named as NURH basis herein) given by Eq. (12.44). Since the NURH is only used for analysis, the NURH with constant weights (or NUH) are used throughout the manuscript. One can see from Fig. 12.16 that the NURH basis have C^1 global continuity. Define the following

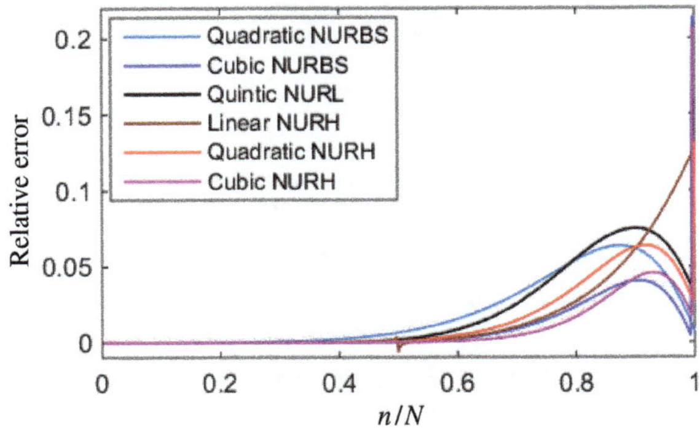

Fig. 12.17. Fixed rod: relative error of discrete spectra using NURBS, NURL and NURH.

vector:

$$\tilde{\boldsymbol{f}} = \{f_1 \ f_1' \ f_2 \ f_2' \ \cdots \ f_m \ f_m'\}^{\mathrm{T}} \tag{12.45}$$

where ξ_j^*, $j = 1, 2, \ldots, n$ are the integration nodes. Then the relation between Eqs. (12.45) and (12.34) is

$$\boldsymbol{f}^{(k)} = \tilde{\boldsymbol{G}}^{(k)} \tilde{\boldsymbol{f}} \tag{12.46}$$

where the weighting matrix $\tilde{G}^{(k)}$ is computed from the basis shown in Eq. (12.44). By using the NURH basis in the physical field, one obtains C^1 continuity between patches. Figure 12.17 compares the NURH results with the NURBS and NURL results. One can see the improved accuracy of the NURH using even linear basis compared with the quintic NURL basis. One can also see from Fig. 12.17 that cubic NURH has the same accuracy as cubic NURBS.

12.4 Isogeometry Analysis of In-Plane Vibrations and Static Deformation by NURL

12.4.1 *Differential and integration rules*

The NURL can also be used for isogeometric analysis like the NURBS. Similar to the NURL surface represented by the NURL, as shown in Eq. (12.26), a 2D function $f(\xi, \eta)$ can be approximated by the NURL basis as

$$f(\xi,\eta) \approx \sum_{i=r}^{r+p}\sum_{j=s}^{s+q} R^{r,s}_{i,j;p,q}(\xi,\eta) f_{ij} \tag{12.47}$$

where $f_{ij} = f(\xi_i, \eta_j)$ are physical points, $\xi_i (i = 0, 1, 2, \ldots, m)$ and $\eta_j (j = 0, 1, 2, \ldots, n)$ are the interpolation nodes of the surface (or the function). Using the chain rule of differentiation, the derivatives of $f[x(\xi,\eta), y(\xi,\eta)]$ with respect to Cartesian coordinates x and y can be given as

$$\frac{\partial f}{\partial x} = \frac{1}{|\boldsymbol{J}|}\left(\frac{\partial y}{\partial \eta}\frac{\partial f}{\partial \xi} - \frac{\partial y}{\partial \xi}\frac{\partial f}{\partial \eta}\right), \quad \frac{\partial f}{\partial y} = \frac{1}{|\boldsymbol{J}|}\left(\frac{\partial x}{\partial \xi}\frac{\partial f}{\partial \eta} - \frac{\partial x}{\partial \eta}\frac{\partial f}{\partial \xi}\right) \tag{12.48}$$

where the determinant $|\boldsymbol{J}|$ of the Jacobian $\boldsymbol{J} = \partial(x, y)/\partial(\xi, \eta)$ is

$$|\boldsymbol{J}| = \frac{\partial x}{\partial \xi}\frac{\partial y}{\partial \eta} - \frac{\partial y}{\partial \xi}\frac{\partial x}{\partial \eta} \tag{12.49}$$

Using Eq. (12.37), the integration of a two-dimensional function defined on a unit square domain can be given as

$$\int_0^1\int_0^1 f(\xi,\eta)\mathrm{d}\xi\mathrm{d}\eta = \sum_{i=1}^{N_\xi}\sum_{j=1}^{N_\eta} C_i^\xi C_j^\eta f(\xi_i^*, \eta_j^*) \tag{12.50}$$

where ξ_i^*, C_i^ξ and η_j^*, C_j^η are respectively integration nodes and weights of ξ- and η-directions shown in Eq. (12.37). N_ξ and N_η are respectively the minimum number of nodes required by Gauss–Lobatto integration for precise integration along ξ- and η-directions. Define the following vectors:

$$\boldsymbol{f} = [f_{11} \ \cdots \ f_{m1} \ \cdots \ f_{mn}]^{\mathrm{T}} \tag{12.51}$$

$$\bar{\boldsymbol{f}} = [\bar{f}_{11} \ \cdots \ \bar{f}_{N_\xi 1} \ \cdots \ \bar{f}_{N_\xi N_\eta}]^{\mathrm{T}} \tag{12.52}$$

where $\bar{f}_{ij} = f(\xi_i^*, \eta_j^*)$. The relationship between the two vectors defined in Eqs. (12.51) and (12.52) is

$$\bar{\boldsymbol{f}} = \boldsymbol{G}\mathbf{f} \tag{12.53}$$

where the weighting coefficient matrix $\boldsymbol{G}$ is determined through the basis $R_{ij}(\xi,\eta)$ in Eq. (12.47) and the integration nodes. Similarly, let

$$\begin{aligned}\bar{\boldsymbol{f}}_x &= [f_x(\xi_1^*,\eta_1^*)\ \cdots\ f_x(\xi_{N_\xi}^*,\eta_1^*)\ \cdots\ f_x(\xi_{N_\xi}^*,\eta_{N_\eta}^*)]^{\mathrm{T}}\\ \bar{\boldsymbol{f}}_y &= [f_y(\xi_1^*,\eta_1^*)\ \cdots\ f_y(\xi_{N_\xi}^*,\eta_1^*)\ \cdots\ f_y(\xi_{N_\xi}^*,\eta_{N_\eta}^*)]^{\mathrm{T}}\end{aligned} \tag{12.54}$$

Using Eqs. (12.47)–(12.49) and the integration nodes, we can get

$$\bar{\boldsymbol{f}}_x = \boldsymbol{G}_x\boldsymbol{f},\quad \bar{\boldsymbol{f}}_y = \boldsymbol{G}_y\boldsymbol{f} \tag{12.55}$$

With the differential quadrature (DQ) rules and integration rules defined above, formulations for in-plane vibration and static deformation of plates will be presented in Section 12.4.2.

12.4.2 *In-plane vibration and static deformation of plates by NURL*

Using the DQ rules of Eq. (12.55), the discretized strain vector for in-plane problem can be given as

$$\begin{bmatrix}\varepsilon_x\\ \varepsilon_y\\ \varepsilon_{xy}\end{bmatrix} = \begin{bmatrix}\boldsymbol{G}_x & \boldsymbol{0}\\ \boldsymbol{0} & \boldsymbol{G}_y\\ \boldsymbol{G}_y & \boldsymbol{G}_x\end{bmatrix}\begin{bmatrix}\boldsymbol{u}\\ \boldsymbol{v}\end{bmatrix} \tag{12.56}$$

or

$$\boldsymbol{\varepsilon} = \boldsymbol{H}\mathbf{U} \tag{12.57}$$

where $\boldsymbol{U} = [\boldsymbol{u};\boldsymbol{v}]$, $\boldsymbol{u}$ and $\boldsymbol{v}$ are respectively the discretized vector of u and v in the form of Eq. (12.51). Using Eqs. (12.49)–(12.53), the integration of u on an arbitrary quadrilateral domain can be given as

$$\int_A u[x(\xi,\eta),y(\xi,\eta)]\mathrm{d}x\mathrm{d}y = \bar{\boldsymbol{C}}^{\mathrm{T}}\boldsymbol{G}\boldsymbol{u} \tag{12.58}$$

where

$$\begin{aligned}\bar{\boldsymbol{C}} &= [\boldsymbol{C}_1^{\mathrm{T}}\ \boldsymbol{C}_2^{\mathrm{T}}\ \cdots\ \boldsymbol{C}_{N_\eta}^{\mathrm{T}}]\\ \boldsymbol{C}_j^{\mathrm{T}} &= C_j^\eta[|\boldsymbol{J}|_{1j}C_1^\xi\ |\boldsymbol{J}|_{2j}C_2^\xi\ \cdots\ |\boldsymbol{J}|_{N_\xi,j}C_{N_\xi}^\xi]\end{aligned} \tag{12.59}$$

where $j = 1, 2, \ldots, N_\eta$. Thus, the strain energy and work potential for in-plane problems can be discretized as

$$U = \frac{1}{2}\boldsymbol{U}^T\boldsymbol{H}^T\boldsymbol{D}\mathbf{HU} \tag{12.60}$$

$$W = \rho h \boldsymbol{U}^T \begin{bmatrix} \boldsymbol{G}^T\boldsymbol{C}\mathbf{G} & \mathbf{0} \\ \mathbf{0} & \boldsymbol{G}^T\boldsymbol{C}\mathbf{G} \end{bmatrix} \ddot{\boldsymbol{U}} - \boldsymbol{U}^T\boldsymbol{R} \tag{12.61}$$

where h is the thickness and

$$\boldsymbol{D} = C\begin{bmatrix} \boldsymbol{C} & \upsilon\boldsymbol{C} & \mathbf{0} \\ \upsilon\boldsymbol{C} & \boldsymbol{C} & \mathbf{0} \\ \mathbf{0} & \mathbf{0} & \boldsymbol{C}(1-\upsilon)/2 \end{bmatrix}, \quad \boldsymbol{R} = h\begin{bmatrix} \boldsymbol{G}_\Gamma^T\boldsymbol{C}_\Gamma\boldsymbol{q}_u \\ \boldsymbol{G}_\Gamma^T\boldsymbol{C}_\Gamma\boldsymbol{q}_v \end{bmatrix} \tag{12.62}$$

where $\boldsymbol{C} = \mathrm{diag}(\bar{\boldsymbol{C}})$ is the diagonal matrix of the integration vector defined by Eq. (12.59); $\boldsymbol{q}_u$ and $\boldsymbol{q}_v$ are the nodal load vectors with respect to x- and y-directions, respectively, $\boldsymbol{G}_\Gamma^T$ and $\boldsymbol{C}_\Gamma$ in Eq. (12.62) are respectively the interpolation and integration matrices on the force boundaries; $C = Eh/(1-\upsilon^2)$ is the stretch rigidity, E is the Young's modulus, υ is the Poisson's ratio, and h is the thickness of the plate. Using the variational principle, one obtains the equilibrium equations as

$$\boldsymbol{K}\mathrm{U} + \boldsymbol{M}\ddot{\boldsymbol{U}} = \boldsymbol{R} \tag{12.63}$$

where the stiffness matrix $\boldsymbol{K}$ and mass matrix $\boldsymbol{M}$ are

$$\boldsymbol{K} = \boldsymbol{H}^T\bar{\boldsymbol{D}}\mathbf{H} \tag{12.64}$$

$$\boldsymbol{M} = \rho h\begin{bmatrix} \boldsymbol{G}^{\mathrm{T}}\boldsymbol{C}\mathbf{G} & \mathbf{0} \\ \mathbf{0} & \boldsymbol{G}^{\mathrm{T}}\boldsymbol{C}\mathbf{G} \end{bmatrix} \tag{12.65}$$

If the force vector $\boldsymbol{R}$ in Eq. (12.63) is taken as $\boldsymbol{0}$, one obtains the equilibrium equation for free in-plane vibration of plates. If the inertial force in Eq. (12.63) is not used, one obtains the equilibrium equation for static in-plane deformation of plates. For static deformations, Eq. (12.63) can be further written as

$$\begin{bmatrix} \boldsymbol{K}_{bb} & \boldsymbol{K}_{bi} \\ \boldsymbol{K}_{ib} & \boldsymbol{K}_{ii} \end{bmatrix}\begin{bmatrix} \boldsymbol{U}_b \\ \boldsymbol{U}_i \end{bmatrix} = \begin{bmatrix} \boldsymbol{R}_b \\ \boldsymbol{R}_i \end{bmatrix} \tag{12.66}$$

where the subscripts b and i respectively refer to the boundary and interior degrees of freedom. If the Dirichlet boundary conditions $\boldsymbol{U}_b$ are given, one obtains

$$\boldsymbol{U}_i = \boldsymbol{K}_{ii}^{-1}(\boldsymbol{R}_i - \boldsymbol{K}_{ib}\boldsymbol{U}_b) \tag{12.67}$$

If IGA based on NURBS is used, the static problems may not be solved so straightforwardly since most degrees of freedom do not have a clear physical meaning.

Now, the NURL results for in-plane and flexural vibration of isotropic thin plates in curvilinear domains are presented to validate its accuracy and convergence properties. In order to show advantages of the NURL in coping with Dirichlet boundary conditions, an example of static in-plane deformation of plate is also presented. The mappings of a square parent domain to the interest domains are carried out via NURL. The same bases as the geometric ones are used for analysis, and this satisfies the isogeometric concept. Unless otherwise specified, the number of sampling points on the edges in the ξ- and η-directions are respectively denoted as M and N. The natural frequencies are expressed as the same frequency parameters as in references, which is done not to compare with round-off errors that are unrealistic. Being different from flexural vibrations, in-plane vibration has two distinct simply supported boundary conditions, where the normal stress and tangential displacement are zeros (S1) or the normal displacement and tangential stress are zeros (S2).

In-plane vibration of a square plate (Fig. 12.18) is analyzed first in Tables 12.1 and 12.2 for the plate with two combinations of simply supported, clamped (C) and free (F) boundaries, since there are plenty of such results in the literatures [334, 335]. A plate with S1–C–S2–F boundary condition means that the plate has S1–S2 at x-direction and C–F at y-direction. The natural frequencies in Tables 12.1 and 12.2 are expressed by the same frequency parameter Ω the one as used in [334, 335]. Quadratic to quintic NURL are used for the two types of plates. One can see from the two tables that the cubic NURL results agree with the exact results for about three significant digits while the quintic NURL results agree with the exact results for about four digits. The quadratic and quartic NURL results are respectively slightly less accurate than the cubic

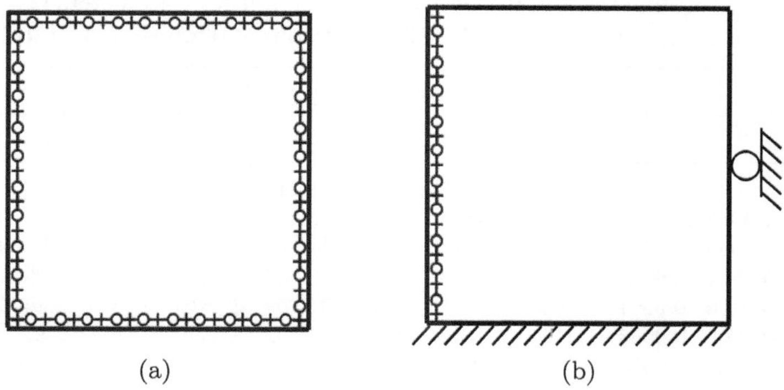

Fig. 12.18. A square plate with two combinations of simply supported, clamped (C) and free (F) boundaries: (a) S1–S1–S1–S1; (b) S1–C–S2–F.

and quintic NURL results. One may have noted from Tables 12.1 and 12.2 that the NURL results converge from the above but not monotonously. The slight variation of convergence should be caused by the searching method used to compute the basis. But we can see that the NURL results are stable.

In order to show the convergence property of the entire spectrum, the S1S1S1S1 plate is further studied. The NURL results with mesh size $M = N = 30$ are compared with those of FEM obtained by the authors using commercial software ANSYS. The material constants of the plate are Young's modulus of 71 GPa, a density of 2700 kg/m^3 and Poisson's ratio of 0.3. The elements used in ANSYS are SHELL181 quadrilateral elements. The element length of FEM is $l = 1/30\,\mathrm{m}$ which is the same as the NURL. The exact results are obtained from the method of [334, 335, 431]. Relative error of FEM, NURBS and NURL are plotted in Fig. 12.19. One can see from Fig. 12.17 that the convergence property of linear NURL is better than that of FEM and the convergence property of the quadratic NURL is similar to that of quadratic NURBS. It is obvious from Fig. 12.19 that about 8% of the converged results of the NURL are smaller than the exact results. The left converged NURL results are larger than the exact results.

In order to show advantages of the NURL in coping with the Dirichlet boundary conditions, static deformation of a thick-walled cylinder under uniform boundary pressure is considered. The stress

Table 12.1. Comparison of the dimensionless frequency parameters $\Omega = (\omega a/\pi)\sqrt{\rho/G}$ for in-plane vibration of a S1–S1–S1–S1 square plate (Fig. 12.18(a)).

	Mode sequence number									
$M = N$	1	2	3	4	5	6	7	8	9	10
					Quadratic NURL					
3	1.0066	1.0066	1.6141	2.5236	4.2875	4.2875	—	—	—	—
4	1.0055	1.0055	1.4869	2.2249	2.2305	2.4504	2.6272	2.6293	3.9160	4.4128
5	0.9964	0.9964	1.3940	1.9722	1.9758	2.2583	2.2589	2.3579	2.9969	3.0004
6	0.9900	0.9901	1.4057	2.0335	2.0336	2.2604	2.2610	2.3598	2.8405	2.9707
7	0.9966	0.9967	1.4142	2.0109	2.0110	2.2492	2.2549	2.3918	2.8754	3.0811
30	0.9954	0.9954	1.4104	1.9860	1.9860	2.2270	2.2270	2.3800	2.8207	2.9774
					Cubic NURL					
4	1.0066	1.0066	1.4192	2.0629	2.0629	2.3932	2.4362	2.4362	3.7043	4.4222
5	1.0005	1.0005	1.4151	2.0623	2.0623	2.2957	2.2964	2.3913	2.9326	3.0897
6	0.9993	0.9993	1.4149	2.0308	2.0308	2.2561	2.2561	2.3894	2.8396	3.0945
7	1.0001	1.0001	1.4133	2.0000	2.0000	2.2391	2.2391	2.3892	2.8416	3.0610
8	0.9999	0.9999	1.4142	2.0022	2.0022	2.2382	2.2382	2.3903	2.8320	3.0333
30	0.9993	0.9993	1.4136	1.9979	1.9979	2.2349	2.2349	2.3888	2.8279	2.9977
					Quartic NURL					
5	1.0000	1.0000	1.4148	2.0629	2.0629	2.2834	2.2834	2.3908	2.8811	3.2168
6	1.0000	1.0000	1.4145	2.0076	2.0076	2.2452	2.2452	2.3907	2.8462	3.2166
7	0.9999	0.9999	1.4141	2.0009	2.0009	2.2373	2.2373	2.3901	2.8311	3.0353
8	1.0000	1.0000	1.4141	1.9987	1.9987	2.2352	2.2352	2.3903	2.8280	3.0000
9	1.0000	1.0000	1.4141	1.9992	1.9992	2.2353	2.2353	2.3903	2.8274	2.9955
30	0.9998	0.9998	1.4140	1.9994	1.9994	2.2357	2.2357	2.3900	2.8283	2.9993
					Quintic NURL					
6	1.0000	1.0000	1.4142	2.0006	2.0006	2.2401	2.2401	2.3905	2.8420	3.2168
7	1.0000	1.0000	1.4142	2.0006	2.0006	2.2367	2.2367	2.3904	2.8296	3.0041
8	1.0000	1.0000	1.4142	2.0001	2.0001	2.2362	2.2362	2.3904	2.8286	3.0012
9	1.0000	1.0000	1.4142	2.0000	2.0000	2.2361	2.2361	2.3904	2.8285	3.0009
10	1.0000	1.0000	1.4142	1.9999	1.9999	2.2360	2.2360	2.3904	2.8285	3.0001
30	0.9999	0.9999	1.4141	1.9998	1.9998	2.2360	2.2360	2.3903	2.8284	2.9998
Exact [334]	1	1	1.4142	2	2	2.2361	2.2361	2.3905	2.8284	3

components of a thick-walled cylinder under the action of uniform internal and external pressure loadings, as shown in Fig. 12.20, are

$$\begin{aligned}\sigma_r &= \frac{a^2b^2(p_2 - p_1)}{b^2 - a^2}\frac{1}{r^2} + \frac{a^2p_1 - b^2p_2}{b^2 - a^2}\\ \sigma_\theta &= -\frac{a^2b^2(p_2 - p_1)}{b^2 - a^2}\frac{1}{r^2} + \frac{a^2p_1 - b^2p_2}{b^2 - a^2}\end{aligned} \tag{12.68}$$

Table 12.2. Comparison of the dimensionless frequency parameters $\Omega = (\omega a/\pi)\sqrt{\rho/G}$ for in-plane vibration of a S1–C–S2–F square plate (Fig. 12.18(b)).

	Mode sequence number									
$M = N$	1	2	3	4	5	6	7	8	9	10
					Quadratic NURL					
3	0.8761	0.9560	1.8602	2.0457	2.9332	3.1035	3.6118	3.6858	—	—
4	0.8674	0.9521	1.5697	1.8550	2.4866	2.7064	3.1320	3.1713	3.3950	3.5083
5	0.8560	0.9420	1.4120	1.7236	2.2374	2.3664	2.4257	2.7774	2.7872	2.8359
6	0.8555	0.9402	1.4176	1.7338	2.2200	2.3315	2.3600	2.7619	2.7703	2.8051
7	0.8601	0.9448	1.4169	1.7241	2.2325	2.3471	2.3653	2.7637	2.8180	2.8418
30	0.8567	0.9439	1.4198	1.7332	2.2127	2.2832	2.3495	2.7406	2.7821	2.8039
					Cubic NURL					
4	0.8606	0.9471	1.4806	1.7894	2.3430	2.5878	2.9776	2.9965	3.3364	3.4930
5	0.8597	0.9467	1.4388	1.7561	2.2409	2.4332	2.5067	2.8146	2.8639	2.8960
6	0.8598	0.9464	1.4275	1.7443	2.2240	2.3622	2.3837	2.7701	2.8162	2.8529
7	0.8599	0.9465	1.4259	1.7405	2.2214	2.3245	2.3674	2.7547	2.7992	2.8356
8	0.8599	0.9467	1.4268	1.7425	2.2196	2.3114	2.3653	2.7543	2.7966	2.8286
30	0.8595	0.9463	1.4250	1.7407	2.2185	2.2955	2.3595	2.7503	2.7918	2.8183
					Quartic NURL					
5	0.8600	0.9468	1.4309	1.7482	2.2306	2.4229	2.4897	2.7977	2.8494	2.9246
6	0.8600	0.9467	1.4268	1.7437	2.2219	2.3423	2.3761	2.7638	2.8084	2.8483
7	0.8600	0.9467	1.4256	1.7413	2.2195	2.3094	2.3637	2.7525	2.7952	2.8302
8	0.8600	0.9467	1.4256	1.7416	2.2188	2.2970	2.3599	2.7511	2.7926	2.8196
9	0.8600	0.9467	1.4257	1.7417	2.2189	2.2957	2.3599	2.7510	2.7924	2.8192
30	0.8600	0.9466	1.4256	1.7417	2.2190	2.2963	2.3606	2.7513	2.7928	2.8200
					Quintic NURL					
6	0.8600	0.9468	1.4261	1.7425	2.2203	2.3295	2.3708	2.7588	2.8209	2.8438
7	0.8600	0.9467	1.4258	1.7419	2.2194	2.3028	2.3625	2.7526	2.7946	2.8253
8	0.8600	0.9468	1.4258	1.7420	2.2193	2.2976	2.3613	2.7520	2.7935	2.8213
9	0.8600	0.9467	1.4258	1.7420	2.2193	2.2970	2.3610	2.7518	2.7933	2.8210
10	0.8600	0.9467	1.4258	1.7420	2.2192	2.2966	2.3609	2.7517	2.7932	2.8205
30	0.8600	0.9467	1.4257	1.7419	2.2192	2.2966	2.3609	2.7516	2.7930	2.8205
Exact [334]	0.8600	0.9468	1.4258	1.7420	2.2193	2.2967	2.3610	2.7517	2.7932	2.8207

Using the strain–displacement relations and Hooke's law, the radial displacement is easily determined as

$$u_r = \frac{1+\upsilon}{E}\left[-\frac{a^2b^2(p_2-p_1)}{b^2-a^2}\frac{1}{r} + (1-2\upsilon)\frac{a^2p_1-b^2p_2}{b^2-a^2}r\right] \tag{12.69}$$

where υ is the Poisson's ratio and E is the Young's modulus. The above solutions are exact. Note that this is a plane strain problem. Therefore, Young's modulus and Poisson's ratio in Eq. (12.62) should

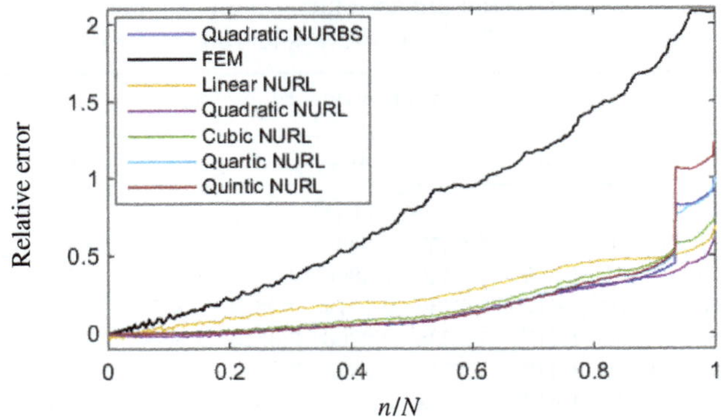

(a)

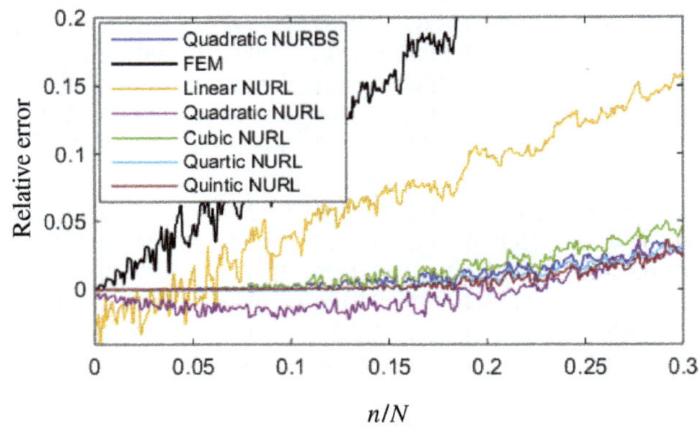

(b)

Fig. 12.19. Simply supported (S1) plate: (a) relative error of discrete spectra using FEM, NURBS and NURL; (b) partial enlargement of (a).

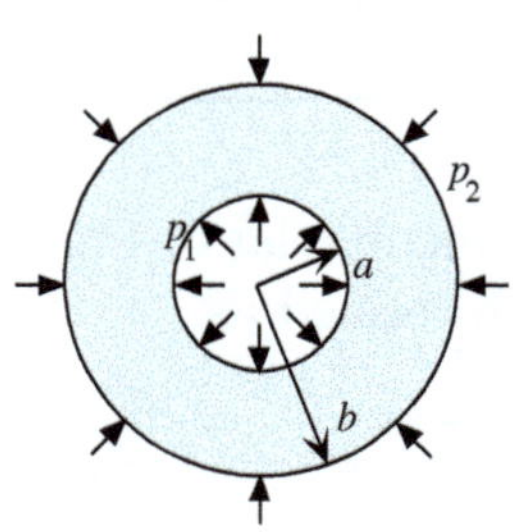

Fig. 12.20. Thick-walled cylinder problem.

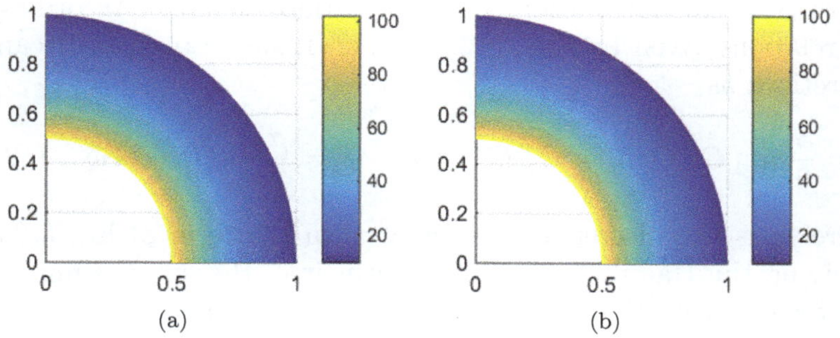

Fig. 12.21. NURL results σ_r for a thick-walled cylinder under boundary pressure $p_1 = -100\,\text{Pa}$ and $p_2 = -10\,\text{Pa}$: (a) inversely; (b) directly.

be replaced by $E = E/(1 - \upsilon^2)$ and $\upsilon = \upsilon/(1 - \upsilon)$. Due to the symmetry of both geometry and loads, the problem can be solved by one NURL element as in Fig. 12.18.

Take $a = 0.5$, $b = 1.0$, $p_1 = -100\,\text{Pa}$, $p_2 = -10\,\text{Pa}$, $E = 72\,\text{GPa}$ and $\upsilon = 0.3$. In order to study the property of the NURL in applying Dirichlet boundary conditions, the above problem will be solved inversely by the NURL, i.e., the exact displacement on the internal and external edges is first obtained from Eq. (12.69), then the stress components are solved from Eq. (12.67) together with the stress–displacement relations to compare with the loading pressures. Quintic NURL with the number of sampling points along the curved edges and the straight edges being $M/2 = N = 15$ are used. One can see from Fig. 12.21(a) that the NURL results agree well with the load conditions. Figure 12.21(b) shows the direct results of NURL obtained from Eq. (12.67) with only homogenous boundary conditions being considered.

12.4.3 *Vibration of membranes by the NURL*

In order to show the process of the derivation above more clearly, vibration of membranes is considered herein. The potential energy U and kinetic energy T of the element are expressed as

$$U = \frac{S}{2}\int_A \left[\left(\frac{\partial w}{\partial x}\right)^2 + \left(\frac{\partial w}{\partial y}\right)^2\right] \mathrm{d}A,$$
$$T = \frac{\rho}{2}\int_A \left(\frac{\partial w}{\partial t}\right)^2 \mathrm{d}A \tag{12.70}$$

where S is surface tension and ρ is surface density. According to the relations given by Eqs. (12.53) and (12.55), Eq. (12.70) can be discretized as

$$U = \frac{S}{2}\boldsymbol{w}^T(\boldsymbol{G}_x^T\boldsymbol{C}\mathbf{G}_x + \boldsymbol{G}_y^T\boldsymbol{C}\mathbf{G}_y)\boldsymbol{w}, \quad T = \frac{\rho}{2}\dot{\boldsymbol{w}}^T\boldsymbol{G}^T\boldsymbol{C}\mathbf{G}\dot{\boldsymbol{w}} \quad (12.71)$$

where $\boldsymbol{w}$ is the discretized vector of w in the form of Eq. (12.51). Applying the Hamilton variational principle, the equilibrium equation for free vibration of membranes can be given as

$$\bar{\boldsymbol{K}}\mathbf{w} + \bar{\boldsymbol{M}}\ddot{\boldsymbol{w}} = 0 \quad (12.72)$$

where

$$\bar{\boldsymbol{K}} = S(\boldsymbol{G}_x^T\boldsymbol{C}\mathbf{G}_x + \boldsymbol{G}_y^T\boldsymbol{C}\mathbf{G}_y), \bar{\boldsymbol{M}} = \rho\boldsymbol{G}^T\boldsymbol{C}\mathbf{G} \quad (12.73)$$

Figure 12.18 plots the four mode shapes of a clamped open sectorial membrane with inner and outer radius 0.5 and 1 respectively. The expression for the frequency parameters Ω in Fig. 12.13 is $\Omega = \omega\sqrt{\rho}/S$. Correctness of the frequency parameters on the caption of Fig. 12.22 can be verified by [49].

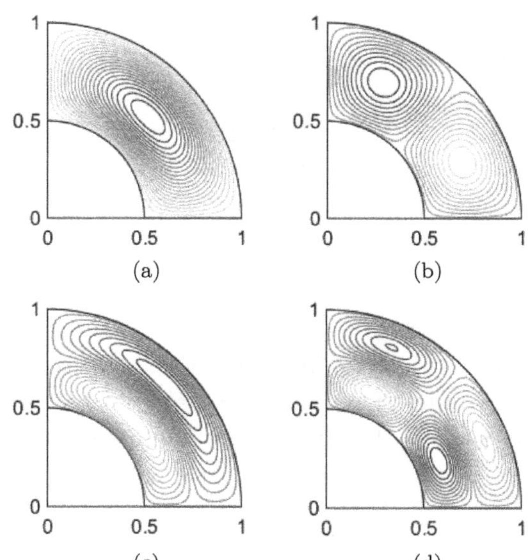

Fig. 12.22. Four mode shapes for vibration of a clamped open sectorial membrane: (a) $\Omega_1 = 6.8138$, (b) $\Omega_2 = 8.2666$, (c) $\Omega_5 = 12.855$, (d) $\Omega_6 = 13.742$.

12.5 Surface Intersection Algorithms

There are at least four typical methods that can be used for surface intersection: (1) the discretization method, where surfaces are discretized as triangular elements and then the intersections are solved through the triangles (Fig. 12.23); (2) the surface subdivision method, where the surfaces are successively discretized according to the intersection of bounding boxes of subdivision surfaces to obtain the intersection (Fig. 12.24); (3) the iterative method, which usually refers to the Newton–Raphson iterative method used to solve the nonlinear system of equations of the intersected surfaces; (4) the

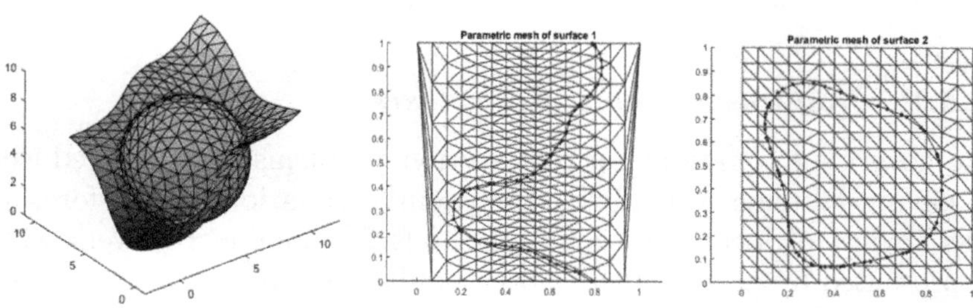

Fig. 12.23. Surface intersection through triangular discretization (left: geometric meshes; middle: parametric domain of the sphere surface; right: parametric domain of the quadric free surface).

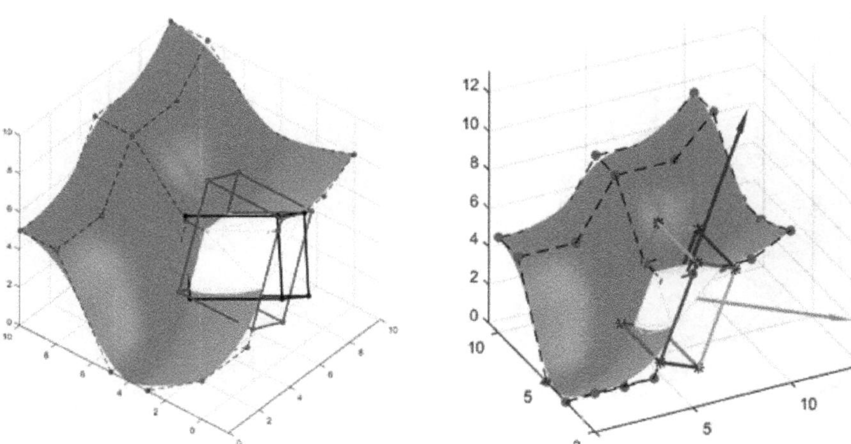

Fig. 12.24. Bounding box of the quadric Bezier patch and its three separated axes.

tracing method, which uses some given intersection points to successively search other intersection points. Practically, none of the methods can be used to tackle all cases of intersections, so this work used them together as usual. To attain results of high accuracy, high efficiency as well as robustness, this work primarily uses the tracing method. Other methods are used as auxiliary methods. In order to solve element singularity and mismatching problems in the process of surface intersection, we developed our own codes of surface trimming as it is hard to find open source codes on surface trimming. Figures 12.25 and 12.26 respectively show the trimmed parametric domain and geometric gird of the sphere and the quadric surfaces shown in Fig. 12.23. Using IGES, one can cope with geometries constructed by commercial CAD software (Fig. 12.27).

12.5.1 *Nonlinear polynomial solvers*

The governing equations for intersection problems typically reduce to solving systems of nonlinear polynomial equations. Therefore, a brief review of local and global methods for nonlinear solvers will be presented first. One of the most popular local methods in solving nonlinear equation systems is the Newton-type method. Advantages of the Newton method are its quadratic convergence and its ease of implementation. Disadvantages are that for each root, a good initial approximation is required, otherwise the method may diverge.

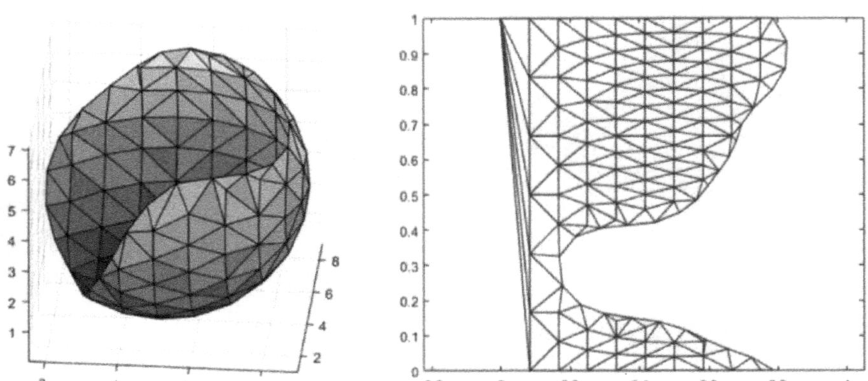

Fig. 12.25. Trimmed sphere surface (left) and its trimmed parametric domain (right).

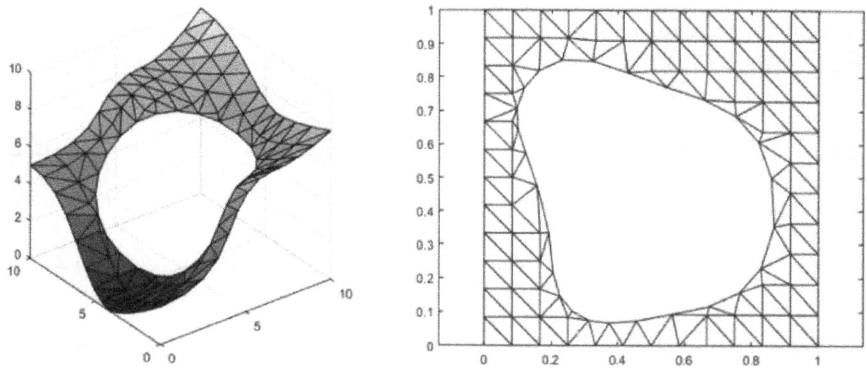

Fig. 12.26. Trimmed quadric surface (left) and its trimmed parametric domain (right).

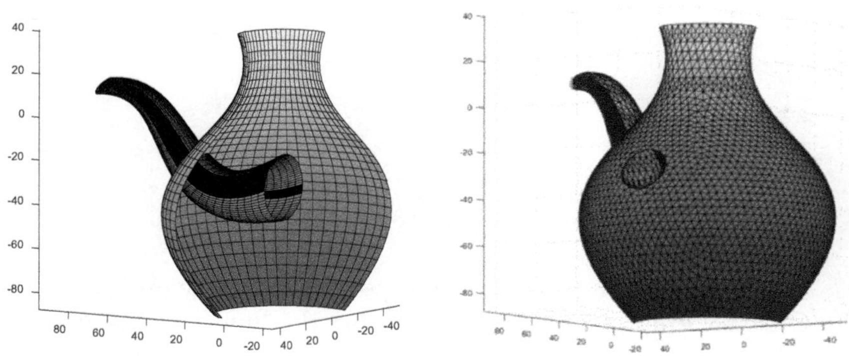

Fig. 12.27. CAD surfaces before (left) and after (right) trimming.

Also, the method cannot by itself provide full assurance that all roots have been found. Global solution methods are designed to compute all roots in some area of interest. Here, we introduce an iterative global root-finding algorithm for n-dimensional nonlinear polynomial equation systems, which belongs to the subdivision class of methods, called projected polyhedron (PP) algorithm [424, 432]. It is easy to visualize and simple in that we only requires two straightforward algorithms in order to implement it: one for subdividing multivariate polynomials in Bernstein form and another for finding the convex hull of a two-dimensional set of points. This algorithm is an extension and generalization of earlier adaptive subdivision algorithms.

For illustration, we enumerate the procedures required by the PP algorithm to find roots of a degree m polynomial equation $f(x) = c_0 + c_1 x + c_2 x^2 + \cdots + c_m x^m$ over the interval $a < x < b$.

(1) Make an affine parameter transformation $x = a + u(b - a)$ such that $0 < u < 1$ as follows:

$$f(u) = \sum_{i=0}^{m} c_i^M u^i, \quad 0 \le u \le 1 \tag{12.74}$$

The transition from the interval $a < x < b$ to the interval $0 < u < 1$ is an affine map, and the polynomials are invariant under affine parameter transformation.

(2) Convert the basis from monomial to Bernstein

$$f(u) = \sum_{i=0}^{n} c_i^B B_{i,m}(u), \quad 0 \le u \le 1 \tag{12.75}$$

where

$$c_i^B = \sum_{j=0}^{i} \frac{\binom{i}{j}}{\binom{m}{j}} c_j^M \tag{12.76}$$

and $B_{i,m}(u)$ is the ith Bernstein polynomial of degree m.

(3) Create a graph of function $f(u)$ using the linear precision property of Bernstein polynomials. Then the graph will become a Bézier curve

$$f(u) = \begin{pmatrix} u \\ f(u) \end{pmatrix} = \sum_{i=0}^{n} \begin{pmatrix} i/m \\ c_i^B \end{pmatrix} B_{i,m}(u) \tag{12.77}$$

where $(i/m, c_i^B)^T$ are control points. Now, the problem of finding roots of the univariate polynomial has been transformed into a geometric problem of finding the intersection of a Bézier curve with the parameter axis.

(4) Construct the convex hull of the Bézier curve.

(5) Intersect the convex hull with the parameter axis.

(6) Discard the regions which do not contain roots by applying the de Casteljau subdivision algorithm and find a sub-region of [0, 1] which may contain the root(s).

(7) If the sub-region is sufficiently small, we conclude that there is a root inside and return it. But when there is more than one root in the sub-region, the sub-region will not be reduced. In such a case, we split the region evenly by applying the de Casteljau subdivision algorithm and we go back to 4.

The algorithm has been generalized to n-dimension. See [424, 432] for more details.

12.5.2 *Surface/surface intersections*

In order to solve general surface–surface (S/S) intersection problems, the following five auxiliary intersection problems need to be considered: point/point (P/P), point/curve (P/C), point/surface (P/S), curve/curve (C/C), curve/surface (C/S). When studying intersection problems, the type of curves and surfaces that we consider can be classified into two types: (1) rational polynomial parametric (RPP); (2) implicit algebraic (IA). non-uniform rational B-spline (NURBS) curves and surfaces can be subdivided into RPP curves and surfaces, and analyzed in a similar manner. The RPP/IA and the RPP/RPP surface intersection that are typical in applications will be discussed here.

(1) Rational polynomial parametric/implicit algebraic (RPP/IA) surface intersection

We start with a rational polynomial parametric surface to implicit algebraic surface intersection problem defined as

$$\begin{aligned} &\boldsymbol{r} = \boldsymbol{r}(u,v) = \left(\frac{X(u,v)}{W(u,v)}, \frac{Y(u,v)}{W(u,v)}, \frac{Z(u,v)}{W(u,v)}\right)^{\mathrm{T}} \cap f(\boldsymbol{r}) = 0, \\ &\quad 0 \le u, v \le 1 \end{aligned} \tag{12.78}$$

This leads to four algebraic equations in five unknowns $\boldsymbol{r}$, u, v (under-constrained system). For the usual low-degree surfaces $f(\boldsymbol{r})$ and low-degree patches $\boldsymbol{r}(u,\ v)$, we can substitute $\boldsymbol{r}(u,\ v)$ into $f(\boldsymbol{r}) = 0$ to obtain an implicit algebraic curve in u, v. Examples of low-order implicit algebraic surfaces in practical use are planes (degree 1), the natural quadrics (cylinder, sphere, cone) (degree 2), and torus (degree 4). In fact, in a survey of mechanical parts (mechanical elements), over 90% of all surfaces involved are of these types.

Now, let us denote the implicit algebraic surface $f(x,y,z)=0$ of total degree m by

$$f(x,y,z)=\sum_{i=0}^{m}\sum_{j=0}^{m-i}\sum_{k=0}^{m-i-j} c_{ijk}x^i y^j z^k = 0 \tag{12.79}$$

By substituting $x = X/W$, $y = Y/W$, $z = Z/W$, where $X(u,v)$, $Y(u,v)$, $Z(u,v)$ and $W(u,v)$ are all of maximum degree p in u and q in v, into Eq. (12.79) and multiplying by $W^m(u,v)$, one obtains an algebraic curve

$$F(u,v)=\sum_{i=0}^{m}\sum_{j=0}^{m-i}\sum_{k=0}^{m-i-j} c_{ijk}X^i(u,v)Y^j(u,v)Z^k(u,v) \\ W^{m-i-j-k}(u,v)=0 \tag{12.80}$$

of maximum degree $M = mp$ and $N = mq$ in u, v, respectively. Consequently, the problem of intersection reduces to the problem of tracing $F(u,v) = 0$ without omitting any special features of the curve, e.g., small loops and singularities, and accurately computing all its branches. The algebraic curve

$$F(u,v)=\sum_{i=0}^{M}\sum_{j=0}^{N} c_{ij}^M u^i v^j = 0 \tag{12.81}$$

can be reformulated in terms of Bernstein polynomials using the method Section 2.2.1 as follows:

$$F(u,v)=\sum_{i=0}^{M}\sum_{j=0}^{N} c_{ij}^B B_{i,M}(u)B_{j,N}(v) = 0 \tag{12.82}$$

where $(u,v) \in [0,1]^2$. In fact, the power basis form of $F(u,v) = 0$ need not be computed at all if polynomial arithmetic for Bernstein polynomials is used.

Given a point on every branch of an algebraic curve, the curve can be traced using differential curve properties. We now consider the intersection curve which lies on the surface represented by the parametric form $\boldsymbol{r}(t) = \boldsymbol{r}(u(t), v(t))$. Differentiating (12.82) with respect to t yields

$$F_u\dot{u} + F_v\dot{v} = 0 \tag{12.83}$$

where u, v are considered as functions of a parameter t. The solution to the differential equation is given by

$$\dot{u} = \xi F_v(u, v), \quad \dot{v} = -\xi F_u(u, v) \tag{12.84}$$

where ξ is an arbitrary non-zero factor. For example, ξ can be chosen to provide arc length parametrization using the first fundamental form of the surface as a normalization condition

$$\xi = \pm\frac{1}{\sqrt{EF_v^2 - 2FF_uF_v + GF_u^2}} \tag{12.85}$$

where E, F and G are first fundamental form coefficients of the parametric surface following standard differential geometry terminology. Note that F in Eq. (12.85) has no relation with function F that appears in other equations above. Equations (12.84) form a system of two first-order nonlinear differential equations which can be solved by the Runge–Kutta or other methods with adaptive step size. For a tracing method to work properly, we must provide all the starting points of all branches in advance. The starting points are identified by looking for characteristic points defined as follows:

(a) *Border points*: The intersections of $F(u, v) = 0$ with all four boundary edges of the parameter space $[0, 1]^2$, e.g., $F(0, v) = 0$, $0 \leq v \leq 1$.
(b) *Turning points:* The u-turning points are the points where the tangent of $F(u, v) = 0$ is parallel to the $u = 0$ axis, which satisfies the simultaneous equations $F = F_v = 0$ (with $F_u \neq 0$). On the other hand, the v-turning points are the points where the tangent of $F(u, v) = 0$ is parallel to the $v = 0$ axis, which satisfies the simultaneous equations $F = F_u = 0$ (with $F_v \neq 0$).
(c) *Singular points*: The points on the curve which satisfy the following three simultaneous equations $F = F_u = F_v = 0$ are

called singular points. Noting that $f(x,y,z)=0$ and $F(u,v)=W^m(u,v)f(x,y,z)=0$, we deduce

$$\begin{aligned} F_u &= mW^{m-1}W_u f + W^m\left(\frac{\partial f}{\partial x}\frac{\partial x}{\partial u}+\frac{\partial f}{\partial y}\frac{\partial y}{\partial u}+\frac{\partial f}{\partial z}\frac{\partial z}{\partial u}\right) \\ &= W^m \nabla f\cdot \boldsymbol{r}_u = 0 \end{aligned} \tag{12.86}$$

and hence at singular points $\nabla f\cdot\boldsymbol{r}_u=\nabla f\cdot\boldsymbol{r}_v=0$. This means that $\nabla f \parallel (\boldsymbol{r}_u\times\boldsymbol{r}_v)$ or that the normals of two surfaces are parallel and since $F(u,v)=0$ at these points, the two surfaces intersect tangentially.

Starting points for tracing algebraic curves could be border points, turning points and singular points. They can be solved through the above definitions together with Eq. (12.82).

(2) Rational polynomial parametric/rational polynomial parametric (RPP/RPP) surface intersection

Rational polynomial parametric surface to rational polynomial parametric surface intersection is defined as

$$\begin{aligned} \boldsymbol{r} = \boldsymbol{r}_1(\sigma,t) &= \left(\frac{X_1(\sigma,t)}{W_1(\sigma,t)},\frac{Y_1(\sigma,t)}{W_1(\sigma,t)},\frac{Z_1(\sigma,t)}{W_1(\sigma,t)}\right)^{\mathrm{T}}, \quad 0\le\sigma,t\le 1 \\ \cap\,\boldsymbol{r} = \boldsymbol{r}_2(u,v) &= \left(\frac{X_2(u,v)}{W_2(u,v)},\frac{Y_2(u,v)}{W_2(u,v)},\frac{Z_2(u,v)}{W_2(u,v)}\right)^{\mathrm{T}}, \quad 0\le u,v\le 1 \end{aligned} \tag{12.87}$$

Formulation can be provided by setting $\boldsymbol{r}_1(\sigma,t)=\boldsymbol{r}_2(u,v)$ which leads to three nonlinear polynomial equations for four unknowns σ, t, u, v. It is an underconstrained system with three equations and four unknowns. One can implicitize [433–436] $\boldsymbol{r}_1(\sigma,t)$ to the form $f(\boldsymbol{r})=0$ and substitution of $\boldsymbol{r}=\boldsymbol{r}_2(u,v)$ into f to reduce the problem to RPP/IA case for a low-degree surface. If this is not possible, one can use the marching methods for RPP/RPP surface intersection.

For the marching methods of RPP/RPP surface intersections, the border and collinear normal points between the two surfaces were found to be more convenient for discovering all connected components of the intersection. Collinear normal points (points on the two parametric surfaces at which the normal vectors are collinear) provide points inside all intersection loops and all singular points.

To simplify the notation, we replace $\boldsymbol{r}_1(\sigma, t)$ by $\boldsymbol{p}(\sigma, t)$ and $\boldsymbol{r}_2(u, v)$ by $\boldsymbol{q}(u, v)$. Then the collinear normal points satisfy the following equations:

$$(\boldsymbol{p}_\sigma \times \boldsymbol{p}_t) \cdot \boldsymbol{q}_u = 0, (\boldsymbol{p}_\sigma \times \boldsymbol{p}_t) \cdot \boldsymbol{q}_v = 0, (\boldsymbol{p} - \boldsymbol{q}) \cdot \boldsymbol{p}_\sigma = 0,$$
$$(\boldsymbol{p} - \boldsymbol{q}) \cdot \boldsymbol{p}_t = 0 \tag{12.88}$$

Equations (12.88) form a system of four nonlinear polynomial equations that can be solved using the methods in Section 2.2.1. Now, we split the patches in (at least) one parametric direction at these collinear normal points. Consequently, starting points are only border points on the boundaries of all subdomains created.

In order to trace the intersection curve, starting points must be located prior to tracing. An intersection curve branch can be traced if its pre-image starts from the parametric domain boundary in either parameter domain. The marching direction coincides with the tangential direction of the intersection curve $\boldsymbol{c}(s)$ which is perpendicular to the normal vectors of both surfaces. Therefore, the marching direction can be obtained as follows:

$$\boldsymbol{c}'(s) = \frac{\boldsymbol{P}(\sigma, t) \times \boldsymbol{Q}(u, v)}{|\boldsymbol{P}(\sigma, t) \times \boldsymbol{Q}(u, v)|} \tag{12.89}$$

where the normalization forces $\boldsymbol{c}(s)$ are to be arc length parametrized and

$$\boldsymbol{P} = \boldsymbol{p}_\sigma \times \boldsymbol{p}_t, \quad \boldsymbol{Q} = \boldsymbol{q}_u \times \boldsymbol{q}_v \tag{12.90}$$

are the normal vectors of $\boldsymbol{p}$ and $\boldsymbol{q}$, respectively. When the two surfaces intersect tangentially, we cannot use Eq. (12.89) since the denominator vanishes. In such cases, we must find the marching direction in an alternate way.

The intersection curve can also be viewed as a curve on the two intersecting surfaces. A curve $\sigma = \sigma(s)$, $t = t(s)$ in the σt-plane defines a curve $\boldsymbol{r} = \boldsymbol{c}(s) = \boldsymbol{p}(\sigma(s), t(s))$ on a parametric surface $\boldsymbol{p}(\sigma, t)$, as well as a curve $u = u(s)$, $v = v(s)$ in the uv-plane defines a curve $\boldsymbol{r} = \boldsymbol{c}(s) = \boldsymbol{q}\{u(s), v(s))$ on a parametric surface $\boldsymbol{q}(u, v)$. We can derive the first derivative of the intersection curve as a curve on the parametric surface using the chain rule:

$$\mathbf{c}'(s) = \boldsymbol{p}_\sigma \sigma' + \boldsymbol{p}_t t', \quad \mathbf{c}'(s) = \boldsymbol{q}_u u' + \boldsymbol{q}_v v' \tag{12.91}$$

Solving Eq. (12.91) together with Eq. (12.89), the solutions are immediately obtained as

$$\begin{aligned} \sigma' &= \frac{\det(\boldsymbol{c}', \boldsymbol{p}_t, \boldsymbol{P}(\sigma,t))}{\boldsymbol{P}(\sigma,t)\cdot\boldsymbol{P}(\sigma,t)}, \quad t' = \frac{\det(\boldsymbol{p}_\sigma, \boldsymbol{c}', \boldsymbol{P}(\sigma,t))}{\boldsymbol{P}(\sigma,t)\cdot\boldsymbol{P}(\sigma,t)} \\ u' &= \frac{\det(\boldsymbol{c}', \boldsymbol{q}_v, \boldsymbol{Q}(u,v))}{\boldsymbol{Q}(u,v)\cdot\boldsymbol{Q}(u,v)}, \quad v' = \frac{\det(\boldsymbol{q}_u, \boldsymbol{c}', \boldsymbol{Q}(u,v))}{\boldsymbol{Q}(u,v)\cdot\boldsymbol{Q}(u,v)} \end{aligned} \tag{12.92}$$

where det denotes the determinant. The points of the intersection curves are computed successively by integrating the initial value problem for a system of nonlinear ordinary differential equations (12.92) using standard numerical techniques.

12.6 Mesh Generation and Optimization

With the intersection curves and trimmed surfaces, the mesh generation methods are considered in this work. For surfaces, three methods are considered here: (1) direct discretization using triangular and quadrilateral elements through the tensor product parametric domain and the trimming curves (Figs. 12.25 and 12.26); (2) iterative discretization according to surface curvature (Fig. 12.28); (3) element optimization through the molecular dynamics method (Fig. 12.29). In order to get meshes dominated by quadrilateral or hexahedral elements, the following techniques are recommended: (1) For B-Rep surface models, direct parametrically discretized quadrilateral elements can be cut by the B-Rep intersection curves to obtain 2D elements. (2) For B-Rep volume models, direct three-dimensional (3D) box with hexahedral elements can be cut by the B-Rep surface model to obtain 3D elements. (3) For CSG volume models, the constructed solids with elements can be intersected with each other to obtain discretized elements for FEA. These procedures mean that the triangularization algorithm is only used for the intersected boundary elements.

For convenience of analyses, mesh mismatching should be considered during the process of mesh generation, or much earlier, in

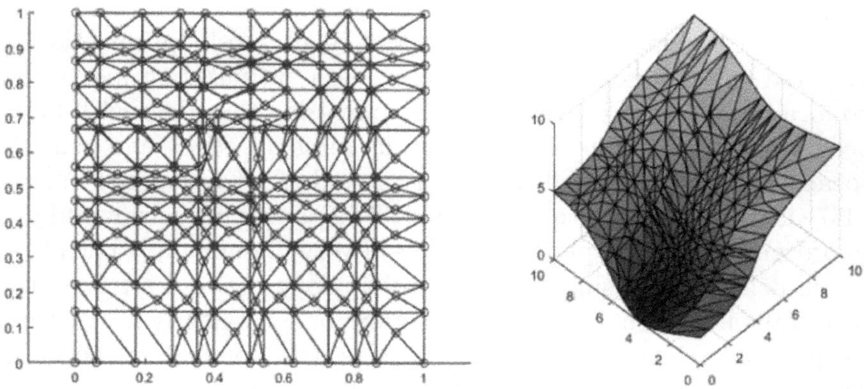

Fig. 12.28. Iterative discretization according to surface curvature (flatness equal to 0.1).

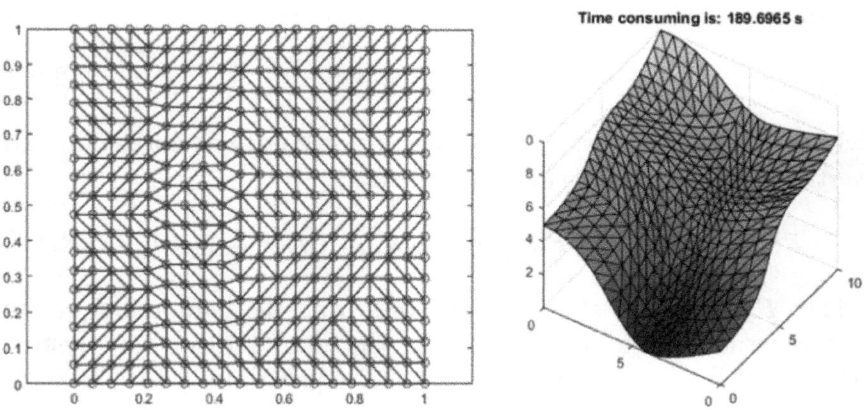

Fig. 12.29. Mesh optimization using the molecular dynamics method.

the process of surface trimming. The discretized elements are not represented by NURBS as expected when IGA was first proposed because this needs additional interpolation that will lose accuracy and NURBS has singularity for triangular and tetrahedral elements. Thus, the elements are represented by a differential quadrature hierarchical (DQH) method proposed by the first author of this work. The DQH method is of high accuracy, is capable of local adaptive analyses, is less sensitive to mesh distortion, and thus all subsequent analyses are carried out by using this method.

Although the DQH method is not sensitive to mesh distortion, much distorted meshes are not expected. Thus, mesh optimization is required to some extent. The simplest mesh optimization method is to locally find distorted elements by comparing the edges with local seed length and then eliminate the large/small edges and remesh locally. One can also use distance-based method to optimize the mesh [437–439]. Especially, the method in [437] is simple in concept and yet is proven to typically produce meshes of very high quality. The algorithm is based on a simple mechanical analogy between a triangular or tetrahedral mesh and a truss structure, or equivalently a structure of springs. In the physical model, the edges (the connections between pairs of points) correspond to bars, and the points correspond to joints of the truss. Each bar has a force–displacement relationship $f(l, l_0)$ depending on its current length l and its unextended length l_0. To solve for the force equilibrium, collect coordinates of all N mesh points into an N-by-m array $\boldsymbol{p}$ (e.g., 2D):

$$\boldsymbol{p} = [\boldsymbol{x}, \boldsymbol{y}] \tag{12.93}$$

where m is the dimension. The force vector $\boldsymbol{F}(\boldsymbol{p})$ has horizontal and vertical components at each mesh point:

$$\boldsymbol{F}(\boldsymbol{p}) = [\boldsymbol{F}_{\text{int},x}(\boldsymbol{p})\ \boldsymbol{F}_{\text{int},y}(\boldsymbol{p})] + [\boldsymbol{F}_{\text{ext},x}(\boldsymbol{p})\ \boldsymbol{F}_{\text{ext},y}(\boldsymbol{p})] \tag{12.94}$$

where $\boldsymbol{F}_{\text{int}}$ contains the internal forces from the bars and $\boldsymbol{F}_{\text{ext}}$ are the external forces (reactions from the boundaries). The system $\boldsymbol{F}(\boldsymbol{p}) = \boldsymbol{0}$ has to be solved for a set of equilibrium positions $\boldsymbol{p}$. A simple approach to solve $\boldsymbol{F}(\boldsymbol{p}) = \boldsymbol{0}$ is to introduce an artificial time dependence. For some $\boldsymbol{p}(0) = \boldsymbol{p}_0$, we consider the system of ODEs (in non-physical units!):

$$\frac{\mathrm{d}\boldsymbol{p}}{\mathrm{d}t} = \boldsymbol{F}(\boldsymbol{p}), \quad t \geq 0 \tag{12.95}$$

If a stationary solution is found, it satisfies our system $\boldsymbol{F}(\boldsymbol{p}) = \boldsymbol{0}$. The system (12.95) is approximated using the forward Euler method. At the discretized (artificial) time $t_n = n\Delta t$, the approximate solution $\boldsymbol{p}_n \approx \boldsymbol{p}(t_n)$ is updated by

$$\boldsymbol{p}_{n+1} = \boldsymbol{p}_n + \Delta t \cdot \boldsymbol{F}(\boldsymbol{p}) \tag{12.96}$$

One can see that this method is similar to the molecular dynamics method [438, 439]. One limitation of the method of [437] is that it requires the surface in implicit algebraic form. However, the method can be easily generalized to NURBS surface as in [438, 439]. Figure 12.29 shows an example of the optimized mesh. Note that for 3D surfaces, one needs to map back the points to the parametric surface after a few iteration steps.

Inspired by the relations between the parametric domain and the physical domain shown in Fig. 12.30, one can take the number of sampling points on one direction to be fixed and vary the number of sampling points on the other direction according to the length of the parametric curves to generate triangular mesh from a NURBS or NURL patch as shown in Fig. 12.31(a). Similar method can be used for three-dimensional mesh generation (Fig. 12.31(b)). The accuracy of the direct usage of linear triangulation in simulation is not good as

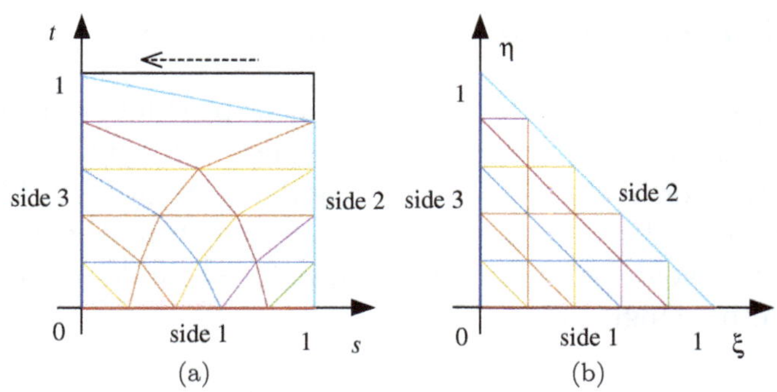

Fig. 12.30. A unit square domain (a) and a unit triangular domain (b).

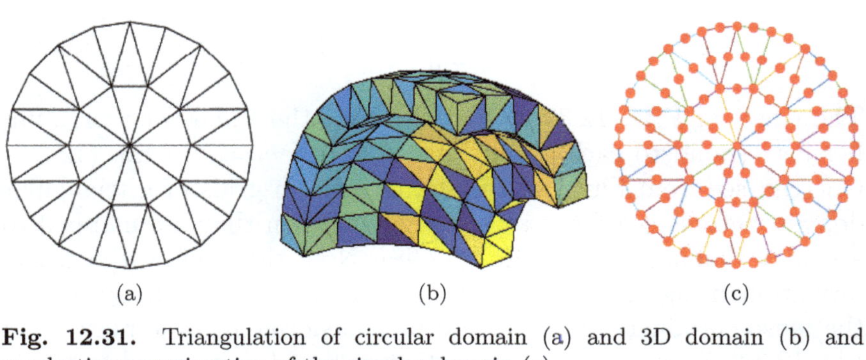

Fig. 12.31. Triangulation of circular domain (a) and 3D domain (b) and quadratic approximation of the circular domain (c).

can be seen from the results section of this work. So, it is necessary to improve the accuracy by using quadratic or higher-order bases for geometries (Fig. 12.31(c)). Elevation of the polynomial order of a geometry needs inverse computation, i.e, one needs to use nodes from the linear triangles to get nodes from the geometry. It is clear that boundary nodes and interior nodes should be computed separately. Due to the nonlinearity of inverse geometric computation and the possibility of the existence of degenerated points, this process should be done carefully. But it can be done efficiently. Since the DQHM is of high accuracy, one can also use the DQH bases for geometries. In this case, geometries and simulations use the same bases, which brings some simplicity in computation. Since quadratic or cubic bases are sufficient for geometries, too much inverse geometric computation may take time, and the order of the bases for simulation may need to be much higher, it is not necessary to remain isoparametric.

12.7 Geometric Mapping of Triangular Patch

Now, we consider how to transform a tensor product surface defined on the parametric coordinates of a unit square domain (Fig. 12.30(a)) to the area coordinates of a unit triangular domain (Fig. 12.30(b)). The unit triangular domain can be transformed from the unit square domain by degenerating the upper edge (for example) of the square domain into a single point:

$$\xi = (1-t)s, \quad \eta = t \tag{12.97}$$

The inverse transformation is

$$s = \frac{\xi}{1-\eta}, \quad t = \eta \tag{12.98}$$

The edges of Figs. 12.30(a) and (b) with the same color or number correspond to each other. Take $s = 0$ if $\eta = 1$ in Eq. (12.98). One can see from Fig. 12.18 that the uniform grids of a triangular domain become scattered to some extent when they are mapped to a square domain through Eq. (12.98). Note that before the transformation using Eq. (12.98), the parametric coordinate system of the tensor product surface should be transformed to the position of Fig. 12.30(a), i.e., the degenerated edge should correspond to $t = 1$.

With the relations in Eq. (12.98), the derivatives of a points on a surface $\boldsymbol{S}(s,t) = \{x(s,t), y(s,t), z(s,t)\}^T$ with respect to area coordinates are

$$\begin{aligned}
\frac{\partial \boldsymbol{S}(s,t)}{\partial \xi} &= \frac{\partial \boldsymbol{S}}{\partial s}\frac{\partial s}{\partial \xi} = \frac{1}{1-\eta}\frac{\partial \boldsymbol{S}}{\partial s} \\
\frac{\partial \boldsymbol{S}(s,t)}{\partial \eta} &= \frac{\partial \boldsymbol{S}}{\partial s}\frac{\partial s}{\partial \eta} + \frac{\partial \boldsymbol{S}}{\partial t} = \frac{\xi}{(1-\eta)^2}\frac{\partial \boldsymbol{S}}{\partial s} + \frac{\partial \boldsymbol{S}}{\partial t}
\end{aligned} \tag{12.99}$$

Let $\boldsymbol{S}_s = \partial \boldsymbol{S}/\partial s$, $\boldsymbol{S}_t = \partial \boldsymbol{S}/\partial t$, and so on. It is obvious that $\boldsymbol{S}_s(s,1) = 0$, and $\boldsymbol{S}_t(s,1)$ changes with the variation of s. Therefore, $\boldsymbol{S}_\xi = 0$ and $\boldsymbol{S}_\eta$ do not have a unique value at area coordinate (0, 1), which is not reasonable. Therefore, $\boldsymbol{S}_\xi$ and $\boldsymbol{S}_\eta$ at the point should be determined through the other way. It is obvious that the NURBS or NURL bases have zero function value and first derivatives at the vertexes. Therefore, the first derivatives at point (0, 1) can be uniquely determined through the two neighboring edges. The point on the triangle at (0, 1) can be interpolated by the two neighboring edges using Lagrange interpolation of two intersected straight lines as

$$\tilde{\boldsymbol{S}}(\xi,\eta) = \frac{1-\xi-\eta}{1-\eta}\boldsymbol{S}(0,t) + \frac{\xi}{1-\eta}\boldsymbol{S}(1,t) \tag{12.100}$$

Therefore, the first derivatives with respect to ξ and η can be given as

$$\begin{aligned}
\frac{\partial \tilde{\boldsymbol{S}}(\xi,\eta)}{\partial \xi} &= \frac{\boldsymbol{S}(1,t) - \boldsymbol{S}(0,t)}{1-\eta} \\
\frac{\partial \tilde{\boldsymbol{S}}(\xi,\eta)}{\partial \eta} &= \frac{\xi}{1-\eta}\frac{\partial \boldsymbol{S}(1,t)}{\partial \eta} + \frac{1-\xi-\eta}{1-\eta}\frac{\partial \boldsymbol{S}(0,t)}{\partial \eta} \\
&\quad + \frac{\xi[\boldsymbol{S}(1,t) - \boldsymbol{S}(0,t)]}{(1-\eta)^2}
\end{aligned} \tag{12.101}$$

Let $(\xi,\eta) \to (0,1)$, then one obtains

$$\begin{aligned}
\frac{\partial \tilde{\boldsymbol{S}}(0,1)}{\partial \xi} &= \boldsymbol{S}_t(0,1) - \boldsymbol{S}_t(1,1) \\
\frac{\partial \tilde{\boldsymbol{S}}(0,1)}{\partial \eta} &= \boldsymbol{S}_t(0,1)
\end{aligned} \tag{12.102}$$

One can degenerate some other edge to make the transformation. The procedure is similar to above. Equations (12.99) and (3.59) are usually sufficient for applications. One only needs to rotate the parametric coordinates of the quadrilateral element to let the edge $t = 1$ to be the edge that degenerates to a point.

12.8 High-Order Mesh Generation through Gmsh and Open CASCADE

This chapter introduces an *a posteriori* approach of high-order mesh generation based on the advanced modeling functions of the open source CAD core Open CASCADE (OCC) and the underlying mathematical algorithms of OCC. The present approach includes the following four steps:

(1) Use the open source mesh solver Gmsh to generate low-order straight-edged meshes;
(2) Curve the edge of the low-order mesh with the aid of the original surface geometry information and the element topology information obtained in step (1);

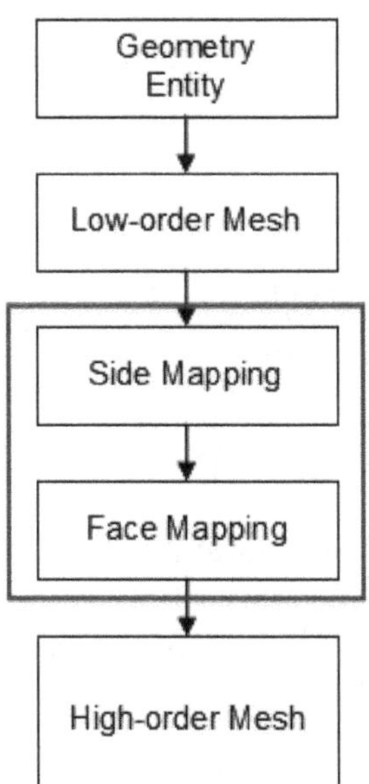

Fig. 12.32. High-order mesh generation process.

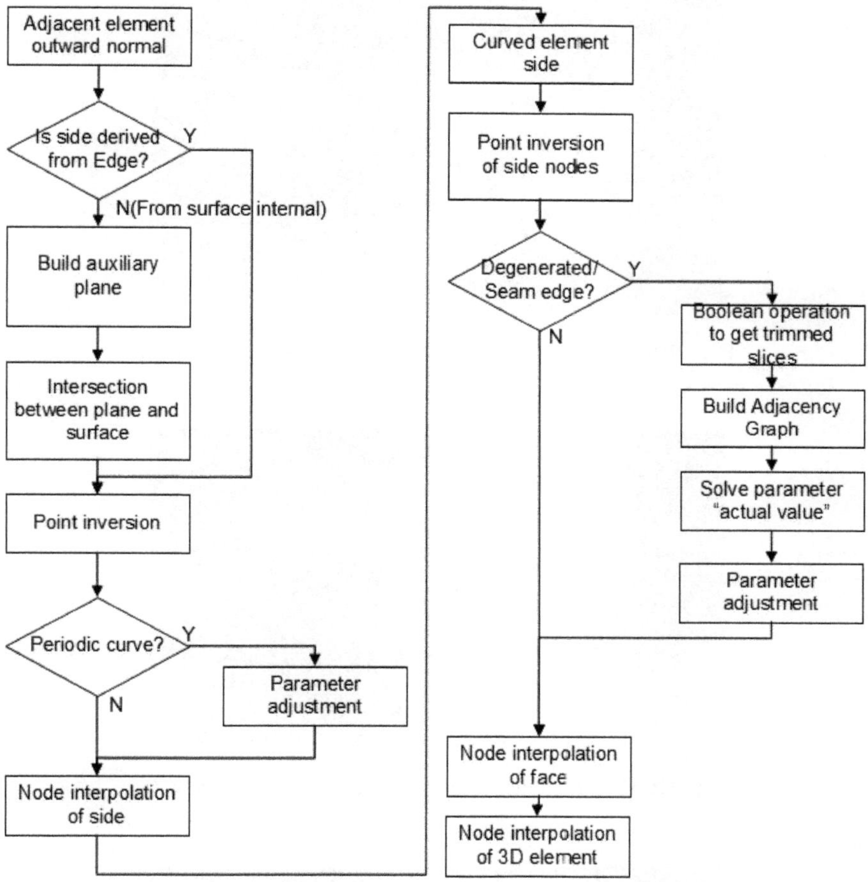

Fig. 12.33. Curving and surfacing process.

(3) Interpolate the interior nodes of the curved elements obtained in Step (2) and project the nodes to the original surface to get more accurate nodes;
(4) Assemble the surface mesh to obtain the final high-order volume mesh.

Figures 12.1 and 12.2 show the flowchart of the high-order mesh generation procedure.

Figures 12.34–12.38 show some examples of the mesh generated through the procedure. Note that the nodes here are non-equally spaced are/are different from Gmsh.

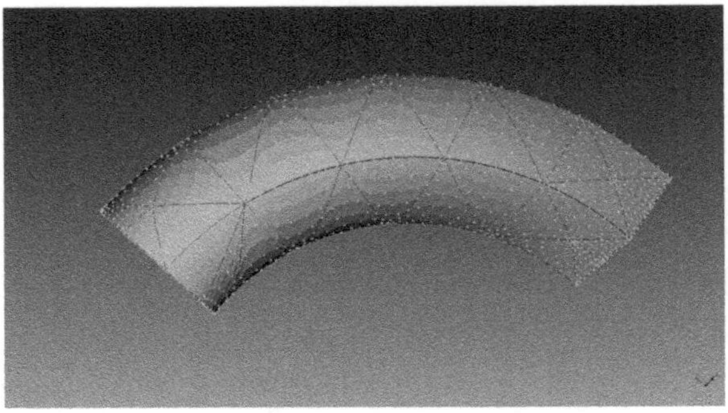

Fig. 12.34. Seventh-order mesh of loft entity: view 1.

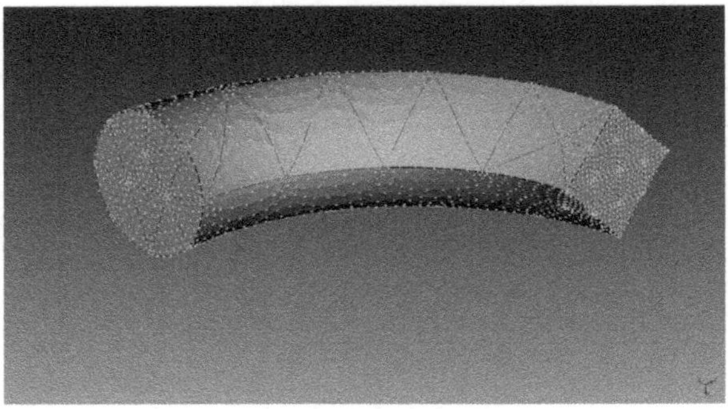

Fig. 12.35. Seventh-order mesh of loft entity: view 2.

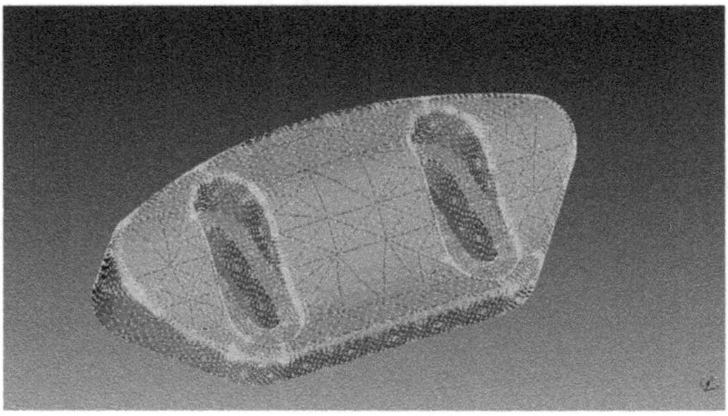

Fig. 12.36. Sixth-order mesh of coverplate entity: view 1.

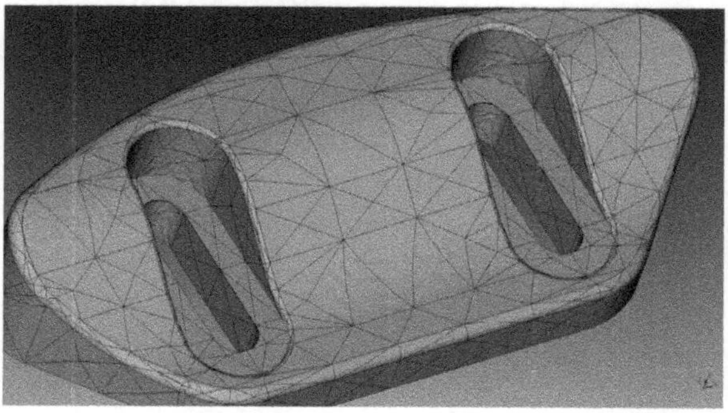

Fig. 12.37. Sixth-order mesh of coverplate entity: hidden nodes.

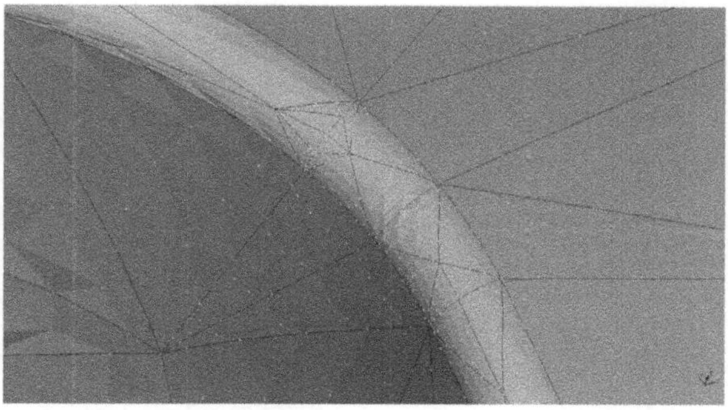

Fig. 12.38. Sixth-order mesh of coverplate entity: mesh on fillet.

12.9 Conclusions

This chapter considers design, mesh generation and analyses. The aim is to achieve a seamless integration of design and analyses as IGA. The process in this chapter is similar to the roadmap of IGA: first discretize CAD models into "exact" NURBS or NURL elements and then carry out IGA. Due to the possible approximation errors for trimmed CAD models and the singularity in computation for triangular and tetrahedral NURBS elements, this chapter transforms the CAD models into DQH elements. Since difficulties of pre-processing in FEA, such as element singularity, element mismatching, etc., originate from surface intersection, this work tries to solve the difficulties of IGA for trimmed surfaces during the process of surface intersection. With the information obtained in surface intersection, mesh optimization is considered in this work. The DQH method is of high accuracy, shows less sensitivity to geometry distortion, and is capable of local adaptive analysis, and thus all subsequent analyses are carried out in this method. It deserves to be noted that throughout Chapters 2–11, NURBS have been extensively used to generate elements for the HQEM.

References

1. Turner, M., J. Peiró, and D. Moxey, Curvilinear mesh generation using a variational framework. *Computer-Aided Design*, 2018. **103**: 73–91.
2. Zhao, Z., *et al.*, High-order curvilinear mesh generation technique based on an improved radius basic function approach. *International Journal for Numerical Methods in Fluids*, 2019. **91**(3): 97–111.
3. Tornabene, F., *et al.*, Strong formulation finite element method based on differential quadrature: A survey. *Applied Mechanics Reviews*, 2015. **67**(2): 020801–020801.
4. Wang, X., Z. Yuan, and C. Jin, Weak form quadrature element method and its applications in science and engineering: A State-of-the-art review. *Applied Mechanics Reviews*, 2017. **69**(3): 030801-030801-19.
5. Zienkiewicz, O.C., *et al.*, Three dimensional stress analysis, in *Proc. IUTAM Symp. High Speed Computing of Elastic Structures*. 1971. 413–432.
6. Zienkiewicz, O.C., R.L. Taylor, and J.Z. Zhu, *The Finite Element Method: Its Basis and Fundamentals*. 6 ed. 2005, New York: Elsevier.
7. Stein, E., R.D. Borst, and T.J.R. Hughes, *The Encyclopedia of Computational Mechanics*. 2004; John Wiley & Sons, Ltd.
8. Petyt, M., *Introduction to Finite Element Vibration Analysis*. 2nd ed. 2010. NewYork: Cambridge University Press.
9. Szabó, B. and I. Babuška, *Introduction to Finite Element Analysis: Formulation, Verification and Validation*. Wiley Series in Computational Mechanics. 2011, New Delhi, India: John Wiley & Sons, Ltd. 364.
10. Ziel, V.S., *et al.*, *High-order 2D mesh curving methods with a piecewise linear target and application to Helmholtz problems. Computer-Aided Design*, 2018. **105**: 26–41.

11. Bert, C.W. and M. Malik, *Differential Quadrature Method in Computational Mechanics: A Review.* Applied Mechanics Reviews, 1996. **49**(1): p. 1–28.
12. Bellman, R. and J. Casti, Differential quadrature and long-term integration. *Journal of Mathematical Analysis and Applications*, 1971. **34**(2): 235–238.
13. Bellman, R., B.G. Kashef, and J. Casti, Differential quadrature: A technique for the rapid solution of nonlinear partial differential equations. *Journal of Computational Physics*, 1972. **10**(1): 40–52.
14. Bert, C.W., S.K. Jang, and A.G. Striz, Two new approximate methods for analyzing free vibration of structural components. *AIAA Journal*, 1988. **26**(5): 612–618.
15. Striz, A. and Y. Loo, Application of differential quadrature to the analysis of static aeroelastic phenomena, in *34th Structures, Structural Dynamics and Materials Conference.* 1993.
16. Xing, Y. and B. Liu, High-accuracy differential quadrature finite element method and its application to free vibrations of thin plate with curvilinear domain. *International Journal for Numerical Methods in Engineering*, 2009. **80**(13): 1718–1742.
17. Mingle, J.O., Computational considerations in non-linear diffusion. *International Journal for Numerical Methods in Engineering*, 1973. **7**(2): 103–116.
18. Quan, J.R. and C.T. Chang, New insights in solving distributed system equations by the quadrature method — I. Analysis. *Computers & Chemical Engineering*, 1989. **13**(7): 779–788.
19. Shu, C. and B.E. Richards, Application of generalized differential quadrature to solve two-dimensional incompressible Navier-Stokes equations. *International Journal for Numerical Methods in Fluids*, 1992. **15**(7): 791–798.
20. Shu, C. and B.E. Richards, Parallel simulation of incompressible viscous flows by generalized differential quadrature. *Computing Systems in Engineering*, 1992. **3**(1): 271–281.
21. Fung, T.C., Imposition of boundary conditions by modifying the weighting coefficient matrices in the differential quadrature method. *International Journal for Numerical Methods in Engineering*, 2003. **56**(3): 405–432.
22. Karami, G. and P. Malekzadeh, Application of a new differential quadrature methodology for free vibration analysis of plates. *International Journal for Numerical Methods in Engineering*, 2003. **56**(6): 847–868.
23. Fung, T.C., Stability and accuracy of differential quadrature method in solving dynamic problems. *Computer Methods in Applied Mechanics and Engineering*, 2002. **191**(13): 1311–1331.

24. Xing, Y. and B. Liu, A differential quadrature analysis of dynamic and quasi-static magneto-thermo-elastic stresses in a conducting rectangular plate subjected to an arbitrary variation of magnetic field. *International Journal of Engineering Science*, 2010. **48**(12): 1944–1960.
25. Mulay, S.S., H. Li, and S. See, On the random differential quadrature (RDQ) method: Consistency analysis and application in elasticity problems. *Computational Mechanics*, 2009. **44**(4): 563–590.
26. Striz, A.G., C. Weilong, and C.W. Bert, Static analysis of structures by the quadrature element method (QEM). *International Journal of Solids and Structures*, 1994. **31**(20): 2807–2818.
27. Striz, A., W. Chen, and C. Bert, High-accuracy plane stress and plate elements in the Quadrature Element Method, in *36th Structures, Structural Dynamics and Materials Conference*, 1995.
28. Chen, W.L., A.G. Striz, and C.W. Bert, High-accuracy plane stress and plate elements in the quadrature element method. *International Journal of Solids and Structures*, 2000. **37**(4): 627–647.
29. Striz, A.G., W.L. Chen, and C.W. Bert, Free vibration of plates by the high accuracy quadrature element method. *Journal of Sound and Vibration*, 1997. **202**(5): 689–702.
30. Zhong, H. and T. Yu, Flexural vibration analysis of an eccentric annular Mindlin plate. *Archive of Applied Mechanics*, 2007. **77**(4): 185–195.
31. Zhong, H. and Z. Yue, Analysis of thin plates by the weak form quadrature element method. *Science China Physics, Mechanics and Astronomy*, 2012. **55**(5): 861–871.
32. Xing, Y., B. Liu, and G. Liu, A differential quadrature finite element method. *International Journal of Applied Mechanics*, 2010. **2**(1): 207–227.
33. Wang, X., Differential Quadrature and Differential Quadrature Based Element Methods, *ed. X. Wang. 2015*: Butterworth-Heinemann.
34. Liu, B., *et al.*, Thickness-shear vibration analysis of circular quartz crystal plates by a differential quadrature hierarchical finite element method. *Composite Structures*, 2015. **131**: 1073–1080.
35. Liu, C., *et al.*, A differential quadrature hierarchical finite element method and its applications to vibration and bending of Mindlin plates with curvilinear domains. *International Journal for Numerical Methods in Engineering*, 2016. **109**(2): 174–197.
36. Jin, C. and X. Wang, Weak form quadrature element method for accurate free vibration analysis of thin skew plates. *Computers & Mathematics with Applications*, 2015. **70**(8): 2074–2086.
37. Hu, Y.C., K.Y. Sze, and Y.X. Zhou, Stabilized plane and axisymmetric Lobatto finite element models. *Computational Mechanics*, 2015. **56**(5): 879–903.

38. Tong, P., Simplex elements of C0 continuity with varying polynomial degrees. *International Journal for Numerical Methods in Engineering*, 1977. **11**(1): 27–38.
39. Peano, A., Hierarchies of conforming finite elements for plane elasticity and plate bending. *Computers & Mathematics with Applications*, 1976. **2**(3–4): 211–224.
40. Szabó, B.A. and A.K. Mehta, P-convergent finite element approximations in fracture mechanics. *International Journal for Numerical Methods in Engineering*, 1978. **12**(3): 551–560.
41. Zienkiewicz, O.C., J.P. De, S.R. Gago, and D.W. Kelly, The hierarchical concept in finite element analysis. *Computers & Structures*, 1983. **16**(1–4): 53–65.
42. Peano, A., *et al.*, Adaptive approximations in finite element structural analysis. *Computers & Structures*, 1979. **10**(1): 333–342.
43. Zienkiewicz, O.C. and R.L. Taylor, *The Finite Element Method: The Basis.* 5 ed. 2000, New York: Elsevier.
44. Zhu, D.-C., Hierarchal finite elements and their application to structural natural vibration problems, in *23rd Structures, Structural Dynamics and Materials Conference.* 1982, American Institute of Aeronautics and Astronautics. 206–211.
45. Zhu, D.C. Development of hierarchical finite element methods at BIAA, in *The International Conference on Computational Mechanics*, 1986. Tokyo.
46. Campion, S.D. and J.L. Jarvis, An investigation of the implementation of the p-version finite element method. *Finite Elements in Analysis and Design*, 1996. **23**(1): 1–21.
47. Beslin, O. and J. Nicolas, A hierarchical functions set for predicting very high order plate bending modes with any boundary conditions. *Journal of Sound and Vibration*, 1997. **202**(5): 633–655.
48. Houmat, A., In-plane vibration of plates with curvilinear plan-forms by a trigonometrically enriched curved triangular p-element. *Thin-Walled Structures*, 2008. **46**(2): 103–111.
49. Houmat, A., Free vibration analysis of arbitrarily shaped membranes using the trigonometric p-version of the finite-element method. *Thin-Walled Structures*, 2006. **44**(9): 943–951.
50. Szabó, B.A. and G.J. Sahrmann, Hierarchic plate and shell models based on p-extension. *International Journal for Numerical Methods in Engineering*, 1988. **26**(8): 1855–1881.
51. Hughes, T.J.R., J.A. Cottrell, and Y. Bazilevs, Isogeometric analysis: CAD, finite elements, NURBS, exact geometry and mesh refinement. *Computer Methods in Applied Mechanics and Engineering*, 2005. **194**(39–41): 4135–4195.

52. Cottrell, J.A., T.J.R. Hughes, and Y. Bazilevs, *Isogeometric Analysis: Toward Integration of CAD and FEA*. 2009, Singapore: John Wiley & Sons, Ltd.
53. Korneev, V.G. and S. Jensen, Domain decomposition preconditioning in the hierarchical p-version of the finite element method. *Applied Numerical Mathematics*, 1999. **29**(4): 479–518.
54. Webb, J.P. and R. Abouchacra, Hierarchal triangular elements using orthogonal polynomials. *International Journal for Numerical Methods in Engineering*, 1995. **38**(2): 245–257.
55. Szabó, B. and I. Babuška, *Finite Element Analysis*. 1991, New York: John Wiley & Sons.
56. Houmat, A., Hierarchical finite element analysis of the vibration of membranes. *Journal of Sound and Vibration*, 1997. **201**(4): 465–472.
57. Houmat, A., An alternative hierarchical finite element formulation applied to plate vibrations. *Journal of Sound and Vibration*, 1997. **206**(2): 201–215.
58. Ribeiro, P., Hierarchical finite element analyses of geometrically non-linear vibration of beams and plane frames. *Journal of Sound and Vibration*, 2001. **246**(2): 225–244.
59. Ribeiro, P. and M. Petyt, Non-linear vibration of composite laminated plates by the hierarchical finite element method. *Composite Structures*, 1999. **46**(3): 197–208.
60. Houmat, A., Free vibration analysis of membranes using the h–p version of the finite element method. *Journal of Sound and Vibration*, 2005. **282**(1–2): 401–410.
61. Zienkiewicz, O.C., *et al.*, Finite element methods in the analysis of reactor vessels. *Nuclear Engineering and Design*, 1972. **20**(2): 507–541.
62. Irons, B.M., A technique for degenerating brick-type isoparametric elements using hierarchical midside nodes. *International Journal for Numerical Methods in Engineering*, 1974. **8**(1): 203–209.
63. Dubiner, M., Spectral methods on triangles and other domains. *Journal of Scientific Computing*, 1991. **6**(4): 345–390.
64. Duffy, M.G., Quadrature over a pyramid or cube of integrands with a singularity at a vertex. *SIAM Journal on Numerical Analysis*, 1982. **19**(6): 1260–1262.
65. Koornwinder, T., Two-variable analogues of the classical orthogonal polynomials, in *Theory and Application of Special Functions*, R.A. Askey, Editor, 1975, Academic Press, 435–495.
66. Carnevali, P., *et al.*, New basis functions and computational procedures for p-version finite element analysis. *International Journal for Numerical Methods in Engineering*, 1993. **36**(22): 3759–3779.

67. Szabó, B. and I. Babuška, *Finite Element Analysis*. Wiley Series in Computational Mechanics. 1991, New Delhi, India: John Wiley & Sons, Ltd.
68. Sherwin, S.J. and G.E. Karniadakis, A new triangular and tetrahedral basis for high-order (hp) finite element methods. *International Journal for Numerical Methods in Engineering*, 1995. **38**(22): 3775–3802.
69. Sherwin, S.J. and G.E. Karniadakis, A triangular spectral element method; applications to the incompressible Navier-Stokes equations. *Computer Methods in Applied Mechanics and Engineering*, 1995. **123**(1–4): 189–229.
70. Gatto, P. and L. Demkowicz, Construction of H1-conforming hierarchical shape functions for elements of all shapes and transfinite interpolation. *Finite Elements in Analysis and Design*, 2010. **46**(6): 474–486.
71. Karniadakis, G.E. and S.J. Sherwin, *Spectral/hp Element Methods for Computational Fluid Dynamics*. 2 ed. 2005, New York: Oxford University Press.
72. Shephard, M.S., S. Dey, and J.E. Flaherty, A straightforward structure to construct shape functions for variable p-order meshes. *Computer Methods in Applied Mechanics and Engineering*, 1997. **147**(3–4): 209–233.
73. Adjerid, S., M. Aiffa, and J.E. Flaherty, Hierarchical finite element bases for triangular and tetrahedral elements. *Computer Methods in Applied Mechanics and Engineering*, 2001. **190**(22–23): 2925–2941.
74. Bittencourt, M.L., Fully tensorial nodal and modal shape functions for triangles and tetrahedra. *International Journal for Numerical Methods in Engineering*, 2005. **63**(11): 1530–1558.
75. Bittencourt, M.L., M.G. Vazquez, and T.G. Vazquez, Construction of shape functions for the h- and p-versions of the FEM using tensorial product. *International Journal for Numerical Methods in Engineering*, 2007. **71**(5): 529–563.
76. Liu, B., *et al.*, A differential quadrature hierarchical finite element method using Fekete points for triangles and tetrahedrons and its applications to structural vibration. *Computer Methods in Applied Mechanics and Engineering*, 2019. **349**: 798–838.
77. Zhong, H., Triangular differential quadrature. *Communications in Numerical Methods in Engineering*, 2000. **16**(6): 401–408.
78. Zhong, H., Triangular differential quadrature and its application to elastostatic analysis of Reissner plates. *International Journal of Solids and Structures*, 2001. **38**(16): 2821–2832.
79. Zhong, H.Z., Free vibration analysis of isosceles triangular Mindlin plates by the triangular differential quadrature method. *Journal of Sound and Vibration*, 2000. **237**(4): 697–708.

80. Zhong, H., Application of triangular differential quadrature to problems with curved boundaries. *Communications in Numerical Methods in Engineering*, 2002. **18**(9): 633–643.
81. Zhong, H., Y. Hua, and Y. He, Localized triangular differential quadrature. *Numerical Methods for Partial Differential Equations*, 2003. **19**(5): 682–692.
82. Zhong, H. and S. Xing, On enforcement of corner conditions in triangular differential quadrature. *Computational Mechanics*, 2004. **33**(6): 413–420.
83. Shen, J., T. Tang, and L.L. Wang, *Spectral Methods: Algorithms, Analysis and Applications.* 2011: Springer.
84. Taylor, M.A., B.A. Wingate, and R.E. Vincent, An algorithm for computing fekete points in the triangle. *SIAM Journal on Numerical Analysis*, 2000. **38**(5): 1707–1720.
85. Bos, L.P., Bounding the Lebesgue function for Lagrange interpolation in a simplex. *Journal of Approximation Theory*, 1983. **38**(1): 43–59.
86. Fekete, M., Über die Verteilung der Wurzeln bei gewissen algebraischen Gleichungen mit ganzzahligen Koeffizienten. *Mathematische Zeitschrift*, 1923. **17**(1): 228–249.
87. Bos, L., On certain configurations of points in Rn which are unisolvent for polynomial interpolation. *Journal of Approximation Theory*, 1991. **64**(3): 271–280.
88. Mavriplis, C. and J.V. Rosendale, triangular spectral elements for incompressible fluid flow, in *11th Computational Fluid Dynamics Conference.* 1993.
89. Chen, Q. and I. Babuška, Approximate optimal points for polynomial interpolation of real functions in an interval and in a triangle. *Computer Methods in Applied Mechanics and Engineering*, 1995. **128**(3): 405–417.
90. Hesthaven, J.S., From electrostatics to almost optimal nodal sets for polynomial interpolation in a simplex. *SIAM Journal of Numerical Analysis*, 1998. **35**(2): 655–676.
91. Pasquetti, R. and F. Rapetti, Spectral element methods on triangles and quadrilaterals: Comparisons and applications. *Journal of Computational Physics*, 2004. **198**(1): 349–362.
92. Blyth, M.G. and C. Pozrikidis, A Lobatto interpolation grid over the triangle. *IMA Journal of Applied Mathematics*, 2005. **71**(1): 153–169.
93. Blyth, M.G., H. Luo, and C. Pozrikidis, A comparison of interpolation grids over the triangle or the tetrahedron. *Journal of Engineering Mathematics*, 2006. **56**(3): 263–272.
94. Luo, H. and C. Pozrikidis, A Lobatto interpolation grid in the tetrahedron. *IMA Journal of Applied Mathematics*, 2006. **71**(2): 298–313.

95. Heinrichs, W., Improved Lebesgue constants on the triangle. *Journal of Computational Physics*, 2005. **207**(2): 625–638.
96. Warburton, T., An explicit construction of interpolation nodes on the simplex. *Journal of Engineering Mathematics*, 2006. **56**(3): 247–262.
97. Briani, M., A. Sommariva, and M. Vianello, Computing Fekete and Lebesgue points: Simplex, square, disk. *Journal of Computational and Applied Mathematics*, 2012. **236**(9): 2477–2486.
98. Zhong, H. and J. Xu, A non-uniform grid for triangular differential quadrature. *SCIENCE CHINA Physics, Mechanics & Astronomy*, 2016. **59**(12): 124611.
99. Liu, C., *et al.*, In-plane vibration analysis of plates in curvilinear domains by a differential quadrature hierarchical finite element method. *Meccanica*, 2016. **52**(4): 1017–1033.
100. Chan, J. and T. Warburton, A Comparison of high order interpolation nodes for the pyramid. *SIAM Journal on Scientific Computing*, 2015. **37**(5): A2151–A2170.
101. O'Malley, B., *et al.*, Discontinuous Galerkin spatial discretisation of the neutron transport equation with pyramid finite elements and a discrete ordinate (SN) angular approximation. *Annals of Nuclear Energy*, 2018. **113**: 526–535.
102. O'Malley, B., *et al.*, Pyramid finite elements for discontinuous and continuous discretizations of the neutron diffusion equation with applications to reactor physics. *Progress in Nuclear Energy*, 2018. **105**: 175–184.
103. Knabner, P. and G. Summ, The invertibility of the isoparametric mapping for pyramidal and prismatic finite elements. *Numerische Mathematik*, 2001. **88**(4): 661–681.
104. Bluck, M.J. and S.P. Walker, Polynomial basis functions on pyramidal elements. *Communications in Numerical Methods in Engineering*, 2008. **24**(12): 1827–1837.
105. Ayala Bravo, C.M.A., *et al.*, Definition of a P-interpolating space of hierarchical bases of finite elements on the pyramid. *Linear Algebra and its Applications*, 2014. **460**: 174–204.
106. Danielson, K.T. and M.D. Adley, Five node pyramid elements for explicit time integration in nonlinear solid dynamics. *Finite Elements in Analysis and Design*, 2018. **141**: 37–54.
107. Yamakawa, S., I. Gentilini, and K. Shimada, Subdivision templates for converting a non-conformal hex-dominant mesh to a conformal hex-dominant mesh without pyramid elements. *Engineering with Computers*, 2011. **27**(1): 51–65.

108. Bedrosian, G., Shape functions and integration formulas for three-dimensional finite element analysis. *International Journal for Numerical Methods in Engineering*, 1992. **35**(1): 95–108.
109. Sherwin, S.J., T.C.E. Warburton, and A.G.E. Karniadakis, Spectral/hp methods for elliptic problems on hybrid grids. *Contemporary Mathematics*, 1998. **218**: 191–216.
110. Meshkat, S. and D. Talmor, Generating a mixed mesh of hexahedra, pentahedra and tetrahedra from an underlying tetrahedral mesh. *International Journal for Numerical Methods in Engineering*, 2000. **49**(1-2): 17–30.
111. Hartmann, R. and T. Leicht, Generation of unstructured curvilinear grids and high-order discontinuous Galerkin discretization applied to a 3D high-lift configuration. *International Journal for Numerical Methods in Fluids*, 2016. **82**(6): 316–333.
112. Chen, J., C.-J. Li, and W.-J. Chen, A 3D pyramid spline element. *Acta Mechanica Sinica*, 2011. **27**(6): 986–993.
113. Lei, N., *et al.*, Quadrilateral and hexahedral mesh generation based on surface foliation theory. *Computer Methods in Applied Mechanics and Engineering*, 2017. **316**: 758–781.
114. Lei, N., *et al.*, Quadrilateral and hexahedral mesh generation based on surface foliation theory II. *Computer Methods in Applied Mechanics and Engineering*, 2017. **321**: 406–426.
115. Owen, S.J. and S. Saigal, H-Morph: an indirect approach to advancing front hex meshing. *International Journal for Numerical Methods in Engineering*, 2000. **49**(1-2): 289–312.
116. Owen, S.J. and S. Saigal, Formation of pyramid elements for hexahedra to tetrahedra transitions. *Computer Methods in Applied Mechanics and Engineering*, 2001. **190**(34): 4505–4518.
117. Yerry, M.A. and M.S. Shephard, Automatic three-dimensional mesh generation by the modified-octree technique. *International Journal for Numerical Methods in Engineering*, 1984. **20**(11): 1965–1990.
118. Bergot, M., G. Cohen, and M. Duruflé, Higher-order finite elements for hybrid meshes using new nodal pyramidal elements. *Journal of Scientific Computing*, 2010. **42**(3): 345–381.
119. Zgainski, F., *et al.*, A new family of finite elements: The pyramidal elements. *IEEE Transactions on Magnetics*, 1996. **32**(3): 1393–1396.
120. Graglia, R.D. and I. Gheorma, Higher order interpolatory vector bases on pyramidal elements. *IEEE Transactions on Antennas and Propagation*, 1999. **47**(5): 775–782.
121. Nigam, N. and J. Phillips, *Higher-order Finite Elements on Pyramids.* arXiv:math/0610206v6, 2007.

122. Ziekiewicz, O., *The Finite Element Method in Engineering Science.* McGrawe Hill, London, 1971.
123. Ainsworth, M. and G. Fu, A lowest-order composite finite element exact sequence on pyramids. *Computer Methods in Applied Mechanics and Engineering*, 2017. **324**: 110–127.
124. Newton, R.E., Degeneration of brick-type isoparametric elements. *International Journal for Numerical Methods in Engineering*, 1973. **7**(4): 579–581.
125. Coulomb, J., F. Zgainski, and Y. Marechal, A pyramidal Element to link hexahedral, prismatic and tetrahedral edge finite elements. *IEEE Transactions on Magnetics*, 1997. **33**(2): 1362–1365.
126. Sherwin, S.J., Hierarchical hp finite elements in hybrid domains. *Finite Elements in Analysis Design*, 1997. **27**(1): 109–119.
127. Nigam, N. and J. Phillips, High-order conforming finite elements on pyramids. *IMA Journal of Numerical Analysis*, 2011. **32**(2): 448–483.
128. Fuentes, F., *et al.*, Orientation embedded high order shape functions for the exact sequence elements of all shapes. *Computers & Mathematics with Applications*, 2015. **70**(4): 353–458.
129. Bergot, M. and M. Duruflé, High-order optimal edge elements for pyramids, prisms and hexahedra. *Journal of Computational Physics*, 2013. **232**(1): 189–213.
130. Bergot, M. and M. Duruflé, Higher-order discontinuous Galerkin method for pyramidal elements using orthogonal bases. *Numerical Methods for Partial Differential Equations*, 2013. **29**(1): 144–169.
131. Holzer, S.M. and Z. Yosibash, *The p-version of the Finite Element Method in Incremental Elasto-plastic Analysis.* 1993, Center for Computational Mechanics, Washington University: St. Louis, MO 63130.
132. Holzer, S.M. and Z. Yosibash, The p-version of the finite element method in incremental elasto-plastic analysis. *International Journal for Numerical Methods in Engineering*, 1996. **39**(11): 1859–1878.
133. Szabó, B.A., R.L. Actis, and S.M. Holzer. *Solution of Elastic-Plastic Stress Analysis Problems by the P-version of the Finite Element Method.* 1995. New York: Springer.
134. Krause, R., R. Mücke, and E. Rank, hp-Version finite elements for geometrically non-linear problems. *Communications in Numerical Methods in Engineering*, 1995. **11**(11): 887–897.
135. Han, W. and M. Petyt, Geometrically nonlinear vibration analysis of thin, rectangular plates using the hierarchical finite element method — I: The fundamental mode of isotropic plates. *Computers & Structures*, 1997. **63**(2): 295–308.

136. Han, W. and M. Petyt, Geometrically nonlinear vibration analysis of thin, rectangular plates using the hierarchical finite element method — II: 1st mode of laminated plates and higher modes of isotropic and laminated plates. *Computers & Structures*, 1997. **63**(2): 309–318.
137. Jeremić, B. and C. Xenophontos, Application of the p-version of the finite element method to elastoplasticity with localization of deformation. *Communications in Numerical Methods in Engineering*, 1999. **15**(12): 867–876.
138. Ribeiro, P. and M. Petyt, Non-linear vibration of beams with internal resonance by the hierarchical finite-element method. *Journal of Sound and Vibration*, 1999. **224**(4): 591–624.
139. Ribeiro, P. and M. Petyt, Nonlinear vibration of plates by the hierarchical finite element and continuation methods. *International Journal of Mechanical Sciences*, 1999. **41**(4–5): 437–459.
140. Ribeiro, P., Nonlinear vibrations of simply-supported plates by the p-version finite element method. *Finite Elements in Analysis and Design*, 2005. **41**(9–10): 911–924.
141. Fonseca, J.R. and P. Ribeiro, Beam p-version finite element for geometrically non-linear vibrations in space. *Computer Methods in Applied Mechanics and Engineering*, 2006. **195**(9–12): 905–924.
142. Lopes Alonso, R. and P. Ribeiro, Flexural and torsional non-linear free vibrations of beams using a p-version finite element. *Computers & Structures*, 2008. **86**(11–12): 1189–1197.
143. Ribeiro, P. and G.H.M. van der Heijden, Elasto-plastic and geometrically nonlinear vibrations of beams by the p-version finite element method. *Journal of Sound and Vibration*, 2009. **325**(1–2): 321–337.
144. Düster, A. and E. Rank, The p-version of the finite element method compared to an adaptive h-version for the deformation theory of plasticity. *Computer Methods in Applied Mechanics and Engineering*, 2001. **190**(15–17): 1925–1935.
145. Düster, A., *et al.*, A numerical investigation of high-order finite elements for problems of elastoplasticity. *Journal of Scientific Computing*, 2002. **17**(1): 397–404.
146. Düster, A. and E. Rank, A p-version finite element approach for two- and three-dimensional problems of the J2 flow theory with non-linear isotropic hardening. *International Journal for Numerical Methods in Engineering*, 2002. **53**(1): 49–63.
147. Suri, M., Analytical and computational assessment of locking in the hp finite element method. *Computer Methods in Applied Mechanics and Engineering*, 1996. **133**(3): 347–371.

148. Düster, A., S. Hartmann, and E. Rank, p-FEM applied to finite isotropic hyperelastic bodies. *Computer Methods in Applied Mechanics and Engineering*, 2003. **192**(47): 5147–5166.
149. Nübel, V., A. Düster, and E. Rank, An rp-adaptive finite element method for the deformation theory of plasticity. *Computational Mechanics*, 2007. **39**(5): 557–574.
150. Yosibash, Z., *et al.*, Axisymmetric pressure boundary loading for finite deformation analysis using p-FEM. *Computer Methods in Applied Mechanics and Engineering*, 2007. **196**(7): 1261–1277.
151. Heisserer, U., *et al.*, On volumetric locking-free behaviour of p-version finite elements under finite deformations. *Communications in Numerical Methods in Engineering*, 2008. **24**(11): 1019–1032.
152. Heisserer, U., *et al.*, p-FEM for finite deformation powder compaction. *Computer Methods in Applied Mechanics and Engineering*, 2008. **197**(6): 727–740.
153. Wang, X. and J. Huang, Elastoplastic buckling analyses of rectangular plates under biaxial loadings by the differential quadrature method. *Thin-Walled Structures*, 2009. **47**(1): 14–20.
154. Zhang, W. and X. Wang, Elastoplastic buckling analysis of thick rectangular plates by using the differential quadrature method. *Computers & Mathematics with Applications*, 2011. **61**(1): 44–61.
155. Andakhshideh, A., S. Maleki, and M.M. Aghdam, Non-linear bending analysis of laminated sector plates using generalized differential quadrature. *Composite Structures*, 2010. **92**(9): 2258–2264.
156. He, R. and H. Zhong, Large deflection elasto-plastic analysis of frames using the weak-form quadrature element method. *Finite Elements in Analysis and Design*, 2012. **50**: 125–133.
157. Xiao, N. and H. Zhong, Non-linear quadrature element analysis of planar frames based on geometrically exact beam theory. *International Journal of Non-Linear Mechanics*, 2012. **47**(5): 481–488.
158. Zhang, R. and H. Zhong, Weak form quadrature element analysis of planar slender beams based on geometrically exact beam theory. *Archive of Applied Mechanics*, 2013. **83**(9): 1309–1325.
159. Zhang, R. and H. Zhong, Weak form quadrature element analysis of spatial geometrically exact shear-rigid beams. *Finite Elements in Analysis and Design*, 2014. **87**: 22–31.
160. Zhong, H., R. Zhang, and N. Xiao, A quaternion-based weak form quadrature element formulation for spatial geometrically exact beams. *Archive of Applied Mechanics*, 2014. **84**(12): 1825–1840.
161. Zhang, R. and H. Zhong, Weak form quadrature element analysis of geometrically exact shells. *International Journal of Non-Linear Mechanics*, 2015. **71**: 63–71.

162. Zhang, R. and H. Zhong, A quadrature element formulation of an energy–momentum conserving algorithm for dynamic analysis of geometrically exact beams. *Computers & Structures*, 2016. **165**: 96–106.
163. Zhang, R. and H. Zhong, A weak form quadrature element formulation for geometrically exact thin shell analysis. *Computers & Structures*, 2018. **202**: 44–59.
164. Fu, C., *et al.*, A modeling and resolution framework for wrinkling in hyperelastic sheets at finite membrane strain. *Journal of the Mechanics and Physics of Solids*, 2019. **124**: 446–470.
165. Schillinger, D., A. Düster, and E. Rank, The hp-d-adaptive finite cell method for geometrically nonlinear problems of solid mechanics. *International Journal for Numerical Methods in Engineering*, 2012. **89**(9): 1171–1202.
166. Schillinger, D., *et al.*, Small and large deformation analysis with the p- and B-spline versions of the finite cell method. *Computational Mechanics*, 2012. **50**(4): 445–478.
167. Abedian, A., *et al.*, The finite cell method for the J2 flow theory of plasticity. *Finite Elements in Analysis and Design*, 2013. **69**: 37–47.
168. Abedian, A., *et al.*, Finite cell method compared to h-version finite element method for elasto-plastic problems. *Applied Mathematics and Mechanics*, 2014. **35**(10): 1239–1248.
169. Taghipour, A., *et al.*, P-version finite elements and finite cells for finite strain elastoplastic problems. *PAMM*, 2016. **16**(1): 243–244.
170. Taghipour, A., *et al.*, The finite cell method for nearly incompressible finite strain plasticity problems with complex geometries. *Computers & Mathematics with Applications*, 2018. **75**(9): 3298–3316.
171. Chapelle, D. and K.J. Bathe, The finite element analysis of shells — fundamentals. *Nuclear Engineering & Design*. 2003, Heidelberg: Springer Verlag.
172. Bischoff, M., *et al.*, Models and finite elements for thin-walled structures, in *Encyclopedia of Computational Mechanics, Vol 2*, E. Stein, R. de Borst, and T.J.R. Hughes, Editors. 2004, John Wiley & Sons, Ltd: New York. pp. 59–137.
173. Reddy, J.N., *Mechanics of Laminated Composite Plates and Shells: Theory and Analysis*. 2004, New York: CRC press.
174. Viola, E., F. Tornabene, and N. Fantuzzi, General higher-order shear deformation theories for the free vibration analysis of completely doubly-curved laminated shells and panels. *Composite Structures*, 2013. **95**: 639–666.
175. Li, X., J. Zhang, and Y. Zheng, Static and free vibration analysis of laminated composite plates using isogeometric approach based on the third order shear deformation theory. *Advances in Mechanical Engineering*, 2014. **6**: 1–16.

176. Tornabene, F., N. Fantuzzi, and M. Bacciocchi, The GDQ method for the free vibration analysis of arbitrarily shaped laminated composite shells using a NURBS-based isogeometric approach. *Composite Structures*, 2016. **154**(1): 190–218.
177. Kapoor, H. and R.K. Kapania, Geometrically nonlinear NURBS isogeometric finite element analysis of laminated composite plates. *Composite Structures*, 2012. **94**(12): 3434–3447.
178. Mantari, J.L., A.S. Oktem, and C. Guedes Soares, A new trigonometric layerwise shear deformation theory for the finite element analysis of laminated composite and sandwich plates. *Computers & Structures*, 2012. **94-95**: 45–53.
179. Thai, C.H., *et al.*, Isogeometric analysis of laminated composite and sandwich plates using a layerwise deformation theory. *Composite Structures*, 2013. **104**: 196–214.
180. Thai, C.H., *et al.*, A generalized layerwise higher-order shear deformation theory for laminated composite and sandwich plates based on isogeometric analysis. *Acta Mechanica*, 2016. **227**(5): 1225–1250.
181. Wu, Y., Y.F. Xing, and B. Liu, Analysis of isotropic and composite laminated plates and shells using a differential quadrature hierarchical finite element method. *Composite Structures*, 2018. **205**: 11–25.
182. Reddy, J. and D. Robbins, Theories and computational models for composite laminates. *Applied Mechanics Reviews* 1994. **47**(6): 147–169.
183. Reddy, J.N., *Theory and Analysis of Elastic Plates and Shells.* 2nd ed. 2006, New York: CRC Press.
184. Carrera, E., Theories and finite elements for multilayered, anisotropic, composite plates and shells. *Archives of Computational Methods in Engineering*, 2002. **9**(2): 87–140.
185. Oñate, E., Structural analysis with the finite element method. *Volume 2. Linear Statics: Beams, Plates and Shells.* Vol. 2. 2013, Netherland: Springer.
186. Babuska, I., B.A. Szabo, and I.N. Katz, The p-version of the finite element method. *Siam Journal on Numerical Analysis*, 1981. **18**(3): 515–545.
187. Stein, E., R.D. Borst, and T.J.R. Hughes, Encyclopedia of computational mechanics. *Volume 1: Fundamentals.* 1st ed. Vol. 1. 2004, Chichester: John Wiley.
188. Királyfalvi, G. and B.A. Szabó, Quasi-regional mapping for the p-version of the finite element method. *Finite Elements in Analysis & Design*, 1997. **27**(1): 85–97.

189. Fortunato, M. and P.-O. Persson, High-order unstructured curved mesh generation using the Winslow equations. *Journal of Computational Physics*, 2016. **307**: 1–14.
190. Szabó, B. and I. Babuška, *Introduction to Finite Element Analysis: Formulation, Verification and Validation*. 2011, Chichester: John Wiley & Sons.
191. Ahmed, N.U. and P.K. Basu, Higher-order finite element modelling of laminated composite plates. *International Journal for Numerical Methods in Engineering*, 1994. **37**(1): 123–139.
192. Rank, E., R. Krause, and K. Preusch, On the accuracy of P-version elements for the Reissner-Mindlin plate problem. *International Journal for Numerical Methods in Engineering*, 1998. **43**(1): 51–67.
193. Woo, K.S., C.H. Hong, and P.K. Basu, Materially and geometrically nonlinear analysis of laminated anisotropic plates by p-version of FEM. *Computers & Structures*, 2003. **81**(16): 1653–1662.
194. Yazdani, S., P. Ribeiro, and J.D. Rodrigues, A p-version layerwise model for large deflection of composite plates with curvilinear fibres. *Composite Structures*, 2014. **108**: 181–190.
195. Peano, A., Hierarchies of conforming finite elements for plane elasticity and plate bending. *Computers & Mathematics with Applications*, 1976. **2**(3): 211–224.
196. Wang, D.W., I.N. Katz, and B.A. Szabó, Implementation of a C1 triangular element based on the p-version of the finite element method. *Computers & Structures*, 1984. **19**(3): 381–392.
197. Chinosi, C., T. Scapolla, and G. Sacchi, A hierarchic family of C1 finite elements for 4th order elleptic problems. *Computational Mechanics*, 1991. **8**(3): 181–191.
198. Bardell, N.S., Free vibration analysis of a flat plate using the hierarchical finite element method. *Journal of Sound & Vibration*, 1991. **151**(2): 263–289.
199. Bardell, N.S., The free vibration of skew plates using the hierarchical finite element method. *Computers & Structures*, 1992. **45**(5): 841–874.
200. Dhatt, G., G. Touzot, and E. Lefrançois, *Finite Element Method*, 2012, Hoboken: John Wiley & Sons.
201. Zhong, H.Z. and Z.G. Yue, Analysis of thin plates by the weak form quadrature element method. *Science China* (*Physics, Mechanics & Astronomy*), 2012. **55**(5): 861–871.
202. Ferreira, L.J.F. and M.L. Bittencourt, Hierarchical high-order conforming C1 bases for quadrangular and triangular finite elements. *International Journal for Numerical Methods in Engineering*, 2017. **109**(7): 936–964.

203. Bernadou, M. and J.M. Boisserie, Curved finite elements of class C1: Implementation and numerical experiments. Part 1: Construction and numerical tests of the interpolation properties. *Computer Methods in Applied Mechanics and Engineering*, 1993. **106**(1): 229–269.
204. Gao, Q. and L. Li, Interpolated boundary conditions in plate bending problems using C1 curved finite elements. *Computer Methods in Applied Mechanics and Engineering*, 1997. **148**(3): 235–255.
205. Hughes, T.J.R., J.A. Cottrell, and Y. Bazilevs, Isogeometric analysis: CAD, finite elements, NURBS, exact geometry and mesh refinement. *Computer Methods in Applied Mechanics and Engineering*, 2005. **194**(39): 4135–4195.
206. Cottrell, J.A., T.J.R. Hughes, and A. Reali, Studies of refinement and continuity in isogeometric structural analysis. *Computer Methods in Applied Mechanics and Engineering*, 2007. **196**(41): 4160–4183.
207. Kiendl, J., *et al.*, Isogeometric shell analysis with Kirchhoff–Love elements. *Computer Methods in Applied Mechanics and Engineering*, 2009. **198**(49): 3902–3914.
208. Kiendl, J., *et al.*, The bending strip method for isogeometric analysis of Kirchhoff–Love shell structures comprised of multiple patches. *Computer Methods in Applied Mechanics and Engineering*, 2010. **199**(37): 2403–2416.
209. Shojaee, S., *et al.*, Free vibration analysis of thin plates by using a NURBS-based isogeometric approach. *Finite Elements in Analysis and Design*, 2012. **61**: 23–34.
210. Guo, Y. and M. Ruess, Nitsche's method for a coupling of isogeometric thin shells and blended shell structures. *Computer Methods in Applied Mechanics and Engineering*, 2015. **284**: 881–905.
211. Guo, Y., *et al.*, Variationally consistent isogeometric analysis of trimmed thin shells at finite deformations, based on the STEP exchange format. *Computer Methods in Applied Mechanics and Engineering*, 2018. **336**: 39–79.
212. Liu, B., *et al.*, Non-uniform rational Lagrange functions and its applications to isogeometric analysis of in-plane and flexural vibration of thin plates. *Computer Methods in Applied Mechanics and Engineering*, 2017. **321**: 173–208.
213. Wu, Y., Y. Xing, and B. Liu, Hierarchical p-version C1 finite elements on quadrilateral and triangular domains with curved boundaries and their applications to Kirchhoff plates. *International Journal for Numerical Methods in Engineering*, 2019. **119**(3): 177–207.
214. Surana, K.S. and R.M. Sorem, P-version hierarchical three dimensional curved shell element for elastostatics. *International Journal for Numerical Methods in Engineering*, 1991. **31**(4): 649–676.

215. Surana, K.S. and R.M. Sorem, Completely hierarchical p-version curved shell element for laminated composite plates and shells. *Computational Mechanics*, 1991. **7**(4): 237–251.
216. Düster, A., H. Bröker, and E. Rank, The p-version of the finite element method for three-dimensional curved thin walled structures. *International Journal for Numerical Methods in Engineering*, 2001. **52**(7): 673–703.
217. Rank, E., *et al.*, High order finite elements for shells. *Computer Methods in Applied Mechanics and Engineering*, 2005. **194**(21): 2494–2512.
218. Muthler, A., *et al.*, High order thin-walled solid finite elements applied to elastic spring-back computations. *Computer Methods in Applied Mechanics and Engineering*, 2006. **195**(41): 5377–5389.
219. Payette, G. and J. Reddy, A seven-parameter spectral/hp finite element formulation for isotropic, laminated composite and functionally graded shell structures. *Computer Methods in Applied Mechanics and Engineering*, 2014. **278**: 664–704.
220. Piegl, L. and W. Tiller, *The NURBS Book*. 1997, Berlin Heidelberg: Springer.
221. Clough, R.W. and C.P. Johnson, A finite element approximation for the analysis of thin shells. *International Journal of Solids & Structures*, 1968. **4**(1): 43–60.
222. Campbell, D.M. and L.R. Herrmann, A finite-element analysis for thin shells. *AIAA Journal*, 1971. **6**: 1842–1847.
223. Bell, K., A refined triangular plate bending finite element. *International Journal for Numerical Methods in Engineering*, 1969. **1**: 101–122.
224. Fraeijs de Veubeke, B., A conforming finite element for plate bending. *International Journal of Solids and Structures*, 1968. **4**(1): 95–108.
225. Argyris, J.H. and N. Lochner, On the application of the SHEBA shell element. *Computer Methods in Applied Mechanics and Engineering*, 1972. **1**(3): 317–347.
226. Argyris, J.H. and D.W. Scharpf, The SHEBA family of shell elements for the matrix displacement method. *The Aeronautical Journal of the Royal Aeronautical Society*, 1968. **72**(694): 873–883.
227. Ivannikov, V., C. Tiago, and P.M. Pimenta. Generalization of the C1 TUBA plate finite elements to the geometrically exact Kirchhoff-Love shell model. *Computer Methods in Applied Mechanics and Engineering*, 2015. **294**: 210–244.
228. Wu, Y., Y. Xing, and B. Liu, Curved p-version C1 finite elements for the finite deformation analysis of isotropic and composite laminated thin shells in irregular shape. *Composite Structures*, 2020: 113459.

229. Cirak, F., M. Ortiz, and P. Schröder, Subdivision surfaces: A new paradigm for thin-shell finite-element analysis. *International Journal for Numerical Methods in Engineering*, 2000. **47**(12): 2039–2072.
230. Cirak, F. and M. Ortiz, *Fully C1-conforming subdivision elements for finite deformation thin-shell analysis. International Journal for Numerical Methods in Engineering*, 2001. **51**(7): 813–833.
231. Kiendl, J., *et al.*, Isogeometric Kirchhoff–Love shell formulations for general hyperelastic materials. *Computer Methods in Applied Mechanics and Engineering*, 2015. **291**: 280–303.
232. Ambati, M., J. Kiendl, and L.D. Lorenzis, *Isogeometric Kirchhoff–love shell formulation for elasto-plasticity. Computer Methods in Applied Mechanics and Engineering*, 2018. **340**: 320–339.
233. Casquero, H., *et al.*, Seamless integration of design and Kirchhoff–Love shell analysis using analysis-suitable unstructured T-splines. *Computer Methods in Applied Mechanics and Engineering*, 2020. **360**: 1–34.
234. Dey, S., M.S. Shephard, and J.E. Flaherty, Geometry representation issues associated with p-version finite element computations. *Computer Methods in Applied Mechanics and Engineering*, 1997. **150**(1): 39–55.
235. Dey, S., R.M. O'Bara, and M.S. Shephard, Towards curvilinear meshing in 3D: The case of quadratic simplices. *Computer-Aided Design*, 2001. **33**(3): 199–209.
236. Sherwin, S.J. and J. Peiró, Mesh generation in curvilinear domains using high-order elements. *International Journal for Numerical Methods in Engineering*, 2002. **53**(1): 207–223.
237. Luo, X.-J., *et al.*, Automatic p-version mesh generation for curved domains. *Engineering with Computers*, 2004. **20**(3): 273–285.
238. Shephard, M.S., *et al.*, Adaptive mesh generation for curved domains. *Applied Numerical Mathematics*, 2005. **52**(2): 251–271.
239. Luo, X.-J., *et al.*, Moving curved mesh adaptation for higher-order finite element simulations. *Engineering with Computers*, 2011. **27**(1): 41–50.
240. Remacle, J.-F., *et al.*, Efficient visualization of high-order finite elements. *International Journal for Numerical Methods in Engineering*, 2007. **69**(4): 750–771.
241. Geuzaine, C. and J.-F. Remacle, Gmsh: A 3-D finite element mesh generator with built-in pre- and post-processing facilities. *International Journal for Numerical Methods in Engineering*, 2009. **79**(11): 1309–1331.

242. Persson, P.O. and J. Peraire, Curved mesh generation and mesh refinement using Lagrangian solid mechanics. *Office of Scientific & Technical Information Technical Reports*, 2009.
243. Hindenlang, F., T. Bolemann, and C.-D. Munz, Mesh Curving Techniques for High Order Discontinuous Galerkin Simulations, in *IDIHOM: Industrialization of High-Order Methods — A Top-Down Approach: Results of a Collaborative Research Project Funded by the European Union, 2010–2014*, N. Kroll, *et al.*, Editors. 2015, Springer International Publishing: Cham, 133–152.
244. Kroll, N., *et al. ADIGMA — A European Initiative on the Development of Adaptive Higher-Order Variational Methods for Aerospace Applications.* 2010. Springer.
245. Kroll, N., *et al. IDIHOM: Industrialization of High-Order Methods — A Top-Down Approach.* 2015. Springer.
246. Johnen, A., J.F. Remacle, and C. Geuzaine, Geometrical validity of curvilinear finite elements. *Journal of Computational Physics*, 2013. **233**: 359–372.
247. Johnen, A., J.-F. Remacle, and C. Geuzaine, Geometrical validity of high-order triangular finite elements. *Engineering with Computers*, 2014. **30**(3): 375–382.
248. Johnen, A. and C. Geuzaine, Geometrical validity of curvilinear pyramidal finite elements. *Journal of Computational Physics*, 2015. **299**: 124–129.
249. Toulorge, T., *et al.*, Robust untangling of curvilinear meshes. *Journal of Computational Physics*, 2013. **254**: 8–26.
250. Toulorge, T., J. Lambrechts, and J.-F. Remacle, Optimizing the geometrical accuracy of curvilinear meshes. *Journal of Computational Physics*, 2016. **310**: 361–380.
251. Xie, Z.Q., *et al.*, *The generation of arbitrary order curved meshes for 3D finite element analysis. Computational Mechanics*, 2013. **51**(3): 361–374.
252. Szałtys, P., *et al.*, High-order 3D anisotropic hybrid mesh generation for high-reynolds number flows, in *IDIHOM: Industrialization of High-Order Methods — A Top-Down Approach: Results of a Collaborative Research Project Funded by the European Union, 2010–2014*; N. Kroll, *et al., Editors.* 2015, Springer International Publishing: Cham, 79–100.
253. Abgrall, R., C. Dobrzynski, and A. Froehly, A method for computing curved meshes via the linear elasticity analogy, application to fluid dynamics problems. *International Journal for Numerical Methods in Fluids*, 2014. **76**(4): 246–266.

254. Poya, R., R. Sevilla, and A.J. Gil, A unified approach for a posteriori high-order curved mesh generation using solid mechanics. *Computational Mechanics*, 2016. **58**(3): 457–490.
255. Lu, Q., *et al.*, Parallel mesh adaptation for high-order finite element methods with curved element geometry. *Engineering with Computers*, 2014. **30**(2): 271–286.
256. Xu, J. and A.N. Chernikov, Automatic curvilinear quality mesh generation driven by smooth boundary and guaranteed fidelity. *Procedia Engineering*, 2014. **82**: 200–212.
257. Gargallo-Peiró, A., *et al.*, Optimization of a regularized distortion measure to generate curved high-order unstructured tetrahedral meshes. *International Journal for Numerical Methods in Engineering*, 2015. **103**(5): 342–363.
258. Gargallo-Peiró, A., *et al.*, Distortion and quality measures for validating and generating high-order tetrahedral meshes. *Engineering with Computers*, 2015. **31**(3): 423–437.
259. Gargallo-Peiró, A., *et al.*, A distortion measure to validate and generate curved high-order meshes on CAD surfaces with independence of parameterization. *International Journal for Numerical Methods in Engineering*, 2016. **106**(13): 1100–1130.
260. Gargallo-Peiró, A., G. Houzeaux, and X. Roca, Subdividing triangular and quadrilateral meshes in parallel to approximate curved geometries. *Procedia Engineering*, 2017. **203**: 310–322.
261. Moxey, D., *et al.*, An isoparametric approach to high-order curvilinear boundary-layer meshing. *Computer Methods in Applied Mechanics and Engineering*, 2015. **283**: 636–650.
262. Moxey, D., *et al.*, *Curvilinear mesh generation for boundary layer problems*, in *IDIHOM: Industrialization of High-Order Methods — A Top-Down Approach: Results of a Collaborative Research Project Funded by the European Union, 2010–2014*, N. Kroll, *et al.*, Editors. 2015, Springer International Publishing: Cham., 41–64.
263. Moxey, D., *et al.*, High-order curvilinear meshing using a thermo-elastic analogy. *Computer-Aided Design*, 2016. **72**: 130–139.
264. Engvall, L. and J.A. Evans, Isogeometric triangular Bernstein–Bézier discretizations: Automatic mesh generation and geometrically exact finite element analysis. *Computer Methods in Applied Mechanics and Engineering*, 2016. **304**: 378–407.
265. Ruiz-Gironés, E., X. Roca, and J. Sarrate, High-order mesh curving by distortion minimization with boundary nodes free to slide on a 3D CAD representation. *Computer-Aided Design*, 2016. **72**: 52–64.
266. Ruiz-Gironés, E., J. Sarrate, and X. Roca, Generation of curved high-order meshes with optimal quality and geometric accuracy. *Procedia Engineering*, 2016. **163**: 315–327.

267. Turner, M., J. Peiró, and D. Moxey, A Variational framework for high-order mesh generation. *Procedia Engineering*, 2016. **163**: 340–352.
268. Eichstädt, J., *et al.*, Accelerating high-order mesh optimisation with an architecture-independent programming model. *Computer Physics Communications*, 2018. **229**: 36–53.
269. Bock, K. and J. Stiller, Optimizing triangular high-order surface meshes by energy-minimization. *Engineering with Computers*, 2018. **34**(4): 659–670.
270. Feng, L., *et al.*, Curved optimal delaunay triangulation. *ACM Transactions on Graphics*, 2018. **37**(4): 1–16.
271. Ims, J. and Z.J. Wang, Automated low-order to high-order mesh conversion. *Engineering with Computers*, 2019. **35**(1): 323–335.
272. Moxey, D., S.P. Sastry, and R.M. Kirby, Interpolation error bounds for curvilinear finite elements and their implications on adaptive mesh refinement. *Journal of Scientific Computing*, 2019. **78**(2): 1045–1062.
273. Shu-Jie, L., Mesh curving and refinement based on cubic Bézier surface for high-order discontinuous Galerkin methods. *Computational Mathematics and Mathematical Physics*, 2019. **59**(12): 2080–2092.
274. Stees, M. and S.M. Shontz, An angular approach to untangling high-order curvilinear triangular Meshes, in *27th International Meshing Roundtable*; X. Roca and A. Loseille, *Editors*. 2019, Springer International Publishing: Cham., 327–342.
275. Stees, M. and S.M. Shontz. High-order mesh generation based on optimal affine combinations of nodal positions, in *Spectral and High Order Methods for Partial Differential Equations ICOSAHOM 2018*. 2020. Cham: Springer International Publishing.
276. Mohammadi, F., *et al.* A Direct high-order curvilinear triangular mesh generation method using an advancing front technique, in *Computational Science – ICCS 2020*. 2020. Cham: Springer International Publishing.
277. Szabó, B., A. Düster, and E. Rank, The p-version of the finite element method, in *Encyclopedia of Computational Mechanics*, 2004.
278. Vincent, P.E. and A. Jameson, Facilitating the adoption of unstructured high-order methods amongst a wider community of fluid dynamicists. *Mathematical Modelling of Natural Phenomercy*, 2011. **6**(3): 97–140.
279. Bert, C.W. and M. Malik, The differential quadrature method for irregular domains and application to plate vibration. *International Journal of Mechanical Sciences*, 1996. **38**(6): 589–606.
280. Malik, M. and C.W. Bert, Vibration analysis of plates with curvilinear quadrilateral planforms by DQM using blending functions. *Journal of Sound and Vibration*, 2000. **230**(4): 949–954.

281. Xing, Y. and B. Liu, High-accuracy differential quadrature finite element method and its application to free vibrations of thin plate with curvilinear domain. *International Journal for Numerical Methods in Engineering*, 2009. **80**(13): 1718–1742.
282. Zhong, H. and T. Yu, A weak form quadrature element method for plane elasticity problems. *Applied Mathematical Modelling*, 2009. **33**(10): 3801–3814.
283. Bardell, N.S., R.S. Langley, and J.M. Dunsdon, On the free in-plane vibration of isotropic rectangular plates. *Journal of Sound and Vibration*, 1996. **191**(3): 459–467.
284. Karunasena, W. and S. Kitipornchai, Free vibration of shear-deformable general triangular plates. *Journal of Sound and Vibration*, 1997. **199**(4): 595–613.
285. Gorman, D.J., A highly accurate analytical solution for free vibration analysis of simply supported right triangular plates. *Journal of Sound and Vibration*, 1983. **89**(1): 107–118.
286. Gorman, D.J., Free vibration analysis of right triangular plates with combinations of clamped-simply supported boundary conditions. *Journal of Sound and Vibration*, 1986. **106**(3): 419–431.
287. Gorman, D.J., Accurate free vibration analysis of right triangular plates with one free edge. *Journal of Sound and Vibration*, 1989. **131**(1): 115–125.
288. Liu, C.F. and Y.T. Lee, Finite element analysis of three-dimensional vibrations of thick circular and annular plates. *Journal of Sound and Vibration*, 2000. **233**(1): 63–80.
289. So, J. and A.W. Leissa, Three-dimensional vibrations of thick circular and annular plates. *Journal of Sound and Vibration*, 1998. **209**(1): 15–41.
290. Zhou, D., *et al.*, Three-dimensional vibration analysis of circular and annular plates via the Chebyshev–Ritz method. *International Journal of Solids and Structures*, 2003. **40**(12): 3089–3105.
291. Leissa, A.W., *Vibration of Plates. NASA SP-160.* 1969, Washington D.C.: Scientific and Technical Information Division of National Aeronautics and Space Administration.
292. Timoshenko, S. and S. Woinowsky-Krieger, *Theory of Plates and Shells.* Second ed. 1959: McGraw-Hill.
293. Muhammad, T. and A.V. Singh, A p-type solution for the bending of rectangular, circular, elliptic and skew plates. *International Journal of Solids and Structures*, 2004. **41**(15): 3977–3997.
294. Liu, B., *et al.*, Thickness-shear vibration analysis of rectangular quartz plates by a numerical extended Kantorovich method. *Composite Structures*, 2014. **107**(0): 429–435.

295. Wang, J., J. Shi, and J.-K. Du. The finite element analysis of thickness-shear vibrations of quartz crystal plates with ANSYS, in *Piezoelectricity, Acoustic Waves, and Device Applications (SPAWDA) and 2009 China Symposium on Frequency Control Technology*. 2009.
296. Gordon, W. and C. Hall, Transfinite element methods: Blending-function interpolation over arbitrary curved element domains. *Numerische Mathematik*, 1973. **21**(2): 109–129.
297. Gordon, W.J. and C.A. Hall, Construction of curvilinear co-ordinate systems and applications to mesh generation. *International Journal for Numerical Methods in Engineering*, 1973. **7**(4): 461–477.
298. Spink, M., *et al. NURBS Toolbox*. 2015; Collection of routines for the creation, and manipulation of non-uniform rational B-splines (NURBS), based on the NURBS toolbox by Mark Spink. Available from: http://octave.sourceforge.net/nurbs/index.html.
299. Zhu, D.C. Development of hierarchical finite element methods at BIAA, in *The International Conference on Computational Mechanics*. 1986. Tokyo.
300. Hinnant, H.E., A fast method of numerical quadrature for p-version finite element matrices. *International Journal for Numerical Methods in Engineering*, 1994. **37**(21): 3723–3750.
301. Liu, C., *et al.*, In-plane vibration analysis of plates in curvilinear domains by a differential quadrature hierarchical finite element method. *Meccanica*, 2016. **52**(4): 1017–1033.
302. Bardell, N.S., An engineering application of the h-p version of the finite element method to the static analysis of a Euler-Bernoulli beam. *Computers & Structures*, 1996. **59**(2): 195–211.
303. Mindlin, R.D., *An Introduction to the Mathematical Theory of Vibrations of Elastic Plates*. 2006, Singapore: World Scientific Publishing Co. Pte. Ltd.
304. Xing, Y., B. Liu, and G. Liu, A differential quadrature finite element method. *International Journal of Applied Mechanics*, 2010. **2**(1): 207–227.
305. Hashemi, S.H. and M. Arsanjani, Exact characteristic equations for some of classical boundary conditions of vibrating moderately thick rectangular plates. *International Journal of Solids and Structures*, 2005. **42**(3–4): 819–853.
306. Liew, K.M., *et al.*, Vibration of thick skew plates based on Mindlin shear deformation plate theory. *Journal of Sound & Vibration*, 1993. **168**(1): 39–69.
307. Shufrin, I. and M. Eisenberger, Stability and vibration of shear deformable plates — first order and higher order analyses. *International Journal of Solids and Structures*, 2005. **42**(3–4): 1225–1251.

308. Liew, K.M., *et al.*, *Vibration of Mindlin Plates: Programming the p-Version Ritz Method.* 1998: Elsevier.
309. Sakata, T., K. Takahashi, and R.B. Bhat, Natural frequencies of orthotropic rectangular plates obtained by iterative reduction of the partial differential equation. *Journal of Sound and Vibration*, 1996. **189**(1): 89–101.
310. Liu, B. and Y. Xing, Thickness-shear vibration analysis of rectangular quartz plates by a differential quadrature finite element method. *AIP Conference Proceedings*, 2014. **1618**(1): 41–44.
311. Wei, G.W., Y.B. Zhao, and Y. Xiang, A novel approach for the analysis of high-frequency vibrations. *Journal of Sound & Vibration*, 2002. **257**(2): 207–246.
312. Cottrell, J.A., *et al.*, Isogeometric analysis of structural vibrations. *Computer Methods in Applied Mechanics and Engineering*, 2006. **195**(41–43): 5257–5296.
313. Liu, F.L. and K.M. Liew, Free vibration analysis of Mindlin sector plates: Numerical solutions by differential quadrature method. *Computer Methods in Applied Mechanics and Engineering*, 1999. **177**(1–2): 77–92.
314. Huang, C.S., O.G. McGee, and A.W. Leissa, Exact analytical solutions for free vibrations of thick sectorial plates with simply supported radial edges. *International Journal of Solids and Structures*, 1994. **31**(11): 1609–1631.
315. McGee, O.G., C.S. Huang, and A.W. Leissa, Comprehensive exact solutions for free vibrations of thick annular sectorial plates with simply supported radial edges. *International Journal of Mechanical Sciences*, 1995. **37**(5): 537–566.
316. Liu, B., *et al.*, Thickness-shear vibration analysis of rectangular quartz plates by a numerical extended Kantorovich method. *Composite Structures*, 2014. **107**(0): 429–435.
317. Liu, B. and Y. Xing, Thickness-shear vibration analysis of rectangular quartz plates by a differential quadrature finite element method. *AIP Conference Proceedings*, 2014. **1618**(1): 41–44.
318. Wang, J., J. Shi, and J.-K. Du. The finite element analysis of thickness-shear vibrations of quartz crystal plates with ANSYS, in *Piezoelectricity, Acoustic Waves, and Device Applications (SPAWDA) and 2009 China Symposium on Frequency Control Technology, Joint Conference of the 2009 Symposium on.* 2009.
319. Bert, C.W., S.K. Jang, and A.G. Striz, Two new approximate methods for analyzing free vibration of structural components. *AIAA Journal*, 1988. **26**(5): 612–618.

320. Bellman, R. and J. Casti, Differential quadrature and long term integration. *Journal of Mathematical Analysis and Applications*, 1971(34): 235–238.
321. Karniadakis, G.E. and S. Sherwin, *Spectral/hp Element Methods for Computational Fluid Dynamics.* 2 ed. 2005, New York: Oxford Science Publications.
322. Sederberg, T.W., *et al.*, Watertight trimmed NURBS. *Acm Transactions on Graphics*, 2008. **27**(3): 15–19.
323. Sederberg, T.W., *et al.*, T-splines and T-NURCCs. *Acm Transactions on Graphics*, 2003. **22**(3): 477–484.
324. Reddy, J.N., *Mechanics of Laminated Composite Plates and Shells: Theory and Analysis.* Second ed. 2003: CRC Press.
325. Liu, B. and Y. Xing, Exact solutions for free in-plane vibrations of rectangular plates. *Acta Mechanica Solida Sinica*, 2011. **24**(6): 556–567.
326. Xing, Y.F. and B. Liu, Exact solutions for the free in-plane vibrations of rectangular plates. *International Journal of Mechanical Sciences*, 2009. **51**(3): 246–255.
327. Liu, B. and Y. Xing, Comprehensive exact solutions for free in-plane vibrations of orthotropic rectangular plates. *European Journal of Mechanics — A/Solids*, 2011. **30**(3): 383–395.
328. Cottrell, J.A., *et al.*, Isogeometric analysis of structural vibrations. Computer Methods in Applied Mechanics and Engineering, 2006. **195**(41–43): 5257–5296.
329. Hughes, T.J.R., A. Reali, and G. Sangalli, Duality and unified analysis of discrete approximations in structural dynamics and wave propagation: Comparison of p-method finite elements with k-method NURBS. *Computer Methods in Applied Mechanics and Engineering*, 2008. **197**(49–50): 4104–4124.
330. Hughes, T.J.R., J.A. Evans, and A. Reali, Finite element and NURBS approximations of eigenvalue, boundary-value, and initial-value problems. *Computer Methods in Applied Mechanics and Engineering*, 2014. **272**: 290–320.
331. Park, C.I., Frequency equation for the in-plane vibration of a clamped circular plate. *Journal of Sound and Vibration*, 2008. **313**(1–2): 325–333.
332. Mohazzab, A.H. and L. Dozio, A spectral collocation solution for in-plane eigenvalue analysis of skew plates. *International Journal of Mechanical Sciences*, 2015. **94–95**(0): 199–210.
333. Liu, C., *et al.*, Micro/macro-mechanical analysis of the interface of composite structures by a differential quadrature hierarchical finite element method. *Composite Structures*, 2016. **154**: 39–48.

334. Liu, B. and Y. Xing, Exact solutions for free in-plane vibrations of rectangular plates. *Acta Mechanica Solida Sinica*, 2011. **24**(6): 556–567.
335. Xing, Y.F. and B. Liu, Exact solutions for the free in-plane vibrations of rectangular plates. *International Journal of Mechanical Sciences*, 2009. **51**(3): 246–255.
336. Zhou, J., *et al.*, The grain size and porosity dependent elastic moduli and yield strength of nanocrystalline ceramics. *Materials Science and Engineering: A*, 2007. **445–446**: 717–724.
337. Dobosz, R., M. Lewandowska, and K.J. Kurzydlowski, The effect of grain size diversity on the flow stress of nanocrystalline metals by finite-element modelling. *Scripta Materialia*, 2012. **67**(4): 408–411.
338. Yang, S. and M. Cho, A scale-bridging method for nanoparticulate polymer nanocomposites and their nondilute concentration effect. *Applied Physics Letters*, 2009. **94**(22): 223104.
339. Khare, R.K., T. Kant, and A.K. Garg, Closed-form thermo-mechanical solutions of higher-order theories of cross-ply laminated shallow shells. *Composite Structures*, 2003. **59**(3): 313–340.
340. Hung, K., *A treatise on three-dimensional vibration of a class of elastic solids*. 1996, Nanyang Technological University.
341. Liew, K.M. and T.M. Teo, Three-dimensional vibration analysis of rectangular plates based on differential quadrature method. *Journal of Sound and Vibration*, 1999. **220**(4): 577–599.
342. Leissa, A.W., The free vibration of rectangular plates. *Journal of Sound and Vibration*, 1973. **31**(3): 257–293.
343. Srinivas, S., C.V.J. Rao, and A.K. Rao, An exact analysis for vibration of simply-supported homogeneous and laminated thick rectangular plates. *Journal of Sound & Vibration*, 1970. **12**(2): 187–199.
344. Cheung, Y.K. and D. Zhou, Three-dimensional vibration analysis of cantilevered and completely free isosceles triangular plates. *International Journal of Solids and Structures*, 2002. **39**(3): 673–687.
345. Kitipornchai, S., *et al.*, Free vibration of isosceles triangular mindlin plates. *International Journal of Mechanical Sciences*, 1993. **35**(35): 89–102.
346. Liew, K.M. and B. Yang, Three-dimensional elasticity solutions for free vibrations of circular plates: A polynomials-Ritz analysis. *Computer Methods in Applied Mechanics and Engineering*, 1999. **175**(1): 189–201.
347. Lim, C.W. and K.M. Liew, A higher order theory for vibration of shear deformable cylindrical shallow shells. *International Journal of Mechanical Sciences*, 1995. **37**(3): 277–295.

348. Liew, K.M., L.X. Peng, and T.Y. Ng, Three-dimensional vibration analysis of spherical shell panels subjected to different boundary conditions. *International Journal of Mechanical Sciences*, 2002. **44**(10): 2103–2117.
349. Liu, C.-F. and Y.-T. Lee, Finite element analysis of three-dimensional vibrations of thick circular and annular plates. *Journal of Sound and Vibration*, 2000. **233**(1): 63–80.
350. Chen, W. and Y. Wang, Minimal degree H(curl) And H(div) conforming finite elements on polytopal meshes. *Mathematics of Computation*, 2017. **86**(307): 2053–2087.
351. Zhou, J., Li, Y., Zhu, R., and Z. Zhang, The grain size and porosity dependent elastic moduli and yield strength of nanocrystalline ceramics. *Materials Science and Engineering: A*. 2007. **445–446**: 717–724.
352. Srinivas, S., C.J. Rao, and A. Rao, An exact analysis for vibration of simply-supported homogeneous and laminated thick rectangular plates. *Journal of sound and vibration*, 1970. **12**(2): 187–199.
353. Huang, K. and A. Dasgupta, A layer-wise analysis for free vibration of thick composite cylindrical shells. *Journal of Sound and Vibration*, 1995. **186**(2): 207–222.
354. Srinivas, S. and A. Rao, Bending, vibration and buckling of simply supported thick orthotropic rectangular plates and laminates. *International Journal of Solids and Structures*, 1970. **6**(11): 1463–1481.
355. Kant, T. and Mallikarjuna, Vibrations of unsymmetrically laminated plates analyzed by using a higher order theory with a C° finite element formulation. *Journal of Sound and Vibration*, 1989. **134**(1): 1–16.
356. Nguyen-Van, H., *et al.*, Buckling and vibration analysis of laminated composite plate/shell structures via a smoothed quadrilateral flat shell element with in-plane rotations. *Computers & Structures*, 2011. **89**(7): 612–625.
357. Garg, A.K., R.K. Khare, and T. Kant, Free vibration of skew fiber-reinforced composite and sandwich laminates using a shear deformable finite element model. *Journal of Sandwich Structures & Materials*, 2006. **8**(1): 33–53.
358. Tornabene, F., N. Fantuzzi, and M. Bacciocchi, The GDQ method for the free vibration analysis of arbitrarily shaped laminated composite shells using a NURBS-based isogeometric approach. *Composite Structures*, 2016. **154**: 190–218.
359. Wang, S., Free vibration analysis of skew fibre-reinforced composite laminates based on first-order shear deformation plate theory. *Computers & Structures*, 1997. **63**(3): 525–538.
360. Srinivas, S., A refined analysis of composite laminates. *Journal of Sound and Vibration*, 1973. **30**(4): 495–507.

361. Ferreira, A., Analysis of composite plates using a layerwise theory and multiquadrics discretization. *Mechanics of Advanced Materials and Structures*, 2005. **12**(2): 99–112.
362. Ferreira, A. and J. Barbosa, Buckling behaviour of composite shells. *Composite Structures*, 2000. **50**(1): 93–98.
363. Ferreira, A., A formulation of the multiquadric radial basis function method for the analysis of laminated composite plates. *Composite Structures*, 2003. **59**(3): 385–392.
364. Pandya, B. and T. Kant, Higher-order shear deformable theories for flexure of sandwich plates — finite element evaluations. *International Journal of Solids and Structures*, 1988. **24**(12): 1267–1286.
365. Ferreira, A., C. Roque, and P. Martins, Analysis of composite plates using higher-order shear deformation theory and a finite point formulation based on the multiquadric radial basis function method. *Composites Part B: Engineering*, 2003. **34**(7): 627–636.
366. Liu, B., *et al.*, Analysis of composite plates using a layerwise theory and a differential quadrature finite element method. *Composite Structures*, 2016. **156**: 393–398.
367. Bhimaraddi, A., Three-dimensional elasticity solution for static response of orthotropic doubly curved shallow shells on rectangular planform. *Composite Structures*, 1993. **24**(1): 67–77.
368. Pagano, N.J., Exact solutions for rectangular bidirectional composites and sandwich plates. *Journal of Composite Materials*, 1970. **4**(1): 20–34.
369. Reddy, T., Three dimensional elastostatic analysis of fibre reinforced composite laminated shells. *M. Tech Dissertation, Department of Civil Engineering*, Indian Institute of Technology Bombay, India, 1992.
370. Ali, J.S.M., K. Bhaskar, and T.K. Varadan, A new theory for accurate thermal/mechanical flexural analysis of symmetric laminated plates. *Composite Structures*, 1999. **45**(3): 227–232.
371. Gordon, W.J., Blending-function methods of bivariate and multivariate interpolation and approximation. *Siam Journal on Numerical Analysis*, 1971. **8**(1): 158–177.
372. Watkins, D.S., On the construction of conforming rectangular plate elements. *International Journal for Numerical Methods in Engineering*, 1976. **10**(4): 925–933.
373. Gordon, W.J. and C.A. Hall, Transfinite element methods: Blending-function interpolation over arbitrary curved element domains. *Numerische Mathematik*, 1973. **21**(2): 109–129.
374. Zienkiewicz, O.C., R.L. Taylor, and J.Z. Zhu, *The Finite Element Method: Its Basis and Fundamentals*. 7 ed. 2013, New York: Elsevier.

375. Pavel, S., S. Karel, and D. Ivo, *Higher-order Finite Element Methods.* 2004, Boca Raton, FL: Chapman & Hall/CRC.
376. Shu, C., *Differential Quadrature and Its Application in Engineering.* 2000, London: Springer-Verlag. 2671–2671.
377. Bos, L., M.A. Taylor, and B.A. Wingate, Tensor product Gauss-Lobatto points are Fekete points for the cube. *Mathematics of Computation*, 2001. **70**(236): 1543–1547.
378. Sevilla, R., S. Fernández-Méndez, and A. Huerta, NURBS-enhanced finite element method (NEFEM). *Archives of Computational Methods in Engineering*, 2011. **18**(4): 441.
379. Taylor, M.A., R.E. Vincent, and B.A. Wingate, An algorithm for computing Fekete points in the triangle. *Siam Journal on Numerical Analysis*, 2000. **38**(5): 1707–1720.
380. Cowper, G.R., *et al.*, Static and dynamic applications of a high-precision triangular plate bending element. *AIAA Journal*, 1969. **7**(10): 1957–1965.
381. Ivannikov, V., C. Tiago, and P.M. Pimenta, Generalization of the C1 TUBA plate finite elements to the geometrically exact Kirchhoff-Love shell model. *Computer Methods in Applied Mechanics & Engineering*, 2015. **294**: 210–244.
382. Tornabene, F., *et al.*, *Laminated composite doubly-curved shell structures: Differential and integral quadrature strong formulation finite element method.* 2016, Società Editrice Esculapio: Bologna.
383. Cui, X.Y., *et al.*, A thin plate formulation without rotation DOFs based on the radial point interpolation method and triangular cells. *International Journal for Numerical Methods in Engineering*, 2011. **85**(8): 958–986.
384. Soedel, W., *Vibrations of Shells and Plates.* 3 ed. 2004, New York: Marcel Dekker, Inc.
385. Qatu, M.S., *Vibration of Laminated Shells and Plates.* 2004, Oxford: Academic Press.
386. Shen, H.S. and Z.X. Wang, Assessment of Voigt and Mori–Tanaka models for vibration analysis of functionally graded plates. *Composite Structures*, 2012. **94**(7): 2197–2208.
387. Reddy, J.N. and C.D. Chin, Thermomechanical analysis of functionally graded cylinders and plates. *Journal of Thermal Stresses*, 1998. **21**(6): 593–626.
388. Shen, H.S. and S.R. Li, Postbuckling of sandwich plates with FGM face sheets and temperature-dependent properties. *Composites Part B Engineering*, 2008. **39**(2): 332–344.

389. Ferreira, A.J.M., Analysis of composite plates using a layerwise theory and multiquadrics discretization. *Mechanics of Advanced Materials & Structures*, 2005. **12**(2): 99–112.
390. Liu, B., *et al.*, Analysis of composite plates using a layerwise theory and a differential quadrature finite element method. *Composite Structures*, 2016. **156**: 393–398.
391. Chandrashekhara, K., Free vibrations of anisotropic laminated doubly curved shells. *Computers & Structures*, 1989. **33**(2): 435–440.
392. Sanders, J.L., Jr, Nonlinear theories for thin shells. Quart.appl. math.xxi, 1961. **21**(1): 21–36.
393. Pandey, S. and S. Pradyumna, A layerwise finite element formulation for free vibration analysis of functionally graded sandwich shells. *Composite Structures*, 2015. **133**: 438–450.
394. Kapuria, S., M. Patni, and M.Y. Yasin, A quadrilateral shallow shell element based on the third-order theory for functionally graded plates and shells and the inaccuracy of rule of mixtures. *European Journal of Mechanics*, 2015. **49**: 268–282.
395. Bo, L., *et al.*, Analysis of functionally graded sandwich and laminated shells using a layerwise theory and a differential quadrature finite element method. *Composite Structures*, 2016. **136**: 546–553.
396. Fazzolari, F.A. and E. Carrera, Free vibration analysis of sandwich plates with anisotropic face sheets in thermal environment by using the hierarchical trigonometric Ritz formulation. *Composites Part B Engineering*, 2013. **50**(50): 67–81.
397. Yang, J. and H.S. Shen, Free vibration and parametric resonance of shear deformable functionally graded cylindrical panels. *Journal of Sound & Vibration*, 2003. **261**(5): 871–893.
398. Wang, Z.X. and H.S. Shen, Nonlinear dynamic response of sandwich plates with FGM face sheets resting on elastic foundations in thermal environments. *Ocean Engineering*, 2013. **57**(2): 99–110.
399. Piegl, L. and W. Tiller, *The NURBS Book*. 2012: Springer Science & Business Media.
400. Rogers, D.F., *An Introduction to NURBS with Historical Perspective*. 2001: Morgan Kaufmann Publishers Inc.
401. Ahmad, S., B.M. Irons, and O.C. Zienkiewicz, Analysis of thick and thin shell structures by curved finite elements. *International Journal for Numerical Methods in Engineering*, 1970. **2**(3): 419–451.
402. Liu, C., *et al.*, A differential quadrature hierarchical finite element method and its applications to vibration and bending of Mindlin plates with curvilinear domains. *International Journal for Numerical Methods in Engineering*, 2017. **109**(2).

403. Srinivas, S., A refined analysis of composite laminates. *Journal of Sound & Vibration*, 1973. **30**(4): 495–507.
404. Oñate, E., *Structural Analysis with the Finite Element Method. Linear Statics: volume 2: Beams, Plates and Shells*. 2013, New Delhi, India: Springer Publishing Company, Incorporated.
405. Owen, D.R.J. and J.A. Figueiras, Anisotropic elasto-plastic finite element analysis of thick and thin plates and shells. *International Journal for Numerical Methods in Engineering*, 1983. **19**(4): 541–566.
406. Qing, G., J. Qiu, and Y. Liu, Free vibration analysis of stiffened laminated plates. *International Journal of Solids & Structures*, 2006. **43**(6): 1357–1371.
407. Olson, M.D. and C.R. Hazell, Vibration studies on some integral rib-stiffened plates. *Journal of Sound & Vibration*, 1977. **50**(1): 43–61.
408. Zeng, H. and C.W. Bert, A differential quadrature analysis of vibration for rectangular stiffened plates. *Journal of Sound & Vibration*, 2001. **241**(2): 247–252.
409. Roh, H.Y. and M. Cho, The application of geometrically exact shell elements to B-spline surfaces. *Computer Methods in Applied Mechanics & Engineering*, 2004. **193**(23): 2261–2299.
410. Scordelis, A. and K. Lo, Computer analysis of cylindrical shells. *Journal of the American Concrete Institute*, 1964. **61**(5): 539–562.
411. Macneal, R.H. and R.L. Harder, A proposed standard set of problems to test finite element accuracy. *Finite Elements in Analysis & Design*, 1985. **1**(1): 3–20.
412. Qu, Y., *et al.*, A unified formulation for vibration analysis of composite laminated shells of revolution including shear deformation and rotary inertia. *Composite Structures*, 2013. **98**(3): 169–191.
413. Viola, E., F. Tornabene, and N. Fantuzzi, General higher-order shear deformation theories for the free vibration analysis of completely doubly-curved laminated shells and panels. *Composite Structures*, 2013. **95**(1): 639–666.
414. Jin, G., *et al.*, A general Fourier solution for the vibration analysis of composite laminated structure elements of revolution with general elastic restraints. *Composite Structures*, 2014. **109**(1): 150–168.
415. De Borst, R., *et al.*, *Nonlinear Finite Element Analysis of Solids and Structures*. 2012: John Wiley & Sons.
416. Simo, J.C. and D.D. Fox, On a stress resultant geometrically exact shell model. Part I: Formulation and optimal parametrization. *Computer Methods in Applied Mechanics and Engineering*, 1989. **72**: 267–304.

417. Sevilla, R., S. Fernández-Méndez, and A. Huerta, NURBS-enhanced finite element method (NEFEM). *Archives of Computational Methods in Engineering*, 2011. **18**(4): 441–484.
418. Liu, B., *et al.*, A differential quadrature hierarchical finite element method using Fekete points for triangles and tetrahedrons and its applications to structural vibration. *Computer Methods in Applied Mechanics & Engineering*, 2019. **349**: 798–838.
419. Kerdid, N. and P.M. Eiroa, Conforming finite element approximation for shells with little regularity. *Computer Methods in Applied Mechanics & Engineering*, 2000. **188**(1-3): 95–108.
420. Wu, Y., Y. Xing, and B. Liu, C1 conforming quadrilateral finite elements with complete second-order derivatives on vertices and its application to Kirchhoff plates. *Science China-technological Sciences*, 2020. **63**(6): 1066–1084.
421. Crisfield, M.A., Nonlinear finite element analysis of solids and structures. Vol. 2: *Advanced Topics*. 1997, Chichester: John Wiley & Sons.
422. Sze, K.Y., X.H. Liu, and S.H. Lo, Popular benchmark problems for geometric nonlinear analysis of shells. *Finite Elements in Analysis & Design*, 2004. **40**(11): 1551–1569.
423. Wang, Z. and Q. Sun, Corotational nonlinear analyses of laminated shell structures using a 4-node quadrilateral flat shell element with drilling stiffness. *Acta Mechanica Sinica*, 2014. **30**(3): 418–429.
424. Farin, G., J. Hoschek, and M.-S. Kim, *Handbook of Computer Aided Geometric Design*. 2002, North Holland: ELSEVIER.
425. Piegl, L. and W. Tiller, *The NURBS Book*. 2 ed. Monographs in Visual Communication 1997: Springer Berlin Heidelberg.
426. Hughes, T.J.R., J.A. Cottrell, and Y. Bazilevs, Isogeometric analysis: CAD, finite elements, NURBS, exact geometry and mesh refinement. *Computer Methods in Applied Mechanics and Engineering*, 2005. **194**(39–41): 4135–4195.
427. Embar, A., J. Dolbow, and I. Harari, Imposing Dirichlet boundary conditions with Nitsche's method and spline-based finite elements. *International Journal for Numerical Methods in Engineering*, 2010. **83**(7): 877–898.
428. Du, X., G. Zhao, and W. Wang, Nitsche method for isogeometric analysis of Reissner–Mindlin plate with non-conforming multi-patches. *Computer Aided Geometric Design*, 2015. **35–36**: 121–136.
429. Schillinger, D., P.K. Ruthala, and L.H. Nguyen, Lagrange extraction and projection for NURBS basis functions: A direct link between isogeometric and standard nodal finite element formulations. *International Journal for Numerical Methods in Engineering*, 2016. **108**(6): 515–534.

430. Sederberg, T.W., *et al.*, T-spline simplification and local refinement. *ACM Trans. Graph.*, 2004. **23**(3): 276–283.
431. Liu, B. and Y. Xing, Comprehensive exact solutions for free in-plane vibrations of orthotropic rectangular plates. *European Journal of Mechanics — A/Solids*, 2011. **30**(3): 383–395.
432. Patrikalakis, N.M. and T. Maekawa, *Shape Interrogation for Computer Aided Design and Manufacturing*. 2002, New York: Springer.
433. Chen, F., L. Shen, and J. Deng, Implicitization and parametrization of quadratic and cubic surfaces by μ-bases. *Computing*, 2007. **79**(2): 131–142.
434. Shen, L. and C. Yuan, Implicitization using univariate resultants. *Journal of Systems Science and Complexity*, 2010. **23**(4): 804–814.
435. Manocha, D. and J.F. Canny, Implicit representation of rational parametric surfaces. *Journal of Symbolic Computation*, 1992. **13**(5): 485–510.
436. Manocha, D. and J.F. Canny, Algorithm for implicitizing rational parametric surfaces. *Computer Aided Geometric Design*, 1992. **9**(1): 25–50.
437. Persson, P.-O. and G. Strang, A simple mesh generator in MATLAB. *SIAM Review*, 2004. **46**(2): 329–345.
438. Zheleznyakova, A.L. and S.T. Surzhikov, Molecular dynamics-based unstructured grid generation method for aerodynamic applications. *Computer Physics Communications*, 2013. **184**(12): 2711–2727.
439. Zheleznyakova, A.L., Molecular dynamics-based triangulation algorithm of free-form parametric surfaces for computer-aided engineering. *Computer Physics Communications*, 2015. **190**: 1–14.

Index

D

J

K

T

U

V

Lightning Source UK Ltd.
Milton Keynes UK
UKHW021042050821
388292UK00001B/47

9 789811 236754